Quantitative

ACCA

Quantitative analysis

M EDDOWES and R STANSFIELD

ACCA
The Chartered Association of Certified Accountants

Longman

PUBLISHED BY LONGMAN GROUP UK LTD IN CO-OPERATION WITH THE CHARTERED ASSOCIATION OF CERTIFIED ACCOUNTANTS

© Longman Group UK Ltd 1989

ISBN 0 85121 4754

Published by

Longman Law, Tax and Finance
Longman Group UK Limited
21–27 Lamb's Conduit Street, London WC1N 3NJ

Associated Offices

Australia	Longman Professional Publishing (Pty) Limited 130 Phillip Street, Sydney, NSW 2000
Hong Kong	Longman Group (Far East) Limited Cornwall House, 18th Floor, Taikoo Trading Estate, Tong Chong Street, Quarry Bay
Malaysia	Longman Malaysia Sdn Bhd No 3 Jalan Kilang A, Off Jalan Penchala, Petaling Jaya, Selangor, Malaysia
Singapore	Longman Singapore Publishers (Pte) Ltd 25 First Lok Yang Road, Singapore 2262
USA	Longman Group (USA) Inc 500 North Dearborn Street, Chicago, Illinois 60610

All rights reserved. No part of this publication may be reproduced, stored in a retrieval system, or transmitted, in any form or by any means, electronic, mechanical, photocopying, recording or otherwise, without either the prior written permission of the copyright holder for which application should be addressed in the first instance to the publishers or a licence permitting restricted copying issued by the Copyright Licensing Agency Ltd, 33–34 Alfred Place, London WC1E 7DP.

A CIP catalogue record for this book is available from the British Library.

Printed in Great Britain by Mackays of Chatham Ltd, Chatham.

For further information and enquiries please contact your local Longman office.

Europe, Latin America, Iran
Please contact our International
Sales Department
Longman House
Burnt Mill
Harlow
Essex CM20 2JE

Arab World
Longman Arab World Centre
Butros Bustani Street
Zokak el Blat
PO Box 11-945
Beirut
Lebanon

New Sphinx Publishing Co. Ltd.
3 Shawarby Street
Kasr el Nil
Cairo
Egypt

Librairie Sayegh
Salhie Street
PO Box 704
Damascus
Syria

Longman Arab World Centre
Al-Hajairi Building
Amir Mohammed Street
PO Box 6587
Amman
Jordan

Longman Arab World Centre
15th Street
PO Box 1391
Khartoum
Sudan

Cameroon
M A W Ngoumbah
BP 537
Limbe
Cameroon

Australia
Longman Cheshire Pty Ltd.
Longman Cheshire House
Kings Gardens
91-97 Coventry Street
South Melbourne
Victoria 3205

Botswana
Longman Botswana (Pty) Ltd.
PO Box 1083
Gaborone

Canada
James D Lang
Marketing Manager
Carswell Legal Publications
2330 Midland Avenue
Agincourt
Ontario
M1S 1P7

Ghana
Sedco Publishing Co. Ltd.
Sedco House
PO Box 2051
Tabon Street
North Ridge
Accra

Hong Kong
Longman Group (Far East) Ltd.
18th Floor Cornwall House
Taikoo Trading Estate
Tong Chong Street
Quarry Bay

India
Orient Longman Limited
5-9-41/1 Bashir Bagh
Hyderabad 500 029

UBS Publishers Distributors
5, Ansari Road
PO Box 7051
New Delhi 110 002

Japan
Longman Penguin Japan Co. Ltd.
Yamaguchi Building
2-12-9 Kanda Jimbocho
Chiyoda-ku
Tokyo 101

Kenya
Longman Kenya Ltd.
PO Box 18033
Funzi Road, Industrial Area
Nairobi

Lesotho
Longman Lesotho (Pty) Ltd.
PO Box 1174
Maseru, 100

Malawi
Dzuka Publishing Co. Ltd.
Blantyre Printing & Publishing
Co. Ltd. PMB 39
Blantyre

Malaysia
Longman Malaysia Sdn, Berhad
No. 3 Jalan Kilang A
Off Jalan Penchala
Petaling Jaya
Selangor

New Zealand
Longman Paul Ltd.
Private Bag
Takapuna
Auckland 9

Nigeria
Longman Nigeria Ltd.
52 Oba Akran Avenue
Private Mail Bag 21036
Ikeja
Lagos

Pakistan
Tahir M Lodhi
Regional Manager
Butterworths
7 Jahangir Street
Islamia Park
Poonch Road
Lahore

Singapore
Longman Singapore
Publishers Pte Ltd.
25 First Lok Yang Road
Off International Road
Jurong Town
Singapore 22

South Africa
Maskew Miller Longman (Pty) Ltd.
PO Box 396
Howard Drive
Pinelands 7405
Cape Town 8000

Swaziland
Longman Swaziland Ltd.
PO Box 2207
Manzini

Tanzania
Ben & Co. Ltd.
PO Box 3164
Dar-es-Salaam

USA
Longman Trade USA.
Caroline House Inc.
520 North Dearborn Street
Chicago
Illinois 60610

Transnational Publishers, Inc.
PO Box 7282
Ardsley-on-Hudson
NY 10503

West Indies
Longman Caribbean (Trinidad) Ltd.
Boundary Road
San Juan
Trinidad

Longman Jamaica Ltd.
PO Box 489
95 Newport Boulevard
Newport West
Kingston 10
Jamaica

Mr Louis A Forde
'Suncrest'
Sunrise Drive
Pine Gardens
St Michael
Barbados

Zimbabwe
Longman Zimbabwe (PVT) Ltd.
PO Box ST 125
Southerton
Harare

The content of this title has been fully approved and endorsed by a specially constituted Advisory Board, whose members are:

Professor Susan Dev, MSc, FCCA, ATII, Professor Emeritus at the London School of Economics and Political Science
Professor Roger Groves, B Com, MSc, Phd, FCA, Professor of Accounting at the Cardiff Business School
Hymie Lipman, BA, MSc (Econ), FCCA, FCA, Head of Accounting and Finance at the Kingston Business School
Anthea Rose, MA, Deputy Secretary of the Chartered Association of Certified Accountants

Contents

Page

Part One: Statistics

1 Basic probability
1.1	Introduction	1
1.2	Basic probability	1
1.2.1	Probability—what do we mean?	2
1.2.2	Properties of the probability measure	2
1.2.3	How to find the value of a probability	2
1.3	The probability of compound events	5
1.3.1	Probability trees	5
1.4	Rules of probability	6
1.4.1	The Addition Rule	6
1.4.2	Conditional probability	8
1.4.3	The Multiplication Rule	9
1.4.4	The probability rules for more than two events	10
1.5	Bayes' Rule	12
1.6	Expected values	15
1.7	Permutations and combinations	16
1.7.1	Using permutations and combinations to calculate probabilities	19
	Summary	19
	Exercises	20

2 Probability distributions 27
2.1	Introduction	27
2.2	Probability distributions for discrete random variables	27
2.2.1	Discrete random variable	27
2.2.2	Probability distribution of a discrete random variable	28
2.2.3	Graphical representation of a discrete probability distribution	29
2.2.4	Expected value and standard deviation of a probability distribution	30
2.3	The binomial probability distribution	32
2.3.1	What is the binomial probability distribution?	32
2.3.2	The binomial probability distribution	34
2.3.3	The expected value and standard deviation for a binomial probability distribution	36
2.4	The Poisson probability distribution	38
2.4.1	What is the Poisson probability distribution?	38
2.4.2	The expected value and variance of a Poisson probability distribution	41
2.5	The Poisson distribution as an approximation to the binomial distribution	41
2.6	Probability distributions for continuous random variables	43
2.6.1	Continuous variables and probability density	43
2.6.2	The uniform distribution	45
2.7	The normal probability distribution for continuous random variables	46
2.7.1	The nature of the normal probability distribution	46
2.7.2	The standard normal probability distribution	47
2.8	Normal distribution as an approximation to the binomial distribution	51
2.9	The normal distribution as an approximation for the Poisson distribution	54
2.10	The combination of variables	55
2.10.1	Independent variables	55
2.10.2	A special case of non-independent variables	56

2.10.3	The nature of the distribution of combined variables	56
	Summary	57
	Exercises	58

3 Decision making under uncertainty 67
3.1	Introduction	67
3.2	Decision rules	68
3.2.1	Decision rules which do not use probability values	68
3.2.2	Decision rules which do use probability values	71
3.2.3	The sensitivity of the decisions to changes in the probabilities	73
3.2.4	The value of perfect information	73
3.3	Expected values and the use of standard deviations to assess risk	74
3.4	Use of utility to incorporate the decision maker's assessment of risk	76
3.4.1	The use of utility values	77
3.5	Decision trees	79
3.5.1	Evaluation of a two stage decision tree	80
3.5.2	Sensitivity analysis and decision trees	87
	Summary	89
	Exercises	90

4 Sampling and sampling distributions 95
4.1	Introduction	95
4.2	Populations and samples: why sample?	95
4.3	The selection of a sample	96
4.3.1	Random sampling	96
4.3.2	Random sample designs	97
4.3.3	Non-random sample designs	98
4.4	Sampling distributions	99
4.4.1	Sampling distribution of sample means	101
4.4.2	Sampling distribution of sample variance	103
4.4.3	Estimate of the standard error of the sampling distribution of sample means	106
4.4.4	Standard sampling distributions—z, t, χ^2, F	107
	Summary	109
	Exercises	111

5 Statistical inference I: estimation—confidence intervals 115
5.1	Introduction	115
5.2	Confidence interval for μ, the population mean	115
5.2.1	When the population variance σ^2 is known	115
5.2.2	The population variance is unknown	117
5.3	Confidence interval for a population proportion, p	119
5.4	Confidence interval for a population variance, σ^2	121
5.5	The choice of appropriate sizes for samples	123
5.5.1	The sample size required to estimate a population mean	123
5.5.2	The sample size required to estimate a population proportion	125
5.5.3	The sample size required to estimate a population variance	125
	Summary	126
	Exercises	127

6 Statistical inference II: hypothesis testing 131
6.1	Introduction	131
6.2	The procedure for any hypothesis test	131
6.2.1	Standard hypothesis tests	133
6.2.2	One and two tailed tests	134
6.3	Hypothesis test on a sample mean—population variance known	135
6.4	Hypothesis test on a sample mean—population variance unknown	140
6.5	Hypothesis test on a sample proportion	142
6.6	Hypothesis test on two population variances	144

6.6.1	Variance ratio, or F, test	144
6.7	Hypothesis test on two sample means—population variance known	147
6.8	Hypothesis test on two sample means—population variances unknown	149
6.9	Hypothesis test on two sample proportions	155
6.10	Hypothesis test on paired data—dependent samples	157
6.11	Non-parametric tests of hypothesis—the chi-squared test	158
6.11.1	To test the association of attributes	158
6.11.2	To test the goodness of fit of an observed frequency distribution to the expected frequency distribution of a discrete variable	164
6.11.3	To test the goodness of fit of an observed frequency distribution to the expected frequency distribution of a continuous variable	168
6.11.4	Does the observed frequency distribution fit the expected frequency distribution too well?	171
	Summary	172
	Exercises	174

7 Linear regression 181

7.1	Introduction	181
7.2	Simple linear regression model	182
7.3	Strength of the linear relationship—the correlation coefficient,	186
7.4	Prediction and estimation using the linear regression model	189
7.4.1	Predictions within the range of the sample data	189
7.4.2	Estimation, error and residuals	189
7.5	Statistical inference in linear regression analysis	192
7.5.1	The underlying assumptions	192
7.5.2	Hypothesis tests to assess the overall linearity of the relationship	194
7.5.3	Confidence intervals in linear regression analysis	197
7.6	Multiple linear regression models	199
7.7	Non-linear relationships	207
7.8	Spearman's rank correlation—r_s	212
	Summary	214
	Exercises	215

8 Time series and forecasting 223

8.1	Introduction	223
8.2	Time series components	223
8.3	The analysis of an additive component model: $A = T + S + E$	226
8.3.1	Calculate the seasonal components for the additive model	226
8.3.2	Deseasonalise the data to find the trend	228
8.3.3	Calculate the errors	229
8.3.4	Forecasting using the additive model	230
8.4	The analysis of a multiplicative component model: $A = T \times S \times E$	230
8.4.1	Calculation of the seasonal components	232
8.4.2	Deseasonalise the data and fit the trend line	233
8.4.3	Calculation of the errors: $A/(T \times S) = E$ or $A - (T \times S) = E$	234
8.4.4	Forecasting with the multiplicative component model	234
8.5	Forecasting using simple exponential smoothing	235
8.5.1	Exponential smoothing of a time series	236
8.5.2	Exponential smoothing with a step change in the time series	237
	Summary	239
	Exercises	240

Part Two: Operational research models

9 Linear programming 247

9.1	Introduction	247
9.2	Problem formulation	247

9.3	Solving the linear programme	252
9.3.1	The graphical solution of a linear programme	253
9.4	Sensitivity analysis	262
9.4.1	How do changes in the provision of a limiting resource affect the solution of a linear programme?	263
9.4.2	How do changes in the non-limiting resources affect the optimum solution?	265
9.4.3	How do changes in the coefficients of the objective function affect the optimum solution?	265
9.5	The simplex solution of multi-variable linear programmes	269
9.6	Sensitivity analysis and the simplex method	275
9.7	The dual linear programming model	279
	Summary	283
	Exercises	283

10 Transportation and assignment models 291

10.1	Introduction	291
10.2	Transportation problem and algorithm	291
10.2.1	The transportation problem	291
10.2.2	The transportation algorithm	293
10.2.3	Finding an initial allocation	294
10.2.4	Testing for optimality	297
10.2.5	Finding the optimum solution	303
10.2.6	Sensitivity analysis	305
10.2.7	Variations in the transportation problem	307
10.3	The assignment problem	311
10.3.1	The assignment algorithm	311
10.3.2	Special cases of the assignment problem	315
	Summary	316
	Exercises	317

11 Network analysis and project scheduling 325

11.1	Introduction	325
11.2	Network diagrams	325
11.2.1	Arrow diagrams	325
11.2.2	Activity-on-node diagrams	328
11.3	Critical path analysis	329
11.3.1	Critical path analysis using activity-on-node diagrams	330
11.3.2	Critical path analysis using an arrow diagram	332
11.4	The costs of a project	335
11.4.1	To minimise the overall project duration with minimum additional cost	335
11.4.2	Completing a project at minimum cost	338
11.5	Uncertainty in activity times	339
11.6	Resource allocation	342
11.6.1	Resource profiles	343
	Summary	345
	Exercises	346

12 Inventory planning and control 353

12.1	Introduction	353
12.2	The basic stock model	353
12.2.1	Assumptions for the basic model	353
12.2.2	The costs of stocking goods	354
12.2.3	The total cost equation	355
12.2.4	The optimum order quantity, q_0	356
12.2.5	The re-order level and re-order interval	357
12.2.6	The economic batch quantity model	358
12.3	Quantity discounts	360
12.4	Further stock models	363

12.4.1	The batch production model	363
12.4.2	The planned shortages model	367
12.5	Uncertainty and the basic stock model	370
12.5.1	A re-order level system	371
12.5.2	A re-order cycle system	377
12.6	Other issues in the control of stock	379
	Summary	380
	Exercises	380

13 Queueing models 387

13.1	Introduction	387
13.2	Components of the queuing system	387
13.2.1	Arrivals and arrival patterns	387
13.2.2	Queue discipline	388
13.2.3	Servers, service time and idleness	389
13.2.4	Criteria for judging a 'good' queue	389
13.3	The Poisson process and the M/M/1 queue	389
13.3.1	The Poisson process	390
13.3.2	The M/M/1 queueing system	392
13.4	Extensions to the M/M/1 model	394
13.4.1	The M/M/2 model	394
13.4.2	Checking the basic Poisson assumptions	395
13.4.3	The costs of a queueing system	396
	Summary	397
	Exercises	397

14 Simulation 403

14.1	Introduction	403
14.2	The principles of the discrete simulation model	403
14.3	The application of a simulation model to a queueing problem	407
14.4	The application of a simulation model to a stock control problem	412
	Summary	414
	Exercises	414

Solutions to exercises	419
Additional exercises	625
December 1988 ACCA Quantitative Analysis Professional Examination	639
Authors' model answers to December 1988 Examination	647
Glossary	663
Appendix One: Mathematical formulae	675
Appendix Two: Statistical tables	683
Index	695

Part One: Statistics

PART ONE

Statistics

The first part of this book is concerned with some aspects of statistics. In the following chapters we discuss basic probability, the binomial, Poisson and normal probability distributions, decision making under uncertainty, sampling, estimation, hypothesis testing, linear regression and time series analysis.

The common threads running through all statistics are variability and uncertainty. It is these factors which make an understanding of statistics important for business and accounting. Product costs, for example, vary, and not simply because we cannot measure them accurately. The quality of the raw material may change from week to week, the production rate may vary or a breakdown occur. All such factors will contribute to the variability of the product cost. For similar reasons, we are uncertain what the demand for a product will be next month or next year.

We hope that the following chapters will help you to understand the importance of statistics in the management of variability and uncertainty in accounting and business generally. The processes and techniques described should enable you to deal with some practical situations immediately, and will lay the foundations for a deeper study of the subject.

Basic probability

1.1 INTRODUCTION

One of the intentions of this book is to help you to make better business decisions. The interest, and the difficulty, of decision making is increased when we do not know precisely what will happen when we have made the decision, that is, there is uncertainty about the outcome. We may know what could happen—what the possible outcomes are—but we are uncertain which of these outcomes will actually occur. **Probability theory** offers a framework for dealing with this uncertainty. That is why it is important for us to understand it.

Students can find probability difficult to grasp initially. There are a number of new concepts, rules and words to understand. The way forward can be smoothed if you begin by concentrating on the practical situation. For a given problem, consider what decisions could possibly be made. For each decision, list all the events that could possibly happen. Sometimes these activities are easy, sometimes they are not, but it is often this first stage in the decision making process that is the most important and the most useful. Once you have a clear statement about the possible outcomes, then with a little practice, probability calculations will become straightforward.

The next section of the chapter is a brief summary of the basic ideas of probability. These ideas are extended to more complex situations in the sections that follow.

1.2 BASIC PROBABILITY

The origins of probability theory lie in the mathematical theory of games. In the mid-seventeenth century, the French mathematicians Pascal and Fermat developed a mathematical model describing the likelihood of certain outcomes in various games of chance, having been prompted by enquiries from a number of prominent gamblers. When tossing dice, spinning a roulette wheel, taking measurements (physical, economic, sociological, etc), drawing a sample from the output of a machine, and so on, the actual results vary from trial to trial, even though the conditions remain the same. The results cannot be predicted exactly and are said to be subject to uncertainty or chance.

Business people make decisions in the face of uncertainty. When a company launches a new product, for example, the marketing manager will be uncertain of the eventual sales potential of that product. When a gambler places a bet, he will be uncertain whether he will win. It is the uncertainty which is common to these two situations. Probability theory is concerned with this uncertainty. A study of probability theory, based on games of chance, provides a valuable analytical tool for measuring and controlling the various forms of uncertainty with which the businessman is faced.

First of all we must explain the terminology which we will be using:

Trial—an activity, the result of which is unknown until the trial has taken place, eg toss a coin; roll a die
Experiment—one or more trials make up an experiment, eg toss a coin 6 times
Outcome—a possible result of an experiment, eg toss a coin 6 times and obtain (head head head tail tail head)
Event—one or more outcomes of an experiment, eg toss a coin 6 times and obtain 4 heads and 2 tails.

Although there may be uncertainty as to the outcome of a single experiment, a sequence of identical experiments often reveals a pattern in the outcomes. This pattern is frequently of use when investigating and analysing a situation in business.

1.2.1 Probability—what do we mean?

We can illustrate the basic idea of probability by the simple experiment of tossing a tack. There are two possible ways in which the tack may land—point up or point down. What is the probability that it lands point down? We will record what happens when we toss the tack ten times. We will then increase the number of tosses, in stages, to 100, 1000, and so on. A possible set of results might be:

TABLE 1.1 **A possible set of results**

Number of tosses	Number point down	Number point up	Proportion point down	Proportion point up
10	6	4	0.6	0.4
100	64	36	0.64	0.36
1000	643	357	0.643	0.357
10000	6431	3569	0.6431	0.3569

As the number of times the tack is tossed increases, the proportion of occasions on which the tack lands point down tends towards a limiting value. In this example the limiting value is 0.643 to 3 decimal places.

On the basis of the figures in TABLE 1.1, if the tack were tossed in the same way for a 10001st time, we can say beforehand that it is more likely to land point down than point up, and we can also measure this likelihood in terms of the previous proportion of 'point down' results, ie we can use the figure 0.643 as a measure of the likelihood that on the 10001st toss of the tack, it will land point down.

In an experiment, if we define the term 'success' as the result we are interested in, then we can write down a simple definition of the term **probability**.

Probability is the 'long-run' proportion of successes obtained when an experiment of chance is carried out identically a large number of times, ie probability measures the likelihood of obtaining a success on the next try of the experiment.

1.2.2 Properties of the probability measure

Since probability is a proportion, two important conclusions follow. If we call the probability of an outcome from an experiment p, then,

1 Individual probabilities must lie between 0 and 1 inclusive, ie:

$$0 \leqslant p \leqslant 1$$

p cannot be negative or greater than 1.

2 The sum of the probabilities of all of the outcomes of an experiment must add up to 1, the probability that something must happen, ie $\sum p = 1$.

Similarly if we call the probability of an event E occurring P(E), then:

$$0 \leqslant P(E) \leqslant 1$$

1.2.3 How to find the value of a probability

Now that we know what the term 'probability' measures, we can consider how to assign numerical values to it in particular situations. There are three ways in which this may be done.

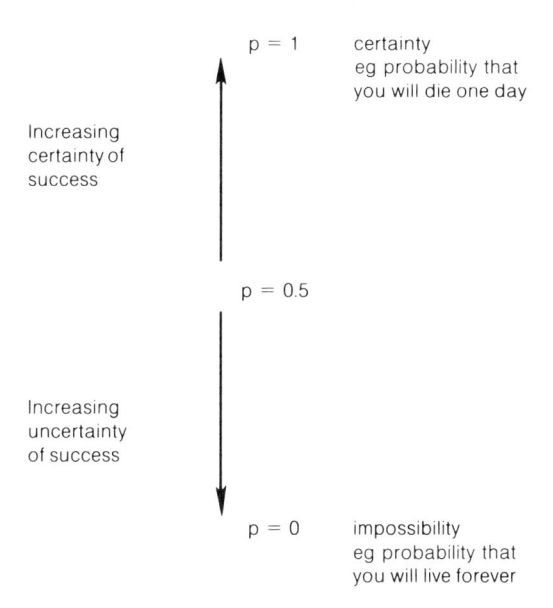

FIGURE 1.1 **The Probability measure**

```
                              p = 1      certainty
                               ↑         eg probability that
                               |         you will die one day
            Increasing         |
            certainty of       |
            success            |
                               |
                               |
                              p = 0.5
                               |
                               |
            Increasing         |
            uncertainty        |
            of success         |
                               ↓
                              p = 0      impossibility
                                         eg probability that
                                         you will live forever
```

PROBABILITY EVALUATED FROM SYMMETRY

There are some situations in which we assign equal probabilities to all possible outcomes because of the symmetry of the problem, eg toss a fair coin once. If the coin is fair, there is no reason why the coin should land heads rather than tails. The two outcomes are equally likely, ie **P(head) = P(tail)** and, since there are only two possible outcomes **P(head) + P(tail) = 1** therefore, **P(head) = P(tail) = 0.5**.

In those trials in which all of the outcomes have the same chance of occurrence, the probability of an event, E, P(E) is:

$$P(E) = \frac{\text{number of equally likely outcomes which comprise E}}{\text{total number of possible outcomes}}$$

Example 1.1: Probabilities derived from symmetry

If a coin is tossed three times, what is the probability of obtaining two tails and one head?
Solution: First we write down all the possible outcomes. These are (hhh, hht, hth, thh, htt, tht, tth, ttt).
Note: To ensure that you write down all of the outcomes, you may find it helpful to use a tree diagram. See Chapter One section **1.3.1**.

There are 8 equally likely outcomes, therefore each has a probability of 1/8 of occurring. The event, E, of two tails and one head, is given by 3 of these outcomes (htt, tht, tth).
Therefore:

$$P(E) = \frac{\text{number of outcomes which give E}}{\text{total number of outcomes}} = \frac{3}{8}$$

Example 1.2: Probabilities derived from symmetry

A fair die is rolled twice. What is the probability that the total score on the two rolls is nine or more?
Solution: Again we begin by writing down all the possible outcomes in a table which contains the total score of the two rolls.

TABLE 1.2 **Total score from two rolls of a fair die**

		|	*Score from first roll*				
		|	1	2	3	4	5	6
	1	|	2	3	4	5	6	7
Score	2	|	3	4	5	6	7	8
from	3	|	4	5	6	7	8	**9**
second	4	|	5	6	7	8	**9**	**10**
roll	5	|	6	7	8	**9**	**10**	**11**
	6	|	7	8	**9**	**10**	**11**	**12**

There are 36 possible outcomes, of which 10 give a total score of 9 or more, therefore:

$$P \text{ (score of 9 or more on two rolls of a die)} = \frac{10}{36}$$

PROBABILITY EVALUATED BY MEASUREMENT—EMPIRICAL

This is the method we used to introduce the meaning of the term probability. The example on page 2 in which a tack is tossed and the number of times it lands point down is recorded, illustrates this method. In general, if a trial is repeated n times and m of these produce a successful outcome, then the probability of a successful outcome is:

$$P(\text{success}) = \frac{m}{n} \quad \text{as long as n is very large}$$

m/n is the relative frequency of occurrence of the particular outcome, in the long-run. These probability values arise from repeated 'experimentation' or from the use of recorded information.

Example 1.3: Repeated experiment to evaluate a probability

A manufacturer tests 500 light bulbs to destruction and finds that 415 last longer than 1000 hours. On the basis of this experiment, he could claim that the probability of any light bulb of this type lasting longer than 1000 hours is:

$$\frac{415}{500} = 0.83$$

Note: The test is destructive so every light bulb cannot be tested. On the other hand, if the manufacturer tested only one light bulb, the probability derived for its lasting more than 1000 hours would be 1 or 0 (it either did or did not), hence the need for many repeated tests.

Example 1.4: The use of recorded information to evaluate a probability

The personnel manager of a packing company extracts the following information from the files concerning the length of service of male employees:

TABLE 1.3 **Lengths of service of male employees**

Length of service, years	*Number of employees*
0 but less than 1	26
1 but less than 2	36
2 but less than 3	16 ⎫
3 but less than 4	20 ⎬ = 38
4 but less than 5	2 ⎭
over 5	0
Total	100

What is the probability that the next male employee to join the company will stay at least two years?
Solution: It can be seen from the table that 38 out of the 100 employees have stayed with the company for two years or more. Therefore, we can say that the empirical probability of the next employee staying for at least two years is:

$$\frac{38}{100} = 0.38$$

This assumes that the next employee to join is 'typical' and that all conditions remain unchanged.

PROBABILITY EVALUATED FROM SUBJECTIVE ESTIMATES

In business, situations frequently arise in which there is neither symmetry nor empirical data. The probability assigned to a particular outcome has to be derived subjectively. The person assigning the probability value uses his or her own judgment and experience in evaluating the likelihood of a particular outcome from the trial.

Example 1.5: The use of subjective probability

The following are examples of the use of subjective probability:

1   An investment analyst judges that a particular investment has a probability of 0.6 of returning a profit within two years.
2   A marketing manager estimates that a new product has a 0.4 chance of 1000 sales in the first month after launch.

1.3 THE PROBABILITY OF COMPOUND EVENTS

Compound events arise when we consider a series of trials and the possible combinations of outcomes. For compound events it is very important that all of the possible outcomes are considered. To do this it is helpful to illustrate the sequence of trials and the resulting outcomes in a diagram. One of the most useful diagrams is the **probability tree**.

1.3.1 Probability trees

The trials are illustrated sequentially by nodes (circles), and each outcome is represented by a 'branch' (line) radiating from the appropriate node. The probability of each outcome is written along the 'branch' and the compound probability of a particular sequence of outcomes is written at the end of each route through the tree.

Example 1.6: To illustrate the use of a probability tree diagram

A die is rolled three times. What are the probabilities of obtaining 0, 1, 2, or 3 fours from the three rolls?
Solution: There are three trials, one after the other, which makes up the experiment. For each trial, there are two outcomes—either a 4 is rolled (4 on the diagram) or a 4 is not rolled (4* on the diagram).
To evaluate the probabilities at the end of each route through the tree, and to determine the probabilities of 0, 1, 2 and 3 fours, we need to know how to calculate the probabilities of compound events. We will look at this in the following section and return to this example later.

QUANTITATIVE ANALYSIS

FIGURE 1.2 **Outcomes when a die is rolled 3 times**

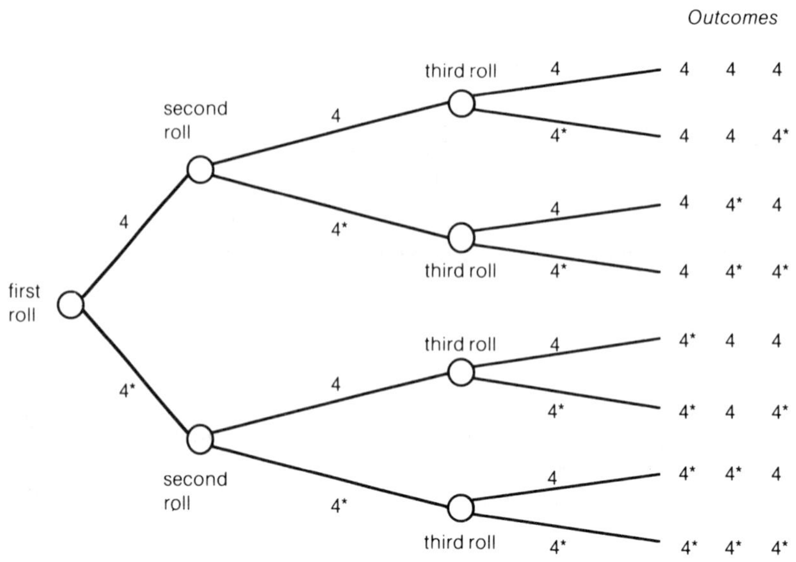

(4 denotes rolling a four; 4* denotes rolling any other score.)

1.4. RULES OF PROBABILITY

Firstly, let us explain the terminology which will be used:

Independent Events A and B are said to be independent if the occurence of event A does not affect the chance of the occurrence of event B and vice versa. For example: a coin is tossed once and a die is rolled once. A is the event of a head on the coin and B is the event of a 6 on the die. Since the outcome on the coin does not influence in any way the outcome on the die, A and B are said to be independent of each other.

Mutually exclusive Events A and B are said to be mutually exclusive if either one or the other may occur but not both. For example: a die is tossed once. A is the event of an even number and B is the event of an odd number. A and B cannot both occur if the die is tossed once only, hence, A and B are mutually exclusive.

There are two basic rules for determining the probability of compound events, the addition rule and the multiplication rule.

1.4.1. The Addition Rule

To illustrate the rule as simply as possible, we will consider only two events; call these events A and B. The Addition Rule is used when we require the probability of A or B or both occurring.

The Addition Rule states:

$$P(A \text{ or } B \text{ or both}) = P(A) + P(B) - P(A \text{ and } B)$$

If A and B are mutually exclusive, the statement reduces to:

$$P(A \text{ or } B) = P(A) + P(B)$$

since mutually exclusive means that events A and B cannot happen together, therefore:

$$P(A \text{ and } B) = 0$$

Example 1.7: To illustrate the Addition Rule

A fair die is rolled once. What is the probability of obtaining a score of 2 or a score which is odd?
Solution: When we roll a die, there are 6 possible outcomes, 1, 2, 3, 4, 5, or 6. Let us call event A a score of 2, and event B an odd score, 1, 3, or 5.

In this example we can find the solution by using either the symmetry of the trial or by using the Addition Rule.

1 Using the symmetry of the trial:

$$P(\text{score of 2 or an odd number}) = \frac{\text{number of successful outcomes (ie 2 or 1, 3, 5)}}{\text{total number of outcomes}}$$
$$= \frac{4}{6}$$

2 Using the Addition Rule:
The two events cannot happen together. They are mutually exclusive, therefore:

$$P(A \text{ and } B) = 0,$$

hence:

$$P(A \text{ or } B) = P(A) + P(B)$$
$$= \frac{1}{6} + \frac{3}{6} = \frac{4}{6}$$

Example 1.8: To illustrate the use of the Addition Rule

We will use the same experiment of rolling the die once. What is the probability of obtaining a score of 2 or any score which is even?
Solution: In this case the two events can happen together. They are not mutually exclusive.

1 Using the symmetry of the trial:
There are 3 outcomes that give us what we want—2, 4 or 6, therefore the probability of obtaining a 2 or any even score is 3/6.
2 Using the Addition Rule gives:

$$P(A \text{ or } B) = P(A) + P(B) - P(A \text{ and } B)$$

since A and B are not mutually exclusive:

$$= 1/6 + 3/6 - 1/6 = 3/6$$

Example 1.9: To illustrate the use of the Addition Rule

A manufactured article can contain any number of flaws—0, 1, 2, 3, 4, etc. From company records the probability that the article has no flaw is 0.9 and the probability that it has one flaw is 0.05. What is the probability that an article has at most one flaw?
Solution: P(at most one flaw) = P(no flaws or one flaw). Again the two events are mutually exclusive, therefore:

$$P(\text{no flaws or one flaw}) = P(\text{no flaws}) + P(\text{one flaw})$$
$$= 0.9 \quad\quad + 0.05$$
$$= 0.95$$

QUANTITATIVE ANALYSIS

Example 1.10: Illustration of the Addition Rule and Venn diagrams

A weather forecaster has estimated the following probabilities for tomorrow:

$$P(\text{Rain}) = 0.4, \ P(\text{Wind}) = 0.7, \ P(\text{Rain and Wind}) = 0.2$$

What is the probability that it will be raining or windy tomorrow?
Solution: Using the Addition Rule:

$$\begin{aligned} P(\text{Rain or Wind or Both}) &= P(\text{Rain}) + P(\text{Wind}) - P(\text{Rain and Wind}) \\ &= 0.4 \quad\quad + 0.7 \quad\quad - 0.2 \\ &= 0.9 \end{aligned}$$

We have used various methods of illustrating the possible outcomes for any particular experiment. We have used lists, tables and tree diagrams. Venn diagrams are another method of presenting a clear picture of what can happen. A rectangle, like the frame of a picture, encloses all the possible outcomes. The events we are interested in are drawn onto the picture as areas, usually circles. In this example one area would represent rain and another wind. The overlap between these two represents rain **and** wind. The rest of the area represents the outcomes when there is no wind or rain. The probabilities are written in the appropriate areas of the diagram. The current example is represented in FIGURE 1.3 below.

FIGURE 1.3 **Venn diagram of tomorrow's forecast weather**

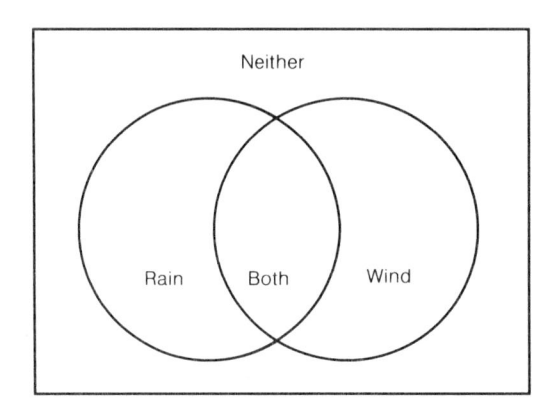

1.4.2 Conditional probability

We will consider two events, E and F, which can occur one after the other, and suppose that E occurs first with probability P(E). Two alternative situations can now arise:

1. F is **independent** of E, ie the probability with which F occurs is not affected by whether or not E has already occurred.
2. E and F are **not independent**, ie the probability with which F occurs depends on whether E has already occurred or not. In this case the probability of event F is said to be **conditional** on the occurrence of event E.

The probability that F will occur, given that E has already occurred, is denoted by:

$$P(F \text{ given } E) \text{ or } P(F|E)$$

If E and F are independent, then:

$$P(F \text{ given } E) = P(F)$$

Example 1.11: Illustration of conditional probability

A box contains 8 red balls and 6 blue balls. If 2 balls are selected at random from the box, what is the probability that the second ball chosen is red?
Solution: There are two possible cases here:

1   the first ball chosen is red, leaving 7 red and 6 blue balls in the box, and
2   the first ball chosen is not red, leaving 8 red and 5 blue.

In case **1**, the probability that the second ball is red is 7/13.
In case **2**, the probability that the second ball is red is 8/13.

1.4.3 The Multiplication Rule

The Multiplication Rule applies when we want to know the probability of events A **and** B occurring at the same time. The Multiplication Rule is:

$$P(A \text{ and } B) = P(A) \times P(B \text{ given } A)$$

If A and B are **independent**, then $P(B \text{ given } A) = P(B)$, and the Multiplication Rule reduces to:

$$P(A \text{ and } B) = P(A) \times P(B)$$

Example 1.12: An illustration of the Multiplication Rule

The experiment is to roll a fair die twice. Event A is that the first roll gives a score of 2, and event B is that the second roll gives an odd score. What is the probability of events A and B occurring?
Solution: Using the Multiplication Rule, the probability of the second roll giving an odd score is not affected by the result of the first throw, hence the two events are independent. Therefore:

$$P(A \text{ and } B) = P(A) \times P(B)$$
$$= \frac{1}{6} \times \frac{3}{6} = \frac{3}{36}$$

A tree diagram can be helpful in answering this type of question. The first step is to illustrate the sequence of outcomes.

FIGURE 1.4   **The sequence of outcomes**

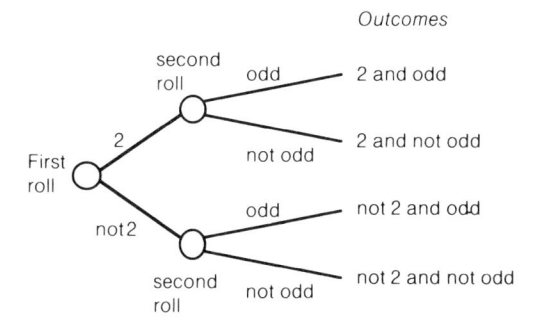

The second step is to add the probabilities to the branches, and the third step is to evaluate the probability of each of the possible sequences of outcomes by multiplying together the probabilities along consecutive branches.

10 QUANTITATIVE ANALYSIS

FIGURE 1.5 **The probability of possible outcomes**

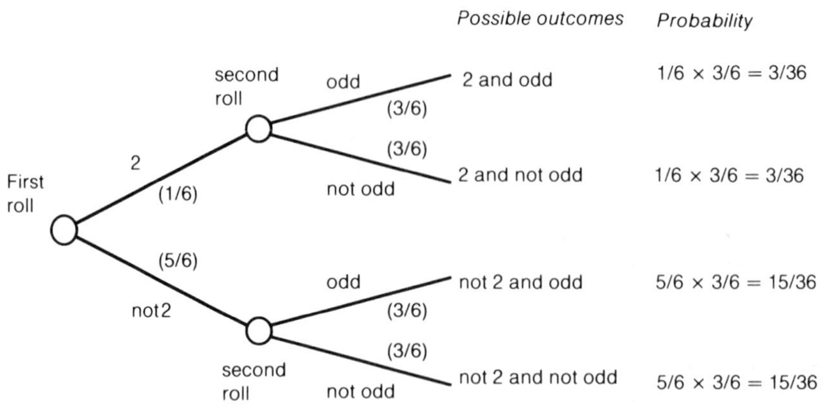

```
                                          Possible outcomes    Probability

                    second    odd       
                    roll     (3/6)      → 2 and odd            1/6 × 3/6 = 3/36
              2    ○
          (1/6)              (3/6)
First                        not odd    → 2 and not odd        1/6 × 3/6 = 3/36
roll  ○
          (5/6)              odd
                             (3/6)      → not 2 and odd        5/6 × 3/6 = 15/36
           not2    ○
                   second    (3/6)
                   roll      not odd    → not 2 and not odd    5/6 × 3/6 = 15/36
```

The top branch gives the solution we require in this example, which is 3/36, exactly as before.

Example 1.13: An illustration of the Multiplication Rule

Suppose a despatch department has 8 large orders to deal with, 5 of which are for the home market and the remainder for export. If 2 of the orders are randomly assigned to a particular group of packers, what is the probability that both of these orders are for the home market?
Solution: Event A: the first order assigned to the packers is for the home market. Event B: the second order assigned is for the home market. We require P(A and B). Using the Multiplication Rule:

$$P(A \text{ and } B) = P(A) \times P(B \text{ given } A)$$

$$= \frac{5}{8} \times \frac{4}{7} = \frac{20}{56}$$

1.4.4 The probability rules for more than two events

The Addition Rule and the Multiplication Rule also apply to situations in which there are more than two events of interest. The Addition Rule, when the events are mutually exclusive, becomes:

$$P(A \text{ or } B \text{ or } C \text{ or } \ldots) = P(A) + P(B) + P(C) + \ldots$$

The Addition Rule cannot be easily generalised if the events are not mutually exclusive.
The Multiplication Rule, when the events are independent, becomes:

$$P(A \text{ and } B \text{ and } C \text{ and } \ldots) = P(A) \times P(B) \times P(C) \times \ldots$$

When the events are not independent, it becomes:

$$P(A \text{ and } B \text{ and } C \text{ and } \ldots) = P(A) \times P(B \text{ given } A) \times P(C \text{ given } A \text{ and } B) \times \ldots$$

Example 1.14: An illustration of the Multiplication Rule and the use of a tree diagram

The Board of Directors of a company consists of 3 accountants, 3 managers and 2 engineers. A planning sub-committee of 3 is chosen at random from the Board. Find the probability that all 3 members of the sub-committee are accountants.
Solution: We can use a tree diagram to illustrate the possible outcomes. To simplify the diagram, A is used when an accountant is selected and A* is used when a production manager or an engineer

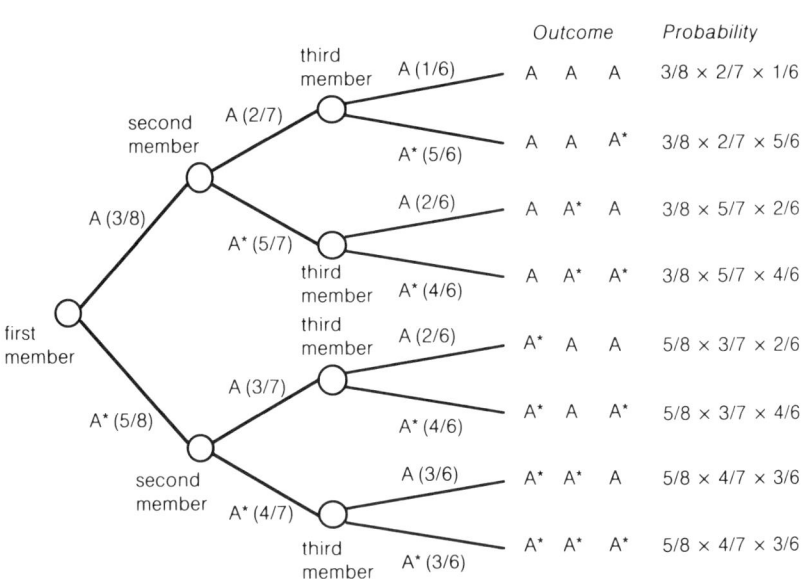

FIGURE 1.6 **Tree diagram for the selection of a sub-committee of 3**

(A = Accountant selected); (A* = Other manager selected)

		Outcome	Probability
third member A (1/6)	A A A		3/8 × 2/7 × 1/6
A (2/7) / A* (5/6)	A A A*		3/8 × 2/7 × 5/6
A (2/6)	A A* A		3/8 × 5/7 × 2/6
A* (5/7) third member A* (4/6)	A A* A*		3/8 × 5/7 × 4/6
third member A (2/6)	A* A A		5/8 × 3/7 × 2/6
A (3/7) A* (4/6)	A* A A*		5/8 × 3/7 × 4/6
A (3/6)	A* A* A		5/8 × 4/7 × 3/6
A* (4/7) third member A* (3/6)	A* A* A*		5/8 × 4/7 × 3/6

is picked. When the tree is complete, we can put on the probabilities. The probabilities change as we go along each branch of the tree because the number of people left unselected reduces from 8 to 7 to 6. Similarly if we have already selected an accountant then the number of accountants has been reduced from 3 to 2. The probabilities are conditional on what has already happened. FIGURE 1.6 shows the selection of a sub-committee of 3.

From the tree diagram the probability that all 3 members of the sub-committee are accountants is (3/8) × (2/7) × (1/6) = 1/56. It is not essential to use a tree diagram to solve this problem; we could use the multiplication rule without the diagram.

P(all members are accountants)
= P(1st chosen is an accountant **and**
2nd chosen is an accountant **and**
3rd chosen is an accountant)
= P(1st chosen is an accountant)
× P(2nd chosen is an accountant, given 1st chosen is an accountant)
× P(3rd chosen is an accountant, given 1st and 2nd chosen are accountants)
= (3/8) × (2/7) × (1/6) = 1/56

Example 1.15: An illustration of the Multiplication Rule

A piece of equipment will work only if three components A, B, C are all working. A, B and C function independently of each other. The probabilities that the components will fail during one year are 0.2, 0.3 and 0.1 respectively. What is the probability that the equipment will fail?
Solution: The tree diagram shows what can happen to each of the three components. The equipment will work only if all three of the components are working. It will fail for any of the other outcomes. For each component the probabilites are:

P(A fails) = 0.2   therefore, P(A does not fail) = 0.8
P(B fails) = 0.3   therefore, P(B does not fail) = 0.7
P(C fails) = 0.1   therefore, P(C does not fail) = 0.9

From the diagram the probability that the component works to the end of the year is:

$$P(A \text{ works and } B \text{ works and } C \text{ works}) = P(A \text{ works}) \times P(B \text{ works}) \times P(C \text{ works})$$
$$= 0.8 \times 0.7 \times 0.9$$
$$= 0.504$$

However, we require the probability that the equipment fails before the end of the year. This probability is the sum of all of the other seven branches of the tree, or, since the total probability is always 1:

$$P(\text{equipment fails}) = 1 - P(\text{equipment does not fail})$$
$$= 1 - 0.504 = 0.496$$

FIGURE 1.7 **Probability of equipment failure**

(F = component failure); (W = component works)

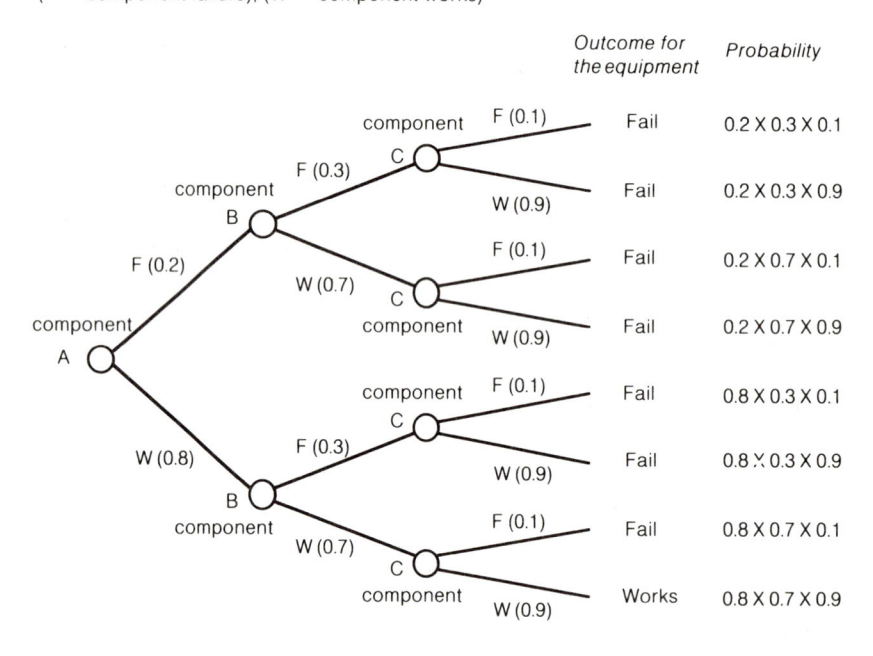

	Outcome for the equipment	Probability
F(0.1)	Fail	0.2 X 0.3 X 0.1
W(0.9)	Fail	0.2 X 0.3 X 0.9
F(0.1)	Fail	0.2 X 0.7 X 0.1
W(0.9)	Fail	0.2 X 0.7 X 0.9
F(0.1)	Fail	0.8 X 0.3 X 0.1
W(0.9)	Fail	0.8 X 0.3 X 0.9
F(0.1)	Fail	0.8 X 0.7 X 0.1
W(0.9)	Works	0.8 X 0.7 X 0.9

1.5 BAYES' RULE

Bayes' Rule is based on the Multiplication Rule and it allows us to revise our estimates of probabilities as we collect more and more information about a particular decision. The Multiplication Rule can be written in two ways:

$$P(A \text{ and } B) = P(A) \times P(B \text{ given } A)$$
$$\text{or } P(A \text{ and } B) = P(B) \times P(A \text{ given } B)$$

If we take $P(A \text{ and } B) = P(B) \times P(A \text{ given } B)$, we may rewrite it as:

$$P(A \text{ given } B) = \frac{P(A \text{ and } B)}{P(B)}$$

This is Bayes' Rule.

BASIC PROBABILITY: CHAPTER ONE

P(A) is our first estimate that the event A will occur when we carry out our experiment. We collect some information before doing the experiment. The information shows that event B has happened. We can now revise our estimate of P(A), replacing it with P(A given that B has occured). The following examples illustrate this process.

Example 1.16: An illustration of Bayes' Rule

A firm is supplied with components from three sources A, B and C. Fifty per cent come from A, 30% come from B and 20% come from C. It is found from experience that 10% of the components supplied by A are defective, 5% of those supplied by B are defective and 6% of those supplied by C are defective.

1. If a component is picked at random, what is the probability that it came from A?
2. If, however, a component picked at random, is first examined and is found to be defective, what then is the probability that it came from A?

Solution:

1. Fifty per cent of the components are supplied by firm A, therefore:

$$P(\text{component supplied by A}) = 50\%, \text{ or } P(A) = 0.5$$

Hence, if a component is picked at random, the probability that it came from A is 0.5.

2. The probability tree is:

FIGURE 1.8 **Probability tree**

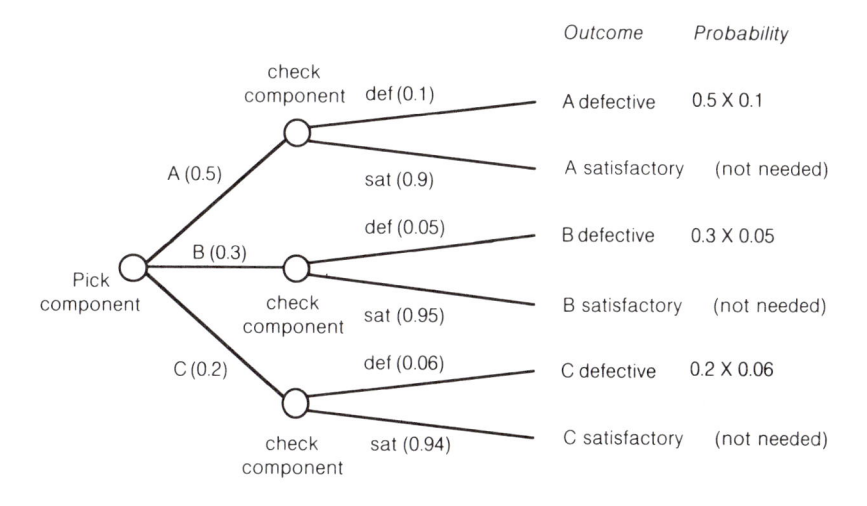

```
                              Outcome         Probability
         check
         component  def (0.1)
                    ────────── A defective    0.5 × 0.1
     A (0.5)
                    sat (0.9)
                    ────────── A satisfactory (not needed)
                    def (0.05)
         B (0.3)    ────────── B defective    0.3 × 0.05
Pick ○
component  check
         component  sat (0.95)
                    ────────── B satisfactory (not needed)
     C (0.2)        def (0.06)
                    ────────── C defective    0.2 × 0.06
         check
         component  sat (0.94)
                    ────────── C satisfactory (not needed)
```

From the diagram the probability of a defective component is:

$$(0.5 \times 0.1) + (0.3 \times 0.05) + (0.2 \times 0.06) = 0.077$$

Hence, the proportion of defective components in the total supply is 0.077. We now need the proportion of these defectives which come from source A.

$$P(\text{component from A, given defective}) = \frac{(0.5 \times 0.1)}{0.077} = 0.65$$

14 QUANTITATIVE ANALYSIS

The probability that the component is from A has changed from 0.5 to 0.65 given the additional information that the component is defective. Using Bayes' Rule this can be written as:

$$P(A \text{ given defective}) = \frac{P(A \text{ and defective})}{P(\text{defective})}$$

$$= \frac{0.5 \times 0.1}{(0.5 \times 0.1) + (0.3 \times 0.05) + (0.2 \times 0.06)}$$

$$= \frac{0.5 \times 0.1}{0.077} = 0.65$$

There is a 0.65 probability that the defective component came from supplier A.

Example 1.17: An illustration of Bayes' Rule

A large company has a central accounting department. Batches of invoices are sent to the department for checking and processing. Ninety per cent of the batches are satisfactory, containing only 1% incorrect invoices. The remaining 10% are however unsatisfactory and contain 5% incorrect invoices.

1 What is the probability that a new batch arriving in the department will be unsatisfactory?
2 One invoice is taken from the batch. It is incorrect. With this new information, what is the probability that the batch is unsatisfactory?
3 A second invoice is taken and is also incorrect. With these two pieces of information, what is the probability that the batch is unsatisfactory?

Solution: We will assume that there is a large number of invoices in each batch so that the probability of an incorrect invoice does not change significantly as we select the invoices.

1 Without any extra information the probability can be based only on past experience, so the probability that this batch is unsatisfactory is 0.1.
2 A tree diagram can be used for the next part of the example.

FIGURE 1.9 **Probability of obtaining an incorrect invoice**

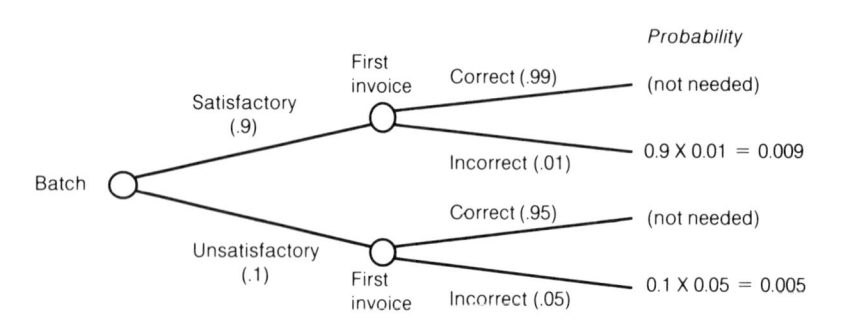

```
                                                            Probability
                              First
                              invoice   Correct (.99) ── (not needed)
              Satisfactory      ○
                 (.9)                   Incorrect (.01) ── 0.9 X 0.01 = 0.009
Batch   ○
              Unsatisfactory            Correct (.95) ── (not needed)
                  (.1)        First ○
                              invoice   Incorrect (.05) ── 0.1 X 0.05 = 0.005
```

The total probability that we obtain an incorrect invoice is:

$$0.009 + 0.005 = 0.014$$

P(batch is unsatisfactory, given the first invoice is incorrect)

$$= \frac{P(\text{batch is unsatisfactory } \textbf{and} \text{ invoice is incorrect})}{P(\text{incorrect invoice})}$$

$$= 0.005/0.014 = 0.357$$

3 The tree for the final part of the example is an extension of the first one:

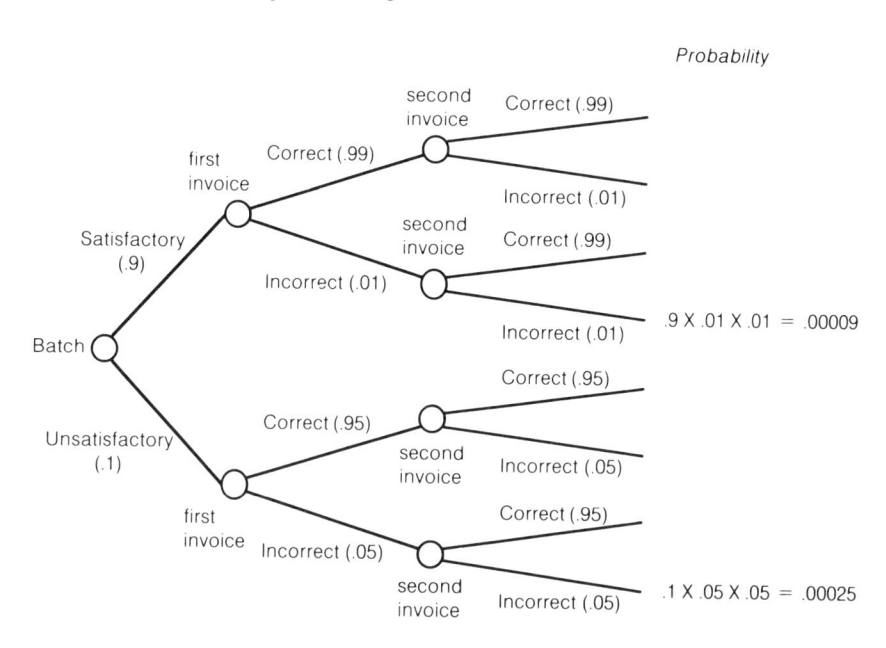

FIGURE 1.10 **Probability of obtaining two incorrect invoices**

[Tree diagram:
- Batch → Satisfactory (.9)
  - first invoice Correct (.99)
    - second invoice Correct (.99)
    - second invoice Incorrect (.01)
  - first invoice Incorrect (.01)
    - second invoice Correct (.99)
    - second invoice Incorrect (.01) → .9 × .01 × .01 = .00009
- Batch → Unsatisfactory (.1)
  - first invoice Correct (.95)
    - second invoice Correct (.95)
    - second invoice Incorrect (.05)
  - first invoice Incorrect (.05)
    - second invoice Correct (.95)
    - second invoice Incorrect (.05) → .1 × .05 × .05 = .00025]

The probability that the first two invoices are incorrect is:

$$0.00009 + 0.00025 = 0.00034$$

P(batch is unsatisfactory given the first 2 invoices are incorrect)

$$= \frac{P(\text{batch is unsatisfactory and 2 invoices are incorrect})}{P(\text{2 incorrect invoices})}$$

$$= 0.00025/0.00034 = 0.735$$

So the probability that this particular batch is unsatisfactory has changed, from 0.10 with historical information only, to 0.36 with one piece of additional information and to 0.74 with two pieces of additional information.

1.6 EXPECTED VALUES

We will now consider only those experiments in which the outcomes have numerical values, eg toss a coin 10 times and record the **number** of heads obtained; take a sample of cartons from the output of a filling machine and note the **number** of defectives in the sample.

If the experiment is repeated a large number of times, and on each occasion the value of the outcome noted, then an **average value** may be calculated. It is this average value (calculated from a large number of experiments) which is known as the **expected value**, $E(x)$. Repeated experiments are time-consuming. The expected value can be found more easily by calculation. The method is similar to that used to find the mean of a frequency distribution. For a frequency distribution the mean is given by:

$$\bar{x} = \frac{\sum fx}{\sum f}$$

If the frequency, f, is replaced by the relative frequency (or probability), p, then the mean or expected value is given by:

$$E(x) = \frac{\sum px}{\sum p}$$

but $\sum p = 1$, hence $E(x) = \sum px$.

Example 1.18: To calculate an expected value

A fair coin is tossed 4 times; what is the expected number of heads?
Solution: The list of all the possible outcomes is:

{hhhh, hhht, hhth, hthh, thhh, hhtt, htth, tthh, thth, htht, thht, httt, thtt, ttht, ttth, tttt}

There are 16 outcomes, each equally likely. Using the symmetry of the problem, the probability of each outcome is 1/16. Therefore:

P(0 heads out of 4 tosses) = 1/16
P(1 head out of 4 tosses) = 4/16
P(2 heads out of 4 tosses) = 6/16
P(3 heads out of 4 tosses) = 4/16
P(4 heads out of 4 tosses) = 1/16
Total probability = 16/16 = 1

Then the expected number of heads from 4 tosses is:

$$E(x) = \sum px = \frac{1}{16} \times 0 + \frac{4}{16} \times 1 + \frac{6}{16} \times 2 + \frac{4}{16} \times 3 + \frac{1}{16} \times 4 = 2$$

The expected number of heads is 2.

Example 1.19: To calculate an expected value

A gambler has a 0.1 chance of winning £1000 and a 0.2 chance of winning £500. If he loses he has to pay £300. What is his expected gain from this game?
Solution: The probability that he loses a game is = 1 − P(Win)
= 1 − (0.1 + 0.2)
= 1 − 0.3 = 0.7

The expected gain per game is:

$$E(x) = \sum px = 0.1 \times 1000 + 0.2 \times 500 + 0.7 \times (-300)$$
$$= -£10 \text{ per game}$$

This figure represents the average loss per game, if the gambler plays the game a very large number of times, under identical conditions. For any individual game he will either gain £1000 or £500 or lose £300, but in the long run he will lose an average of £10 per game.

The expected value is used to help assess whether a particular course of action is worth undertaking. If the expected value is large, then the activity may be worthwhile. See Chapter Three.

1.7 PERMUTATIONS AND COMBINATIONS

If we are dealing with experiments in which selections are being made from a group of items, then it is useful to be able to calculate quickly how many possible selections there are. To do this we must first decide whether the order in which the items are selected is important or not.

Permutations: the items in the group are identifiable and the order in which the items are selected is important.

Example 1.20: To illustrate a permutation

How many arrangements of 3 letters can we make from the group of letters A, B, C, D, E?
Solution: The first letter in the group may be chosen in any one of 5 ways. This leaves 4 letters remaining in the group. The second letter can be chosen in any one of 4 ways. This leaves 3 letters in the group. The third letter can be chosen in any one of 3 ways. Each of the 5 possible first choices could be combined with any one of the 4 second choices which, in turn, could be combined with any one of the 3 possible third choices. Hence, the total number of arrangements of 3 letters is:

$$5 \times 4 \times 3 = 60$$

We can say that the number of permutations of 5 items taken 3 at a time is $5 \times 4 \times 3 = 60$. These arrangements are:

```
ABC ACB BAC BCA CAB CBA
ABD ADB BAD BDA DAB DBA
ABE AEB BAE BEA EAB EBA
ADC ACD DAC DCA CAD CDA
AEC ACE EAC ECA CAE CEA
ADE AED DAE DEA EAD EDA
DBC DCB BDC BCD CDB CBD
EBC ECB BEC BCE CEB CBE
DEC DCE EDC ECD CDE CED
DBE DEB BDE BED EDB EBD
```

We must now produce a general expression for a permutation. We will then be able to evaluate any permutation given the size of the group and the number of items to be selected.

Let us consider the number of permutations of n items taken three at a time. The first item can be chosen in any one of n ways. The second item can be chosen in any one of (n − 1) ways. The third item can be chosen in any one of (n − 2) ways. The number of permutations is n × (n − 1) × (n − 2). This direct method is easy to use if the size of the group being chosen is small, but the expression for the permutation becomes very unwieldy if the group size is large. The introduction of a notation and mathematical shorthand is helpful. A permutation of n items taken r at a time can be denoted by ^nP_r. The symbol ! when placed after a number is used to denote a factorial. For example:

$$6! = 6 \times 5 \times 4 \times 3 \times 2 \times 1$$

and:

$$10! = 10 \times 9 \times 8 \times 7 \times 6 \times 5 \times 4 \times 3 \times 2 \times 1$$

In general, $n! = n \times (n-1) \times (n-2) \times (n-3) \times \cdots \times 1$.

A special point to note is that $1! = 1$ and, by definition, $0! = 1$. With a little manipulation the factorial notation can be used to represent a permutation.

Consider **Example 1.20** again. The number of permutations of 5 items taken 3 at a time. We found that:

$$^5P_3 = 5 \times 4 \times 3$$

We can re-write this as follows without changing the value:

$$^5P_3 = 5 \times 4 \times 3 \times \frac{2 \times 1}{2 \times 1}$$

$$= \frac{5 \times 4 \times 3 \times 2 \times 1}{2 \times 1}$$

$$= \frac{5!}{2!}$$

Similarly:

$$^nP_3 = n \times (n-1) \times (n-2) \times \frac{(n-3) \times (n-4) \times \cdots \times 1}{(n-3) \times (n-4) \times \cdots \times 1}$$

$$= \frac{n!}{(n-3)!}$$

We will not always want to choose the items three at a time. We must generalise further to allow for any number of choices. The number of permutations of n items taken r at a time is:

$$^nP_r = \frac{n!}{(n-r)!}$$

Combinations: a combination refers to the number of selections which may be made from a group of items when the order of the selection is not important.

Example 1.21: To illustrate a combination

How many selections of 3 letters can we make from the group of letters A, B, C, D, E? The order of selection is not important.

Solution: Referring back to the solution for **Example 1.20**, we can see that for each row in the list of the permutations, the actual letters used are the same. Each row in this list represents one combination of 3 letters. Hence, the number of combinations of 5 letters taken 3 at a time is 10. We also note that the number of permutations of 3 letters taken together is:

$$^3P_3 = \frac{3!}{(3-3)!} = \frac{3!}{0!} = \frac{3!}{1} = 6$$

This can be used to produce an expression for the number of combinations of 5 items taken 3 at a time, denoted by $^5C_3$:

$$^5C_3 = \frac{^5P_3}{^3P_3} = \frac{5!/2!}{3!}$$

$$= \frac{5!}{3! \times 2!}$$

$$= 10$$

If we generalise in stages as before, the overall pattern emerges. What is the number of combinations of n items taken 3 at a time?

$$^nC_3 = \frac{^nP_3}{^3P_3} = \frac{n!/(n-3)!}{3!}$$

$$= \frac{n!}{3! \times (n-3)!}$$

Now consider what is the number of combinations of n items taken r at a time?

$$^nC_r = \frac{^nP_r}{^rP_r} = \frac{n!/(n-r)!}{r!}$$

$$= \frac{n!}{r! \times (n-r)!}$$

BASIC PROBABILITY: CHAPTER ONE

1.7.1 Using permutations and combinations to calculate probabilities

Example 1.22: To illustrate the use of combinations to calculate a probability

Consider again **Example 1.14:** the Board of Directors of a company consists of 3 accountants, 3 managers and 2 engineers. A planning sub-committee of 3 is chosen at random from the Board. Find the probability that all 3 members of the sub-committee are accountants.
Solution: The number of combinations of 3 accountants from 3 accountants is:

$$^3C_3 = \frac{3!}{3!(3-3)!} = 1$$

the number of combinations of 3 people from 8 is:

$$^8C_3 = \frac{8!}{3!(8-3)!} = 56$$

Hence, the probability that all 3 committee members will be accountants is:

$$\frac{^3C_3}{^8C_3} = \frac{1}{56}$$

as before.

Example 1.23: To illustrate the use of combinations to calculate a probability

A committee of 10 is to be selected at random from 3 accountants, 8 managers and 6 scientists. What is the probability that the committee will consist of 1 accountant, 5 managers and 4 scientists?
Solution:

The number of combinations of 1 accountant is $^3C_1 = \dfrac{3!}{1!(3-1)!} = 3$

The number of combinations of 5 managers is $^8C_5 = \dfrac{8!}{5!(8-5)!} = 56$

The number of combinations of 4 scientists is $^6C_4 = \dfrac{6!}{4!(6-4)!} = 15$

The number of combinations of 10 people is $^{17}C_{10} = \dfrac{17!}{10!(17-10)!} = 19448$

Therefore the probability of a committee with 1 accountant, 5 managers and 4 scientists is:

$$\frac{^3C_1 \times {}^8C_5 \times {}^6C_4}{^{17}C_{10}} = \frac{3 \times 56 \times 15}{19448} = 0.130$$

SUMMARY

Probability theory is concerned with uncertainty. We carry out an experiment. We know all the possible outcomes, but not which of them will actually occur. Probability tells us how likely it is that each of the outcomes will occur.

An event is one or more outcomes which are of interest to us. Probability values lie between 0 and 1 inclusive. The sum of the probabilities of all the outcomes of an experiment must equal 1. Probability can be evaluated by using the symmetry of the experiment, or by repeated measurement, or by subjective estimation.

QUANTITATIVE ANALYSIS

There are two basic rules of probability. The addition rule gives the probability of event A or event B or both occurring:

$$P(A \text{ or } B \text{ or both}) = P(A) + P(B) - P(A \text{ and } B)$$

For mutually exclusive events this becomes:

$$P(A \text{ or } B) = P(A) + P(B)$$

The Multiplication Rule gives the probability of two events happening together:

$$P(A \text{ and } B) = P(A) \times P(B \text{ given } A)$$

For independent events this becomes:

$$P(A \text{ and } B) = P(A) \times P(B)$$

Before considering probabilities or rules, always illustrate the possible outcomes of an experiment. Lists, tables, tree diagrams or Venn diagrams can be used. When you are clear about the outcomes, only then should the probabilities be added.

Bayes' Rule is used to modify probability values as new information becomes available. It is based on the Multiplication Rule and may be written as:

$$P(A \text{ given } B) = \frac{P(A \text{ and } B)}{P(B)}$$

The expected value of an experiment is:

$$E(x) = \sum px$$

When we have to decide whether to carry out an experiment or not, it is often useful to calculate this average value.

Permutations and combinations are used to determine the number of ways of selecting r items from a group of m items.

$$\text{Permutations: } {}^nP_r = \frac{n!}{(n-r)!} \quad \text{—the order of selection is important.}$$

$$\text{Combinations: } {}^nC_r = \frac{n!}{r!(n-r)!} \quad \text{—the order of selection is not important.}$$

EXERCISES

Exercise 1.1

A coin is tossed 4 times. What is the probability of getting 2 heads and 2 tails,

(i) in that order
(ii) in any order?

Answers on page 421.

BASIC PROBABILITY: CHAPTER ONE

Exercise 1.2

A die is thrown twice. The two outcomes are multiplied together. What is the probability of getting

(i) 23
(ii) 12?

Answers on page 422.

Exercise 1.3

The employees of Tutial plc are split by department and sex in the following table:

Department	Female	Male
Production	6	20
Maintenance	3	10
Stores	5	5
Transport	2	8
Sales	5	10

One employee is selected at random. What is the probability that this employee is:

(i)   a female
(ii) from maintenance
(iii) a male from stores
(iv) a female from stores or transport
(v)   from production or sales?

The same company, Tutial plc, want to select 2 employees at random to serve on a consultative committee. What is the probability that these employees are:

(vi)   both female
(vii) both from production
(viii) one from sales and one from transport
(ix)   one a female from maintenance and the other a male from stores
(x)   both females from stores or one person from production and the other a male from sales?

Answers on page 422.

Exercise 1.4

During a course, trainee accountants are learning to check invoices for errors. The tutor has selected 10 invoices, 4 of which contain errors. He selects 2 of the 10 invoices at random and asks a trainee to check them. What is the probability that the 2 invoices will be:

(i) both in error
(ii) one in error, the other satisfactory?

If the trainee has a 0.8 probability of picking out an erroneous invoice correctly, and a 0.9 probability of picking out a good invoice correctly, what is the probability that the trainee will correctly identify 2 invoices given that they are:

(iii) both erroneous
(iv) one erroneous, the other satisfactory?

Answers on page 423.

Exercise 1.5

A shop receives clothes in batches of 100 items. They inspect each batch by taking a random sample of 5 items. If all 5 items are satisfactory, the batch of 100 is accepted. A particular batch contains 8 defective items. What is the probability that this batch will be accepted?

Answers on page 424.

Exercise 1.6

Ten stages are involved in the production of the documentation for the manufacture of an assembly. At each stage, the probability of error is 0.002. What is the probability that the documentation will be:

(i) completely accurate
(ii) contain one error?
(iii) A clerical reorganisation could reduce the process to 5 stages, but it is estimated that the probability of error at each stage would increase to 0.003. How would this alter the probabilities in (i) and (ii)?

Answers on page 424.

Exercise 1.7

An electronic system has 3 components, R, S and T. The probability that the components work for a year are 0.95 for R, 0.9 for S and 0.93 for T.

(i) If the system is designed so that all 3 components must work for the system to function, what is the probability that the system will work all year?
(ii) If the system is redesigned so that it functions as long as any 2 of the components are working, what now is the probability that the system will work all year?
(iii) A further redesign means that the system will function as long as at least 1 component is working. Now what is the probability that the system will work all year?

Answers on page 425.

Exercise 1.8

A similar system has 2 components, G and H, one of which must work for the system to function. The distribution for the lifetimes of the components, in years, are given below.

Lifetime (years)	1	2	3	4	5
Probability for Component G	0.1	0.2	0.2	0.3	0.2
Component H	0.2	0.3	0.3	0.2	0.0

(i) What is the probability distribution for the lifetime of the system?
(ii) What is the expected lifetime of each of the components and of the system as a whole?

Answers on page 426.

Exercise 1.9

Drivers are classified by an insurance company as low, average or high risk drivers. The company estimates that at present they have 25% low, 60% average, and 15% high risk drivers on their books. The probability that such drivers have a given number of accidents during a year are given below:

Driver risk	Low	Average	High
1 accident/year	0.01	0.03	0.10
2 accidents/year	0	0.01	0.05
3 accidents/year	0	0	0.01
4 accidents/year	0	0	0

Required:
(i) If Mr. A has no accidents in the year, what is the probability that he is a high risk driver?
(ii) If Ms. B has no accidents for 4 years, what is the probability that she is a low risk driver?

Answers on page 427.

Exercise 1.10

A company is considering launching a new product. It estimates the probability of high sales for the product as 0.6 and of low sales as 0.4. The company decides to carry out a market survey, which from past experience it knows has a probability of 0.8 of being correct. If the survey predicts low sales, how does this change the initial probabilities of the sales levels?

Answers on page 428.

Exercise 1.11

An insurance company wants to calculate the annual premium it should charge for building insurance on £60,000 houses. Its records show that 2 in every 100 such houses are likely to be damaged each year. Of the damaged houses, 5% are completely destroyed, 25% suffer loss of £8000 and the remainder a loss of £4000.

Required:
Ignoring other costs, what premium should the company charge to break even?

Answers on page 429.

Exercise 1.12

There are 25 members of a society at its annual general meeting. All of them are willing to be elected chairman, secretary, treasurer or to one of the 4 other committee posts.

Required:
(i) How many different ways of selecting the chairman, secretary and treasurer are there, out of the 25 members?
(ii) When these officials have been elected, how many different ways are there of selecting the 4 other committee members?
(iii) How many different committees can therefore be chosen from the 25?

Answers on page 429.

Exercise 1.13

A transport firm employs 10 drivers.

(i) On a particular day it needs to send a driver to each of 5 factories. How many different ways are there of doing this?

(ii) On another day the firm needs to send 2 drivers to each of these 5 factories. How many different ways are there of doing this?

Answers on page 430.

Exercise 1.14

Troophit plc manufacture nuts and bolts. To satisfy their various customers, the nuts and bolts are made in both imperial and metric sizes. One day in the stores, a box containing fifteen 5mm bolts is accidentally tipped into a bin containing thirty 3/16" bolts, and a box of fifteen 5mm nuts into a bin of thirty 3/16" nuts.

Required:

(i) If 1 nut and 1 bolt are picked out at random from the bins, what is the probability that they fit each other?

(ii) If 2 nuts and 2 bolts are picked at random from the bins, what is the probability that they will be 2 matching pairs?

Answers on page 430.

Exercise 1.15

An investor has £10,000 available to purchase shares in either chemical or brewing companies. His broker has provided him with the following estimates of the likely returns over the next 12 months:

Possible annual return £	−2000	−1000	0	1000	2000	3000	4000
Probability from chemicals	0.05	0.1	0.2	0.2	0.2	0.2	0.05
brewers	0.05	0.2	0.3	0.2	0.1	0.1	0.05

Required:
What are the expected annual returns for:

(i) chemicals
(ii) brewers?

Answers on page 431.

Exercise 1.16

The World Life Assurance Company Ltd uses recent mortality data in the calculation of its premiums. The following table shows, per thousand of population, the number of persons expected to survive to a given age:

Age	0	10	20	30	40	50	60	70	80	90	100
Number surviving to given age	1000	981	966	944	912	880	748	525	261	45	0

Required:
1   Use the table to determine the probability that:
    (a)   a randomly chosen person born now will die before reaching the age of 60,
    (b)   a randomly chosen person who is aged 30 now will die before reaching the age of 60,
    (c)   a randomly chosen person who is aged 50 now will die before reaching the age of 60.
    Comment on the order of magnitude of your 3 answers.
2   The company is planning to introduce a life insurance policy for persons aged 50. This policy requires a single payment paid by the insured at the age of 50 so that if the insured dies within the following 10 years the dependent will receive a payment of £10,000; however if this person survives, then the company will not make any payment.

Required:
Ignoring interest on premiums and any administrative costs, calculate the single premium that the company should charge to break even in the long run.
3   If twelve persons each take out a policy as described in **2** and the company charges each person a single premium of £2000, find the probability that the company will make a loss.
4   The above table was based on the ages of all people who died in 1986. Comment on the appropriateness to the company when calculating the premiums it should charge.
5   The table can be expanded to include survival numbers, per thousand of population, of other ages:

Age	50	52	54	56	58	60
Number surviving to given age	880	866	846	822	788	748

Required:
    (a)   Given that a person aged 50 now dies before the age of 60, use this new information to estimate the expected age of death.
    (b)   Calculate a revised value for the single premium as described in **2**, taking into account the following additional information: the expected age of death before 60 as estimated at (a); a constant interest rate of 8% pa on premiums; an administration cost of £100 per policy; a cost of £200 to cover profit and commissions.

(ACCA, December 1987)

Answers on page 432.

Exercise   1.17

Every Friday morning Minos Security Ltd delivers wages from a local bank to each of 5 firms in a certain area. Having collected the necessary money from the bank, a security van visits the 5 firms in turn but, as a precaution against robbery, the same route is not used on each occasion. Immediately prior to each Friday's delivery, the van driver chooses at random, according to a procedure established by the operations manager, the order in which the 5 firms will be visited.

Required:
1   (a)   What is the probability of the van following a particular route (ie visiting the firms in a particular order) on any specific occasion?
    (b)   What is the probability of the van following a different route on each of 4 consecutive weeks?
    (c)   What is the probability of 2 particular firms, A and B, being visited consecutively?
2   In order to reduce the delivery time, the operations manager has decided to modify the routing procedure. Firms A, B and C are close together, and so are firms D and E. Accordingly the van visits the first 3 firms in a randomly chosen order, and then visits the other 2 firms, again in random order. How does this modify each of the 3 probabilities in **1**?

3 As an additional precaution to the procedure in **2**, the van driver is instructed never to follow the same route on 2 successive weeks. What now is the probability of the van following a different route on each of 4 consecutive weeks?

(ACCA, December 1983)

Answers on page 433.

CHAPTER TWO

Probability distributions

2.1 INTRODUCTION

In the previous chapter experiments of chance were considered. The term **probability** was given to a measure of the likelihood of obtaining particular outcomes from an experiment. The probabilities of compound events were then obtained using the rules for combining individual probabilities.

In this chapter we are concerned with probability distributions. First we will discuss probability distributions of discrete random variables and then move on to examine those of continuous random variables.

We are now dealing with those experiments of chance in which the outcomes may be described by numerical values. These outcomes are represented by what is called a **random variable**. Random variables are either **discrete** (take integer values only) or **continuous** (take any value within a specified range). We will discuss probability distributions for both types of variable in this chapter.

The probability distributions for discrete random variables can be of various kinds. Two of the most common are the **binomial** and the **Poisson** probability distributions. They are particularly important in auditing procedures, for example when checking the proportion of accounts in error. There are also many probability distributions for continuous random variables. The most commonly used is the **normal probability distribution**. This will be discussed in detail in section **2.7** of this chapter. The normal distribution becomes particularly important when we consider the distribution of the mean values of random variables rather than the individual variable values. See Chapter Four.

Some of the calculations using the binomial and Poisson distributions can be long and tedious. There are various ways of using approximations which reduce the calculations to much simpler levels and give very similar results. Section **2.5** shows how to use the Poisson distribution as an approximation to the binomial distribution. Sections **2.8** and **2.9** show how to use the normal distribution as an approximation to the binomial and the Poisson distributions.

2.2 PROBABILITY DISTRIBUTIONS FOR DISCRETE RANDOM VARIABLES

2.2.1 Discrete random variable

If the outcomes of an experiment of chance can be described by whole numbers then these outcomes may be represented by a discrete random variable, as shown in TABLE 2.1 below:

TABLE 2.1 **Discrete random variables**

	Experiment of chance	Random variable	Values of the random variable
(a)	Toss a coin twice and note the number of heads obtained.	The number of heads.	0, 1, 2
(b)	Roll a die once and note the score.	The score on the die.	1, 2, 3, 4, 5, 6
(c)	Select a sample of 5 items from the output of a mass production process and note the number of defectives.	Number of defectives in the sample.	0, 1, 2, 3, 4, 5

28 QUANTITATIVE ANALYSIS

TABLE 2.1 **Discrete random variables**

Experiment of chance	Random variable	Values of the random variable
(d) Record the number of accidents per week on a given stretch of road.	Number of accidents which occur per week.	0, 1, 2, 3, …
(e) Record the demand per day for a hire car.	The number of requests for a car per day.	0, 1, 2, 3, …
(f) The number of large cakes sold per day in a baker's shop, if the maximum stock level for the cakes is five.	The number of cakes sold per day.	0, 1, 2, 3, 4, 5

Let us denote the random variable by R and the values which R can take by r. Hence, in TABLE 2.1 (f) above, R is the number of cakes sold per day, and:

$$r = 0 \text{ or } 1 \text{ or } 2 \text{ or } 3 \text{ or } 4 \text{ or } 5.$$

2.2.2 Probability distribution of a discrete random variable

Consider an experiment in which we check a sample of 10 invoices. We record whether each invoice is satisfactory or not. There are 11 possible outcomes to this experiment, as shown in TABLE 2.2:

TABLE 2.2 **Number of satisfactory invoices**

Outcome number	1	2	3	4	5	6	7	8	9	10	11
Number of satisfactory invoices	0	1	2	3	4	5	6	7	8	9	10
Number of unsatisfactory invoices	10	9	8	7	6	5	4	3	2	1	0

The number of satisfactory invoices is a discrete random variable. We do not know which of the 11 possible outcomes will actually occur when we begin the experiment, but we can assign a probability of occurence to each outcome. This list of probabilities is what is referred to as a probability distribution for the number of satisfactory invoices in the sample of 10.

If we consider an experiment in which the outcomes may be described by a discrete random variable, R, the set of probabilities which correspond to the set of values for R is called the probability distribution of R. The probability that the random variable takes a particular value, r, is denoted by $P(R = r)$

Example 2.1: Probability distributions of discrete random variables

(a) Experiment: toss a coin twice and note the number of heads.
Possible outcomes:
$$\{TT, HT, TH, HH\}$$

Values of the random variable:
$$r = 0, 1, 2$$

Probability Distribution:
$$P(R = r) = \{1/4, 2/4, 1/4\}$$

(b) Experiment: roll a die once and note the score on the die.
Possible outcomes:
$$\{1, 2, 3, 4, 5, 6\}$$

Values of random variable:

$$r = 1, 2, 3, 4, 5, 6$$

Probability Distribution:

$$P(R = r) = \{1/6, 1/6, 1/6, 1/6, 1/6, 1/6\}$$

(c) Experiment: an investment analyst assesses the probability of achieving specific returns, in £'000, on an investment.

Possible outcomes, in £'000:

$$1, 2, 3, 4, 5$$

Values of random variable:

$$r = 1, 2, 3, 4, 5$$

Probability Distribution:

$$P(R = r) = \{0.4, 0.3, 0.2, 0.1, 0.0\}$$

In the above examples the probability distributions are represented by a list of probabilities. This is an inconvenient format, particularly if there are many possible outcomes. In each of the examples, there is a relationship between the members of the probability distribution. This means that the list of individual probabilities may be replaced by a single mathematical function which describes the relationship.

In **Example 2.1(a)**, $P(R = r) = {}^2C_r \times (1/4)$ $r = 0, 1, 2$
In **Example 2.1(b)**, $P(R = r) = 1/6$ $r = 1, 2, \ldots, 6$
In **Example 2.1(c)**, $P(R = r) = 0.1(5 - r)$ $r = 1, 2, \ldots, 5$

Individual members of the probability distribution can be calculated as and when required. Whenever possible, the probability distribution of a discrete random variable is represented by a mathematical function, $f(r)$, where $P(R = r) = f(r)$ (over the appropriate range of r).

2.2.3 Graphical representation of a discrete probability distribution

The probability distribution of a discrete random variable may be represented by a line graph.

Example 2.2: Graphical representation of the probability distribution of a discrete random variable

(a) For **Example 2.1(a)**, in which a coin is tossed twice, the probability distribution is illustrated by FIGURE 2.1 below:

FIGURE 2.1   **Probability distribution of the number of heads obtained when a coin is tossed twice**

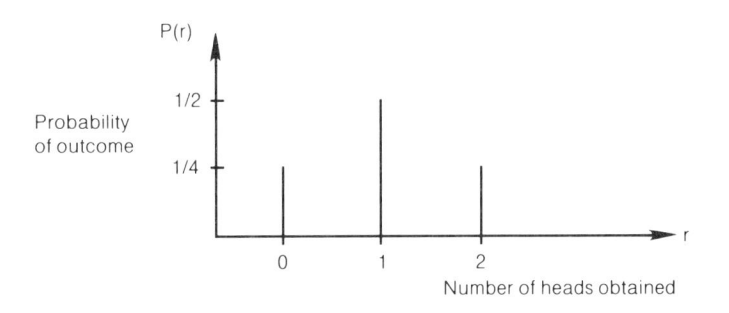

30 QUANTITATIVE ANALYSIS

(b) For **Example 2.1 (b)**, in which a die is rolled once, the probability distribution may be illustrated by FIGURE 2.2:

FIGURE 2.2 **Probability distribution of the outcomes when a die is rolled once**

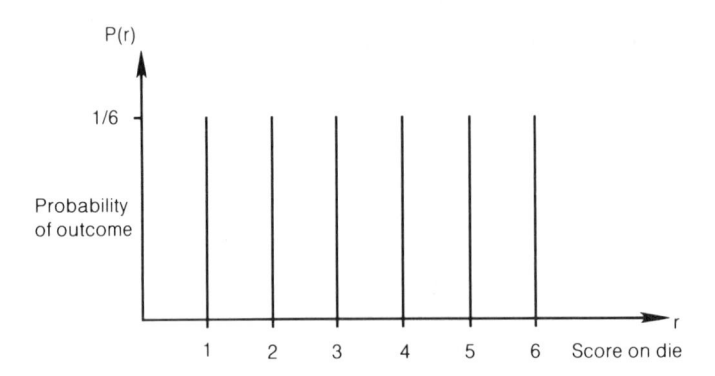

In each case, all of the usual rules governing probabilities apply. Each probability in the distribution must lie in the range

$$0 \leqslant p \leqslant 1$$

The sum of all of the probabilities in the distribution must equal 1. In **Example 2.1(a)** above, the sum of the probabilities is:

$$\frac{1}{4} + \frac{2}{4} + \frac{1}{4} = 1$$

In **Example 2.1 (b)** above, the sum of the probabilities is:

$$6 \times (1/6) = 1$$

2.2.4. Expected value and standard deviation of a probability distribution

Consider the number of cars sold per day by a particular salesman over a period of 30 days:

TABLE 2.3 **Number of cars sold per day**

Number of cars sold per day	Number of days	Calculations	
r	f	fr	fr^2
0	5	0	0
1	4	4	4
2	8	16	32
3	4	12	36
4	6	24	96
5	3	15	75
Totals	30	71	243

The mean number of cars sold per day is:

$$\bar{r} = \frac{\sum fr}{\sum f} = \frac{(5 \times 0) + (4 \times 1) + (8 \times 2) + (4 \times 3) + (6 \times 4) + (3 \times 5)}{5 + 4 + 8 + 4 + 6 + 3}$$

$$\bar{r} = \frac{71}{30} = 2.37 \text{ cars sold per day}$$

The variance of the number of cars sold per day is:

$$\sigma^2 = \frac{\sum fr^2}{\sum f} - \bar{r}^2 = \frac{243}{30} - 2.367^2 = 2.50$$

The standard deviation = $\sqrt{\text{variance}}$

Therefore:

$$\sigma = \sqrt{\frac{243}{30} - 2.367^2} = 1.58 \text{ cars sold per day}$$

The frequency distribution may be used to estimate the probability of selling a particular number of cars per day as follows:

TABLE 2.4 **Probability of number of cars sold per day**

Number of cars sold per day, r	Number of days, f	Probability of selling r cars per day, P(r)	Calculations rP(r)	r^2P(r)
0	5	5/30	0	0
1	4	4/30	4/30	4/30
2	8	8/30	16/30	32/30
3	4	4/30	12/30	36/30
4	6	6/30	24/30	96/30
5	3	3/30	15/30	75/30
Totals	30	1	71/30	243/30

The probability is the relative frequency of occurrence of each value of the discrete random variable. The mean and standard deviation may be found using the probability distribution in exactly the same way as using the frequency distribution, but relative frequency (probability) replaces frequency. For a probability distribution:

$$\text{Mean} = \frac{\sum rP(r)}{\sum P(r)} = \frac{(0 \times 5/30) + (1 \times 4/30) + (2 \times 8/30) + (3 \times 4/30) + (4 \times 6/30) + (5 \times 3/30)}{1}$$

$$= 2.37 \text{ cars sold per day}$$

$$\text{Variance} = \frac{\sum r^2 P(r)}{\sum P(r)} - \left(\frac{\sum rP(r)}{\sum P(r)}\right)^2 = \frac{243/30}{1} - 2.37^2 = 2.50$$

The mean using a probability distribution is usually referred to as the **expected value**. It represents the average value which the random variable will take if the experiment of chance is repeated a very large number of times. The expected value of a discrete random variable may be denoted by $E(r)$ or by μ. Since $\sum P(r) = 1$, the expected value is given by:

$$E(r) = \sum rP(r)$$

QUANTITATIVE ANALYSIS

The variability or dispersion of the probability distribution may be measured using the standard deviation or the variance of the values of the discrete random variable:

$$\text{Variance} = (\text{standard deviation})^2$$

Again, since $\sum P(r) = 1$:

$$\sigma^2 = \sum r^2 P(r) - (E(r))^2$$

Example 2.3: Expected value and standard deviation of a probability distribution

In an experiment a coin is tossed three times and the number of heads is noted. The possible outcomes are:

$$\{ttt, tth, tht, htt, thh, hth, hht, hhh\}$$

The discrete random varible is the number of heads obtained and its possible values are:

$$r = 0 \text{ or } 1 \text{ or } 2 \text{ or } 3$$

From the list of possible outcomes the probability distribution is:

TABLE 2.5 **Probability distribution for number of heads obtained**

Number of heads, r	0	1	2	3
P(r)	1/8	3/8	3/8	1/8

(Note: $\sum P(r) = 1$)

The expected value of the number of heads obtained is:

$$E(r) = \sum rP(r) = (0 \times 1/8) + (1 \times 3/8) + (2 \times 3/8) + (3 \times 1/8) = 12/8$$
$$= 1.5$$

The variance of the number of heads is:

$$\sigma^2 = \sum r^2 P(r) - (E(r))^2 = \{0 \times 1/8 + 1 \times 3/8 + 4 \times 3/8 + 9 \times 1/8\} - 1.5^2$$
$$= 3.0 - 2.25 = 0.75$$

Therefore the standard deviation of the number of heads is given by:

$$[\sigma = \sqrt{0.75} = 0.87]$$

2.3 THE BINOMIAL PROBABILITY DISTRIBUTION

2.3.1 What is the binomial probability distribution?

The binomial probability distribution may be illustrated by the following example. Let us consider an experiment in which a die is rolled 4 times. We are interested in those outcomes which produce a 6. Each trial of rolling the die is either a 'success', if a 6 is obtained, or it is a 'failure' if a score of 1 to 5 is obtained. The probability of a 'success' is 1/6 and the probability of a 'failure' is 5/6.

FIGURE 2.3 **Possible outcomes of the experiment**

6 denotes rolling a six; 6* denotes rolling any score other than six.

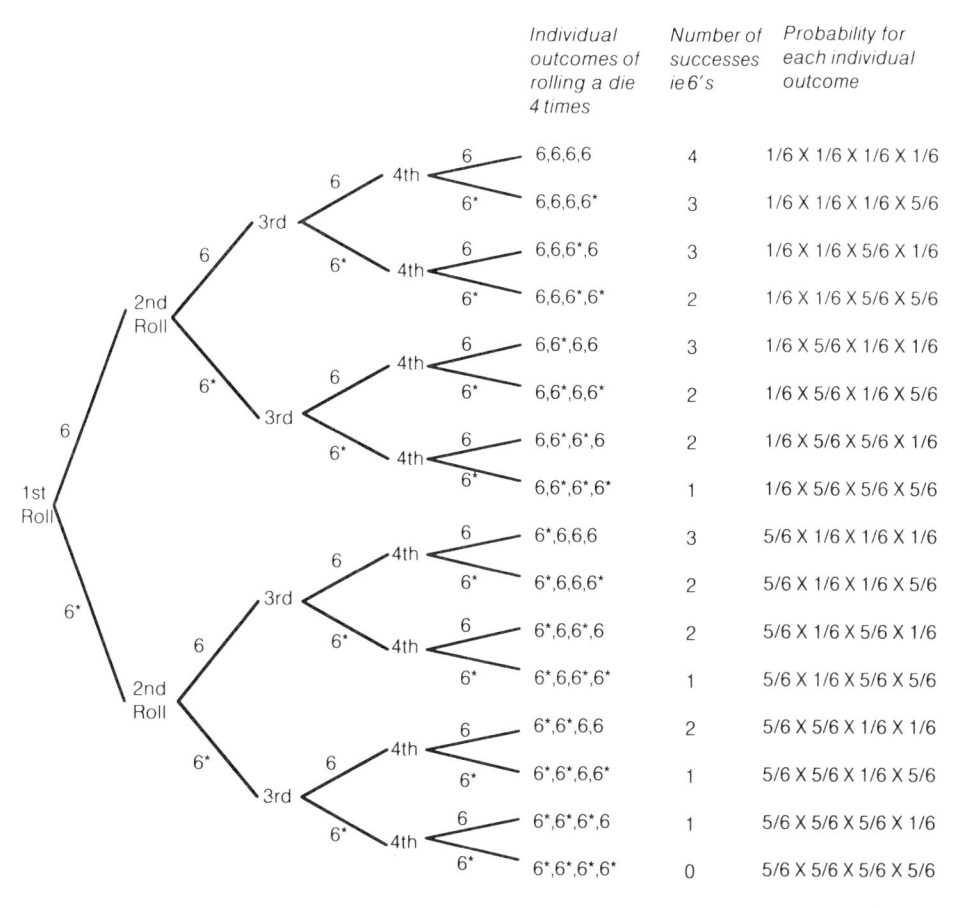

	Individual outcomes of rolling a die 4 times	Number of successes ie 6's	Probability for each individual outcome
	6,6,6,6	4	1/6 X 1/6 X 1/6 X 1/6
	6,6,6,6*	3	1/6 X 1/6 X 1/6 X 5/6
	6,6,6*,6	3	1/6 X 1/6 X 5/6 X 1/6
	6,6,6*,6*	2	1/6 X 1/6 X 5/6 X 5/6
	6,6*,6,6	3	1/6 X 5/6 X 1/6 X 1/6
	6,6*,6,6*	2	1/6 X 5/6 X 1/6 X 5/6
	6,6*,6*,6	2	1/6 X 5/6 X 5/6 X 1/6
	6,6*,6*,6*	1	1/6 X 5/6 X 5/6 X 5/6
	6*,6,6,6	3	5/6 X 1/6 X 1/6 X 1/6
	6*,6,6,6*	2	5/6 X 1/6 X 1/6 X 5/6
	6*,6,6*,6	2	5/6 X 1/6 X 5/6 X 1/6
	6*,6,6*,6*	1	5/6 X 1/6 X 5/6 X 5/6
	6*,6*,6,6	2	5/6 X 5/6 X 1/6 X 1/6
	6*,6*,6,6*	1	5/6 X 5/6 X 1/6 X 5/6
	6*,6*,6*,6	1	5/6 X 5/6 X 5/6 X 1/6
	6*,6*,6*,6*	0	5/6 X 5/6 X 5/6 X 5/6

In order to determine a function which will enable us to generate the probability distribution, let us consider the particular event of rolling 2 sixes, ie 2 successes and 2 failures. Two sixes may arise in any of the following 6 ways:

$$6\ 6\ 6^*\ 6^*\ ;\ 6\ 6^*\ 6\ 6^*\ ;\ 6\ 6^*\ 6^*\ 6\ ;\ 6^*\ 6^*\ 6\ 6\ ;\ 6^*\ 6\ 6^*\ 6\ ;\ 6^*\ 6\ 6\ 6^*$$

The probability of each of these combinations is the same:

$$\frac{1}{6} \times \frac{1}{6} \times \frac{5}{6} \times \frac{5}{6} = \left(\frac{1}{6}\right)^2 \times \left(\frac{5}{6}\right)^2$$

and hence, the probability of obtaining exactly 2 sixes is:

$$\left(\frac{1}{6}\right)^2 \times \left(\frac{5}{6}\right)^2 \times \text{(the number of ways of obtaining 2 sixes from 4 rolls of the die)}$$

$$P(\text{exactly 2 sixes from 4 throws}) = \left(\frac{1}{6}\right)^2 \times \left(\frac{5}{6}\right)^2 \times 6 = \frac{150}{1296}$$

We can use the tree diagram to determine the probability of all of the other events:

$$P(0 \text{ sixes out of 4 trials}) = \left(\frac{1}{6}\right)^0 \times \left(\frac{5}{6}\right)^4 \times 1 = \frac{625}{1296}$$

$$P(1 \text{ six out of 4 trials}) = \left(\frac{1}{6}\right)^1 \times \left(\frac{5}{6}\right)^3 \times 4 = \frac{500}{1296}$$

$$P(2 \text{ sixes out of 4 trials}) = \left(\frac{1}{6}\right)^2 \times \left(\frac{5}{6}\right)^2 \times 6 = \frac{150}{1296}$$

$$P(3 \text{ sixes out of 4 trials}) = \left(\frac{1}{6}\right)^3 \times \left(\frac{5}{6}\right)^1 \times 4 = \frac{20}{1296}$$

$$P(4 \text{ sixes out of 4 trials}) = \left(\frac{1}{6}\right)^4 \times \left(\frac{5}{6}\right)^0 \times 1 = \frac{1}{1296}$$

$$\text{Total probability} = \frac{1296}{1296} = 1$$

We can now derive the function which describes the probability distribution for the number of sixes obtained when a die is rolled 4 times:

$$P(r \text{ sixes from 4 trials}) = \left(\frac{1}{6}\right)^r \times \left(\frac{5}{6}\right)^{4-r} \times (\text{the number of events with r sixes})$$

From Chapter One, we know that the number of ways of obtaining r successes from 4 trials is given by the number of combinations of r items from 4 items, ie:

$$^4C_r = \frac{4!}{r!(4-r)!}$$

Hence:

$$P(r \text{ sixes from 4 trials}) = \left(\frac{1}{6}\right)^r \times \left(\frac{5}{6}\right)^{4-r} \times {}^4C_r$$

The essential features of this binomial experiment are:

1 There are 4 identical trials.
2 The trials are independent.
3 For each trial there are 2 possible outcomes, ie 6 or not 6, success or failure.
4 For each trial, the probability of a success is the same, ie $\frac{1}{6}$.

These 4 features define the conditions which are necessary for the binomial probability distribution to apply. We are now in a position to experiment with any number of trials.

2.3.2 The binomial probability distribution

In any experiment in which there are n identical, independent trials with 2 possible outcomes (denoted success and failure), and the probability of a success on any one trial is constant, then the probability of obtaining r successes from the n trials is given by:

$$P(r) = (\text{probability of r successes}) \times (\text{probability of } (n-r) \text{ failures})$$
$$\times (\text{number of ways this can be done})$$

$$P(r) = \frac{n!}{r!(n-r)!} p^r q^{n-r} \qquad r = 0, 1, 2, \ldots, n$$

where p is the probability of a success on any one trial, and q is the probability of a failure and $p + q = 1$.

Example 2.4: The binomial distribution

In a group of 5 machines, which run independently of each other, the chance of a breakdown on each machine in any day is 0.2. If a machine breaks down it is not used again for the rest of that day. What is the probability of 0, 1, 2, 3, 4, 5 machines breaking down in any day?

Solution: The situation satisfies the conditions necessary for the binomial distribution to be appropriate, since:

1 The 5 machines represent 5 identical trials.
2 The 5 machines run independently of each other.
3 There are 2 possible outcomes for each machine—it breaks down in a given day, or it does not.
4 For each machine, the probability that it breaks down on any day is the same, 0.2. The probability that it does not break down in any day is therefore 0.8.

Hence, the probability of r breakdowns in any one day is given by:

$$P(r) = {}^5C_r(0.2)^r(0.8)^{5-r} \quad r = 0, 1, \ldots, 5$$

TABLE 2.6 **Probability of breakdowns in any one day**

Number of machines breaking down, r	Number of machines not breaking down, $5-r$	Number of ways this can happen ${}^5C_r$		P(r machines breakdown/day)
0	5	$\frac{5!}{0!(5-0)!}$ = 1		$1 \times (0.2)^0 \times (0.8)^5 = 0.3277$
1	4	$\frac{5!}{1!(5-1)!}$ = 5		$5 \times (0.2)^1 \times (0.8)^4 = 0.4096$
2	3	$\frac{5!}{2!(5-2)!}$ = 10		$10 \times (0.2)^2 \times (0.8)^3 = 0.2048$
3	2	$\frac{5!}{3!(5-3)!}$ = 10		$10 \times (0.2)^3 \times (0.8)^2 = 0.0512$
4	1	$\frac{5!}{4!(5-4)!}$ = 5		$5 \times (0.2)^4 \times (0.8)^1 = 0.0064$
5	0	$\frac{5!}{5!(5-5)!}$ = 1		$1 \times (0.2)^5 \times (0.8)^0 = 0.0003$
			Total	1.0000

The discrete probability distribution of P(r) against r may be illustrated by a line graph:

FIGURE 2.4 **Probability distribution for the number of machines out of 5 which break down per day**

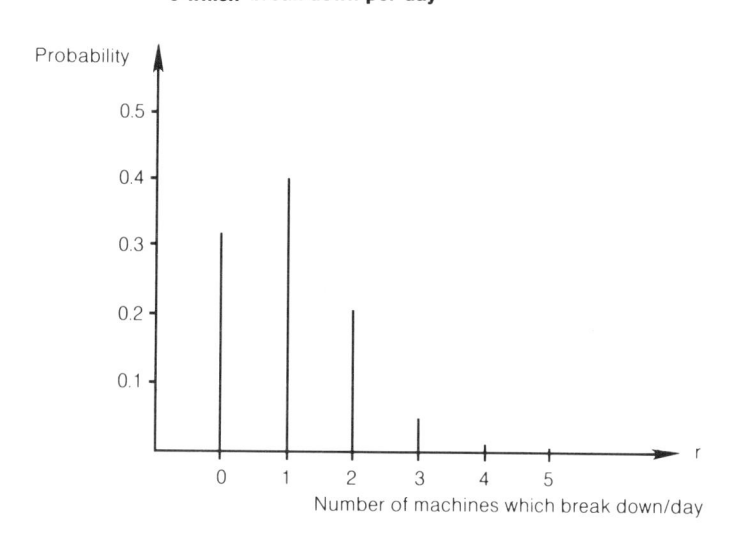

2.3.3 The expected value and standard deviation for a binomial probability distribution

For any discrete random variable the expected value of the variable is:

$$E(r) = \sum rP(r)$$

It can be shown that for a variable with a binomial probability distribution:

$$E(r) = \sum rP(r) = np$$

where n is the number of trials, p is the probability of a success on any trial and P(r) is the binomial probability. Similarly, for any discrete random variable, the standard deviation of the variable values is:

$$\sigma = \sqrt{\sum r^2 P(r) - (\sum rP(r))^2}$$

Again for a variable with a binomial probability distribution, it may be shown that:

$$\sigma = \sqrt{\sum r^2 P(r) - (\sum rP(r))^2} = \sqrt{np(1-p)} = \sqrt{npq}$$

and, the variance is:

$$\sigma^2 = npq$$

where q is the probability of a failure on any trial. These properties are illustrated in **Example 2.5**, below.

Example 2.5: Expected value and standard deviation for a binomial probability distribution

The data from **Example 2.4,** on 5 machines, is used to calculate the expected number of machines breaking down per day, and the standard deviation of the number of machines breaking down per day. The calculation is done using the general formulae for E(r) and σ, and then again using the reduced formulae which are applicable only to the binomial distribution.

TABLE 2.7 **Probability of r machines breaking down per day**

Number of machines breaking down/day r	Probability of r machines breaking down/day, P(r)	rP(r)	r²P(r)
0	0.3277	0	0
1	0.4096	0.4096	0.4096
2	0.2048	0.4096	0.8192
3	0.0512	0.1536	0.4608
4	0.0064	0.0256	0.1024
5	0.0003	0.0015	0.0075
Total	1.0000	0.9999	1.7995

Note: The probabilities are rounded to 4 decimal places, leading to small errors in the fourth decimal place of the totals. The totals should be 1.0000 and 1.8000.

The expected number of machine breakdowns in a day, using the general formula is:

$$E(r) = \sum rP(r)$$

$$E(r) = 1.0$$

and, for the binomial distribution:

$$E(r) = np = 5 \times 0.2 = 1.0$$

The expected number of breakdowns per day is 1 machine. Similarly, the standard deviation of the number of breakdowns per day using the general formula is:

$$\sigma = \sqrt{\sum r^2 P(r) - (\sum rP(r))^2} \quad \text{therefore}$$

$$\sigma = \sqrt{1.8 - 1.0^2} = 0.8944 \quad \text{(to 4 decimal places)}$$

Again, using the formula specific to the binomial distribution gives:

$$\sigma = \sqrt{npq} = \sqrt{5 \times 0.2 \times 0.8} = 0.8944 \quad \text{(to 4 decimal places)}$$

and variance:

$$\sigma^2 = npq = 5 \times 0.2 \times 0.8 = 0.8$$

We sometimes want to know the proportion of successes out of n trials rather than the number of successes. Modifying the above formulae gives:

$$E(\text{proportion of successes in n trials}) = \frac{np}{n} = p$$

and the standard deviation of the proportion of successes is given by:

$$\sigma(\text{proportion of successes}) = \frac{\sqrt{npq}}{n} = \sqrt{\frac{pq}{n}}$$

Therefore in **Example 2.5**, the expected proportion of machine breakdowns per day is:

$$\frac{np}{n} = \frac{1}{5} = 0.2 = p$$

and the standard deviation of the proportion of machines breaking down per day is:

$$\sigma = \sqrt{\frac{0.2 \times 0.8}{5}} = 0.179 \quad \text{(to 3 decimal places)}$$

The variance of the proportion of machines breaking down per day is:

$$\sigma^2 = \frac{0.2 \times 0.8}{5} = 0.032$$

Example 2.6: Expected value and standard deviation for a binomial probability distribution

A company produces a large number of springs, 10% of which are defective. Samples of 100 springs are selected at random for quality control inspection purposes. Find the expected number of defective springs and the standard deviation of the number of defective springs per sample of 100. Write down the probability that one such sample contains at least 15 defectives.
Solution: The binomial distribution is appropriate since:

1   there are 100 identical trials;
2   the trials are independent since the springs are selected randomly;
3   for each trial there are 2 possible outcomes—the spring is defective or it is not;
4   the probability of any one spring being defective is virtually constant at 0.1. There is a large number of springs from which to select the sample of 100, therefore the proportion of defective springs will not vary significantly as the sample is removed from the total output of springs.

QUANTITATIVE ANALYSIS

The expected number of defective springs is:

$$E(r) = np = 100 \times 0.1 = 10 \text{ springs per sample}$$

Standard deviation of the number of defectives, $\sigma = \sqrt{npq}$:

$$\sigma = \sqrt{100 \times 0.1 \times 0.9} = 3 \text{ springs per sample}$$

The probability that there are r defectives in a sample is given by:

$$P(r \text{ defectives out of } 100) = {}^{100}C_r (0.1)^r (0.9)^{100-r}$$

$$r = 0, 1, 2, \ldots, 100$$

$$P(r \geq 15) = P(15) + P(16) + P(17) + \cdots + P(100)$$

$$= 1 - P(0) + P(1) + \cdots + P(14)$$

This is a long and tedious calculation and methods of approximately evaluating this probability are found in sections **2.5** and **2.8**.

2.4 THE POISSON PROBABILITY DISTRIBUTION

2.4.1 What is the Poisson probability distribution?

Consider a situation in which the number of accidents occurring per week on a particular stretch of road is observed. We may regard the number of accidents occurring per week as a random variable which may take the values 0, 1, 2, 3, ... (there is no upper limit to the value of the variable). The number of trials, ie the number of potential accidents, is unlimited. If we consider a small interval of time within the week, a minute, say, then within this minute an accident will either happen or not happen. The probability of an accident occurring within a particular minute is very small and it may be assumed to be the same for all individual minutes.

The probability distribution of the number of accidents occurring per week is given by the function:

$$P(r \text{ accidents per week}) = \frac{m^r e^{-m}}{r!}$$

$$r = 0, 1, 2, 3, \ldots$$

where m is the average number of accidents per week on the stretch of road and e is the constant 2.718 ... which may be found by using a calculator which has statistical functions or by using mathematical tables. This is a Poisson probability distribution of the number of accidents occurring per week. The essential features of this situation which make the Poisson probability distribution appropriate are:

1 Each small interval of time may be regarded as a trial. During the trial there is either an accident (success) or there is not (failure). The intervals are so small that there can be only one success during the interval and the probability of success is very small and constant.
2 The number of successes in a large interval of time is independent of the number of successes in any other large interval, ie the number of successes are randomly scattered through time.
3 The average number of successes is constant through time.

The random variables may refer to intervals other than time. For example, we might be interested in the number of typing errors made per page or the number of defects per mile in the surface of a road. The Poisson probability distribution can also be used to model these situations. All we need to know is the average number of successes per interval, ie the mean number of errors per page or the mean number of defects per mile of road.

In general the Poisson probability distribution is given by:

$$P(r \text{ successes per interval}) = \frac{m^r e^{-r}}{r!} \quad r = 0, 1, 2, 3, \ldots$$

where m is the mean number of successes per unit. Poisson probability tables are available in which values of $P(r)$ are tabulated for some values of m and r.

Example 2.7: Poisson probability distribution

It is known that the average demand for a telephone line is 3 per 5 minute interval. Calculate the probability that the demand will be for 0, 1, 2, 3, 4 or more than 4 lines per 5 minutes.
Solution: This situation may be represented by a Poisson probability distribution since:

1   There is an unlimited number of trials, ie small intervals of time in which there may or may not be a single demand for a telephone line, and the probability of a success in any one small interval of time is very small and may be assumed to be constant.
2   The demand for a telephone line is assumed to be randomly scattered through time.
3   The average number of demands for a line in any 5 minute period may be assumed to be constant over time.

In this example the mean number of demands for a line = 3 demands per 5 minutes. Therefore the Poisson probability distribution is:

$$P(r \text{ demands per 5 minutes}) = \frac{3^r e^{-3}}{r!} \quad r = 0, 1, 2, 3, 4, \ldots$$

$$P(0 \text{ demands/5 mins}) = P(0) = \frac{3^0 e^{-3}}{0!} = \frac{1}{1} \times 0.0498 = 0.0498$$

$$P(1 \text{ demands/5 mins}) = P(1) = \frac{3^1 e^{-3}}{1!} = \frac{3}{1} \times P(0) = 3 \times 0.0498 = 0.1494$$

$$P(2 \text{ demands/5 mins}) = P(2) = \frac{3^2 e^{-3}}{2!} = \frac{3}{2} \times P(1) = \frac{3}{2} \times 0.1494 = 0.2240$$

$$P(3 \text{ demands/5 mins}) = P(3) = \frac{3^3 e^{-3}}{3!} = \frac{3}{3} \times P(2) = \frac{3}{3} \times 0.2240 = 0.2240$$

$$P(4 \text{ demands/5 mins}) = P(4) = \frac{3^4 e^{-3}}{4!} = \frac{3}{4} \times P(3) = \frac{3}{4} \times 0.2240 = 0.1680$$

$$P(\text{More than 4 demands/5 mins}) = \{P(5) + P(6) + P(7) + P(8) + \cdots\}$$

$$= 1 - \{P(0) + P(1) + P(2) + P(3) + P(4)\}$$

$$= 1 - \{0.0498 + 0.1494 + 0.2240 + 0.2240 + 0.1680\}$$

$$= 1 - 0.8152 = 0.1848$$

The conditions necessary for the Poisson probability distribution mean that if we know the average number of successes per 5 minutes, m, say, as in **Example 2.7,** but we want to know the probability of r successes per hour, then we simply set up a Poisson probability distribution with a new mean of m × 12. In **Example 2.7** the mean number of demands for a telephone line per hour is $3 \times 12 = 36$ demands per hour. Similarly, if we wanted the probability distribution of the demands per minute, the mean would be $m = 3/5 = 0.6$ demands per minute.

Example 2.8: Poisson probability distribution

On average a production line develops 3.4 faults per 5 day working week. What is the probability that 2 faults develop on any day of production?

Solution: This situation may be described by a Poisson probability distribution because:

1 There is an unlimited number of trials, ie small intervals of time, in which there may or may not be a single fault on the production line. The probability of a success in any one small interval of time is very small and may be assumed to be constant.
2 The faults are assumed to be randomly scattered through time.
3 The average number of faults in any 5 day period is assumed to be constant over time.

The mean number of faults = 3.4 per 5 days. Therefore the number of faults per day is:

$$\frac{3.4}{5} = 0.68 = m$$

Hence:

$$P(r \text{ faults per day}) = \frac{0.68^r e^{-0.68}}{r!} \quad r = 0, 1, 2, 3, \ldots$$

Therefore:

$$P(2 \text{ faults per day}) = \frac{0.68^2 e^{-0.68}}{2!} = 0.1171$$

Example 2.9: Poisson probability distribution

A car hire firm has 2 cars which it hires out by the day. The daily demand for the cars follows a Poisson distribution with a mean of 1.3 requests per day. If both cars are used equally, calculate the probabilities that on any day:

1 no car is hired;
2 a particular car is not hired;
3 both cars are hired.

Solution: The discrete random variable is the number of requests per day for a car. The probability of r requests per day is given by:

$$P(r) = \frac{1.3^r e^{-1.3}}{r!} \quad r = 0, 1, 2, 3, \ldots$$

1 P(no car is hired on a given day) = P(demand per day is 0)

$$P(r = 0) = \frac{1.3^0 e^{-1.3}}{0!} = 0.2725$$

2 P(a particular car is not hired) = P(demand is 0 **or** demand is 1 **and** the other car is used)

$$= P(r = 0) + P(r = 1) \times P(\text{other car used})$$

$$= 0.2725 + \frac{1.3 e^{-1.3}}{1!} \times \frac{1}{2}$$

$$= 0.4497$$

Note: P(other car is used) = $\frac{1}{2}$, since the cars are used equally.

3 P(both cars are hired) = P(demand per day ≥ 2)

$$= 1 - P(r \leqslant 2)$$
$$= 1 - \{P(r = 0) + P(r = 1)\}$$
$$= 1 - 0.6268 = 0.3732$$

2.4.2 The expected value and variance of a Poisson probability distribution

The expected value or mean number of successes per interval must be determined from the data for any particular situation. Once the expected value is known, then the variance is also known, since it is one of the properties of a Poisson probability distribution that:

$$\text{Expected value} = \text{variance}$$

Therefore the standard deviation of the number of successes per interval is:

$$\text{Standard deviation} = \sqrt{\text{variance}} = \sqrt{\text{expected value}}$$

This is a useful property if we have a set of observations for a variable and we want to know whether or not a Poisson probability distribution would be appropriate.

2.5 THE POISSON DISTRIBUTION AS AN APPROXIMATION TO THE BINOMIAL DISTRIBUTION

In certain circumstances the Poisson probability distribution may be used as an approximation to the binomial probability distribution. This is of value in situations such as the one described in **Example 2.6**, section **2.3.3**, in which the calculation is long and tedious using the binomial probability distribution. If the necessary conditions are satisfied then the calculation using the Poisson probability distribution is much simpler and gives very similar results. The Poisson approximation is good if the binomial situation has:

1 a large number of trials; ie n is large, preferably $n \geqslant 30$;
2 a small probability of a success in each trial, ie p is small, preferably $p \leqslant 0.10$;
3 an expected number of successes less than 5, ie $np \leqslant 5$.

In these circumstances the Poisson distribution with a mean, $m = np$ gives a reasonable approximation to the binomial distribution. The bigger n is and the smaller p is, the better the approximation will be. The values given above are guidelines only. In section **2.8**, we will consider how to deal with difficult binomial calculations when the value of p is large. In these cases the normal distribution may be used as an approximation.

Example 2.10: The Poisson distribution as an approximation to the binomial distribution

The manufacturer of a pocket calculator knows from experience that 1% of all the calculators manufactured and sold by his company are defective and will have to be replaced under the guarantee. A large accounting firm purchases 500 calculators from the manufacturer. Find the probability that 5 or more of the calculators will have to be replaced.
Solution: This situation is properly described by a binomial probability distribution:

$$n = 500, \ p = 0.01, \ q = 0.99 \text{ and } np = 5$$

Since n is large, p is small and $np \leqslant 5$, a Poisson probability distribution can be used as an approximation to the binomial.

42 QUANTITATIVE ANALYSIS

(a) The calculation using the binomial distribution; probability of r defectives in the sample is given by:

$$P(r) = {}^{500}C_r \times 0.01^r \times 0.99^{500-r} \quad r = 0, 1, 2, \ldots, 500$$

$$P(r \geqslant 5) = 1 - P(r \leqslant 5) = 1 - \{P(r=0) + P(r=1) + P(r=2) + P(r=3) + P(r=4)\}$$

$$P(0) = 0.01^0 \times 0.99^{500} \times \frac{500!}{500!0!} = 0.99^{500} \qquad\qquad = 0.00657$$

$$P(1) = 0.01^1 \times 0.99^{499} \times \frac{500!}{499!1!} = 0.01 \times 0.99^{499} \times 500 \qquad = 0.03318$$

$$P(2) = 0.01^2 \times 0.99^{498} \times \frac{500!}{498!2!} = 0.01^2 \times 0.99^{498} \times \frac{500 \times 499}{2 \times 1} = 0.08363$$

$$P(3) = 0.01^3 \times 0.99^{497} \times \frac{500!}{497!3!} = 0.01^3 \times 0.99^{497} \times \frac{500 \times 499 \times 498}{3 \times 2 \times 1} = 0.14023$$

$$P(4) = 0.01^4 \times 0.99^{496} \times \frac{500!}{496!4!} = 0.01^4 \times 0.99^{496} \times \frac{500 \times 499 \times 498 \times 497}{4 \times 3 \times 2 \times 1} = 0.17600$$

Total 0.43961

Therefore:

$$P(r \geqslant 5) = 1 - \{0.43961\}$$

$$= 0.56039 \quad \text{(to 5 decimal places)}$$

The probability that 5 or more calculators will have to be replaced is 0.560 (to 3 decimal places)

(b) the calculation using the Poisson distribution:

$$m = np = 500 \times 0.01 = 5$$

The probability of r defectives per sample is approximately given by:

$$P(r) = \frac{5^r e^{-5}}{r!} \quad r = 0, 1, 2, 3, \ldots$$

Therefore:

$$P(r \geqslant 5) = 1 - \{P(r=0) + P(r=1) + P(r=2) + P(r=3) + P(r=4)\}$$

$$P(0) = e^{-5} \qquad\qquad\qquad\qquad = 0.00674$$

$$P(1) = e^{-5} \times 5^1 = P(0) \times 5 = 0.03369$$

$$P(2) = e^{-5} \times \frac{5^2}{2!} = P(1) \times \frac{5}{2} = 0.08422$$

$$P(3) = e^{-5} \times \frac{5^3}{3!} = P(2) \times \frac{5}{3} = 0.14037$$

$$P(4) = e^{-5} \times \frac{5^4}{4!} = P(3) \times \frac{5}{4} = 0.17547$$

Total 0.44049

Therefore:

$$P(r \geqslant 5) = 1 - 0.44049$$

$$= 0.55951 \quad \text{(to 5 decimal places)}$$

The probability that 5 or more calculators will have to be replaced is 0.560 to 3 significant figures. To this accuracy, it is the same value as that obtained in part (a) using the correct binomial distribution.

2.6 PROBABILITY DISTRIBUTIONS FOR CONTINUOUS RANDOM VARIABLES

2.6.1 Continuous variables and probability density

The earlier sections of this chapter considered probability distributions for discrete random variables. The rest of the chapter deals with probability distributions for continuous random variables, in particular the uniform and normal distributions. If our experiment of chance results in a random variable which is continuous, then the variable can take any value within a specified range, compared with the specific values taken by a discrete variable. We therefore have to modify the way in which the variable values are specified and also the way in which the probability distribution is described.

We may, for example, be carrying out an experiment to measure the height or the weight of students, or the time to make a journey, or the volume of a plastic bottle. If the plastic bottles are made in a mass production process, the actual volume of the bottles will vary about the specified volume. Suppose the volume of the bottles is meant to be 200 ml, the actual volumes may vary between 190 ml and 210 ml for example. If the volumes of a batch of bottles are measured, the continuous random variable, the volume of the bottles, may take any value over the range 190 ml to 210 ml. There is an unlimited number of possible values for the variable. The probability function from which the probability distribution is obtained is now a continuous function of the random variable. Suppose we have a continuous random variable X which can take values x, then the probability function is a continuous function of x for values of x between x_1 and x_2, ie;

$$\text{probability function} = f(x) \quad x_1 < x < x_2$$

If the probability function is now plotted against x the resulting graph will be a curve rather than the line diagram for discrete variables.

Example 2.11: Continuous probability functions

(a) Suppose the probability function is:

$$f(x) = 0.125(5 - x) \quad \text{for} \quad 1 < x < 5$$

Then the probability graph looks like FIGURE 2.5:

FIGURE 2.5 **Probability density function, f(x) = 0.125 (5 − x)**

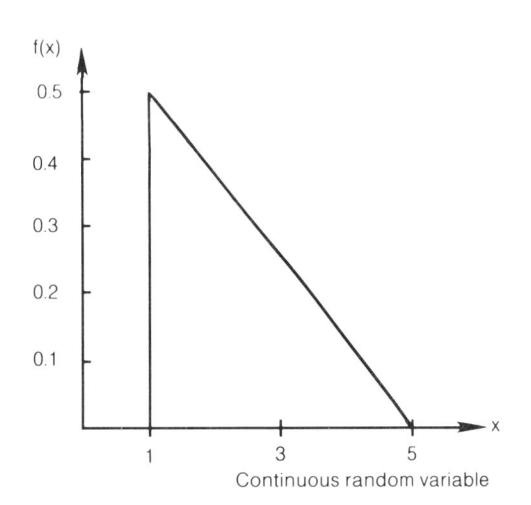

Continuous random variable

(b) Suppose the probability function is:

$$f(x) = 6x(1 - x) \quad 0 \leqslant x \leqslant 1$$

Then the probability graph looks like FIGURE 2.6:

FIGURE 2.6 **Probability density function, f(x) = 6x(1 − x)**

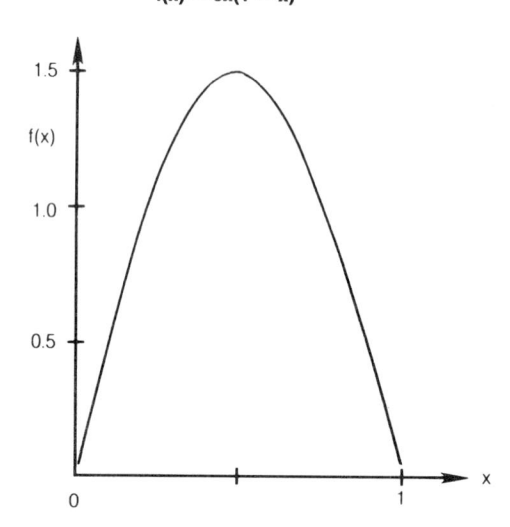

When dealing with a discrete variable, we have specific values of the variable to which a probability may be attached. This is clearly not possible with a continuous variable. In order to solve this difficulty, we think of the probability as being associated with a range of values of the continuous variable. For example, we may want to know the probability that the volume of a bottle lies in the range 195 ml to 197 ml.

The probability is then represented graphically by an area rather than by the height of a line. This is a similar procedure to that of representing a frequency by the area of a rectangle when drawing a histogram. In a histogram, the total area of the rectangles represents the total frequency. In the same way, the total area under a probability function represents the total probability, 1. In **Example 2.11**(a) above, we may represent the probability of X taking a value between 2 and 2.5 by the shaded area of the probability function graph.

FIGURE 2.7 **Probability that X takes a value between 2 and 2.5**

Area gives the probability that x takes a value between 2 and 2.5

Continuous random variable

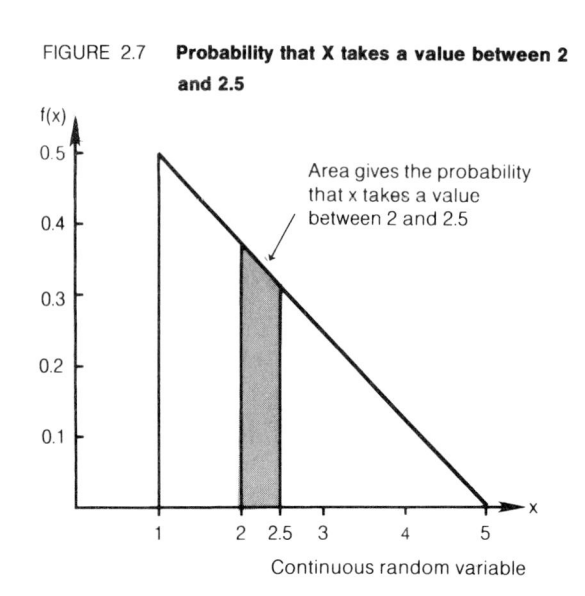

In this context the probability function, f(x), is called the **probability density**. Hence:

$$\text{probability density, } P(x) = f(x) \quad x_1 \leqslant x \leqslant x_2$$

The probability that the continuous variable takes a value within a given range is then the area under the probability density function for that range. With continuous random variables, the probability always refers to a range of values of the variable, eg less than 5 minutes, 21.0 to 23.5 m or more than 3.6 kg. The probability never refers to a specific value of the variable, eg the probability that x is 5 minutes does not have any meaning for a continuous variable.

2.6.2 The uniform distribution

The uniform distribution is a simple example of a probability distribution of a continuous random variable. We will use it as an example to illustrate the points made in the previous section.

Example 2.12: The uniform distribution

Take two accountants and measure the time it takes them to travel from home to work. The first accountant's journey takes between 20 and 25 minutes. The second accountant's journey takes between 20 and 30 minutes. Any journey time between these limits is equally likely. FIGURES 2.8 and 2.9 below illustrate these two distributions. They are examples of the uniform distribution.

FIGURE 2.8 **Probability distribution of journey time to work for the first accountant**

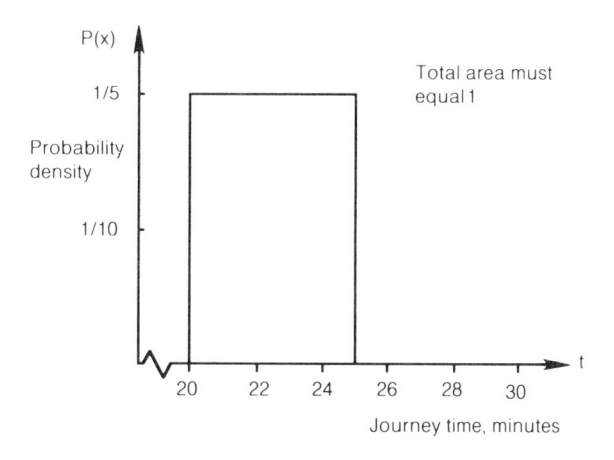

FIGURE 2.9 **Probability distribution of journey time to work for the second accountant**

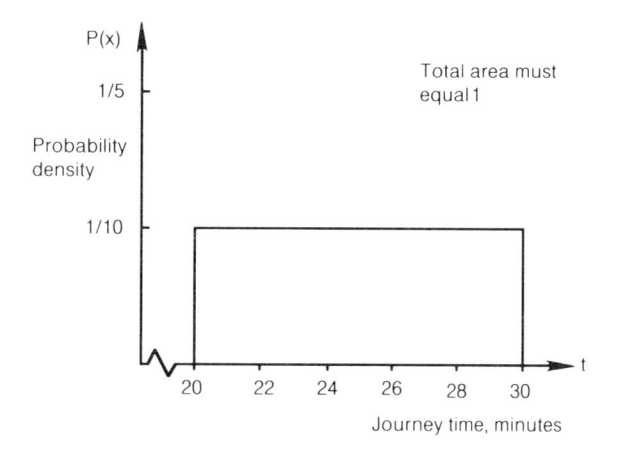

If we remember that the total probability must always equal 1, then the area under the curve must equal 1. This makes it easy to calculate the probability density for the uniform distribution. For the first accountant, the total time interval is 5 minutes. The width of the rectangle is 5, therefore the height (the probability density) must be 1/5. The total area under the curve is (1/5) × 5 = 1. Similarly for the second accountant, his total time interval is 10 minutes, therefore the probability density must be 1/10 so that the total area under the curve is (1/10) × 10 = 1, the total probability. We can now calculate the probability for each accountant that his journey will take between 20.5 minutes and 22.8 minutes. The probability is the area under each curve between these two limits.

First accountant:

$$P(\text{journey takes between 20.5 and 22.8 mins}) = (1/5) \times (22.8 - 20.5)$$
$$= (1/5) \times 2.3$$
$$= 0.46$$

Second accountant:

$$P(\text{journey takes between 20.5 and 22.8 mins}) = (1/10) \times (22.8 - 20.5)$$
$$= (1/10) \times 2.3$$
$$= 0.23$$

2.7 THE NORMAL PROBABILITY DISTRIBUTION FOR CONTINUOUS RANDOM VARIABLES

2.7.1 The nature of the normal probability distribution

The most important probability distribution for continuous variables is the normal probability distribution. It is appropriate in many situations where measurement is involved, eg the weight or volume of a product delivered by a filling machine on a packaging line in a factory, the heights of adult males, the lifetimes of electrical goods. The essential features of the uniform distribution discussed in the last section also apply to the normal distribution:

1 The area under the normal curve gives the probability that the continuous variable takes a value within a given range.
2 The total area under the normal curve is equal to the total probability, 1.
3 The probability that the continuous variable takes a specific value has no meaning.

The **normal probability distribution** is a symmetrical distribution, centred on the mean value of the variable and existing theoretically for values of the variable from minus infinity to plus infinity, so that the values of the continuous variable are unlimited in both the positive and negative directions. In practice, however, the normal distribution is used for variables which take values over a limited range. A typical normal probability distribution is illustrated in FIGURE 2.10 below.

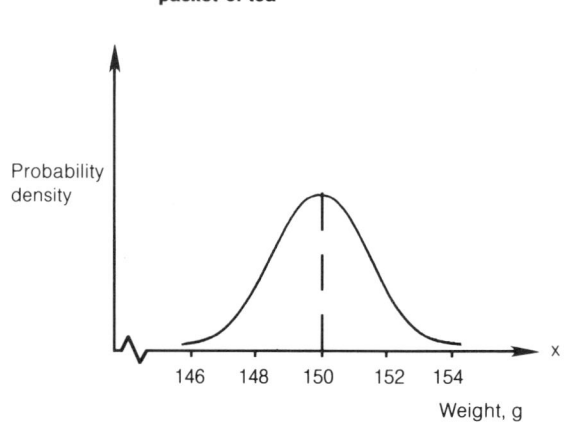

FIGURE 2.10   **Probability distribution for the weight of a packet of tea**

The probability density function for the normal distribution is a complicated mathematical function which depends on the values of the mean (μ) and variance (σ^2) of the variable. Fortunately, this property enables all normal probability distributions to be reduced to just one distribution, the standard normal probability distribution. Standard tables have been constructed for this distribution which give the area under the curve for various values of the continuous variable, enabling us to determine the probabilities we require.

In the standard normal distribution the mean of the variable is zero, the variance is 1 and, therefore, the standard deviation is 1. The process of converting any normal probability distribution to the standard normal distribution simply involves re-writing the values of the variable (whether they are number of centimetres or minutes or grams) in terms of the number of standard deviations away from the mean value of the variable. Once the conversion from real units to standard deviation units (z units) has been done, the required probabilities may be found from the tables.

Normal probability distribution tables can be arranged in a number of ways, all of which give the same solution, but by slightly different procedures. The tables in Appendix Two give the probability of the variable value being more than z standard deviations above the mean.

2.7.2 The standard normal probability distribution

The standard normal probability distribution may be represented as follows:

FIGURE 2.11 **Standard normal probability distribution**

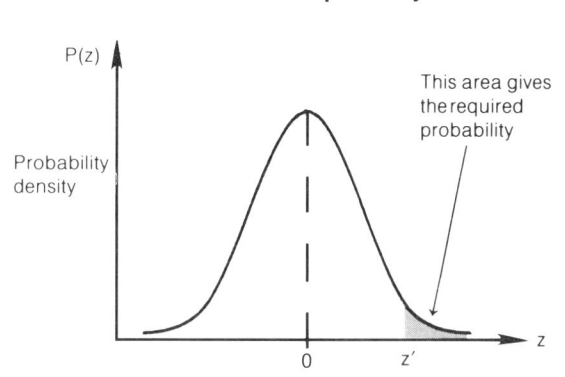

The probabilities are tabulated for values of z from $z = 0$ to $z = 3.5$ approximately. Since the distribution is symmetrical the same figures can be used to give the probability of the variable value being z or more standard deviations below the mean.

FIGURE 2.12 **Standard normal probability distribution**

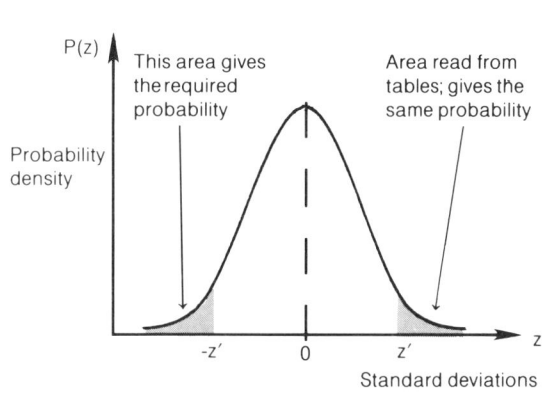

QUANTITATIVE ANALYSIS

The values of a continuous variable may be converted to the number of standard deviations from the mean using:

$$z = \frac{x - \mu}{\sigma}$$

where x is the value of the variable, μ is the mean of the variable values and σ is the standard deviation. The numerical example which follows illustrates the use of the normal distribution tables.

Example 2.13: Use of the standard normal distribution tables

A firm manufactures light bulbs for which the length of life is normally distributed with a mean lifetime of 600 hours and a standard deviation of lifetime of 40 hours. What is the probability that a bulb will have a lifetime:

1 less than 700 hours;
2 less than 550 hours;
3 between 550 and 700 hours.
4 The 2% of bulbs with the shortest lifetimes, have lifetimes below which value?

Solution: The probability density function for the lifetimes of the bulbs is illustrated below in FIGURE 2.13:

FIGURE 2.13 **Probability function for the lifetime of the bulbs**

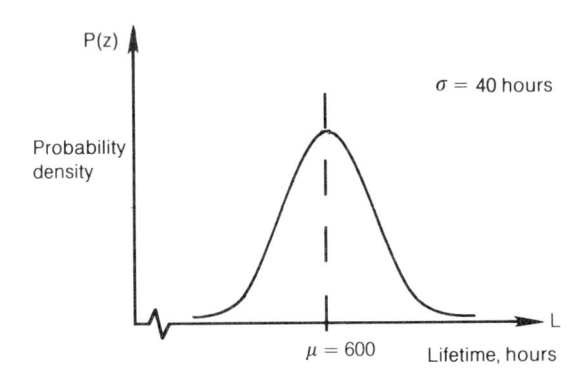

1 The probability that a bulb will have a lifetime less than 700 hours is given by the shaded area in FIGURE 2.14 below:

FIGURE 2.14 **Probability that the bulb will have a lifetime of less than 700 hours**

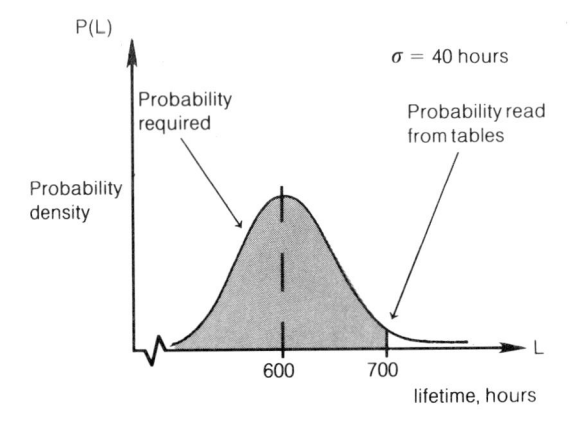

Standardise this normal distribution to one with mean = 0 and standard deviation = 1. Calculate how many standard deviations away from the mean, z, 700 hours represents.

$$z = \frac{L - \mu}{\sigma}$$

$$z = \frac{700 - 600}{40} = 2.5 \text{ standard deviations above the mean}$$

The normal probability tables give:

$$P(z > 2.5) = 0.0062$$

Since the total probability is 1:

$$P(z < 2.5) = 1 - 0.0062$$
$$= 0.9938$$

Therefore, the probability that a bulb has a lifetime less than 700 hours is 0.9938. This may also be interpreted as, the proportion of bulbs which have lifetimes less than 700 hours is 0.9938, or 99.38%

2   The probability that a bulb will have a lifetime less than 550 hours is given by the shaded area in FIGURE 2.15 below:

FIGURE 2.15   **Probability that the bulb will have a lifetime of less than 550 hours**

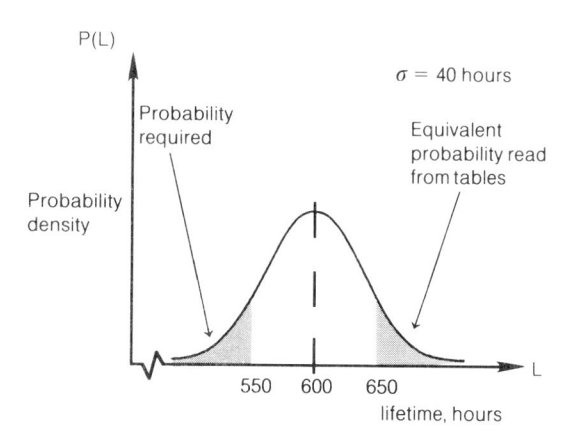

Standardise this normal distribution to one with mean = 0 and standard deviation = 1. Calculate how many standard deviations away from the mean, z, 550 hours represents.

$$z = \frac{L - \mu}{\sigma} \text{ standard deviations}$$

$$z = \frac{550 - 600}{40} = -1.25 \text{ standard deviations below the mean}$$

Remember that the normal distribution is symmetrical, therefore:

$$P(z \leq -1.25) = P(z \geq 1.25)$$

The normal probability tables give:

$$P(z \geq 1.25) = 0.1056$$

Therefore:

$$P(z \leq -1.25) = 0.1056$$

The probability of a bulb having a lifetime below 550 hours is 0.1056. This may also be interpreted as, 10.56% of bulbs have lifetimes less than 550 hours.

3 The probability that a bulb will have a lifetime between 550 hours and 700 hours is given by the shaded area in FIGURE 2.16:

FIGURE 2.16 **Probability that the bulb will have a lifetime between 550 and 700 hours**

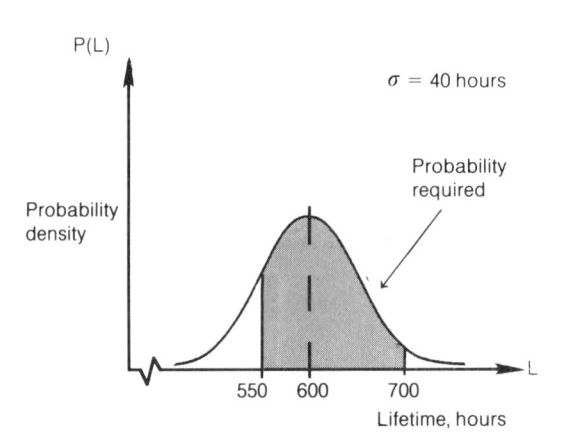

$$P(550 < L < 700 \text{ hours}) = P(L < 700 \text{ hours}) - P(L < 550 \text{ hours})$$

We have already calculated the necessary probabilities in parts **1** and **2** Therefore:

$$P(550 < L < 700 \text{ hours}) = 0.9938 - 0.1056 = 0.8882$$

The probability of a bulb having a lifetime between 550 hours and 700 hours is 0.8882, or 88.82% of bulbs have lifetimes between 550 and 700 hours.

4 This question is slightly different. We are starting with a percentage of bulbs and are required to find the corresponding lifetime. Let L be the value of lifetime which we are trying to find. The situation is illustrated in FIGURE 2.17.

Firstly, we find the probability 0.02, in the standard normal tables. The corresponding value of z is 2.055 approximately. This means that the probability of a value more than 2.055 standard deviations above the mean is 0.02. Since the normal distribution is symmetrical, we can also say that the probability of a value which is less than 2.055 standard deviations below the mean is 0.02. Therefore, the required lifetime is 2.055 × 40 hours = 82.2 hours below the

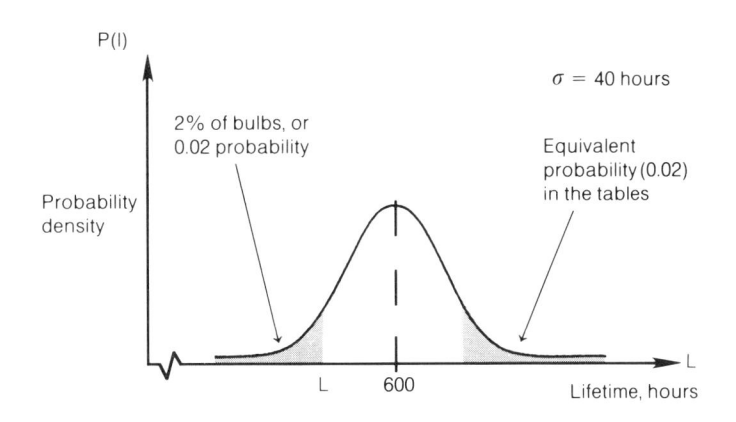

FIGURE 2.17 **Two per cent of light bulbs with the shortest lives**

mean. The value of L is 600 − 82.2 = 517.8 hours. The 2 % of light bulbs with the shortest lives have lifetimes below 517.8 hours.

2.8 NORMAL DISTRIBUTION AS AN APPROXIMATION TO THE BINOMIAL DISTRIBUTION

In some situations the amount of arithmetic required to evaluate a probability from the binomial or Poisson distributions is considerable. It is therefore convenient to use the normal distribution to provide an approximation of the required probability more quickly.

A minor problem arises, however, since the normal distribution assumes that the variable is continuous and the binomial and Poisson distributions assume that the variable is discrete. This problem can be solved if a correction is applied when using the approximation. The correction is called the **continuity correction**.

From the binomial distribution we can calculate, for example, the probability of exactly 2 defectives in a sample of n items. To use the normal distribution as an approximation, we pretend that this discrete variable value, 2, is a value of a continuous variable in the range 1.5 to 2.5. This is the continuity correction. The normal distribution used has the same mean and standard deviation as the correct binomial distribution. The area under the normal curve from 1.5 to 2.5 gives an approximate value for the discrete probability of exactly 2 defectives.

The approximation is used only if the correct binomial calculations would take too long, and only then if certain conditions are met. In section **2.5** we used the Poisson distribution as an approximation to the binomial. This worked well as long as n was large, p was small and np ⩽ 5. The normal distribution can be used as an approximation to the binomial if both np and nq are greater than 5. This generally means that n must be large and p bigger than 0.1, preferably near to 0.5. All of these figures are guidelines only. The bigger n, np and nq, the better will be the approximation.

The mean and standard deviation using the binomial are given by:

$$E(r) = np[= \mu] \quad \text{and} \quad \sigma = \sqrt{npq}$$

These values are used to calculate the z values for the normal distribution as explained in section **2.7**.

Example 2.14: Normal approximation to the binomial

A factory produces microchips in large numbers each day. The defective rate is very high at 40% of the production. A sample of 20 is drawn at random from the day's output. What is the probability that 14 or more out of the 20 are defective?

Solution: The calculation may be done using the correct binomial distribution:

$$P(r \text{ defectives in a sample of } 20) = (0.4)^r \times (0.6)^{20-r} \times {}^{20}C_r$$

$$r = 0, 1, \ldots, 20$$

$$P(14 \text{ or more defectives}) = P(14) + P(15) + P(16) + P(17) + P(18) + P(19) + P(20)$$

$$P(14) = 0.4^{14} \times 0.6^6 \times \frac{20!}{14!6!} = 0.004854$$

$$P(15) = 0.4^{15} \times 0.6^5 \times \frac{20!}{15!5!} = 0.001294$$

$$P(16) = 0.4^{16} \times 0.6^4 \times \frac{20!}{16!4!} = 0.000270$$

$$P(17) = 0.4^{17} \times 0.6^3 \times \frac{20!}{17!3!} = 0.000042$$

$$P(18) = 0.4^{18} \times 0.6^2 \times \frac{20!}{18!2!} = 0.000005$$

$$P(19) = 0.4^{19} \times 0.6^1 \times \frac{20!}{19!1!} = 0.000000$$

$$P(20) = 0.4^{20} \times 0.6^0 \times \frac{20!}{20!0!} = 0.000000$$

$$\overline{0.006465}$$

$$P(14 \text{ or more chips are defective from a sample of } 20) = 0.006465$$

The calculation using the normal approximation is very much easier. Firstly we check that the conditions apply for using the approximation:

$$np = 20 \times 0.4 = 8$$
$$nq = 20 \times 0.6 = 12$$

The approximation should be reasonably close.

We must now consider the application of the continuity correction. The discrete variable value 14 is taken to mean a continuous variable value in the range 13.5 to 14.5. Instead of finding the discrete probability that 14 or more chips are defective, we find the continuous probability of more than 13.5 defective chips. The mean for the normal distribution is $np = 8$ and the standard deviation is:

$$\sqrt{npq} = \sqrt{20 \times 0.4 \times 0.6} = 2.19$$

Calculate the z value at 13.5:

$$z = \frac{13.5 - 8}{2.19} = 2.51 \text{ standard deviations above the mean}$$

The standard normal tables give:

$$P(z \geqslant 2.51) = 0.0060$$

Therefore:

$$P(\text{number of defectives} \geqslant 13.5) = 0.0060$$

FIGURE 2.18   **The continuous probability of more than 13.5 defective chips**

Compare this figure from the short normal calculation with:

$$P(14 \text{ or more defectives in a sample of } 20) = 0.0065$$

from the long binomial calculation.

Example 2.15: Normal approximation to the binomial—proportions

From past experience, the auditors of ABC & Co Ltd have found that the mean number of incorrect entries in the company's accounts is 35 per 1000 entries in any one financial year. What is the probability that the proportion of incorrect entries found during the current audit is more than 0.05?
Solution:

$$\text{mean proportion} = p = \frac{35}{1000} = 0.035$$

$$\text{standard deviation} = \sqrt{\frac{pq}{n}} = \sqrt{\frac{0.035 \times 0.965}{1000}} = 0.0058$$

In this case the proportion of errors is assumed to be continuous and the normal approximation is applied without the use of the continuity correction. (The effect of the continuity correction in this type of problem is illustrated in **Exercise 6.16** at the end of Chapter Six.)

FIGURE 2.19   **Probability of the proportion of errors**

Calculate the z value at p = 0.05:

$$z = \frac{0.05 - 0.035}{0.0058} = 2.586 \text{ standard deviations above the mean}$$

The standard normal tables give:

$$P(z > 2.586) = 0.0048$$

Therefore:

$$P(\text{proportion of errors} > 0.05) = 0.0048$$

2.9 THE NORMAL DISTRIBUTION AS AN APPROXIMATION FOR THE POISSON DISTRIBUTION

The normal distribution can be used in a similar way, as an approximation to the Poisson distribution. The same continuity correction is used to convert from a discrete to a continuous variable. The mean, m, and the standard deviation, $\sqrt{m}$, are used in the normal approximation. The larger m is, the more reliable the approximation. $m \geq 10$ is a general guideline, but as the example below shows, smaller values of the mean can give reasonable approximations.

Example 2.16: Normal approximation to the Poisson distribution

On average, a computer installation suffers 2 breakdowns per week. Estimate the probability that in any given month (4 weeks) there will be more than 10 breakdowns.
Solution: The average is m = 8 breakdowns per month. The number of breakdowns per month is Poisson distributed with:

$$P(r \text{ breakdowns/month}) = \frac{8^r e^{-8}}{r!} \quad r = 0, 1, 2, 3, \ldots$$

We require:

$$P(r \geq 10) = 1 - \{P(0) + P(1) + P(2) + P(3) + \cdots + P(10)\}$$

Using the Poisson distribution:

$$P(r \geq 10) = 1 - 0.816 = 0.184$$

The probability of more than 10 breakdowns in 4 weeks is 0.184. Even though the mean is below the suggested level of 10 we will use the normal approximation.

The continuity correction means that we require the probability of more than 10.5 breakdowns in the 4 week period. For the normal approximation the mean is 8 and the standard deviation is $\sqrt{8}$, therefore the z value for 10.5 breakdowns is:

$$z = \frac{10.5 - 8}{\sqrt{8}} = 0.884 \text{ standard deviations above the mean}$$

From the standard normal tables:

$$P(z \geq 0.884) = 0.1894$$

The probability of more than 10.5 breakdowns in a month is 0.1894 using the normal approximation. This compares with the figure of 0.184 using the Poisson distribution.

2.10 THE COMBINATION OF VARIABLES

2.10.1 Independent variables

Occasions arise in which we wish to combine existing variables to form new variables. For example, in a certain factory a piece of equipment is assembled in 2 stages. The times taken at each stage are independent of each other, therefore we may assume that the time taken to assemble the equipment, t_E, is given by:

$$t_E = t_1 + t_2 \text{ where } t_1 \text{ and } t_2 \text{ are the times taken at the two stages.}$$

In another factory, cartons are filled with orange juice and then 4 randomly selected cartons are packed into trays. If the weight of a filled carton of orange juice is denoted by w and the weight of a tray of 4 cartons is denoted by T, then:

$$T = w_1 + w_2 + w_3 + w_4$$

At this stage, we do not need to define the nature of the distributions of the original variables. Providing that the existing variables are independent of each other, we can deduce the mean and variance of the combined variable. If we have 2 independent variables, with values x and y, and we add them to form a variable with values z, then:

$$z = x + y$$

Irrespective of the nature of the distributions of x and y, the mean of z is the sum of the means of x and y, hence:

$$\mu_z = \mu_x + \mu_y$$

and the variance of z is the sum of the variances of x and y, that is:

$$\text{Var}(z) = \text{Var}(x) + \text{Var}(y)$$

It should be noted that standard deviations cannot be combined directly. Variances must be used.

Example 2.17: The mean and variance of the sum of independent variables

The daily demand for a product has a mean of 100 items per day and a standard deviation of 12 items per day. We wish to use the daily mean sales to calculate the mean sales for a week.
Solution: If the demand for each day is independent of the next, the calculations are straightforward. We are taking 7 random samples from the distribution of daily sales and adding them to give the weekly figure. The mean sales per week is:

$$\text{weekly mean demand} = 7 \text{ days} \times 100 = 700 \text{ items/week}$$

In order to determine the standard deviation of the weekly demand, we must first work in terms of the variance. For the daily demand, the variance $= 12^2 = 144$. For the weekly demand, the variance $= 7 \times 12^2 = 7 \times 144$, giving a standard deviation of $\sqrt{7 \times 144} = 31.75$ items per week.

In other situations we may wish to form new variables from the difference between existing independent variables. The basic principle is the same. If we have 2 independent variables, with values x and y, and we subtract them to form a variable with values z, then:

$$z = x - y$$

Irrespective of the nature of the distributions of x and y, the mean of z is the difference between the means of x and y, hence:

$$\mu_z = \mu_x - \mu_y$$

However, the variance of z is still the sum of the variances of x and y, that is:

$$\mathrm{Var}(z) = \mathrm{Var}(x) + \mathrm{Var}(y)$$

Example 2.18: The mean and variance of the difference between independent variables

A timber yard has ready-cut planks of wood (L1), which have a mean length of 200 cm with a standard deviation of 1.7 cm. Lengths (L2) are cut from these planks, with mean 70 cm and standard deviation 0.3 cm. What will be the mean and standard deviation length of the pieces (L3) that are left?

Solution: The original length and the length cut off are independent of each other, hence for the piece remaining:

$$\mathrm{mean\ length} = \mathrm{Mean}(L1) - \mathrm{Mean}(L2) = (200 - 70) = 130\ \mathrm{cm}$$

and

$$\mathrm{variance} = \mathrm{Var}(L1) + \mathrm{Var}(L2) = 1.7^2 + 0.3^2 = 2.98$$

Hence the standard deviation of L3 = $\sqrt{2.98}$ = 1.73 cm.

Whether we add or subtract independent variables, we always add their variances. The relationships described above may be extended to any number of independent variables. We will not consider the situation in which the variables are not independent, except for the special case described in the next section.

2.10.2 A special case of non-independent variables

If the new variable, z, is derived by multiplying variable x by a number, a, so that $z = ax$, then we are no longer dealing with the combination of independent variables. The mean of z is given by:

$$\mu_z = a\mu_x$$

and the variance of z is given by:

$$\mathrm{Var}(z) = a^2 \mathrm{Var}(x)$$

Example 2.19: The mean and variance of a combination of non-independent variables

The product mentioned in **Example 2.17** contains 5 identical components. What is the mean and variance of the daily demand for the components?

Solution: Superficially this looks like the weekly demand calculation, but the situation is quite different. We do not have 5 independent random values of the variable. We have one value for the one day, which is multiplied by 5 to give the number of components required.

Daily demand for components:

$$\begin{aligned}\mathrm{mean} &= 5 \times 100 = 500\ \mathrm{components\ per\ day} \\ \mathrm{variance} &= 5^2 \times 12^2\end{aligned}$$

Therefore, the standard deviation = 5 × 12 = 60 components per day.

2.10.3 The nature of the distribution of combined variables

If the independent variables, which are to be combined, are all normally distributed, then the combination will itself be normally distributed. For example, if $z = x + y$, and x and y are both independent and normally distributed, then z will also be normally distributed. The same follows if $z = x - y$

Similarly, if $z = ax$ where a is a constant, then z will have a normal distribution if x is normally distributed. This property is not generally true for any distribution. For example, if x and y were independent and each binomially distributed, then z would not be binomially distributed. These relationships hold for any number of independent variables. For example, refer to the situation described earlier in which cartons are filled with orange juice and then 4 randomly selected cartons are packed into trays. If the weight of a filled carton is normally distributed with a mean weight of $\bar{w}$ g and a standard deviation of σ_w then the weight of a tray of 4 cartons is also normally distributed, with a mean weight of:

$$\bar{T} = \bar{w} + \bar{w} + \bar{w} + \bar{w} = 4 \times \bar{w}$$

and a variance of:

$$\sigma_T^2 = \sigma_w^2 + \sigma_w^2 + \sigma_w^2 + \sigma_w^2 = 4 \times \sigma_w^2$$

SUMMARY

A random variable represents the numerical outcomes of an experiment. Random variables are either discrete (integer values only) or continuous (take any value). For a discrete variable the probability that the variable takes a value r is found from the discrete probability function:

$$P(r) = f(r) \qquad \text{over a specified range of r}$$

The expected value of the variable is:

$$E(r) = \sum rP(r)$$

and the standard deviation is:

$$\sigma = \sqrt{\sum r^2 P(r) - (E(r))^2}$$

For the binomial distribution we have n independent, identical trials. Each trial has two possible outcomes—success or failure. The probability of success is the same for all trials. The probabilities for a binomial situation are given by:

$$P(r) = \frac{n!}{r!(n-r)!} p^r q^{n-r} \quad r = 0, 1, 2, \ldots, n$$

$$E(r) = np \quad \text{and} \quad \sigma = \sqrt{npq}$$

For the Poisson distribution there is an unlimited number of identical, independent trials, each with a small, constant probability of success. The probabilities for a Poisson situation are given by:

$$P(r) = \frac{m^r e^m}{r!} \quad r = 0, 1, 2, 3, \ldots$$

$$E(r) = m \quad \text{and} \quad \sigma = \sqrt{m}$$

The Poisson distribution can be used as an easy way of calculating binomial probabilities if $n \geqslant 30$, $p \leqslant 0.1$ and $np \geqslant 5$. These are general guidelines.

For a continuous variable, probability is represented by the area under the probability density curve. Probability applies only to ranges of the variable value, not to exact values.

All normal distributions depend on the mean and standard deviation of the variable, and can be reduced to one distribution. This is the standard normal distribution. Variable values are converted to the number of standard deviations away from the mean, z. The probabilities are tabulated for values of z. The normal can be used to approximate the binomial distribution if np and nq are more than 5. The normal can also be used to approximate the Poisson distribution if $m \geqslant 10$.

If independent normal variables are combined by adding or subtracting them, then the combined variable is also normally distributed. The mean of the combined variable is the sum or difference of the means of the individual variables. The variance of the combined variable is the sum of the variances of the individual variables.

If x is a normal variable and z = ax, where a is a constant, then z is also normally distributed with a mean of ax and a variance of $a^2 Var(x)$.

EXERCISES

Exercise 2.1

Which of the following situations could be modelled by a binomial distribution? In each case give a brief reason for your answer.

1. Sparks Garage Ltd sell 10 cars of a particular type during a given month. Sparks' manager wishes to determine the probability that 2 of those customers will file a claim against the warranty on their car, when it is known from past records that 5% of all purchasers of cars of this type file warranty claims.
2. Estimation of the probability that the petrol consumption of a new car will exceed 30 mpg, if an AA report suggests that the average consumption for such a car is 33 mpg.

Answers on page 435.

Exercise 2.2

Experience indicates that 7% of the invoices waiting to be checked in the accounts department of a large firm, are faulty in some way. A random sample of 20 invoices is selected: what is the probability that:

1. exactly 3 are faulty;
2. at least 3 are faulty?

Answers on page 435.

Exercise 2.3

An old-established family firm has decided to go public. It has been reported that 80% of the stockbroking firms are recommending the issue to their clients. Assume that this report is true.

Required:
Suppose 6 stockbrokers are contacted at random. Find the probability that at least 4 of them are recommending the purchase of this issue to their clients.

Answers on page 435

Exercise 2.4

A fruit and vegetable wholesaler buys bananas in large consignments directly from the docks. In view of the perishable nature of bananas, he budgets to accept up to 10% unusuable bananas in each consignment. As the wholesaler cannot check all of the bananas individually, he has set up an acceptance sampling scheme on which to base his decision of whether to buy the consignment or not.

He takes a random sample of 30 hands of bananas from the consignment. If no more than 2 contain bad bananas, he accepts the consignment outright. If more than 2 hands contain bad bananas, he rejects the whole consignment.

Required:
If a particular consignment actually contains 5% of hands with bad bananas, what is the probability that it will be rejected?

Answers on page 436.

Exercise 2.5

A manufacturer wishes to produce a transistor with no more than 1% defective. Quality is checked regularly by taking a random sample of 10 transistors from the production line and determining the number of defectives in the sample.

Required:
Name the probability distribution which may be used to describe the distribution of the number of defectives per sample.

Answers on page 437.

Exercise 2.6

During a normal day, the average number of lorries arriving to unload at a depot is 3 per hour.

Required:
1   Name the probability distribution which may be used to describe the distribution of the number of lorries arriving per hour.
2   What is the probability that in any hour, more than 4 lorries will arrive?

Answers on page 437.

Exercise 2.7

Pearsons Engineering Ltd mass produce bolts and other items. The machine shop contains a number of machines. The average number of machine breakdowns per hour is 2. One machine maintenance engineer is attached permanently to the shop but calls for assistance when there are more than 2 breakdowns in any one hour. During a 120 hour week, how often on average would he call for assistance?

Answers on page 437.

Exercise 2.8

Star Holdings plc use a computer for all of their record keeping. The computer manager wishes to archive all files which were created more than 5 years ago. In order to assess the demand for access to these files, he consults the computer log for the last 100 working days and draws up the following table:

Number of old files accessed	Number of days
0	10
1	30
2	27
3	17
4	6
5	5
6	4
7	1

Required:
1 Calculate the mean number of old files accessed per day and the standard deviation of the number of old files accessed per day.
2 Explain why it might be reasonable to expect that the number of old files accessed per day would follow a Poisson distribution.
3 Assuming that the variable above does follow a Poisson distribution with the mean calculated in **1**, find the expected frequencies for this distribution.

Answers on page 438.

Exercise 2.9

The accounting department of Star Holdings plc uses an average of 4 boxes of floppy disks per month. Assuming that the demand for floppy disks follows a Poisson distribution, find the number of boxes of disks which the department should stock at the beginning of each month so that the chance of running out during the month is less than 4%.

Answers on page 439.

Exercise 2.10

Oranges are packed in crates, each containing 250. On average 0.6% of the oranges are found to be bad when the crates are opened.

Required:
What is the probability that in a given crate there will be no more than 2 bad oranges?

Answers on page 440.

Exercise 2.11

A machine dispenses soap powder into packets. The amount dispensed is normally distributed with a mean of 930 g and a standard deviation of 20 g.

Required:
1 What proportion of the packets will contain less than 900 g?
2 It is required that no more than 2.5% of the packets should contain less than 900 g. How should the machine be reset in order to meet this requirement?

Answers on page 441

Exercise 2.12

The lifetimes of a certain type of electrical component are approximately normally distributed with a mean of 80 hours and a standard deviation of 30 hours. If the manufacturer replaces all components which fail before the guaranteed minimum lifetime of 45 hours, what proportion of his output will he have to replace? If the manufacturer now decides that he wishes to replace only the 10% shortest lived components, what value should he use as the guaranteed minimum lifetime?

Answers on page 442.

Exercise 2.13

HJ Woolley & Sons Ltd manufacture knitting needles. Three of the most popular sizes for fine knitting are:

	Diameter, mm	Tolerance, mm
Number 10	3.25	±0.125
Number 11	3.00	±0.125
Number 12	2.75	±0.125

The steel rod from which the needles are made is first cut, shaped and buffed by Woolley's. A given batch of rod emerges from these processes with a mean diameter of 3.10 mm and a standard deviation of diameter of 0.10 mm. Assume the diameters are normally distributed.

Required:
Determine the proportion of these prepared rods which are suitable for finishing as No. 11 needles. The finishing process does not alter the diameter of the rods in any way.

Answers on page 443.

Exercise 2.14

The amount of sugar dispensed by a packaging machine is normally distributed with a standard deviation of 20 g. It is possible to adjust the machine to set the mean amount dispensed at any required level (to the nearest gram). The packets are labelled 1000 g.

In order to comply with the Weights and Measures Act 1979 the amount of sugar dispensed must satisfy 3 conditions:

1 the average amount of sugar dispensed must be at least 1000 g;
2 no more than 2.5% of the packets must contain less than 975 g;
3 no more than 1 in 10,000 packets contain less than 950 g.

At present the machine is set to dispense a mean amount of 1010 g.

Required:
1 What proportion of packets contain less than 975 g?
2 What proportion of packets contain less than 950 g?
3 What is the minimum new setting that can be chosen so that the packer complies with the above regulations? The standard deviation is unchanged.

Answers on page 443.

Exercise 2.15

EG Mersoe & Company is a small firm which mills flour. At the moment the machine which bags the flour is in need of replacement. Mr Mersoe has been offered two possible models. For the basic model, the weight of flour delivered to the bags can be expected to follow a normal distribution with a standard deviation of 15 g. Alternatively, Mr Mersoe can pay £50,000 extra for a more sophisticated machine which would function in the same way but the standard deviation of the weights delivered to the bags would be reduced to 5 g.

When filling the flour bags with a nominal weight of 1 kg (1000 g) the machine must satisfy all of the following 3 weight regulations:

1 The average weight must be at least 1000 g.
2 At most 2.5% of the bags may weigh less than 985 g.
3 At most 0.01% of the bags may weigh less than 970 g.

The company sells 4,000,000 of the 1 kg bags per year. The variable cost of the flour is 15p per kg.

QUANTITATIVE ANALYSIS

Required:
How much money would be saved per year if Mr Mersoe buys the more sophisticated machine?

Answers on page 444.

Exercise 2.16

A canvasser for a political party claims that 45% of voters in a given area intend to vote for his party. The whole area has 10,800 registered voters. Assuming that all who are eligible will vote and that the canvasser's claim is based on an unbiased sampling method, what is the probability that the party in question will receive less than 5000 votes?

Answers on page 446.

Exercise 2.17

When appointing computer operators, a firm requires candidates to pass a written examination. The paper contains 100 multiple choice questions, each with 3 answers, only one of which is correct. A pass is obtained by answering 40 or more questions correctly.

Required:
Estimate the probability that a candidate who chooses to answer each question randomly will pass the examination.

Answers on page 447.

Exercise 2.18

The Achilles Art Company sells special pens (for use on overhead projector transparencies) in boxes of 5. For various reasons, principally incorrect storage of the finished goods stock, a certain number of defective pens is discovered in the boxes sold to customers. The quality control manager is of the opinion that the percentage defective is approximately constant. Recently, however, there seem to have been more customer complaints than usual and he has been asked to investigate the situation. Accordingly he has instigated a sampling inspection scheme, which has produced the following results from a sample of 500 boxes chosen randomly from the finished goods stock:

Number of defectives per box	Number of boxes
0	392
1	73
2	30
3	3
4	2
5	0

Required:
1   (a)   From the above sample, estimate the proportion of defective pens.
    (b)   Conditions suggest that the number of defectives per box should follow a binomial distribution. Assuming that this is the case, determine the number of boxes out of 500 that you would expect to contain respectively 0 defectives, exactly 1 defective and exactly 2 defectives.

(ACCA, December 1982)

Answers on page 447.

PROBABILITY DISTRIBUTIONS: CHAPTER TWO

Exercise 2.19

As a result of recent analysis of labour costs throughout all the branches of the Olympus Assurance Company, it has been decided that the total number of employees should be reduced by 3%. The usual retirement age in the company is 65 and it is estimated that the desired reduction can be achieved by offering early retirement to all employees aged 62 and above. For this purpose, an early retirement scheme has been drawn up, based on negotiations with the unions concerned and the informal reactions of employees. It has been estimated that the probability of employees accepting the scheme depends on their age as follows:

Age	Probability of accepting early retirement
62	0.6
63	0.7
64	0.8

In one particular branch of the company, there are 10 employees aged 62 and above as detailed in the following table:

Age	62	63	64
Number	5	3	2

Required:
1. What is the expected number of employees accepting early retirement at this branch? What is the probability of early retirement being accepted by all 10 employees?
2. What is the probability of exactly 9 employees accepting early retirement? *Note*: The one person who does not accept early retirement can be aged 62, 63 or 64. These 3 cases must be considered individually.
3. Given that the variance of a binomial distribution is $np(1-p)$, determine the variance of the total number of employees accepting early retirement from this branch. Using a normal distribution approximation, estimate the probability of at least 8 employees accepting the scheme.

(ACCA, December 1985)

Answers on page 448.

Exercise 2.20

Over a period of time a certain branch of the London and South Eastern Bank has analysed its daily note issue and found that the demand for 1 pound, 5 pound and 10 pound notes on any day of the week has an approximately normal distribution with the following parameters:

	Number of notes	
Denomination	Mean	Standard deviation
£ 1	1,200	250
£ 5	600	100
£10	50	5

You may ignore all other denominations. Furthermore, the 3 denominations are independent of each other in this respect.

Required:
1. What is the probability that demand for cash exceeds £5000 on any one day?

2 At the beginning of a particular day, the branch finds that it has £1600 in 1 pound notes, £3500 in 5 pound notes and £600 in 10 pound notes. Assuming that no cash will be paid in during the day, what is the probability that the branch will be able to meet all its demands for cash during the day? (Your answer should assume that the appropriate combinations of lower denominations may not be used instead of higher denominations so that, for example, two 5 pound notes cannot be issued instead of one 10 pound note).

Which of the three denominations is most critical in respect of the bank being able to meet its demands for cash?

3 If appropriate combinations of lower denomination notes could be issued instead of higher denomination, explain (but do not calculate) how your answer would have to be amended to allow for this.

(ACCA, June 1982)

Answers on page 449.

Exercise 2.21

Classic Cars Ltd is a car rental firm which has 15 cars for hiring purposes. Each day 10 of these cars are rented to regular customers and the daily demand for the remaining cars may be assumed to have a Poisson distribution with mean 4.

Required:
1 Determine the probabilities that 10,11,12,13, 14 and 15 cars are used on any one day. Hence calculate the expected number of cars rented each day.
2 The daily holding cost to the firm of possessing a car is considered to be £6 whether or not it is used, whilst maintenance costs for each car total £2 for each day the car is used. If the daily hiring charge is £20, find the firm's expected daily profit.
3 Determine whether it would be advisable for the firm to change its number of cars from 15 to either 14 or 16, assuming that the daily demand distribution does not alter.

For what range of values of the daily holding cost would the situation of possessing 15 cars be optimal?

(ACCA, June 1988)

Answers on page 452.

Exercise 2.22

Electra Ltd is a small electronics company specialising in the assembly of a range of items from bought-in components. Due to recent changes in its product range, the company has found itself with a stock of 2000 solenoids. Solenoids are classified by their electrical resistance which is measured in ohms, and the solenoids in stock can now only be used in one product with an electrical resistance in the range of 210–220 ohms. As the solenoids were originally purchased for use in other products with different designs specifications, there is some doubt as to how much of the existing stock can still be used. It is known, however, that the mean resistance of the solenoids in stock is somewhere between 205 ohms and 225 ohms and, from previous experience, it may reasonably be assumed that the distribution of the resistance is approximately normal with a standard deviation of about 5 ohms.

Bob Watt, the works accountant, has been asked to advise on what should be done with the stock as there seem to be two possible courses of action, each involving certain associated costs. The company could sell its entire stock of 2000 solenoids for scrap at £0.50 each and order new supplies to the required specification. Alternatively, all 2000 solenoids could be inspected, measured and separated into 2 groups: those which comply with the required specification of 210–220 ohms, and those which do not. This second group would then have to be sold for scrap at £0.50 each. The variable cost of inspecting and measuring each solenoid has been estimated to amount to £1.40 whereas the unit cost of new solenoids is £4.00.

Required:

1. If the mean resistance of the solenoids in stock is 210 ohms, estimate how many of these will fall within the range of 210–220 ohms. In this case, which is the more economical policy: to scrap all the solenoids or to test each one and scrap only those that do not fall within the specification?

2. Determine the return from the two available courses of action for mean resistances of 205, 210, 215, 220 and 225 ohms and represent the two sets of returns graphically. From your graph, estimate the range within which the mean resistance of the 2000 solenoids must lie for it to be more economical to test each one.

3. If the mean resistance of all solenoids is equally likely to be any one of the 5 values given in **2**, which course of action would you recommend? Would your recommendation change if a mean resistance of 215 ohms were considered twice as likely as each of the other 4 values?

(ACCA, June 1985)

Answers on page 453.

Decision making under uncertainty

3.1 INTRODUCTION

The making of the 'right' decision at the 'right' time is essential for the successful management of a company. Wrong or unwise decisions may be costly, possibly even fatal for an organisation. It is important, therefore, that those involved in management should make use of every tool available in order to arrive at the 'best' decision.

One of the first things to be decided is what is meant by the 'best' decision. 'Best' for what or for whom? The objectives to be achieved must be carefully thought out before any decision is made. This is often difficult because of the conflicting objectives within any organisation and the way the activity of one part of the company affects other parts. For example, the control of stock in a manufacturing company may seem relatively straightforward, but when the inter-relationships of the various functions of the company are considered, numerous problems may arise. The manufacturing department will look for long, uninterrupted production runs that will reduce time and set up costs. However, such long runs may result in large stocks of raw materials and work in progress and produce finished goods in only a few product lines. This situation is likely to conflict with the marketing department. Marketing will probably prefer a large and varied stock of finished goods as well as a flexible manufacturing system which can produce special orders at short notice. In addition, the accountant will prefer to minimise investment in stock so that the capital can be used elsewhere.

How then do we decide what is the optimum stock control policy for the company? Do we consider the company as a whole and optimise overall? Or, do we consider one or another of the individual functions of the company and choose a stock control policy which optimises costs for this function without regard to the other functions? This is a process called **sub-optimisation**. Decision making is a broad, difficult and interesting subject. There are numerous types of decisions and numerous ways of dealing with each type. There are subjective approaches. How, for instance, do you decide what kind of play, film or book you like? In business there are also subjective decisions. How should a firm be organised? Should it be centralised or de-centralised? What style of management should be adopted? These are not the kinds of decision which we will be considering in this chapter.

Accountants are frequently concerned with the types of decision which lend themselves, at least in part, to a quantitative approach to the decision making. In this chapter we will illustrate some possible approaches to the quantitative analysis of a decision problem. You should remember, however, that usually decisions are made by a mixture of quantitative and subjective approaches.

For successful decision making it is always useful to take the following steps:

Step 1: Define the objective for the decision.
Step 2: Define the possible decisions that could be taken.
Step 3: Define the possible outcomes for each decision.
Step 4: Evaluate each outcome.
Step 5: Choose the optimum decision on the basis of the stated objective.

In the following, all of our procedures will start by listing the possible decisions and the possible outcomes. Each outcome will then be evaluated for each decision. This is a common base for all quantitative decision making. These basic steps are the most important whether the decision is very simple or extremely complicated. Although this chapter deals with only a small part of a very complicated and wide-ranging topic, we will consider a number of possible objectives. The 'best' one will depend on the decision maker and on the circumstances.

Example 3.1: Illustrating decisions, outcomes and payoffs

Singles plc is trying to set the price for one of its computer products. Three prices are under consideration. The marketing department has provided information about the expected sales volumes at each price. These figures and the associated costs are as follows:

TABLE 3.1 **Estimated sales volume at each price**

Possible price per unit	£8.00	£8.60	£8.80
Estimated sales volume (units/year)			
Best possible	16000	14000	12500
Most likely	14000	12500	12000
Worst possible	10000	8000	6000

Fixed costs are £40,000 per year and variable costs of sales are £4.00 per unit.

The decision the company has to make is which price to charge. Notice that for simplicity we are considering only 3 prices, ie there are only 3 possible decisions. For each decision, again for simplicity, there are only 3 possible outcomes—the different sales volumes.
Solution: For each outcome we will calculate the payoff—in this case, the payoff is the annual contribution.

TABLE 3.2 **Calculation of contribution per year**

Price per unit	£8.00	£8.60	£8.80
Variable cost per unit	£4.00	£4.00	£4.00
Contribution per unit	£4.00	£4.60	£4.80
	Total contribution (£/year)		
Best possible	64,000	64,400	60,000
Most likely	56,000	57,500	57,600
Worst possible	40,000	36,800	28,800

We will use these figures to illustrate the difficulty that uncertainty has introduced. We can develop sensible arguments that lead to any of the 3 possible decisions. The highest contribution, based on the most likely sales volume, would be £57,600. This would be achieved by making the decision to set the price at £8.80 per unit. However, it might be felt that a price of £8.60 would be better for the company, since the most likely contribution is almost as good, but the contributions for the other 2 outcomes are higher than for a price of £8.80. However, if we now take the fixed costs into consideration, the unit price of £8.00 is the only one which does not lead Singles to a possible loss overall, ie here the contribution is never less than the fixed costs of £40,000.

Hence, there are arguments to justify any one of the 3 decisions. The decision which is finally made depends on what the objective of the decision is and what the attitudes towards risk are of the decision maker. A cautious management may well prefer a unit price of £8.00 to either of the other 2—the possible gains are less but so are the possible losses. Management has to consider, amongst other things, its attitude to the risks involved in each decision. This will be discussed later in the chapter. We will now see how the decision rules can be applied to a particular decision.

3.2 DECISION RULES

Decisions may be made using a **decision rule**. There are several of these rules and they can be used to define the objective which is Step 1 of the decision making procedure. The decision maker still has to decide which rule to use. Not all of the rules require the use of the probability of the outcomes of the decisions, hence, we will discuss the rules in two groups.

3.2.1 Decision rules which do not use probability values

1. **Maximax rule**—maximise the maximum payoff.
2. **Maximin rule**—maximise the minimum payoff.
3. **Minimax rule**—minimise the maximum opportunity loss.

Example 3.2: Illustrating decision rules without probabilities

Suppose that you are the owner of a cake shop, the Cake Box. At the beginning of each day, you have to decide how many fresh cream gateaux to stock in order to meet demand during the day. Each gateau costs £0.70 and sells for £1.30 each. The gateaux cannot be stored overnight and if any remain at the end of the day they are sold off for £0.30 each. An analysis of past sales data gives the following information:

TABLE 3.3 **Demand for gateaux**

Number of gateaux demanded per day	1	2	3	4	5
Frequency	5	10	15	15	5
Relative frequency/probability	0.1	0.2	0.3	0.3	0.1

The decision you have to make is how many gateaux should be bought at the beginning of each day.

Solution: Your possible decisions are to buy 1, 2, 3, 4 or 5 gateaux each day. The possible outcomes are that demand will be for 1, 2, 3, 4 or 5 gateaux in a day. In this case the decisions and the outcomes look the same, but remember that you control the decisions. You will decide which decision to make. You have no control over the outcomes. Your customers, the outside world, impose one of the outcomes on you. For this reason the outcomes are also referred to as 'states of nature'. When we have listed the decisions and the corresponding outcomes, we evaluate each outcome. A payoff table is set out below. This table gives the payoff, or monetary gain, for each combination of decision and outcome.

TABLE 3.4 **Payoff table**

Possible outcomes: demand in gateaux per day	Payoff-profit, £/day Number of gateaux bought per day (possible decisions)				
	1	2	3	4	5
1	0.60	0.20	−0.20	−0.60	−1.00
2	0.60	1.20	0.80	0.40	0.0
3	0.60	1.20	1.80	1.40	1.00
4	0.60	1.20	1.80	2.40	2.00
5	0.60	1.20	1.80	2.40	3.00

We will consider the decision problem 'how many gateaux should the Cake Box buy for the beginning of each day?', using each of the decision rules set out at the beginning of this section.

1 Maximax rule—maximise the maximum payoff. For each possible decision, each column in the table above, the maximum payoffs are as follows:

TABLE 3.5 **Maximum payoffs**

Decision: number of gateaux bought per day	Maximum payoff: profit, £/day
1	0.60
2	1.20
3	1.80
4	2.40
5	3.00 ← **maximum**

Under this rule, you would decide to buy 5 gateaux for the beginning of each day. This is the gambler's approach—you go for the possibility of the biggest gain and ignore what you might lose.

2   Maximin rule—maximise the minimum payoff. For each possible decision, the minimum payoffs from the payoff table are as follows:

TABLE 3.6 **Minimum payoffs**

Decision: number of gateaux bought per day	Minimum payoff: profit, £/day
1	0.60 ← maximum
2	0.20
3	−0.20
4	−0.60
5	−1.00

Under this rule, you would make the decision which maximises the minimum payoff. You decide to buy 1 gateau per day. This is a very cautious decision criterion.

3   Minimax rule—minimise the maximum opportunity loss. Your decision could be analysed by looking at opportunity loss rather than payoff. An opportunity loss table shows, for each possible outcome, the 'shortfall' in profit which would arise from making the wrong decision. For example, if 2 gateaux are demanded (the second row in the payoff table), the best decision would be to buy 2, giving a payoff of £1.20. If, however, you had decided to buy 3 gateaux, the payoff would be £0.80, which is £0.40 less than you could have received for this outcome. This £0.40 is what is called the **opportunity loss** or **regret**. The opportunity loss table is derived from the rows of the payoff table by finding the largest payoff for each outcome and then calculating the opportunity loss for all of the payoffs associated with that outcome.

TABLE 3.7 **Opportunity loss table**

Possible outcomes: demand in gateaux per day	\multicolumn{5}{c}{Opportunity loss, £/day Number of gateaux bought per day (possible decisions)}
	1	2	3	4	5
1	0.0	0.40	0.80	1.20	1.60
2	0.60	0.0	0.40	0.80	1.20
3	1.20	0.60	0.0	0.40	0.80
4	1.80	1.20	0.60	0.0	0.40
5	2.40	1.80	1.20	0.60	0.0

The decision rule to use with the opportunity loss table is the minimax rule which is also called the **minimax regret rule** (regret = opportunity loss). For each decision, the maximum opportunity loss is selected. The decision which leads to the minimum value of these maximum regrets is chosen.

TABLE 3.8 **Maximum opportunity losses**

Decision: number of gateaux bought per day	Maximum opportunity loss (£/day) (from table above)
1	2.40
2	1.80
3	1.20 ← ⎫
4	1.20 ← ⎬ minimum
5	1.60 ⎭

The minimum value of these maximum opportunity losses is £1.20 which arises from ordering 3 or 4 gateaux per day. Hence, you would choose either of these decisions on the basis of the minimax rule.

Each of the criteria for decision making which we have considered so far, has led to a different decision. It is up to you to choose at the outset, the criterion which you think is the 'best'. The application of that decision rule will then lead to your best decision.

3.2.2 Decision rules which do use probability values

In the previous section we did not use the information given about the probability of each outcome occurring. We will now consider how these probabilities may be used with the decision rules.

1. Maximum likelihood rule—maximise the most likely payoffs. Look at the relative frequencies/probabilities of the daily demand for gateaux.

TABLE 3.9 **Relative frequencies/probabilities of the daily demand for gateaux**

Number of gateaux demanded per day	1	2	3	4	5
Frequency	5	10	15	15	5
Relative frequency/probability	0.1	0.2	0.3	0.3	0.1

The largest probability (0.3) occurs for daily demands of 3 and 4 gateaux per day. For each of these outcomes, the payoffs associated with each decision are examined and the maximum selected.

TABLE 3.10 **Maximum payoff for each decision**

Decision: number of gateaux bought per day	Maximum payoff (£/day)
3	1.80 when the outcome is 3 or more
4	2.40 when the outcome is 4 or more ← **maximum**

Under this rule, the Cake Box decides to buy 4 gateaux for the beginning of each day.

2. Optimise the expected values. The most commonly used way of incorporating the probabilities into the decision rule is to calculate expected values. This is done for either the payoffs, or the opportunity losses, for each decision. The decision is selected which gives the largest expected payoff, or the smallest opportunity loss.

   (a) Maximise expected payoff. The expected payoff for a decision is:

$$E(\text{payoff for a decision}) = \sum (\text{probability} \times \text{payoff})$$

(summed for all of the outcomes of that decision)

Using the Cake Box example, the expected payoff for the decision to buy 5 gateaux per day is:

$E(\text{payoff if decide to buy 5 gateaux})$
$= (0.1 \times -1.0) + (0.2 \times 0.0) + (0.3 \times 1.0) + (0.3 \times 2.0) + (0.1 \times 3.0)$
$= £1.1 \text{ per day}$

The expected payoff means that in the long run, this will be the **average** profit to be gained per day, if you decide to buy 5 gateaux each day. The payoff table for the Cake Box is repeated below with probabilities added. This is followed by a table which shows the calculation of the expected payoff for each decision.

QUANTITATIVE ANALYSIS

TABLE 3.11 **Payoff table**

	Payoff/profit, £ per day Number of gateaux bought per day (possible decisions)					
Possible outcomes: demand in gateaux per day	1	2	3	4	5	Probability
1	0.60	0.20	−0.20	−0.60	−1.00	0.1
2	0.60	1.20	0.80	0.40	0.0	0.2
3	0.60	1.20	1.80	1.40	1.00	0.3
4	0.60	1.20	1.80	2.40	2.00	0.3
5	0.60	1.20	1.80	2.40	3.00	0.1

TABLE 3.12 **Calculation of expected payoff**

(Probability × payoff from the table above)

	Number of gateaux bought per day (possible decisions)				
Possible outcomes: demand in gateaux per day	1	2	3	4	5
1	0.06	0.02	−0.02	−0.06	−0.10
2	0.12	0.24	0.16	0.08	0.0
3	0.18	0.36	0.54	0.42	0.30
4	0.18	0.36	0.54	0.72	0.60
5	0.06	0.12	0.18	0.24	0.30
Expected payoff (£/day) = total of column	0.60	1.10	1.40	1.40	1.10

The maximum value of the expected payoff is £1.40 per day. Hence, using the criterion of maximising the expected payoff, the Cake Box should make the decision to buy 3 or 4 gateaux each day. In this type of example, where the decision is repeated many times, the expected value criterion for decision making is usually considered to be the most suitable one.

(b) Minimise expected opportunity loss. The same procedure may be carried out using the opportunity loss table and the probability of the occurrence of each outcome. In this case we choose the decision which gives the minimum expected opportunity loss rather than the maximum expected payoff.

TABLE 3.13 **Opportunity loss**

	Opportunity loss, £ Number of gateaux bought per day (possible decisions)					
Possible outcomes: demand in gateaux per day	1	2	3	4	5	Probability
1	0.0	0.40	0.80	1.20	1.60	0.1
2	0.60	0.0	0.40	0.80	1.20	0.2
3	1.20	0.60	0.0	0.40	0.80	0.3
4	1.80	1.20	0.60	0.0	0.40	0.3
5	2.40	1.80	1.20	0.60	0.0	0.1

TABLE 3.14 **Calculation of expected opportunity loss**
(probability × opportunity loss)

	Number of gateaux bought per day (possible decisions)				
Possible outcomes: demand in gateaux per day	1	2	3	4	5
1	0.0	0.04	0.08	0.12	0.16
2	0.12	0.0	0.08	0.16	0.24
3	0.36	0.18	0.0	0.12	0.24
4	0.54	0.36	0.18	0.0	0.12
5	0.24	0.18	0.12	0.06	0.0
Expected opportunity loss, (£/day) = total of column	1.26	0.76	0.46	0.46	0.76

The minimum expected opportunity loss is £0.46 per day. Hence, your decision would be to buy 3 or 4 gateaux each day. This is the same decision as the one made when we used the criterion of maximising the expected payoff. This is always the case.

3.2.3 The sensitivity of the decisions to changes in the probabilities

The probabilities we use are based either on historical information or on subjective estimates. The values are uncertain in themselves, and it is useful to see how sensitive our choice of decision is to alterations in the probabilities.

Sensitivity analysis is an important theme which will recur throughout the book. The procedure is to estimate some sensible changes in the probabilities and to repeat the calculations which led to the choice of decision. We will use the criterion of maximising the expected payoff to illustrate the effect. The original and one alternative set of probabilities are given below. In practice you would check more than one alternative, depending on the importance of the decision.

TABLE 3.15 **Sensitivity of decision to changes in probabilities**

	Possible decisions: number of gateaux bought per day				
	1	2	3	4	5
Original probabilities	0.1	0.2	0.3	0.3	0.1
Expected payoffs £/day	0.60	1.10	1.40	1.40	1.10
Alternative probabilities	0.2	0.2	0.2	0.2	0.2
Expected payoffs* £/day	0.60	1.00	1.20	1.20	1.00

* These expected payoffs are calculated in the same way as those in section **3.2.2**.

In this case the decision does not change. The decision to buy 3 or 4 gateaux gives the maximum expected payoff. There is, however, a reduction in the average profit which the shop will make per day from £1.40 with the original probabilities to £1.20 with the alternative probabilities. In this example the decision is not sensitive to small changes in the probabilities. Small variations in the probabilities of the outcomes do not cause a change in the choice of decision.

3.2.4 The value of perfect information

We can sometimes reduce the uncertainty involved in making a decision by collecting more information. However, we will usually have to pay for this additional information. The **value of perfect information** tells us the maximum amount it is worth paying for it. If we know, in advance, which one of the outcomes will actually occur, then we can choose the decision which will lead to the maximum payoff. This does not mean that we can control the choice of outcome.

For example, the Cake Box could take orders for gateaux to be delivered the following day. You cannot control how many orders you will receive, but you can then buy the right number of gateaux each day to maximise the payoff. The number bought will now change from day to day depending on the number ordered.

The expected payoff now becomes:

$$E = \sum (\text{payoff for given order size} \times \text{probability of that order size})$$
$$E = (0.60 \times 0.1) + (1.20 \times 0.2) + (1.80 \times 0.3) + (2.40 \times 0.3) + (3.00 \times 0.1) = £1.86$$

The difference between this figure and the maximum expected payoff without perfect information, is called 'the value of perfect information'.

The value of perfect information for the Cake Box is:

$$£1.86 - £1.40 = £0.46 \text{ per day}$$

which is equal to the minimum expected opportunity loss.

The value of perfect information tells us the maximum we should pay for additional information about the likelihood of each outcome arising. The Cake Box could afford to pay up to £0.46 per day to operate an ordering system, or for 'market research' information to enable the daily demand to be predicted more accurately than at present.

3.3 EXPECTED VALUES AND THE USE OF STANDARD DEVIATIONS TO ASSESS RISK

The criterion of maximum expected payoff (or minimum expected opportunity loss) provides one number for us to use when choosing the 'best' decision. The payoff table gives the actual values of the payoffs for individual outcomes. It shows how diverse the payoffs may be. An inspection of the payoff table and the spread of the payoffs gives a good indication of the risk involved in each decision. An alternative approach to the assessment of risk is to calculate the standard deviation of the payoffs, just as we would for any other distribution. The example below compares two investments in this way. This example and the gateaux example are treated arithmetically in the same way. There is, however, a fundamental difference between the two. The gateaux decision will be repeated day after day many times and the idea of an expected (average) payoff is easy to grasp. The investment decision will be taken once only, making the practical meaning of the expected payoff less obvious.

Example 3.3: Expected payoff and the use of standard deviations

Two investments are available at the same cost. The possible net returns (discounted to the present day) are shown below, together with their associated probabilities.

TABLE 3.16 **Possible net returns and associated probabilities**

Comparison of investment decisions

Net returns £'000	−3	−2	−1	0	1	2	3	4
Probabilities:								
Investment 1	0	0	0.1	0.2	0.3	0.2	0.2	0
Investment 2	0.1	0.1	0.1	0.1	0.1	0.1	0.2	0.2

For Investment 1, the expected return is:

$$E(\text{Investment 1}) = \sum (\text{return} \times \text{probability})$$

Hence:

$$E(\text{Investment 1}) = (-3 \times 0) + (-2 \times 0) + (-1 \times 0.1) + (0 \times 0.2) + (1 \times 0.3)$$
$$+ (2 \times 0.2) + (3 \times 0.2) + (4 \times 0)$$

Therefore:

$$E(\text{Investment 1}) = £1200$$

Similarly, for Investment 2:

$$E(\text{Investment 2}) = (-3 \times 0.1) + (-2 \times 0.1) + (-1 \times 0.1) + (0 \times 0.1) + (1 \times 0.1)$$
$$+ (2 \times 0.1) + (3 \times 0.2) + (4 \times 0.2)$$

Therefore:

$$E(\text{Investment 2}) = £1100$$

On the basis of the expected returns only, Investment 1 is marginally better. If the investment decision could be carried out many times under the same conditions, then this investment would return £1200 on average each time.

This decision rule does not take into account the risk associated with the investment, ie the variability in the possible returns. This risk may be assessed in terms of the variance or the standard deviation of the returns.

From Chapter Two, we know that the variance of a probability distribution may be written as:

$$\text{Variance} = \sum px^2 - (E(x))^2$$
$$\text{where } E(x) = \sum px$$

In this example, x is the return on the investment and p is the probability of receiving that return.

TABLE 3.17 **Calculation of mean and variance for the investments problem**

Return £'000 x	Investment 1 p	px	px²	Investment 2 p	px	px²
−3	0	0	0	0.1	−0.3	0.9
−2	0	0	0	0.1	−0.2	0.4
−1	0.1	−0.1	0.1	0.1	−0.1	0.1
0	0.2	0	0	0.1	0	0
1	0.3	0.3	0.3	0.1	0.1	0.1
2	0.2	0.4	0.8	0.1	0.2	0.4
3	0.2	0.6	1.8	0.2	0.6	1.8
4	0	0	0	0.2	0.8	3.2
Total	1.0	1.2	3.0	1.0	1.1	6.9

Investment 1:

$$\text{Variance} = 3.0 - 1.2^2 = 1.56 \; (£'000)^2$$

Therefore:

$$\text{the standard deviation of the returns} = \sqrt{1.56} \; £'000$$
$$= £1250$$

Investment 2:

$$\text{Variance} = 6.9 - 1.1^2 = 5.69 \; (\pounds'000)^2$$

Therefore:

$$\text{the standard deviation of the returns} = \sqrt{5.69} \; \pounds'000$$
$$= \pounds 2385$$

Investment 1 is the less risky scheme since the variance of the returns is much lower than that of Investment 2.

TABLE 3.18 **Expected value and standard deviation for two investment schemes**

	Expected return, £	Standard deviation, £
Investment 1	1200	1250
Investment 2	1100	2385

Both expected return and the variability of the returns lead us to choose Investment 1. Investment 1 will yield the higher expected return and has the lower risk as measured by the standard deviation.

3.4 USE OF UTILITY TO INCORPORATE THE DECISION MAKER'S ASSESSMENT OF RISK

So far we have concentrated on decision rules. The decision maker chooses the rule he prefers and, when he has done that, the rule leads him to the best decision. Once the rule is chosen, it makes no difference who the decision maker is. The same decision will be the 'best' whether the decision maker is a millionaire or a penniless student, whether he likes risk or likes safety, although the decision maker's attitudes and circumstances are partially reflected in his choice of rule. Utility theory is a way of allowing the decision maker to impose his attitudes on a further stage in the process. He not only selects the rule, he also changes all the monetary outcomes to his own utility values. The same rule no longer leads to the same decision for all decision makers. Each person can fit the process to his own circumstances.

A typical example is to consider the investment of £1000 in two possible schemes. The first (Scheme 1) will give a 10% return per annum with no risk. Hence, the outcome will be £1100 at the end of the year. With the second (Scheme 2) you either lose all your money or you double it by the year end. The payoff table is:

TABLE 3.19 **Payoff in one year's time**

	Possible decision invest £1000 in:	
Possible outcomes	Payoff, £	
for Scheme 2	Scheme 1	Scheme 2
Success	1100	2000
Failure	1100	0

On a monetary scale the £1100 payoff is 0.55 of the way from £0 to £2000. The idea of utility is to create a new scale for each individual decision maker with the minimum payoff (£0) at 0 and the maximum (£2000) at 100. In this example, the decision maker has now to position £1100 on this utility scale.

We will look at two decision makers. The first is a student. The original £1000 is all the money she has, so its loss would be extremely serious. The risks for her are high. Therefore, the utility value of £1100 is high. We need a number for the utility. This is achieved by asking her to estimate what probability, p, the maximum payoff (£2000) would have to have before she would take the gamble and invest in Scheme 2. If the probability of success for Scheme 2 is low, say 0.1, then in her situation she would not take the risk. She would take the risk-free decision of Scheme 1. If, however, the probability of success was 1, she would invest in Scheme 2 because there is no longer any risk involved in the scheme. Somewhere between p = 0.1 and p = 1 there will be a point where she would move from investing in the risk-free Scheme 1 and instead invest in Scheme 2 and accept the risk. This value of p, where the change occurs, represents her utility value of £1100. Let us say it is p = 0.95 and the utility will be 0.95 × 100 = 95. The student's scale has changed from a monetary scale of £0–£1100–£2000 to a utility scale of 0–95–100.

Our second decision maker has £500,000 in capital. The loss of £1000 would have a very marginal effect on her. The risks are small. Let us also assume that she enjoys a small gamble. For both reasons the utility of £1100 to her is small. She has to estimate a value for the same probability as the student. For what value of p, the probability of success, would she change from Scheme 1, £1100 risk-free, to Scheme 2, the gamble? She estimates p = 0.2. If the probability of success is less than 0.2 she would stick to the risk-free investment and take the £1100 payoff. If p were more than 0.2 she would take the gamble of Scheme 2. Her utility value for £1100 would be 0.2 × 100 = 20. Her scale of values has changed from the monetary scale of £0–£1100–£2000 to the utility scale of 0–20–100.

The monetary scale has been converted into two quite different utility scales, because of the two decision makers' different situations and attitudes.

The example in the next section uses the decision rule which maximises expected values to illustrate the use of the utility scale compared with the normal monetary scale.

3.4.1 The use of utility values

In **Example 3.4** we illustrate the use of utility values, compared with monetary payoffs, using the decision rule which maximises expected values. The example first uses the criterion of maximising the expected payoff. The payoffs are then converted into utility values and the new criterion of maximising the expected utility is used.

Example 3.4: The calculation of a utility value

Suppose you have saved £5000 as a deposit for a house which you hope to buy next year. A friend then offers you a stake in a business venture which he is planning. If the venture fails, you will lose your £5000 and the chance of buying your house. If it succeeds, you will be paid £30,000 at the end of a year. A business analyst estimates the probability of the success of the venture as 0.3. The alternative decision is to put your money in a bank account at a fixed interest rate of 9% per annum, and with no risk of losing it. The payoff table is given below:

TABLE 3.20 **Payoff**

Possible outcome	Payoff, £ Possible decision— invest £5000 in:		Probability
	Venture	Bank	
Venture succeeds	30,000	5450	0.3
Venture fails	0	5450	0.7
Expected payoff, £	9000 ↑ choose on basis of maximum expected payoff	5450	

Using the monetary scale, the business venture has the larger expected payoff. Hence, using the criterion, you would elect to risk your money by putting it into the venture, in the hope of a large payoff. However, this may not be a realistic decision since the loss of your money would mean that you could not buy your house.

We assign a utility value of 0 to the smallest payoff (which is £0 in this example) and a utility value of 100 to the largest payoff (£30,000 in this example), ie:

$$U(0) = 0 \text{ and } U(30\,000) = 100$$

Utility values can range from 0 to 100, or 0 to 1, or any other convenient scale. The actual utility value is not important, it is its relative value which matters.

A utility value is now required for the payoff of £5450. We need to consider what the probability, p, would have to be before you would consider £5,450 with certainty, to be equally as attractive as £30,000 with probability p and £0 with probability $(1 - p)$.

Suppose you decide that you need at least a 60% chance of the business venture succeeding, ie $p = 0.6$, then, the utility of £5,450 is given by:

$$U(5450) = p \times 100$$
$$= 0.6 \times 100 = 60$$

The table of utility values for this example is then:

TABLE 3.21 **Utility table**

Possible outcome	Possible decision—invest £5000 in: Venture	Bank	Probability
Venture succeeds	100	60	0.3
Venture fails	0	60	0.7
Expected utility	30	60	

choose because it is the higher expected utility.

The decision with the higher expected utility is the investment in the bank. This reverses the decision based on the criterion of expected payoff. The assessment of this decision maker's risk involved in the business venture has changed the decision. If we graph the utility values against the payoffs, we can assess the decision maker's attitude towards risk. We plot the utility values and then join U(0) and U(100) by a straight line. If the utility value for £5450 falls above this line, the decision maker is a risk avoider. If it falls below the line, the decision maker is a risk taker.

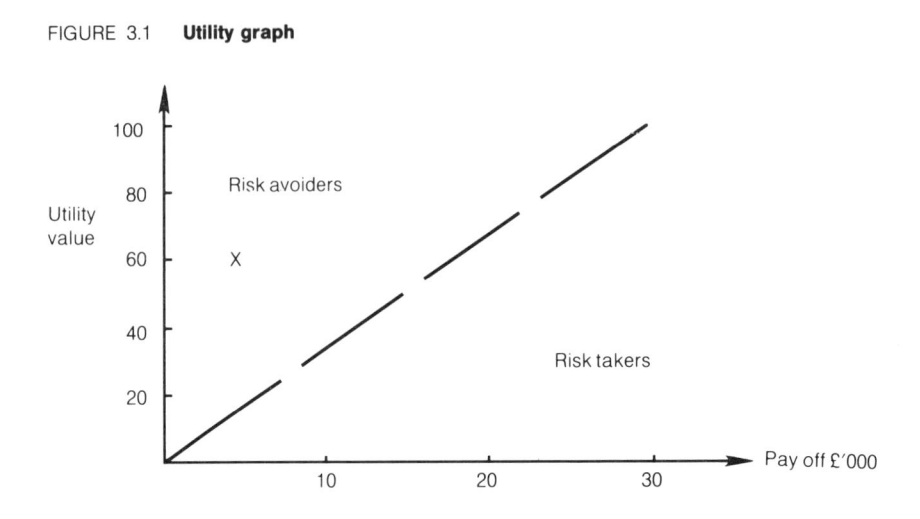

FIGURE 3.1 **Utility graph**

We can see from the graph, that in this case, the decision maker is a risk avoider. The idea of utility can be developed to cater for problems in which there are several possible decisions, but we will not do this here.

3.5 DECISION TREES

The examples we have used so far in this chapter have all involved single decisions. However, we are often in situations in which the outcome from one decision leads us to a second decision, and so on. This process cannot be represented by a payoff table and some other decision making process must be used.

A **decision tree** is a branching diagram which is similar to a probability tree. It is used to represent problems in which a series of decisions has to be made under conditions of uncertainty. Any one of the decisions may be dependent on the outcome of preceding decisions or the outcomes of a trial.

The first step is to draw a diagram which shows the structure of the problem. Decision trees are constructed from left to right. The branches represent the possible alternative decisions which could be made and the various possible outcomes which might arise. It is helpful to distinguish between the two types of branch. In this text, decision branches will be drawn as broken lines emerging from square nodes, and the outcomes of a trial as solid lines emerging from round nodes.

The square nodes, from which the decision branches are drawn, represent the points at which the decision maker selects his decision. The round nodes represent the points at which the outcome of the decision arises. The decision maker has no control over the outcome and can only estimate the probability of the various outcomes actually occurring.

When all of the decisions and outcomes have been represented on the tree, each of the possible routes through the tree is considered and the monetary payoff is shown at the end of each route. Any costs incurred by the decisions are indicated along the appropriate branches.

Example 3.5: A single decision illustrated by a decision tree

A business man needs to borrow £15,000 for a year in order to finance a capital project. He approaches a bank who would charge him interest of 15% per annum. If the bank decides not to make the loan, it would invest the £15,000 at a guaranteed return of 9% per annum. From past experience, the bank estimates that 4% of this type of client default on the repayment of loans. What should the bank decide to do? Should it grant the client a loan or not? This is a single decision which can be represented by either a payoff table or a decision tree. Both approaches are given below.

Solution (using a payoff table): We will maximise the expected net payoff at the end of the year. The net payoff is the sum received at the end of the year less the amount initially invested. Hence, if the loan is granted and repaid, the net payoff will be:

$$\text{Net payoff} = £((15{,}000 + 15\% \text{ of } 15{,}000) - 15{,}000) = £2250$$

TABLE 3.22 **Net payoff at year end £**

Possible outcomes	Payoff, £ Possible decision: Grant loan	Do not grant	Probability
Client repays	2250	1350	0.96
Client defaults	−15,000	1350	0.04
Expected net payoff, £	1560	1350	

The total expected net payoff will be a maximum (£1560) if the bank makes the decision to grant the loan.

Solution (using a decision tree): We will use the same criterion of maximising the expected net payoff at the end of the year.

FIGURE 3.2 **Decision tree for Example 3.5**

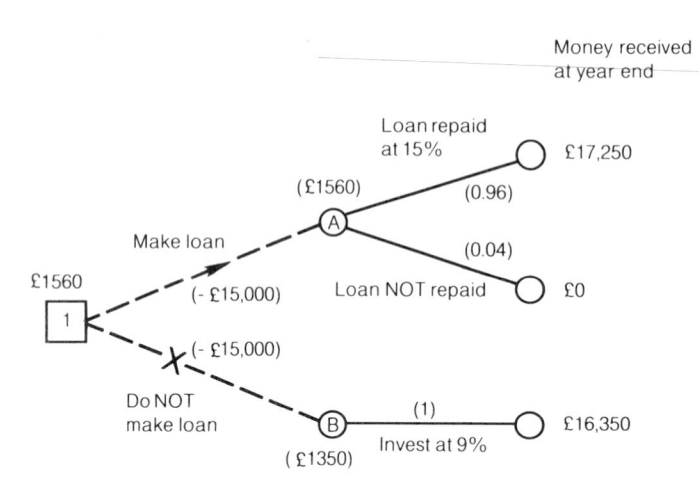

The tree is now evaluated in a similar way to the payoff table. The expected net payoffs at nodes A and B are as follows. At Node A the expected net payoff is:

$$E(\text{Make Loan}) = \{17{,}250 \times 0.96 + 0 \times 0.04\} - 15{,}000$$
$$= 16{,}560 - 15{,}000$$
$$= £1560$$

At Node B the expected net payoff is:

$$E(\text{Do not Make Loan}) = \{16{,}350 \times 1.0 - 15{,}000\}$$
$$= £1350$$

The higher net expected value is at Node A. Hence, the decision is made to make the loan to the client.

3.5.1 Evaluation of a two stage decision tree

Example 3.6: Evaluating a two stage decision process using a decision tree

We will extend the previous example. The bank is considering whether to check the credit worthiness of a client before making the decision about a loan. A credit rating agency charges the bank £80 for each credit check made. The bank now has two decisions to make. First it has to decide whether or not to carry out a credit check on the client. Then it has to decide whether or not to make the loan.

To use the credit checking system sensibly the bank has to estimate how accurate it might be. It has taken a random sample of 1000 people to whom it has already given loans. A retrospective

DECISION MAKING UNDER UNCERTAINTY: CHAPTER THREE

check has been made on the credit worthiness of the people in the sample. The following data are now available:

TABLE 3.23 **Recommendations of credit agency**

Recommendation from credit check	Actual outcome from the loan		Total
	Client repaid	Client defaulted	
Make loan	735	15	750
Do not make loan	225	25	250
Total	960	40	1000

Which decisions should the bank make?

Solution:

Step 1: Construct the decision tree as shown below. The probabilities shown are derived in *Step 2* below.

FIGURE 3.3 **Decision tree for bank loan with credit check**

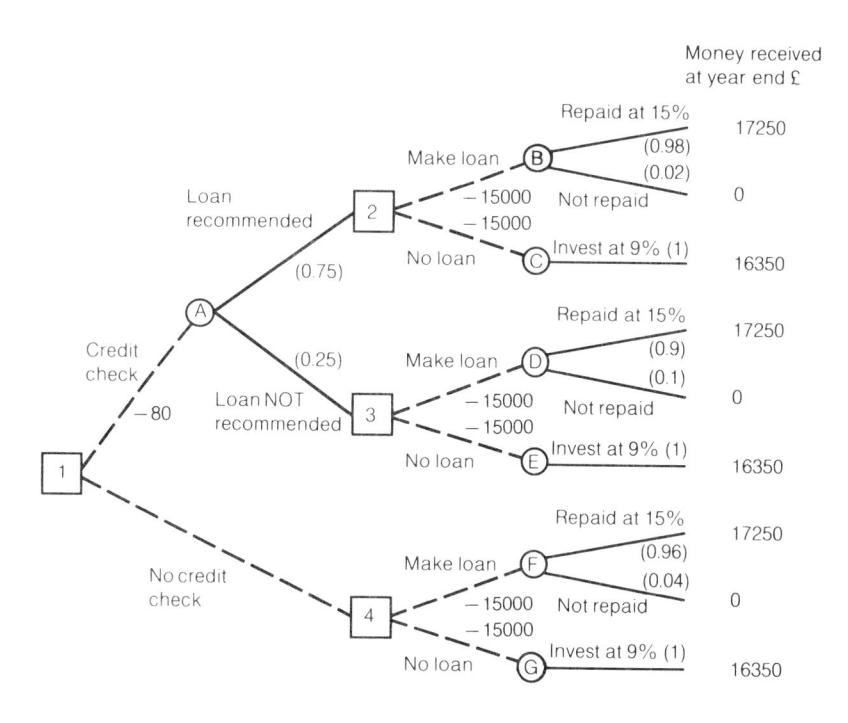

Step 2: The probabilities of each outcome must now be calculated. Using TABLE 3.23:

P(client will repay loan, given credit check recommended a loan) = 735/750 = 0.98

P(client will not repay loan, given credit check recommended a loan) = 15/750 = 0.02

P(client will repay loan, given credit check did not recommend a loan) = 225/250 = 0.9

P(client will not repay loan, given credit check did not recommend loan) = 25/250 = 0.1

Step 3: Evaluate the tree. The evaluation of the tree is carried out from right to left. The estimated monetary values listed at the extreme right of the tree are used to calculate the expected payoff at the first nodes to the left. Any costs incurred are deducted from the expected values. This process is continued through the tree using the expected values already calculated as the estimated monetary values for the next set of calculations. When decision nodes are reached, the decision branch is chosen which leads to the highest expected payoff at that point. Once the decision has been made, the other decision branches are crossed out and the expected payoff is transferred to the decision node.

First we will look at the outcome nodes B and C which lead back to the decision node 2 (should the client be granted the loan?). The expected payoff at outcome node B is:

$$E(B) = £17,250 \times 0.98 + £0 \times 0.02 = £16,905$$

and the net expected payoff:

$$NE(B) = £16,905 - £15,000 = £1905$$

The expected payoff at outcome node C is:

$$E(C) = £16,350 \times 1.0 = £16,350$$

and the net expected payoff:

$$NE(C) = £16,350 - £15,000 = £1350$$

Looking at the tree we now assume that we are at decision node 2. The maximum net expected payoff at this point is £1905 at node B and we would therefore make the decision to give the client the loan.

Having made the decision, we modify the tree. The net expected value of £1905 is written at decision node 2. The decision branch for not making the loan is crossed out. See the final tree at the end of the example. We now do the same for outcome nodes D and E which lead back to decision node 3. The expected payoff at outcome node D is:

$$E(D) = (£17,250 \times 0.9) + (0 \times 0.1) = £15,525$$

and the net expected payoff:

$$NE(D) = £15,525 - £15,000 = £525$$

Similarly at outcome node E:

$$E(E) = £16,350 \times 1.0 = £16,350$$

and the net expected payoff:

$$NE(E) = £16,350 - £15,000 = £1350$$

If we were at decision node 3, the maximum net expected payoff would be £1350 and we would make the decision not to grant the loan. We now modify this part of the diagram. We write the net expected payoff of £1350 at decision node 3 and cross out the decision branch for making the loan.

The final calculation of this first stage begins with outcome nodes F and G and leads back to decision node 4.

$$E(F) = £17,250 \times 0.96 + £0 \times 0.04 = £16,560$$

and the net expected payoff:

$$NE(F) = £16,560 - £15,000 = £1560$$

$$E(G) = £16,350 \times 1.0 = £16,350$$

and the net expected payoff:

$$NE(G) = £16,350 - £15,000 = £1350$$

DECISION MAKING UNDER UNCERTAINTY: CHAPTER THREE

At decision node 4 the maximum expected net payoff is £1560 and we would make the decision to grant the client the loan. £1560 is written at decision node 4 on the modified tree and the alternative decision branch crossed out.

We now step back along the tree to the remaining nodes A and 1. Following the same procedure, we calculate the expected value at node A, using the expected net payoffs we have written at decision nodes 2 and 3. The expected payoff at outcome node A is:

$$E(A) = (£1905 \times 0.75) + (£1350 \times 0.25) = £1766$$

and the net expected payoff:

$$NE(A) = £1766 - £80 = £1686$$

since the credit check costs £80.

We can now evaluate the first decision, decision node 1. Should the bank use the credit check? At this node the maximum expected value is £1686 at decision node 1 and cross out the alternative decision branch.

Summary: The sequence through the tree which generates the maximum expected net return is shown by the arrowed lines in FIGURE 3.4. At decision node 1, use the credit check. If the loan is recommended, then, at decision node 2, make the loan. If the loan is not recommended, then, at decision node 3, do not make the loan but invest the money at a guaranteed 9% per annum instead. The complete decision tree for **Example 3.6** is:

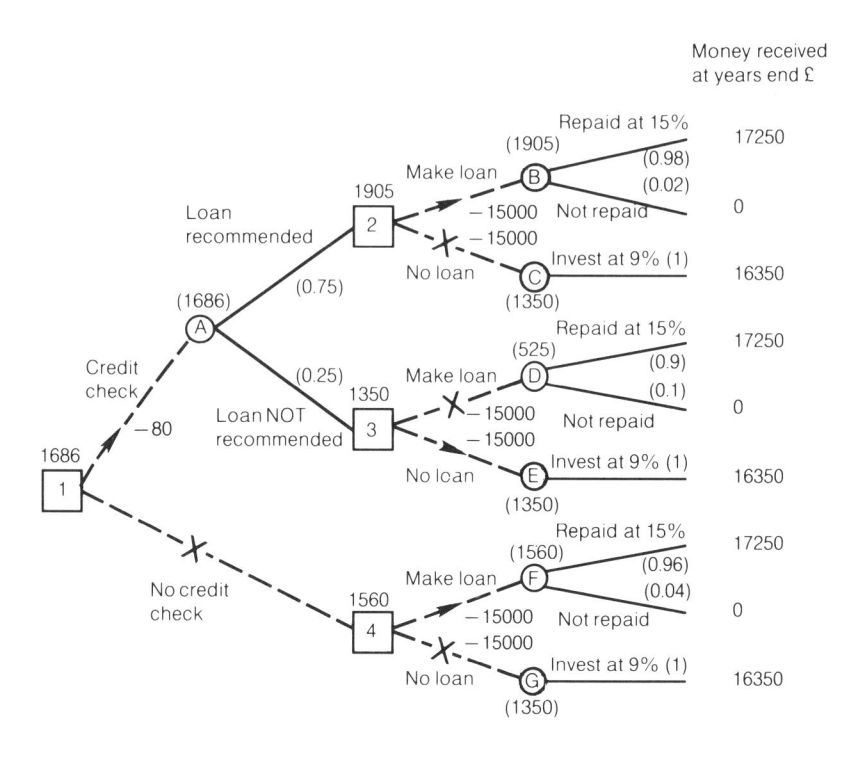

FIGURE 3.4   **Complete decision tree for Example 3.6**

Example 3.7: A decision tree which involves the time value of money and two-stage outcomes

Tranda plc, a market research bureau, expects to be able to expand its business by providing personal computers for its data collection staff. The company has to decide whether to buy or lease these machines. The growth of the business over the next four years cannot be predicted exactly, but it can be classified as high, average or low. In the first year after the installation of the computers, it is estimated that the probability of the growth being high is 0.6. The corresponding

84 QUANTITATIVE ANALYSIS

probabilities for average and low growth in this first year are 0.3 and 0.1 respectively. In the following three year period, the prediction of expected growth is reduced to just two categories—high and low. It is estimated that, if the growth is high in the first year, there is a 0.75 probability of the growth remaining high over the next three year period. A first year average growth has a 0.5 chance of deteriorating to low in years 2 to 4. An initially low growth has a 0.9 chance of remaining low throughout the period. The net cash revenues generated have been estimated as:

TABLE 3.24 **Net cash revenue,**

Growth	Net cash revenue, £/year arising at year end
High	£20,000
Average	£14,000
Low	£11,000

The initial purchase cost of the machine is £35,000. Tranda's alternative leasing arrangement is an immediate payment of £15,000 plus 25% of the net cash revenue paid at the end of each year. The company expects 12% per annum return on capital.

We will use a decision tree to advise Tranda whether or not it should buy or lease the computers. The decision criterion is to maximise the expected net present value using 12% per annum rate of return.

Solution:

Step 1: Draw a decision tree for the purchase or lease situation:

FIGURE 3.5 **Decision tree for the purchase or lease situation**

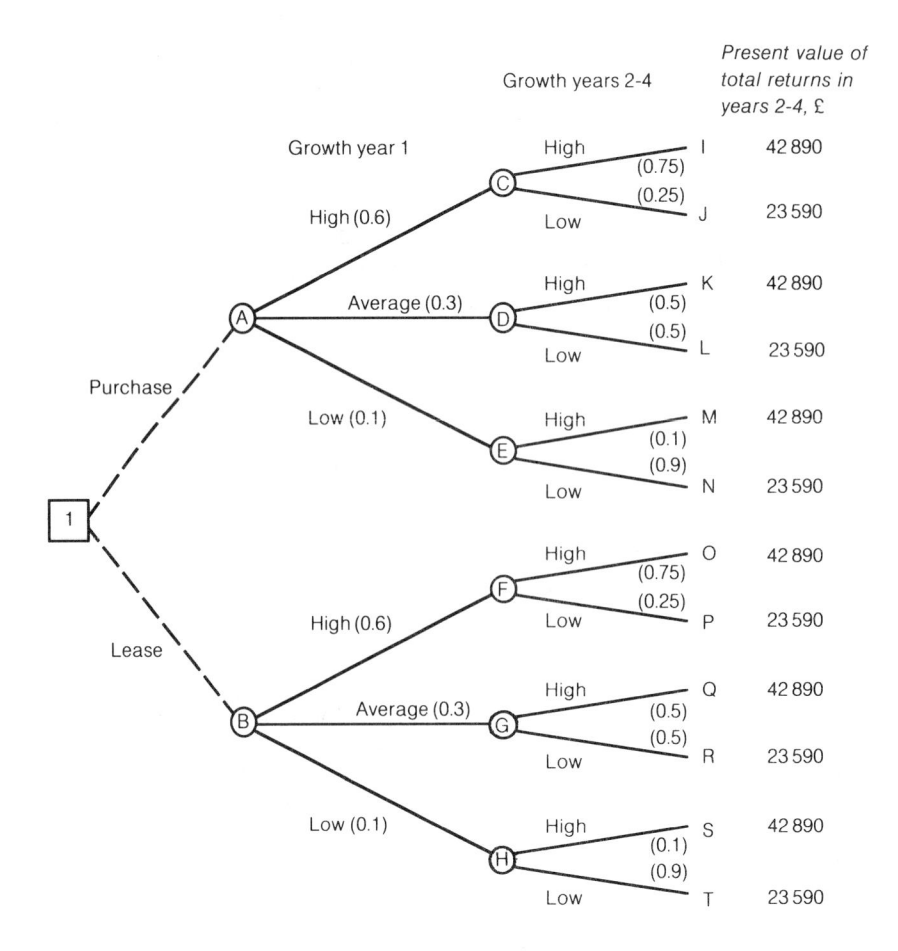

	Growth years 2-4	Present value of total returns in years 2-4, £
Growth year 1 — High (0.6) → C: High (0.75)	I	42 890
Low (0.25)	J	23 590
Average (0.3) → D: High (0.5)	K	42 890
Low (0.5)	L	23 590
Low (0.1) → E: High (0.1)	M	42 890
Low (0.9)	N	23 590
Lease — High (0.6) → F: High (0.75)	O	42 890
Low (0.25)	P	23 590
Average (0.3) → G: High (0.5)	Q	42 890
Low (0.5)	R	23 590
Low (0.1) → H: High (0.1)	S	42 890
Low (0.9)	T	23 590

Note that the two halves of the tree, for the purchase and lease options, are identical apart from the initial costs. The cost of the lease cannot be calculated until an estimate is obtained of the expected returns. This is done in the final step.

Step 2: Calculate the present value of the sums due in years 2 to 4. The monetary values on the extreme right of the decision tree are the total revenues for years 2 to 4 inclusive, discounted at 12% per annum to the present time, ie the present values of these revenues. The details of these calculations are set out below. If £A is received at the end of each year and the company requires 12% per annum on its capital, then the present value of £A per year for years 2 to 4 is:

$$\text{Present value} = \frac{A}{(1 + 0.12)^2} + \frac{A}{(1 + 0.12)^3} + \frac{A}{(1 + 0.12)^4}$$

$$= A\left(\frac{1}{1.12^2} + \frac{1}{1.12^3} + \frac{1}{1.12^4}\right)$$

$$= A \times 2.1445$$

Hence at node I, where A, the return per year, is £20,000, the present value of the total return for years 2 to 4 at 12% per annum is:

$$PV_I = £20,000 \times 2.1445 = £42,890$$

The corresponding figure at node J is:

$$PV_J = £11,000 \times 2.1445 = £23,590$$

These present value figures are repeated alternately at nodes K to T.

Step 3: Evaluate the tree. Using these present values, the expected values at outcome nodes C to H can be calculated. At outcome node C, the expected present value of the returns in years 2 to 4 is:

$$EPV_C \text{ (years 2–4)} = (£42,890 \times 0.75) + (£23,590 \times 0.25) = £38,065$$

On this route, a return of £20,000 is allocated to year 1. The present value of this sum is:

$$\frac{£20,000}{1.12} = £17,857$$

Hence, the expected present value at node C for years 1 to 4 is:

$$EPV_C \text{ (years 1–4)} = £38,065 + £17,857 = £55,922$$

At outcome node D, the expected present value of the returns in years 1 to 4 at 12% per annum is:

$$EPV_D = ((£42,890 \times 0.5) + (£23,590 \times 0.5)) + \frac{£14,000}{1.12} = £45,740$$

At outcome node E, the expected present value for years 1 to 4 is:

$$EPV_E = ((£42,890 \times 0.1) + (£23,590 \times 0.9)) + \frac{£11,000}{1.12} = £35,341$$

From the symmetry of the problem, the expected present values at nodes F, G, and H are £55,922, £45,740 and £35,341, respectively. This completes the calculation for all the extreme right hand nodes. We can now move through the tree to the left, to evaluate the expected present value at nodes A and B. Both nodes take the same value.

$$EPV_A = EPV_B = £55,922 \times 0.6 + £45,740 \times 0.3 + £35,341 \times 0.1 = £50,809$$

QUANTITATIVE ANALYSIS

The net expected present value at outcome node A, if the computers are purchased, is:

$$\text{Net EPV}_A = \text{the expected present value} - \text{the cost of purchase}$$
$$= £50,809 - £35,000 = £15,809$$

To find the net expected present value at outcome node B, we have to calculate the cost of the leasing arrangement. £15,000 is payable immediately plus 25% of the net cash revenue per annum. We require the expected present value of all these payments. The expected present value of the net cash revenues is £50,809. Hence the expected present value of the cost of the leasing arrangement is:

$$£15,000 + 25\% \text{ of } £50,809 = £15,000 + £12,702 = £27,702$$

Therefore, the net expected present value at outcome node B, if the computers are leased, is:

$$£50,809 - £27,702 = £23,107$$

We are now back to the decision node 1, at the far left of the tree. On the basis of maximising the net expected present value of the returns, we compare the value at outcome node A (£15,809 for purchasing) with the value at node B (£23,107 for leasing). The company should make the decision to lease the computers. The final diagram for **Example 3.7** is:

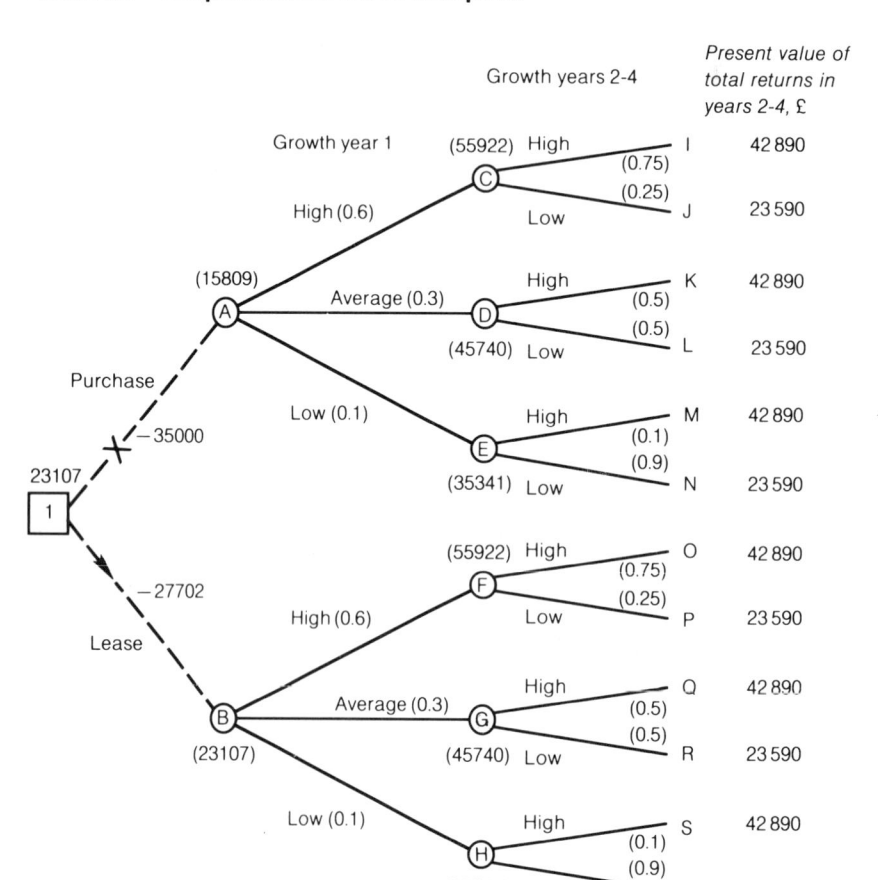

FIGURE 3.6 **Complete decision tree for Example 3.7**

Notice that no account has been taken of any scrap value for the computers at the end of the four years if they were purchased. This item could reverse the decision.

3.5.2 Sensitivity analysis and decision trees

The decisions made using a decision tree depend on the probabilities assigned to the various outcomes. The size of the changes necessary to alter the decisions is a measure of the sensitivity of the problem. This is an important aspect to consider since the probabilities will almost certainly be estimates and thus subject to unreliability. We need to know how dependent the decision is on the estimates of probability.

Example 3.8: Sensitivity analysis of a decision tree

The Cacus Chemical Company has recently developed a new product for which a substantial market is likely to exist in one year's time. Due to the highly unstable nature of this product, a new production process must be set up at a cost of £2.5 million to cope with the anticipated high temperature reactions. This process will take one year to develop, but it is estimated that there is only a 0.55 probability that it will provide adequate standards of safety. In the light of this, the company is considering the additional development of a computerised control system (CCS) which will detect and warn against dangerous reaction conditions. Research on the CCS will take one year and cost £1 million and Cacus estimate that there is a 0.75 probability that the CCS can be developed successfully.

Development of the CCS can either begin immediately or be postponed for one year until the safety of the new process is known. If the CCS is developed immediately and the new process proves to have an adequate standard of safety, then the CCS will be unnecessary and the £1 million expenditure will have been wasted. On the other hand, if the CCS is postponed and the new process turns out to be unsafe, a subsequently successful development of the CCS will have delayed the product by one year. If neither the new process nor the CCS are successful, there is no way in which the product can be safely manufactured and the project will have to be abandoned.

If sales of the new product can commence in one year's time, it has been estimated that the discounted profit would amount to a total of £10 million before any allowance is made for depreciation on the new process or the CCS. If the launch of the product is delayed by one year, however, the total return is expected to fall to £8.5 million due to the possibility of other manufacturers entering the market. For simplicity, you may ignore the effects of discounting on the expenditure on the CCS.

Required:
1 Draw a decision tree to represent the various courses of action open to the company.
2 Which course of action would you recommend to the management of Cacus?
3 By how much would the probability of a successful new process development (currently estimated to be 0.55) have to change before you would alter your recommendation in **2**? Does the decision in question appear particularly sensitive to the value of this probability?

(ACCA, December 1986)

Solution:
1 The decision tree for this problem is shown in FIGURE 3.7 on page 88.
2 To evaluate the tree calculate the net expected returns at the nodes. At outcome node D, expected return is:

$$8.5 \times 0.75 + 0 \times 0.25 = £6.375m$$

and the net expected return is:

$$£6.375m - £1m = £5.375m$$

At outcome node E, net expected return is zero.
Therefore, at decision node 2, we would decide to develop the CCS and transfer £5.375m to this node. At outcome node A, the net expected return is:

$$(10 \times 0.55 + 5.375 \times 0.45) - 2.5 = £5.419m$$

88 QUANTITATIVE ANALYSIS

FIGURE 3.7 **Decision tree for Example 3.8**

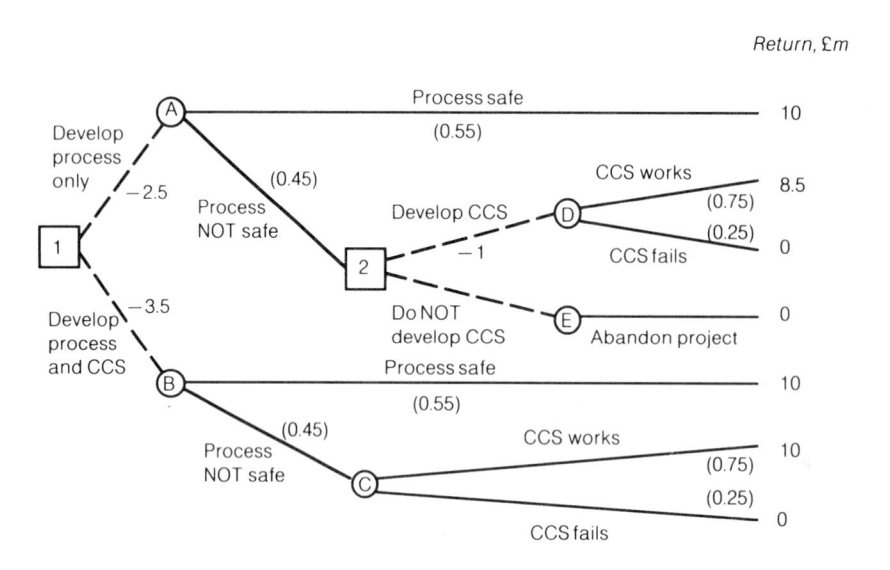

At outcome node B, the net expected return is:

$$(10 \times 0.55 + (10 \times 0.75 + 0 \times 0.25) \times 0.45) - 3.5 = £5.375m$$

Therefore at decision node 1, we would choose to develop the processing plant only, initially, and to develop the CCS a year later, only if the process proves to be unsafe. The net expected return will be £5.419m. The final tree for **Example 3.8** is:

FIGURE 3.8 **Complete decision tree for Example 3.8**

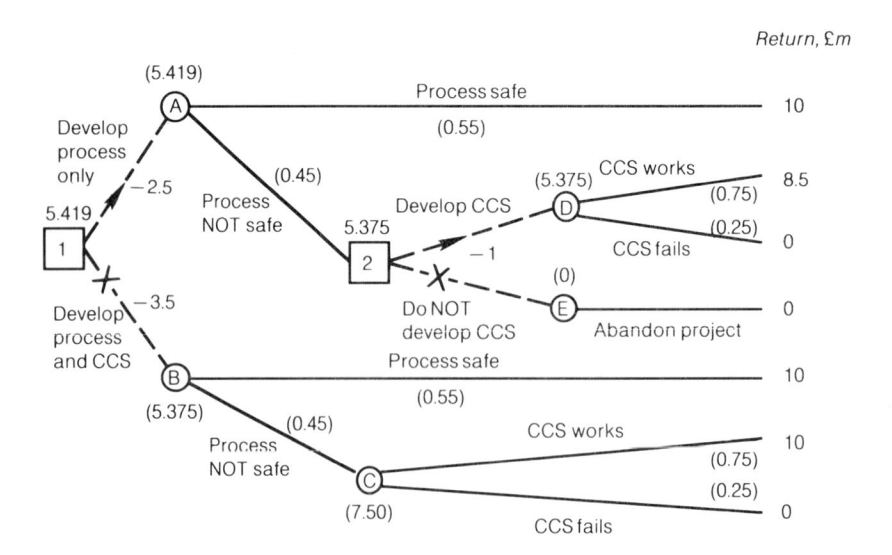

3 The sensitivity of the decision. The net expected returns at nodes A and B are almost the same, £5.419m and £5.375m. The decision is strongly dependent on the estimates of the probabilities. Sensitivity analysis allows us to calculate the values of the probabilities which would change the current decision.

In this question we are concerned only with the probability that the new process will be safe. In practice, the probability that the CCS works, also affects the expected values. A full sensitivity analysis would consider both of these. Let the probability that the production process is safe be p. Currently p = 0.55. The net expected return at A is:

$$10 \times p + 5.375 \times (1 - p) - 2.5 \quad (£m)$$

$$= 4.625p + 2.875 \quad (£m)$$

The net expected return at B is:

$$10 \times p + (10 \times 0.75 + 0 \times 0.25)(1 - p) - 3.5$$

$$= 2.5p + 4.0 \ (£m)$$

Equating the net expected returns at A and B gives:

$$4.625p + 2.875 = 2.5p + 4.0$$

Hence, 2.125p = 1.125 and p = 0.529.

Therefore, if the probability that the new process is safe is 0.529, the 2 alternative decisions will yield the same net expected returns. Hence, if the probability is less than 0.529, the decisions to develop both the process and the CCS immediately will yield the larger net expected return, ie the initial decision will be reversed.

Since p = 0.529 is close to p = 0.55, the decision is very sensitive to the value assigned to this probability. A small error in this estimate can reverse the decision taken. This shows the importance of sensitivity analysis in the decision making process.

SUMMARY

This chapter is about decision making where there is uncertainty. It deals with some of the quantitative approaches available for decision making but you should note that there is usually a mixture of quantitative and subjective elements in all practical decision making.

In every decision situation there are 3 fundamental steps which are always required:

1 define all the possible decisions that can be taken;
2 for each decision, define all the possible outcomes;
3 for each decision and outcome, calculate the payoff.

The payoff table is a convenient way of illustrating these steps. The decision maker then has to choose the decision rule that fits his own situation. The rule will indicate which decision he should make.

Some decision rules do not use probabilities. The maximax rule (for gamblers) and the maximin rule (for the cautious) use the payoffs directly. The minimax rule uses opportunity losses which are derived from the payoffs. More complicated decision rules use probability to calculate either the expected payoff or the expected opportunity loss for each decision. It is important to check the sensitivity of these rules to changes in the probabilities of each outcome. Additional information can lead to the making of better decisions. The 'value of perfect information' gives the maximum amount of money it is worth paying for this additional information.

The rules considered so far do not reflect the variability of the payoffs, nor the decision maker's attitude to risk. The standard deviation of payoffs for each decision can be used to show how variable the results may be. Utility is used to reflect the decision maker's attitude to risk. Each payoff is converted into a utility value by the decision maker. The expected utility for each decision is based on these values and their probabilities. The decision rule then uses the utility values to replace the payoffs. Situations often involve a chain of decisions, where one decision leads to another. Decision trees are a good way of illustrating this type of situation. The tree consists of decision nodes and decision branches, followed by outcome nodes and branches. The tree shows all the possible decisions and outcomes in the same way that the payoff table does for a single decision. The tree is evaluated in a stepwise fashion from the ends of the branches back towards the first decision. It gives the sequence of decisions which should be taken as the outcome of each

QUANTITATIVE ANALYSIS

decision becomes known. Again it is important to use sensitivity analysis to check how the decisions depend on changes in the probabilities.

EXERCISES

In **Exercises 3.1** to **3.4** you are given four decision situations. In each situation use each of the following rules to decide what should be done:

(a) Maximax payoff
(b) Maximin payoff
(c) Minimax opportunity loss
(d) Maximum expected payoff
(e) Minimum expected opportunity loss

Also, calculate the value of perfect information.

Exercise 3.1

A bakery produces bread for sale to shops. Each loaf costs 30p and is sold to the shops for 40p. The demand over a recent period of 50 days is given below:

Demand per day in '000 s	10	12	14	16	18
Number of days	5	10	15	15	5

If the bakery produces a loaf but cannot sell it to a shop, it loses 20p per loaf. Use each of the above rules in turn to decide how many loaves the bakery should produce per day.

Answers on page 456.

Exercise 3.2

A theatre has to decide how many programmes to produce for a run of performances. The production cost of the programmes is made up of a fixed cost of £200 plus 30p for each copy. The programmes are sold for 60p each, and, in addition, there is advertising revenue of £300. From previous experience, it is estimated that audience attendance will be:

Total audience	4000	4500	5000	5500	6000
Probability	0.1	0.3	0.3	0.2	0.1

It is expected that 40% of the audience will buy a programme.

Required:
1 Use each of the above rules in turn to decide how many programmes the theatre should order.
2 The advertisers offer to increase their payment to £400 if the audience is over 5250, and the demand for programmes is satisfied. How does this offer affect your recommendations in **1**?
3 Sensitivity: if the probability of each audience size was 0.2, how would the decision using the rule of maximising expected payoff be affected?

Answers on page 458.

Exercise 3.3

Kilroy's produce a very special non-alcoholic beverage. It is produced in 40 pint batches. A number of batches are brewed together during a week and each Monday they are ready for drinking. However, by the following Sunday, the drink has deteriorated so badly that any full batches remaining are thrown away. The secret pseudo-alcoholic additive used in the brew is purchased from a small laboratory. The laboratory has to set up its equipment well in advance and can produce only a fixed amount per week for a six month period. The additive is rushed to Kilroy's where it must be used immediately.

The beverage sells for £1.50 per pint and has variable costs of 70p per pint. However, Kilroy's feel that if they fail to meet demand, they lose custom equivalent, in the long term, to 30p per pint of demand not met. Demand has shown no clear trends over the last 50 weeks and has been as follows:

Number of batches demanded/week	3	4	5	6	7
Number of weeks	5	10	15	10	10

Required:

1 Use each of the rules above, in turn, to decide what Kilroy's should do during the next six month period.
2 Sensitivity: if Kilroy's increase the selling price to £1.75 per pint, would this change any of the decisions?

Answers on page 461.

Exercise 3.4

A publisher is bringing out a book on quantitative analysis for accountants. Its marketing department has produced the following sales forecast, covering the next three years:

Books demanded during three years	2000	3000	4000	5000
Probability	0.1	0.5	0.2	0.2

The contribution to the publisher's fixed costs and profit from every book sold is £9. If a book is not sold, the publisher loses £4 per book. If the publisher cannot meet demand, he estimates he will lose £1 per book of demand not met, to cover loss of goodwill and future sales.

Use each of the above rules, in turn, to decide how many books the publisher should print to cover this three year period.

Answers on page 463.

Exercise 3.5

The publisher in **Exercise 3.4** has suggested that you should use the further step of asking the marketing director and the finance director to estimate their utilities for various sums of money from £0 to £50,000. Their responses are given below:

Monetary value £'000	0	10	20	30	40	50
Utilities:						
Marketing director's	0	10	20	35	55	100
Finance director's	0	40	70	85	95	100

Required:

1 Plot the two utility curves and comment on the risk characteristics of the two directors.
2 Use the utility curve of each director in turn to estimate the utilities of the payoffs in **Exercise 3.4**. Reconsider the publishing decisions in that exercise using the criterion of maximising expected utility. What would each director recommend?

Answers on page 464.

Exercise 3.6

You have to make a journey from Stafford to Nottingham and then to Cambridge. For each leg of the journey you have decided that there are two possible routes. You have estimated possible times for each leg of the journey, together with the probability of achieving these times. The data are given below:

Route	from	to	time hours	probability	time hours	probability
1	Stafford	Nottingham	1.0	0.4	1.5	0.6
2	Stafford	Nottingham	1.25	0.5	1.5	0.5
3	Nottingham	Cambridge	2.25	0.2	2.75	0.8
4	Nottingham	Cambridge	2.5	0.6	3.0	0.4

Required:

1. Use a decision tree to illustrate your choices.
2. Decide which route you should take to minimise total travel time.

Answers on page 466.

Exercise 3.7

Hetros Ltd is a small specialist chemical company whose major product is Hetrozone, an expensive industrial solvent. Because Hetrozone deteriorates rapidly, production can only be scheduled on a monthly basis and stocks of Hetrozone cannot be kept from one month to the next. At the start of each month, a production figure for that month is decided upon and the necessary raw materials are obtained. Unfortunately, demand for Hetrozone varies randomly from month to month and, if demand exceeds the monthly production figure, sales are lost because demand must be satisfied more or less immediately. If, on the other hand, production exceeds demand in any month, the surplus is wasted and has no value. The selling price (p) of Hetrozone is £2400 per ton and the variable production costs (v) amount to £1500 per ton.

In an analysis of demand over the last few months, the sales manager finds that the monthly demand for Hetrozone is between 10 and 20 tons. In order to simplify his analysis of the demand situation, he decides to assume three levels of demand, 'low' (10 tons), 'medium' (15 tons) and 'high' (20 tons), with the following probabilities:

Demand (Tons)	Probability
10	0.3
15	0.6
20	0.1

Required:
1. In terms of the demand levels specified, draw a decision tree to represent the production alternatives open to the company and their various outcomes.
2. Assuming that the demand distribution does not change, advise the company on a production level that will maximise contribution in the long term.
3. It may be shown that the optimum production level, Q, occurs where the probability of demand exceeding Q is equal to v/p (the ratio of the unit variable production cost to selling price). Assuming, more realistically, that the distribution of demand is normal rather than the simple distribution used in **1** and **2**, estimate the mean and standard deviation of monthly demand and find the optimum production level Q.

(ACCA, June 1983)

Answers on page 467.

DECISION MAKING UNDER UNCERTAINTY: CHAPTER THREE

Exercise 3.8

Cadmus, Carna and Clytie, a firm of the investment consultants, have been approached by one of their clients with regard to the investment of a sum of £100,000 over a period of two years. After a thorough survey of the available opportunities, two alternatives (A and B) are proposed, one involving a small amount of risk, the other being risk free.

Investment A will lead to a return of either 8%, 10% or 12% in each year but, due to the nature of the investment, there will be some correlation between year 1 and year 2 returns. This is shown by the following table which gives the probability of various returns in year 2 given the returns in year 1.

	Year 2		
Year 1	8%	10%	12%
8%	0.6	0.3	0.1
10%	0.2	0.5	0.3
12%	0.1	0.2	0.7

At this stage, the three different returns in year 1 are considered to be equally likely. Investment B will produce a certain return of 9.5% per year. You may ignore the effects of taxation, and you may assume that the interest earned in year 1 is re-invested for the second year.

Required:
1. Assuming that whichever alternative is chosen, the investment will be made for the full two year period:
   (a) Draw a decision tree to represent the alternative courses of action and outcomes.
   (b) On the basis of the expected value of returns, which investment would you recommend?
   (c) What is the probability that investment B produces a greater return than investment A?
2. Indicate how your decision tree should be modified if it is possible to switch from investment A to investment B at the end of year 1. In this case, what investment strategy will produce maximum expected return?

(ACCA December 1984)

Answers on page 469.

Sampling and sampling distributions

4.1 INTRODUCTION

We will now turn our attention to an important activity in statistics—sampling.

The term **population** refers to the entire group of people or items to which the statistical investigation relates. The term **sample** refers to a small group selected from that population. In the same way, we use the term **parameter** to refer to a population measure and the term **statistic** to refer to the corresponding sample measure. For example, if we consider our population to be the current membership of ACCA, the mean salary of the full membership is a **population parameter**. However, if we take a sample of 100 members, their mean salary is referred to as a **sample statistic**. The purpose of a statistical investigation may be to measure a population parameter, for example the mean height of professional accountants in the UK, or the spread of wages paid to manual workers in the steel industry in France. Alternatively, the purpose may be to verify a belief held about the population, for example, the belief that violence on television contributes to the increase in violence in society, or that jogging keeps you fit.

The use of data from samples inevitably leads us into the realms of statistical inference, since it becomes necessary to know what may be inferred about the population from the sample. What does the sample statistic tell us about the population parameter, or what does the evidence of the sample allow us to conclude about our belief with respect to the population? How does the mean height of a sample of professional accountants relate to the mean height of the population of professional accountants? If, in a sample of adults, the joggers are fitter, can we claim for the population as a whole that jogging keeps you fit? With an appropriately chosen sample, it is possible to estimate population parameters from sample statistics and to use sample evidence to test beliefs held about the population.

Statistical inference is a large and important aspect of statistics. Information is gathered from a sample and this information is used to make deductions about some aspect of the population. For example, an auditor may check a sample of a company's transactions and, if the sample is satisfactory, he will assume, or infer, that all the company's transactions are satisfactory. He uses a sample because it is cheaper, quicker and more pratical than checking all the transactions carried out in the company.

The first part of this chapter is concerned with different ways of selecting samples. If a sample is not selected correctly, we cannot use statistical inference to draw reliable conclusions about the population.

4.2 POPULATIONS AND SAMPLES: WHY SAMPLE?

In a survey of student opinion about the catering services provided by a college, the target population is all students registered with that college. However, if the survey was concerned with catering services generally in colleges, the target population would be all students in all colleges. Since it is unlikely that in either case the survey team could collect data from the whole population, a small group of students would be selected to represent the population. The survey team would then draw appropriate inferences about the population from the evidence produced by the sample. It is very important to define the population at the beginning of an investigation in order to ensure that any inferences made are meaningful.

Populations such as those referred to above are limited in size and are known as **finite** populations. Populations which are not limited in size are referred to as **infinite** populations. In practice, if a population is sufficiently large that the removal of one member does not appreciably alter the probability of selection of the next member, then the population is treated statistically as if it were infinite. For example, a machine produces nails, a certain proportion of which are defective. If the output of nails in a day is 100,000 and we select a small sample from this output, the probability of selecting a defective nail does not change noticeably as the sample members are removed. The finite population of 100,000 items can be treated as an infinite population. On first consideration, we might feel that it would be better to use the whole population for our statistical investigation if at all possible. However, in practice we often find that the use of a sample is better.

The advantages of using a sample rather than the population are:

1. **Practicality** The population could be very large, possibly infinite. Hence, it would be physically impossible to collect data from the whole population.
2. **Time** If the data are needed quickly, then there may not be enough time to cover the whole population. For example, if we are concerned with checking the quality of goods produced by a mass production process, the delay in delivery while every item is checked will be unacceptable.
3. **Cost** The cost of collecting data from the whole population may be prohibitively high. In the example above, the cost of checking every item could make the mass produced item excessively expensive.
4. **Errors** If data were collected from a large population, then the actual task of collecting, handling and processing the data would involve a large number of people and the risk of error increases rapidly. Hence, the use of a sample with its smaller data set will often result in fewer errors.
5. **Destructive tests** The collection of the data may involve destructive testing, in which case, it is obviously undesirable to deal with the entire population. For example, a manufacturer wishes to make a claim about the durability of a particular type of battery. He runs tests on some batteries until they fail to determine what would be a reasonable claim about all the batteries.

The advantages of using the population rather than a sample are:

1. **Small populations** If the population is small, so that any sample taken would be large relative to the size of the population, then the time, cost and accuracy involved in using the population rather than the sample will not be significantly different.
2. **Accuracy** If it is essential that the information gained from the data is accurate, then statistical inference from sample data may not be sufficiently reliable. For example, it is necessary for a shop to know exactly how much money has been taken over the counter in the course of a year. It is not sufficient for the owner to record takings on a sample of days out of the year. The problem of errors is still relevant here but any errors in the data will be ones of arithmetic rather than unreliability of statistical estimates.

4.3 THE SELECTION OF A SAMPLE

It is extremely important that the members of a sample are selected so that the sample is as representative of the population as possible, given the constraints of availability, time and money. A biased sample will give a misleading impression about the population.

There are several methods of selecting sample members. These methods may be divided into two categories—**random** sample designs and **non-random** sample designs.

4.3.1 Random sampling

Statistical inference can be used only with this kind of sample. Random sampling means that every member of the population has an equal chance of being selected for the sample. If the population consists of groups of members with different characteristics, which are important in the investigation, then random sampling should result in a sample which contains members from each of these groups. The larger the group, the more likely it is that its members will be selected for the sample, hence, the larger its representation in the sample. The size of the group is related to the probability that its members will be selected, therefore, random sampling has a tendency to lead towards a sample which is representative of the population.

The first step in selecting a random sample from a finite population is to establish a **sampling frame**. This is a list of all members of the population. It does not matter what form the list takes as long as the individual members can be identified. Each member of the population is given a number, then some random method is used to select numbers and the sample members are thus identified. The representativeness of the sample depends on the quality of the sampling frame. It is important that the sampling frame possesses the following properties:

1 **Completeness**—all the population members should be included in the sampling frame. Incompleteness can lead to defects in the sample, especially if the members which are excluded belong to the same group within the population.
2 **Accuracy**—the information for each member should be accurate and there should be no duplication of members.

This list of the members of the population should not be confused with a census of the population. The purpose of the sampling frame is solely to enable individuals to be identified for selection purposes. The data required have then to be collected from the sample members. If the data are collected from all members of the sample frame, ie the whole population, then this is a census. The random selection method can be manual, such as picking numbered balls from a bag, or using random number tables. Alternatively, a computer can be used to generate random numbers. These are then used to select the sample.

4.3.2 Random sample designs

SIMPLE RANDOM SAMPLE DESIGN

The basic principle of random sampling has already been described in the previous section and the procedure for the selection of a simple random sample was outlined. A simple random sample is most suitable when the members of the population are similar for the purposes of the investigation. For example, a simple random sample would be suitable for selecting a sample of 20 employees of a company to take part in a survey if we are interested only in the fact that the individuals are employees of this company. In the later chapters on statistical inference it will be assumed that this is the kind of sample that has been chosen. All the random sampling selection methods that follow are amenable to statistical analysis, but in slightly different, more complicated, ways.

STRATIFIED RANDOM SAMPLING

If the populations members have different characteristics which are of interest, then the simple random sampling method may not give the most representative sample for a given sample size. For example, in the survey of employees of a company, it may be important to distinguish between male and female employees. A simple random sample design could result in too many members of one sex, unless a large sample is used. The stratified random sampling method enables us to produce a sample which more accurately reflects the composition of the population. The procedure requires the sampling frame to be subdivided into the groups of interest. These groups are referred to as **strata.** In our example, the males and females should be identified. We need to know how many members of the population fall into each category. We then use a simple random sampling method to select sample members from each stratum separately in proportion to their number in the population. If the population contains 60% males and 40% females, and we are selecting a sample of 100 members, then we select 60 men from the male stratum and 40 women from the female stratum. The main advantage of stratified random sampling is that we can use smaller sample sizes to get the same results as with a simple random sample.

SYSTEMATIC SAMPLE DESIGN

The advantage of systematic sampling is that the selection of the sample is much simpler than the previous methods we have considered. If the population contains 1000 members and we require a sample of 100, then every tenth population member (1000/100) is selected from the sampling frame. The random element is introduced by selecting the starting point at random. For example, if we have set of invoices, numbered 1 to 1000, we use random numbers to select the first invoice number, and then pick out every following tenth invoice. If the random number were 3, for example, we pick out invoice numbers 3, 13, 23, 33, etc. This gives the required sample. We have used only one random number, compared with 100 for the other methods.

This is a satisfactory procedure as long as there are no periodic effects or odd groupings in the invoices. If for some reason one customer always has invoices ending in 3, the sample would be biased and the method unacceptable.

MULTI-STAGE SAMPLE DESIGN

As the name implies, the selection of the sample by this method requires several stages. The method is most commonly used when the population is distributed over a wide geographical area. For example, the population might be the world-wide membership of the ACCA. The first stage is to divide the population into a few clearly defined areas. In our example, these could be individual countries. The proportion of the sample allocated to these areas is determined by the proportion of the population in each area. Hence, if 80% of the ACCA membership were in England and Wales, and we were looking for a sample of 1000 members, then we allocate 800 sample members to England and Wales, as with stratified sampling. The next stage is to define some smaller areas—these might be local government districts, and then companies within these districts. A sample is taken of the local government districts. Within the chosen districts, a sample of the companies is selected. Finally individual members are sampled from the chosen companies in the selected districts. With this method the actual sampling is quicker and more convenient than simple random sampling from widely dispersed populations.

CLUSTER SAMPLING DESIGN

In previous designs we have selected items one at a time. In cluster sampling clusters of items are formed, which it is assumed are reasonably representative of the whole population. Clusters are then randomly selected and all the items in the cluster are included in the sample. For example, suppose a large firm stores its invoices in batches of 50. If, in a year, there are 10,000 invoices generated, then there will be 200 batches. These batches can be used as clusters. Suppose the firm wants a sample of 300 invoices. This could be achieved by selecting 300 individual invoices randomly from the 10,000. Alternatively, the cluster sample design allows 6 clusters to be randomly selected from the 200 batches. This is a much easier and quicker method, but we must be sure that there is no bias within the batches.

There are a number of other random sample designs, but those described above are the most commonly quoted ones.

4.3.3 Non-random sample designs

For many surveys, especially in the area of market research, sampling frames do not exist. For example, if we wish to investigate housewives' views of a new product, it would be difficult to draw up a sampling frame for housewives.

Quota sampling is a commonly used sampling method in this situation. Initially the important characteristics of the target population are identified as for stratified sampling, for example male/female, age groups, social class, etc. The sample is divided proportionately into these groups as far as it is possible, but from then on, it is left to the individual field workers to decide how to obtain the sample members. There is no question of identifying individuals first or choosing them randomly.

In auditing, so called **judgement sampling** can be used, where the accountant uses a mixture of hunch, prior knowledge and judgement to select his sample. There is no attempt at stratification and randomness.

4.4 SAMPLING DISTRIBUTIONS

One of the purposes of sampling is to obtain estimates of various population parameters such as the mean, standard deviation or proportion. The sample statistics are used to estimate the unknown values of these parameters. We need to know, therefore, how the statistics can best be used for this estimation and how reliable the estimates are likely to be.

It will be assumed at all times that the samples used are simple random samples. Although the ideas apply to all random sampling methods, the detailed calculations are different. We are helped in this investigation by knowing a little about the probability distributions of the sample statistics we are using. These distributions can be constructed, in principle, as described below.

We select a sample of a particular size, called n, from a population of size N. For this sample we determine the value of the statistic corresponding to the parameter in which we are interested. We take a second sample of the same size and determine the statistic as before. We expect the numerical values of the statistic from the two samples to be different. If we go on to take a third and fourth sample, of the same size, we would expect to obtain still further values of the statistic. However, as we continue to take further samples, some values of the statistic will be repeated and, by the time we have taken all possible samples of size n from our population, we can draw up a frequency distribution for the values obtained of the statistic. The corresponding relative frequency distribution is the probability distribution for the statistic. This frequency distribution of the sample statistic is called a **sampling distribution.**

For example, if we take all possible samples of size n = 5 from a population of 50 items, we will have 2,118,760 different samples. (This is calculated using a combination, because we are not interested in the order in which the items are picked.)

$$^{50}C_5 = \frac{50!}{5!45!} = 2118760$$

For each sample of 5, we calculate the sample mean and the sample standard deviation. The frequency distribution of all 2,118,760 sample means gives the sampling distribution of means of samples of size 5 from this population. In the same way, the plot of all the sample standard deviations gives the sampling distribution of the standard deviations of samples of size 5 from the population. If the 50 items can be divided into good and bad items, for example, then we can calculate the proportion of good items in each sample of 5 and plot the sampling distribution of the sample proportions of good items. Each of these distributions will be different. The shapes will depend on the population, the sample size and the statistic we are measuring. The example below illustrates this idea for a much smaller population.

Example 4.1: To illustrate a sampling distribution

We have a finite population of six numbers:

$$4, 8, 12, 16, 20, 24$$

We will draw up a sampling distribution using samples of size 2, taking the arithmetic mean as our statistic. There are:

$$^6C_2 = 15$$

different samples.

Solution:

TABLE 4.1 **Sample means, n = 2**

Possible samples size $n = 2$	Sample mean, $\bar{x}$ $\bar{x} = (x_1 + x_2)/2$
4, 8	6
4, 12	8
4, 16	10
4, 20	12
4, 24	14
8, 12	10
8, 16	12
8, 20	14
8, 24	16
12, 16	14
12, 20	16
12, 24	18
16, 20	18
16, 24	20
20, 24	22

The sampling distribution for the means of these samples is:

TABLE 4.2 **Sampling distribution**

Sample mean, $\bar{x}$ sample size $n = 2$	Frequency, f
6	1
8	1
10	2
12	2
14	3
16	2
18	2
20	1
22	1
Total	15

Although a sampling distribution could be constructed for any statistic, the discussion below will be confined to the two most useful ones—the sampling distribution of sample means and the sampling distribution of sample variances.

The statistical procedure, which is developed in the following chapters, depends on establishing the relationships between a sampling distribution and the parent population. In this book we do not attempt to provide the full theoretical development of these general relationships but simply state the conclusions which we wish to use. Once the connection between the sampling distribution and the population has been established in general, then we may take a particular single sample and use the relationships to draw conclusions about the unknown population from which the sample was drawn. In all cases we assume that the population is normal, or approximately normal, and that the required sample statistics are known. In this way we can, for example, determine what the population mean (μ) is likely to be, or within what limits we expect its variance (σ^2) to lie.

4.4.1 Sampling distribution of sample means

Suppose we are interested in the mean value of some characteristic of the population, for example the mean height of ten year old children, the mean wage of clerical workers in the textile industry, or the mean diameter of steel washers produced by a machine. These population means can be estimated from a sample. If the original, or parent, population is normally distributed, then the sampling distribution of sample means will also be normally distributed. Even for non-normal parent populations, if the sample size is large (n ⩾ 30), the sampling distribution of sample means will be approximately normally distributed. This very important property is referred to as the **Central Limit Theorem**. This allows us to use here, all of the ideas in Chapter Two section **2.7** on the normal distribution, the standardised tables and z values.

From the sampling distribution we can calculate the average value of all of the sample means. This is the expected value of the sample mean:

$$E(\bar{x}) = \frac{\sum f\bar{x}}{\sum f}$$

If the parent population is normal, the expected value of the sample mean is the same as the population mean μ. Hence:

$$E(\bar{x}) = \mu$$

This result is true only if the method of selecting the sample is random. The mean of the sampling distribution is said to be an **unbiased estimator** of the population mean. **Example 4.2** shows that this can also be true for a non-normal population.

Example 4.2: To illustrate the calculation of E($\bar{x}$)

Refer back to **Example 4.1** above. The sampling distribution for the means of samples with n = 2 is:

TABLE 4.3 **Calculations of E($\bar{x}$)**

Sample mean, $\bar{x}$	Frequency, f	f$\bar{x}$
6	1	6
8	1	8
10	2	20
12	2	24
14	3	42
16	2	32
18	2	36
20	1	20
22	1	22
Total	15	210

The expected value of the sampling distribution is:

$$E(\bar{x}) = \frac{\sum f\bar{x}}{\sum f} = \frac{210}{15} = 14.0$$

The population is:

$$4, 8, 12, 16, 20, 24$$

and the population mean is:

$$\mu = (4 + 8 + 12 + 16 + 20 + 24)/6 = 84/6 = 14.0$$

Hence, in this example:

$$E(\bar{x}) = \mu$$

However, as has been already mentioned, in practice we would not actually construct the sample distribution in this way, hence, $E(\bar{x})$ cannot be calculated. Usually we have information about a single sample only. However, since we know that:

$$E(\bar{x}) = \mu$$

we can use this single sample mean as the unbiased estimate of the population mean:

$$\bar{x} = \hat{\mu}$$

where $\hat{}$ denotes an estimated value.

The reliability of the estimate will be discussed in detail later, but it may be assessed in terms of the dispersion of the sampling distribution. The standard deviation of a sampling distribution is referred to as the standard error of the sampling distribution, denoted by SE. In this particular case, the standard error of the sample means would be denoted by $SE_{\bar{x}}$ where:

$$SE_{\bar{x}} = \sqrt{\frac{\sum f(\bar{x} - E(\bar{x}))^2}{\sum f}} = \sqrt{\frac{\sum f\bar{x}^2}{\sum f} - (E(\bar{x}))^2}$$

FIGURE 4.1 **Normal parent population**

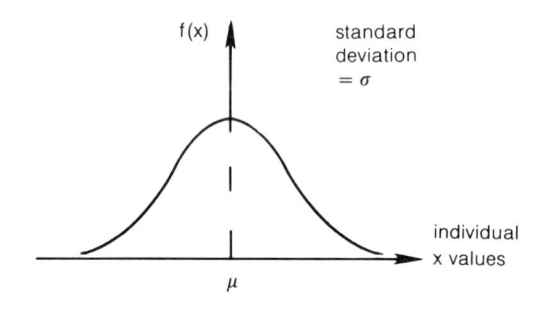

FIGURE 4.2 **Sampling distribution of means, $\bar{x}$ of samples of size n = 4**

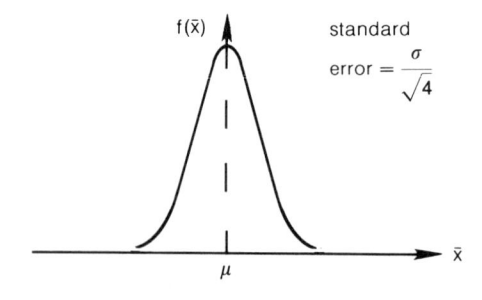

For a normal parent population, we find that the standard error of the sampling distribution of sample means is given by:

$$SE_{\bar{x}} = \sqrt{\frac{(N - n)\sigma^2}{(N - 1)n}}$$

where σ^2 is the population variance.

If the population is very large compared to the sample size (usually this is taken to be n/N ⩽ 0.05), then:

$$\sqrt{\frac{(N-n)}{(N-1)}} = 1 \text{ approximately}$$

and the standard error becomes:

$$SE_{\bar{x}} = \sqrt{\frac{\sigma^2}{n}} = \sqrt{\frac{\text{population variance}}{\text{sample size}}}$$

If we vary the sample size, we find that the mean of the sampling distribution does not change, that is $E(\bar{x}) = \mu$ the unbiased estimator is independent of the size of the sample, but the $SE_{\bar{x}}$ decreases as the sample size increases.

FIGURE 4.3 **Normal population individual values**

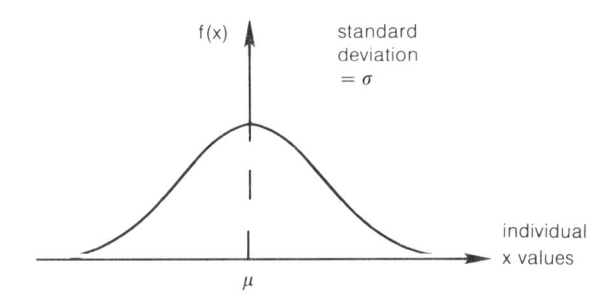

FIGURE 4.4 **Sampling distribution of means, $\bar{x}$, for sample sizes n = 4, 10, 25**

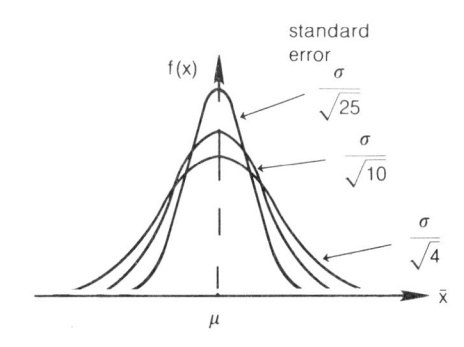

In calculating the standard error of the sample means we have assumed that we know σ^2, the population variance. In practice this is unlikely to be the case, and we need some way of estimating the variance of the population from the sample.

4.4.2 Sampling distribution of sample variance

The sampling distribution of sample variances can be set up using the method described above, but this time, the variance of each sample is recorded. However, unlike the sampling distribution of sample means, the sampling distribution of sample variances is not normally distributed. If the parent population is normal, the χ^2 (chi-squared) distribution represents the sampling distribution of the sample variance.

FINITE PARENT POPULATION

If the parent population is finite, then the expected value of the sample variances is given by:

$$E(s^2) = \frac{N}{(N-1)} \frac{(n-1)\sigma^2}{n}$$

hence:

$$\sigma^2 = \frac{(N-1)n}{N(n-1)} E(s^2)$$

Therefore, if the expected value of the sample variances is known, the variance of the population (σ^2) may be determined. $E(s^2)$ is an unbiased estimator of the population variance. Again, it is more likely that the variance of one sample only will be available to us. This must then be used to estimate the population variance. Hence:

$$\hat{\sigma}^2 = \frac{(N-1)}{N} \frac{ns^2}{(n-1)}$$

where:

$$s^2 = \frac{\sum (x - \bar{x})^2}{n}$$

This gives us an unbiased estimate of the population variance.

Example 4.3: To illustrate the sampling distribution of sample variances

We will look at **Example 4.1** again, where we have a finite population of numbers:

4, 8, 12, 16, 20, 24.

We will draw up a sampling distribution, using samples of size 2 and taking the sample variance as our statistic. The sample variance:

$$s^2 = \frac{\sum (x - \bar{x})^2}{n}$$

Solution:

TABLE 4.4 **Sample variance, n = 2**

Possible samples	Sample variance, s^2
4, 8	4
4, 12	16
4, 16	36
4, 20	64
4, 24	100
8, 12	4
8, 16	16
8, 20	36
8, 24	64
12, 16	4
12, 20	16
12, 24	36
16, 20	4
16, 24	16
20, 24	4

The sampling distribution for the variances of these samples is:

TABLE 4.5 **Calculations of E (s²)**

Sample variance, s^2	Frequency, f	fs^2
4	5	20
16	4	64
36	3	108
64	2	128
100	1	100
Total	15	420

The expected value of the sampling distribution is given by:

$$E(s^2) = \frac{\sum fs^2}{\sum f} = \frac{420}{15} = 28.0$$

Therefore:

$$\sigma^2 = \frac{(N-1)}{N} \frac{n}{(n-1)} E(s^2)$$

$$= \frac{(6-1)}{6} \frac{2}{(2-1)} \times 28.0$$

$$= 46.6$$

Taking the parent population, we can calculate the value of σ^2 directly. We know that the mean, $\mu = 14.0$, hence:

TABLE 4.6 **Calculations of σ^2**

Variable value, x	$(x - \mu)^2$
4	100
8	36
12	4
16	4
20	36
24	100
Total	280

Therefore:

$$\sigma^2 = \frac{\sum (x - \mu)^2}{n} = \frac{280}{6} = 46.6$$

This is the same value as that derived from the sampling distribution of sample variances.

LARGE (INFINITE) POPULATIONS

If the parent population is large, then:

$$\sqrt{\frac{N}{(N-1)}} = 1 \text{ approximately}$$

The unbiased estimator of the population variance then becomes:

$$E(s^2) = \frac{(n-1)\sigma^2}{n} \quad \text{or} \quad \sigma^2 = \frac{n}{(n-1)} E(s^2)$$

and the unbiased estimate of the population variance is given by:

$$\hat{\sigma}^2 = \frac{n}{(n-1)} s^2$$

where:

$$s^2 = \frac{\sum (x - \bar{x})^2}{n}$$

hence, $\hat{\sigma}^2$ may be written as:

$$\hat{\sigma}^2 = \frac{n}{(n-1)} \frac{\sum (x - \bar{x})^2}{n} = \frac{\sum (x - \bar{x})^2}{(n-1)}$$

This is the unbiased estimator of the population variance. You should note that the unbiased estimator of the population variance is frequently quoted as:

$$s^2 = \frac{\sum (x - \bar{x})^2}{n-1}$$

(s is used not $\hat{\sigma}$).

It is vital not to confuse this with the sample variance:

$$s^2 = \frac{\sum (x - \bar{x})^2}{n}$$

You must be clear which statistic you are actually calculating. In this book we will denote the unbiased estimator of the population variance by $\hat{\sigma}^2$ to avoid confusion.

The tabulated distribution we use with sample variances from normal populations is the χ^2 distribution where the sampling distribution of $\frac{ns^2}{\sigma^2}$ is a χ^2 distribution with $(n-1)$ degrees of freedom. (See section **4.4.4**.)

4.4.3 Estimate of the standard error of the sampling distribution of sample means

In section **4.4.1** the standard error of the sampling distribution of sample means was quoted for large populations:

$$SE_{\bar{x}} = \sqrt{\frac{\sigma^2}{n}}$$

To evaluate the standard error from this equation, we must know σ^2, the population variance. If we do not know σ^2 we use its best estimate which is given by:

$$\hat{\sigma}^2 = \frac{ns^2}{(n-1)}$$

Hence, the best estimate of the standard error is:

$$\widehat{SE}_{\bar{x}} = \sqrt{\frac{(ns^2)/n}{(n-1)}} = \sqrt{\frac{s^2}{(n-1)}} \quad \text{where} \quad s^2 = \frac{\sum (x - \bar{x})^2}{n}$$

or:

$$\widehat{SE}_{\bar{x}} = \sqrt{\frac{\hat{\sigma}^2}{n}} = \frac{\hat{\sigma}}{\sqrt{n}} \quad \text{where} \quad \hat{\sigma}^2 = \frac{\sum (x - \bar{x})^2}{n - 1}$$

4.4.4 Standard sampling distributions—z, t, χ^2, F

There are four standard sampling distributions which will be referred to frequently in the following chapters. They are the normal (z), t, χ^2, and F distributions. In this section the assumptions behind each distribution are outlined, together with the use of the distributions in statistical inference.

THE STANDARD NORMAL (z) DISTRIBUTION FOR SAMPLE MEANS

In Chapter Two section **2.7.** we discussed the normal distribution. We described how to transform any normal distribution into the standard normal distribution which has mean, $\mu = 0$, and variance, $\sigma^2 = 1$. The variable for this standard normal distribution is called z, where:

$$z = \frac{x - \mu}{\sigma}$$

These values of z are used with the standard normal tables to find the required probabilities. The sampling distribution for sample means is a normal distribution if the samples are simple random samples chosen from a normal population. We can treat this normal distribution in the same way as that summarised above, but the value of the variable z, is:

$$z = \frac{\bar{x} - \mu}{SE_{\bar{x}}}$$

since the sampling distribution of sample means is a distribution of $\bar{x}$ (not x) which has a mean of μ and a standard deviation or standard error denoted by $SE_{\bar{x}}$. z measures the number of standard errors that the sample mean is away from the population mean.

Since $SE_{\bar{x}} = \sigma/\sqrt{n}$ for large populations, the z value may be written as:

$$z = \frac{\bar{x} - \mu}{\sigma/\sqrt{n}}$$

To use this equation we must know the population variance (σ^2). If we do not know σ^2, we estimate it, using the sample variance, by:

$$\hat{\sigma}^2 = \frac{ns^2}{(n - 1)}$$

The standard variable becomes:

$$\frac{\bar{x} - \mu}{\hat{\sigma}/\sqrt{n}}$$

but its distribution is no longer normal. The standard normal (z) distribution is replaced by the t distribution.

THE t DISTRIBUTION FOR SAMPLE MEANS

If the simple random sample is taken from a normal population, for which we do not know σ^2, the standardised distribution for the sample means is a t distribution, where:

$$t = \frac{\bar{x} - \mu}{s/\sqrt{(n - 1)}} = \frac{\bar{x} - \mu}{\hat{\sigma}/\sqrt{n}}$$

The t distribution is symmetrical about the population mean, μ, but unlike the normal distribution, the exact shape of the t distribution varies for each value of n, the sample size. When n is small the t distribution is much more spread out than the z distribution. As the sample size increases the t distribution merges into the standard normal distribution; hence it is possible to use the normal distribution as an approximation to the t distribution for large sample sizes. In this case a sample size is large if $n \geq 30$.

The dependence of the t distribution on the sample size is not a simple dependence. The t distribution actually varies with the number of **degrees of freedom** in the problem. For example, if we are dealing with the mean of a single sample of size n, the number of degrees of freedom will be $(n - 1)$, but if we are considering the means of two samples, size n_1 and n_2, then the number of degrees of freedom will be $(n_1 + n_2 - 2)$.

The idea behind the name, degrees of freedom, can be illustrated with a simple example. If we are given the mean of 5 numbers then we have the freedom to choose any 4 numbers, but the value of the fifth number is determined by the given mean. For example, if the mean of 5 numbers is 6, we can choose 2, 7, 9 and 3 as the first 4 numbers. The fifth number, (y), is then determined because the mean is:

$$6 = (2 + 7 + 9 + 3 + y)/5 = (21 + y)/5, \quad \text{ie y must equal 9}$$

We have no freedom to choose the final number and so we have 4 degrees of freedom.

Once the t value has been calculated, standard probability tables—t tables—are used to determine the required probability as for the standard normal distribution. However, since these t tables must accommodate different degrees of freedom as well as different t values, it is necessary to condense the information provided. In fact, these tables are usually arranged to give the t values associated with specific probabilities for various degrees of freedom. (See the table in Appendix Two.) The use of the table will be explained in more detail in Chapter Five.

THE χ^2 SAMPLING DISTRIBUTION FOR THE SAMPLE VARIANCE

The same assumptions apply to the χ^2 distribution, ie a simple random sample is assumed to be taken from a normal population. The statistic:

$$\chi^2 = n \frac{s^2}{\sigma^2}$$

follows the χ^2 sampling distribution with $(n - 1)$ degrees of freedom. Like the t distribution, the shape of this distribution depends on the number of degrees of freedom.

FIGURE 4.5 **χ^2 distributions for different degrees of freedom**

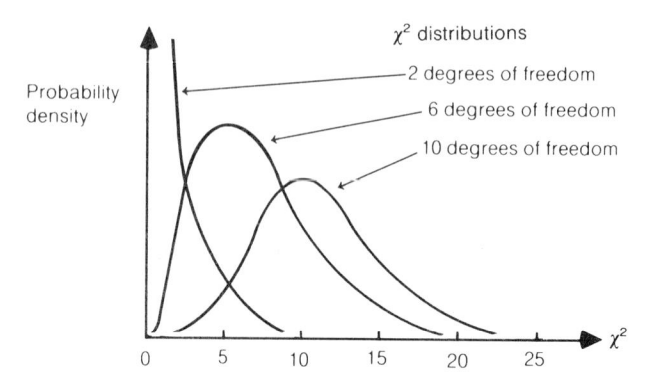

There is a separate χ^2 distribution for each degree of freedom. The distribution is not symmetrical and it changes as the sample size increases.

The χ^2 tables are summarised in a similar way to the t tables. Particular values are given for most degrees of freedom. This will be considered in more detail in sections **5.4** and **6.11**.

THE F SAMPLING DISTRIBUTION FOR TWO SAMPLE VARIANCES

If we have two samples which we can assume are randomly selected from normal populations, then in order to compare the two sample variances, a new sampling distribution, the F distribution, is required. The statistic:

$$F = \frac{n_1 s_1^2}{(n_1 - 1)\sigma_1^2} \bigg/ \frac{n_2 s_2^2}{(n_2 - 1)\sigma_2^2}$$

follows the F distribution. Like the t tables, the F tables are restricted to a few useful values of probability. They give values of the F statistic for combinations of the degrees of freedom in the two samples. The F distribution is considered in more detail in Chapter Six.

WHAT IF THE POPULATION IS NOT NORMAL?

All these standard distributions, z, t, χ^2 and F, assume that the sample is a simple random sample taken from a normal population. We control the choice of the sample and so we can ensure that it is a simple random sample, but we have no control over the normality of the distribution of the population. The easiest way to check the population assumption is to plot the sample data. If the plot looks reasonably symmetrical then we can make the required assumption. There are more formal statistical tests for normality but these are not necessary in this text. If the sample is obviously skewed, we must be careful how we proceed. There are alternative courses of action available.

If we are concerned with means, the Central Limit Theorem allows us to use the z distribution, if the sample size is at least 30. The Central Limit Theorem states that, if we take large enough samples from any distribution, with mean μ and standard deviation σ, however skewed that distribution may be, the distribution of sample means will be approximately normal. The larger the sample the nearer to the normal the distribution will be. The general rule is that n should be 30 or more. When sample sizes are at least 30, and we are concerned with means, the normality assumption may be ignored.

The second alternative is to transform the variable. A variable which has a non-normal distribution may be transformed into a normal distribution by, for example, taking logarithms. The third alternative is to use completely different non-parametric statistics, which do not need the normality assumption.

SUMMARY

A population consists of all of the items with which a particular study is concerned. A sample is a much smaller number of items chosen from this population. A statistic is calculated from data collected from the sample. This is used to draw inferences about the corresponding population parameter.

To use a statistical analysis the sample must be chosen randomly. This means that every item must have an equal chance of being selected for the sample.

There are a number of random sampling procedures. Simple random sampling is the most basic. The population members are numbered. This establishes the sampling frame. Random numbers are taken from tables or from a computer. These random numbers are used to identify the appropriate items for the sample. Other sampling methods are used to reduce the necessary sample size which ensures representation of the population characteristic in the sample, or to simplify the sampling procedure. Amongst these are stratified, systematic, multi-stage and cluster sampling.

If we select all possible samples of size n from a normal population and calculate the sample statistic for each one, we can then establish the sampling distribution for that statistic. This enables us to see how the sample statistics are related to the population parameters. The plot of, for example, all the sample means of samples of size n is called the sampling distribution of sample means.

For a normal population, the sampling distribution of sample means (sample size n) has an expected value:

$$E(\bar{x}) = \mu \quad \text{the population mean}$$

and a standard error:

$$SE_{\bar{x}} = \sqrt{\frac{(N-n)}{(N-1)} \frac{\sigma^2}{n}}$$

where σ^2 is the population variance, when the sample size n is large compared with the number of items in the population (N).

If the population size is infinite or $n/N \leqslant 0.05$, then:

$$SE_{\bar{x}} = \sqrt{\frac{\sigma^2}{n}}$$

The generalisations made from the sampling distributions enable us to use the statistics from single samples to estimate the corresponding population parameters:

$$\hat{\mu} = \bar{x}$$

and

$$\hat{\sigma} = \sqrt{\frac{n}{n-1}} s \quad \text{where} \quad s = \sqrt{\frac{\sum (x - \bar{x})^2}{n}}$$

or

$$\hat{\sigma} = \sqrt{\frac{\sum (x - \bar{x})^2}{n-1}}$$

Four important standard distributions are used in the following chapters. All of them assume a simple random sample. The z (normal) distribution describes the distribution of sample means, if the parent population is normal:

$$z = \frac{\bar{x} - \mu}{\sigma/\sqrt{n}}$$

This distribution is also appropriate if the population is not normal but the sample size is large, $n \geqslant 30$, from the Central Limit Theorem.

If σ^2 is not known, but the population is normal, the standardised distribution of the sample means is the t distribution where:

$$t = \frac{\bar{x} - \mu}{s/\sqrt{(n-1)}} = \frac{\bar{x} - \mu}{\hat{\sigma}/\sqrt{n}}$$

The t distribution depends on the number of degrees of freedom in the problem.

The χ^2 distribution represents sample variances, so long as the population is normal:

$$\chi^2 = n \frac{s^2}{\sigma^2}$$

with $(n - 1)$ degrees of freedom.

The F distribution describes the ratio of two sample variances from two normal populations:

$$F = \frac{n_1 s_1^2}{(n_1 - 1) \sigma_1^2} \bigg/ \frac{n_2 s_2^2}{(n_2 - 1) \sigma_2^2}$$

This distribution depends on the number of degrees of freedom in each sample.

These standard distributions are tabulated in different ways and will be considered in detail in later chapters.

EXERCISES

Exercise 4.1

Are the populations given below finite or infinite?

1 The orders processed by a laundry last week.
2 The emergency calls that could be made to the fire brigade.
3 The voters on the electoral register for a town.
4 The orders that could be processed by an engineering works.
5 All the computers that could be manufactured in the UK.

Answers on page 472.

Exercise 4.2

What would the population be for a survey of:

1 The public's attitude to smoking.
2 The basic wages of hourly paid workers in the shoe manufacturing industry.
3 A company's accounts for an audit of the year ending April 19X9.
4 The average lifetime of a particular electronic component.

Answers on page 472.

Exercise 4.3

Describe how you would obtain a sampling frame for surveys of:

1 The households in Ayetown, postal district 7.
2 The students at present studying at the University of Beeton.
3 The solicitors in Ceetown.
4 The retail sales outlets for electrical goods in Deeton.

Answers on page 472.

Exercise 4.4

Explain why the following methods of collecting data may lead to bias and non-randomness in a survey:

1 To obtain the public's views on vivisection, print a tear-out questionnaire in a wild life magazine.
2 To obtain the public's reaction to a particular method of taxation, carry out house-to-house interviews during the day time.
3 To find out the public's attitude to the problems of living in the inner city, telephone a sample of households, randomly selected from the telephone directories of the major cities.

Answers on page 472.

Exercise 4.5

The staff in a large office get their drinks from a vending machine in a nearby corridor. The manager feels that too much time is being spent by certain staff in using the machine. How would you gather data to decide whether this was true or not?

Answers on page 473.

Exercise 4.6

A local authority wishes to collect information on residents' attitudes to a proposed leisure centre.

Required:
Comment on each of the following methods which have been suggested for obtaining a suitable sample of residents.

1. Interview 500 people at the local football match.
2. Set up an exhibition about the leisure centre in the Town Hall, and ask visitors to make comments in a book.
3. Put an article in the local paper and ask people to write in with comments.
4. Ring up 500 names from the telephone book.
5. Send batches of questionnaires to the secretaries of local sports clubs, and ask them to circulate the questionnaires to their members.

Answers on page 473.

Exercise 4.7

The main methods of data collection are:

- (a) Postal questionnaire.
- (b) Direct observation with on-the-spot recording.
- (c) Personal interviewing.
- (d) Telephone interviewing.
- (e) Abstracting data from published statistics.

Required:
Discuss the advantages and disadvantages of each of these methods in the following situations, and suggest which would be the most appropriate to gather data for a survey of:

1. Traffic flow along a particular stretch of road.
2. Sales of a brand of cat food.
3. Reaction to a new television series.
4. Reaction of catalogue agents to a mail order company's new method of payment.
5. Response of consumers to a new food product.

Answers on page 473.

Exercise 4.8

The following is a small population of 24 weights, measured in grammes. The weights are numbered from 1 to 24 for identification.

Number	1	2	3	4	5	6	7	8	9	10	11	12
Weight g	4	6	12	11	12	35	31	7	3	13	13	7
Number	13	14	15	16	17	18	19	20	21	22	23	24
Weight g	14	5	33	14	4	5	6	6	30	12	7	5

Required:

1. Calculate the mean of the population of weights.
2. Use the random numbers given at the end of the question to select random samples, using in turn each of the methods below. Calculate the sample mean weight for each of the samples. Calculate the standard deviation of the weights for each of the samples.

(a) Select 3 simple random samples of 6 items each. Use columns 1, 2, 3 of the random numbers table.
(b) Select 3 systematic random samples of 6 items each. Use columns 4, 5, 6 of the random numbers table.
(c) Select 3 stratified random samples of 6 items each. To do this split the population into its three strata of light, medium and heavy weights. Use columns 7, 8, 9 of the random numbers table.
(d) Select 3 cluster samples of 6 items each. To do this, split the population into 4 clusters, each containing 3 light, 2 medium and 1 heavy weight. Use columns 10, 1, 2 of the random numbers table.

3 Compare the resulting 12 sample means with the population mean and comment on how well each sampling method estimates the population mean.

When using the random numbers, start at the top of each indicated column, and continue to the next column if you need to.

Column Number	1	2	3	4	5	6	7	8	9	10
	01	15	04	19	08	09	02	09	18	18
	17	22	15	02	04	10	16	03	24	03
	07	24	21	08	19	12	11	15	13	17
	20	07	14	09	13	18	19	12	05	02
	21	10	17	15	16	05	16	20	14	16
	23	14	01	07	03	10	23	04	02	05
	02	21	24	23	06	14	21	23	19	08
	24	09	12	12	11	04	01	16	15	19
	23	08	09	21	14	02	03	21	07	22
	04	07	13	20	19	16	20	22	14	01

Answers on page 473.

Exercise 4.9

In section **4.4** of this chapter, the idea of a sampling distribution was illustrated by an example in which we took all samples of 2 items from a small population of 6 numbers, 4, 8, 12, 16, 20, 24. We will extend this example as follows:

Required:

1 Take all 15 samples containing 4 items from this population.
2 For each of these samples of 4 items, calculate the sample mean, $\bar{x}$, and sample standard deviation, s.
3 Plot the frequency diagram for the sample means of all samples of 2. The values are listed in section **4.4.1**. On the same axes, plot the frequencies of the sample means of all the samples of 4 using the values which you have just calculated in **2**. Compare the diagrams.
4 Plot the frequency diagram for the sample standard deviations of samples of 2. The values are listed in section **4.4.2**. On the same axes, plot the frequencies of the sample standard deviations of all the samples of 4 using the values which you have just calculated in **2**. Compare the diagrams.

Answers on page 476.

Statistical inference I: estimation—confidence intervals

5.1 INTRODUCTION

In Chapter Four we discussed how a sample might be used to produce an estimate of a population parameter. We also mentioned that in any situation in which an estimate is used, it is necessary to have some idea of how reliable this estimate is. The standard error of the sampling distribution will give us an indication of reliability, since the smaller the standard error, the less dispersed is the sampling distribution, and, hence, the less variable the sample statistic. However, it is simpler to appreciate the reliability of an estimate if we set up a range of values within which we can be reasonably sure that the population parameter lies. This range of values is called a **confidence interval**. Confidence intervals can be set up for any population parameter, although they are most commonly used for means and proportions.

5.2 CONFIDENCE INTERVAL FOR μ, THE POPULATION MEAN

5.2.1 When the population variance σ^2 is known

If the parent population is normal then we know that the sampling distribution of sample means will also be normal. If the population has a mean represented by μ and a standard deviation represented by σ, the sampling distribution of sample means will have a mean given by $E(\bar{x}) = \mu$ and a standard error given by $SE_{\bar{x}} = \sigma/\sqrt{n}$. From the Central Limit Theorem, we know that the above is also true for non-normal populations if the sample size, n, is at least 30.

If we draw a sample of size n from the population and we find that it has a mean of $\bar{x}$, $\bar{x}$ can be used to estimate μ, but how reliable is this estimate?

The figure below illustrates the sampling distribution of sample means. The values $\bar{x}_1$ and $\bar{x}_2$ are symmetrically placed with respect to the mean and enclose 95% of the sampling distribution. The area under the curve, above $\bar{x}_2$ is 2.5%, and the area under the curve below $\bar{x}_1$ is 2.5%. Hence, we can say that 95% of all samples of size n taken from our parent population will have mean values between $\bar{x}_1$ and $\bar{x}_2$. Putting this another way—there is a 95% probability that a sample of size n, taken from the population, will have a mean between $\bar{x}_1$ and $\bar{x}_2$.

FIGURE 5.1 **Sampling distribution of sample means, $\bar{x}$, with sample size n, for a normal population, with mean μ and variance σ^2**

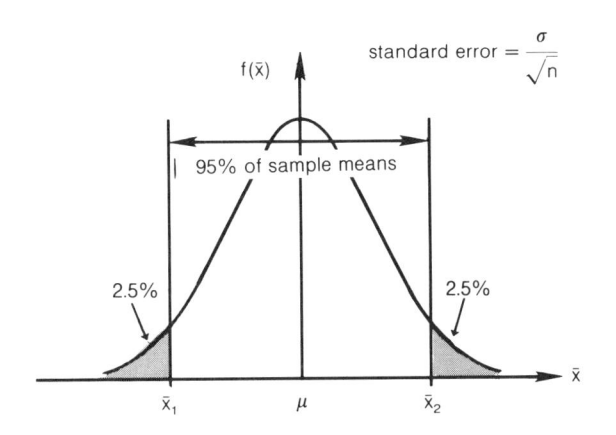

$\bar{x}_1$ is $(\bar{x}_1 - \mu)/SE_{\bar{x}}$ standard errors below the mean and $\bar{x}_2$ is the same distance above the mean. Since $\bar{x}_1$ and $\bar{x}_2$ enclose 95% of the distribution, we can determine from the standard normal tables that $\bar{x}_2$ is 1.96 standard errors above the mean and $\bar{x}_1$ is 1.96 standard errors below the mean.

$$\bar{x}_1 = \mu - 1.96 \, SE_{\bar{x}}$$

and:

$$\bar{x}_2 = \mu + 1.96 \, SE_{\bar{x}}$$

Therefore the range from $\bar{x}_1$ to $\bar{x}_2$ may be written as:

$$\mu \pm 1.96 \, SE_{\bar{x}}$$

We now take one sample from a population, calculate its sample mean and, using this sample mean, we draw inferences about the mean (μ) of the population from which the sample is taken. We are 95% sure that our single $\bar{x}$ lies between $\bar{x}_1$ and $\bar{x}_2$. If $\bar{x}$ does lie between $\bar{x}_1$ and $\bar{x}_2$, then μ must lie somewhere in the range:

$$\bar{x} \pm 1.96 \, SE_{\bar{x}}$$

We say we are 95% confident of this. Hence, $\bar{x} \pm 1.96 \, SE_{\bar{x}}$ is the 95% confidence interval for the population mean. If, for example, $\bar{x}$ exactly equals $\bar{x}_1$, then μ is right at the top of the range. If $\bar{x}$ is less than $\bar{x}_1$, μ is not in the confidence interval. We have picked one of the 5% of samples for which the inference is wrong. In the above discussion we chose to enclose 95% of the sampling distribution. This was an entirely subjective choice. Any size of range can be used depending on how confident you wish to be that you are enclosing the population mean within the range. Typical values are 90%, 95% or 99%. Whatever the value chosen, the format of the confidence interval remains the same. The only difference is the value of the standard normal variable, z. Hence, the general form of a confidence interval for a population mean is:

$$\bar{x} \pm z_{\alpha/2} \, SE_{\bar{x}}$$

where $z_{\alpha/2}$ is the value of the standard normal variable above which $(\alpha/2) \, 100\%$ of the values lie. This gives the $(1 - \alpha) \, 100\%$ confidence interval.

For example, if we require a 95% confidence interval, then $\alpha = 0.05$ and the confidence interval may be written as:

$$\bar{x} \pm z_{0.025} \, SE_{\bar{x}}$$

If the population standard deviation is known, then the standard error of the sampling distribution of sample means is:

$$SE_{\bar{x}} = \sigma/\sqrt{n}$$

and the $(1 - \alpha) \, 100\%$ confidence interval for the population mean becomes:

$$\bar{x} \pm z_{\alpha/2} \frac{\sigma}{\sqrt{n}}$$

Example 5.1: To calculate a confidence interval for a population mean, when the population standard deviation is known

An importer packs tea into packets containing a nominal 125 g. His filling machine is known to work to a standard deviation (σ) of 10 g. A sample of 50 packets (n) has a mean weight ($\bar{x}$) of 128.5 g.

What is the 95% confidence interval for the population mean weight (μ) delivered by the machine? Assume that the weight of tea delivered to the packets is normally distributed.
Solution: The 95% confidence interval for the population mean is:

$$\bar{x} \pm 1.96 \frac{\sigma}{\sqrt{n}}$$

where 1.96 is the number of standard errors above and below the mean which encloses 95% of the normal distribution. Hence, the 95% confidence interval is:

$$128.5 \pm 1.96 \times \frac{10}{\sqrt{50}} = 128.5 \pm 2.77 \text{ g}$$

There is a 95% probability that the population mean weight of tea (μ) delivered to the packets is between 125.73 g and 131.27 g. The range of ± 2.77 g represents approximately $\pm 2\%$ of the sample mean fill of 128.5 g. A 2% variability is probably not large in this context, hence, the sample mean may be regarded as a reliable estimate of the population mean. Remember however that there is a 5% chance that we are wrong and μ lies outside this range.

Example 5.2: To calculate a confidence interval for a population mean when the population standard deviation is known

A machine which packs sugar has for a long time given a normal distribution of weights of filled packets. The standard deviation (σ) of weight has been 2.5 g. The machine is adjusted to give a new metric size pack. A random sample of 20 (n) of the new packets is weighed. The mean weight of the sample ($\bar{x}$) is 1002 g. Assuming that the adjustment has not affected the variability of the filling process, what is the 99% confidence interval for the population mean weight of the metric packs?
Solution: The 99% confidence interval for the population mean is:

$$\bar{x} \pm 2.576 \frac{\sigma}{\sqrt{n}}$$

where 2.576 is z, the number of standard errors above and below the mean which encloses 99% of the normal distribution. Hence, the 99% confidence interval is:

$$1002 \pm 2.576 \times \frac{2.5}{\sqrt{20}} = 1002 \pm 1.44 \text{ g}$$

There is a 99% probability that the population mean weight of sugar (μ) delivered to the packets is between 1000.56 g and 1003.44 g. The range of ± 1.44 g represents approximately $\pm 0.1\%$ of the sample mean fill of 1002 g. A 0.1% variability is small, hence, the sample mean may be regarded as a reliable estimate of the population mean. In this case there is a 1 in 100 chance that we are wrong and that μ lies outside this range.

5.2.2 When the population variance is unknown

If the population variance is not known, then the standard error of the sampling distribution of sample means must be estimated as described in Chapter Four:

$$\widehat{SE}_{\bar{x}} = \frac{s}{\sqrt{n-1}} = \frac{\sigma}{\sqrt{n}}$$

where:

$$s = \sqrt{\frac{\sum (x - \bar{x})^2}{n}}$$

is the sample standard deviation and:

$$\hat{\sigma} = \sqrt{\frac{\sum (x - \bar{x})^2}{(n-1)}}$$

is the unbiased estimate of the population standard deviation. The distribution of the appropriate standardised variable is no longer the normal. Instead a t distribution must be used. The $(1 - \alpha)$ 100% confidence interval for the population mean becomes:

$$\bar{x} \pm t_{\alpha/2,\,(n-1)} \frac{s}{\sqrt{(n-1)}} \quad \text{or} \quad \bar{x} \pm t_{\alpha/2,\,(n-1)} \frac{\hat{\sigma}}{\sqrt{n}}$$

where:

$$t_{\alpha/2,\,(n-1)}$$

is the standard t variable for $(n - 1)$ degrees of freedom, above which $(\alpha/2)$ 100% of the t distribution lies.

As explained in Chapter Four, if the sample size, n, is at least 30, then the t distribution may be approximated by the normal distribution.

Example 5.3: To calculate a confidence interval for a population mean, when the population standard deviation is unknown

A random sample of 25 packs (n) of apples was found to have a mean weight ($\bar{x}$) of 1020 g with a standard deviation (s) of 12 g. Find an approximate 95% confidence interval for the population mean weight of the apples. It is assumed that the population is normal.
Solution: Since the population standard deviation is not known, the standard error of the sample can be estimated only. The appropriate sampling distribution to use is the t distribution. The 95% confidence interval for the population mean is:

$$\bar{x} \pm t_{0.025,24} \frac{s}{\sqrt{(n-1)}}$$

where $t_{0.025,24}$ is the standard t variable above which 2.5% of the t distribution lies for (25–1) degrees of freedom.

Using the standard t tables in Appendix Two, we find that:

$$t_{0.025,24} = 2.064$$

Hence, the 95% confidence interval is:

$$1020 \pm 2.064 \times \frac{12}{\sqrt{24}} = 1020 \pm 5.06 \text{ g}$$

There is a 95% probability that the population mean weight (μ) of apples in the packs is between 1015 g and 1025 g.

The range of ±5.06 g represents approximately ±0.5% of the sample mean fill of 1020 g. A 0.5% variability is probably not large in this context, and hence, the sample mean may be regarded as a reliable estimate of the population mean.

Example 5.4: To calculate a confidence interval for a population mean when the population standard deviation is unknown

A manufacturer of car tyres is interested in obtaining an estimate of the average lifetime of one particular model. He draws a random sample of 10 tyres (n) and subjects them to a forced-life test. The sample mean lifetime ($\bar{x}$) is 22,500 miles with a standard deviation (s) of 3000 miles. What is the approximate 99% confidence interval for the mean lifetime of his whole production of tyres of this type? As usual we assume that the population is normal.

Solution: Since the population standard deviation is not known, the standard error of the sampling distribution can be estimated only. This means that the appropriate standard distribution to use is the t distribution. The 99% confidence interval for the population mean is:

$$x \pm t_{0.005,9} \frac{s}{\sqrt{(n-1)}}$$

where $t_{0.005,9}$ is the standard t variable above which 0.5% of the t distribution lies, for (10–1) degrees of freedom.

Using the t tables in Appendix Two, we find that:

$$t_{0.005,9} = 3.25$$

Hence, the 99% confidence interval is:

$$22500 \pm 3.25 \times \frac{3000}{\sqrt{9}} = 22500 \pm 3250 \text{ miles}$$

There is a 99% probability that the population mean lifetime (μ) of this type of tyre is between 19,250 miles and 25,750 miles.

The range of ±3250 miles represents approximately ±14% of the sample mean life of 22,500 miles. A 14% variability is probably quite large in this context, and hence, the sample mean may be doubtful as a reliable estimate of the population mean.

5.3 CONFIDENCE INTERVAL FOR A POPULATION PROPORTION, p

We are often interested in the **proportion** of occasions in which a characteristic occurs within a population. For example, auditors are interested in the proportion of errors in a set of accounts. From Chapter Two, we know that the proportion of occasions in which a characteristic arises follows a binomial probability distribution. However, if we consider only large samples and cases in which the proportion of occurrences is neither small nor large, then we can assume that the sampling distribution of sample proportions is approximately normal. This means that we can set up confidence intervals for the population proportion based on z, the standard normal variable. We use the same guidelines for this approximation as in Chapter Two section **2.8**, on the normal approximation to the binomial, which is valid when:

$$np \geq 5 \quad \text{and} \quad n(1-p) \geq 5$$

We will use p to denote the proportion of occurrences in the population and $\hat{p}$ to denote the proportion in a sample. The standard error of the sampling distribution of sample proportion is:

$$SE_{\hat{p}} = \sqrt{\frac{p(1-p)}{n}}$$

(As discussed in section **2.3.3** this is the standard deviation of a binomial probability distribution.)

However, since the population proportion, p, is likely to be unknown, the standard error will usually be estimated by using $\hat{p}$ as an estimate of p. Therefore:

$$\widehat{SE}_{\hat{p}} = \sqrt{\frac{\hat{p}(1-\hat{p})}{n}}$$

120 QUANTITATIVE ANALYSIS

We can now write down the $(1 - \alpha)100\%$ confidence interval for the population proportion as follows:

$$\hat{p} \pm z_{\alpha/2}\sqrt{\frac{\hat{p}(1 - \hat{p})}{n}}$$

where $z_{\alpha/2}$ is the value of the standard normal variable above which $(\alpha/2)100\%$ of the sampling distribution lies.

Example 5.5: To calculate a confidence interval for a population proportion

A random sample of transactions in a set of accounts is taken for investigation by a firm of auditors. Out of the sample of 500 (n) entries, 10 reveal some error of entry or procedure. Estimate the 95% confidence interval for the proportion of errors in the whole population of transactions.

Solution: As np = 10 and n(1 − p) = 490 are both bigger than 5, the normal approximation to the binomial can be used. As the population proportion, p, is not known, the standard error of the sampling distribution must be estimated using the sample proportion, $\hat{p}$. The 95% confidence interval for the population proportion is:

$$\hat{p} \pm 1.96\sqrt{\frac{\hat{p}(1 - \hat{p})}{n}}$$

where ±1.96 is the value of the standard variable, z, which encloses 95% of the standard normal distribution.

$$\hat{p} = 10/500 = 0.02$$

Therefore:

$$\widehat{SE}_{\hat{p}} = \sqrt{0.02 \times 0.98/500} = 0.0063$$

Hence, the 95% confidence interval is:

$$0.02 \pm 1.96 \times 0.0063 = 0.02 \pm 0.012$$

There is a 95% probability that the proportion, p, of transaction errors in the population lies between 0.008 and 0.032.

The range of ±0.012 in the proportion represents approximately 60% of the sample proportion of 0.02. A 60% variability is probably large in this context, and, hence, the sample proportion may be doubtful as a reliable estimate of the population proportion.

Example 5.6: To calculate a confidence interval for a population proportion

A random sample is taken from the personal loan accounts at a bank. Of the 1000 accounts (n) in the sample, 60 are found to have been in arrears for at least three months. What are the 90% confidence limits for the proportion, p, of accounts in the population which have been in arrears for at least three months?

If the bank has a total of 30,000 personal loan accounts, what is the 90% confidence interval for the number of accounts which have been in arrears for at least three months?

Solution: Confidence limits are the values at the ends of the confidence interval. As np = 60 and n(1 − p) = 940 are both bigger than 5, the normal approximation to the binomial can be used. As the population proportion, p, is not known, the standard error of the sampling distribution must be estimated using the sample proportion, $\hat{p}$. The sample proportion of accounts in which we are interested is:

$$\hat{p} = 60/1000 = 0.06$$

and also:

$$\widehat{SE}_{\hat{p}} = \sqrt{\hat{p}(1 - \hat{p})/n} = \sqrt{0.06 \times 0.94/1000} = 0.0075$$

The 90% confidence interval for the population proportion is given by:

$$\hat{p} \pm 1.645 \, \widehat{SE}_{\hat{p}}$$

where ± 1.645 are the values of the standard normal variable, z, which encloses 90% of the distribution. Hence:

$$0.06 \pm 1.645 \times 0.0075 = 0.06 \pm 0.012$$

The confidence limits are, therefore:

$$0.06 + 0.012 = 0.072$$

and

$$0.6 - 0.012 = 0.048$$

If there is a total of 30,000 accounts, then the 90% confidence interval for the number of accounts of interest is:

$$30{,}000(0.06 \pm 0.012) = 1800 \pm 360$$

Hence, we are 90% confident that the total number of accounts, which have been in arrears for at least three months, lies between 1440 and 2160.

5.4 CONFIDENCE INTERVAL FOR A POPULATION VARIANCE, σ^2

A confidence interval for the population variance is also useful in situations in which it is important to control the spread of values of a variable. The following example illustrates a typical case. A company produces different components which fit together to form a car body. The mean dimensions of the population of components may be exactly as required, but if the variation in dimension is too large some of the components will not fit together. The company has to control both the mean and the variance of the population. From a sample of components, estimates of both the population mean and the population variance are needed. Both must be acceptable for the body assembly plant to work satisfactorily.

The χ^2 sampling distribution is used to calculate confidence intervals for the population variance. This assumes that the population is normal. In Chapter Four we stated that:

$$\chi^2 = n \frac{s}{\sigma^2} \quad \text{where} \quad s^2 = \frac{\sum (x - \bar{x})^2}{n}$$

or

$$\chi^2 = (n - 1) \frac{\hat{\sigma}^2}{\sigma^2} \quad \text{where} \quad \hat{\sigma}^2 = \frac{\sum (x - \bar{x})^2}{(n - 1)}$$

is distributed with a χ^2 distribution with $(n - 1)$ degrees of freedom. Thus:

$$\sigma^2 = n \frac{s^2}{\chi^2} = (n - 1) \frac{\hat{\sigma}^2}{\chi^2}$$

and the $(1 - \alpha) \, 100\%$ confidence interval for σ^2 is:

$$\frac{n \, s^2}{\chi^2_{\alpha/2, (n-1)}} \quad \text{to} \quad \frac{n \, s^2}{\chi^2_{(1 - \alpha/2), (n-1)}}$$

where $\chi^2_{\alpha, (n-1)}$ is the χ^2 value with $(n - 1)$ degrees of freedom, above which a proportion α of the χ^2 distribution lies. Because we are dividing by χ^2, notice that the larger χ^2 value $(\chi^2_{\alpha/2})$ gives the smaller, lower limit.

122 QUANTITATIVE ANALYSIS

The assumption that the parent population is normal is crucial when assessing population variances. Care is needed when checking that the sample looks normal.

Example 5.7: To calculate a confidence interval for a population variance

A sample of 19 components is measured, giving a sample mean ($\bar{x}$) of 125 mm and a sample standard deviation (s) of 1.5 mm. What is the 95% confidence interval for the population variance σ^2? We assume that the dimension is normally distributed.

Solution: The diagram illustrates the χ^2 values we use for the 95% confidence interval:

FIGURE 5.2 χ^2 **sampling distribution for sample variances for 18 degrees of freedom**

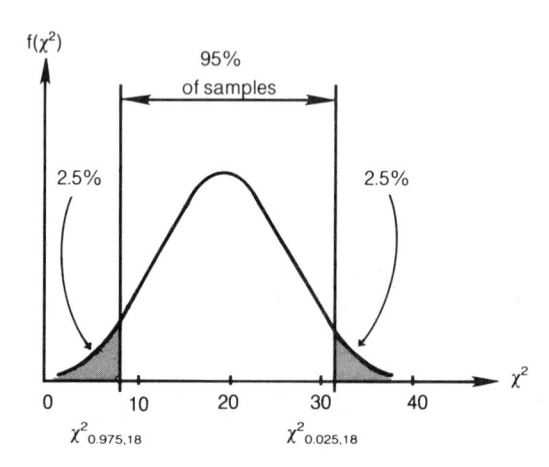

From the χ^2 table in Appendix Two, we need $\chi^2_{0.975,18}$ below which 2.5% of the distribution lies, and $\chi^2_{0.025,18}$, above which 2.5% of it lies. The values are:

$$\chi^2_{0.975,18} = 8.23$$

and

$$\chi^2_{0.025,18} = 31.53$$

The 95% confidence interval for the population variance σ^2 is:

$$\frac{19 \times 1.5^2}{31.53} \quad \text{to} \quad \frac{19 \times 1.5^2}{8.23}$$

which is 1.36 to 5.19 mm².

The confidence interval for the population standard deviation (σ) is:

$$\sqrt{1.36} \quad \text{to} \quad \sqrt{5.19}$$

which is 1.17 mm to 2.28 mm.

Notice that the confidence interval is not symmetrical about:

$$\hat{\sigma} = \sqrt{\frac{n}{(n-1)}}\, s = 1.54 \text{ mm}$$

Example 5.8: To calculate a confidence interval for a population variance

In **Example 5.3** we took a random sample from a normal population of packs of apples. The sample size (n) is 25, the sample mean ($\bar{x}$) 1020 g and the sample standard deviation (s) 12 g. What is the 90% confidence interval for the population standard deviation, σ?

Solution: The 90% interval for the variance is given by:

$$\frac{n s^2}{\chi^2_{0.05,24}} \text{ to } \frac{n s^2}{\chi^2_{0.95,24}}$$

Using the χ^2 table in Appendix Two this is:

$$\frac{25 \times 12^2}{36.42} \text{ to } \frac{25 \times 12^2}{13.85}$$

which is:

$$98.9 \text{ to } 259.9$$

The 90% confidence interval for the population standard deviation about 12 g is 9.9 g to 16.1 g.

5.5 THE CHOICE OF APPROPRIATE SIZES FOR SAMPLES

Throughout this chapter we have assumed that the sample size is given. We have not considered how we would actually decide on a suitable value. The confidence interval formulae may be used to determine an appropriate size for our sample. But first we must make some decisions. We must decide for example what level of confidence we should use—90%, 95%, or 99%?

We also need to know, given the level of confidence, what width of confidence interval we are willing to accept—± 2 g for a mean, $\pm 0.3\%$ for a proportion or 5 mm width for a standard deviation, for example?

If we already know σ or p, we can then calculate the required sample size. Usually however this is not the case, and we have to estimate σ or p. To do this, we have either to guess the values, using some prior knowledge, or we have to take a preliminary sample and use this to estimate σ or p. In the sections that follow, we will use the samples taken from earlier examples in the chapter as our preliminary samples.

5.5.1. The sample size required to estimate a population mean

Example 5.9: To calculate a sample size for the confidence interval of a population mean, μ, when σ is known

In **Example 5.1** a sample of 50 packets (n) of tea has a mean weight ($\bar{x}$) of 128.5 with a population standard deviation (σ) of 10 g. The 95% confidence interval was 128.5 $\pm$ 2.77 g. What should the sample size (n) be to give a 95% confidence interval of ± 2.0 g for the population mean μ?
Solution: The range of the confidence interval is:

$$\pm z_{\alpha/2} \frac{\sigma}{\sqrt{n}}$$

For the 95% interval in this question we want:

$$z_{\alpha/2} \sigma / \sqrt{n} \text{ to be } \leqslant 2 \text{ g}$$

Hence:

$$1.96 \times \frac{10}{\sqrt{n}} \leqslant 2$$

Therefore:

$$\sqrt{n} \geqslant \frac{1.96 \times 10}{2} = 9.8$$

and:

$$n \geqslant 96$$

The sample size, n, should be increased from 50 to at least 96 packets of tea to reduce the 95% confidence interval from ± 2.77 g to ± 2.0 g.

Example 5.10: To calculate a sample size for the confidence interval of a population mean, μ, when σ is not known

In **Example 5.3** we had 25 packs (n) of apples with a sample mean ($\bar{x}$) of 1020 g and a sample standard deviation (s) of 12 g. This gave a 95% confidence interval of ± 5.06 g. What must the sample size be if we want a 99% confidence interval of ± 5.0 g for the population mean?
Solution: The range of the confidence interval is:

$$\pm t_{\alpha/2, (n-1)} \frac{s}{\sqrt{(n-1)}} \quad \text{which must be less than } \pm 5 \text{ g}$$

We use the preliminary sample standard deviation of 12 g to estimate s, the standard deviation of the final sample. Unfortunately the t value we want to use, $t_{0.005, (n-1)}$, depends on the sample size, and we do not know n until we have done the calculation. We will ignore this problem for the moment, simply noting that our estimate of n is likely to be high, and we will use $t_{0.005, 24} = 2.797$ in the formula. Substituting in the values we have:

$$\frac{2.797 \times 12}{\sqrt{(n-1)}} \leqslant 5.0$$

Therefore:

$$\sqrt{(n-1)} \geqslant (2.797 \times 12)/5 = 6.71$$

Hence:

$$n - 1 \geqslant 45.1$$

And:

$$n \geqslant 46.1$$

The estimated sample size should be at least 47 to give ± 5 g for the 99% confidence interval. If we wish to be more accurate, we can now re-calculate n with the new degrees of freedom, or, since we now know that the sample size is likely to be greater than 30, we could use the z value in the equation instead of t. Replacing t by $z_{0.005} = 2.576$ gives:

$$\frac{2.576 \times 12}{\sqrt{(n-1)}} \leqslant 5$$

Therefore:

$$n - 1 \geqslant 38.2$$

Hence:

$$n \geqslant 39.2$$

So the required sample size is at least 40 when we make this correction, compared with 47 without the correction.

5.5.2. The sample size required to estimate a population proportion

Example 5.11: To calculate the sample size required to estimate a population proportion

Referring to **Example 5.5**, in which an auditor took a sample of 500 transactions to estimate the error rate, let us find out how big a sample the auditor would have to take if he wanted to be within 0.005 of the population proportion with 95% confidence. The error rate in the original sample was 0.02 with a 95% confidence interval of ± 0.012.

Solution: The range for the confidence interval for a proportion is:

$$\pm z_{\alpha/2} SE_{\hat{p}}$$

This is estimated by:

$$\pm z_{\alpha/2} \sqrt{\frac{\hat{p}(1-\hat{p})}{n}}$$

The auditor requires that the range of the 95% confidence interval about the sample proportion shall be at most ± 0.005. Therefore, we require:

$$1.96 \sqrt{\frac{\hat{p}(1-\hat{p})}{n}} \leqslant 0.005$$

Using the previous sample, we can now put $\hat{p} = 0.02$ and solve the inequality for n:

$$1.96 \times \sqrt{\frac{0.02 \times 0.98}{n}} \leqslant 0.005$$

Therefore:

$$\frac{1.96^2 \times 0.02 \times 0.98}{n} \leqslant 0.005^2$$

Hence:

$$\frac{1.96^2 \times 0.02 \times 0.98}{0.005^2} \leqslant n$$

Therefore:

$$n \geqslant 3012$$

If the auditor wishes to be within 0.005 of the population proportion at 95% confidence, his sample size should be at least 3012.

If we do not take a preliminary sample, we would have to guess a sensible value for the error rate, $\hat{p}$. If we estimate it to be 0.03, this will change the required sample size to n = 4500. Because the observed error rate is so small, a small absolute change, from 0.02 to 0.03, makes a very big difference to the required sample size.

5.5.3. The sample size required to estimate a population variance

Since the confidence interval for a population variance is not symmetrical, there are additional difficulties in estimating n for a given range. However, the problem can be overcome if we specify the upper limit only. Further problems arise since the sample standard deviation must be estimated either from a preliminary sample or by guessing, and the χ^2 value required has an unknown number of degrees of freedom. The following example illustrates how to deal with these problems.

126 QUANTITATIVE ANALYSIS

Example 5.12: To calculate the sample size for the confidence interval of a population variance, σ^2

In **Example 5.8** the 25 packs (n) of apples had a sample standard deviation (s) of 12 g. We calculated that the 90% confidence interval for the population standard deviation, σ, is 9.9 g to 16.1 g. What would the sample size have to be to reduce the upper 90% limit for the standard deviation to 15 g?

Solution: Remember that the upper confidence limit is given by the lower, smaller, χ^2 value. The upper 90% limit for the variance is

$$\frac{n s^2}{\chi^2_{0.95, (n-1)}}$$

which we want to be less than or equal to 15^2. Therefore:

$$\frac{n \, 12^2}{\chi^2_{0.95, (n-1)}} \leqslant 15^2$$

Hence:

$$n \leqslant \frac{15^2}{12^2} \chi^2_{0.95, (n-1)}$$

Therefore:

$$n \leqslant 1.56 \chi^2_{0.95, (n-1)}$$

In this inequality as n increases so does χ^2. We can use the χ^2 tables in Appendix Two to find the value of n by trial and error.

TABLE 5.1 Calculation of sample size, to give the upper confidence limit for σ

n	$\chi^2_{0.95, (n-1)}$	$\chi^2_{0.95, (n-1)} \times 1.56$	Comment
25	13.85	21.61	less than n — try larger n
31	18.49	28.84	less than n — try larger n
41	26.51	41.35	more than n — use this n

A sample size increased from the original 25 to at least 41 will give a 90% upper confidence limit of at most 15 g for the population standard deviation.

SUMMARY

The information from a simple random sample taken from a normal population is used to draw inferences about the mean, variance or proportion of the parent population. The 95% confidence interval tells us that we are 95% sure that the population parameter lies in this range. Conversely, there is a 5% chance that we are wrong and it lies outside the range. The $(1 - \alpha)$ 100% confidence intervals are calculated from the following formulae (using the standard z, t and χ^2 sampling distributions tabulated in Appendix Two):

1. Confidence interval for the population mean μ, when σ^2 is known:

$$\bar{x} \pm z_{\alpha/2} \frac{\sigma}{\sqrt{n}}$$

2 Confidence interval for the population mean μ, when σ^2 is not known:

$$\bar{x} \pm t_{\alpha/2,(n-1)} \frac{s}{\sqrt{(n-1)}} \quad \text{or} \quad \bar{x} \pm t_{\alpha/2,(n-1)} \frac{\hat{\sigma}}{\sqrt{n}}$$

which, for samples of at least 30, can be approximated by:

$$\bar{x} \pm z_{\alpha/2} \frac{s}{\sqrt{(n-1)}}$$

From the Central Limit Theorem, this formula also applies to any non-normal populations, for samples with $n \geqslant 30$.

3 Confidence interval for the population proportional p, when both $np \geqslant 5$ and $n(1-p) \geqslant 5$:

$$\hat{p} \pm z_{\alpha/2} \sqrt{\frac{\hat{p}(1-\hat{p})}{n}}$$

4 Confidence interval for the population variance, σ^2:

$$\text{from} \quad \frac{n s^2}{\chi^2_{\alpha/2,(n-1)}} \quad \text{to} \quad \frac{n s^2}{\chi^2_{(1-\alpha/2),(n-1)}}$$

Except where stated, all these formulae assume that the parent population is normal. This assumption is most critical for the variance. The sample size, n, required to give a confidence interval of a specified width and level of confidence can also be calculated using these formulae, if σ^2 is known or we are able to estimate s or p from a preliminary sample.

EXERCISES

Exercise 5.1

A machine dispenses liquid chocolate to moulds to make chocolate bars. Over a long period of time the weights of the chocolate dispensed have followed a normal distribution with a standard deviation of 2.5 g. As part of the routine quality control process, a random sample of 15 of the bars is taken from the output of a production run, and weighed. The sample mean weight is 99.5 g. Set up a 95% confidence interval for the true mean weight of chocolate dispensed to the moulds.

Answers on page 478.

Exercise 5.2

A particular component in a transistor circuit has a lifetime which is known to follow an approximately normal distribution. A random sample of 50 components from a week's production has an average lifetime of 840 hours with a standard deviation of 22 hours. Find an approximate 99% confidence interval for the mean lifetime of the population of these components.

Answers on page 478.

Exercise 5.3

The quality control manager of Britalite Bulbs plc wishes to know the average lifetime of a particular type of bulb made by the company. A random sample of 75 bulbs is taken and tested to destruction. The average lifetime of the sample is 2920 hours with a standard deviation of 400 hours.

Required:
1 How good an estimate of the population mean is this sample mean likely to be?
2 What should the manager do if he wishes to estimate the population mean lifetime to within 50 hours at the 95% confidence level?

Answers on page 479.

Exercise 5.4

A random sample of 800 housewives in a town centre one morning, revealed that 480 would like to see the shopping centre become traffic-free.

Required:
What are the 90% confidence limits for the proportion of all housewives in the town who would like the shopping centre to be traffic-free?

Answers on page 480.

Exercise 5.5

Ms Sally Briggs is the sales manageress for a confectionery manufacturer. In a survey of a random sample of 200 outlets in Wales, she found that 50 would be willing to stock a proposed new product.

Required:
1 Find a 95% confidence interval for the proportion of outlets in Wales which will stock this product.
2 Ms Briggs intends to conduct a similar survey in Scotland. However, she wishes to obtain an estimate of the proportion of outlets willing to stock the new product to within ± 4%. How large a sample should be used in Scotland? (Assume that she uses a 95% confidence interval.)

Answers on page 480.

Exercise 5.6

Jones plc is a company retailing fashionwear for men and women through a nationwide chain of 20 high street stores in England and Wales.

Jones is considering introducing its own credit card. As part of a feasibility study, Jones conducted a survey of a random sample of its customers at its Oxford Street store in London. From a sample of 80 customers, 32 indicated that they would apply for such a credit card if it should become available. These 32 people were then asked to anticipate how much they would borrow on the credit card. For the 32 replies, the mean amount was £450 with a standard deviation of £150.

Required:
1 Find a 95% confidence interval for the proportion of all customers who would apply for a credit card.
2 Find a 95% confidence interval for the mean amount of anticipated loans on the credit card.

After this pilot survey, the company intends to repeat the survey throughout England and Wales.

Required:
3 How large a sample should be selected if the company wishes to be able to estimate, to within 5%, the proportion of customers who would apply for the card, and also, wishes to estimate the mean amount of loans to within £20? (In both cases use a 95% confidence interval.)

4 Give brief reasons why the results of the nationwide survey might differ from the results in the Oxford Street store.

Answers on page 481.

Exercise 5.7

As part of an annual company audit, the 12,143 sales invoices issued by Ixion plc during the last year have been sampled and checked for accuracy. The required sample of 200 invoices was drawn randomly and the results were classified according to the number of items that each invoice contains, giving rise to the following data:

		Number of invoices with		
Items per invoice	Number sampled	0 errors	1 error	2 errors
1– 5	64	60	2	2
6–10	90	74	8	8
10 +	46	32	8	6
	200			

Required:
Estimate the proportion of all invoices which contain at least one error, complete with a 95% confidence interval.

(ACCA, December 1985, alternative paper)

Answers on page 482.

Exercise 5.8

Nemesis Products plc is in the process of determining the value of the company's consumable stocks for the annual statement of accounts. In order to reduce the amount of time spent physically counting stock, the chief accountant has developed a sampling scheme for use in one section of the store which deals with a limited range of low value items which are stocked in large quantities. From the full range of 55 stock items, 6 items are chosen at random and the selected items are then physically counted. As a result of applying this procedure, the following data are obtained:

Item number	Unit cost (£)	Number in stock
1	0.6	74
2	2.6	10
3	1.2	27
4	0.8	49
5	0.4	32
6	1.6	6

Required:
1. Calculate the average value of stock held per item for the given sample of 6 items and from this estimate the total value of this group of 55 stock items.
2. Determine a 95% confidence interval for the total value of this group of 55 stock items. Explain carefully the meaning of your confidence interval and also indicate any assumptions which you have made.

3 On the evidence of the sample of data which has been obtained, how many items should be sampled in future years if the estimate of total value is to be correct to within £150 (with 95% probability)?

(ACCA, December 1985)

Answers on page 482.

Exercise 5.9

Martin Electronic Controls Ltd is a small manufacturer of electronic components, specialising in industrial control equipment. Prior to an external audit, the chief accountant of this company decided to undertake a preliminary spotcheck and sampled 18 of the 1200 sales items from the previous month in order to have some general idea as to their overall total value. The values, in pounds, of these sales items are given in the following table:

82	30	98	116	80	150	200	88	70
90	160	100	86	76	90	140	76	68

(Note that, using the usual terminology, $\sum x = 1,800$ and $\sum x^2 = 207,200$.)

Required:
1. On the basis of the sample values obtain a point estimate of the total value of all items. What basic assumption are you making when finding this estimate?
2. Provide 95% confidence limits for the total value of all items. What further assumption are you making when providing these confidence limits?
3. How many items should the external auditor investigate if he wishes his estimate of the total value of the 1200 items to be within £5000 of the true value?

(ACCA, December 1987)

Answers on page 484.

Statistical inference II: hypothesis testing

6.1 INTRODUCTION

To comply with packing regulations, a filling machine must be set so that on average it delivers a particular weight to each container. Once the machine has been set, the manufacturer knows that he cannot assume that this average fill will remain constant. During normal operation there will be an inherent variation in the amount delivered by the machine. The problems will increase as the machine wears. To keep a check on the filling process, samples of the output could be taken and the average fill determined. As we have seen in Chapter Five, one course of action available to the manufacturer is to use the sample average to construct a confidence interval for the population average fill.

Alternatively, he can use a **hypothesis test** to assess whether his machine is performing satisfactorily. In this case, the manufacturer holds a belief, or hypothesis, about the average fill. He then uses the sample average as evidence to support his belief or not. On the basis of the sample evidence, the manufacturer decides how likely it is that the belief, or hypothesis, is correct.

If a die were thrown 102 times, then, on average, 17 of the outcomes will be sixes, ie the expected number of sixes is 17. Suppose, however, that we roll a die 102 times and actually obtain 20 sixes, could we automatically assume that the die is biased in favour of the six? Since the outcome when a die is rolled is subject to chance, 17 sixes will not be obtained in every set of 102 trials. The question is, therefore, how many sixes can we obtain before we say that the result is due to bias, rather than to chance? Clearly, if all 102 outcomes were sixes, we would be confident that the die was biased. Where do we draw the line between a result attributable to the chance nature of an experiment with a fair die, and a result attributable to a biased die? An appropriate hypothesis test, or **test of significance** as it is often called, will enable us to come to a decision about these kinds of problems. These tests essentially use a set of decision rules, and hence, are based on the judgement of the decision maker. We are never in a situation in which we absolutely prove or disprove an hypothesis; we simply make an assessment of how likely it is that the hypothesis is true.

6.2 THE PROCEDURE FOR ANY HYPOTHESIS TEST

To evaluate the sample evidence, we must formulate our hypothesis so that a known probability distribution can be used. This initial hypothesis is called the **null hypothesis** and is denoted by H_0. The null hypothesis is always formulated to state that the sample statistic is consistent with the assumed population parameter. Having set up the null hypothesis, we investigate the sample to see if the evidence is consistent with this hypothesis. However, in order to preserve as much objectivity as possible, it is important that the hypothesis is formulated before the data are collected. There is a complete spectrum of possible results, but in order to make a decision, we usually divide the results into three categories:

1 the evidence is consistent with the null hypothesis;
2 the evidence is not consistent with the null hypothesis;
3 the evidence is inconclusive and more data are needed before a decision can be made.

If the result falls into category **1** then the decision is made to accept the null hypothesis as probably being true. It is assumed that the difference between the value of the sample statistic and the population parameter is explained by the chance variation inherent in sampling.

If the result falls into category **2** the decision is made to reject the null hypothesis as probably being untrue. It is assumed that the difference between the sample statistic and the population parameter is not explained by the inherent sampling variation. It is common practice in this case to adopt an **alternative hypothesis**. The hypothesis test does not test the sample evidence for consistency with this alternative hypothesis. If the alternative hypothesis is adopted, all we are saying is, since the null hypothesis has not been supported, we will assume that the alternative hypothesis is true instead. We have not provided any statistical evidence for its correctness.

The alternative hypothesis is usually denoted by H_1 and like H_0, should be formulated at the outset of the investigation. For example, a machine produces metal discs. It is set to give a mean diameter of 2.0 cm. A sample of discs from its output has a mean diameter of 2.3 cm. The question then arises as to whether the machine is still set correctly. The null hypothesis assumes that the machine is still set correctly and the sample mean is consistent with the sample having been drawn from a normal population with a mean of 2.0 cm. If when we carry out the hypothesis test we find that the sample evidence is not consistent with the null hypothesis, then we must decide what alternative conclusion to draw. The alternative hypothesis could simply be that the population mean is not 2.0 cm, or, the alternative hypothesis could be that the population mean is greater than 2.0 cm. The alternative hypothesis will determine the exact conditions which we use for the test. These two cases can be symbolised as follows:

Case 1: $\qquad H_0: \mu = 2.0$ cm

$\qquad\qquad\qquad H_1: \mu \neq 2.0$ cm

Case 2: $\qquad H_0: \mu = 2.0$ cm

$\qquad\qquad\qquad H_1: \mu \geq 2.0$ cm

If the result falls into category **3**, then no decision can be made until more data are available and the test is carried out again. It should be noted, however, that the definition of these three categories is made on a subjective basis by the decision maker. Whichever result is obtained, we never definitely prove or disprove the null hypothesis. The best we can do is to say that the null hypothesis is almost certainly true, or, that the null hypothesis is very unlikely to be true. The following example will help to clarify why this is so.

Example 6.1: To illustrate the principle of a hypothesis test

Consider again the experiment of rolling a die 102 times. Twenty sixes are obtained. Does the evidence suggest that the die is biased in favour of the sixes?

Solution: If the die is not biased then the number of sixes obtained, when the die is rolled 102 times, follows the binomial distribution. The null hypothesis is that the number of sixes is binomially distributed with p = 1/6, ie the die is not biased in favour of the six.

The alternative hypothesis is that the die is biased towards the six—the number of sixes is not binomially distributed with p = 1/6.

The probability of obtaining r sixes from 102 rolls, if the die is unbiased, is given by:

$$P(r) = {}^{102}C_r \times (1/6)^r \times (5/6)^{102-r}$$

$$r = 0, 1, 2, 3, \ldots, 102$$

Therefore the probability of the observed number of sixes is:

$$P(r = 20) = {}^{102}C_{20} \times (1/6)^{20} \times (5/6)^{82}$$

$$= 0.073$$

which is a small probability, and the probability of the expected number of sixes is:

$$P(r = 17) = {}^{102}C_{17} \times (1/6)^{17} \times (5/6)^{85}$$

$$= 0.105$$

which is also a small probability.

How do we assess whether the difference between these individual probabilities is important or not? Some standard measure against which we can compare the probabilities is required. The simplest standard to use is the total probability of all of the outcomes which is 1, ie:

P(all possible outcomes) = 1

is used as the basis for comparison.

Since the probability of any individual value is likely to be small compared to 1, the procedure which is used in practice, is to compare P(20 or more sixes) with 1. If P(20 or more sixes) is 'small' compared to 1, we say that, if the null hypothesis is true, this event is unlikely to occur. We do not attribute the occurrence of the event to statistical fluctuation or to chance. We assume there is a cause. We do not know what the cause is but the most reasonable assumption is that the die is biased. Consequently we reject the null hypothesis since the evidence does not support it. If P(20 or more sixes) is not 'small' compared to 1, then the event could reasonably have occurred by chance. The null hypothesis is not contradicted by the evidence and therefore should not be rejected. This result does not prove that the null hypothesis is true; we are saying only that there is no evidence to reject it.

The only remaining problem is to decide what we mean by 'small'. It is at this point that the decision maker has to use his/her judgment with respect to the level of certainty required for a particular decision. The decision maker has to choose a boundary value for P(20 or more sixes). If the actual probability is larger than this value, then the null hypothesis is accepted. If the actual probability is smaller than the boundary value, the decision is made to reject the null hypothesis. The decision maker can choose any value for the boundary, but generally the following guidelines are used:

1. Boundary value for the probability = 0.05. If the probability is less than this, we say that the null hypothesis could very well be untrue. It is in doubt and is rejected.
2. Boundary value for the probability = 0.01. If the probability is less than this, we say that the null hypothesis is most likely untrue. It is in serious doubt and is rejected.
3. Boundary value for the probability = 0.001. If the probability is less than this, we say that the null hypothesis is almost certainly untrue and is rejected.

When the decision is being made, the term **significant** is used to describe the result. For example, if the decision is to be made at the 0.05 boundary and the actual probability is less than 0.05, then we say that the result is significant at the 5% level. This means that the probability of obtaining, by chance, a sample statistic greater than or equal to the observed value is less than 5%, hence the null hypothesis is in doubt. If, however, the actual probability is more than 0.05, the result is not significant at the 5% level. The probability of obtaining that sample statistic or larger by chance is greater than 5%, and there is no reason to reject the null hypothesis.

It must be emphasised that the failure of any result to reject the null hypothesis does not in any way prove conclusively that the null hypothesis is true. There may be many ways of looking at the sample; many statistics which could be calculated. Some of these statistics will be consistent with the null hypothesis and others will not. Some tests will provide supportive evidence, others will provide contradictory evidence. The decision maker must evaluate all of the evidence available and exercise his/her personal judgment.

Let us return to the example about the die. We will choose to make our decision at the 5% level of significance. The next step is to calculate the probability of obtaining 20 or more sixes. This is a somewhat time-consuming calculation, but eventually we find that P(20 or more sixes) = 0.2480.

We now compare P(20 or more sixes) with the boundary value of 0.05. P(20 or more sixes) is much bigger than 0.05, therefore, we say that the result is not significant at the 5% level. The evidence is consistent with the null hypothesis. We have no reason to reject this hypothesis. The probability of obtaining 20 or more sixes by chance with a fair die is almost 25% and is therefore quite likely to occur. We have no reason to suppose that the die is biased.

6.2.1. Standard hypothesis tests

In principle, we can test the significance of any statistic related to any probability distribution. However, we are usually concerned with a few standard cases. The sample statistics, mean, proportion and variance, are related to the normal, t, F or chi squared distributions.

1. **Normal test** This test is used to test a sample mean ($\bar{x}$) against a population mean (μ). It is used with samples of any size (n) when the population variance (σ^2) is known. In addition, if we wish to test a sample proportion, p, then the normal test may be used if the sample size is large, $np > 5$ and $n(1 - p) > 5$, since in this case the normal distribution can be used to approximate the binomial distribution.
2. **t test** This is used to test a sample mean against a population mean. It is used with samples of any size when the population variance is unknown. For large samples, the t distribution approaches the normal distribution, and the normal test can then be used to approximate the t test.
3. **Variance ratio test or F test** This test is used to compare population variances. It is used with samples of any size, drawn from normal populations.
4. **Chi squared test** This is a non-parametric test, that is, values of sample statistics are not required. The test uses the frequency of occurrence of the variable values. It can be used to test the association between attributes or the goodness of fit of an observed frequency distribution to a standard distribution.

6.2.2 One and two tailed tests

As has been said already, the choice of the significance level at which the decision is made is up to the decision maker. One aspect which is taken into consideration is the nature of the alternative hypothesis. It is customary for this alternative hypothesis to influence the choice of decision boundary.

Refer again to the example above in which a machine produces metal discs to a mean diameter of 2.0 cm. A random sample of discs, from the output, has a mean diameter of 2.3 cm. If the decision maker is interested simply in whether the machine is still set correctly, then it does not matter to him whether the sample mean is larger or smaller than 2.0 cm. The crucial issue is whether that sample mean could reasonably be expected by random sampling from that population. Hence, the null hypothesis and alternative hypothesis will be:

$$H_0: \mu = 2.0 \text{ cm}$$

$$H_1: \mu \neq 2.0 \text{ cm}$$

If the decision maker is carrying out his test at the 5% significance level, then he will assume that boundaries are placed symmetrically on the sampling distribution, as shown below:

FIGURE 6.1 **Five per cent two tail test**

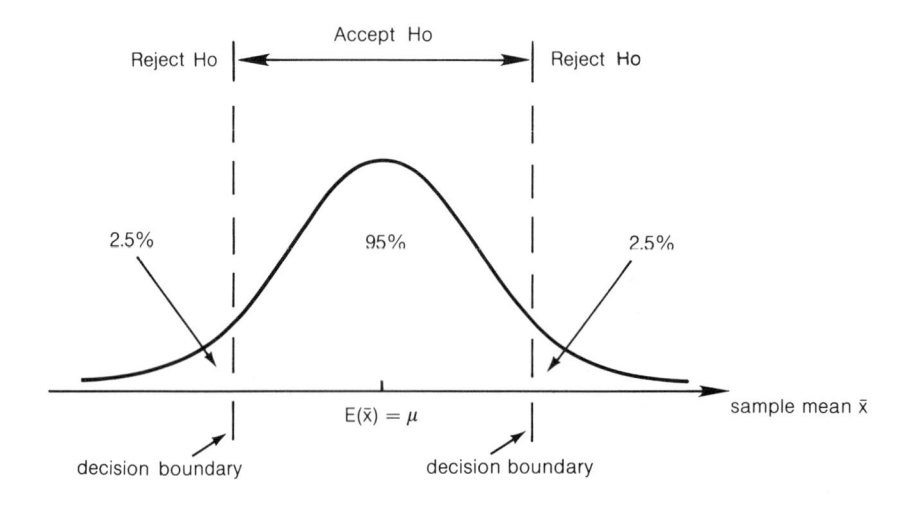

In this case, the decision maker is using a **two tail test**. He is indifferent about whether the sample mean is actually larger or smaller than the supposed population mean. He simply wants to know if there has been any change.

However, if the decision maker is concerned that the machine's mean has increased, then he would adopt a different alternative hypothesis:

$$H_0: \mu = 2.0 \text{ cm}$$

$$H_1: \mu > 2.0 \text{ cm}$$

In this case, the decision maker is looking for significant differences in a specific direction. He may still wish to make his decision at the 5% level of significance, but this time he chooses to put a single boundary on the sampling distribution as shown below:

FIGURE 6.2 **Five percent one tail test**

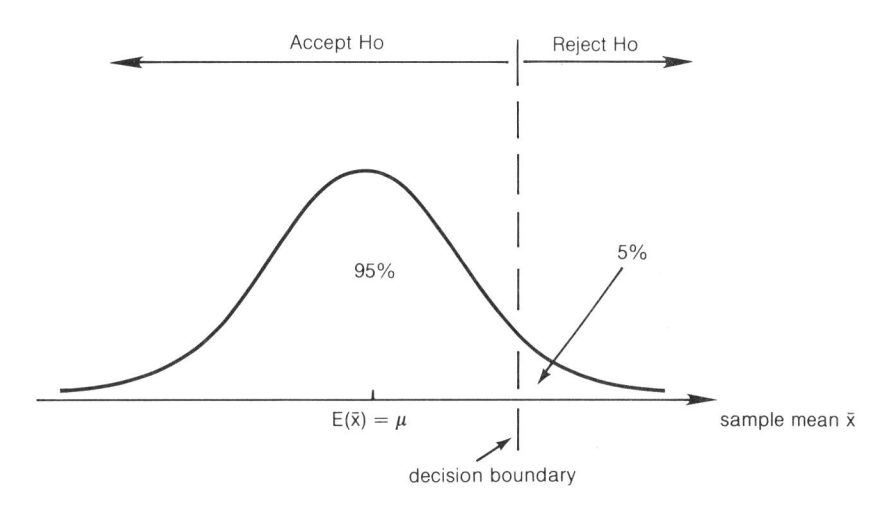

This is referred to as a **one tailed test**.

The essential effect of moving between a one and two tailed test is to alter the level of significance at which the decision is made. It is not a crucial issue, since we will be confident about our decisions only if the results are clear cut one way or the other. We will now look at specific types of test, remembering that the basic principles are the same in each case.

6.3 HYPOTHESIS TEST ON A SAMPLE MEAN—POPULATION VARIANCE KNOWN

The precise way in which a hypothesis test is carried out can vary. The method described below is one which is compatible with the way in which the standard probability distribution tables are usually printed.

Example 6.2: To illustrate a hypothesis test on a sample mean ($\bar{x}$) when the population variance σ^2 is known

A sugar refiner packs sugar into bags weighing, on average, 1.0 kg (μ) with a standard deviation (σ) of 0.01 kg. A sample of 16 bags (n) is taken at random from the production and the mean weight ($\bar{x}$) is found to be 1.01 kg. Is there any evidence to suggest that the filling machine has drifted from its correct setting?
Solution: The output from a filling machine can be expected to follow approximately a normal probability distribution. The null hypothesis is that the machine has not drifted from its correct setting:

H_0: The sample mean is consistent with the sample having been drawn from a normal population with a mean of 1.0 kg, ie $\mu = 1.0$ kg.

Is there evidence that the setting of the machine has remained at its proper level or has it changed? If the evidence from the sample leads us to reject the null hypothesis, the logical

136 QUANTITATIVE ANALYSIS

assumption is that the machine has drifted from its correct setting, therefore the alternative hypothesis is:

H_1: The sample was not drawn from a normal distribution with a mean of 1.0 kg, ie $\mu \neq 1.0$ kg.

It follows from H_0 that the sampling distribution of sample means is also a normal distribution with a mean of 1.0 kg and a standard error of $(0.01/\sqrt{16})$ kg.

We will choose to make our decision of whether or not to accept the null hypothesis at the 5% level of significance, using a two tail normal test. The boundary values on the sampling distribution are shown below:

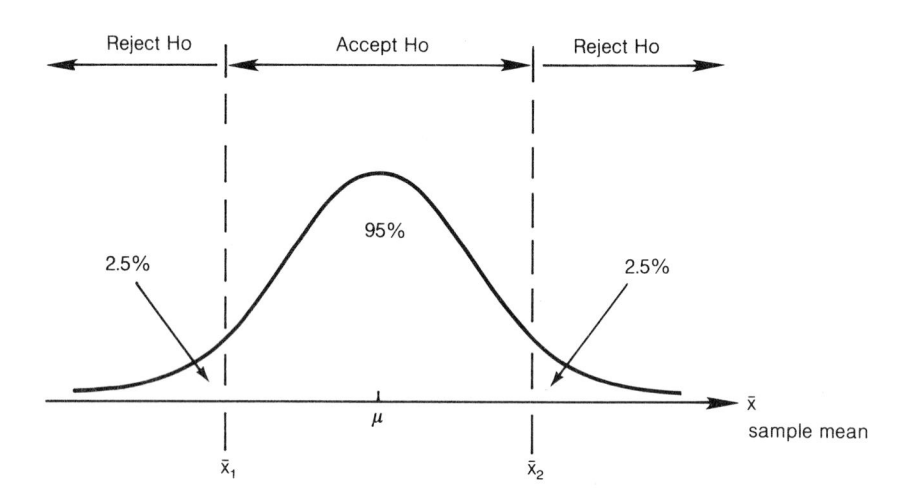

FIGURE 6.3 **Boundary values on the sampling distribution for 5% level of significance**

Using the standard normal distribution tables, we find that $\bar{x}_1$ and $\bar{x}_2$ are both 1.96 standard errors away from the mean, and the diagram can be redrawn as follows:

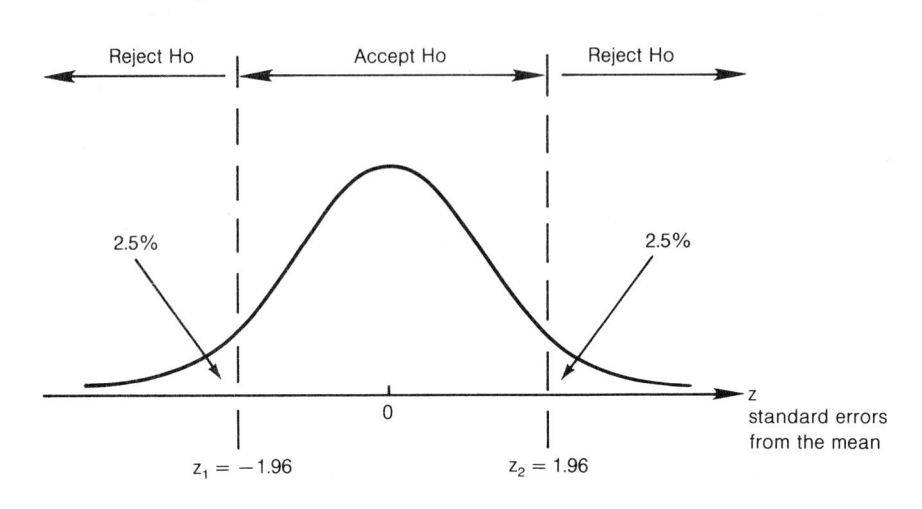

FIGURE 6.4 **Boundary values on the standard normal distribution for 5% level of significance**

If we now consider the sample mean ($\bar{x}$) of 1.01 kg, we can calculate the number of standard errors by which this is above the mean, using:

$$z = \frac{\bar{x} - \mu}{SE_{\bar{x}}} = \frac{\bar{x} - \mu}{\sigma/\sqrt{n}}$$

This is the same equation as that used in Chapter Two, but the population standard deviation is replaced by the standard error of the sampling distribution. Hence:

$$z = \frac{1.01 - 1.0}{0.01/4} = 4.0 \text{ standard errors above } \mu$$

In terms of the standardised variable, 4.0 is greater than the boundary value of 1.96 which we have already found. This means that:

$$P(\text{standardised variable} \geq 4.0) < 0.025$$

and the result is significant at the 5% level of significant. If we plot $z = 4.0$ on the diagram, it falls well within the 'Reject H_0' region. Since the result is significant at the 5% level, we conclude that there is reasonable evidence that the mean sample is not consistent with the null hypothesis. We reject this hypothesis in favour of the alternative hypothesis. The probability of a sample mean of 1.01 kg or greater occurring from sampling fluctuation in a random sample of 16, drawn from a normal population with a mean of 1.0 kg, is less than 5%. We choose to believe that this sample was taken from a population for which the mean was not 1.0 kg. We conclude that the machine has drifted from its correct setting. Let us now look at another example of the procedure and try to streamline it a little.

Example 6.3: To illustrate a hypothesis test on a sample mean ($\bar{x}$) when the population variance (σ^2) is known

A punch machine makes the holes in metal washers. The machine is intended to operate so that the hole is, on average, 4.00 mm (μ) with a standard deviation (σ) of 0.20 mm. A sample of 25 (n) washers is selected at random and the mean diameter is found to be 3.88 mm ($\bar{x}$). Is there any evidence to suggest that the machine has drifted from its correct setting?
Solution: The diameter of the holes made by the punch can be expected to follow approximately a normal probability distribution.
   The null hypothesis is that the machine has not drifted from its correct setting.

H_0: The sample mean is consistent with the sample having been drawn from a normal population with a mean of 4.0 mm, ie $\mu = 4.0$ mm.

Is there evidence that the setting of the machine has remained at its proper level, or has it changed? If the evidence from the sample leads us to reject the null hypothesis, the logical assumption is that the machine has drifted from its correct setting. The alternative hypothesis is:

H_1: The sample was not drawn from a normal distribution with a mean of 4.0mm, ie $\mu \neq 4.0$ mm.

It follows from H_0 that the sampling distribution of sample means is also a normal distribution with a mean of 4.0 mm and a standard error of $(0.2/\sqrt{25})$ mm. It follows from H_1 that we will use a two tail test.
   We will choose to make our decision on whether or not to accept the null hypothesis at the 1% level of significance. Using the standard normal distribution tables, the boundary values are both 2.576 standard errors from the mean μ. This is illustrated on the diagram overleaf:

FIGURE 6.5   **Boundary values for 1% level of significance**

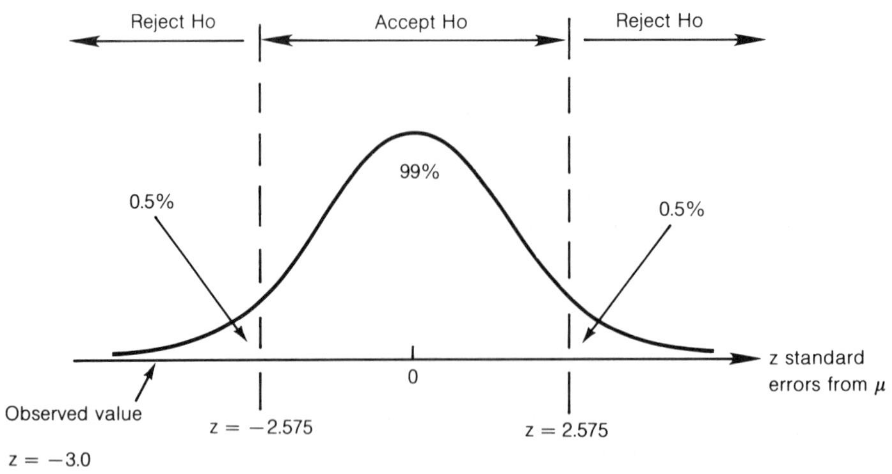

If we now consider the sample mean of 3.88 mm, we can calculate the number of standard errors which this is below the mean using:

$$z = \frac{\bar{x} - \mu}{SE_{\bar{x}}} = \frac{\bar{x} - \mu}{\sigma/\sqrt{n}}$$

Hence:

$$z = \frac{3.88 - 4.0}{0.20/5} = -3.0$$

z is referred to as the **test statistic**.

This test statistic, −3.0, is less than the boundary value of −2.575. This means that:

$$P(\text{standardised variable} \leq -3.0) < 0.005$$

The result is significant at the 1% level. This is also clear from the above diagram. We do not actually need to know what the probability is in order to make our decision.

Since the result is significant at the 1% decision level, we conclude that there is strong evidence that the sample is not consistent with the null hypothesis. We reject this hypothesis in favour of the alternative hypothesis. The probability of a sample mean of 3.88 mm or less occurring from sampling fluctuation in a random sample of 25, drawn from a normal population with a mean of 4.0 mm, is less than 1%. We believe that this sample was taken from a population for which the mean was not 4.0 mm. We conclude that the machine has drifted from its correct setting.

Example 6.4: To illustrate a hypothesis test on a sample mean ($\bar{x}$) when the population variance (σ^2) is known

The height to which a variety of plant grows in a nurseryman's plantation is approximately normally distributed with a mean (μ) of 53 cm and a variance (σ^2) of 12 cm². Last year, a box of 15 (n) of these plants received, by mistake, double the usual fertiliser. The mean height of the plants in this box was 55 cm ($\bar{x}$). Is there any evidence to suggest that the extra fertiliser has had a beneficial effect?

Solution: The null hypothesis is that the extra fertiliser has had no effect.

H_0: The sample mean is consistent with the sample having been drawn from a normal population with a mean of 53 cm, ie $\mu = 53$ cm.

Is there evidence that the extra fertiliser has had no effect or that it has increased the average height of the plants? If we decide to reject the null hypothesis, the logical assumption is that the extra fertiliser has had a beneficial effect, and the plants are taller. The alternative hypothesis is:

H_1: The sample was not drawn from a normal distribution with a mean of 53 cm but from a population with a mean larger than 53 cm, ie $\mu > 53$ cm

It follows from H_0 that the sampling distribution of sample means is also a normal distribution with a mean of 53 cm and a standard error of $\sqrt{(12/15)}$cm. It follows from H_1 that we will use a one tail test. We will make our decision to accept the null hypothesis at the 0.1% level of significance. Using the standard normal distribution tables, we find that the boundary value, z is 3.09 standard errors above the mean. The boundary value on the sampling distribution is shown below:

FIGURE 6.6   **Boundary value for 0.1% level of significance**

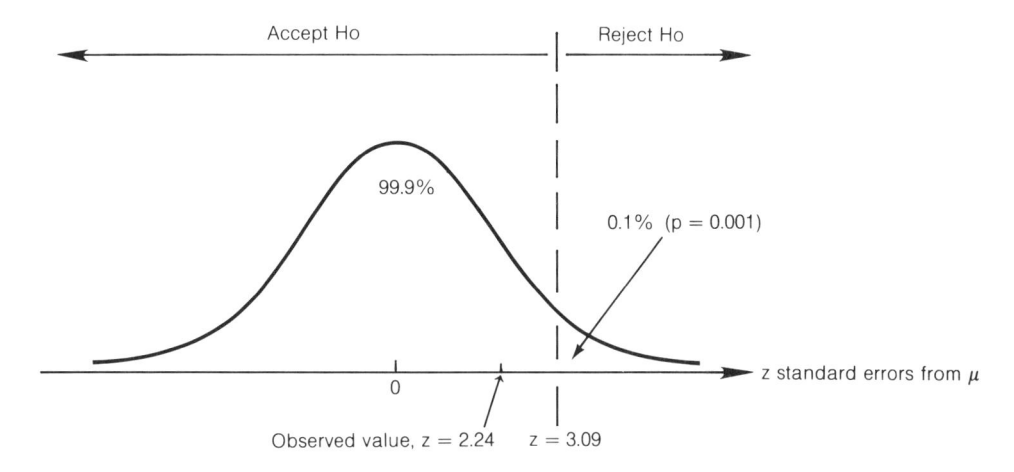

We calculate the test statistic as before:

$$z = \frac{\bar{x} - \mu}{SE_{\bar{x}}} = \frac{\bar{x} - \mu}{\sigma/\sqrt{n}}$$

Hence:

$$z = \frac{55 - 53}{\sqrt{(12/15)}} = 2.24$$

The test statistic, 2.24, is less than the boundary value of 3.09. This means that:

P (standardised variable $\geqslant$ 2.24) > 0.001

as can be seen from the diagram. The result is not significant at the 0.1% level of significance. Since the result is not significant, we conclude that the evidence is consistent with the null hypothesis which we accept. The probability of a sample mean of 55 cm or more occurring from sampling fluctuation in a random sample of 15 drawn from a normal population with a mean of 53 cm, is greater than 0.1%. We believe that this sample was taken from a population for which the mean was 53 cm. We conclude that the extra fertiliser has not had a beneficial effect.

6.4 HYPOTHESIS TEST ON A SAMPLE MEAN—POPULATION VARIANCE UNKNOWN

We saw in Chapters Four and Five that if the population variance, σ^2, is unknown then we can estimate it using the sample standard deviation (s). The appropriate standardised distribution then becomes the t distribution with (n − 1) degrees of freedom.

Example 6.5: To illustrate a hypothesis test on a sample mean when the population variance is unknown

Britelite plc produce light bulbs. For one particular type of bulb, it has always been claimed that the expected lifetime is 1500 hours (μ). In order to test a new batch, a sample of 10 (n) was taken. The mean lifetime of the sample was 1410 hours ($\bar{x}$) with a standard deviation of 90 hours (s). Does the evidence suggest that the expected lifetime has changed from 1500 hours?
Solution: The null hypothesis is that the sample has been drawn from the given population.

H_0: The sample mean is consistent with the sample having been drawn from a normal population with a mean of 1500 hours, ie $\mu = 1500$ hours.

H_1: The sample was not drawn from a normal distribution with a mean of 1500 hours, ie $\mu \neq 1500$ hours.

It follows from H_1, that we will use a two tailed test, and from H_0 that the sampling distribution of sample means is also a normal distribution with a mean of 1500 hours and a standard error of $(\sigma/\sqrt{10})$ hours. Since σ is unknown, the hypothesis test uses the standard t distribution with $(10 - 1) = 9$ degrees of freedom.

We will make our decision on the null hypothesis at the 5% level of significance. Using the t distribution tables in Appendix Two, $t_{0.05/2,9}$ is ± 2.26. The boundary values on the standard distribution are shown below:

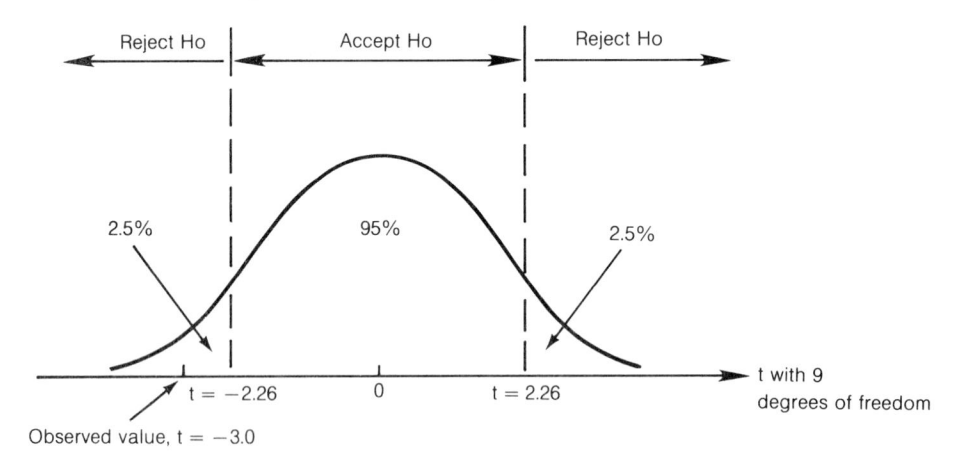

FIGURE 6.7 **t boundary values at 5% level of significance**

The test statistic is now:

$$t = \frac{\bar{x} - \mu}{\widehat{SE}_{\bar{x}}} = \frac{\bar{x} - \mu}{\hat{\sigma}/\sqrt{n}}$$

From section **4.4.3**, we know that:

$$\hat{\sigma} = \sqrt{\frac{n}{n-1}} \, s$$

Therefore:

$$\widehat{SE}_{\bar{x}} = \frac{\hat{\sigma}}{\sqrt{n}} = \sqrt{\frac{n}{n-1}} \frac{s}{\sqrt{n}} = \frac{s}{\sqrt{n-1}}$$

Hence:

$$t = \frac{1410 - 1500}{90/\sqrt{(10-1)}} = -3.0$$

The test statistic, -3.0, is less than the boundary value of -2.26. This means that:

$$P(\text{standardised variable} \leq -3.0) < 0.025$$

The result is significant at the 5% level.

Since the result is significant, we conclude that there is reasonably strong evidence that the sample is not consistent with the null hypothesis. We reject this hypothesis. The probability of a sample mean of 1410 hours or less, occurring by sampling fluctuation in a random sample of 10 drawn from a normal population with a mean of 1500 hours, is less than 5%. We believe that this sample was not taken from such a population. The mean lifetime of the bulbs has changed.

Example 6.6: To illustrate a hypothesis test on a sample mean when the population variance is unknown

A company packs its product (sugar crystals) into cartons labelled 250 g. The company claims that the average contents of the packets is at least 250 g (μ). A Trading Standards Officer visits the company's factory with the intention of testing this claim. The officer selects a random sample of 8 (n) packets and weighs the contents. The following weights were obtained:

TABLE 6.1 **Weights of a random sample of 8 packets**

Packet Number	1	2	3	4	5	6	7	8
Weight, g	247	252	248	253	246	245	248	250

What conclusions can the Trading Standards Officer draw from this sample? (Assume the weights of the packets of sugar are normally distributed).
Solution:

H_0: The sample mean is consistent with the sample having been drawn from a normal population with a mean (μ) of 250 g, ie $\mu = 250$ g.

H_1: The sample was drawn from a distribution with a mean of less than 250 g, ie $\mu < 250$ g.

It follows from H_0 that the sampling distribution of sample means is a normal distribution with a mean of 250 g and a standard error of $(\sigma/\sqrt{8})$g. It follows from H_1 that we will use a one sided test. Using the sample data we can calculate the sample mean ($\bar{x}$) and standard deviation (s):

$$\bar{x} = \frac{\sum x}{n} = \frac{1989}{8} = 248.6 \text{ g}$$

$$s = \sqrt{\frac{\sum x^2}{n} - (\bar{x})^2} = 2.643 \text{ g}$$

Since σ is unknown, the hypothesis test is carried out using the standard t distribution with $(8-1) = 7$ degrees of freedom. We will make our decision to accept the null hypothesis at the 1% level of significance. Using the t distribution tables in Appendix Two we find that $t_{0.01,7} = -3.00$. The boundary values are shown in FIGURE 6.8 overleaf.

FIGURE 6.8 t boundary values at the 1% level of significance

The test statistic is calculated as before:

$$t = \frac{\bar{x} - \mu}{\widehat{SE}_{\bar{x}}} = \frac{\bar{x} - \mu}{s/\sqrt{n-1}}$$

Hence:

$$t = \frac{248.6 - 250}{2.643/\sqrt{(8-1)}} = -1.40$$

The test statistic, -1.40, is greater than the boundary value of -3.00. (See the diagram.) This means that:

$$P(\text{standardised variable} \leq -1.40) > 0.01.$$

The result is not significant at the 1% level. Since the result is not significant, we accept the null hypothesis at this level. The probability of a sample mean of 248.6 g or less occurring due to sampling fluctuation in a random sample of 8, drawn from a normal population with a mean of 250 g, is more than 1%. We choose to believe that this sample was taken from such a population. The company's claim is correct.

6.5 HYPOTHESIS TEST ON A SAMPLE PROPORTION

The normal hypothesis test can also be used to test a sample proportion of occurrences. If the proportion is actually binomially distributed but the sample size is large, then the normal distribution may be used as an approximation to the binomial.

Example 6.7: To illustrate a hypothesis test on a sample proportion when the population proportion is known

A supplier of electronic components makes a product which sometimes fails immediately it is used. The supplier tries to control his manufacturing process so that the proportion of faulty products is less than 4%. A batch of 500 components is supplied and 28 prove to be faulty. Is there any evidence that the manufacturing process has gone out of control and too many faulty products are being produced?
Solution:

H_0: The proportion of faulty components produced is still 4%, ie $p = 0.04$.

H_1: The proportion of faulty components produced has increased, ie $p > 0.04$.

It follows from H₁ that we will use a one tailed test. The sample of 500 components is large, therefore we will approximate the binomial distribution by a normal distribution. The sampling distribution of sample proportions will be approximately normal with a mean proportion p = 0.04. The standard error of the sampling distribution is:

$$SE_{\hat{p}} = \sqrt{p(1-p)/n} = \sqrt{0.04 \times 0.96/500} = 0.00876$$

The proportion of defectives in the sample is:

$$\hat{p} = 28/500 = 0.056$$

We will test at the 1% level of significance. The boundary value for the standardised normal variable is 2.33.

FIGURE 6.9 **z boundary values at 1% level of significance**

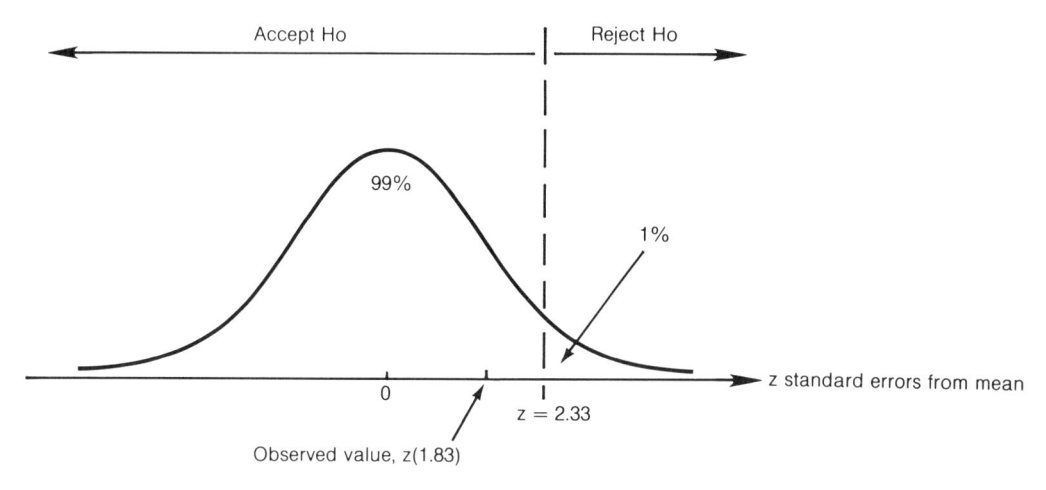

Test statistic:

$$z = \frac{\hat{p} - p}{SE_{\hat{p}}}$$

therefore:

$$z = \frac{0.056 - 0.04}{0.00876} = 1.83$$

The test statistic is less than the boundary value at 1%. (See the diagram.) The result is not significant at the 1% level:

$$P(\text{standardised variable} \geq 1.83) > 0.01.$$

The sample proportion of defective components, 0.056, can be explained in terms of sampling fluctuation.

We accept the null hypothesis at the 1% level of significance. There is no evidence to suggest that the production process has gone out of control and is producing more than 4% defectives.

So far we have considered a single sample and compared a sample statistic with a supposed population parameter. We must now consider those situations in which we have two samples which we wish to compare. How can we adapt the hypothesis tests we have used until now, to deal with the comparison of two sample statistics? We will examine this in the following sections.

6.6 HYPOTHESIS TEST ON TWO POPULATION VARIANCES

We mentioned in section **5.4** that there are many situations where the variability in the data is just as important as the average value. When we are evaluating an investment portfolio, the expected return is important but so also is the risk associated with the investment. This risk may be evaluated from the variance of the possible returns on the investment (see section **3.3**). In Chapter Five, we demonstrated how to construct a confidence interval for a population variance on the basis of the variance of a single sample. Suppose, however, that we have two independent samples and we wish to know whether they come from normal populations with the same variance. For example, suppose a company produces a particular component on two separate production lines, A and B. The specification is the same on the two lines. How can we determine whether the two outputs are equally variable? One way would be to draw a random sample from each machine and compare the sample variances, using an appropriate hypothesis test. The same procedure may be used to compare the risk involved in two different investment portfolios. A comparison of the variances of the actual returns in previous years, will enable a judgment to be made.

6.6.1 Variance ratio, or F, test

We stated in section **4.4**, that the ratio of two sample variances may be compared using the distribution of the F statistic, where:

$$F = \frac{\hat{\sigma}_1^2}{\hat{\sigma}_2^2}$$

and since the best estimate of the population variance is given by:

$$\hat{\sigma}^2 = \frac{n}{n-1} s^2$$

Then:

$$F = \frac{n_1 s_1^2}{(n_1 - 1)} \frac{(n_2 - 1)}{n_2 s_2^2}$$

The null hypothesis assumes that the two samples are independent and are drawn from normal populations with the same variance, $\sigma_1^2 = \sigma_2^2$. The expected value for F is 1. We know from our previous discussion, that even if the null hypothesis is true, σ_1^2 is very unlikely to have exactly the same value as σ_2^2, because of sampling fluctuation. Hence, the F statistic is unlikely to be exactly 1. The decision which we have to make, using the hypothesis test, is whether the actual value of F is sufficiently close to 1 for us to say that most likely the samples are drawn from normal populations with the same variance. In that case, the difference in the values of σ_1^2 and σ_2^2 may be attributed to the fluctuations inherent in sampling.

As was mentioned in section **4.4.4**, the exact shape of the F distribution depends on the number of degrees of freedom of both samples. When we are estimating a single population parameter from a sample, we lose one degree of freedom. For each sample there are $(n - 1)$ degrees of freedom remaining.

To make the standard table for the F distribution more manageable, values of F greater than or equal to 1 only, are given, ie they are one tail tables. In order to use these tables, when we calculate F we divide the larger variance by the smaller. The test statistic is:

$$F = \text{(larger estimated variance)}/\text{(smaller estimated variance)}$$

Example 6.8: To illustrate the use of the F hypothesis test

A stockbroker is examining two alternative investments, A and B, on behalf of a client. Investment A has been available for 10 years and has produced an expected annual return over this period of 17.8%. Investment B has been available for 8 years and it has also produced an expected annual

return of 17.8%. The variances of the annual returns on the two investments are 3.21%^{2} and 7.14%^{2}. Is there any evidence that the risks associated with investments A and B are unequal? Assume that the distributions of the annual returns on the investments are approximately normal.
Solution: The variance of the annual returns may be used to assess risk. We wish to know whether the two samples of annual returns from the two investments are drawn from normal populations with the equal variances. Therefore:

$$H_0: \sigma_A^2 = \sigma_B^2$$

And:

$$H_1: \sigma_A^2 \neq \sigma_B^2$$

We will test the null hypothesis using a two tail F test at the 5% level of significance. This is equivalent to a 2.5% one tail test, therefore, we use the $\alpha = 0.025$ lines in the F table.

The best estimates of the two population variances may be obtained from the sample variances as follows:

$$\hat{\sigma}_A^2 = \frac{n_A}{n_A - 1} s_A^2 = \frac{10}{9} \times 3.21 = 3.567 \text{ with 9 degrees of freedom}$$

$$\hat{\sigma}_B^2 = \frac{n_B}{n_B - 1} s_B^2 = \frac{8}{7} \times 7.14 = 8.160 \text{ with 7 degrees of freedom}$$

Since:

$$\hat{\sigma}_B^2 > \hat{\sigma}_A^2,$$

$$F = \frac{\text{larger estimated variance}}{\text{smaller estimated variance}} = \frac{\hat{\sigma}_B^2}{\hat{\sigma}_A^2}$$

Remember, to use the standard tables, we require $F \geq 1$.

The F tables are printed to give the degrees of freedom of the larger variance ($v_1 = 7$ dfs) across the page and the degrees of freedom of the smaller variance ($v_2 = 9$ dfs) down the page. Using the 2.5% F tables in Appendix Two, which is equivalent to the 5% two tail level of significance, with 7 and 9 degrees of freedom, the boundary value is:

$$F_{0.05/2, 7, 9} = 4.197$$

Using the sample data, the test statistic is:

$$F = \frac{8.160}{3.567} = 2.29$$

FIGURE 6.10 **F boundary values at 5% level of significance, two tail**

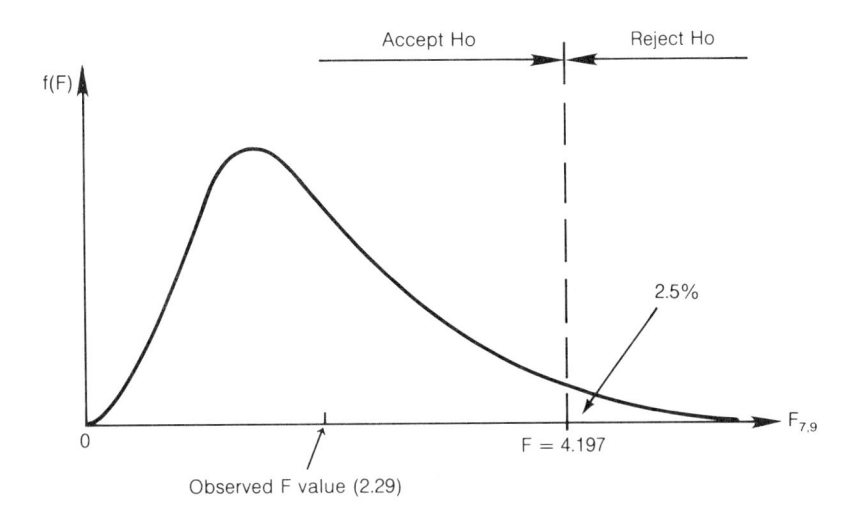

146 QUANTITATIVE ANALYSIS

Since:

$$2.29 < F_{0.05/2,7,9}$$

the result is not significant at the 5% level. The evidence supports the null hypothesis at this level. We have no reason to suppose that the risks (as measured by the variance of the annual returns) in the two investments are unequal.

Example 6.9: To illustrate the use of the F hypothesis test

POP Chemicals plc manufacture chemical X using a batch process. There are two different processes available and the management of POP wish to use the process which produces the more consistent product. The consistency of chemical X can be assessed in terms of its density. The more consistent the process is, the smaller will be the variation in the density.

The production manager at POP carries out 10 (n_1) trials with process 1 and achieves a standard deviation of 4.1 units (s_1) for the density of the chemical. He then carries out 12 (n_2) trials with process 2. This produces a standard deviation of 2.0 units (s_2) for the density of the chemical. Is process 2 more consistent than process 1? Assume that the density of the chemical is normally distributed.

Solution: The null hypothesis is that the two samples are drawn from normal populations with equal variances. The consistency of the two processes is the same:

$$H_0: \sigma_1^2 = \sigma_2^2$$

and:

$$H_1: \sigma_1^2 > \sigma_2^2$$

Process 2 is more consistent, giving a smaller variance. (If the samples show that $\hat{\sigma}_2$ is bigger than $\hat{\sigma}_1$, we know, without any statistical test, that H_1 is certainly not true.) H_1 indicates that we must use a one tail test. We will test the null hypothesis at the 1% level of significance. The test statistic is:

$$F = \frac{\hat{\sigma}_1^2}{\hat{\sigma}_2^2}$$

The best estimates of the two population variances are obtained from the sample variances as follows:

$$\hat{\sigma}_1^2 = \frac{n_1}{n_1 - 1} s_1^2 = \frac{10}{9} \times 4.1^2 = 18.68 \text{ with 9 degrees of freedom}$$

$$\hat{\sigma}_2^2 = \frac{n_2}{n_2 - 1} s_2^2 = \frac{12}{11} \times 2.0^2 = 4.36 \text{ with 11 degrees of freedom}$$

Using the 1% one tail F tables, ie the 0.01 lines in Appendix Two, with 9 degrees of freedom (v_2) across the page and 11 degrees of freedom (v_2) down the page, the boundary value is:

$$F_{0.01,9,11} = 4.632$$

Using the sample data, the F statistic is:

$$F = \frac{18.68}{4.36} = 4.28$$

Since:

$$4.28 < F_{0.01,9,11} = 4.632$$

the result is not significant at the 1% level. The evidence supports the null hypothesis at this level and we have no reason to suppose that process 2 is more consistent than process 1.

6.7 HYPOTHESIS TEST ON TWO SAMPLE MEANS—POPULATION VARIANCE KNOWN

In sections **6.3** to **6.5**, we were dealing with a single sample and assessing whether it was drawn from a specified normal population. We will now look at situations in which we have two samples. We wish to determine whether the two samples are drawn from normal populations with equal means. For example, if an auditor satisfies himself that a company's invoicing system is working satisfactorily, he may take a sample of invoices to estimate the error rate in the population. If the system continues to work correctly, a second sample should give an estimate of the population error rate which is not significantly different from the first.

Similarly, in a production process, if a machine is set to fill bottles with sauce, a random sample of bottles from the production line will enable an estimate to be made of the population average fill. If the process continues to function correctly, subsequent samples should not give estimates of the average fill which are signficantly different. This is an important aspect of quality control.

In both of these examples, it might reasonably be assumed that the population variance remains the same. However, if we consider two separate production lines filling bottles with sauce, we may take a sample from each line to see if there is any significant difference between the population mean fills. There is no reason to assume in this case that the two population variances are equal.

In the particular case of unknown population variances, the type of hypothesis test used depends on whether the variances may be assumed to be equal or not. The form of the null hypothesis is, however, the same in all cases. For a hypothesis test on two sample means, the null hypothesis assumes that the two samples are drawn from population with equal means.

$$H_0: \mu_1 = \mu_2, \text{ the population means are equal.}$$

A new variable is created which is the difference between the sample means $(\bar{x}_1 - \bar{x}_2)$ and compared with the assumed difference between the population means, ie $\mu_1 - \mu_2 = 0$. If the difference between the sample means is not significantly different to zero, then we assume that the null hypothesis is supported. If the difference is significantly different to zero, then we assume that the null hypothesis is not supported.

If σ_1^2 and σ_2^2 are known, the test statistic follows a normal distribution and is:

$$z = \frac{(\bar{x}_1 - \bar{x}_2) - (\mu_1 - \mu_2)}{SE_{\bar{x}_1 - \bar{x}_2}}$$

where:

$$SE_{\bar{x}_1 - \bar{x}_2} = \sqrt{\frac{\sigma_1^2}{n_1} + \frac{\sigma_2^2}{n_2}}$$

Example 6.10: To illustrate the normal hypothesis test for two sample means when the population variances are known

The Sweet Tooth Sugar Company has two production lines for the filling of 1 kg bags (μ_1 and μ_2) with sugar. Using data collected over a long period of time, the production manager estimates that the population standard deviation of the weight delivered for line 1 is 0.02 kg (σ_1) and for line 2 is 0.04 kg (σ_2). A random sample of 10 (n_1) bags is taken from line 1 and the mean weight in the bags is found to be 1.018 kg ($\bar{x}_1$). A similar random sample of 12 (n_2) bags is taken from line 2 and the mean fill is 0.989 kg ($\bar{x}_2$). Is there any evidence to suggest that the two production lines are dispensing different mean amounts of sugar?

Solution: The null hypothesis is that the two sample means are consistent with the samples having been drawn from normal populations with the same mean.

$$H_0: \mu_1 = \mu_2, \text{ ie } \mu_1 - \mu_2 = 0$$

and:

$$H_1: \mu_1 \neq \mu_2, \text{ ie } \mu_1 - \mu_2 \neq 0$$

The choice of a two tail test follows from H_1.

Since the population variances (σ_1^2 and σ_2^2) are known, we test the difference between the sample means using a normal test. Let us choose to test at the 1% level of significance. From the standard normal distribution tables in Appendix Two, the boundary value for z are ± 2.576.

FIGURE 6.11 **z boundary values for 1% level of significance**

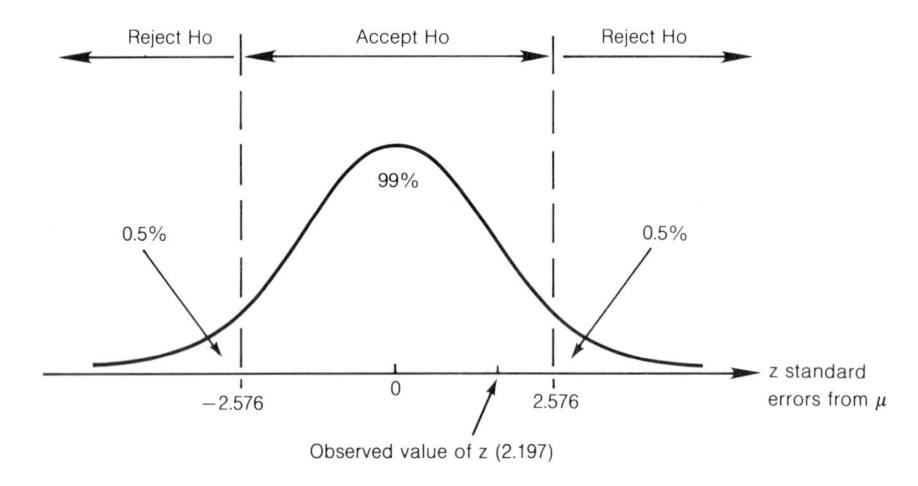

The test statistic is:

$$z = \frac{(\bar{x}_1 - \bar{x}_2) - (\mu_1 - \mu_2)}{SE_{\bar{x}_1 - \bar{x}_2}}$$

Where:

$$SE_{\bar{x}_1 - \bar{x}_2} = \sqrt{\frac{\sigma_1^2}{n_1} + \frac{\sigma_2^2}{n_2}} = \sqrt{\frac{0.02^2}{10} + \frac{0.04^2}{12}} = 0.0132 \text{ g}$$

Therefore:

$$z = \frac{(1.018 - 0.989) - 0}{0.0132} = 2.197$$

Since:

$$2.197 < z_{0.01} = 2.576$$

the result is not significant at the 1% level. We have no reason to reject H_0. We assume that the two production lines are filling the bags to the same average weight.

Example 6.11: To illustrate the normal hypothesis test for two sample means when the population variances are known

For several seasons, a fruit grower has grown two similar varieties of gooseberries. The yield per plant has been virtually the same for both varieties with a variance of 1.26 kg^2 (σ_A^2) for variety A and a variance of 1.20 kg^2 (σ_B^2) for variety B. He now proposes to use a new field for his gooseberry production and he does not know whether the new soil conditions will affect both varieties equally. As an experiment, the grower plants 30 bushes (n_A and n_B) of each variety in the new field. Variety A yields an average of 3.0 kg ($\bar{x}_A$) per bush, whereas variety B yields 3.5 kg ($\bar{x}_B$) per bush on average. Is there any evidence that variety B has a greater mean yield than variety A in this field?
Solution: The null hypothesis is that the two samples are drawn from normal distributions with the same mean:

$$H_0: \mu_A = \mu_B, \text{ ie } \mu_A - \mu_B = 0$$

and:

$$H_1: \mu_A < \mu_B, \quad \text{ie variety B has a larger mean yield.}$$

This means we must use a one tail test.

Since the population variances are known, we test the difference between the sample means using a normal test. Let us test at the 5% level of significance. From the standard normal distribution tables in Appendix Two, the boundary value for $z_{0.05}$ is -1.645. The test statistic is:

$$z = \frac{(\bar{x}_A - \bar{x}_B) - (\mu_A - \mu_B)}{SE_{\bar{x}_A - \bar{x}_B}}$$

Where:

$$SE_{\bar{x}_A - \bar{x}_B} = \sqrt{\frac{\sigma_A^2}{n_A} + \frac{\sigma_B^2}{n_B}} = \sqrt{\frac{1.26}{30} + \frac{1.20}{30}} = 0.286 \text{ kg}$$

Therefore:

$$z = \frac{(3.0 - 3.5) - 0}{0.286} = -1.746$$

FIGURE 6.12 **z boundary values at the 5% level of significance**

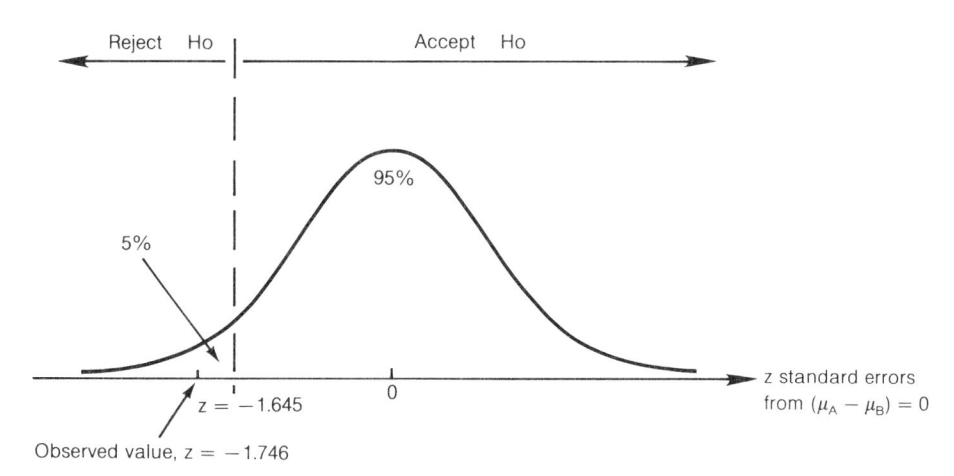

Since:

$$-1.746 < -z_{0.05} = -1.645$$

the result is significant at the 5% level. There is reasonably strong evidence that the samples are not consistent with H_0. We reject H_0 and accept H_1. The probability of obtaining a difference in sample means of 0.5 kg or more from sampling fluctuation, is less than 5%. We assume that Variety B has a bigger yield than Variety A in this field.

6.8 HYPOTHESIS TEST ON TWO SAMPLE MEANS—POPULATION VARIANCES UNKNOWN

The standard error used in the tests in this section depends on whether it may be assumed that the two population variances are equal. The assumption is investigated using the F test which was described in section **6.6**. The standard error of the difference between two sample means is:

$$SE_{\bar{x}_1 - \bar{x}_2} = \sqrt{\frac{\sigma_1^2}{n_1} + \frac{\sigma_2^2}{n_2}}$$

If σ_1^2 and σ_2^2 are unknown then they must be estimated by the sample standard deviations. Two possible cases arise:

1 If the population variances are equal, then $\sigma_1^2 = \sigma_2^2 = \sigma^2$ and:

$$SE_{\bar{x}_1 - \bar{x}_2} = \sqrt{\frac{\sigma^2}{n_1} + \frac{\sigma^2}{n_2}} = \sigma \sqrt{\frac{1}{n_1} + \frac{1}{n_2}}$$

There are two samples available from which σ could be estimated. The best estimate is obtained by pooling the two sample standard deviations (s_1^2 and s_2^2), rather than by using one or the other alone. The best estimate of the population standard deviation is given by:

$$\hat{\sigma} = \sqrt{\frac{(n_1 s_1^2 + n_2 s_2^2)}{(n_1 + n_2 - 2)}}$$

Therefore the best estimate of the required standard error is:

$$\widehat{SE}_{\bar{x}_1 - \bar{x}_2} = \sqrt{\frac{(n_1 s_1^2 + n_2 s_2^2)}{(n_1 + n_2 - 2)} \left(\frac{1}{n_1} + \frac{1}{n_2} \right)}$$

Where:

$$s^2 = \frac{\sum (x - \bar{x})^2}{n}$$

Alternatively, we may write:

$$\widehat{SE}_{\bar{x}_1 - \bar{x}_2} = \sqrt{\frac{((n_1 - 1)\hat{\sigma}_1^2 + (n_2 - 1)\hat{\sigma}_2^2)}{(n_1 + n_2 - 2)} \left(\frac{1}{n_1} + \frac{1}{n_2} \right)}$$

Where:

$$\hat{\sigma}^2 = \frac{\sum (x - \bar{x})^2}{n - 1}$$

The test statistic for the hypothesis test on the two sample means is no longer normal, but follows a standard t distribution, with $(n_1 + n_2 - 2)$ degrees of freedom. It is:

$$t = \frac{(\bar{x}_1 - \bar{x}_2) - (\mu_1 - \mu_2)}{\sqrt{\frac{(n_1 s_1^2 + n_2 s_2^2)}{(n_1 + n_2 - 2)} \left(\frac{1}{n_1} + \frac{1}{n_2} \right)}}$$

We therefore treat problems of this type in the same way as those in section **6.4**.

2 If the population variances are not equal, each population variance must be estimated by the corresponding sample variance:

$$\hat{\sigma}^2 = \frac{n}{n - 1} s^2$$

Hence:

$$\widehat{SE}_{\bar{x}_1 - \bar{x}_2} = \sqrt{\frac{s_1^2}{n_1 - 1} + \frac{s_2^2}{n_2 - 1}}$$

Where:

$$s^2 = \frac{\sum (x - \bar{x})^2}{n}$$

Or, alternatively:

$$\widehat{SE}_{\bar{x}_1 - \bar{x}_2} = \sqrt{\frac{\hat{\sigma}_1^2}{n_1} + \frac{\hat{\sigma}_2^2}{n_2}}$$

Where:

$$\hat{\sigma}^2 = \frac{\sum (x - \bar{x})^2}{n - 1}$$

The test statistic for the hypothesis test on the two sample means is now:

$$\frac{(\bar{x}_1 - \bar{x}_2) - (\mu_1 - \mu_2)}{\sqrt{\frac{s_1^2}{n_1 - 1} + \frac{s_2^2}{n_2 - 1}}}$$

This statistic is no longer either normally or t distributed. The appropriate distribution is approximately t, but the dependence on the number of degrees of freedom is more complex. Fortunately, however, if the sample sizes are large, $n \geq 30$, the distribution of this new statistic is approximately normal, as described by the Central Limit Theorem. In this book, we will discuss problems in this category only where the sample sizes are large.

To select the appropriate test statistic for problems in which the population variances are not known, we must know which assumption to make. Are the population variances equal or not? We use the F test to make this decision.

Example 6.12: To illustrate the normal hypothesis test for two sample means when the population variances are unknown

A certain type of polymer is made by batch production. Ten samples (n_1 and n_2) were taken from each of two consecutive batches and the percentage of chemical X in each sample was measured. For batch 1 the mean percentage per sample was 68.2% ($\bar{x}_1$) with a standard deviation of 0.70% (s_1). For batch 2, the mean content of chemical X was 67.0% ($\bar{x}_2$) with a standard deviation of 0.74% (s_2). Is there any evidence to suggest that the two batches contain different percentages of chemical X?

Solution: The null hypothesis is that the sample means are consistent with the two sets of samples having been drawn from normal populations with the same mean.

$$H_0: \mu_1 = \mu_2$$
$$H_1: \mu_1 \neq \mu_2$$

The alternative hypothesis is that the yields of the two processes are different. A two tail test is therefore appropriate.

Since the population variances are unknown, we must use the F test to determine whether it is reasonable to assume that the two population variances are equal. For the F test:

$$H_0: \sigma_1^2 = \sigma_2^2$$
$$H_1: \sigma_1^2 \neq \sigma_2^2$$

We will test the null hypothesis at the 5% level using a two tail test. This means that we use the 0.025 lines in the F tables in Appendix Two with 9 and 9 degrees of freedom.

$$F_{0.05/2, 9, 9} = 4.026$$

$$\hat{\sigma}_1^2 = \frac{n_1}{n_1 - 1} s_1^2 = \frac{10}{9} \times 0.70^2 = 0.544$$

$$\hat{\sigma}_2^2 = \frac{n_2}{n_2 - 1} s_2^2 = \frac{10}{9} \times 0.74^2 = 0.608$$

152 QUANTITATIVE ANALYSIS

Since σ_2^2 is the larger variance, the F statistic is:

$$F = \frac{0.608}{0.544} = 1.12$$

Since:

$$1.12 < F_{0.05/2,9,9} = 4.026$$

the result is not significant at the 5% level. The evidence is consistent with the null hypothesis. We can reasonably make the assumption that the two population variances are equal and use a t test on the means.

We will now continue with the hypothesis test on the two sample means. We will test at the 5% level using a two tail t test with:

$$10 + 10 - 2 = 18$$

degrees of freedom. From the tables in Appendix Two, we find that:

$$t_{0.05/2,18} = 2.10$$

Since we assume:

$$\sigma_1^2 = \sigma_2^2$$

$$\widehat{SE}_{\bar{x}_1 - \bar{x}_2} = \sqrt{\frac{(n_1 s_1^2 + n_2 s_2^2)}{(n_1 + n_2 - 2)} \left(\frac{1}{n_1} + \frac{1}{n_2}\right)} = \sqrt{\frac{(10 \times 0.70^2 + 10 \times 0.74^2)}{(10 + 10 - 2)} \left(\frac{1}{10} + \frac{1}{10}\right)}$$

$$= \sqrt{0.1153} = 0.3395$$

The test statistic is:

$$t = \frac{(\bar{x}_1 - \bar{x}_2) - 0}{\widehat{SE}_{\bar{x}_1 - \bar{x}_2}} = \frac{68.2 - 67.0}{0.3395} = 3.53$$

Since:

$$3.53 > t_{0.05/2,18} = 2.10$$

the result is significant at the 5% level. The evidence is very probably not consistent with the null hypothesis. We reject H_0 at this decision level, and we therefore assume that H_1 is true. The two batches contain different amounts of chemical X on average.

Example 6.13: To illustrate the normal hypothesis test for two sample means when the population variances are unknown

JBO plc manufacture batteries and claim that on average their batteries last longer than batteries from their competitor Sparky plc. A consumer organisation took a random sample of 35 (n_1) batteries from JBO's production and tested the batteries to destruction. The mean life was found to be 198.0 hours ($\bar{x}_1$) with a standard deviation of 8.7 hours (s_1). A similar test was done on a random sample of 30 (n_2) batteries from Sparky's output. The mean life was found to be 193.8 hours ($\bar{x}_2$) with a standard deviation of 5.8 (s_2) hours. Does the evidence substantiate JBO's claim?

Solution: The null hypothesis is that the two samples are drawn from normal populations with the same mean.

$$H_0: \mu_1 = \mu_2$$

$$H_1: \mu_1 > \mu_2$$

The alternative hypothesis is that JBO's batteries last longer, therefore a one tail test will be used.

Since the population variances are unknown, we must use the F test to determine whether to assume that the two population variances are equal, and therefore, which statistic to use to test the means. For the F test:

$$H_0: \sigma_1^2 = \sigma_2^2$$
$$H_1: \sigma_1^2 \neq \sigma_2^2$$

We will test the null hypothesis at the 5% level, using a two tail test:

$$\hat{\sigma}_1^2 = \frac{n_1}{n_1 - 1} s_1^2 = \frac{35}{34} \times 8.7^2 = 77.92$$

$$\hat{\sigma}_2^2 = \frac{n_2}{n_2 - 1} s_2^2 = \frac{30}{29} \times 5.8^2 = 34.80$$

Since $\hat{\sigma}_1^2$ is the larger variance, the F statistic is:

$$F = \frac{77.92}{34.80} = 2.24$$

From the 0.025 one tail lines in the F tables in Appendix Two, we take 34 degrees of freedom across the page and 29 down the page. $F_{0.05/2, 34, 29}$ lies between $F_{0.025, 30, 29} = 2.092$ and $F_{0.025, 50, 29} = 1.987$.

Our F statistic is bigger than both of these, therefore the result is significant at the 5% level. The evidence is not consistent with the null hypothesis. We cannot make the assumption that the two population variances are equal. We must assume:

$$\sigma_1^2 \neq \sigma_2^2.$$

A hypothesis test using the t statistic is no longer appropriate. However, since the sample sizes are large, n_1 and $n_2 \geqslant 30$, we can proceed by using the normal test as an approximation to the correct statistic for the two sample means. We will test at the 5% level, using a one tail normal test. From the table in Appendix Two, we find that $z_{0.05} = 1.645$. Since we assume:

$$\sigma_1^2 \neq \sigma_2^2$$

Then:

$$\widehat{SE}_{\bar{x}_1 - \bar{x}_2} = \sqrt{\frac{s_1^2}{(n_1 - 1)} + \frac{s_2^2}{(n_2 - 1)}} = \sqrt{\frac{8.7^2}{34} + \frac{5.8^2}{29}} = 1.84$$

The test statistic is:

$$z = \frac{(\bar{x}_1 - \bar{x}_2) - 0}{\widehat{SE}_{\bar{x}_1 - \bar{x}_2}} = \frac{198.0 - 193.8}{1.84} = 2.28$$

Since:

$$2.28 > z_{0.05} = 1.645$$

The result is significant at the 5% level. The evidence is very probably not consistent with the null hypothesis. We reject H_0 at this decision level. We therefore assume that the alternative hypothesis, H_1, is true. The mean lifetime for the JBO batteries is greater than for the Sparky batteries. We accept that the evidence substantiates JBO's claim.

154 QUANTITATIVE ANALYSIS

Example 6.14: To illustrate the normal hypothesis test for two sample means when the population variance are unknown

Electra plc manufacture electrical components. One assembly operation requires a one-month training period for a new employee to reach maximum efficiency. Since the training programme is costly, the management of Electra are anxious to reduce the time needed as much as possible. The training manager has devised a new method of training and management have asked you to carry out a test to determine whether the new method reduces the training period on average.

Two groups of 10 (n_1 and n_2) new employees have been trained for three weeks. One group has used the new method and one group the standard procedure. At the end of the three week period, each new employee assembled a component and the time taken was recorded:

TABLE 6.2 **Time taken to assemble one component, minutes**

Normal Training Group 1	New Training Group 2
32	35
37	31
35	29
28	25
41	34
44	40
35	27
31	32
34	31
30	33

Advise the management of Electra on whether the data provides evidence that the new procedure reduces the average training period required.

Solution: The null hypothesis assumes that the two samples are drawn from normal populations with the same mean:

$$H_0: \mu_1 = \mu_2$$
$$H_1: \mu_1 > \mu_2$$

The normal training gives a longer assembly time, therefore a one tail test will be used.

In order to decide which statistic to use to test the sample means, we must first test the variances. For the F test:

$$H_0: \sigma_1^2 = \sigma_2^2$$
$$H_1: \sigma_1^2 \neq \sigma_2^2$$

We will test the null hypothesis at the 5% level, using a two tail test. Using the data given:

$$\bar{x}_1 = 34.7 \text{ mins} \quad s_1 = 4.691 \text{ minutes}$$
$$\bar{x}_2 = 31.7 \text{ mins} \quad s_2 = 4.026 \text{ minutes}$$

Therefore:

$$\hat{\sigma}_1^2 = \frac{n_1}{n_1 - 1} s_1^2 = \frac{10}{9} \times 4.691^2 = 24.45$$

$$\hat{\sigma}_2^2 = \frac{n_2}{n_2 - 1} s_2^2 = \frac{10}{9} \times 4.026^2 = 18.01$$

Since $\hat{\sigma}_1^2$ is the larger variance, the F statistic is:

$$F = \frac{24.45}{18.01} = 1.36$$

From the 0.025 lines of the one tail F tables in Appendix Two:

$$F_{0.05/2,9,9} = 4.026$$

$$1.36 < F_{0.05/2,9,9}$$

The result is not significant at the 5% level. The evidence is consistent with the null hypothesis. We therefore make the assumption that the two population variances are equal. We will now continue with the hypothesis test on the two sample means.

We choose to test at the 1% level, using a one tail t test with:

$$10 + 10 - 2 = 18 \text{ degrees of freedom}$$

From the tables in Appendix Two, we find that:

$$t_{0.01,18} = 2.552$$

Since we assume:

$$\sigma_1^2 = \sigma_2^2$$

Then:

$$\widehat{SE}_{\bar{x}_1 - \bar{x}_2} = \sqrt{\frac{(n_1 s_1^2 + n_2 s_2^2)}{(n_1 + n_2 - 2)}\left(\frac{1}{n_1} + \frac{1}{n_2}\right)} = \sqrt{\frac{(10 \times 4.691^2 + 10 \times 4.026^2)}{(10 + 10 - 2)}\left(\frac{1}{10} + \frac{1}{10}\right)}$$

$$= \sqrt{4.246}$$

$$= 2.06$$

The test statistic is:

$$t = \frac{(\bar{x}_1 - \bar{x}_2) - 0}{\widehat{SE}_{\bar{x}_1 - \bar{x}_2}} = \frac{34.7 - 31.7}{2.06} = 1.46$$

Since:

$$1.46 < t_{0.01,18} = 2.552$$

the result is not significant at the 1% level. The evidence is consistent with the null hypothesis. We accept H_0 at this decision level. There is no evidence to suppose that the new training method has reduced the average time required for training.

6.9 HYPOTHESIS TEST ON TWO SAMPLE PROPORTIONS

In section **6.5**, we stated that if we draw a large random sample from a population in which the proportion of occurrences, p, of a certain characteristic, follows a binomial distribution, then the sampling distribution of the sample proportion, $\hat{p}$, is approximately normally distributed. In the same way we find that if two large samples are drawn independently from two binomial populations, then the statistic $(\hat{p}_1 - \hat{p}_2)$ is normally distributed with mean $(p_1 - p_2)$ and standard error:

$$SE_{\hat{p}_1 - \hat{p}_2} = \sqrt{\frac{p_1(1 - p_1)}{n_1} + \frac{p_2(1 - p_2)}{n_2}}$$

156 QUANTITATIVE ANALYSIS

where $\hat{p}$ refers to the sample statistic, p refers to the population parameter and n_1, n_2 are both large, that is, 30 or more. We are usually interested in whether or not the two samples are drawn from binomial populations with the same proportion of occurrences, ie $p_1 = p_2$. The test statistic is approximately normally distributed when the sample sizes are large:

$$z = \frac{(\hat{p}_1 - \hat{p}_2) - (p_1 - p_2)}{SE_{\hat{p}_1 - \hat{p}_2}}$$

Example 6.15: To illustrate a hypothesis test of two sample proportions

The internal auditors of a large company are interested in the system by which incoming invoices are processed. They take a random sample of 50 (n_1) completed invoices and check them in detail for correctness of procedure. Four are found to be defective in some way. The auditors then suggest some modifications to the procedure and these are implemented. Allowing time for the clerks to adjust to the new procedures, the auditors then take a random sample of 60 (n_2) completed invoices. This time they find three incorrect ones. Is there any evidence that the new procedures have improved the error rate?

Solution: The null hypothesis is that the two samples are randomly drawn from two binomial populations with equal proportions of defectives:

$$H_0: p_1 = p_2 = p$$
$$H_1: p_1 > p_2$$

ie the new procedures have reduced the error rate, therefore a one tail test is appropriate.

We will make the decision at the 5% level of significance. The normal test is appropriate since the sample sizes are both large. Using the standard normal tables in Appendix Two:

$$z_{0.05} = 1.645$$
$$\hat{p}_1 = 4/50 = 0.08$$

And:

$$\hat{p}_2 = 3/60 = 0.05$$

Assuming that H_0 is true, the best estimate of the proportion of defective invoices in the population is obtained by combining the proportions from the two samples. In total there are 7 defectives out of 110 invoices. Therefore, the best estimate of the population proportion is:

$$\bar{p} = 7/110 = 0.0636$$

Therefore:

$$\widehat{SE}_{\hat{p}_1 - \hat{p}_2} = \sqrt{\frac{\bar{p}(1-\bar{p})}{n_1} + \frac{\bar{p}(1-\bar{p})}{n_2}}$$

$$\widehat{SE}_{\hat{p}_1 - \hat{p}_2} = \sqrt{\frac{0.0636 \times 0.9364}{50} + \frac{0.0636 \times 0.9364}{60}}$$

$$= 0.0467$$

The test statistic is:

$$z = \frac{(\hat{p}_1 - \hat{p}_2) - (p_1 - p_2)}{\widehat{SE}_{\hat{p}_1 - \hat{p}_2}} = \frac{0.08 - 0.05}{0.0467} = 0.64$$

Since:

$$0.64 < z_{0.05} = 1.645$$

the result is not significant at the 5% level. The evidence is consistent with H₀ at this decision level. We have no reason to suppose that the modifications to the invoice system have reduced the error rate.

6.10 HYPOTHESIS TEST ON PAIRED DATA—DEPENDENT SAMPLES

In some situations a straightforward hypothesis test on two sample means will not be valid because the samples are not independent. There are factors involved which affect both samples in some unknown way. In these cases, it may be possible to overcome the problem by pairing individual members of one sample with the members of the other sample. A hypothesis test may then be carried out on the average difference between the paired measurements. The procedure is illustrated in the following example.

Example 6.16: To illustrate the use of paired data in a hypothesis test

A consumer organisation wishes to compare the wearing quality of a type of car tyre which is produced by two different manufacturers, X and Y. The tyres were put, as full sets, onto different cars. Each car had all X tyres or all Y. Unfortunately, it was then argued that any differences between X and Y were caused by the car or the driver, not by the tyres. To avoid this confusion, the organisation decides to choose one tyre from manufacturer X and one from manufacturer Y at random and to fit them to the rear wheels of six cars. Each pair of tyres will then receive exactly the same treatment. The results, for the mileage to replacement, are given in the table.

TABLE 6.3 **Results of tests—mileage to replacement**

Car Number	1	2	3	4	5	6
Mileage '000 for X tyre	50.1	47.0	48.6	48.8	50.2	48.0
Mileage '000 for Y tyre	53.9	50.3	48.5	51.3	49.7	51.0

Is there any evidence to suggest that the two makes of tyre have different lives?
Solution: Since the amount of wear the tyres receive will depend on a variety of factors such as the driver, the car and the road conditions, the two samples are deliberately not independent. A hypothesis test on the two sample means is not appropriate, since we would not know whether the result arose from the tyres or from the other factors. We can eliminate this latter effect by using paired data to produce one sample. For each car we calculate:

TABLE 6.4 **Difference in mileage between tyre X and tyre Y**

Car number	1	2	3	4	5	6
Mileage '000 for X tyre	50.1	47.0	48.6	48.8	50.2	48.0
Mileage '000 for Y tyre	53.9	50.3	48.5	51.3	49.7	51.0
Difference, d (= X − Y)	−3.8	−3.3	0.1	−2.5	0.5	−3.0

The mean value of d, $\bar{x}_d = -2.0$ miles '000.
The standard deviation of d, $s_d = 1.675$ miles '000.
The null hypothesis is:

$$H_0: \mu_d = 0: \text{there is no difference between the makes}$$

and:

$$H_1: \mu_d \neq 0: \text{the makes of tyre have different lives}$$

It follows from H_1 that a two tail test is required. We will make our decision at the 5% level of significance, using a t distribution, with (n − 1) = 5 degrees of freedom. From the standard t tables in Appendix Two:

$$t_{0.05/2, 5} = \pm 2.571$$

The test statistic is:

$$t = \frac{\bar{x}_d - 0}{\widehat{SE}_d} = \frac{-2.0}{1.675/\sqrt{5}} = -2.67$$

Since:

$$-2.67 < -t_{0.05/2, 5} = -2.571$$

the result is just significant at the 5% level. We reject H_0 and accept H_1. It is reasonable to assume that there is a difference in the mean lives of the tyres produced by the two manufacturers.

6.11 NON-PARAMETRIC TESTS OF HYPOTHESIS—THE CHI-SQUARED TEST

The tests in the preceding sections were concerned with comparing sample statistics with the corresponding population parameters. Except for large samples, it was necessary to assume that the populations were normal or approximately normal. We will now discuss examples of hypothesis tests which require neither this assumption nor the use of population parameters. This group of tests is referred to as **non-parametric tests**. The general procedure for the hypothesis test is the same as that used for parametric tests. The calculation of the test statistic is different.

In this section we will consider the most commonly used non-parametric test—the **chi-squared test**. It is a method of comparing a set of observed frequencies with the frequencies expected if the null hypothesis is true. There are two main areas of application which we will consider in turn.

6.11.1 To test the association of attributes

An attribute is a characteristic of a variable. The characteristics can usually be assigned to a category. For example, eye colour is an attribute of a person, and eye colour can be assigned to the categories brown, blue or green. The state of a customer's bank account can be categorised by 'always in credit', 'usually in credit', 'frequently overdrawn', 'permanently in debt'. The monthly sales of a product can be described as 'high', 'medium' or 'low'. Suppose that we are interested in two different characteristics of a variable and wish to know whether there is any relationship between them. For example, we have the grades achieved by a group of students in an accountancy examination and in a mathematics examination. We are interested in whether there is any relationship between the grades achieved in the accountancy examination and whether the students passed or failed in mathematics. The following categories could be established:

TABLE 6.5 **Example of a contingency table**

Mathematics examination	Grade in accountancy examination
	A B C Fail
Passed	
Failed	

The number or frequency of students who passed mathematics and achieved a grade A in accountancy is entered in the top left hand part of the table. The number of students who failed mathematics and achieved a grade A in accountancy is entered in the bottom left hand part, and so on. This type of table is called a **contingency table**. The above table has two rows and four columns and is referred to as a **(2 by 4) table**.

Using an appropriate null hypothesis, we can calculate the number of students we would expect to find in each cell. If the null hypothesis is correct, then the discrepancies between the observed and expected frequencies should be small and explainable in terms of fluctuation due to sampling. If the null hypothesis is not true, then the discrepancies will be large and indicative of a real difference between the observed and expected frequencies. We use the same decision rules as in the previous tests. The test statistic is obtained from the differences between the observed and expected frequencies for all the cells in the contingency table.

If f_O denotes the observed frequency of an event and f_E denotes the expected frequency, $(f_O - f_E)$ is the discrepancy between observation and expectation. The test statistic is:

$$\sum \left[\frac{(f_O - f_E)^2}{f_E} \right]$$

It is necessary to square the deviation, $(f_O - f_E)$, to avoid cancelling effects from positive and negative values. In addition, to make the statistic independent of the size of the frequencies used, the square of the deviation is divided by the expected frequency. This standardises all the values. This statistic follows a χ^2 distribution provided that the expected frequencies are not too small. The guideline which is usually used is:

$$f_E \geqslant 5$$

If one or more of the expected frequencies is less than 5, then categories must be combined until the limit is exceeded.

For small contingency tables, usually only two by two, and small samples ($n \leqslant 30$), a further correction is sometimes used. This is called **Yates' correction**. The test statistic is:

$$\chi^2 = \sum \left[\frac{(|f_O - f_E| - 0.5)^2}{f_E} \right]$$

where $|f_O - f_E|$ denotes the size only of the discrepancy, $(f_O - f_E)$, ignoring its sign. This correction is made because χ^2 is a continuous distribution and the sample data are discrete. In Chapter Two, we discussed the need for such a correction when using the normal distribution to approximate discrete distributions. For large samples, the difference between the corrected and uncorrected values of χ^2 is small and therefore the correction is not needed.

As we stated in Chapter Four, the shape of the χ^2 distribution depends on the number of degrees of freedom in the problem. When using a contingency table, the number of degrees of freedom is:

$$(r-1)(c-1)$$

where r and c are the number of rows and columns in the contingency table. If the contingency table has only one row, the number of degrees of freedom is $(c - 1)$.

Example 6.17: To illustrate the use of the χ^2 hypothesis test

Autosure plc is a large insurance company, specialising in car insurance. It has always been company policy to charge different premiums according to the size of the car being insured; the bigger the car, the bigger the premium. However, this policy is under review since branch managers have been reporting a higher rate of claims for personal injury for accidents involving small cars. One of the company's analysts has examined data from 566 recent claims. The data collected are shown below:

TABLE 6.6 **Analysis of data from 566 recent claims**

	Size of car insured		
Type of claim	Small	Medium	Large
Personal injury	120	57	42
No personal injury	149	105	93

Do these data indicate that the frequency of personal injury claims is associated with the size of the car insured?

Solution: We must first set up the null hypothesis. If there is no association between the type of claim and the size of the car, we would expect the frequency of claims in the contingency table to be in proportion to the total in each category:

H_0 There is no association between the type of claim and the size of the car insured.

H_1: There is some association between the claim and the size of car.

We will test the hypothesis at the 5% level of significance using a χ^2 test with $(2-1)(3-1) = 2$ degrees of freedom. From the tables in Appendix Two, we find that:

$$\chi^2_{0.05,2} = 5.991$$

To calculate the test statistic χ^2 we must determine the expected frequencies, from the totals in each category.

TABLE 6.7 **Observed frequencies**

Observed frequencies	Size of car insured			
Type of claim	Small	Medium	Large	Total
Personal injury	120	57	42	219
No personal injury	149	105	93	347
Total	269	162	135	566

There are 566 claims, 219 of which involved personal injury, therefore the proportion involving personal injury is 219/566. There are 269 small cars involved, and if there is no connection between the two factors, we would expect 219/566 of these 269 to fall into the first category on average. The expected frequency in the first cell of the table is:

$$(219/566) \times 269 = 104.08$$

Similarly, the expected number of claims in the other categories can be calculated. The complete set of figures is shown in the tables below. The expected frequencies are left as decimals. Since they are mean values, they should not be rounded to the nearest whole number.

TABLE 6.8 **Calculation of expected frequencies**

Basic calculation	Size of car insured			
Type of claim	Small	Medium	Large	Total
Personal injury	269 × 219/566	162 × 219/566	135 × 219/566	219
No personal injury	269 × 347/566	162 × 347/566	135 × 347/566	347
Total	269	162	135	566

The expected frequencies are given in the table below.

TABLE 6.9 **Expected frequencies**

Expected frequencies	Size of car insured			
Type of claim	Small	Medium	Large	Total
Personal injury	104.1	62.7	52.2	219
No personal injury	164.9	99.3	82.8	347
Total	269	162	135	566

The χ^2 test statistic is:

$$\chi^2 = \sum \left[\frac{(f_O - f_E)^2}{f_E} \right]$$

χ^2 is calculated below:

TABLE 6.10 **To calculate χ^2**

f_O	f_E	$(f_O - f_E)$	$(f_O - f_E)^2$	$(f_O - f_E)^2/f_E$
120	104.1	15.9	252.81	2.43
57	62.7	−5.7	32.49	0.52
42	52.2	−10.2	104.04	1.99
149	164.9	−15.9	252.81	1.53
105	99.3	5.7	32.49	0.33
93	82.8	10.2	104.04	1.26

check = 0 $\chi^2 = 8.06$

Therefore: $\chi^2 = 8.06$, and this is illustrated in the following diagram:

FIGURE 6.13 **χ^2 boundary value at 5% level of significance, 2 degrees of freedom**

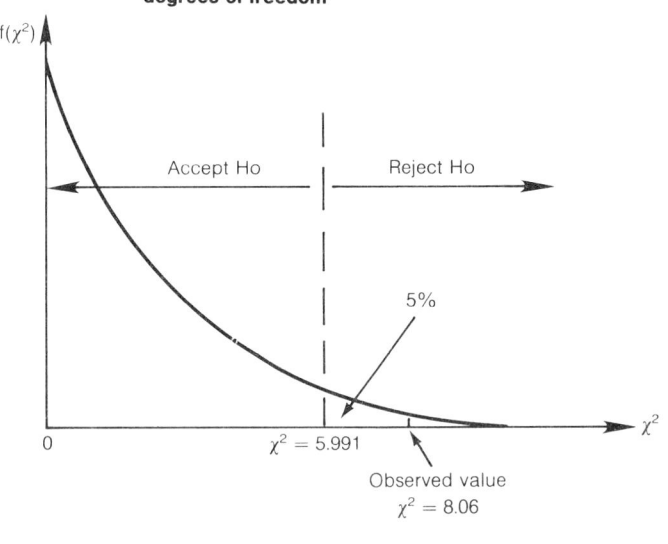

Since:

$$8.06 > \chi^2_{0.05,2} = 5.991$$

the result is significant at the 5% level. We reject H_0 at this level, and accept H_1. We can be reasonably sure that the data indicates an association between the claims involving personal injury and the size of the car insured. At this stage we do not know what kind of association there is. The company think that they are getting more personal injury claims than they would expect for small cars. The χ^2 value could have come from exactly the opposite situation. We must inspect the discrepancies in the data to find the nature of the association:

TABLE 6.11 **Discrepancies between observed and expected frequencies**

Discrepancies $(f_O - f_E)$	Size of car insured		
Type of claim	Small	Medium	Large
Personal injury	+15.9	−5.7	−10.2
No personal injury	−15.9	+5.7	+10.2

162 QUANTITATIVE ANALYSIS

The table confirms the company's suspicions. The number of personal injury claims is bigger than expected for small car owners. It is also lower for both medium and large cars. In view of the extra expense of personal injury claims, the company should reconsider its charging policy.

Example 6.18: To illustrate the use of the χ^2 hypothesis test

An international firm of accountants takes 150 school leavers onto its accounting technicians training scheme on the basis of a personal interview with each candidate. The training manager wishes to compare the trainees' performance during the first year of training with their final school report to see if there is any association between the two. The data collected are shown below:

TABLE 6.12 **Observed frequencies**

Performance in training	Final school report		
	Good	Average	Poor
Good	18	12	5
Average	39	34	18
Poor	6	11	7

Solution: We must first set up a suitable null hypothesis. H_0 must be chosen in such a way that we can calculate the expected frequencies. If we suppose that there is no association between the trainees' performance in their first year and their final report from school, we would expect the frequency of trainees in the contingency table to be in proportion to the total numbers in each category:

H_0: There is no association between the trainees' performance in their first year and their final report from school.

H_1: There is some association between the trainees' performance and their final report.

We will test the hypothesis at the 5% level of significance using a χ^2 test with:

$$(3 - 1)(3 - 1) = 4 \text{ degrees of freedom}$$

provided that all of the expected frequencies are at least 5.

To calculate the test value of χ^2, we determine the total number of trainees in each category and use these to find the expected frequencies.

TABLE 6.13 **Total number of trainees in each category**

Observed frequencies	Final school report			
Performance in training	Good	Average	Poor	Total
Good	18	12	5	35
Average	39	34	18	91
Poor	6	11	7	24
Total	63	57	30	150

There are 150 trainees, 35 of whom give a good performance during their first year. Therefore the proportion giving a good performance is 35/150. We will use this to calculate the expected frequencies for the top row of the table.

There are 63 trainees with good school reports, and if there is no connection between the two factors, we would expect 35/150ths of these 63 to fall into the first category on average. The expected frequency in the first cell of the table is:

$$(35/150) \times 63 = 14.7$$

Similarly, the expected number of trainees giving a good performance following an average school report is:

$$(35/150) \times 57 = 13.3$$

whilst the expected number giving a good performance following a poor school report is:

$$(35/150) \times 30 = 7.0$$

It should be noted that these three expected frequencies must add up to the total for the row:

$$14.7 + 13.3 + 7.0 = 35$$

We deal with the other rows in the table in a similar way. The proportion of trainees giving an average performance during the first year is 91/150, therefore the expected number who would have had a good school report is:

$$(91/150) \times 63 = 38.2$$

The remaining expected frequencies are calculated in the same way. The complete distribution is shown in the table below:

TABLE 6.14 **Expected frequencies**

Expected frequencies: performance in training	Final school report			
	Good	Average	Poor	Total
Good	14.7	13.3	7	35
Average	38.2	34.6	18.2	91
Poor	10.1	9.1	4.8*	24
Total	63	57	30	150

All of the rows and columns of the expected frequency table must give the same totals as the original contingency table. We have one expected frequency which is less than 5 in cell (3,3). To use the χ^2 distribution, we must amalgamate two of the categories. It does not matter from the point of view of the test, whether we reduce the number of categories for the school report or for the performance in training. The choice is made in terms of which is most sensible for the problem. Suppose in this case, we re-define the school reports as either 'Good' or 'Not good' and add together the Average and Poor columns of the table. The contingency table then becomes:

TABLE 6.15 **Amended observed frequencies**

Amended observed frequencies: performance in training	Final school report		
	Good	Not good	Total
Good	18	17	35
Average	39	52	91
Poor	6	18	24
Total	63	87	150

The corresponding expected frequencies are:

TABLE 6.16 **Amended observed frequencies**

Amended expected frequencies: performance in training	Final school report		
	Good	Not good	Total
Good	14.7	20.3	35
Average	38.2	52.8	91
Poor	10.1	13.9	24
Total	63	87	150

All expected frequencies now exceed 5, and the χ^2 test may be continued as before, but with:

$$(3 - 1)(2 - 1) = 2 \text{ degrees of freedom}$$

instead of 4.

From the tables in Appendix Two:

$$\chi^2_{0.05,2} = 5.991$$

$$\chi^2 = \sum \left[\frac{(f_O - f_E)^2}{f_E} \right]$$

This is calculated in the table below:

TABLE 6.17 **Calculation of χ^2**

f_O	f_E	$(f_O - f_E)$	$(f_O - f_E)^2$	$(f_O - f_E)^2/f_E$
18	14.7	3.3	10.89	0.74
17	20.3	−3.3	10.89	0.54
39	38.2	0.8	0.64	0.02
52	52.8	−0.8	0.64	0.01
6	10.1	4.1	16.81	1.66
18	13.9	−4.1	16.81	1.21

check = 0 $\chi^2 = 4.18$

Since:

$$4.18 < \chi^2_{0.05,2} = 5.991$$

the result is not significant at the 5% level. We are reasonably sure that the evidence is consistent with H_0 and we accept it at this level. We assume that there is no association between the performance of the trainees during their first year and their last school report.

6.11.2 To test the goodness of fit of an observed frequency distribution to the expected frequency distribution of a discrete variable

In previous sections it has usually been assumed that the samples are drawn from standard probability distributions—normal, binomial or Poisson. For example, we assume that the weight of sugar delivered by a filling machine is approximately normal, or the number of defectives per sample from a production process is binomially distributed. If we can test a sample for consistency with the assumption about the population, we have a method of checking that the process is working correctly. In addition, it is much easier to analyse data if it can be assumed that the underlying distribution is a standard one. We require a method of testing such assumptions.

The χ^2 hypothesis test can be used to determine whether a sample is taken from a particular population distribution. The procedure is the same no matter which distribution we assume for the population. On the basis of the null hypothesis, we calculate the expected frequency for each value (or range of values) of the variable. The observed and expected frequencies are compared using the χ^2 test. The number of degrees of freedom is determined by:

$$v = n - 1 - k$$

where n is the number of pairs of observed and expected frequencies and k is the number of population parameters estimated from the sample. One additional degree of freedom is always lost because the total frequency is fixed.

As in the case of the association of attributes, the expected frequencies should exceed 5. If this is not the case, the frequencies are combined until the limit is exceeded. This will reduce the number of degrees of freedom in the problem.

Example 6.19: To illustrate the use of the χ^2 test to examine the goodness of fit of an observed distribution to a binomial distribution.

Steel washers are made on a mass production basis. The company has established quality control checks. A random sample of 30 washers is taken and the number of defectives is recorded. During a week, 320 samples were taken. The distribution of the number of defectives is given in the table below:

TABLE 6.18 **Number of defectives in a sample of 30 washers**

Number of defectives per sample	0	1	2	3	4	5
Number of samples per week	13	49	87	109	56	6

Does the distribution of the number of defectives per sample follow a binomial distribution? The binomial distribution implies that defectives are produced at random throughout the week with a constant probability. The above process will work in this way if it is properly set up and operated. In quality control terms, the process is in control. If the process goes out of control, the defective rate will change. It may become erratic, with runs of defectives followed by runs of good washers, so that the defectives are not random through the week. On the other hand, the defective rate may get steadily worse as the machine wears. In either case, the distribution of defectives will not be binomial and the χ^2 test should show this. In practice, a quality control system will be designed to pick out these changes as they occur, not at the end of the week.

Solution: We will set up the null hypothesis to assume that the distribution is binomial and then use this to calculate the expected number of samples, out of 320, which would contain 0, 1, ..., 30 defectives. To do this we require the probability that any one item is defective. Since we do not know the proportion of defectives in the population, we must estimate it using the sample. The proportion of defectives in the samples, $\hat{p}$, is given by:

$$\frac{\text{total number of defectives found}}{\text{total number of washers examined}} = \frac{0 \times 13 + 1 \times 49 + 2 \times 87 + 3 \times 109 + 4 \times 56 + 5 \times 6}{30 \times 320}$$

$$\hat{p} = 0.08375$$

For convenience, we will round this figure to 0.084. (This is a very high defective rate for such a basic product. It should lead to immediate action.)

H_0: The number of defective washers per sample follows a binomial distribution with $n = 30$ and $p = 0.084$.

H_1: The number of defectives does not follow this binomial distribution.

TABLE 6.19 **Calculation of expected frequencies**

Number of defectives, r	P(r)	Expected frequency, f_E	Observed frequency, f_O
0	$^{30}C_0(0.084)^0(0.916)^{30} = 0.07192$	23.0	13
1	$^{30}C_1(0.084)^1(0.916)^{29} = 0.19787$	63.3	49
2	$^{30}C_2(0.084)^2(0.916)^{28} = 0.26310$	84.2	87
3	$^{30}C_3(0.084)^3(0.916)^{27} = 0.22519$	72.1	109
4	$^{30}C_4(0.084)^4(0.916)^{26} = 0.13939$	44.6	56
5	$^{30}C_5(0.084)^5(0.916)^{25} = 0.06647$	21.3	6
6 or more	(1 − sum of above) = 1 − 0.96394 = 0.03606	11.5	0
	Total 1.00	320.0	320

We will test at the 1% level of significance using a χ^2 test. The probability of r defectives per sample is given by:

$$P(r) = {}^{30}C_r \times (0.084)^r \times (0.916)^{30-r} \quad r = 0, 1, 2, \ldots, 30$$

We can now calculate the expected frequencies, including a category of 6 or more. (See TABLE 6.19 on page 165.)

$$\text{Expected frequency, } f_E = P(r) \times 320$$

All of the expected frequencies exceed 5. The number of degrees of freedom is:

$$v = 7 - 1 - 1 = 5$$

There are 7 pairs of observed and expected frequencies. One degree of freedom is lost because the total frequency is fixed and one more because the population proportion was estimated from the sample. From the tables in Appendix Two:

$$\chi^2_{0.01,5} = 15.086$$

To calculate the test statistic:

$$\chi^2 = \sum \left[\frac{(f_O - f_E)^2}{f_E} \right]$$

$$\chi^2 = \frac{(13 - 23.0)^2}{23.0} + \frac{(49 - 63.3)^2}{63.3} + \cdots + \frac{(6 - 21.3)^2}{21.3} + \frac{(0 - 11.5)^2}{11.5}$$

$$= 51.96$$

Since:

$$51.96 > \chi^2_{0.01,5} = 15.086$$

the result is significant at the 1% level. There is strong evidence that the sample data is not consistent with H_0, which we reject at this level. We choose to accept the alternative hypothesis that the number of defectives per sample does not follow a binomial distribution with p = 0.084. The implication of this for the production process is that the defectives are not being produced at random.

The data show there are more occasions on which three or four defectives are produced than we would expect, but there are fewer samples with very low or very high defective rates. In practice the production department would investigate the causes of this behaviour, when the quality control system picked up the evidence during the week.

Example 6.20: To illustrate the use of the χ^2 test to examine the goodness of fit of an observed distribution to a Poisson distribution

A company auditor would like to determine whether invoice errors per day are random or occur with some pattern. He decides that, if they are random, they should follow a Poisson distribution. This would indicate a random, but steady, error rate from the people involved in preparing the invoices. If the distribution is not Poisson, it could mean that some of the invoice clerks are behaving erratically or having problems with certain types of invoice. A sample of 197 consecutive days produced the results shown in the table below:

TABLE 6.20 **Invoice errors per day**

Number of errors per day	0	1	2	3	4	5	6	7	8	9	10	11
Number of days	6	14	25	43	36	30	21	9	8	2	2	1

Is there any evidence that the number of invoice errors per day follows a Poisson distribution?

Solution: The null hypothesis assumes that the distribution is Poisson. This will enable us to calculate the expected number of days with 0, 1, ... errors. We require the mean number of errors per day in the population. Since this is unknown, we must estimate it from the sample. The mean number of errors per day is given by:

$$m = \frac{\sum ((\text{number of errors per day}) \times (\text{number of days}))}{\text{total number of days}} = \frac{789}{197}$$

$$= 4.01 \text{ to 2 decimal places}$$

H_0: The number of invoice errors per day follows a Poisson distribution with a mean of 4.01 errors per day.

H_1: The number of errors does not follow this Poisson distribution.

We will test at the 5% level of significance, using a χ^2 test. The probability of r defectives per sample is given by:

$$P(r) = \frac{m^r}{r!} e^{-m} \quad r = 0, 1, 2, \ldots$$

The expected frequencies are:

$$f_E = P(r) \times 197$$

TABLE 6.21 **Calculation of expected frequencies**

Number of defectives, r	P(r) $(m^r e^{-m})/r!$	Expected frequency, f_E	Observed frequency, f_O
0	$(4.01^0 e^{-4.01})/0! = 0.01813$	3.6 } = 17.9	6 } = 20
1	$(4.01^1 e^{-4.01})/1! = 0.07271$	14.3	14
2	$(4.01^2 e^{-4.01})/2! = 0.14579$	28.7	25
3	$(4.01^3 e^{-4.01})/3! = 0.19488$	38.4	43
4	$(4.01^4 e^{-4.01})/4! = 0.19536$	38.5	36
5	$(4.01^5 e^{-4.01})/5! = 0.15668$	30.9	30
6	$(4.01^6 e^{-4.01})/6! = 0.10472$	20.6	21
7	$(4.01^7 e^{-4.01})/7! = 0.05999$	11.8	9
8	$(4.01^8 e^{-4.01})/8! = 0.03007$	5.9 } = 10.2	8 } = 13
9 or more	1 − above = 1 − 0.97833 = 0.02167	4.3	5
Total	1.00	197.0	197

Since the first and the last expected frequencies do not exceed 5, the first two and the last two categories must be combined. The degrees of freedom is then:

$$v = 8 - 1 - 1 = 6$$

There are 8 pairs of observed and expected frequencies. One degree of freedom is lost because the total frequency is fixed and one more because the population mean was estimated from the sample. From the tables in Appendix Two:

$$\chi^2_{0.05, 6} = 12.592$$

The test statistic is:

$$\chi^2 = \sum \left[\frac{(f_O - f_E)^2}{f_E} \right]$$

$$\chi^2 = \frac{(20 - 17.9)^2}{17.9} + \frac{(25 - 28.7)^2}{28.7} + \cdots + \frac{(9 - 11.8)^2}{11.8} + \frac{(13 - 10.2)^2}{10.2}$$

$$= 2.90$$

Since:
$$2.90 < \chi^2_{0.05,6} = 12.592$$

the result is not significant at the 5% level. The evidence is consistent with H_0, which we accept at this level. The errors per day do follow a Poisson distribution with a mean of 4.01.

6.11.3 To test the goodness of fit of an observed frequency distribution to the expected frequency distribution of a continuous variable

In much of the statistical work with sample data, we have assumed that the population is normal. The χ^2 test can be used to test this assumption, using a similar method to that described in the two previous examples. However, since we are now working with a continuous variable instead of a discrete one, we must consider ranges of values, not individual values. As before, we make a decision whether the differences between the sample frequencies and the expected frequencies can be attributed to sampling variation or whether they indicate the unsuitability of the model.

Example 6.21: To illustrate the use of the χ^2 test to examine the goodness of fit of an observed distribution to a normal distribution

MR Confectionery plc produce sweets and chocolate. One particular line makes and wraps chocolate bars of 250 g nominal weight. A sample of 100 bars was taken from the output of one day and each bar was weighed. The weights are given in the tables below:

TABLE 6.22 **Weights of a sample of 100 chocolate bars**

Weight, g	Number of bars
240 but < 245	4
245 but < 250	12
250 but < 255	30
255 but < 260	36
260 but < 265	15
265 but < 270	3
Total	100

Is this sample from a normal population?

Solution: The first task is to use the sample to estimate the population mean and standard deviation.

TABLE 6.23 **To estimate population mean and standard deviation**

Mid-point of weight, g x	Number of bars f	fx	fx²
242.5	4	970.0	235225.0
247.5	12	2970.0	735075.0
252.5	30	7575.0	1912687.5
257.5	36	9270.0	2387025.0
262.5	15	3937.5	1033593.5
267.5	3	802.5	214668.75
Total	100	25525.0	6518275.0

Mean:
$$\bar{x} = 25525.0/100 = 255.25 \text{ g}$$

Standard deviation:

$$s = \sqrt{(6518275.0/100) - (255.25)^2} = 5.494 \text{ g}$$

The best estimate of the population mean is:

$$\hat{\mu} = 255.25 \text{ g}$$

and the best estimate of the population standard deviation is:

$$\hat{\sigma} = \sqrt{\frac{n}{n-1}}\, s = \sqrt{\frac{100}{99}} \times 5.494 = 5.522 \text{ g}$$

The null hypothesis is:

H_0: The sample is taken from a normal population with a mean of 255.25 g and a standard deviation of 5.522 g.

H_1: The sample is not from this normal population.

We will test the hypothesis at the 5% level of significance, using a χ^2 test. The expected frequencies are the number of chocolate bars in each of the ranges, assuming a normal distribution with the given mean and standard deviation.

FIGURE 6.14 **Assumed normal distribution**

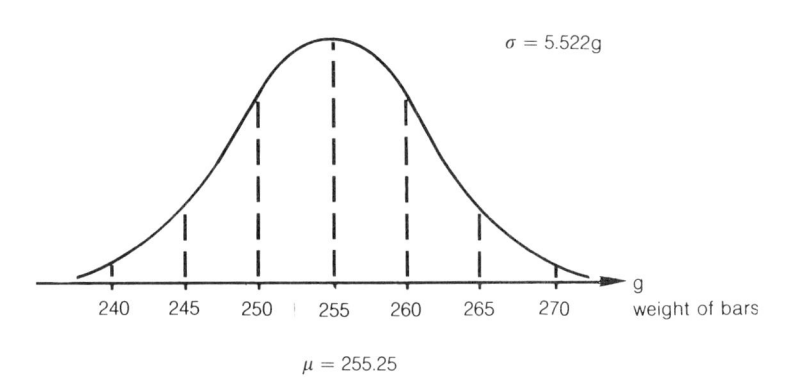

We must calculate the probability of a chocolate bar having a weight in each of the bands. For the first band, we require:

$$P(\text{weight} < 240 \text{ g})$$

and for the second band, we require:

$$P(240 \text{ g} < \text{weight} < 245 \text{ g})$$

and so on. If we determine the number of standard deviations each boundary is away from the mean, we can use the standard normal tables to find the required probabilities. The standard normal variable:

$$z = \frac{x - 255.25}{5.522}$$

For the first band:

$$z = \frac{240 - 255.25}{5.522} = -2.76$$

170 QUANTITATIVE ANALYSIS

From the standard normal tables in Appendix Two:

$$P(z < -2.76) = 0.00289$$

The expected frequency of bars less than 240 g is $0.0029 \times 100 = 0.29$. Similarly, for the second band, if z_l is the lower limit:

$$z_l = -2.76$$

from above. If z_u is the upper limit:

$$z_u = \frac{245 - 255.25}{5.522} = -1.86$$

From the standard normal tables:

$$P(z < -1.86) = 0.0314$$

Therefore:

$$P(240 \text{ g} < \text{weight} < 245 \text{ g}) = 0.0314 - 0.0029 = 0.0285$$

The expected frequency of bars between 240 g and 245 g is:

$$0.0285 \times 100 = 2.85$$

The calculations are summarised in the TABLE 6.24.

TABLE 6.24 **Calculation of expected frequencies**

x g	z	$P(\text{stand var} < z)$	$P(z_l < z < z_u)$	Expected frequency $P(z_l < z < z_u) \times 100$	Observed frequency
240	−2.76	0.0029	0.0029	0.29	0
245	−1.86	0.0314	0.0285	2.85	4
250	−0.95	0.1711	0.1397	13.97	12
255	−0.05	0.4801	0.3090	30.90	30
260	0.86	0.8051	0.3250	32.50	36
265	1.77	0.9616	0.1565	15.65	15
270	2.67	0.9962	0.0346	3.46	3
and $P(z > 2.67) =$		0.0038	0.0038	0.38	0
			Total	100.00	100

Since some of the expected frequencies are <5, we must combine the first three and the last three categories.

TABLE 6.25 **Amended frequencies and χ^2 calculation**

Observed f_O	Expected f_E	Discrepancy $(f_O - f_E)$	$\dfrac{(f_O - f_E)^2}{f_E}$
16	17.11	−1.11	0.07
30	30.90	−0.90	0.03
36	32.50	+3.50	0.38
18	19.49	−1.49	0.11
Totals 100	100.00	0	$0.59 = \chi^2$

There are:

$$v = 4 - 1 - 2 = 1 \text{ degree of freedom}$$

Two population parameters, the mean and standard deviation, were estimated, therefore 2 degrees of freedom are lost. The other degree of freedom is lost because the total frequency is fixed. From the χ^2 tables in Appendix Two:

$$\chi^2_{0.05,1} = 3.841$$

The test statistic from the table above is:

$$\chi^2 = \sum \left[\frac{(f_O - f_E)^2}{f_E} \right] = 0.59$$

Since:

$$0.59 < \chi^2_{0.05,1} = 3.841$$

the result is not significant at the 5% decision level. The evidence is consistent with the null hypothesis at this level. We assume that the sample is from a normal population with mean 255.25 g and standard deviation 5.522 g.

6.11.4 Does the observed frequency distribution fit the expected frequency distribution too well?

In the previous examples we have considered χ^2 values which cut off the top 5% or 1% of the distribution. We have accepted the null hypothesis if the actual value of χ^2 has been less than the boundary value. We must now consider the implications of a χ^2 value which is very small. As far as our current test is concerned, a very small value of χ^2 results in us assuming that the observed frequency distribution does fit the proposed model. However, a very small value of χ^2 means that there is virtually no difference between the corresponding values of the observed and expected frequencies. This immediately raises suspicions about the observed distribution, since we expect some differences to arise due to sampling. The occurrence of small values of χ^2 should lead us to a test using the left hand tail of the χ^2 distribution. We will determine the likelihood of obtaining by sampling fluctuations a χ^2 value which is less than the boundary value.

Example 6.22: To illustrate the use of the χ^2 hypothesis test when an observed distribution appears to fit the model too well

Suppose in **Example 6.20**, where the internal auditor was checking invoices, the value of χ^2 had been 0.290, instead of 2.90. Since this is a small value, we wish to question whether it indicates that the observed frequency distribution fits the expected Poisson distribution better than we expect for a randomly selected sample.
Solution: The null hypothesis is the same as that for the initial test.

H₀: The number of invoice errors per day follows a Poisson distribution with a mean of 4.01.

H₁: The errors per day do not follow this Poisson distribution.

We will test at the 99% level of significance using a χ^2 test with 6 degrees of freedom. The 99% level of significance means that, as a result of sampling fluctuations, there is only a 1% chance of obtaining a χ^2 value less than the boundary value.

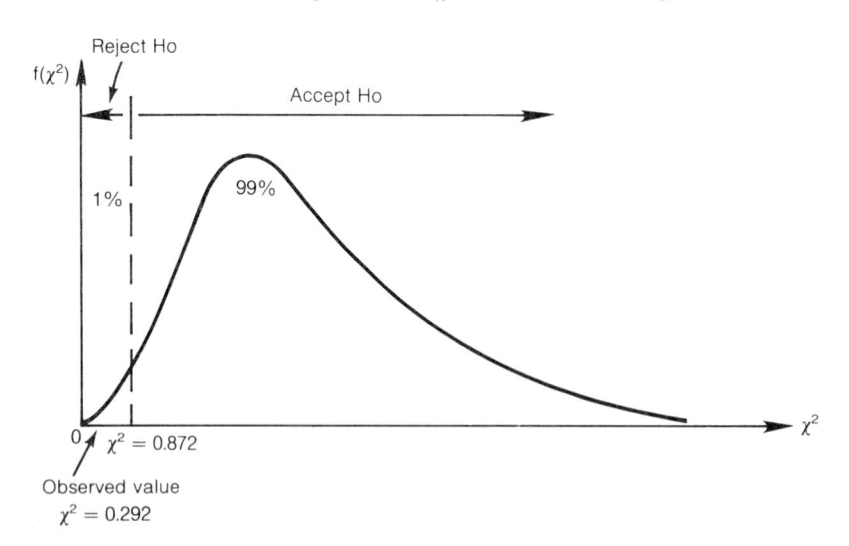

FIGURE 6.15 **Lower boundary value for a χ^2 distribution with 6 degrees of freedom**

From the table in Appendix Two, $\chi^2_{0.99,6} = 0.872$.

Since the supposed value of χ^2 is 0.290, the result is significant at the 99% level. The evidence is not consistent with the null hypothesis and we reject it in favour of the alternative hypothesis. The sample does not represent a random selection from a Poisson distribution with a mean of 4.01 errors per day. The observed frequencies fit the Poisson distribution too closely to be a random sample. Someone may have fixed the data!

SUMMARY

The procedure for all hypothesis tests is the same. We decide which question we wish to answer. We set up null and alternative hypotheses, H_0 and H_1. These clarify the conclusions which we may draw and indicate the test statistics to use. We choose the level of significance and, together with the hypotheses, this sets the boundary values, either one or two tail, for the test. We gather a random sample of data, calculate the test statistic, compare it with the boundary values and draw our conclusions. The main tests in the chapter are summarised below.

There are two tests for a single sample mean ($\bar{x}$) against a population mean (μ). If we know the population variance, σ^2, the test statistic is:

$$z = \frac{\bar{x} - \mu}{\sigma/\sqrt{n}}$$

If we do not know σ^2, we estimate it using the sample variance, s^2, and the test statistic is:

$$t = \frac{\bar{x} - \mu}{s/\sqrt{(n-1)}}$$

with $(n - 1)$ degrees of freedom.

To test a single sample proportion, $\hat{p}$, against a population proportion, p, if the sample size is at least 30, and np and $n(1 - p) \geq 5$, the test statistic is:

$$z = \frac{\hat{p} - p}{\sqrt{\dfrac{p(1-p)}{n}}}$$

The situation becomes more complicated when we wish to test two independent samples.

To test two sample variances (s_1^2 and s_2^2) we use:

$$F_{(n_1-1),(n_2-1)} = \frac{n_1 s_1^2}{(n_1-1)} \bigg/ \frac{n_2 s_2^2}{(n_2-1)}$$

with the larger variance on the top.

To test two sample means ($\bar{x}_1$ and $\bar{x}_2$), if we know the population variances (σ_1^2 and σ_2^2), the statistic is:

$$z = \frac{(\bar{x}_1 - \bar{x}_2) - (\mu_1 - \mu_2)}{\sqrt{\dfrac{\sigma_1^2}{n_1} + \dfrac{\sigma_2^2}{n_2}}}$$

To test two sample means, if we do not know the population variances, we must first test the sample variances, (s_1^2 and s_2^2) using the F test above. If the F test shows that we may assume $\sigma_1^2 = \sigma_2^2$, then to test the means we use the test statistic:

$$t = \frac{(\bar{x}_1 - \bar{x}_2) - (\mu_1 - \mu_2)}{\sqrt{\dfrac{(n_1 s_1^2 + n_2 s_2^2)}{(n_1 + n_2 - 2)}\left(\dfrac{1}{n_1} + \dfrac{1}{n_2}\right)}}$$

with $(n_1 + n_2 - 2)$ degrees of freedom.

If the F test shows that we must assume $\sigma_1^2 \neq \sigma_2^2$, then as long as both sample sizes are at least 30, the statistic is:

$$z = \frac{(\bar{x}_1 - \bar{x}_2) - (\mu_1 - \mu_2)}{\sqrt{\dfrac{s_1^2}{(n_1 - 1)} + \dfrac{s_2^2}{(n_2 - 1)}}}$$

To test two sample proportions, $\hat{p}_1$ and $\hat{p}_2$, if both samples are at least 30, the statistic is:

$$z = \frac{(\hat{p}_1 - \hat{p}_2)}{\sqrt{\bar{p}(1 - \bar{p})\left(\dfrac{1}{n_1} + \dfrac{1}{n_2}\right)}}$$

where $\bar{p}$ is the proportion from the two samples combined. (We have not covered the above two situations when the sample size is less than 30.)

If we have two dependent samples, we take the differences between the dependent pairs and treat these as a single sample. The test statistic is:

$$t = \frac{\bar{x}_d - 0}{s_d/\sqrt{(n - 1)}}$$

with $(n - 1)$ degrees of freedom.

If we wish to compare frequencies, we use the non-parametric χ^2 test. The test statistic is always calculated in the same way using observed and expected frequencies. It is:

$$\chi^2 = \sum \left[\frac{(f_O - f_E)^2}{f_E}\right]$$

Two applications of this statistic were considered.

If we have two attributes, we use the test to see if there is any association between them. A contingency table is constructed with the observed frequencies. This table and the null hypothesis are used to calculate the expected frequencies. The number of degrees of freedom is:

$$(\text{rows} - 1) \times (\text{columns} - 1)$$

In the second situation, the observed frequencies are compared with the expected frequencies from an assumed probability distribution, such as the normal or Poisson. The number of degrees of freedom is:

$$(n - 1 - k)$$

where n is the number of pairs and k is the number of population parameters estimated from the sample. The expected frequencies in both cases must be greater than 5.

Finally, the χ^2 test was used to check whether the data was too good a fit to the expected distribution, showing a lack of randomness.

EXERCISES

Exercise 6.1

Betta Bolt plc produce a variety of industrial fasteners. One of their product range is a non-magnetic bolt. The bolts are manufactured to a mean length of 5 cm with a known standard deviation of 0.05 cm. During the latest production run, a random sample of 25 was taken. The mean length of the bolts in the sample was 5.025 cm. What conclusions can be drawn about the average length of the bolts in this production run?

Answers on page 486.

Exercise 6.2

Shush plc manufacture soft drinks. At the moment they are producing half litre bottles of lemonade. The filling machine is set to fill to 500 ml. Every hour 30 bottles are randomly selected and checked to see whether the machine is underfilling on average. The mean volume of the latest sample is 495.6 ml and the standard deviation of the volumes is 8.3 ml. Is there evidence to suggest that the machine needs to be reset?

Answers on page 486.

Exercise 6.3

A company packs its product, freeze dried coffee, into cartons labelled 500 g. The company claims that the average contents of the packets is at least 500 g. A Trading Standards Officer visits the company's factory with the intention of testing this claim. The officer selects a random sample of 10 packets and weighs the contents. The following weights were obtained:

Packet Number	1	2	3	4	5	6	7	8	9	10
Weight, g	497	498	502	503	495	496	497	500	501	496

Required:
1   Write down clearly the null hypothesis and the alternative hypothesis for an appropriate test in this situation.
2   Conduct the test, using the above data, at the 1% significance level. State clearly your conclusions.

Note: You may assume that the weights of the packets are normally distributed.

Answers on page 487.

Exercise 6.4

A pottery company normally expects 10% of its output of coffee mugs to be defective in some way and to have to be sold off as seconds. Recently the company has introduced some new quality control procedures. In order to assess the effectiveness of these procedures, the line manager takes a random sample of 100 of the mugs and examines them. He finds 7 defectives. On the basis of this evidence alone, has there been a significant improvement in the quality of the product?

Answers on page 487.

Exercise 6.5

A machine is used to fill packets with jelly sweets. When the machine is working correctly, it fills with a known variance of:

$$\sigma^2 = 0.42 \text{ g}^2$$

The machine is set, using the variance, so that the packing regulations are met for the weight of the packets. In order to monitor the variability of the weights delivered to the packets, 10 packets are removed at random from the line every hour, weighed and the sample variance calculated. One such check gave a sample variance of 0.516 g^2. Does the sample provide any evidence to suggest that the variability of the filling machine has increased?

Answers on page 488.

Exercise 6.6

A large supermarket chain stocks a well-known brand of breakfast cereal. This cereal is sold under the brand name and also in a plain box as the supermarket's own brand. The manufacturer of the cereal fills both types of box. However, there is some suspicion that the supplier is putting less, on average, into the own brand packets. Random samples were taken from the output of the two types of packet and the following statistics were calculated:

$$\text{Normal brand—25 packets—} \bar{x}_1 = 503.6 \text{ g } s_1 = 5.03 \text{ g}$$
$$\text{Own brand \quad —30 packets—} \bar{x}_2 = 498.2 \text{ g } s_2 = 4.39 \text{ g}$$

Required:
Is there any evidence to justify the suspicion?

Answers on page 488.

Exercise 6.7

Yield is a critical factor affecting the total costs of a chemical process. COM plc are constantly trying to improve the yield from their plant which produces the chemical ABC1. The plant ran for 8 weeks before the latest adjustments were made, and gave the following weekly yields:

Week	1	2	3	4	5	6	7	8
% Yield	91.2	90.4	90.8	91.4	90.0	89.8	91.6	90.6

The plant settings were then adjusted and the plant has now run for a further 6 weeks, giving the following yields:

Week	9	10	11	12	13	14
% Yield	91.0	90.8	91.2	91.7	90.9	91.6

Required:
Is there any evidence to suggest that the adjustments have been successful in increasing the yield of ABC1?

Answers on page 490.

Exercise 6.8

COM plc also produce chemical ABC2 using two different processes. The company wishes to compare the yield from the two processes. It is known that the yields can be seriously affected by the purity of the raw material. The production manager selects eight batches of the raw material and uses some of each batch in the two processes. The yields are:

Batch number	1	2	3	4	5	6	7	8
Yield from process 1	70.1	67.0	68.6	68.8	70.2	68.0	69.5	68.4
Yield from process 2	73.9	70.3	68.5	71.3	69.7	71.0	69.8	68.2

What conclusions can you draw about the yield of chemical ABC2 from the two processes?

Answers on page 491.

Exercise 6.9

An external auditor is inspecting the accounting system of a small company. His first inspection of 100 transactions reveals 56 to be in error in some way. The company is allowed a month to improve its systems. When the auditor carries out his second inspection, he finds that 28 transactions are in error out of the 75 he checks. Is there any evidence that the error rate has improved between the inspections?

Answers on page 492.

Exercise 6.10

The Domestic Chemical Corporation plc commissioned a market research survey designed to test consumer preferences for five brands of detergent which they produce. A sample of 200 respondents gave the following responses:

Brand	A	B	C	D	E
% Preferring brand	19	20	15	25	21

Do these data show any preference for a particular brand?

Answers on page 493.

Exercise 6.11

Refer to **Exercise 6.10**. The Domestic Chemical Corporation decided to launch a new advertising campaign for Brand E. They then commissioned a similar survey with a new sample of 100 people. The second survey yielded the following results:

Brand	A	B	C	D	E
% Preferring brand	20	19	14	25	22

Required:
1 Do these data show any preference between the brands?
2 Has the market share of Brand E changed following the advertising campaign?
3 Has the pattern of preferences changed following the advertising campaign?

Answers on page 493.

Exercise 6.12

In a market survey concerned with home loans, a random sample of 250 people who had taken out a first mortgage during the last year was interviewed. As a result of the interview, the following data were collected:

Source of loan	Less than £10,000	£10,000 and over
Bank	25	25
Building Society	125	75

Required:
Carry out an appropriate statistical test to see if there is any evidence to show that the source and the size of the home loan are related.

Answers on page 495.

Exercise 6.13

Chips plc market a range of telephones and associated equipment. One particular product is a compact telephone answering system. The Sales Director thinks that the number of units sold per week follows a Poisson distribution. The data for the last 50 weeks is as follows:

Units sold per week	0	1	2	3	4	5	6	7	8	$\geqslant 8$
Number of weeks	1	3	6	11	10	7	5	3	4	0

Do the data support the Sales Director's view?

Answers on page 496.

Exercise 6.14

The Achilles Art Company sells special pens (for use on overhead projector transparencies) in boxes of 5. For various reasons—principally incorrect storage of the finished goods stock—a certain number of defective pens is discovered in the boxes sold to customers. The quality control manager is of the opinion that the percentage defective is approximately constant. Recently, however, there seem to have been more customer complaints than usual and he has been asked to investigate the situation. Accordingly he has instigated a sampling inspection scheme, which has produced the following results from a sample of 500 boxes chosen randomly from the finished goods stock:

Number of defectives per box	Number of boxes
0	392
1	73
2	30
3	3
4	2
5	0

Required:
Test whether the observed distribution of defectives per box is significantly different from a binomial distribution. Comment on your conclusions, indicating why any differences might have occurred.

(ACCA, December 1982)

Answers on page 497.

Exercise 6.15

The London Midland and Scottish Bank is at present reviewing its policy regarding the issue of statements to current account holders. Under the present system all customers receive a statement each quarter (ie every three months) as well as having the option of requesting a statement at any other time. It has been proposed, however, that all current account customers should be allowed to specify how frequently they wish to receive a statement by choosing one of the following four options:

- (a) No regular statement.
- (b) Every month.
- (c) Every three months.
- (d) Every six months.

In addition, customers would still have the freedom to request a statement at any other time to avoid having to wait for the arrival of their next regular statement. It has been estimated that there would be at most 600,000 irregular statement requests per year, which should occur more or less evenly throughout the year. At the present time, the bank has approximately 1.2 million current accounts.

To assess the acceptability of the new scheme, a random sample of 150 current account customers was selected and their preferences as regards the four statement options have been ascertained. The results are as follows:

Preferred frequency	No statement	Every 1 month	Every 3 months	Every 6 months
Number of customers	11	65	61	13

Required:

1. Construct a frequency distribution of the number of regular statements received per year for the sample of 150 customers and determine the mean and standard deviation of your distribution.
2. Determine a 95% confidence interval for the true mean number of regular statements per year and explain carefully the meaning of this interval.
3. All statements are despatched daily from the bank's computer centre. Under present conditions, there is capacity for 33,000 statements to be issued each day. Assuming that the computer centre works for 300 days a year, comment on whether there will be sufficient capacity to cope with the likely number of statements which will have to be issued. Your answer should incorporate an appropriate test of significance and any assumptions that you make should be clearly stated.

(ACCA, December 1984)

Answers on page 498.

Exercise 6.16

Martin Electronic Controls Ltd is a small manufacturer of electronic components, specialising in industrial control equipment.

Required:

1. The chief accountant made an inspection of 300 recent sales invoices, to make sure that the number of minor irregularities—such as the wrong date, incorrect addition, not rubber-stamped, or not signed by the recommended person—lies within the company's accepted error rate of 2%. If 9 of the 300 sample items contain a minor irregularity, is there statistical evidence that the stipulated error rate has been violated?

2 How many of the 300 sample items described in **1** would need to contain a minor irregularity in order for there to be statistical evidence, at the 5% level of significance, that the stipulated error rate has been exceeded?

(ACCA, December 1987)

Answers on page 500.

Exercise 6.17

As part of an investment portfolio analysis, two similar companies which operate in the same market are being compared in terms of their earnings per share. The companies' annual results over the last six years have provided the following information:

Year	Earnings per share (p)	
	Company A	Company B
19X0	14.3	13.8
19X1	15.6	14.6
19X2	17.2	16.4
19X3	16.4	16.8
19X4	14.9	15.0
19X5	17.6	16.4

The person performing the analysis calculated the mean and standard deviation of each company's results and then computed a t statistic as follows:

Company	Mean (p)	Standard deviation (p)
A	16.0	1.2977
B	15.5	1.2050

$$t = \frac{16.0 - 15.5}{\sqrt{\frac{1.2977^2}{6} + \frac{1.2050^2}{6}}} = 0.69 (10df)$$

He then concluded that the result was insignificant on both a one and two tailed basis, and consequently decided that the two companies have the same average earnings per share.

Required
1 Critically evaluate the analysis which has been performed and give three reasons why the procedure used and the conclusion reached are inappropriate and invalid.
2 Perform an appropriate analysis of the data and state your conclusions.

(ACCA, June 1986)

Answers on page 501.

Exercise 6.18

Croesus Construction Ltd has recently been analysing the unit cost associated with different types of building. It has obtained the following data relating to a random sample of 7 schools and 5 office blocks that were completed during the period 1971 to 1980. The data show the cost per square metre of floor area (in terms of 1982 prices) for each building involved:

Cost per square metre	
Schools	Offices
28	37
31	42
26	34
27	37
23	35
38	
37	

Required:

1. Do the data support the hypothesis that during the period in question the cost per square metre for office blocks was greater than that for schools?
2. If it were known that during the period 1971–1980, the company had completed 24 school buildings and 27 office blocks, would this cause you to alter the conclusion that you have reached in **1** and, if so, how?
3. Discuss briefly some of the limitations to the analysis you have performed and suggest two other factors that you might take into account in a more detailed comparison of unit costs.

(ACCA, December 1982)

Answers on page 503.

Exercise 6.19

1. Explain briefly the meaning of the term 'significantly different' in statistical hypothesis testing.
2. The four production plants of Zeus Company Ltd are based at Aybridge, Beedon, Crambourne and Deepool. A random sample of employees at each of these four plants have been asked to give their views on a productivity-based wage deal that the company is proposing. The table below summarises these views:

	Production plant			
View	Aybridge	Beedon	Crambourne	Deepool
In favour	80	40	50	60
Against	35	30	40	25

Required

(a) A chi-squared analysis of the contingency table gave a χ^2 value of 7.34. Test the hypothesis that there is no significant difference in views between the production plants. Explain what action might have been taken to reach a clearer decision.

(b) The employees at Ayebridge, Beedon Crambourne and Deepool were also asked to indicate whether they were under 40 years of age. The numbers aged under 40 for the four production plants were 75, 48, 54 and 57 respectively, of which 60, 28, 30 and 45 respectively were in favour of the wage proposal. Analyse the two resulting 2 × 4 contingency tables to test separately for the two age groups the hypothesis that there is no significant difference in views between the production plants.

(c) Estimate the overall percentage of the employees in favour of the wage deal for each age group. Are these percentages 'significantly different'?

(ACCA, June 1988)

Answers on page 505.

Linear regression

7.1 INTRODUCTION

The statistical analyses in the previous chapters have been concerned with the behaviour of single variables. We will now turn our attention to an analysis which involves two or more variables and the relationships between them.

For example, consider a company which regularly places advertisements for one of its products in a local newspaper. The company keeps records, on a monthly basis, of the amount of money spent on advertising and the corresponding sales of this product. If advertising is effective at all, then we can see intuitively that there is likely to be a relationship between the amount of money spent on advertising and the corresponding monthly sales. We would expect that the larger the sum spent on advertising, the greater the sales, at least within certain limits. There is, however, no theoretical basis from which we could write down an equation which would exactly link sales to expenditure on advertising. There are a number of factors which will work together to determine the exact value of sales each month—factors such as the price of the item, the price of a competitor's product, perhaps the time of year, or the weather conditions. Nevertheless, if the expenditure on advertising is thought to be a major factor in determining sales, knowledge of the relationship between the two variables would be of great use for the estimation of sales, and related budgeting and planning activities.

The term **association** is used to refer to the relationship between variables. For the purposes of the statistical analysis two aspects of the problem are defined. The term **regression** is used to describe the nature of the relationship, and the term **correlation** is used to describe the strength of the relationship. We need to know, for example, whether the monthly advertising expenditure is strongly related to the monthly sales and therefore will provide a reliable estimate of sales, or whether the relationship is weak and will provide a general indicator only.

The general procedure in the analysis of the relationships between variables is to use a sample of corresponding values of the variables to establish the nature of any relationship. We can then develop a mathematical equation, or **model**, to describe this relationship. From the mathematical point of view, linear equations are the simplest to set up and analyse. Consequently, unless a linear relationship is clearly out of the question, we would normally try to describe the relationship between the variables by means of a linear model. This procedure is called **linear regression**. A measure of the fit of the linear model to the data is an indicator of the strength of the linear relationship between the variables and, hence, the reliability of any estimates made using this model. For example:

FIGURE 7.1 **A linear relationship**

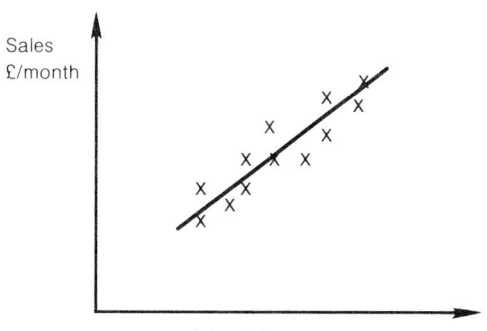

This graph indicates that a linear model might be a suitable way of describing the relationship between sales and advertising expenditure.

FIGURE 7.2 **A non-linear relationship**

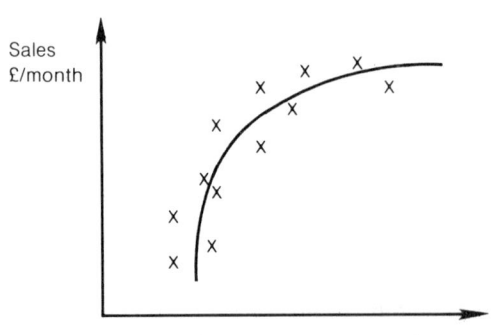

This graph indicates that a linear model would not be a suitable way of describing the relationship between sales and advertising expenditure.

Linear regression is our first example of the use of mathematical models. The purpose of a model is to help us to understand a particular situation, possibly to explain it and then to analyse it. We may use the model to make forecasts or predictions. A model is usually a simplification of the real situation. We have to make simplifying assumptions so that we can construct a manageable model but the model must still be sufficiently realistic to be worthwhile. Linear regression models are some of the most frequently used. They range from the simple two variable model, with which we will be mainly concerned, to the complicated multivariable models, which we will mention briefly. These models are widely used because there are many easy-to-use computer programmes available to carry out the required calculations.

Unfortunately the validation and interpretation of the computer output is much more difficult. You must be very careful when using such packages to ensure that you understand thoroughly what the output is about and how to evaluate it.

This chapter will cover all of the steps in the analysis of a simple linear regression model. We will take one sample of data and use it to illustrate each of the steps. The chapter concludes with sections on multiple regression models, the treatment of some examples of non-linear relationships and, finally, with an example of a non-parametric measure of correlation, Spearman's rank correlation coefficient.

7.2. SIMPLE LINEAR REGRESSION MODEL

Simple linear regression is concerned with what are called **bivariate distributions**, that is, distributions of two variables. We are interested in whether or not there is any linear relationship between the two variables. The population for bivariate distributions is made up of pairs of variable values rather than values of a single variable. For example, we may be interested in the heights and weights of a number of people, the price and quantity sold of a product, employees' age and salary, chickens' age and weight, weekly departmental costs and hours worked or distance travelled and the time taken.

The first step in the analysis is to consider the variables qualitatively. What are the factors with which we are concerned? How do we think they will affect one another?

As an example, let us say that a poultry farmer wishes to predict the weights of the chickens he is rearing. Weight is the variable which we wish to predict, therefore weight is the **dependent variable**. We will plot the dependent variable on the **y axis**. It is suggested to the farmer that weight depends on the chicken's age. Age is said to be the **independent variable**. This is the variable, the value of which we assume is known, which we can use to estimate weight. The independent variable will be plotted on the **x axis**. If we can establish the nature of the relationship between the age and weight of chickens, then we can predict the weight of a chicken at a given age. Any chicken for which the weight differs significantly from the prediction can be investigated.

How would we expect weight to change with age? At first, we expect the weight to increase with age, but when the chicken is fully grown, we would expect its weight to level off with small fluctuations depending on the feed or the time of year. Its weight gain and its adult weight will also depend on the breed of the chicken and the way it is fed and housed. There are clearly a lot of factors besides age, which will affect the weight. This process of thinking through the possible relationship, deciding on the dependent (y) and independent (x) variables, and thinking of other factors which could affect the relationship, is a very important part of all modelling. Our purpose is not 'to do some linear regression', but, rather, to try to explain and understand weight variations through the modelling. Can we explain a chicken's weight simply by looking at its age?

The probable conclusion, from the above discussion, is that there will be several factors working together to determine the precise weight of a particular chicken, but a general picture is:

FIGURE 7.3 **Possible relationship between weight and age of chickens of a particular breed**

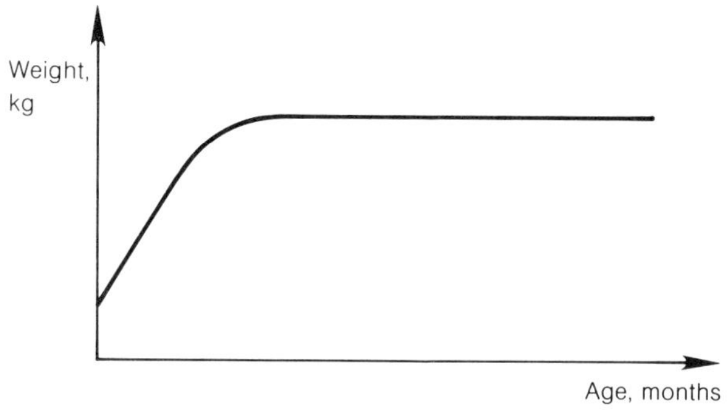

We are now at the stage of collecting a sample data to test our ideas.

Example 7.1: To illustrate the setting up of a simple linear regression model

The example we will analyse is concerned with the time it takes to make deliveries. We are running a special delivery service for short distances in a city. We wish to cost the service and to do this we must estimate the time for deliveries of any given distance.

There are factors, other than the distance travelled, which will affect the times taken—traffic congestion, time of day, road works, the weather, the road system, the driver, his transport. However, the initial investigation will be as simple as possible. We will consider distance only, measured as the shortest practical route in miles, and the time taken in minutes.

The relevant population is all of the possible journeys, with their times, which could be made in the city. It is an infinite population and we require a random sample from this population. For simplicity, let us use a systematic sample design for this preliminary sample. We will measure the time and distance for every tenth journey starting from a randomly selected day and a randomly selected hour of next week. The firm works a 6 day week, excluding Sundays. The random number, chosen by throwing a die, is 2, so next Tuesday is the chosen day. The service runs from 8am to 6pm. A random number, between 0 and 9, is chosen from random number tables to select the starting time. The number chosen is 6, so the first journey chosen is the first one after 1pm (ie this is the sixth hour, beginning at 8am), then we take every tenth delivery, after that.

The sample data for the first 10 deliveries will be used for the analysis.

TABLE 7.1 **Sample data for delivery distances and times**

Distance, miles	Time, minutes
3.5	16
2.4	13
4.9	19
4.2	18
3.0	12
1.3	11
1.0	8
3.0	14
1.5	9
4.1	16

184 QUANTITATIVE ANALYSIS

We wish to explain variations in the time taken, the dependent variable (y), by introducing distance as the independent variable (x). Generally we would expect the time taken to increase as distance increases. Hence, we first plot the sample data to obtain an impression of any relationship which exists between the variables.

FIGURE 7.4 **Plot of delivery times against distance for a random sample of deliveries**

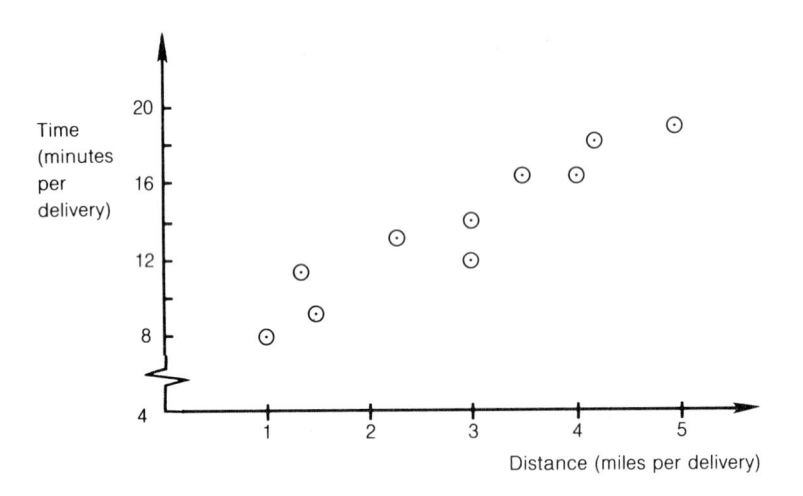

The plot does show a general increase of time with distance. It also looks as though the plotted points cluster around a straight line. This means that we could use a linear model to describe the relationship between these two variables. The points are not exactly on a line. It would be surprising if they were, in view of all the other factors which we know can affect journey time. A linear model will be an approximation only to the true relationship between journey time and distance, but the evidence of the plot is that it is the best available.

In the population from which we have taken our sample, for each distance, there are numerous different journeys and numerous different times for each of the journeys. In fact, for any distance, there is a distribution of possible delivery times. Our sample of 10 journeys is, in effect, a number of different samples, each taken from these different distributions. We have taken a sample of size 1 from the 1.0 mile deliveries and the 1.3 mile deliveries, a sample of size 2 from the 3.0 mile deliveries and we have taken no samples from the distributions for distances not on our list.

FIGURE 7.5 **An illustration of the distribution of journey times against distance in the population**

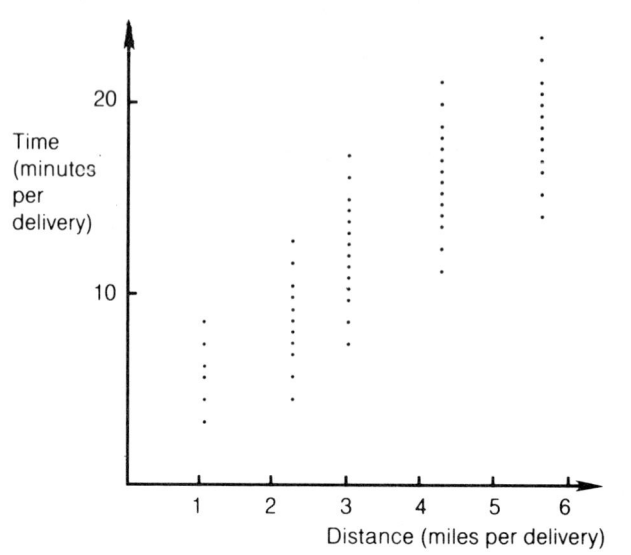

This idea of the underlying population of points is important for the subsequent analysis. We will return to the assumptions we are making about this population later in the chapter.

We have decided that the best model to describe the relationship between journey time and distance is probably linear. We now require a method of finding the most suitable line to fit through our sample of points. This line is referred to as the line of **best fit**. We must next define what we mean by 'best fit'. Look at the diagram below.

FIGURE 7.6   **Plot of delivery times against distance for a random sample of deliveries**

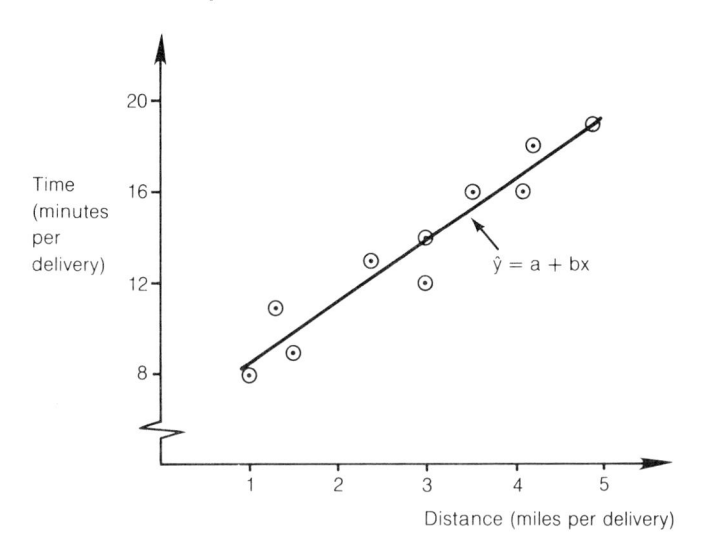

The line drawn through the data is a possible linear model to describe the relationship between the variables. The equation of this line may be described by:

$$\hat{y} = a + bx$$

where a is the intercept on the y axis and b is the slope of the line. The slope of the line is often called the **regression coefficient**.

Let us consider a particular value of the distance travelled, which we will call x_1. For x_1, the actual time taken is y_1, whereas the time taken as predicted by the linear model is:

$$\hat{y}_1 = a + bx_1$$

The difference between the two values:

$$e_1 = y_1 - \hat{y}_1$$

is called the **error** or **deviation** or **residual**. We can determine the error for all of the plotted points. The linear model which best fits the data is the one for which the total error has its smallest value. To avoid positive and negative values cancelling each other, all of the errors are made positive by squaring. The line of 'best fit' which we will use is the line which minimises the squares of the differences between the observed y values and the corresponding y values calculated from the line of best fit. This line is called the **least squares regression line**. Other criteria could be chosen to define the 'best' line, but in a situation where the values of the independent variable may be assumed known and the amount of scatter about the line is independent of the value of x, the least squares method is the most commonly used.

Using differential calculus, it is possible to determine the slope and intercept of the 'least square' regression line. We will not give the detailed working here since it is common practice simply to use the results.

The equations for the slope and the intercept of the least squares regression line are:

$$\text{slope, } b = \frac{n \sum xy - \sum x \sum y}{n \sum x^2 - (\sum x)^2}$$

where n is the sample size.

$$\text{intercept, } a = \frac{\sum y}{n} - \frac{b \sum x}{n}$$

The calculations for our sample of size n = 10 are given below. The linear model is:

$$\hat{y} = a + bx.$$

TABLE 7.2 **Calculations for the regression line**

Distance, x, miles	Time, y, mins	xy	x^2	y^2
3.5	16	56.0	12.25	256
2.4	13	31.2	5.76	169
4.9	19	93.1	24.01	361
4.2	18	75.6	17.64	324
3.0	12	36.0	9.0	144
1.3	11	14.3	1.69	121
1.0	8	8.0	1.0	64
3.0	14	42.0	9.0	196
1.5	9	13.5	2.25	81
4.1	16	65.6	16.81	256
Totals 28.9	136	435.3	99.41	1972
($\sum x$)	($\sum y$)	($\sum xy$)	($\sum x^2$)	($\sum y^2$)

(The final column is used in later calculations)

$$\text{The slope, } b = \frac{10 \times 435.3 - 28.9 \times 136}{10 \times 99.41 - 28.9^2} = \frac{422.6}{158.9} = 2.66$$

$$\text{and the intercept, } a = \frac{136}{10} - \frac{2.66 \times 28.9}{10} = 5.91$$

We now insert these values in the linear model, giving:

$$\hat{y} = 5.91 + 2.66x$$

or:

$$\text{delivery time (mins)} = 5.91 + 2.66 \text{ delivery distance (miles)}$$

The slope of the regression line (2.66 minutes per mile) is the estimated number of minutes per mile needed for a delivery. The intercept (5.91 miles) is the estimated time to prepare for the journey and to deliver the goods, that is, the time needed for each journey other than the actual travelling time. The intercept gives the average effect on journey time of all of the influential factors except distance which is explicitly included in the model. It is important to remember that these values are based on a small sample of data. We must determine the reliability of the estimates, that is, we must calculate the confidence intervals for the population parameters.

7.3 STRENGTH OF THE LINEAR RELATIONSHIP—THE CORRELATION COEFFICIENT, r

In the above example, the plot of the data showed that an underlying linear model looked reasonable. We therefore calculated the least squares regression line. However, we do not have an objective assessment of how well the linear model fits the data. The fit looks strong from our graph but this could be deceptive since the spread of the plotted points on the page will be determined in part by the scales chosen. We require an objective measure of the strength of the linear relationship.

Let us consider two variables, x and y. We consider that a relationship is likely between these variables. A sample of data yields the scatter diagram shown below with the least squares regression line superimposed. The line y = ȳ has also been added to the graph.

FIGURE 7.7 **Scatter diagram of y against x**

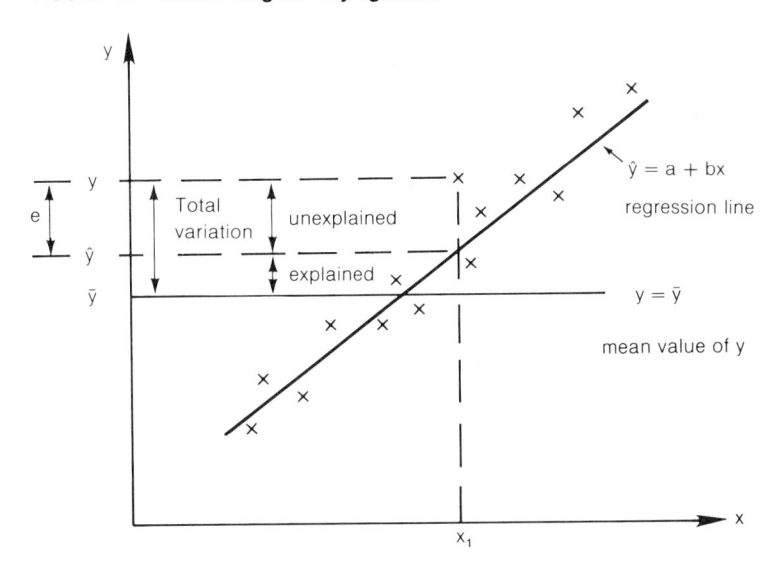

If we consider a particular value of x, say x_1, any point in the sample with this value of x will have a corresponding value of y. There may, in fact, be several points with the same value of x and different values of y, but in each case the actual y value may be split into two components. This can be written as:

actual value of y = value due to the linear relationship + value due to other factors

$$y = \hat{y} + e$$

where e is the residual, that is, the difference between the actual value of y and the value of y on the line.

The linear relationship only partly explains the variation in the value of y. The unexplained portion is the residual, e. If the relationship between x and y was perfectly linear, all the e's would be zero. As the strength of the linear relationship diminishes, the residuals increase. This procedure forms the basis from which we can develop a measure of the strength of the linear relationship. We must consider all of the points in the sample, not just one or two. Using y as a reference point, the total variation in the value of y may be written as:

$$\sum (y - \bar{y})^2$$

This is the total variation in the value of y, independent of the value of x. The total variation in the value of y due to the linear relationship may be written as:

$$\sum (\hat{y} - \bar{y})^2$$

This is referred to as the variation in y which is explained by the regression, ie by the introduction of the independent variable x. As x varies, y varies and y is related to x by $\hat{y} = a + bx$. The variation in y which is not explained by the linear relationship is written as:

$$\sum (y - \hat{y})^2$$

This variation arises from the other factors not included in the linear model. This variation is not explained by the regression.

The ratio of the explained variation to the total variation is used as a measure of the strength of the linear relationship. The stronger the linear relationship, the closer this ratio will be to one. The ratio is called the **coefficient of determination** and is given the symbol r^2 where:

$$r^2 = \frac{\sum (\hat{y} - \bar{y})^2}{\sum (y - \bar{y})^2}$$

The coefficient of determination is frequently expressed as a percentage and tells us the amount of the variation in y which is explained by the introduction of x into the model. A perfect linear relationship between x and y would result in $r^2 = 1$ or 100%. No linear relationship at all means $r^2 = 0$ or 0%. The coefficient of determination does not indicate whether y increases or decreases as x increases. This information may be obtained from the **Pearson product moment correlation coefficient, r**. This coefficient is the square root of the coefficient of determination:

$$r = \sqrt{\frac{\sum (\hat{y} - \bar{y})^2}{\sum (y - \bar{y})^2}}$$

For the purposes of calculation it is useful to re-arrange this expression algebraically to give:

$$r = \frac{n \sum xy - \sum x \sum y}{\sqrt{(n \sum x^2 - (\sum x)^2)(n \sum y^2 - (\sum y)^2)}}$$

This is the sample correlation coefficient.

The value of r always lies between -1 and $+1$. The sign of r is the same as the sign of the slope, b. If b is positive, showing a positive relationship between the variables, then the correlation coefficient, r, will also be positive. If the regression coefficient, b, is negative, then the correlation coefficient, r, is also negative.

As the strength of the linear relationship between the variables increases, the plotted points will lie more closely along a straight line and the magnitude of r will be closer to 1. As the strength of the linear relationship diminishes, the value of r is closer to zero. When r is zero, there is no linear relationship between the variables. This does not necessarily mean that there is no relationship of any kind. FIGURES 7.8 and 7.9 below will both give values of the correlation coefficient which are close to zero.

FIGURE 7.8 **No relationship of any kind between the variables**

FIGURE 7.9 **A strong non-linear relationship between the variables**

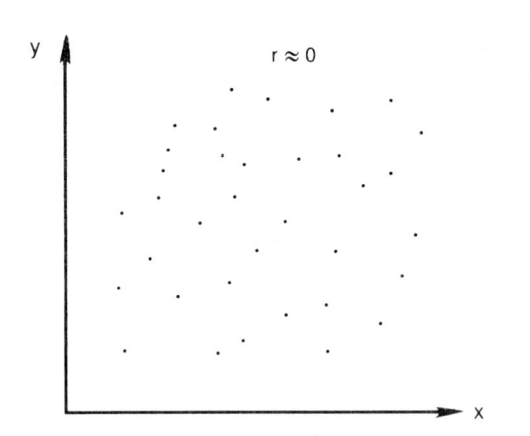

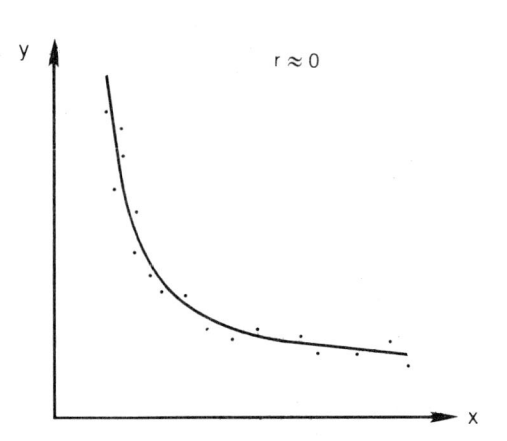

Return to **Example 7.1** above, in which a model was set up to predict delivery times for journeys of a given distance within a city. The correlation coefficient, r, is calculated as follows:

$$r = \frac{n \sum xy - \sum x \sum y}{\sqrt{(n \sum x^2 - (\sum x)^2)(n \sum y^2 - (\sum y)^2)}}$$

Notice that we have already evaluated the top line and part of the bottom line in the calculation for the slope of the regression line, b.

$$r = \frac{10 \times 435.2 - 28.9 \times 136}{\sqrt{(10 \times 99.41 - 28.9^2)(10 \times 1972 - 136^2)}} = \frac{422.6}{\sqrt{158.9 \times 1224}}$$

$$r = 0.958$$

This value of the correlation coefficient is very close to +1 which indicates a very strong linear relationship between the delivery distance and the time taken. This conclusion confirms the subjective assessment made from the scatter diagram.

The coefficient of determination ($r^2 \times 100\%$) gives the percentage of the total variability in delivery time which we have explained in terms of the linear relationship with the delivery distance. In this example, the coefficient of determination is high:

$$r^2 = 0.958^2 \times 100 = 91.8\%$$

The sample model:

$$\text{time (minutes)} = 5.91 + 2.66 \text{ (distance in miles)}$$

has explained 91.8% of the variability in the observed times. It has not explained 8.2% of the variability in the journey times. This 8.2% of the variability is caused by all of the other factors which influence the journey time but have not been included in the model.

7.4 PREDICTION AND ESTIMATION USING THE LINEAR REGRESSION MODEL

7.4.1 Predictions within the range of the sample data

We can use the model to predict the mean journey time for any given distance. If the distance is 4.0 miles, then our estimated mean journey time is:

$$\hat{y} = 5.91 + 2.66 \times 4.0 = 16.6 \text{ minutes}$$

An important word of caution is required here. It is not good practice to use the model to make predictions for values of the independent variable which are outside the range of the data used to set up the model. In our sample, the distances range from 1.0 mile to 4.9 miles. We have no evidence that the model is valid outside this range. The relationship between time and distance may change as distances increase. For example, a longer journey might include the use of a high-speed motorway, but the sample data was drawn only from the slower city journeys. In the same way, longer journeys may have to include meal or rest stops which will considerably distort the time taken. The information from which we have been working is a sample from the population of journey times within the distance range 1 to 4.9 miles. If we wish to extrapolate to distance outside this range, we should collect more data. If we are unable to do this, we must be very careful using the model to predict the journey times. These predictions are likely to be unreliable.

7.4.2 Estimation, error and residuals

How accurate are our predictions likely to be? In the next section we will consider this question in terms of the familiar idea of confidence intervals. However, it is also useful to assess the reliability of the predictions by means of the differences between the observed values of the dependent variable, y, and the predicted values, ŷ, for each value of the independent variable, x. These errors, or residuals, e, are the unexplained part of each observation and are important for two reasons. Firstly, they allow us to check that the model and its underlying assumptions are sound. Secondly, we can use them to give a crude estimate of the likely errors in the predictions made using the line.

The table below gives the residuals for **Example 7.1**.

TABLE 7.3 Calculation of residuals (y − ŷ)

Distance in miles, x	Observed time, mins, y	Estimated times = 5.91 + 2.66x Mins, ŷ	Residual e = (y − ŷ)
3.5	16	15.22	+0.78
2.4	13	12.29	+0.71
4.9	19	18.94	+0.06
4.2	18	17.08	+0.92
3.0	12	13.89	−1.89
1.3	11	9.37	+1.63
1.0	8	8.57	−0.57
3.0	14	13.89	+0.11
1.5	9	9.90	−0.90
4.1	16	16.82	−0.82

We can examine the suitability of the model by plotting the residuals on the y axis, against either the calculated values of ŷ or, in bivariate problems, the x values. This procedure is particularly important in multiple regression, when the original data cannot be plotted initially on a scatter diagram so that the linearity of the proposed relationship may be assessed. If the linear model is a good fit, the residuals will be randomly and closely scattered about zero. There should be no pattern apparent in the plot. The residual plot for the current example is given below.

FIGURE 7.10 **Plot of residuals (y − ŷ) against calculated ŷ values**

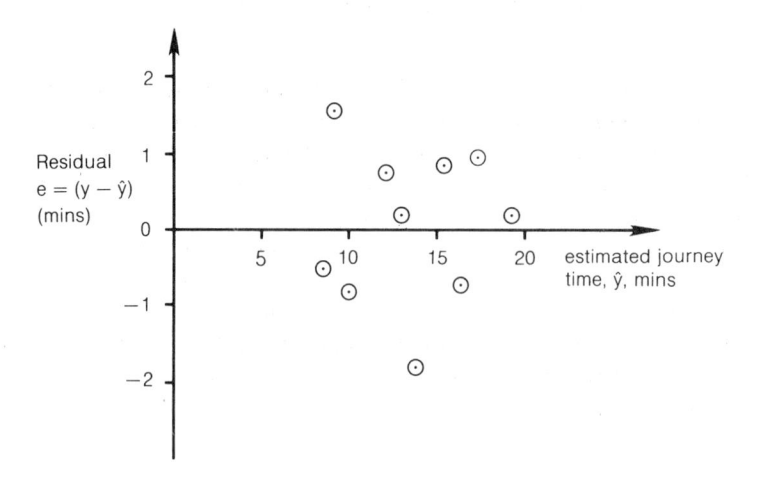

If the underlying relationship had in fact been a curve, then the residual pattern would have shown this very clearly. An example of the effect of fitting a linear model when the relationship between the variables is actually curvilinear is shown in FIGURES 7.11 and 7.12.

FIGURE 7.11 **Original data with calculated regression line**

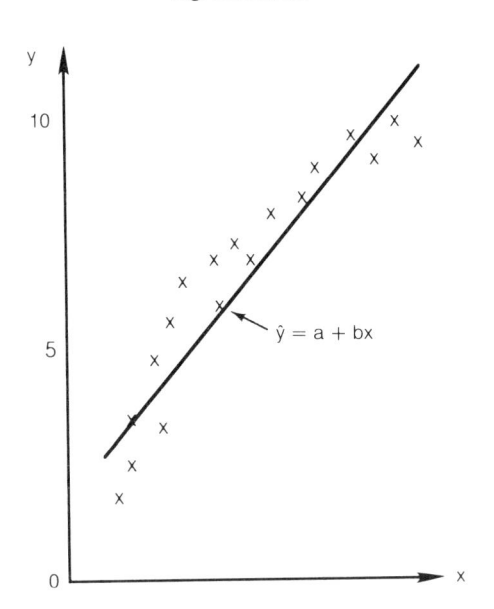

FIGURE 7.12 **Residual plot showing curved relationship**

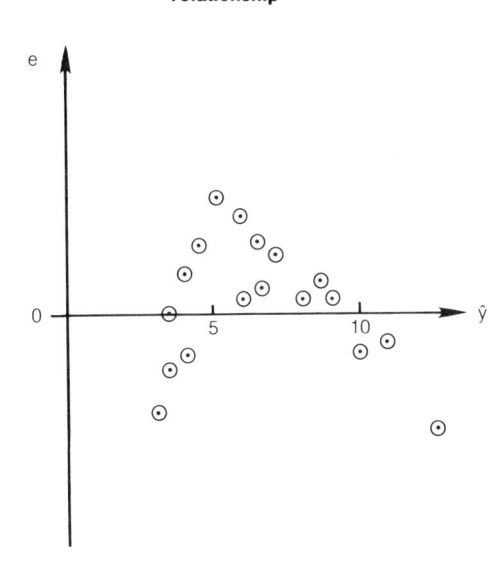

The residuals also allow us to assess the spread of the errors. One of the basic assumptions behind the least squares method is that the spread of data about the line is the same for all of the values of x, that is, the amount of variability in the data is the same across the range of x. See FIGURES 7.13 and 7.14 below:

FIGURE 7.13 **Constant variability for all x**

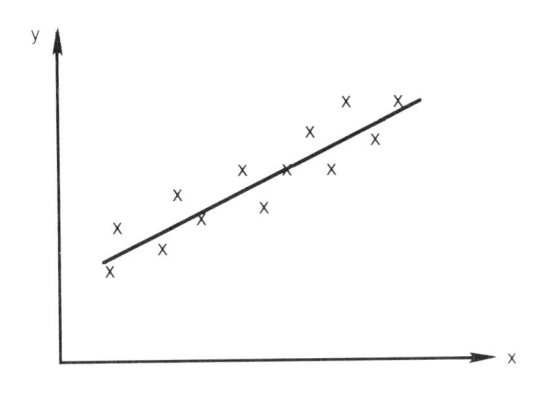

FIGURE 7.14 **Variability changing with x**

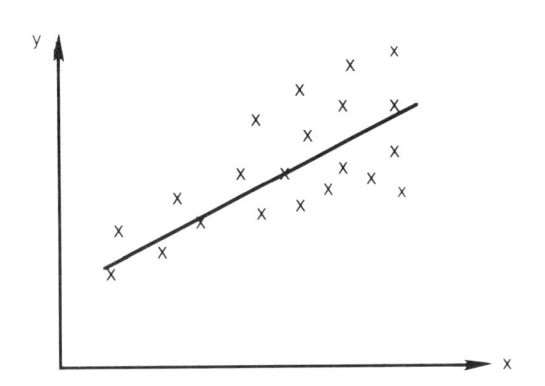

The variability in the data is constant across the range, hence the least squares method may be used to fit the 'best' line.

The variability in the data is not constant across the range, hence the least squares method is unsuitable for fitting the 'best' line.

The residual pattern for the journey time example shows two large residuals (-1.89 and $+1.63$). These may indicate that the data used do not conform to the assumption of uniform spread. The consequences of this will be that the confidence limits, described in the next section, will be unusable.

If the residuals give a pattern of the type shown in FIGURE 7.15 overleaf, then we know that the variance is changing as x changes.

The only way to continue with the statistical analysis of confidence intervals and hypothesis testing in this case, is to transform the data (often by taking logs of the x values) until the residual plot gives a random scatter of points about $e = 0$, with no large values.

The setting up and evaluation of a linear regression model can be a lengthy task, especially if the data set is moderately large. Fortunately there are many computer packages available which will take the drudgery out of the work by performing all of the arithmetic tasks to produce the required statistics. Unfortunately, however, it is all too easy to collect data and, without any further thought, to

FIGURE 7.15 **Residual plot for data which do not have a fixed variance for all x**

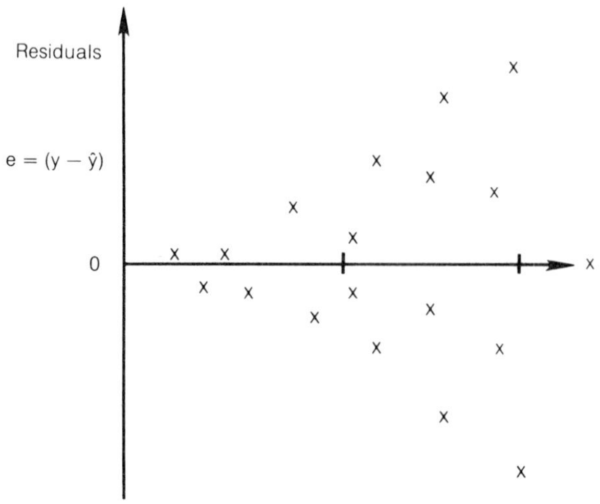

enter it into a computer package. The programme will produce a linear model, no matter how unsuitable that may be. In the following section we will examine how to assess and fully evaluate a linear regression model.

7.5 STATISTICAL INFERENCE IN LINEAR REGRESSION ANALYSIS

7.5.1 The underlying assumptions

In this section, we will discuss some of the necessary assumptions underlying the further analysis of the linear regression model. The data, from which a linear regression model is constructed, is a sample of the population of pairs of x and y values. Essentially we are using the sample to build a model which we hope will represent the relationship in the population as a whole. The relationship between the dependent variable, y, and the independent variable, x, is described by:

$$y = \alpha + \beta x + \varepsilon$$

where ε is the deviation of the actual value of y from the line:

$$y = \alpha + \beta x$$

for a given value of x.

$$y = \alpha + \beta x$$

is the linear model we would set up if we had all of the population data.

For any given x, the population of y values is assumed to be normally distributed about the population line, with the same variance for all x's. (See FIGURE 7.16.)

y is the mean of all the y values for a given x. As before, Greek letters refer to the population parameters, such as μ and σ. ε is the error, or residual, the difference between the actual y value and the mean value from the line. If the least squares method is used to determine the line of best fit, then we are minimising $\sum e^2$. The linear model which we calculate from the sample is:

$$\hat{y} = a + bx$$

where $\hat{y}$ is an estimate of the population mean y for a given value of x, and a and b are the sample statistics used to estimate the population parameters α and β.

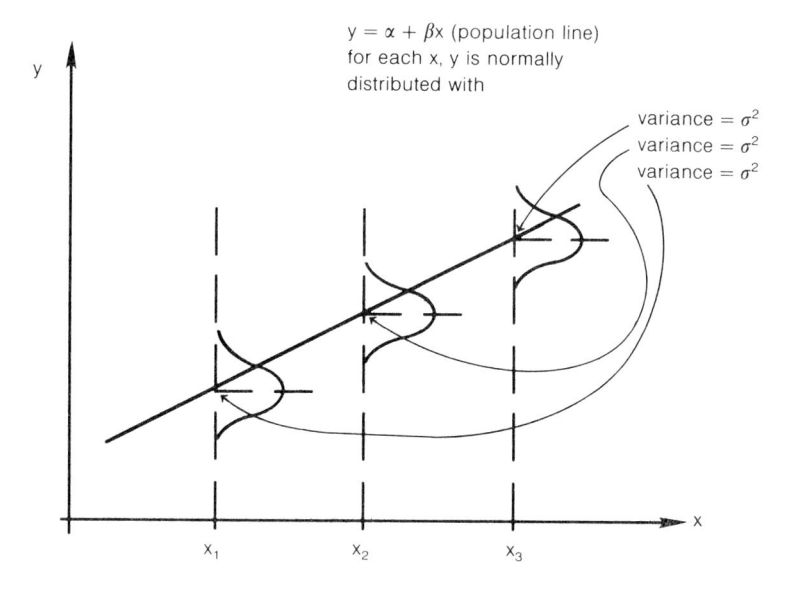

FIGURE 7.16 **Distribution of y values in the population**

$y = \alpha + \beta x$ (population line) for each x, y is normally distributed with variance $= \sigma^2$, variance $= \sigma^2$, variance $= \sigma^2$

As in any sampling situation, if we take a second sample, different values of a and b will arise. There is an exact analogy between the use of $\bar{x}$ to estimate μ and the use of a to estimate α. By making assumptions about the sampling distribution of $\bar{x}$, we can find confidence intervals for the value of the population mean μ. Exactly the same procedure may be used for α and β by making inferences from the sample values of a and b. Our basic model is:

$$y = \alpha + \beta x + \varepsilon$$

Assumptions:

1. The underlying relationship is linear.
2. The independent variable, x, is assumed to be known and is used to predict the dependent variable, y.
3. The errors or residuals, ε, are normally distributed.
4. For any given x, the expected value of ε is zero, ie $E(\varepsilon) = 0$.
5. The variance of ε is constant for all values of x, ie the variance of $\varepsilon = \sigma^2$.
6. The errors are independent.

If these assumptions hold, then the distribution of the population of y values, for a given x, is normal, with mean:

$$\mu_{y|x} = \alpha + \beta x$$

where $\mu_{y|x}$ denotes the mean of y for a given x, and variance $= \sigma^2$.

The line, set up from the sample data, is the best estimate of this population line, with a as the best estimate of α and b as the best estimate of β. Since there are many possible samples which could be drawn from a given population, it is not possible to be sure that a particular set of sample data is actually drawn from the given population. Hypothesis tests may be used on the sample to assess the compatibility of the sample with the population. The tests show how confident we may be about the linearity of the parent population. If there is no linear relationship in the population ρ, the population correlation coefficient, will be zero, and β, the slope of the regression line, will also be zero. Once we have tested the overall linearity, we may wish to calculate confidence intervals for the slope β, for the intercept α, for the mean value of y given a value of x, or for individual values of y for a given x. As in previous chapters, we will use a random sample to calculate sample statistics and to estimate the corresponding population parameters.

7.5.2 Hypothesis tests to assess the overall linearity of the relationship

We are using sample data which has been drawn at random from a population in order to estimate a suitable linear relationship for the population. We do not actually know that the underlying relationship in the population is linear. The random sampling process could, quite legitimately, result in a sample which exhibits linear properties but which was actually drawn from a population in which the underlying relationship is not linear. These possibilities are illustrated below:

FIGURE 7.17 **A random sample drawn from a population in which the underlying relationship is linear**

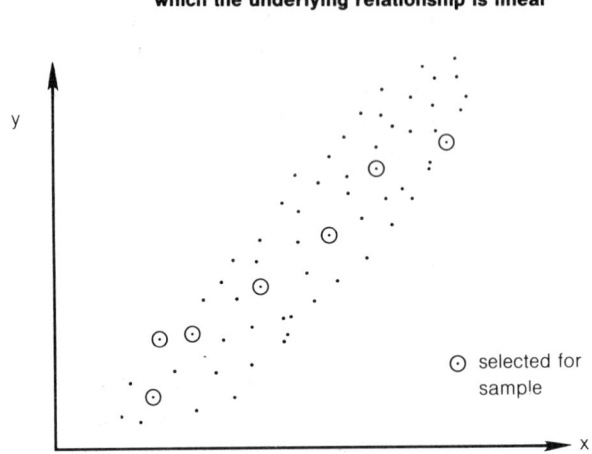

FIGURE 7.18 **A random sample drawn from a population in which the underlying relationship is not linear**

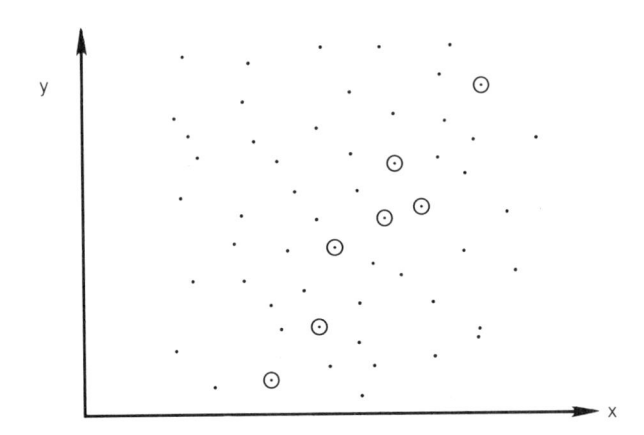

We require some means of assessing the likelihood that a linear relationship in the sample implies a linear relationship in the population. Hypothesis tests help with this assessment. As in any situation in which hypothesis tests are used, we can never prove beyond all doubt that the population relationship is compatible with the relationship derived from the sample. We simply determine the consistency, or otherwise, of the sample evidence with the given null hypothesis. Linear regression generates several statistics and it is possible to perform separate hypothesis tests on these. We therefore build up an accumulative picture of the evidence for or against the basic hypothesis of a linear relationship in the population. We will now look at these hypothesis tests in turn. The null hypothesis is essentially the same for all of the tests. It is that there is no linear relationship between the dependent and independent variables in the population.

TESTING THE POPULATION CORRELATION COEFFICIENT ρ

The evaluation of the Pearson product moment correlation coefficient, r, depends on the size of the sample. The interpretation of the value of r is independent of size from the point of view of the sample, but the implications for the population relationship are different for different sample sizes. A different inference will be drawn when considering a correlation coefficient of, say, 0.90, which arises from a sample of 6 items, compared to the same value which arises from a sample of 20 items. We can feel more confident that the underlying relationship is linear in the second case, since the chance of obtaining a sample which exhibits linearity, from a population which does not, decreases as the size of the sample increases. The correlation coefficient is assessed using a t test:

H_0: There is no linear relationship between the y and x variables. The independent variable, x, does not help in predicting the values of y, ie $\rho = 0$.

H_1: $\rho \neq 0$, ie there is some linear relationship between the x and y variables. x does help to predict the y values.

Using this alternative hypothesis, we have a two sided test. If we had decided that only a positive value for ρ would be sensible, then H_1: $\rho > 0$ and we would now use a one sided test. The test statistic is:

$$t = \sqrt{\frac{r^2(n-2)}{(1-r^2)}}$$

The number of degrees of freedom is $(n-2)$, because we have calculated $\bar{x}$ and $\bar{y}$ to find r, using up two degrees of freedom. n is the number of pairs of values in the sample. If we wish to test at the 5% level using a two tail test, the test statistic would be compared with $t_{0.025,(n-2)}$, found from the tables in Appendix Two.

To illustrate the procedure, we will return to **Example 7.1** which was concerned with the estimation of journey times from the journey distance. Previously we have found that r = 0.958. Therefore the test statistic is:

$$t = \sqrt{\frac{0.958^2 \times 8}{(1-0.958^2)}} = \sqrt{\frac{7.342}{0.082}}$$

$$= 9.45$$

The number of degrees of freedom is:

$$(10 - 2) = 8$$

From the tables in Appendix Two.

$$t_{0.025,8} = 2.306$$

The test statistic (9.45) is greater than 2.306, therefore we reject H_0 at the 5% level of significance and choose to accept H_1. The evidence is not consistent with the null hypothesis at this level. We assume that the correlation coefficient in the population is not zero and that there is a linear relationship between journey time and distance. This is the result we would expect with such a high value of the sample correlation coefficient, r.

HYPOTHESIS TEST ON THE SLOPE OF THE SIMPLE REGRESSION LINE

In simple linear regression, the hypothesis test on the slope of the line does exactly the same job as the test on the correlation coefficient. We do either one test or the other, but not both. In multiple regression, however, where we have a regression coefficient for each of the independent variables, the two tests fulfil different functions and both are needed.

H_0: There is no linear relationship between the variables. x does not help in predicting y, ie $\beta = 0$.

H_1: $\beta \neq 0$, ie there is a linear relationship and x does help to predict the y values.

In this case a two sided test is used. However, as in the test on ρ, we can alter this to a one sided test if we think that $\beta > 0$ or $\beta < 0$ is a more appropriate alternative hypothesis. The test statistic for β

is very similar to the one we used for μ and p in Chapter Six. When the population variance is unknown, the test statistic for a sample mean is:

$$t = \frac{\text{(sample statistic—parameter assumed in } H_0)}{\text{best estimate of standard error of statistic}} = \frac{(\bar{x} - \mu)}{s/\sqrt{n-1}}$$

The test statistic for the regression coefficient, b, is:

$$t = \frac{b - 0}{\text{estimated standard error of b}}$$

The estimated standard error of b is:

$$se_b = \frac{\hat{\sigma}_e}{\sqrt{\sum (x - \bar{x})^2}}$$

where σ_e^2 is the variance of the distribution of the residuals about the population regression line. Remember, we are assuming that this variance is the same for all values of x. $\hat{\sigma}_e^2$ is the best estimate of this population variance σ_e^2:

$$\hat{\sigma}_e^2 = \frac{\sum e^2}{(n-2)} = \frac{\sum (y - \hat{y})^2}{(n-2)}$$

For computational purposes, this expression may be re-arranged algebraically as:

$$\hat{\sigma}_e^2 = \frac{(\sum y^2 - a \sum y - b \sum xy)}{(n-2)}$$

Again, to illustrate the procedure, refer to **Example 7.1** about the journey times and distances. Using the first expression for $\hat{\sigma}_e^2$:

$$\hat{\sigma}_e^2 = \frac{0.78^2 + 0.71^2 \cdots (-0.82)^2}{8} = \frac{10.01}{8} = 1.25$$

Therefore:

$$\hat{\sigma}_e = 1.12 \text{ minutes}$$

and:

$$\sqrt{\sum (x - \bar{x})^2} = \sqrt{\left(\sum x^2 - \frac{(\sum x)^2}{n}\right)} = \sqrt{15.889} = 3.99$$

Therefore:

$$se_b = \frac{1.12}{3.99} = 0.281$$

The test statistic for β is:

$$t = \frac{2.66}{0.281} = 9.47$$

If allowance is made for rounding errors, this value of t is the same as the t statistic obtained in the test on the correlation coefficient, 9.47 compared with 9.45.

To test at the 5% level using a two sided test, we compare the test statistic with the boundary value found from the tables in Appendix Two:

$$t_{0.025, 8} = 2.306$$

Since $9.47 > 2.306$, we reject H_0 and choose to accept H_1. At the 5% decision level, the evidence is not consistent with the null hypothesis. This is the same conclusion as before. We choose to assume that there is a linear relationship between the journey time and distance, ie $\beta \neq 0$, x does help to explain the variability in y.

7.5.3 Confidence intervals in linear regression analysis

CONFIDENCE INTERVAL FOR THE SLOPE OF THE POPULATION REGRESSION LINE, β

The $(1 - p)\,100\%$ confidence interval for the slope, β, defines a range of values about b, the sample estimate of β, within which we may be $(1 - p)\,100\%$ confident that the actual value of β lies. Put another way, for $(1 - p)\,100\%$ of the samples, the true value of β will lie within the confidence interval. The confidence interval for β has the same format as the ones we met in Chapter Five:

$$b \pm t_{(p/2),(n-2)}\, se_b$$

From the above, we know that:

$$se_b = \frac{\hat{\sigma}_e}{\sqrt{\sum (x - \bar{x})^2}}$$

Let us calculate the 95% confidence interval for the slope of the regression model derived in **Example 7.1** about journey times and distances:

$$b \pm t_{0.025,8}\, se_b = 2.66 \pm 2.31 \times 0.281 = 2.66 \pm 0.65$$

We are 95% confident that the population slope, β, lies between 2.01 and 3.31 minutes per mile. There is a 5% chance that β lies outside this range.

CONFIDENCE INTERVAL FOR THE MEAN VALUE OF y FOR A GIVEN VALUE OF x

We now return to the basic assumption in regression analysis that for a given value of x, which we will call x_0, the possible values of y are normally distributed. The mean value of these normal distributions is the value of y on the population regression line. We will call this mean value $\mu_{y|x}$. The $(1 - p)\,100\%$ confidence interval for $\mu_{y|x}$ is:

$$\hat{y} \pm t_{(p/2),(n-2)}\, \hat{\sigma}_e \sqrt{\frac{1}{n} + \frac{(x_0 - \bar{x})^2}{\sum (x - \bar{x})^2}}$$

where $\hat{y}$ is the estimated y value, calculated from the sample regression, $\hat{y} = a + bx$.

Notice that this confidence interval depends on the value of x_0. The width of the interval about the regression line, therefore, varies as x varies. The interval is at its narrowest when $x_0 = \bar{x}$, the sample mean of the x's. The interval is then the more familiar:

$$\hat{y} \pm t_{(p/2),(n-2)}\, \hat{\sigma}_e / \sqrt{n}$$

As x_0 moves further away from $\bar{x}$, in either direction, the width increases.

In the **Example 7.1** which we have been following, the 95% confidence interval for $\mu_{y|x}$ is:

$$\hat{y} \pm 2.31 \times 1.12 \sqrt{\frac{1}{10} + \frac{(x_0 - 2.89)^2}{15.89}}$$

where:

$$\hat{y} = 5.91 + 2.66x_0$$

Values for this interval will be calculated in the next section.

CONFIDENCE INTERVAL FOR INDIVIDUAL VALUES OF y FOR A GIVEN VALUE OF x

A further assumption of the model is that y values are distributed about the regression line with variance, σ_e^2, which is the same for all values of x. Since we are using a sample, there are two elements of variability for individual y values. One arises from the estimated position of the mean, $\mu_{y|x}$, and the other arises from the variability of the individual values about this mean.

These two elements are different in that the first is due to the fluctuations inherent in sampling and the effect can be reduced if the sample is increased. The second is there by the nature of the variables and is unavoidable. It can be argued, therefore, that the confidence interval for individual values of y is not like other confidence intervals, which arise entirely due to sampling fluctuation. Some books refer to this interval as a 'prediction' interval rather than a confidence interval. Whatever name is used, it is important to understand the distinction between the $(1 - p)$ 100% interval for $\mu_{y|x}$ and that for the individual y values, when x is given. The confidence interval expressions look very similar. The only difference is that the variance for individual y's, given x, is increased by $\hat{\sigma}_e^2$.

The $(1 - p)$ 100% confidence interval for individual y values, given $x = x_0$, is:

$$\hat{y} \pm t_{(p/2),(n-2)} \hat{\sigma}_e \sqrt{1 + \frac{1}{n} + \frac{(x_0 - \bar{x})^2}{\sum (x - \bar{x})^2}}$$

where:

$$\hat{y} = a + bx_0$$

The 95% confidence interval in **Example 7.1** is:

$$\hat{y} \pm 2.31 \times 1.12 \sqrt{1 + \frac{1}{10} + \frac{(x_0 - 2.89)^2}{15.89}}$$

The table below and FIGURE 7.19 on page 199 illustrate the behaviour of the two confidence intervals, as x_0 changes.

TABLE 7.4   Calculation of confidence intervals for $\mu_{y|x}$ and for y given x_0, for Example 7.1

Distance x_0 miles	Estimated time $\hat{y} = 5.91 + 2.66x_0$ minutes	95% confidence intervals for $\mu_{y|x} \pm$ mins	y given $x_0 \pm$ mins
1.0	8.57	1.47	2.98
2.0	11.23	1.00	2.77
2.89($\bar{x}$)	13.60	0.82	2.71
3.0	13.89	0.82	2.71
4.0	16.55	1.09	2.81
4.9	18.94	1.54	3.01

These are long and complicated calculations for a sample of only 20 values. Much of the work can be taken away by a computer package but it is important to understand what the package is doing and how to interpret its output. Unfortunately, different packages use slightly different terms and symbols. If you have a regression analysis package available to you, it will help if you work through a simple example, like the one in this book, by hand, then use the package. Compare the computer output with the hand calculation until you thoroughly understand it. A clear understanding of the two variable linear models will also help considerably when we tackle multiple regression,

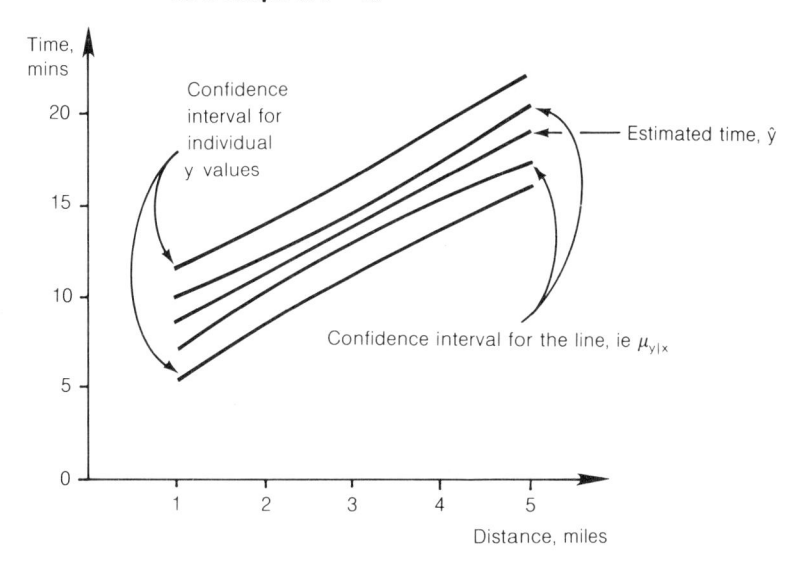

FIGURE 7.19 **Plot of 95% confidence intervals for $\mu_{y|x}$ and y given x_0, for a sample of n = 10**

the calculations for which are always done by computer. We discuss multiple regression in the following section.

7.6 MULTIPLE LINEAR REGRESSION MODELS

In the previous sections it was mentioned that the chosen independent variable is unlikely to be the only factor which affects the dependent variable. There will be many situations in which we can identify more than one factor which we feel must influence the dependent variable. For example, we wish to predict costs per week for a production department. It is reasonable to suppose that departmental costs will be affected by production hours worked, raw material used, number of items produced, and maintenance hours worked. It seems sensible to use all of the factors we have identified to predict the departmental costs. For a sample of weeks, we can collect the data on costs, production hours, raw material usage, etc, but we will no longer be able to investigate the nature of the relationship between costs and the other variables by means of a scatter diagram. In the absence of any evidence to the contrary, we begin by assuming a linear relationship, and, only if this proves to be unsuitable, will we try to establish a non-linear model. The linear model for multiple linear regression is:

$$y = \alpha + \beta_1 x_1 + \beta_2 x_2 + \beta_3 x_3 \ldots \beta_n x_n + \varepsilon$$

The variation in y is explained in terms of the variation in a number of independent variables, which, ideally, should be independent of each other. For example, if we decide to use 5 independent variables, the model is:

$$y = \alpha + \beta_1 x_1 + \beta_2 x_2 + \beta_3 x_3 + \beta_4 x_4 + \beta_5 x_5 + \varepsilon$$

As with simple linear regression, the sample data is used to estimate α, β_1, β_2, etc. The line of 'best fit' for the sample data is:

$$y = a + b_1 x_1 + b_2 x_2 + \cdots + b_n x_n$$

The intercept, a, and the regression coefficients, b, are evaluated using the criterion of minimising the sums of squares of the errors, $\sum (y - \hat{y})^2$. For the further analysis of the regression model, the following assumptions are made about the errors, for any given x:

1   $E(\varepsilon) = 0$
2   variance of $\varepsilon = \sigma_e^2$ is the same for all x's, and
3   the errors are independent of one another.

These assumptions are the same as those made for the simple regression case. However, here they lead to very complex calculations. Fortunately, a computer package will perform these calculations, leaving us free to concentrate on the interpretation and evaluation of the multiple linear regression model. In the following section, we will indicate the steps which should be taken when setting up and using a multiple regression model, but at all stages, we assume that the actual calculations will be done by computer.

STEP 1: LOOK AT THE DATA

The first step, as usual, is to look at the data and to think how the dependent variable might be related to each of the independent variables. Remember, we wish to explain variations in y in terms of the variations in the independent variables. There is no point in introducing additional independent x variables if they do not improve our ability to explain the variations in y. We cannot tell at the outset of our analysis how the variables are related since we can no longer produce a scatter diagram. We should, however, calculate the correlation coefficients, r, of all of the pairs of variables. This will enable us to determine whether each x is linearly related to y and also, whether the x's are independent of each other. This is important in multiple regression. We can test each of these correlation coefficients, as shown in **section 7.5**, to see whether the values are significantly different from zero. We are looking for high correlations between y and the x's, but low correlations between the x's themselves. If we find high correlations between the x's, for example, between x_1 and x_5, then it is unlikely that both x_1 and x_5 should be included in the final model. However, at this stage we just note that we should be wary of these particular highly correlated independent variables.

STEP 2: DETERMINE ALL OF THE STATISTICALLY VALID MODELS

We can set up linear models between y and any combination of the independent variables, but the model is valid only if there is a significant overall linear relationship and if each of the regression coefficients, b, is significantly different from zero. We can assess the overall significance by using an F test. We must then use a t test for each of the regression coefficients, b, to determine whether they are significantly different from zero. If a regression coefficient is not significantly different from zero, then that independent variable is not helping to predict y and that model is not valid.

The full procedure is to set up the multiple linear regression model for all combinations of the independent variables. Evaluate each model using the F test on the whole model and the t test for each regression coefficient. If the F test or any of the t tests are not significant, then the model is not valid and it cannot be used. In this way, invalid models are excluded until only the set of valid models remains. This process can involve a lot of work. For example, if we have five independent variables, there will be 31 possible models: one model with all five variables, five models containing four of the variables, ten with three variables, ten with two variables and five models with one variable.

Stepwise regression is a method of short circuiting this process. The method may start from either of the two extremes. In the forward method we begin by regressing y against each of the independent variables separately. We choose the best of these models, ie the one with the highest correlation coefficient. A second variable is added to this best variable. y is then regressed against all of these pairs of variables. This procedure substantially reduces the number of models which must be tested.

The backward method starts from the full model containing all of the independent variables, five in the example above. The variable which makes least contribution to the overall model is eliminated, leaving four variables. The linear model for these four variables is established. If this is not valid, the variable making the least contribution is eliminated, leaving three variables. This process is then repeated to eliminate the next variable. At each elimination there is a check to see that a significant variable has not been eliminated. This procedure has to be carried out with care, because it is possible to miss good, valid models by trying to reduce the work involved. No matter which method is used, a number of valid models may be produced, each of which is significant overall, and for which all of the regression coefficients are significant.

STEP 3: CHOOSING THE BEST MODEL FROM THE GROUP OF VALID MODELS

This procedure may be illustrated by looking at an example in which three valid models have been identified. Initially there were five independent x variables, x_1, x_2, x_3, x_4, x_5 but three of these, x_2, x_4

and x_5, have been excluded from all of the valid models. These variables do not help to predict y. The three valid models are therefore:

Model 1: y predicted by x_1 alone
Model 2: y predicted by x_3 alone
Model 3: y predicted by x_1 and x_3 together

In order to choose between the models, we examine the values of the multiple correlation coefficient, r, and the standard deviation of the residuals, $\hat{\sigma}_e$. The multiple correlation coefficient is the ratio of the 'explained' variance to the total variance and is calculated in the same way as the correlation coefficient for simple two variable regression. A model which describes well the relationship between y and the x's has a multiple correlation coefficient, r, which is close to ± 1 and a value of $\hat{\sigma}_e$ which is small. The coefficient of determination, r^2, which is often given by computer packages, describes the percentage of the variability in y which is explained by the model. A suitable model has a value of r^2 which is close to 100%.

In the example mentioned above, to decide between the single variable Models 1 and 2, we simply choose the model with the higher value of r^2 and the lower value of σ_e. Let us say that the preferred model is Model 1. The next step is to compare Models 1 and 3. The difference between these models is the introduction of x_3 into Model 3. The question is whether it significantly improves the prediction of y or whether its effect is marginal? As you might now guess there is a further test to help us answer this question. It is called the **partial F test**. We will now look at an example to illustrate the whole procedure.

Example 7.2: To illustrate the setting up of an appropriate multiple linear regression model

A large sweet manufacturer is trying to develop a model to predict the sales of one of its long established brands. The following data have been collected. All money values are deflated to 19X0.

TABLE 7.5   To determine a model to predict sales

Date	Sales per 6 mths £millions	Brand advertising £m	Brand price, p/item	Weighted competitor price, p/item	Index of consumer spending
19X0 Jan–Jun	126	4.0	15.0	17.0	100.0
Jul–Dec	137	4.8	14.8	17.3	98.4
19X1 Jan–Jun	148	3.8	15.2	16.8	101.2
Jul–Dec	191	8.7	15.5	16.2	103.5
19X2 Jan–Jun	274	8.2	15.5	16.0	104.1
Jul–Dec	370	9.7	16.0	18.0	107.0
19X3 Jan–Jun	432	14.7	18.1	20.2	107.4
Jul–Dec	445	18.7	13.0	15.8	108.5
19X4 Jan–Jun	367	19.8	15.8	18.2	108.3
Jul–Dec	367	10.6	16.9	16.8	109.2
19X5 Jan–Jun	321	8.6	16.3	17.0	110.1
Jul–Dec	307	6.5	16.1	18.3	110.7
19X6 Jan–Jun	331	12.6	15.4	16.4	110.3
Jul–Dec	345	6.2	15.7	16.2	111.8
19X7 Jan–Jun	364	5.8	16.0	17.7	112.3
Jul–Dec	384	5.7	15.1	16.2	112.9

Determine the 'best' model for the prediction of sales.
Solution: Step 1 is to look at the data. Sales per 6 months is the dependent variable, y. There are 5 independent variables, x. Four of these are advertising expenditure, brand price, competitors' price and the index of consumer spending. The fifth is time, which can be represented by calling the first period, January—June 19X0, time = 1, the next period time = 2, and so on, to time = 16 for the last period, July—December 19X7.

A computer is used to calculate the correlation coefficients, r, for all of the 6 variables. We test these correlation coefficients in turn using the hypothesis:

$H_0: \rho = 0$: the population correlation coefficient is 0. There is no linear relationship between this pair of variables.

Ideally this should be true for all pairs of the independent variables.

$H_1: \rho \neq 0$: the population correlation coefficient is not 0. There is a linear relationship between this pair of variables.

This should be true for the pairings of the dependent variable with the independent variables.

We will test at the 5% and 1% decision levels using a two-sided test. From the standard t distribution tables, at the 5% level of significance:

$$t_{0.05/2, 14} = 2.145$$

and the 1% level:

$$t_{0.01/2, 14} = 2.977$$

The test statistic is:

$$t = \sqrt{\frac{r^2(n-2)}{(1-r^2)}}$$

with $(n-2)$ degrees of freedom.

The correlation coefficients and the corresponding level of significance are given below:

TABLE 7.6 **Correlation coefficients r (significance level)**

	Dependent variable	Independent variables			
	Sales	Time	Advertising	Price	Competitor price
Time	0.68(1%)	—			
Advertising	0.64(1%)	0.10	—		
Price	0.23	0.17	−0.01	—	
Competitor price	0.23	−0.05	0.21	0.70(1%)	—
Index	0.82(1%)	0.96(1%)	0.27	0.23	0.03

The dependent variable, sales, has an encouragingly strong linear relationship with time, brand advertising expenditure and index of consumer spending. Unfortunately, the independent variables, time and index of consumer spending, are very highly correlated, (0.96). It is unlikley that both variables should be included in the final model. The same is true for the two price variables with a correlation coefficient of 0.70. We will bear these comments in mind during Step 2.

Step 2: The backward stepwise method will be used to find the valid models. We start with all of the variables in the model and step backwards from 5 to 4 variables, and so on, until all of the valid models have been identified. The 5 variable model is:

Sales = $a + b_1 x$(time) $+ b_2 x$(advertising) $+ b_3 x$(price) $+ b_4 x$(competitor price) $+ b_5 x$(index)

We start by establishing the overall validity of the model using an F test.

A computer is used to produce an **analysis of variance table**, which divides the total variation in the sales into two parts, the part explained by the model and the part which is not explained by the model, that is, the variation due to the regression and the unexplained, or residual, variation. The computer will calculate two variances:

$$\text{Mean squares due to regression} = \frac{\sum (\hat{y} - \bar{y})^2}{df_{reg}}$$

which measures the variation explained by the regression model, and:

$$\text{Mean squares due to residuals} = \frac{\sum (y - \hat{y})^2}{df_{resid}}$$

which measures the variation not explained by the regression model.

Note: The total number of degrees of freedom = n − 1, where n is the number of data sets; in this example, n = 16. df_{reg} is the number of degrees of freedom due to the regression which is given by the number of independent variables, k. In this model k = 5 = df_{reg}. df_{resid} is the number of degrees of freedom due to the residuals which is given by:

$$df_{resid} = df_{total} - df_{reg} = (n - 1) - \text{number of independent variables}$$

In this example, df_{resid} = 16 − 1 − 5 = 10.

If the model describes well the relationship between y and all of the independent variables, x, then this latter variance will be small. For the model as a whole:

H_0: There is no linear relationship between any of the independent variables and sales, ie $\beta_1 = \beta_2 = \beta_3 = \beta_4 = \beta_5 = 0$

H_1: There is a linear relationship between one or more of the independent variables and sales, ie at least one $\beta_i \neq 0$.

In order to have a useful, valid model we must be able to reject H_0 and accept H_1. The F test statistic is the ratio of the two variances, described above:

$$F = \frac{\text{Mean squares due to regression}}{\text{Mean squares due to residual}}$$

This is a one tail test because the mean squares due to regression must be the larger for us to accept H_1. In the previous sections when F tests were used, the tests were two sided, since we put the bigger variance on top, whichever it was. In regression analysis, there is no choice, the variance due to regression is always on the top. If it is less than the variation due to the residual, the model does not explain the variation in y and we accept H_0. This F test statistic is compared with:

$$F_{0.05, k, (n - 1 - k)}$$

From the standard F distribution tables:

$$F_{0.05, 5, 10} = 3.326$$

In the example the test statistic is:

$$F = 68196/1717 = 39.7$$

39.7 > $F_{0.05, 5, 10}$, therefore, we have a highly significant result. The evidence is not consistent with H_0 at the 5% decision level, hence we choose to reject H_0 and accept H_1. We may be confident in assuming that there is a strong linear relationship between at least one of the independent variables and sales.

We now test each individual regression coefficient. We will assume that the computer has calculated the necessary t statistics. For the first coefficient:

H_0: Time does not help to explain the variation in sales, given that the other variables are in the model, ie $\beta_1 = 0$

H_1: Time does contribute to the model and should be included, ie $\beta_1 \neq 0$

We will test at the 5% level using a two sided t test with:

$$(n - 1 - k) = 10 \text{ degrees of freedom}$$

The boundary values at the 5% level are:

$$t_{0.025, 10} = \pm 2.228.$$

The test statistic is:

$$t = \frac{b_1 - 0}{s_{b1}}$$

The computed t values must lie outside these boundaries for us to take the decision to reject H_0. H_0 must be rejected before we can assume that the model is valid.

Since all of the independent variables are treated in the same way, the results for the 5 variable in the current model are summarised below:

TABLE 7.7 **Testing the b coefficients of the 5 variable model**

Independent variable	Regression coefficient, b	t statistic	Significance at the 5% decision level
Time	−13.4	−1.29	not significant
Advertising	6.6	2.21	almost significant at 5%
Price	−6.4	−0.41	not significant
Competitor price	12.1	0.84	not significant
Index	30.5	2.65	significant at 5%

We can now see that the current model is not valid because 4 of the regression coefficients are not significantly different from zero. We must now decide which variable to leave out of the model. This is done one at a time. Do not jump to the conclusion that we can automatically leave out all of the variables which have non-significant regression coefficients.

TABLE 7.8 summarises the steps taken as we reduce the model from 5 to 4 to 3 to 2 and finally to one independent variable. The dashes indicate that a variable is not included in the model. At

TABLE 7.8 **Testing different regression models**

Variables in model	Overall model significant at 5%? (F)	Time	Advert Exp.	Price	Compet. Price	Index	Model valid?	$\hat{\sigma}_e$	r
5	yes	no	?	no	no	yes	no	42	0.94
4	yes	—	yes	no	no	yes	no	43	0.93
3	yes	—	yes	—	no	yes	no	41	0.93
2	yes	—	yes	—	—	yes	yes	41	0.93
2	yes	yes	yes	—	—	—	yes	50	0.89
1	yes	—	—	—	—	yes	yes	62	0.82

each stage, the tests on the overall regression and on the individual regression coefficients are carried out. Using the results of these tests, it is decided which variable to exclude. In addition, for each model, we have recorded in the table the standard deviation of the residuals, $\hat{\sigma}_e$, and the multiple correlation coefficient, r. If the model is a good fit, $\hat{\sigma}_e$ should be small and r should be as close to 1 as possible.

Step 3: Which of the valid models should we use? There are no valid models until we reduce to two independent variables. We decide between the two valid models with two independent variables by comparing the standard deviations of residuals. The standard deviation of residuals should be as small as possible. The first model, with advertising expenditure and the index of consumer spending as the independent variables, is the better model because $\hat{\sigma}_e = 41$, compared with 50 for the model with advertising expenditure and time as the independent variables.

The final stage is to compare the best two variable model with the best one variable model. From the correlation table, we know that the index of consumer spending gives the best single variable model with $r = 0.82$. We carry out a partial F test to see if the addition of the extra independent variable, advertising expenditure, has significantly improved the model. For this test:

H_0: advertising expenditure does not contribute to the explanation of the variability in sales, hence the one variable model should be used with the index of consumer spending as the only independent variable, ie $\beta_2 = 0$

H_1: advertising expenditure does contribute, and the two variable model should be used with index and advertising expenditure included, ie $\beta_2 \neq 0$

The test statistic is:

$$F = \frac{(r^2_{larger} - r^2_{smaller})}{(1 - r^2_{larger})/(n - 1 - k_{larger})}$$

Where r_{larger} is the correlation coefficient of the model with the larger number of variables, n is the number of data sets and k_{larger} is the number of independent variables in the larger model. In this example, $n = 16$ and $k_{larger} = 2$. As in the earlier F test on the model as a whole, the partial F test is a one sided test. The number of degrees of freedom are 1 and $(n - 1 - k_{larger})$.

The test statistic is compared with the boundary value $F_{0.05,1,(n-1-k)}$. From the 5% standard F distribution tables:

$$F_{0.05,1,13} = 4.667$$

The test statistic is, using unrounded values of r:

$$F = \frac{0.859 - 0.666}{(1 - 0.859)/13} = 17.8$$

which is much bigger than the boundary value at the 5% decision level. The evidence is not consistent with H_0 at this level. We choose to reject H_0 and to accept H_1. The introduction of advertising expenditure does improve the model and we should use the two variables, index of consumer spending and advertising expenditure, in the regression model. The final model is:

Estimated Sales = $-1476 + 9.54 \times$(advertising) $+ 15.8 \times$(index)   £m/6 months

The regression coefficients of advertising expenditure and the index of consumer spending are positive as we would expect. The constant at -1476 (£m) sales looks absurd, but remember the model is valid only for the ranges of advertising expenditure and index of consumer spending covered by the sample data. The advertising expenditure ranges from £3.8m to £19.8m and the index of consumer spending from 98.4 to 112.9.

This is a good example of the dangers and difficulties of trying to interpret, separately, each individual constant of a multiple regression model. The aim of the statistical models is to explain the variation in sales. The aim is not to provide specific information on the individual effects of

206 QUANTITATIVE ANALYSIS

advertising, or the index, on sales. Within the range of the sample data, the model gives some guidance on these effects, but we are using a sample of data, which often contains interactions between the different variables. The individual constants should be used with caution and judgment.

Finally in the analysis, we must examine the pattern and size of the residuals. We know that the standard deviation of the residuals, $\hat{\sigma}_e$, is £41m for sales which vary from £126m to £445m. We should, therefore expect some large errors. The errors are plotted below, where:

residual = actual sales − estimated sales

TABLE 7.9 **The size of the residuals**

Actual sales, £m	Estimated sales*, £m	Residuals, £m
126	142	−16
137	125	12
148	159	−11
191	242	−51
274	247	27
370	307	63
432	361	71
445	417	28
367	424	−57
367	350	17
321	346	−25
307	335	−28
331	387	−56
345	350	−5
364	354	10
384	362	22

* using 'best' regression model

FIGURE 7.20 **Residual plot for two variable model**

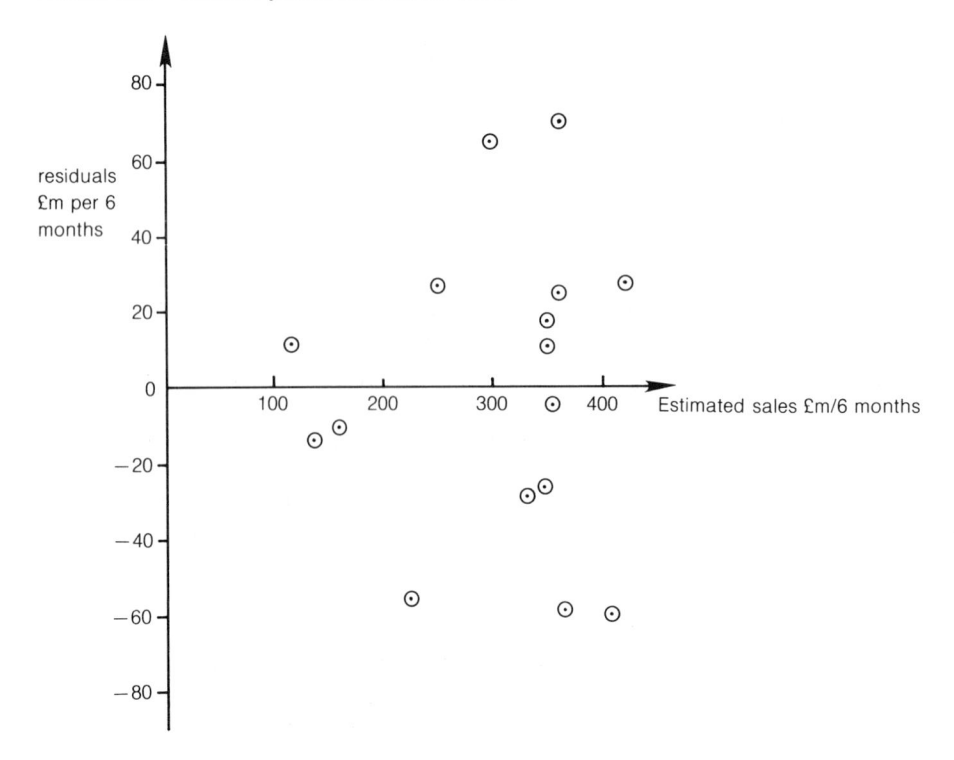

There are 8 errors which are 10% or more of the actual sales values. The largest error is 27% of the actual sales. Will this size of error be acceptable to a company trying to plan for the next 6 months? The answer to this question will depend on the other methods which are available and whether they are any more reliable.

7.7 NON-LINEAR RELATIONSHIPS

Let us now return to a situation in which there are just two variables but the relationship between them is not linear. This is a common occurrence. In practice, many of the underlying relationships between variables are curvilinear. For example, the relationship may be:

$$y = ax^2 + bx + c$$

or:

$$y = ax^3 + bx^2 + cx + d$$

or:

$$y = a + \frac{b}{x}$$

or:

$$y = ae^{bx}$$

The scatter diagram and correlation coefficient will indicate to us that a linear model would not be appropriate.

If the relationship between the variables is strong, that is, there is relatively little scatter about the curvilinear model, then we may be able to guess the nature of the best model. It is, however, difficult to fit any non-linear model directly to a sample of data. The process is easier if we can manipulate the non-linear model into a linear form. In the first two cases, given above, the functions x^2 and x^3 can be given different names, and a multiple linear regression procedure used. For example, if the model:

$$y = ax^3 + bx^2 + cx + d$$

is thought to best describe the relationship between y and x, we would use x, x^2 and x^3 as the independent variables, re-writing the model as:

$$y = aZ + bX + cx + d$$

where:

$$Z = x^3 \quad \text{and} \quad X = x^2$$

These variables are treated in the analysis as though they were ordinary independent variables, even though we know that Z, X, and x cannot be independent of one another. The best model is still chosen in the way indicated in the previous section.

The third and fourth cases, given above, are treated differently. These are dealt with by what is called a **linear transformation**. For example, if the underlying relationship between y and x is:

$$y = 1 + \frac{4}{x}$$

then a plot of y against x will produce a curvilinear pattern. If, however, we plot y against (1/x), the resulting scatter diagram will indicate a straight line. The process is illustrated below:

TABLE 7.10 **Calculations of y = 1 + 4/x and x = 1/x**

x	y = 1 + 4/x	X = 1/x
0	infinity	infinity
1	5	1
2	3	0.5
3	2.33	0.33
4	2	0.25

FIGURE 7.21 **Plot of y against x**

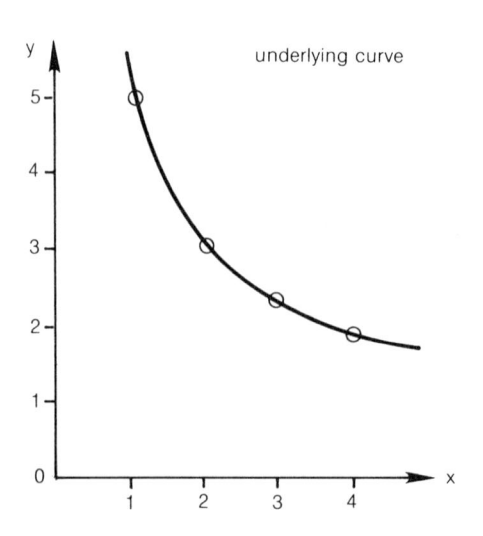

FIGURE 7.22 **Plot of y against X**

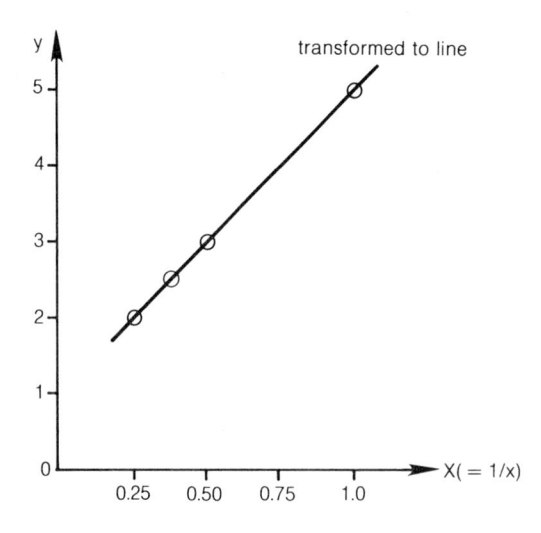

The linear model corresponding to the transformed relationship is:

$$y = 1 + 4X$$

where:

$$X = 1/x$$

In general, if the initial scatter diagram of y against x indicates that the underlying relationship is of the form $y = \alpha + \beta/x$, then a plot of y against X, where $X = 1/x$, will indicate a straight line. We can now use the procedures of simple linear regression to establish the model $\hat{y} = a + bX$. The calculated values of a and b are the best estimates of α and β.

The fourth relationship, in the above list, involves the transformation of the y values using logarithms to the base e:

$$y = a\, e^{bx}$$

Taking logarithms to the base e of both sides gives:

$$\ln(y) = \ln(a) + \ln(e^{bx})$$

Therefore,

$$\ln(y) = \ln(a) + bx \quad \text{where} \quad \ln = \log_e \quad \text{and} \quad \ln(e) = 1$$

If $Y = \ln(y)$ and $A = \ln(a)$, then, $Y = A + bx$ is a linear relationship between Y and x.

If we think that $y = a\ e^{bx}$ is the underlying relationship between y and x, then we must transform each y by taking logarithms to the base e. We do a simple linear regression of $Y(=\ln(y))$ on x to find the values of A and b. The antilogarithm of A yields the value of a and the original non-linear model can then be written down.

The method of linear regression can, therefore, be applied to non-linear relationships. However, some algebraic skill is required in order to write down a suitable initial model, and even more care must be taken when interpreting and using the model.

Example 7.3: To illustrate the setting up of a non-linear regression model

The following table shows the total annual production of tableware in the country of Blueland for the period 19X0–X7:

TABLE 7.11 **Annual production of tableware, 19X0–X7**

Year	Annual Production, ('000 tonnes)
19X0	740
19X1	804
19X2	879
19X3	961
19X4	1042
19X5	1137
19X6	1242
19X7	1357

Required:

1. Plot a scatter diagram of total annual production against year. Comment on the scatter diagram.
2. Plot a scatter diagram of log(total annual production) against year. Comment on the scatter diagram.
3. It has been suggested that the relationship between the total annual production and the year may be described by:

$$y = ab^x$$

   where y is the total annual production (in '000 tonnes) and x is the number of years since 19X0.
   Use the sample data given to estimate a and b. Interpret the value of b.
4. Use the model fitted in **3** to predict the total Blueland production for 19X8 and 19X9. Comment on the value of these predictions.

(ACCA, June 1988)

Solution:

1. The scatter diagram of total annual production, y, against the year is shown is FIGURE 7.23 overleaf.

210 QUANTITATIVE ANALYSIS

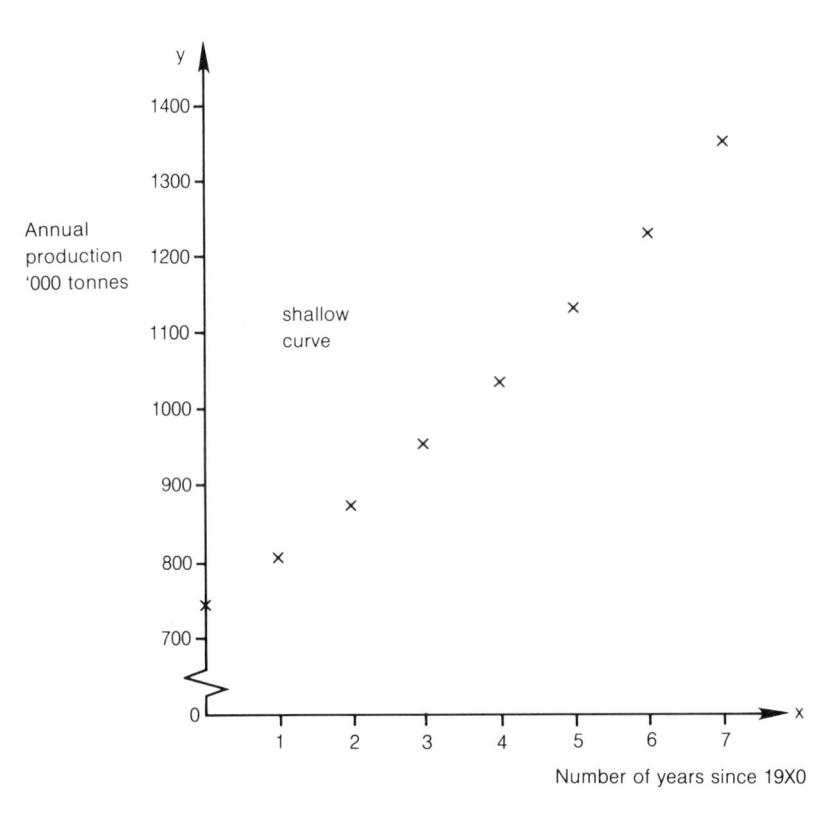

FIGURE 7.23 **Scatter diagram of total annual production**

[Scatter diagram: y-axis labelled "Annual production '000 tonnes" with values from 700 to 1400; x-axis labelled "Number of years since 19X0" from 0 to 7. Points show a shallow curve rising from about 750 at x=0 to about 1350 at x=7.]

The scatter diagram indicates that there is a relationship between y and x, but the relationship may be curvilinear.

2 We must first transform each total annual production into its logarithm.

TABLE 7.12 **Logarithm for each total annual production**

Year	Log(total annual production) $Y = \log(y)$
19X0	2.869
19X1	2.905
19X2	2.944
19X3	2.983
19X4	3.018
19X5	3.056
19X6	3.094
19X7	3.133

The linear relationship between log(y) and x is clearly seen from the scatter diagram. The underlying non-linear relationship between y and x must be one which will transform into:

$$Y = \log(y) = A + Bx$$

where:

$$Y = \log(y)$$

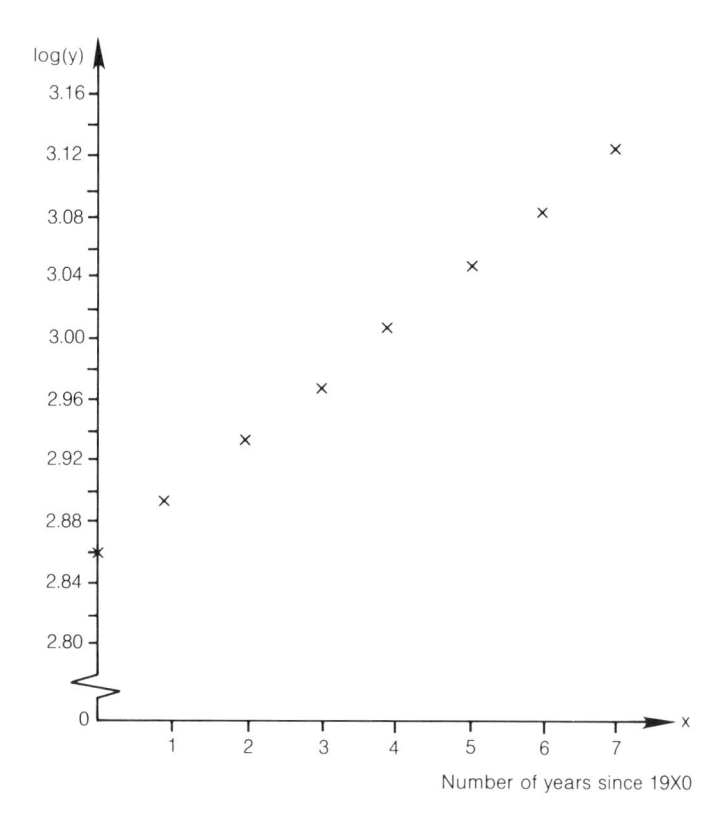

FIGURE 7.24 **The scatter diagram of log(y) against x**

Number of years since 19X0

3 The suggested model is:

$$y = ab^x$$

Taking logs of both sides:

$$\log(y) = \log(ab^x) = \log(a) + \log(b^x)$$

Hence:

$$\log(y) = \log(a) + x \log(b)$$

Let $A = \log(a)$ and $B = \log(b)$: we now have the expected linear model:

$$Y = \log(y) = A + Bx$$

We will use the technique of simple linear regression to evaluate A and B.

TABLE 7.13 **Calculations for regression model**

Year	Log(total annual production) $Y = \log(y)$	Number of years since 19X0 x	x^2	$x \log(y)$
19X0	2.869	0	0	0
19X1	2.905	1	1	2.905
19X2	2.944	2	4	5.888
19X3	2.983	3	9	8.949
19X4	3.018	4	16	12.072
19X5	3.056	5	25	15.280
19X6	3.094	6	36	18.564
19X7	3.133	7	49	21.931
Total	24.002	28	140	85.589

The 'least squares' method is used to fit the 'best' line to the sample data:

$$B = \frac{n \sum (x \log(y)) - \sum x \sum \log(y)}{n \sum x^2 - (\sum x)^2}$$

$$B = \frac{8 \times 85.589 - 28 \times 24.002}{8 \times 140 - (28)^2} = 0.03767$$

The intercept, A is given by:

$$nA = \sum \log(y) - B \sum x$$

ie,

$$8A = 24.002 - 0.03767 \times 28$$

$$A = 2.8684$$

The linear model is therefore:

$$\log(y) = 2.8684 + 0.03767x$$

Since $A = \log(a)$, then:

$$a = 10^A = 738.6$$

and $B = \log(b)$, then:

$$b = 10^B = 1.0906$$

The relationship between the total annual production and the number of years since 19X0 may now be described by:

$$y = 739 \times (1.091)^x$$

Interpretation of b: if we re-write this model as $y = 739(1 + 9.1/100)^x$, the nature of the relationship becomes more obvious. The annual production is 739 ('000 tonnes) in 19X0 when $x = 0$. The annual production then grows at a rate of 9.1% per year thereafter. b is the ratio of the total annual production in a given year to the total annual production in the previous year.

4 Prediction of future production levels:

$$\text{In 19X8, } x = 8, y = 739(1.091)^8 = 1483 \text{ thousand tonnes}$$

$$\text{In 19X9, } x = 9, y = 748(1.091)^9 = 1618 \text{ thousand tonnes}$$

Extreme care must be taken when extending a model beyond the range of the data used to build it. We are assuming that the conditions which have existed between 19X0 and 19X7 continue to exist. This is perhaps not unreasonable for the 19X8 forecast but as we move further into the future the forecasts will become rapidly less reliable.

7.8 SPEARMAN'S RANK CORRELATION—r_s

In the analysis discussed in the preceding sections of the this chapter, it has been assumed that the data involved measurement of some kind. If the only data available is some kind of ranking, then a non-parametric procedure must be used. A typical example arises in market research, such as food taste tests. If we are asked to taste four soups, we can rank them in order of preference from 1 to 4, but we will not be able to give a precise numerical value to the preference. This kind of data is called **ordinal data**. The data we have used until now are called **interval data**.

Suppose two people are asked to taste the four soups mentioned above. It is likely that there will be some variation in the two orders of preference. The two sets of ranks constitute two ordinal variables. We can compare the two sets of ranks for the **degree of agreement**, using **Spearman's rank correlation coefficient**.

The Spearman's rank correlation coefficient, r_s, is the non-parametric statistic which corresponds to the Pearson product moment correlation coefficient:

$$r_s = 1 - \frac{6 \sum d^2}{n(n^2 - 1)}$$

where d is the difference in rank for each item. Like the Pearson coefficient, the value of r_s varies from -1 to $+1$, but it does not measure the linear association between the two sets of ranks, just overall association. The significance of the Spearman coefficient may be tested in a similar way to the Pearson coefficient.

Example 7.4: Calculation of Spearman's rank correlation coefficient

In this example, two people have ranked ten samples of new soup mixes. We wish to know what level of agreement there is between the two tasters.

TABLE 7.14 **To determine level of agreement between two tasters**

Soup	A	B	C	D	E	F	G	H	I	J	Total
Rank given by
First person	1	3	4	9	8	10	2	7	6	5	
Second person	3	2	5	7	9	10	1	6	4	8	
Difference (d)	−2	1	−1	2	−1	0	1	1	2	−3	0
d^2	4	1	1	4	1	0	1	1	4	9	26

Therefore:

$$r_s = 1 - \frac{6 \times 26}{10 \times 99}$$

$$= 1 - 0.158 = 0.842$$

In order to assess the significance of this value, we will carry out a hypothesis test at the 5% level.

H_0: There is either no association or a negative association (disagreement) between the rankings of the 2 tasters, ie $\rho_s \leqslant 0$

H_1: There is a positive association (agreement) between the rankings ie $\rho_s > 0$

This is a one sided test. A two sided test would be used if we were interested in either a positive or negative association.

The test statistic, for $n \geqslant 10$, is approximately:

$$z = \frac{(r_s - 0)}{1/\sqrt{(n-1)}} = \frac{0.842}{1/3} = 2.53$$

From the standard normal tables:

$$z_{0.05} = 1.645$$

The test statistic is greater than 1.645. The result is significant at the 5% decision level. The evidence is not consistent with the null hypothesis at this level. We choose to reject H_0 and accept H_1 at the 5% level. There is evidence of a positive agreement between the two tasters. The normal approximation is true only for $n \geq 10$. Special tables are available for sample sizes of less than 10.

SUMMARY

Linear regression is a modelling process. We wish to explain the variability in the dependent variable, y, by introducing independent variables, x, into the model.

The building of the model always begins by thinking qualitatively about which factors are likely to influence the dependent variable. The main ones are then selected. In simple linear regression, there is only one independent variable. The model is:

$$y = \alpha + \beta x + \varepsilon$$

where ε is the residual, the part of y which is not explained by the model. α and β are usually estimated from a sample of data using the least squares method of fitting the 'best' straight line. For this line the slope is:

$$b = \frac{n \sum xy - \sum x \sum y}{n \sum x^2 - (\sum x)^2}$$

and the intercept is:

$$a = \frac{\sum y}{n} - \frac{b \sum x}{n}$$

The strength of the linear relationship is measured by the Pearson product moment correlation coefficient, r, where:

$$r = \frac{n \sum xy - \sum x \sum y}{\sqrt{(n \sum x^2 - (\sum x)^2)(n \sum y^2 - (\sum y)^2)}}$$

The nearer r is to -1 or $+1$, the better the linear model describes the relationship between x and y. The coefficient of determination, r^2, measures the proportion of the variability in y explained by the introduction of x.

To carry out statistical tests on the regression model, it is assumed that the residuals are normally distributed with a mean of zero and a standard deviation σ_e, which is constant for all x values. Confidence intervals may be set up for the population parameters ρ, α and β, using the sample statistics r, a and b. Hypothesis tests are used to assess the compatibility of the sample statistics with assumed values of the population parameters. For all of the tests, the null hypothesis is the assumption that there is no linear relationship between the variables in the population. The test statistic for ρ is:

$$t = \sqrt{\frac{r^2(n-2)}{(1-r^2)}}$$

with $(n - 2)$ degrees of freedom, and for β:

$$t = \frac{b - 0}{\hat{\sigma}_e / \sqrt{\sum (x - \bar{x})^2}}$$

with $(n - 2)$ degrees of freedom.

For a given value of x, say x_0, we can calculate confidence intervals for the mean value of y and for individual values of y about the mean. The width of these confidence intervals varies with x_0 and they form curves on either side of the estimated regression line.

Multiple linear regression has more than one independent x variable included in the model. The calculations are normally done using a computer package. Hypothesis tests are used to check that the overall relationship is significant and that each regression coefficient is significant. If any of these tests are not significant, that particular model must not be used. If a large number of independent variables are included in the original investigation, there are short-cut methods available for eliminating invalid models and establishing the best model.

Non-linear models may be manipulated algebraically into linear forms. Simple or multiple linear regression methods are then used to determine the values of the coefficients and to evaluate the suitability of the model.

Spearman's rank correlation coefficient is used in a similar way to Pearson's coefficient, when we have ordinal data, such as preferences or positions, rather than direct measurements.

EXERCISES

Exercise 7.1

1 A wholesaler wishes to explain the variation in the quantity of lettuces sold each day in a wholesale vegetable market.

Required:
Identify the factors which might affect this quantity.

2 The following data on quantity and price are available:

| Quantity ('000/day) | 28 | 29 | 34 | 35 | 37 | 37 | 41 | 46 |
| Price (p/item) | 30 | 31 | 25 | 26 | 22 | 24 | 16 | 12 |

Required:
(a) Plot the data, calculate the correlation coefficient, r, and test its significance at the 5% level.
(b) Set up the linear regression model and explain the meaning of the coefficients.
(c) If the price of the lettuce was 45p each, estimate how many would be sold? Comment on your assumptions.

Answers on page 509.

Exercise 7.2

A study is being carried out on the amount of money people save, compared with the amount they earn. The following data have been collected from an initial small random sample of 9 people.

| Income, £'000/year | 15 | 6 | 9 | 3 | 20 | 11 | 14 | 10 | 12 |
| Savings £/year | 2000 | 200 | 500 | 500 | 2500 | 1800 | 1500 | 1500 | 1600 |

Required:
1 Plot the data. Measure and assess the strength of any linear relationship.
2 Set up the regression model and give the meaning of the coefficients.
3 What are the other factors which you might consider?

Answers on page 510.

Exercise 7.3

The tourist office of a large holiday town is interested in the relationship between the number of holiday makers staying at hotels in the town, and the expenditure on hotel advertising. It has randomly chosen 6 hotels of similar size and gathered the following information for the current season:

Hotel	1	2	3	4	5	6
Advertising, £	9000	6000	10000	8000	7000	4000
Number of Guests	1100	1200	1600	1300	1100	800

Required:
1. Establish a model to explain the variation in the number of guests and test its validity.
2. Comment on the likely accuracy of forecasts made using this model.

Answers on page 511.

Exercise 7.4

As a basis for setting standard costs, the accountant of a firm has collected the following data from one of the production departments.

Output '000 units/day	12	19	12	17	15	15	17	16	18	19
Costs £'000/day	3.2	4.1	2.9	3.8	3.6	3.5	3.9	3.7	4.0	4.2

Required:
1. Set preliminary standards.
2. Comment on how accurate these might be and say what other factors you might include in the standards to improve their accuracy.

Answers on page 512.

Exercise 7.5

Garden Groceries own a small chain of 12 shops. There is a large spread in the size of the shops. The financial director of the group is considering the amalgamation of a number of the smaller shops with a view to improving profitability. He has assumed that he will retain at least the total turnover of the shops concerned, after amalgamation. He now wishes to relate profit to turnover. The data for each shop for the last financial year are given below:

Shop	Annual profit, £'000	Turnover, £'000
1	2	50
2	4	60
3	11	85
4	17	85
5	18	100
6	28	120
7	34	140
8	36	155
9	48	180
10	55	210
11	71	250
12	85	300

Required:
1. Set up a model for the relationship between profit and turnover. Interpret the constants in the model. Comment on the suitability of the model.
2. Advise the financial director on amalgamation.

Answers on page 513.

Exercise 7.6

Fifteen students have taken an accountancy and a mathematics examination. Their rankings in the two examinations are given below.

Required:
Is there any association between the two results?

Student	Rank in Examination Accountancy	Mathematics
1	10	13
2	5	4
3	12	10
4	1	1
5	6	11
6	2	2
7	7	8
8	11	9
9	15	14
10	3	5
11	9	7
12	14	12
13	13	15
14	4	3
15	8	6

Answers on page 514.

Exercise 7.7

The management accountant of Atlas Stores plc is analysing the profitability of one of the company's larger shops. From the figures for the latest quarter, he extracts the following data which give the number of customers and the shop turnover for each week during the quarter:

Week	Number of customers ('000)	Turnover (£000)
1	5.2	10.7
2	6.1	11.9
3	4.8	9.6
4	5.6	11.1
5	5.0	10.4
6	6.2	12.1
7	4.7	9.6
8	5.8	11.4
9	4.9	9.7
10	4.6	9.5
11	5.8	11.6
12	5.7	11.2
13	5.8	11.8

Using linear regression analysis to examine the relationship between the number of customers and the total turnover gives rise to the following equation:

$$\text{Turnover} = 1.60 + 1.71 \times \text{customers}$$

In this situation, however, it appears that the intercept (1.60) should really be zero since, without customers, there would be no turnover. Instead of the above regression equation, therefore, he decides to use a 'zero-intercept' regression of the form $y = bx$, where the least squares estimate of the gradient b is given by:

$$b = \sum xy / \sum x^2$$

Required:
1 Fit a 'zero-intercept' line of the form $y = bx$ to the data and explain the meaning of the regression coefficient b.
2 Plot a scatter diagram of the data and draw on your diagram the two regression lines. Comment on the apparent fit of these two lines.
3 Give two possible reasons why a non-zero intercept (such as the value 1.60) may arise when fitting a general line of the form $y = a + bx$ to a sample of data, even though the true relationship must pass through the origin. What relationship would you advise the management accountant to use in this situation and what precautions, if any, would you suggest?

(ACCA, June 1987)

Answers on page 515.

Exercise 7.8

The management accountant of Pan Products has been analysing the manufacturing times of all the company's major products. He has obtained the following data for one particular item which is produced on a batch basis to satisfy specific customer orders:

Batch number	Batch size	Manufacturing time (hours)
1	32	21.4
2	24	17.0
3	30	20.4
4	45	29.6
5	15	12.6
6	26	19.1
7	50	34.2
8	18	15.2
9	20	16.3
10	40	29.2

The data show, in chronological order, the manufacturing times of all batches of this item produced in the last year. Having obtained these times, the accountant developed a regression relationship between manufacturing time (y) and batch size (x) of the form:

$$y = 3.5 + 0.6x$$

with a correlation coefficient of 0.99. In the light of the high correlation, he concluded that batch size was a reliable and effective predictor of manufacturing time.

Required:
1 Explain the meaning of the calculated regression coefficients in the context of this example.
2 Calculate the deviations of the observed manufacturing times from the regression line and show that these deviations sum to zero. Explain why the deviations will always sum to zero.
3 Calculate the sum of squared deviations and explain the importance of this value.
4 Plot the deviations against both the batch size and the batch number. What do you deduce from the form of your graphs? Would you agree with the view that batch size is a 'reliable and effective predictor of manufacturing time'?

(ACCA, December 1986)

Answers on page 516.

Exercise 7.9

Pandora plc operates a regionalised distribution system with a number of delivery vans attached to each of the company's 7 distribution depots. When a decision has been made at each depot as to the number of deliveries to be made on any particular day, each driver is allocated a convenient group of customers. The driver makes one journey from the depot to deliver cartons to each customer in turn, following the most appropriate route, before finally returning to the depot. As part of an exercise to assess the effectiveness of the distribution system, data have been collected from a random sample of drivers relating to the time taken for a particular group of deliveries, the number of customers visited and the overall size of the load. The data are as follows:

Group of deliveries	Time taken (hours)	Number of customers	Total load (cartons)
1	6.3	6	24
2	4.7	3	15
3	5.0	4	17
4	8.5	7	30
5	3.8	4	14
6	4.1	6	20

Required:
1 Plot the data in a suitable form and comment on the relationships which are apparent.
2 Determine the product moment correlation coefficient between time taken and each of the other two variables. Comment on the two variables as possible predictors of total delivery time.
3 If total delivery time is regressed on both the number of customers and the total load, this produces a multiple correlation coefficient of 0.99. Explain carefully the meaning of this value and compare it with the two individual correlation coefficients which you determined in **2**.

(ACCA, December 1985)

Answers on page 518.

Exercise 7.10

Themis Processing Ltd is a wholly owned subsidiary of Tantalus Products plc and was formed in 19X6 to develop a new process for recovering iron from the waste slag of steel mills. Since that time, Themis has built 10 new slag processing plants of varying sizes which all operate successfully and run at a profit. Currently Themis is considering two additional plants and, because the average processing cost is a critical factor in overall profitability, it is important to estimate future processing costs early in the planning stage.

The chief accountant has suggested that some of the key factors influencing processing costs are the economies of scale inherent in different plant sizes. To substantiate this view, he has obtained the following cost data relating to the existing 10 plants:

Plant	First began operations	Capacity (tons/month)	Average cost (£ per ton)
1	19X6	900	51.95
2	19X6	500	57.18
3	19X8	1,750	46.90
4	19X9	2,000	45.37
5	19X0	1,400	46.03
6	19X0	1,500	48.15
7	19X1	3,000	44.22
8	19X2	1,100	48.72
9	19X2	2,600	45.40
10	19X3	1,900	44.69

The average cost per ton for each plant was obtained by summing all direct and indirect expenses and then dividing the total cost by the number of tons actually produced in 19X5 by the plant concerned.

Required:
1. If capacity is represented by x and average cost per ton by y, plot scatter diagrams of y against x, and y against 1/x. Comment on the form of your diagrams.
2. The results of performing a linear regression analysis of y against 1/x are as follows:

Coefficient	Value	Standard error
Intercept (a)	41.75	0.602
Gradient (b)	7,925.0	667.2

Residual Standard Error $(s) = 0.984$

Variability Explained $(r^2) = 0.946$

Plot the appropriate form of the above regression line on both of your diagrams in **1**. Explain the meaning of the various pieces of information given above.

3. On your scatter diagram of y and 1/x, draw 95% limits based on the residual standard error. Explain why these limits do not provide a prediction of confidence interval when forecasting the average cost per ton for a given value of 1/x.

(ACCA, December 1985)

Answers on page 520.

Exercise 7.11

Odin Chemicals Ltd is aware that its power costs are a semi-variable cost and over the last 6 months these costs have shown the following relationship with a standard measure of output:

Month	Output (standard units)	Total power costs (£'000s)
1	12	6.2
2	18	8.0
3	19	8.6
4	20	10.4
5	24	10.2
6	30	12.4

Required:
1   Using the method of least squares, determine an appropriate linear relationship between total power costs and output.
2   If total power costs are related to both output and time (as measured by the number of the month) the following least squares regression equation is obtained:

$$\text{Power Costs} = 4.42 + 0.82 \times \text{Output} + 0.10 \times \text{Month}$$

where the regression coefficients (ie 0.82 and 0.10) have t values of 2.64 and 0.60 respectively, and the coefficient of multiple correlation amounts to 0.976.

Compare the relative merits of this fitted relationship with the one you determined in **1**. Explain (without doing any further analysis) how you might use the data to forecast total power costs in month 7.

(ACCA, June 1982)

Answers on page 522.

Exercise 7.12

Venus Tableware Ltd is an important tableware producer in the country of Blueland. In this country the amount of imports and exports of ceramic tableware is insignificant compared to local production and so it can reasonably be concluded that the total Blueland production in any one year is essentially equal to the total sales. The following table shows the total Blueland production, the Venus Tableware sales and the consequent market share that Venus Tableware commanded within Blueland for the 8 year period 19X0 to 19X7.

Year	Total Blueland ('000 tonnes)	Venus Tableware ('000 tonnes)	Venus Tableware, (%)
19X0	744	113	15.2
19X1	773	108	14.0
19X2	828	131	15.8
19X3	900	144	16.0
19X4	936	146	15.6
19X5	977	157	16.1
19X6	1,007	163	16.2
19X7	1,066	175	16.4

Required:
1   Plot on a scatter diagram the logarithm of total Blueland production against the year. Interpret your scatter diagram.
2   Express the total Blueland production time series in the form:

$$Y = ab^x$$

where Y = total Blueland production (in '000 tonnes).

$$X = \text{year} - 19X0$$

Estimate a and b and interpret your value of b.

3   Use the model fitted in part **2** to predict the total Blueland production for the next three years: 19X8, 19X9 and 19X0.
4   The chief accountant of Venus Tableware wishes to use the total Blueland production time series to forecast the company sales. As Venus Tableware had increased its market share from 15.2% to 16.4%, he felt it might be wise to produce two sets of forecasts:
    (a)   a pessimistic forecast, assuming that their market share remains at 16.4%

(b) an optimistic forecast, where the company's market share increases by 0.2% over each of the next three years, to reach 17.0% by 19X0.

Obtain a set of optimistic and pessimistic forecasts for Venus Tableware annual sales for each of the next three years.

5 Outline the main shortcomings of regression-based forecasting techniques.

(ACCA, June 1988)

Answers of page 523.

CHAPTER EIGHT

Time series and forecasting

8.1 INTRODUCTION

All businesses have to plan their future activities. When making both short and long term plans, managers will have to make forecasts of the future values of important variables such as sales, interest rates, costs, etc. In this chapter we will look at ways of using past data to make these forecasts.

In the previous chapter, on regression, we used independent variables to explain the variation in a dependent variable. In this chapter, we are doing a similar job using time as the independent variable in all cases. For example, we may wish to explain the variability in sales by looking at the way in which they have changed with time, ignoring other factors. If we can explain the past pattern, we can use it to forecast future values. A data set in which the independent variable is time, is referred to as a **time series**.

Care is required since the historical pattern is not always relevant for particular forecasts. A company may deliberately plan to change its pattern if, for example, it has been making a loss. There may be large external factors which completely modify the pattern. There may be a major change in raw material prices, world inflation may suddenly increase or a natural disaster may affect the business unpredictably.

In section **8.2**, we begin by looking at time series which contain components, such as trend, seasonal variation and cyclical variation. The components can be combined in a number of ways. We will look at two specific models: the **additive component model** and the **multiplicative component model**. As the names imply, the components are added or multiplied, respectively. For each of these models, there are different ways of calculating the trend component. We will use a combination of moving averages and linear regression.

The model considered in the final section uses simple exponential smoothing, giving a different type of forecasting model. The method attempts to track the movement of a time series. Simple models can be modified to cope with trend and seasonal patterns but we will look only at the basic version.

You should note that the techniques described in this chapter are not the only, nor necessarily the best, forecasting methods for any particular forecasting situation. There are many more sophisticated statistical techniques available. As well as the quantitative techniques, there are qualitative methods which must be used when there is little or no past data. The **Delphi technique**, which uses experts to guess what might happen, and the **scenario writing method**, are examples of these.

8.2 TIME SERIES COMPONENTS

The value of a variable, such as sales, will change over time due to a number of factors. For example, a company may be expanding the market for a new product, hence there will be an upward movement in the sales of the product as time progresses. The general change in the value of a variable over time is referred to as the **trend, T**. The examples in the following sections are ones in which the trend is linear. This means that we can easily model the trend by using the technique of regression to calculate the line of best fit. The model is then used to forecast future trend. In practice, there may be no trend at all, with demand fluctuating about a fixed value, or more likely, there will be a non-linear trend. A non-linear trend is more difficult to model and it is more difficult to obtain a forecast of trend. The graphs below represent the trend demand at different stages of a product's life cycle.

There is an underlying upward trend for the newly launched product and a dying curve for the old product reaching the end of its economic life. It is difficult to fit an equation to these trend curves.

The **moving average technique**, described in the following sections, can be used to separate the trend from the seasonal pattern. The technique uncovers the historical trend by smoothing away the seasonal fluctuations. However, the moving average trend is not used to forecast future trend values because there is too much uncertainty involved in extending such a series.

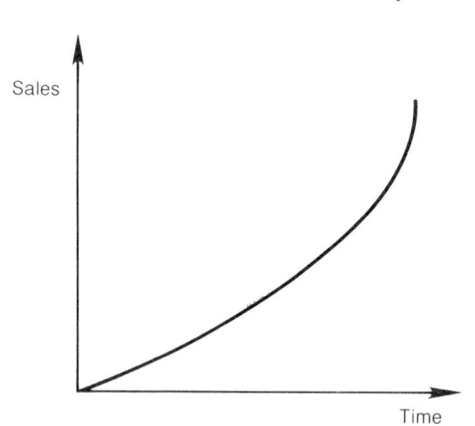

FIGURE 8.1 **Sales of a successful new product**

FIGURE 8.2 **Sales of a product nearing the end of its life**

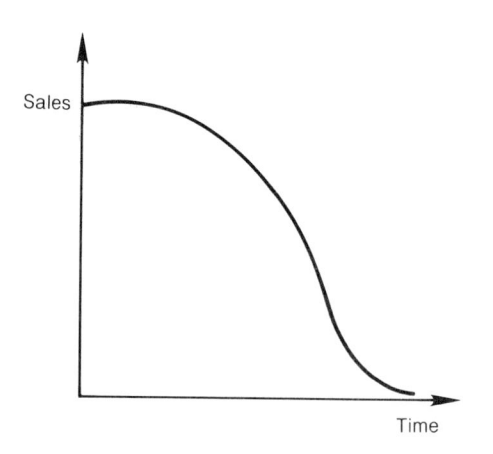

Generally it is found that variable values do not indicate a trend only. There is frequently a cyclical variation apparent in the values. If these regular fluctuations occur in the short term, they are referred to as **seasonal variation**, S. Longer-term fluctuation is called simply **cyclical variation**. The seasonal patterns in examples used in this chapter refer to the traditional seasons, but in forecasting generally, the term 'season' is applied to any systematic pattern. It may be the pattern of retail sales during the week, in which case the 'season' is a day. We may be interested in a seasonal pattern of traffic flow during the day and during the week. This will give us an hourly 'seasonal' pattern, superimposed on a daily 'seasonal' pattern, which both fluctuate about a daily trend. If we use annual data, we cannot identify a seasonal pattern. Any fluctuation about the annual trend data would be described by the cyclical component. This cyclical factor will not be included in our examples. It is seen only in long-term data, covering ten, fifteen or twenty years, where large scale economic factors cause additional fluctuations about the trend.

These cyclical factors were apparent in economic data from about 1960 to 1975. This was the time when many of these forecasting ideas were being developed, but since then, the overall economic pattern has changed. We will concentrate on short-term models which exclude the cyclical component.

The final term in our model also arises in the linear regression model. It is the **error** or **residual**, the part of the actual observation that we cannot explain using the model. We can use the errors to give us a measure of how well our model fits the data. Two measures are usually used. These are, the **mean absolute deviation**:

$$\text{MAD} = \frac{\sum |\text{Actual} - \text{Forecast}|}{n} = \frac{\sum |E_t|}{n}$$

which is the sum of all the errors, ignoring their sign, divided by the number of forecasts, and the **mean square error**:

$$\text{MSE} = \frac{\sum (E_t)^2}{n}$$

which is the sum of the squares of the errors, divided by the number of forecasts. This second measure emphasises the large errors.

In the analysis of a time series, we attempt to identify the factors which are present and to build a model which combines them in an appropriate way.

Example 8.1: To illustrate the choice of an appropriate time series model

The data below are the quantities of a product sold by Lewplan plc during the last 13 three monthly periods.

TABLE 8.1 **Quantities of product sold over last 13 three monthly periods**

Date	Quantity sold, '000
Jan–Mar 19X6	239
Apr–Jun	201
Jul–Sep	182
Oct–Dec	297
Jan–Mar 19X7	324
Apr–Jun	278
Jul–Sep	257
Oct–Dec	384
Jan–Mar 19X8	401
Apr–Jun	360
Jul–Sep	335
Oct–Dec	462
Jan–Mar 19X9	481

We wish to analyse this data set to see if we can find a historical pattern. If there is a consistent pattern, we will use it to forecast the quantity sold in subsequent three monthly periods.
Solution: The figures are plotted below. In time series diagrams it is customary to join the points with straight lines so that any pattern can be seen more clearly:

FIGURE 8.3 **Lewplan plc, sales per 3 months**

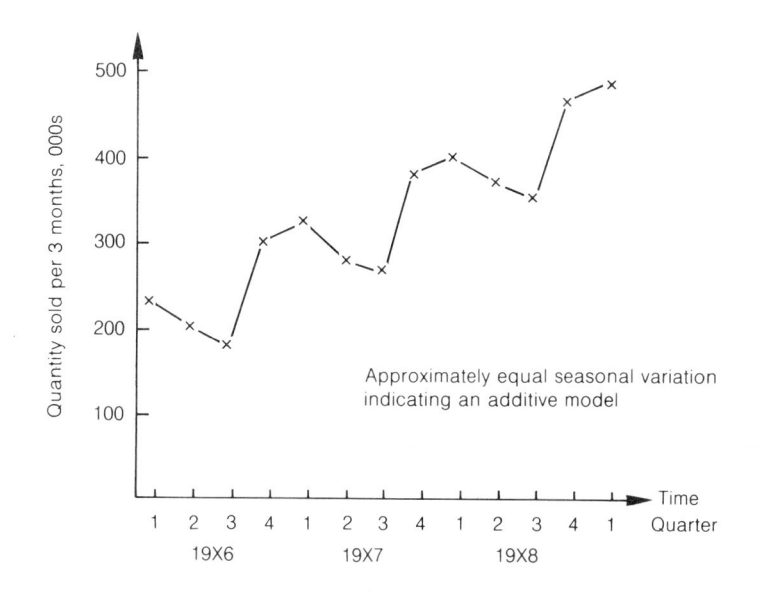

226 QUANTITATIVE ANALYSIS

The diagram suggests that there may be an increasing trend, overlaid by seasonal fluctuations. The sales in the winter seasons, 1 and 4, are consistently higher than those in the summer seasons, 2 and 3. The seasonal component appears to be fairly constant over the three years. The trend is for the sales to increase overall from around 230 in 19X6 to 390 in 19X8, but the seasonal fluctuations have not increased. This indicates that the additive component model should be more suitable. See section **8.3**.

8.3 THE ANALYSIS OF AN ADDITIVE COMPONENT MODEL: A = T + S + E

The additive component model is one in which the variation of the value of the variable over time can best be described by adding the relevant components. Assuming that cyclical variations are not included, the actual value of the variable, A, may be modelled by:

$$\text{Actual value} = \text{trend} + \text{seasonal variation} + \text{error}$$

that is:

$$A = T + S + E$$

In both the additive and multiplicative component models, the general analysis procedure is the same:

Step 1: Calculate the seasonal components.
Step 2: Remove the seasonal component from the actual values. This is called deseasonalising the data. Calculate the trend from these deseasonalised figures.
Step 3: Deduct the trend figures from the deseasonalised figures to leave the errors.
Step 4: Calculate the mean average deviation (MAD) or the mean square error (MSE) to judge whether the model is reasonable, or to select the best from different models.

8.3.1 Calculate the seasonal components for the additive model

Example 8.2: Setting up the additive component model for a time series

Refer to **Example 8.1** in the previous section which relates to the quarterly sales of Lewplan plc. We have already decided that an additive model is appropriate for these data, therefore the actual sales may be modelled by:

$$A = T + S + E$$

To eliminate the seasonal components, we will use the method of moving averages. If we add together the first 4 data points, we obtain the total sales for 19X6: dividing this by 4 gives the quarterly average for 19X6, ie:

$$(239 + 201 + 182 + 297)/4 = 229.75$$

This figure contains no seasonal component because we have averaged them out over the year. We are left with an estimate of the trend value for the middle of the year, that is, for the mid-point of quarters 2 and 3. If we move on for 3 months, we can calculate the average quarterly figure for April 19X6–March 19X7 (251), for July 19X6–June 19X7 (270.25), and so on. This process generates the 4 point moving averages for this set of data. The set of moving averages represent the best estimate of the trend in the demand.

We now use these trend figures to produce estimates of the seasonal components. We calculate:

$$A - T = S + E$$

Unfortunately, the estimated values of the trend given by the 4 point moving averages are for points in time which are different from those for the actual data. The first value, 229.75, represents a point in the middle of 19X6, exactly between the April–June and the July–September quarters. The second value, 251, falls between the July–September and the October–December actual figures. We require a deseasonalised average which corresponds with the figure for an actual quarter. The position of the deseasonalised averages is moved by reaveraging pairs of values. The first and second values are averaged, centring them on July–September 19X6, ie:

$$(229.75 + 251)/2 = 240.4$$

is the deseasonalised average for July–September 19X6. This deseasonalised value, called the **centred moving average**, can be compared directly with the actual value for July–September 19X6 of 182. Notice that this means that we have no estimated trend figures for the first 2 or last 2 quarters of the time series. The results of these calculations are shown in the table below:

TABLE 8.2 **Calculation of the centred 4 point moving average trend values for the model A = T + S + E**

Date	Quantity '000s A	4 quarter total	4 quarter moving average	Centred moving average, T	Estimated seasonal component A − T = S + E
Jan–Mar 19X6	239		—		
Apr–Jun	201		—		
		919	229.75		
Jul–Sep	182			240.4	−58.4
		1004	251		
Oct–Dec	297			260.6	+36.4
		1081	270.25		
Jan–Mar 19X7	324			279.6	+44.4
		1156	289		
Apr–Jun	278			299.9	−21.9
		1243	310.75		
Jul–Sep	257			320.4	−63.4
		1320	330		
Oct–Dec	384			340.3	+43.8
		1402	350.5		
Jan–Mar 19X8	401			360.2	+40.8
		1480	370		
Apr–Jun	360			379.8	−19.8
		1558	389.5		
Jul–Sep	335			399.5	−64.5
		1638	409.5		
Oct–Dec	462		—		
Jan–Mar 19X9	481		—		

For each quarter of the year, we have estimates of the seasonal components, which include some error or residual. There are two further stages in the calculations before we have usable seasonal components. We average the seasonal estimates for each season of the year. This should remove some of the errors. Finally we adjust the averages, moving them all up or down by the same amount, until their total is zero. This is done because we require the seasonal components to average out over the year. The correction factor required is: (the sum of the estimated seasonal values)/4. The estimates from the last column in TABLE 8.2 are shown under their corresponding quarter numbers. The procedure is shown in the table below:

TABLE 8.3 **Calculation of the average seasonal components**

	Year	Quarter of the year				
		1	2	3	4	
	19X6	—	—	−58.4	+36.4	
	19X7	+44.4	−21.9	−63.4	+43.8	
	19X8	+40.8	−19.8	−64.5	—	
Total		+85.2	−41.7	−186.3	+80.2	
Average		85.2 ÷ 2	−41.7 ÷ 2	−186.3 ÷ 3	80.2 ÷ 2	
Estimated seasonal component		+42.6	−20.8	−62.1	+40.1	Sum = −0.2
Adjusted seasonal component*		+42.6	−20.7	−62.0	+40.1	Sum = 0

* In this example, two of the seasonal components have been rounded up, and two have been rounded down, so that the sum equals zero.

228 QUANTITATIVE ANALYSIS

The seasonal components confirm our comments on the diagram at the end of section **8.2**. Both winter quarters are above the trend by about 40 thousand units and the two summer quarters are below the trend by 21 and 62 thousand units, respectively.

A similar procedure is used for seasonal variation over any time period. For example, if the seasons are days of the week, take a 7 point moving average to remove the daily 'seasonal' effect, rather than a 4 point moving average. This average will represent the trend at the middle of the week, that is, on day 4, therefore it is not necessary to centre these moving averages.

8.3.2 Deseasonalise the data to find the trend

Step 2 is to deseasonalise the basic data. This is shown below by deducting the appropriate seasonal component from each quarter's actual sales, ie A − S = T + E.

TABLE 8.4 **Calculation of the deseasonalised data**

Date	Quarter number	Quantity sold, '000 A	Seasonal component S	Deseasonalised quantity, '000 A − S = T + E
Jan–Mar 19X6	1	239	(+42.6)	196.4
Apr–Jun	2	201	(−20.7)	221.7
Jul–Sep	3	182	(−62.0)	244.0
Oct–Dec	4	297	(+40.1)	256.9
Jan–Mar 19X7	5	324	(+42.6)	281.4
Apr–Jun	6	278	(−20.7)	298.7
Jul–Sep	7	257	(−62.0)	319.0
Oct–Dec	8	384	(+40.1)	343.9
Jan–Mar 19X8	9	401	(+42.6)	358.4
Apr–Jun	10	360	(−20.7)	380.7
Jul–Sep	11	335	(−62.0)	397.1
Oct–Dec	12	462	(+40.1)	421.9
Jan–Mar 19X9	13	481	(+42.6)	438.4

These re-estimated trend values, with errors, can be used to set up a model for the underlying trend. The values are plotted on the original diagram, which now shows a clear linear trend:

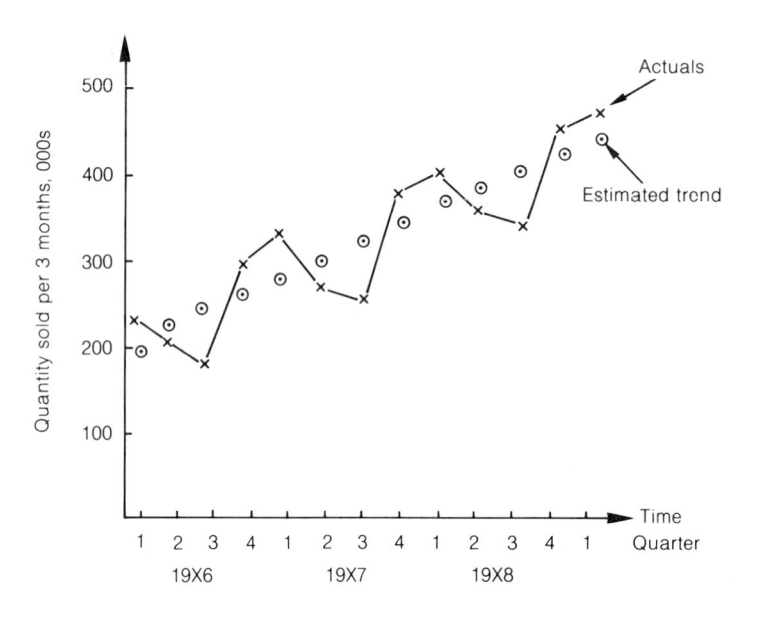

FIGURE 8.4 **Lewplan plc, actual and deseasonalised sales per 3 months**

The equation of the trend line is:

$$T = a + b \times \text{quarter number}$$

where a and b represent the intercept and slope of the line. The least squares method can be used to determine the line of 'best fit', therefore the equations for a and b, from the previous chapter on linear regression are:

$$b = \frac{n \sum xy - \sum x \sum y}{n \sum x^2 - (\sum x)^2} \quad \text{and} \quad a = \frac{\sum y}{n} - \frac{b \sum x}{n}$$

where x is the quarter number and y is (T + E) in the above table. Using a calculator, we find:

$$\sum x = 91 \quad \sum x^2 = 819 \quad \sum y = 4158.7 \quad \sum xy = 32747.1 \quad n = 13$$

It follows by substitution that:

$$b = 19.978 \text{ and } a = 180.046$$

Hence the equation of the trend model may be written:

$$\text{Trend quantity ('000s)} = 180.0 + 20.0 \times \text{quarter number}$$

8.3.3 Calculate the errors

Step 3 in the procedure, before using the model to forecast, is to calculate the errors or residuals. The model is:

$$A = T + S + E$$

We have calculated S in section **8.3.1** and T in section **8.3.2**. We can now deduct each of these from A, the actual quantity, to find the errors in the model.

TABLE 8.5 **Calculation of errors for the additive component model**

Date	Quarter number	Quantity sold, '000 A	Seasonal component, '000 S	Trend component, '000 T	Error, '000 A − S − T = E
Jan–Mar 19X6	1	239	(+42.6)	200	−3.6
Apr–Jun	2	201	(−20.7)	220	+1.7
Jul–Sep	3	182	(−62.0)	240	+4.0
Oct–Dec	4	297	(+40.1)	260	−3.1
Jan–Mar 19X7	5	324	(+42.6)	280	+1.4
Apr–Jun	6	278	(−20.7)	300	−1.3
Jul–Sep	7	257	(−62.0)	320	−1.0
Oct–Dec	8	384	(+40.1)	340	+3.9
Jan–Mar 19X8	9	401	(+42.6)	360	−1.6
Apr–Jun	10	360	(−20.7)	380	+0.7
Jul–Sep	11	335	(−62.0)	400	−3.0
Oct–Dec	12	462	(+40.1)	420	+1.9
Jan–Mar 19X9	13	481	(+42.6)	440	−1.6

The final column in the table can be used for *Step 4*, the calculation of the mean absolute deviation (MAD) or the mean square error (MSE) of the errors.

$$\text{MAD} = \frac{\sum |E_t|}{n} = \frac{28.7}{13} = 2.2$$

$$\text{MSE} = \frac{\sum (E_t)^2}{n} = \frac{78.85}{13} = 6.1$$

The errors are small at about 1 or 2%. The historical pattern is highly consistent and should give a good short-term forecast.

8.3.4 Forecasting using the additive model

The forecast for this additive component model is:

$$F = T + S \quad \text{('000 units per quarter)}$$

where the trend component $T = 180 + 20 \times$ quarter number, and the seasonal components, S, are $+42.6$ for January–March, -20.7 for April–June, -62.0 for July–September and $+40.1$ for October–December.

The quarter number for the next three monthly period, April–June 19X9, is 14, therefore the forecast trend is:

$$T_{14} = 180 + 20 \times 14 = 460 \quad \text{('000 units per quarter)}$$

The appropriate seasonal component is -20.7 ('000 units). Therefore the forecast for this quarter is:

$$F(\text{April–June 19X9}) = 460 - 20.7 = 439.3 \text{ ('000 units)}$$

It is important to remember that the further ahead the forecast, the more unreliable it becomes. We are assuming that the historical pattern continues uninterrupted. This assumption may hold for short periods but is less and less likely to be true the further we go into the future.

8.4 THE ANALYSIS OF A MULTIPLICATIVE COMPONENT MODEL: A = T × S × E

In some time series, the seasonal component is not a fixed amount each year. Instead it is a percentage of the trend values. As the trend increases, so does the seasonal variation.

Example 8.3: Setting up a multiplicative component model for a time series.

CD plc sell a range of products. The quarterly sales of one product for the last 13 quarters are given in TABLE 8.6:

TABLE 8.6 **CD plc quarterly sales**

Date	Quarter number	Quantity sold, '000 A
Jan–Mar 19X6	1	70
Apr–Jun	2	66
Jul–Sep	3	65
Oct–Dec	4	71
Jan–Mar 19X7	5	79
Apr–Jun	6	66
Jul–Sep	7	67
Oct–Dec	8	82
Jan–Mar 19X8	9	84
Apr–June	10	69
Jul–Sep	11	72
Oct–Dec	12	87
Jan–Mar 19X9	13	94

The scatter diagram for these data is:

FIGURE 8.5 **CD plc sales per 3 months**

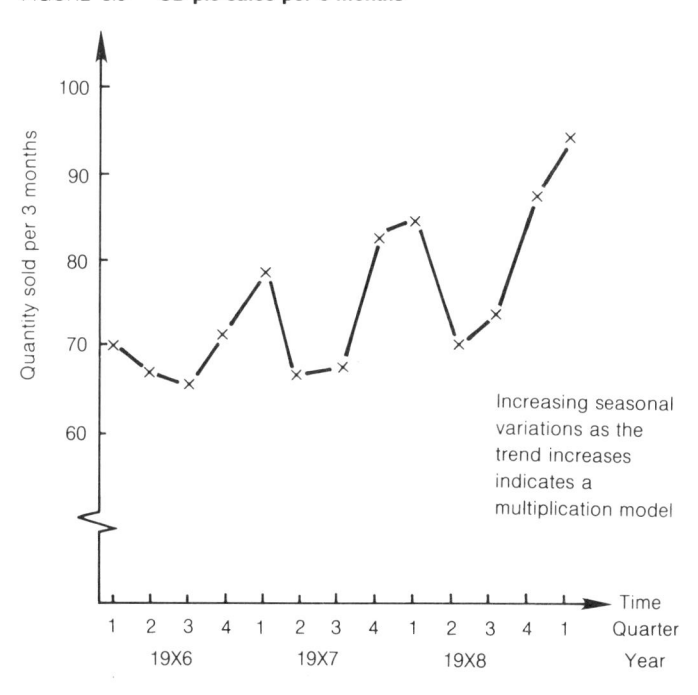

Increasing seasonal variations as the trend increases indicates a multiplication model

This product has a similar seasonal pattern to the previous example, with high winter values and low summer values, but the size of variations about the trend line are increasing. A multiplicative component model should be suitable for these data.

Actual values = trend × seasonal variation × error

that is:

$$A = T \times S \times E$$

In this example the trend looks linear but this will become clearer when we have smoothed the series.

8.4.1 Calculation of the seasonal components

Initially the same procedure is followed as for the additive model. The centred moving average trend values are calculated but the estimated seasonal components are ratios derived from A/T = S × E. The calculations are shown in the table below:

TABLE 8.7 **Calculation of the seasonal components, CD plc.**

Date	Quarter number	Quantity sold '000 A	4 point moving average	Centred 4 point moving average T	Seasonal ratio A/T = S × E
Jan–Mar 19X6	1	70			
Apr–Jun	2	66			
			68.00		
Jul–Sep	3	65		69.13	0.940
			70.25		
Oct–Dec	4	71		70.25	1.011
			70.25		
Jan–Mar 19X7	5	79		70.50	1.121
			70.75		
Apr–Jun	6	66		72.13	0.915
			73.50		
Jul–Sep	7	67		74.13	0.904
			74.75		
Oct–Dec	8	82		75.13	1.092
			75.50		
Jan–Mar 19X8	9	84		76.13	1.103
			76.75		
Apr–Jun	10	69		77.38	0.892
			78.00		
Jul–Sep	11	72		79.25	0.909
			80.50		
Oct–Dec	12	87		—	—
Jan–Mar 19X9	13	94		—	—

The seasonal components ratios are derived from the quarterly estimates in a similar way to those for the additive model. Since the seasonal values are ratios and there are 4 seasons, we require the seasonal components to total to 4 rather than zero. (If the data comprised 7 daily seasons in each week, then the seasonal components would be required to total to 7.) If the total is not 4, the values are adjusted as before. The estimates from the last column above are shown under their corresponding quarter numbers below:

TABLE 8.8 **Calculation of seasonal components, CD plc**

		Quarter of the year				
	Year	1	2	3	4	
	19X6	—	—	0.940	1.011	
	19X7	1.121	0.915	0.904	1.092	
	19X8	1.103	0.892	0.909	—	
Total		2.224	1.807	2.753	2.103	
Average		2.224 ÷ 2	1.807 ÷ 2	2.753 ÷ 3	2.103 ÷ 2	
Estimated seasonal component ratio		1.112	0.903	0.918	1.051	Sum = 3.984
Adjusted seasonal component ratio*		1.116	0.907	0.922	1.055	Sum = 4

* The adjusted value is obtained by multiplying each estimated seasonal component ratio by (4/3.984).

The seasonal effect on the sales of the January-March quarter is estimated to increase sales by 11.6% of the trend value (1.116). Similarly the seasonal effect of the October-December quarter is to raise sales by 5.5% of the trend. For the other 2 quarters, the seasonal effect is to depress the sales below the trend values to 90.7% and 92.2% of trend respectively.

8.4.2 Deseasonalise the data and fit the trend line

We have now found estimates of the seasonal component and can deseasonalise the data by calculating $A/S = T \times E$. These estimated trend values are calculated below:

TABLE 8.9 **Calculation of the trend for CD plc**

Date	Quarter number	Quantity sold '000 A	Seasonal component ratio S	Deseasonalised quantity '000 $A/S = T \times E$
Jan–Mar 19X6	1	70	1.116	62.7
Apr–Jun	2	66	0.907	72.8
Jul–Sep	3	65	0.922	70.6
Oct–Dec	4	71	1.055	67.3
Jan–Mar 19X7	5	79	1.116	70.8
Apr–Jun	6	66	0.907	72.8
Jul–Sep	7	67	0.922	72.7
Oct–Dec	8	82	1.055	77.7
Jan–Mar 19X8	9	84	1.116	75.2
Apr–Jun	10	69	0.907	76.1
Jul–Sep	11	72	0.922	78.2
Oct–Dec	12	87	1.055	82.4
Jan–Mar 19X9	13	94	1.116	84.2

The trend values are superimposed on the original scatter diagram:

FIGURE 8.6 **CD plc actual and deseasonalised sales per 3 months**

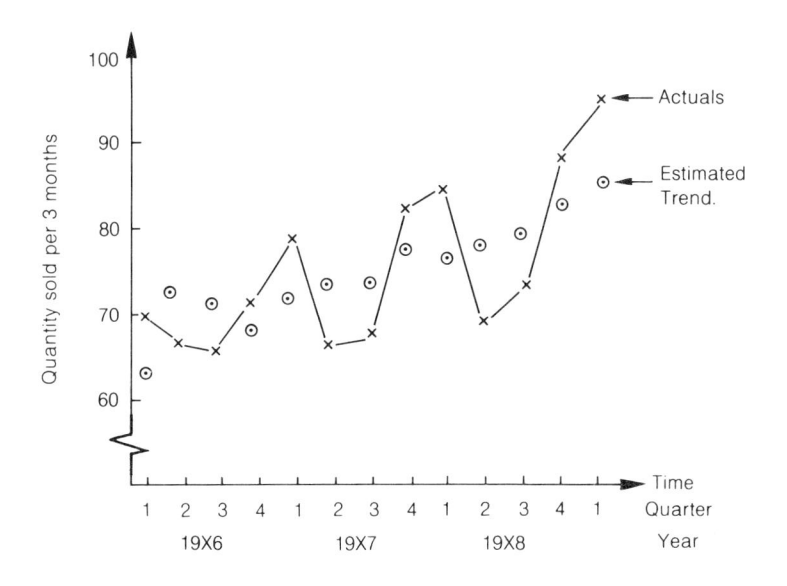

The trend which emerges is erratic. The sales values in this time series are not as consistent as the ones in the first example for Lewplan plc. CD plc is probably a more realistic example.

We now have to decide how to model the trend. It is not a curve but looks roughly linear, even though the values are erratic, particularly in 19X6. For simplicity, we will assume that the trend is linear and use the least squares method to fit the 'best' line to the data. The trend line, using the same procedure as in section **8.3.2**, is:

$$T = 64.6 + 1.36 \times \text{quarter number} \qquad (\text{'000 units per 3 months})$$

We use this equation to estimate the value of the trend sales for each of the periods.

8.4.3 Calculation of the errors: A/(T × S) = E or A − (T × S) = E

We have now calculated the trend and seasonal components. We can use these to find the errors between the observed sales, A, and the sales which are forecast by the model, T × S. The table below gives the errors, both as a proportion, E = A/(T × S), and as absolute values, A − (T × S).

TABLE 8.10 **Errors for CD plc**

Date	Quarter number	Quantity sold '000 A	Seasonal component, '000 S	Trend component, '000 T	T × S	Errors A/(T × S)	A − (T × S)
Jan–Mar 19X6	1	70	1.116	66.0	73.7	0.95	−3.7
Apr–Jun	2	66	0.907	67.3	61.0	1.08	+5.0
Jul–Sep	3	65	0.922	68.7	63.3	1.03	+1.7
Oct–Dec	4	71	1.055	70.0	73.9	0.96	−2.9
Jan–Mar 19X7	5	79	1.116	71.4	79.7	0.99	−0.7
Apr–Jun	6	66	0.907	72.8	66.0	1.00	0
Jul–Sep	7	67	0.922	74.1	68.3	0.98	−1.3
Oct–Dec	8	82	1.055	75.5	79.7	1.03	+2.3
Jan–Mar 19X8	9	84	1.116	76.8	85.7	0.98	−1.7
Apr–Jun	10	69	0.907	78.2	70.9	0.97	−1.9
Jul–Sep	11	72	0.922	79.6	73.3	0.98	−1.3
Oct–Dec	12	87	1.055	80.9	85.4	1.02	+1.6
Jan–Mar 19X9	13	94	1.116	82.3	91.9	1.02	+2.1

The errors are high in the first year, as we could have guessed from the plot of the deseasonalised figures. However, from the January–March 19X7 quarter, the errors are all 2 or 3% of the actual values and the model looks reasonably satisfactory.

8.4.4 Forecasting with the multiplicative component model

Forecasting with either model assumes that we can fit an equation to the trend values. In both of the examples used we have been lucky. The trend has been clearly linear. If the trend had been a curve, we would have had to guess the relationship, and use some of the techniques mentioned in the previous chapter for dealing with non-linear relationships. Once we have established the trend equation, the calculation of the forecast is straightforward. The forecast is:

$$F = T \times S$$

where:

$$T = 64.6 + 1.36 \times \text{quarter number} \qquad (\text{'000 units per 3 months})$$

and the seasonal component ratios are 1.116 for quarter 1, 0.907 for quarter 2, 0.922 for quarter 3, and 1.055 for quarter 4. The next quarter is April-June 19X9, which is quarter 14 in the series and quarter 2 in the year. The forecast sales are:

$$F = T \times S$$
$$= (64.6 + 1.36 \times 14) \times 0.907$$
$$= 83.64 \times 0.907 = 75.9 \text{ ('000 units per quarter)}$$

Given the errors for the model, we would hope that this estimate will be within 2 or 3% of the actual value. Similarly, the forecast for October-December 19X9 is found using quarter number 16 and the seasonal component for quarter 4:

$$F = T \times S$$
$$= (64.6 + 1.36 \times 16) \times 1.055$$
$$= 86.36 \times 1.055 = 91.1 \text{ ('000 units per quarter)}$$

We expect the error on this forecast to be larger than the previous one because it is further into the future.

8.5 FORECASTING USING SIMPLE EXPONENTIAL SMOOTHING

If the methods described in the previous sections produce unsatisfactory models for a time series, we can consider a smoothing technique, similar to moving averages, for forecasting one period ahead. This technique is called **exponential smoothing**. The exponential smoothing method can be extended to deal with time series which contain trends and seasonal patterns, but we will look only at the simplest model, which is intended for time series without obvious trend or seasonal patterns. The forecasting formula for simple exponential smoothing is:

$$F_t = a A_t + (1 - a)F_{t-1}$$

where:

$$0 \leqslant a \leqslant 1$$

F_t is the forecast made at time t for the following period. A_t is the observed value at time t, and a is the smoothing constant which must lie between 0 and 1.

The technique is called exponential smoothing because of the weights which are attached to each past observation. In the moving average method, used in sections **8.3** and **8.4**, we gave equal weight to each of the four observations and completely ignored any earlier values. The effect of the exponential smoothing formula is to give a weight to all of the observations in the time series. The weights are a, for A_t, $a(1-a)$ for A_{t-1}, $a(1-a)^2$ for A_{t-2}, $a(1-a)^3$ for A_{t-3}, and so on. This is demonstrated below:

$$F_t \quad = a A_t \quad + (1 - a)F_{t-1} \tag{1}$$

$$F_{t-1} = a A_{t-1} + (1 - a)F_{t-2} \tag{2}$$

$$F_{t-2} = a A_{t-2} + (1 - a)F_{t-3} \tag{3}$$

Substituting (3) into (2) gives:

$$F_{t-1} = a A_{t-1} + a(1-a)A_{t-2} + (1-a)^2 F_{t-3} \tag{4}$$

Substituting (4) into (1) gives:

$$F_t = a A_t + a(1-a)A_{t-1} + a(1-a)^2 A_{t-2} + (1-a)^3 F_{t-3}$$

All of the past values of the series are contained in the forecast. The relative weight given to each past value depends on a. If a is near to 1, then a heavy weight is put on the most recent observation and little on the past values. If a is near 0, more weight is put on past values and little on the latest data. If a is small, it smooths out fluctuations but will not respond to sudden changes in the series. If a is large, it will not smooth the series but will follow sudden changes more quickly. The choice of smoothing constant depends on the criterion which we decide to use to judge the model, and on the kind of series we are dealing with. If the data are stable and unlikely to jump, a small value of a will be suitable. If the series is erratic, we may prefer a larger value of the smoothing constant.

Computationally, the exponential smoothing method is very useful because it is not necessary to retain the past data. It is all contained in the one figure, F_{t-1}. The storage and calculations needed to run the system are small compared with our previous models. This property makes the method useful for stock control forecasting, where the system may have to deal with thousands of items.

8.5.1 Exponential smoothing of a time series

Our first example is a time series which has no trend or other obvious pattern.

Example 8.4: Forecasting using simple exponential smoothing

The data below are the daily demands for component Z at the stores of Modem plc. The plot given towards the end of the section shows an erratic time series without any obvious trend.

TABLE 8.11   **Daily demands for component Z**

Day	1	2	3	4	5	6	7	8	9
Demand for component Z	15	6	11	7	17	7	12	15	8

We will use the simple exponential smoothing model to forecast, from day to day, the next day's demand. The forecast is:

$$F_t = a A_t + (1 - a)F_{t-1}$$

We have to choose two values to start the procedure. We must estimate the first forecast, F_0. From past experience we will put $F_0 = 10$. This is the forecast made on day 0 for day 1, and is compared with the actual value, $A_1 = 15$. This start value is not as important as the choice of a, since its effect soon becomes small as the forecasts advance. a is usually chosen by trying out a number of values on the sample data and using the one which gives the lowest MAD or MSE for the forecasting errors. In practice, this is done by computer. We will use $a = 0.1$ and $a = 0.9$ in our calculations to demonstrate the effect of two extreme values.

TABLE 8.12   **Daily demand for component Z at Modem plc—forecasting a time series, with $F_0 = 10$**

Day	Actual demand A	$a = 0.1$ $F_t = 0.1A_t + 0.9F_{t-1}$ F	Errors A − F	$a = 0.9$ F	Errors A − F
1	$A_1 = 15$	F_0 = 10	+5.0	10	+5.0
2	$A_2 = 6$	$F_1 = 0.1A_1 + 0.9F_0 = 10.5$	−4.5	14.5	−8.5
3	$A_3 = 11$	$F_2 = 0.1A_2 + 0.9F_1 = 10.05$	+0.95	6.85	+4.15
4	$A_4 = 7$	$F_3 = 0.1A_3 + 0.9F_2 = 10.15$	−3.15	10.59	−3.59
5	$A_5 = 17$	$F_4 = 0.1A_4 + 0.9F_3 = 9.83$	+7.17	7.36	+9.64
6	$A_6 = 7$	$F_5 = 0.1A_5 + 0.9F_4 = 10.55$	−3.55	16.04	−9.04
7	$A_7 = 12$	$F_6 = 0.1A_6 + 0.9F_5 = 10.19$	+1.81	7.90	+4.10
8	$A_8 = 15$	$F_7 = 0.1A_7 + 0.9F_6 = 10.37$	+4.63	11.59	+3.41
9	$A_9 = 8$	$F_8 = 0.1A_8 + 0.9F_7 = 10.84$	−2.84	14.66	−6.66

The actual data and the two forecasts are plotted below:

FIGURE 8.7   Component Z—Modem plc—actual demand with two forecasts

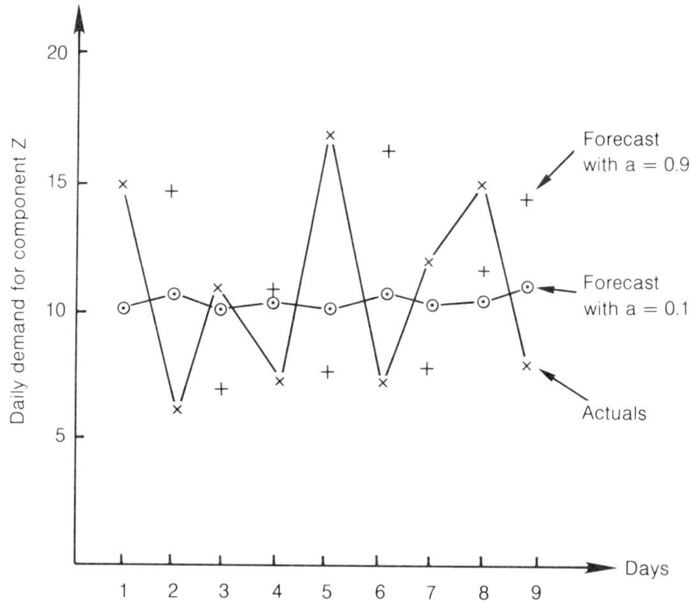

The mean absolute deviation (MAD) and the mean square errors (MSE) for each model are give below:

TABLE 8.13   Component Z—comparison of the two forecasting models

Smoothing constant	MAD	MSE
a = 0.1	3.7	17.0
a = 0.9	6.0	41.6

The small smoothing constant gives the better model. This is because the time series has large erratic fluctuations which are best represented by its average value of between 10 and 11. The model with the large smoothing constant, a = 0.9, tries to follow the fluctuations in the patttern. It cannot do this because the fluctuations are so erratic, therefore it is a less suitable model.

8.5.2   Exponential smoothing with a step change in the time series

One of the difficulties for all forecasting methods is a step change in the time series. How quickly does the exponential smoothing method deal with this type of change? We will illustrate the problem using different values of a in the simple exponential smoothing model.

Example 8.5: Forecasting with simple exponential smoothing when there is a step change in the time series

We will use the data from **Example 8.4**, but from day 5 onwards, we will add 10 units to all of the values. This is the step change in the demand.

TABLE 8.14   Data from day 5 onwards, with a step change of +10

Day	1	2	3	4	5	6	7	8	9
Modified demand	15	6	11	7	27	17	22	25	18

Step change of +10

238 QUANTITATIVE ANALYSIS

We will use the same start forecast, $F_0 = 10$, and produce forecasts for $a = 0.1$ and $a = 0.9$. The forecasts, together with the values of the MAD and MSE, are given in the table below:

TABLE 8.15 **Daily demand for component Z at Modem plc—time series with step change, $F_0 = 10$**

Day	Demand A	$a = 0.1$ Forecast	Error	$a = 0.9$ Forecast	Error
1	15	10.00	+5.00	10.00	+5.00
2	6	10.50	−4.90	14.50	−8.50
3	11	10.05	+0.95	6.85	+4.15
4	7	10.15	−3.15	10.59	−3.59
		------ step change ----			
5	27	9.83	+17.17	7.36	+19.64
6	17	11.55	+5.45	25.04	−8.04
7	22	12.09	+9.91	17.80	+4.20
8	25	13.08	+11.92	21.58	+3.42
9	18	14.28	+3.72	24.66	−6.66
	MAD		6.9		7.0
	MSE		70.5		72.4

The suitability of these forecasts can be seen in the following plot:

FIGURE 8.8 **Actual demand with two forecasts for component Z—Modem plc—time series with a step change at day 5**

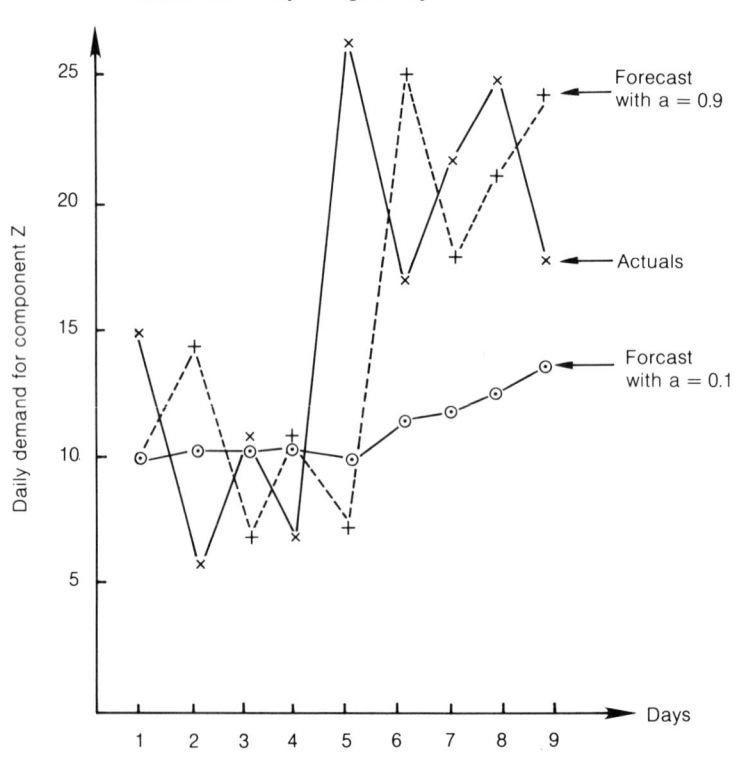

The important point in this example is the speed with which the model reacts and adjusts to the new level of demand after the step. When $a = 0.1$, the forecast is very slow to increase to the new value. In fact, in this small sample, the forecast does not reach the new level at all. The forecasts after the step are much too small. When a = 0.9, the model reacts almost immediately. It still has large errors after the step but they are both positive and negative.

The values of the MAD and the MSE for the two models are almost identical even though they react in such different ways. We may prefer the characteristics of one model rather than another for reasons other than the ones considered so far. For example, the criterion may be to find the model which gives the smallest mean square deviation. In this case, we repeat the calculations described above, for different values of the smoothing constant. This is usually done in steps of 0.1 using a computer. For this example, a = 0.4 gives the minimum value of the MSE of 53.4. This 'best' model responds in between the two extremes. It jumps much more quickly that the a = 0.1 model, but does not over-react to all the erratic movements in the data as the a = 0.9 model does.

SUMMARY

A time series is any set of data which is recorded over time. It may, for example, be annual, quarterly, monthly or weekly data. Models use the historical pattern of the time series to forecast how the variable will behave in the future. The short-term forecasts will be more accurate than the longer-term ones. The further ahead we forecast, the less likely it is that the historical pattern will remain unchanged, and the larger will be the errors.

There are two basic models which can be used to represent a time series, the additive model and the multiplicative model. In both cases, it is assumed that the value of the variable is made up of a number of components. The series may contain: trend—general movement in the value of the variable; seasonal variation—short-term periodic fluctuations in the variable values; cyclical variation—long-term periodic fluctuations in the variable values; and error components—residual term. Data sets large enough to include the cyclical component were not considered in this text.

The component models are:

$$\text{Additive } A = T + S + E$$

$$\text{Multiplicative } A = T \times S \times E$$

In both cases, moving averages are used to deseasonalise the time series. These deseasonalised data are used to set up a model to describe the trend. The model is used to forecast future trend. If a linear model is suitable, we use the least squares method to fit the 'best' line. The fitting of curves is much more difficult.

The simple exponential smoothing model is used to make forecasts for time series which do not have obvious components. This technique tracks the series without any attempt to explain the movements. The forecast in period t for the following period is:

$$F_t = a A_t + (1 - a) F_{t-1}$$

where:

$$0 \leqslant a \leqslant 1$$

This model uses all of the past data which is contained in the previous forecast, F_{t-1}. The modeller selects a value for a, the smoothing constant, which best suits the situation.

Statistical methods are not normally used to assess the validity or accuracy of the forecasting model, in contrast to the situation for linear regression models. The forecaster chooses the best model. Two measures give guidance on how well a model fits the past data. These are:

$$\text{Mean absolute deviation (MAD)} = \frac{\sum |E_t|}{n}$$

and

$$\text{Mean square error (MSE)} = \frac{\sum (E_t)^2}{n}$$

EXERCISES

Exercise 8.1

Amada plc had the following turnover, corrected for inflation, in the last 11 three month periods.

Year	1				2				3		
Quarter	1	2	3	4	1	2	3	4	1	2	3
Turnover	22	28	34	27	31	43	43	41	46	53	56

Required:
1. Fit an additive component model to the data, assuming a linear trend.
2. Forecast the next three periods. Comment on the likely accuracy of your forecasts.

Answers on page 525.

Exercise 8.2

The demand for chairs from Peace Retailers has been as follows:

Year	1		2				3			
Quarter	3	4	1	2	3	4	1	2	3	4
Demand	157	137	156	151	153	141	154	152	154	142

Required:
1. Fit a suitable additive component model.
2. Forecast the first two quarters of year 4.

Answers on page 527.

Exercise 8.3

The quarterly outputs of Cobournes plc are given below.

Year	1				2				3		
Quarter	1	2	3	4	1	2	3	4	1	2	3
Output	24	50	56	63	79	89	79	80	93	100	88

Required:
1. Analyse the quarterly outputs using an additive component model.
2. Comment on the trend.

Answers on page 529.

Exercise 8.4

Doble-Flood have not been performing well. Their profits (corrected for inflation) for the last 10 three monthly periods are given below.

Year	1				2				3	
Quarter	1	2	3	4	1	2	3	4	1	2
Profit	146	106	123	89	97	74	80	53	56	35

Required:
1. Fit a multiplicative component model to the time series.
2. Forecast the next two quarters of year 3.

Answers on page 530.

Exercise 8.5

Banham and Barsey's output has been improving year by year. The figures are given below.

Year	1			2				3		
Quarter	2	3	4	1	2	3	4	1	2	3
Output	400	715	600	585	560	975	800	765	720	1235

Required:
1. Fit a suitable component model to the data.
2. Forecast the next two quarters.

Answers on page 532.

Exercise 8.6

Using the Peace Retailers data in **Exercise 8.2**:

Required:
1. Calculate the errors, the mean absolute deviation (MAD) and the mean square errors (MSE), for the additive component model calculated in **Exercise 8.2**.
2. Use a simple exponential smoothing model, with $F_0 = 150$ and $a = 0.2$, to calculate forecasts, errors and the corresponding MAD and MSE.
3. Compare the values in **1** and **2** and decide which is the more appropriate model.

Answers on page 534.

Exercise 8.7

Purchase plc records the weekly demand for each of its products. Data are given below for its printer systems.

Week	1	2	3	4	5	6	7	8	9	10
Demand	15	16	14	18	12	14	10	11	13	12

Required:
1. Use simple exponential smoothing with $F_0 = 15$, and $a = 0.1$, to forecast the series. Calculate the MSE and comment on the suitability of the smoothing constant.
2. Repeat for $a = 0.5$.

Answers on page 535.

Exercise 8.8

Droco plc have just taken over Cindalites. As part of its rationalisation programme, Droco have closed the stores at its new acquisition and insist that everything is ordered through Droco's central stores. Unfortunately, the store's computer is still using its normal forecasting system, which is based on simple exponential smoothing with $a = 0.2$. The demand on both stores separately, before the closure, and the total demand on Droco's central store after the closure, is given below, for one common spare part.

Week	1	2	3	4	5	6	7	8	9	10	11
At Droco	30	32	35	28	33	42	40	39	45	38	43
At Cindalite	10	12	9	11	8	closed					

Required:
1 What forecasts will the Droco store's computer make for these weeks, assuming $F_0 = 30$?
2 Suggest how the Droco computer system should have been changed. How would your suggestion affect the forecasts?

Answers on page 536.

Part Two: Operational research models

PART TWO

Operational research models

The second part of this book is concerned with operational research models. In the following chapters, we cover linear programming, the transportation and assignment problems, inventory control, queueing and simulation.

There has never been a generally accepted definition of operational research. However, in the context in which we are involved, we may take as a working definition that operational research is the application of the scientific method to business problems. Operational research often, but not always, involves mathematical and statistical models. The models normally deal with complex situations, where there may be many interacting objectives and variables. Different people within an organisation may view the same situation in quite different ways. Problems change with time, as do the people involved with them.

We have already looked at problems of variability and uncertainty in decision making, and at some of the statistical methods of dealing with the effects. In the chapters on regression and time series, in particular, we developed models to help explain variability. The OR models, however, have been developed to deal with many different types of complex situation. There is a standard set of techniques with which operational researchers are expected to be familiar. However, it is rare in practice that a problem fits neatly into the framework of these models. It is more usual that the techniques give the modeller ideas about possible ways of approaching new challenges. The model must be custom made for each situation.

Linear programming

9.1 INTRODUCTION

There are many activities in an organisation which involve the allocation of resources. These resources include labour, raw materials, machinery and money. The allocation of these resources is sometimes referred to as **programming**. Problems arise because the resources are usually in limited supply. If a company makes several products, using the same machines and labour force, management must decide how many of each product to produce. The decision will be made so that a management objective is satisfied. Management may wish to arrange production in order to maximise the total contribution made each month, or to maximise the utilisation of the machinery each week, or to minimise the cost of labour each week. The decision variables are the amounts of each product to be made in a given time period.

Similarly, if the company has an amount of capital to invest in a number of projects, the money allocated to each project will be governed by some objective. It may be necessary to minimise risk, or maximise capital growth. The decision variables are the amounts of money allocated to each project.

In general, the objective is to determine the most efficient method of allocating these resources to the variables so that some measure of performance is optimised. Modelling methods can often be used to aid in this allocation process. Mathematical programming is the use of mathematical models and techniques to solve programming problems. There are a number of different techniques which fall under the general heading of mathematical programming but we will consider only the one which is most commonly used—**linear programming**.

Linear programming is a suitable method for modelling an allocation problem if the objective and the constraints on the resources can all be expressed as linear relationships of the variables. The technique has a number of distinct steps:

1. The linear programme must first be formulated mathematically. This means that the variables over which we have control and the objective must be identified. The objective and the constraints on the resources are then written down as linear relationships in terms of the variables.
2. Once the linear programme is complete, all of the feasible combinations of the variables are identified. The combination which optimises the objective may then be selected. If only two variables are involved, a graphical solution is possible. If, however, we have a multivariable problem, we must resort to an algebraic method for which a computer package will be used.
3. Once the optimum solution has been identified, it must be evaluated. This will include a sensitivity analysis.

As with any other mathematical aid to decision making, the solution from the linear programme is just one piece of management information which contributes to the final decision. We will begin by considering the formulation of a linear programme.

9.2 PROBLEM FORMULATION

The basic procedure is the same for the formulation of all linear programmes:
Step 1: Identify the variables in the problem for which the values can be chosen, within the limits of the constraints.
Step 2: Identify the objective and the constraints on the allocation.
Step 3: Write down the objective in terms of the variables.
Step 4: Write down the constraints in terms of the variables.

248 QUANTITATIVE ANALYSIS

The same procedure is used no matter how many variables there are, but we will look first at a two variable problem.

Example 9.1: To formulate a two variable linear programme

A small family firm produces two old-fashioned non-alcoholic drinks, 'Pink Fizz' and 'Mint Pop'. The firm can sell all that it produces but production is limited by the supply of a major ingredient and by the amount of machine capacity available. The production of 1 litre of 'Pink Fizz' requires 0.02 hours of machine time, whereas the production of 1 litre of 'Mint Pop' requires 0.04 hours of machine time. 0.01 kg of a special ingredient is required for 1 litre of 'Pink Fizz'. 'Mint Pop' requires 0.04 kg of this ingredient per litre. Each day the firm has 24 machine hours available and 16 kg of the special ingredient. The contribution is £0.10 on 1 litre of 'Pink Fizz' and £0.30 on 1 litre of 'Mint Pop'. How much of each product should be made each day, if the firm wishes to maximise the daily contribution?

Solution:

Step 1: Identify the variables. Within the limits of the constraints, the firm can decide how much of each type of drink to make. Let p be the number of litres of 'Pink Fizz' produced per day. Let m be the number of litres of 'Mint Pop' produced per day.

Step 2: Identify the objective and the constraints. The objective is to maximise the daily contribution. Let £P per day be the contribution. This is maximised within constraints on the amounts of machine time and the special ingredient available.

Step 3: Express the objective in terms of the variables:

$$P = 0.10p + 0.30m \quad (\text{£/day})$$

This is the objective function—the quantity we wish to optimise.

Step 4: Express the constraints in terms of the variables. The contribution is maximised subject to the following constraints on production:

(a) Machine time: to produce p litres of 'Pink Fizz' and m litres of 'Mint Pop' requires $(0.02p + 0.04m)$ hours of machine time each day. There is a maximum of 24 machine hours available each day, therefore the production must be such that the number of machine hours required is less than or equal to 24 hours per day. Therefore:

$$0.02p + 0.04m \leqslant 24 \text{ hours/day}$$

(b) Special ingredient: to produce p litres of 'Pink Fizz' and m litres of 'Mint Pop' requires $(0.01p + 0.04m)$ kg of the ingredient each day. There is a maximum of 16 kg available each day, therefore the production must be such that the amount of the special ingredient required is at most 16 kg per day. Therefore:

$$0.01p + 0.04m \leqslant 16 \text{ kg/day}$$

There are no further constraints on the production, but it is sensible to assume that the firm cannot make negative amounts of drink, therefore.

(c) Non-negativity:

$$p \geqslant 0$$
$$m \geqslant 0$$

The complete linear programme is as follows. Maximise:

$$P = 0.10p + 0.30m \quad (\text{£/day})$$

subject to:

$$\text{Machine time: } 0.02p + 0.04m \leqslant 24 \text{ hours/day}$$

$$\text{Special ingredient: } 0.01p + 0.04m \leqslant 16 \text{ kg/day}$$

$$p, m \geqslant 0$$

Example 9.2: To formulate a two variable linear programme

A manufacturer of high precision machined components produces two different types, X and Y. In any given week there are 4000 man-hours of skilled labour available. Each component X requires 1 man-hour to produce it and each component Y requires 2 man-hours. The manufacturing plant has the capacity to produce a maximum of 2250 components of type X each week as well as 1750 components of type Y. Each component X requires 2 kg of metal rod and 5 kg of metal plate, whereas each component Y needs 5 kg of rod and 2 kg of plate. Each week there are 10,000 kg each of rod and plate available. The company supplies a car manufacturer with 600 components of type X each week on a regular basis. In addition, there is an agreement with the unions that at least 1500 components will be produced each week in total.

If the unit contribution for component X is £30 and for component Y is £40, how many of each type should be made in order to maximise the total contribution per week?
Solution: First the linear programme must be formulated.

Step 1: Choose the variables: produce x components of type X per week and y components of type Y per week.
Step 2: What is the objective? What are the constraints on the production? The objective is to maximise the total weekly contribution. The production is constrained by the amount of:

(a) labour—maximum available is 4000 hours per week.
(b) machine capacity—there is a separate limit for each product. The machines can make at most 2250 components of type X each week and at most 1750 components of type Y each week.
(c) rod—maximum available is 10,000 kg per week.
(d) plate—maximum available is 10,000 kg per week.

In addition, minimum amounts of each product are required:

(a) Regular orders—the number of component X made must be at least enough to satisfy the regular orders;
(b) Union agreement—the total number of components (x + y) must at least satisfy the agreement.

Step 3: The objective function. Let £P be the total contribution per week, where:

$$P = 30x + 40y \quad (\text{£/week})$$

Step 4: The constraints on production. For each resource constraint, the amount of resource required each week to produce x components of type X and y components of type Y is given below, together with the maximum amount of the resource available.

$$\text{Labour required: } 1x + 2y \leqslant 4000 \text{ hours/week}$$
$$\text{Machine capacity required: } x \leqslant 2250 \text{ components/week}$$
$$y \leqslant 1750 \text{ components/week}$$
$$\text{Rod required: } 2x + 5y \leqslant 10,000 \text{ kg/week}$$
$$\text{Plate required: } 5x + 2y \leqslant 10,000 \text{ kg/week}$$

In addition:

$$\text{Regular orders: } x \geqslant 600 \text{ components/week}$$
$$\text{Union agreement: } x + y \geqslant 1500 \text{ components/week}$$
$$\text{Non-negativity: } x, y \geqslant 0$$

The complete linear programme is:

Produce x components of type X and y components of type Y each week. Maximise:

$$P = 30x + 40y \quad (\text{£/week})$$

subject to the constraints:

$$\begin{aligned}
\text{Labour: } 1x + 2y &\leq 4000 \text{ hours/week} \\
\text{Machine capacity: } x &\leq 2250 \text{ components/week} \\
y &\leq 1750 \text{ components/week} \\
\text{Rod: } 2x + 5y &\leq 10000 \text{ kg/week} \\
\text{Plate: } 5x + 2y &\leq 10000 \text{ kg/week} \\
\text{Regular orders: } x &\geq 600 \text{ components/week} \\
\text{Union agreement: } x + y &\geq 1500 \text{ components/week} \\
\text{Non-negativity } x, y &\geq 0
\end{aligned}$$

We will look next at a problem involving more than two variables. The procedure is identical.

Example 9.3: To formulate a multivariable linear programme

Electra plc produce personal computers and word processors. Currently four models are being produced:

(a) the Jupiter—512K memory, single disk drive;
(b) the Venus—512K memory, double disk drive;
(c) the Mars—640K memory, double disk drive;
(d) the Saturn—640K memory, hard disk.

Each computer passes through three departments in the factory—sub-assembly, assembly and test. The details of the times required in each department for each model are given in TABLE 9.1, together with the maximum capacities of the departments. The marketing department has assessed the demand for each model. The maximum forecast demands are also given in the table, together with the unit contribution for each model:

TABLE 9.1 **Times required in each department for each model**

Department	Jupiter	Venus	Mars	Saturn	Maximum available hours/month
Sub-assembly	5	8	20	25	800
Assembly	2	3	8	14	420
Test	0.1	0.2	2	4	150
Maximum forecast demand/month	100	45	25	20	
Contribution £/unit	15	30	120	130	

Construct the linear programme for this product mix problem, if the objective is to maximise the total contribution per month.

Solution:
Step 1: Choose the variables. Produce:

j units of the Jupiter per month,
v units of the Venus per month,
m units of the Mars per month and
s units of the Saturn per month.

Step 2: What is the objective? What are the constraints on the production? The objective is to maximise the total contribution each month. The production is constrained by the number of labour hours available in the three departments and by the number of each model which can be sold.
Step 3: The objective function. Let £P be the total contribution per month, where:

$$P = 15j + 30v + 120m + 130s \quad (£/month)$$

Step 4: The constraints on production. For each department, the amount of time required to produce j, v, m and s units of the respective models is linked to the maximum time available.

$$\begin{aligned}
\text{Sub-assembly:} \quad & 5j + 8v + 20m + 25s \leqslant 800 \text{ hours/month} \\
\text{Assembly:} \quad & 2j + 3v + 8m + 14s \leqslant 420 \text{ hours/month} \\
\text{Test:} \quad & 0.1j + 0.2v + 2m + 4s \leqslant 150 \text{ hours/month} \\
\text{Demand for Jupiter:} \quad & j \leqslant 100 \text{ units/month} \\
\text{Demand for Venus:} \quad & v \leqslant 45 \text{ units/month} \\
\text{Demand for Mars:} \quad & m \leqslant 25 \text{ units/month} \\
\text{Demand for Saturn:} \quad & s \leqslant 20 \text{ units/month} \\
\text{Non-negativity:} \quad & j,v,m,s \geqslant 0
\end{aligned}$$

The complete linear programme is: each month produce j, v, m and s models, respectively of the Jupiter, Venus, Mars and Saturn. Maximise:

$$P = 15j + 30v + 120m + 130s \quad (£/month)$$

subject to the constraints given above.

Example 9.4: To formulate a multivariable linear programme

A portfolio manager wishes to invest up to £100,000 to maximise the total annual interest income. She has narrowed her choices to four possible investments, A, B, C and D. Investment A yields 6% per annum, investment B yields 8% per annum, investment C 10% per annum and investment D 9% per annum. The four investments have varying risks and conditions attached to them. For safety, the manager feels that at least half of the funds must be placed in A and B. For liquidity, at least 25% of the funds must be placed in investment D. Volatile government policies indicate that no more than 20% of the investment should be in C, whereas tax considerations require at least 30% of the funds to be placed in A. Formulate a linear programming model for this investment problem.
Solution: Invest £a in investment A, £b in investment B, £c in investment C and £d in investment D. The objective is to maximise the total interest income per year. The investment is constrained by the considerations of safety, liquidity, government policy and taxation. Let £R be the total interest income per year, where:

$$R = 0.06a + 0.08b + 0.10c + 0.09d \quad (£/year)$$

Maximise subject to:

$$\text{Total investment: } a + b + c + d \leq 100{,}000 \text{ £ available}$$

$$\text{Safety: } a + b \geq 0.5(a + b + c + d) \text{ £}$$

$$\text{Liquidity: } d \geq 0.25(a + b + c + d) \text{ £}$$

$$\text{Government policy: } c \leq 0.2(a + b + c + d) \text{ £}$$

$$\text{Tax: } a \geq 0.3(a + b + c + d) \text{ £}$$

$$\text{Non-negativity: } a, b, c, d \geq 0$$

To solve a linear programme, it is conventional to arrange the constraints so that the variables appear only on the left hand side of the inequalities. This is done below. The complete linear programme is:

Invest £a in investment A,

£b in investment B,

£c in investment C

and £d in investment D.

Maximise the total interest income per year, where:

$$R = 0.06a + 0.08b + 0.10c + 0.09d \quad \text{(£/year)}$$

subject to the constraints:

$$\text{Total investment: } a + b + c + d \leq 100{,}000 \text{ £}$$

$$\text{Safety: } 0.5a + 0.5b - 0.5c - 0.5d \geq 0 \text{ £}$$

$$\text{Liquidity: } -0.25a - 0.25b - 0.25c + 0.75d \geq 0 \text{ £}$$

$$\text{Government policy: } -0.2a - 0.2b + 0.8c - 0.2d \leq 0 \text{ £}$$

$$\text{Tax: } 0.7a - 0.3b - 0.3c - 0.3d \geq 0 \text{ £}$$

$$\text{Non-negativity: } a, b, c, d \geq 0$$

The objectives in the four examples above have all required a quantity to be maximised. The procedure is identical at this stage if the objective is to minimise some measure. See the examples at the end of the chapter.

9.3 SOLVING THE LINEAR PROGRAMME

We are now at the stage of considering how to find the values of the variables which will satisfy all of the constraints simultaneously and optimise the objective. We are more accustomed to dealing with equations rather than inequalities. It is a simple matter to change the inequalities into equations. We add an additional variable to the left hand side of the inequality. The purpose of this variable is to represent the difference in value between the two sides of the inequality. To demonstrate this procedure, let us refer to **Example 9.2** which concerned the manufacture of machined components X and Y. We will include an additional variable in each constraint to produce a set of equations. This variable is denoted by s, hence, s_1 is included in the first constraint, s_2 in the second constraint and so on. We will also impose the condition that the values of these variables cannot be negative ie $s_i \geq 0$. This means that the variable is added to the left hand side for all $\leq$ constraints and subtracted from all $\geq$ constraints. The linear programme becomes: produce x components of type X and y components of type Y each week. The objective is to maximise the total weekly contribution. Maximise:

$$P = 30x + 40y \quad \text{(£/week)}$$

Subject to the constraints:

$$\begin{aligned}
\text{Labour: } 1x + 2y + s_1 &= 4000 \text{ hours/week} \\
\text{Machine capacity: } x \quad\quad + s_2 &= 2250 \text{ components/week} \\
y + s_3 &= 1750 \text{ components/week} \\
\text{Rod: } 2x + 5y + s_4 &= 10{,}000 \text{ kg/week} \\
\text{Plate: } 5x + 2y + s_5 &= 10{,}000 \text{ kg/week} \\
\text{Regular orders: } x \quad\quad - s_6 &= 600 \text{ components/week} \\
\text{Union agreement: } x + y - s_7 &= 1500 \text{ components/week} \\
\text{Non-negativity: } x, y &\geq 0
\end{aligned}$$

These additional variables are called **slack variables** in the $\leq$ constraints. They represent the amount of the resource not used, that is, the difference between the resource used and the maximum available. For example, look at the labour constraint above. Suppose 1000 components of each type were made each week, then the number of hours used is $1 \times 1000 + 2 \times 1000 = 3000$ hours. Since 4000 hours are available, the spare capacity, or slack, is $(4000 - 3000) = 1000$ hours. Therefore, for this combination of x and y, s_1 takes the value 1000.

In the $\geq$ constraints, the slack variables are referred to as **surplus variables** since they represent the amount of resource being used over and above the minimum requirement. For example, look at the 'regular orders' constraint, when 1000 components of type X are being produced. The minimum number of type X required by this constraint is 600, hence a production level of 1000 gives a surplus of 400 components above the minimum. Therefore, s_6 takes the value 400.

We now have a set of simultaneous equations. However, we cannot solve these by the usual algebraic methods to produce a set of unique values for the variables because the number of variables is greater than the number of equations. A unique set of solutions will arise only if the number of variables and the number of equations are the same. The best we can do is to identify a set of feasible solutions to the equations. This set of feasible solutions gives all combinations of the variables which satisfy all of the constraints. We will then select from this set the particular solution or solutions which optimise the objective.

How do we set about identifying the set of feasible solutions? This can be done graphically if the problem involves only two variables. However, we must resort to an algebraic method if the problem is multivariate.

9.3.1 The graphical solution of a linear programme

A linear equation represents a set of points which lie on a straight line. A linear inequality represents an area of a graph. For example, $x \leq 7$ says that x takes a value which is less than 7 or is equal to 7. The situation can be illustrated graphically as follows. Draw the line $x = 7$. See left hand graph in FIGURE 9.1. This divides the graph into three sets of points: those for which $x = 7$, the line itself; those for which $x < 7$, the area to the left of the line; and those for which $x > 7$, the area to the right of the line. We do not require this last set. It is usual to shade the area not required. See the right hand graph in FIGURE 9.1:

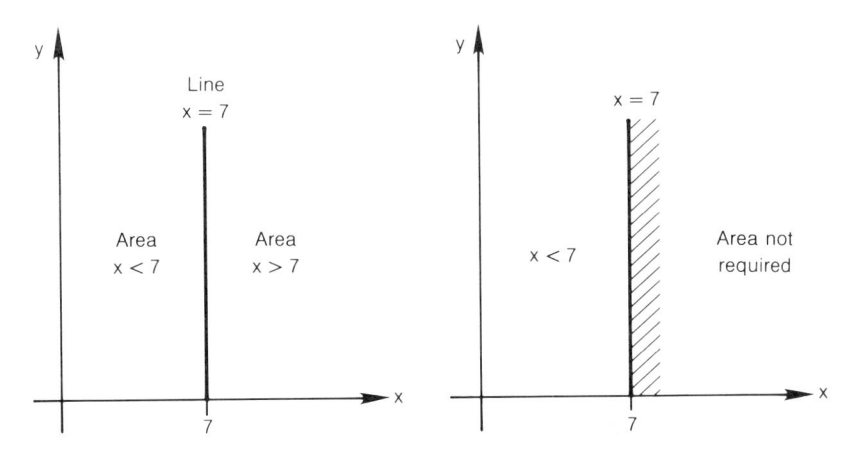

FIGURE 9.1 **Graphical representation of the inequality $x \leq 7$**

Suppose x + y ⩽ 10. Which area does this represent? The procedure is the same as in the previous example. First of all we draw the line x + y = 10. See the left hand graph below. Again the line divides the graph into three sets of points: those for which x + y = 10, the line; those for which x + y < 10, the area below the line; and those for which x + y > 10, the area above the line.

A useful technique for deciding which is the rejected area on the graph is to take any point on the graph away from the line and substitute its values into the inequality. If the inequality still makes sense, then that point is a feasible solution. If the inequality is untrue, then the point is infeasible and lies in the rejected region. The origin is a convenient point to use. Substitute x = y = 0 into the inequality x + y ⩽ 10. We have 0 + 0 ⩽ 10 which is a true statement, therefore the origin is a feasible solution and we should reject the other side of the line. See the right hand graph in FIGURE 9.2.

FIGURE 9.2 **Graphical representation of the inequality x + y ⩽ 10**

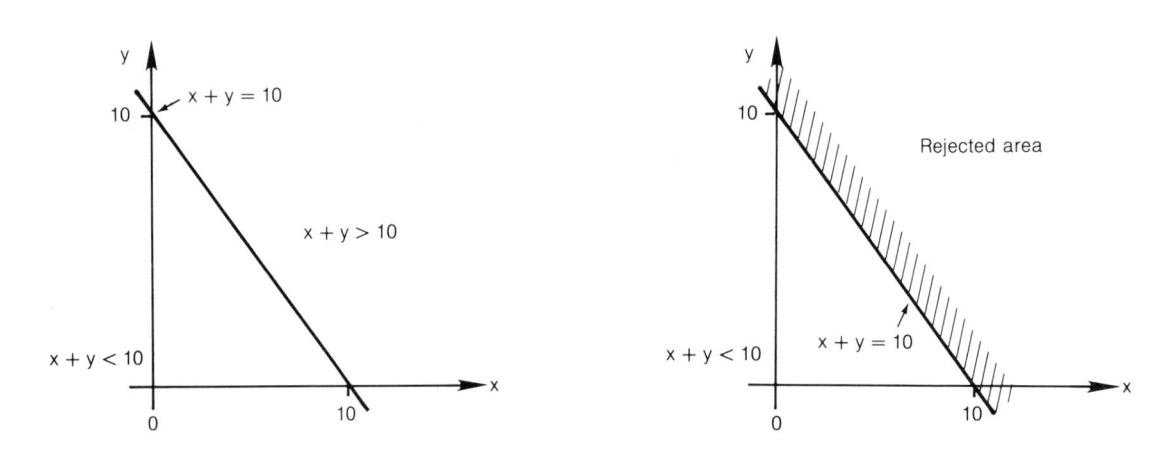

Each constraint in the linear programme can be drawn in this way and the rejected area shaded. If all of the constraints are drawn on the same graph, the area which remains unshaded is the set of points which satisfies all of the constraints simultaneously. This area is called the **feasible region**. For a linear programme, it does not matter which variable is plotted on which axis. The origin should always be included on the graph. False zeros must not be used.

Let us now apply this procedure to the linear programme for **Example 9.1** about the production of the two types of soft drink. We can illustrate the constraints graphically.

$$\text{Machine time: } 0.02p + 0.04m \leqslant 24 \text{ hours/day}$$

Plot the line 0.2p + 0.04m = 24. An easy way of plotting the line is to find the points where the line crosses the p and m axes. Put p = 0 into the equation and calculate m, ie when p = 0, m = 600. Put m = 0 into the equation and calculate p, ie when m = 0, p = 1200. Plot these two points and join them to give the line. This method always works unless the line passes through the origin. In that case revert to the alternative procedure of substituting any other value of p and finding the equivalent value of m.

To find which side of the line to shade put p = 0 and m = 0 in the inequality:

$$0.02 \times 0 + 0.04 \times 0 < 24$$

This statement is true, so the origin is included in the feasible area.

FIGURE 9.3 **Graphical representation of the inequality 0.02p + 0.04m ⩽ 24**

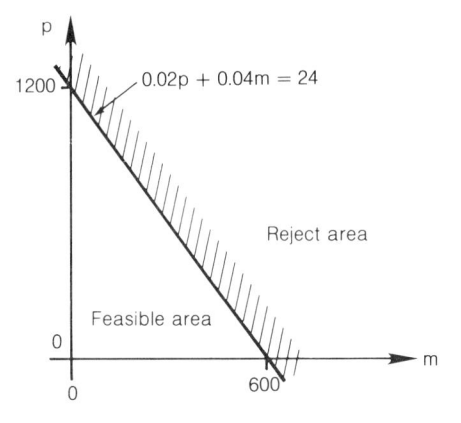

Special Ingredient: $0.01p + 0.04m \leqslant 16$

Plot the line:

$$0.01p + 0.04m = 16$$

Again the origin is included in the feasible region so we shade out the area above the line.

FIGURE 9.4 **Graphical representation of the inequality 0.01p + 0.04m ⩽ 16**

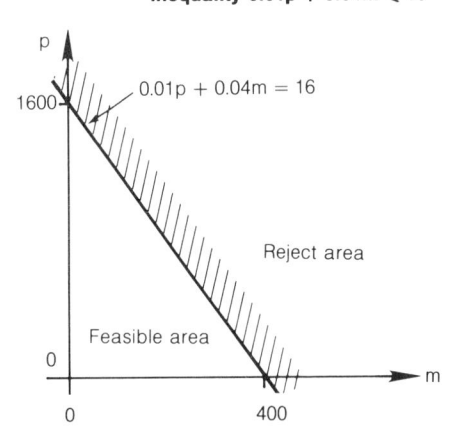

Non-negativity: $p \geqslant 0$ and $m \geqslant 0$

Shade out negative values of each variable.

FIGURE 9.5 **Graphical representation of the non-negativity constraints**

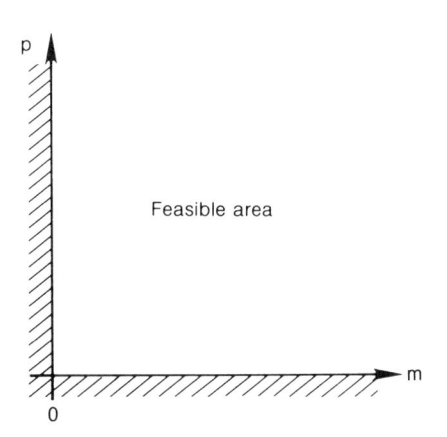

256 QUANTITATIVE ANALYSIS

Putting these four constraints together on one graph gives:

FIGURE 9.6 **Graphical representation of the constraints for Example 9.1**

The area left unshaded by all of the constraints is the **feasible region** and this contains all of the possible combinations of production which will satisfy the given constraints. The co-ordinates of any point within the feasible region represents a possible combination of soft drink production for this firm.

We must now consider how to choose the production which will maximise the firm's daily contribution. The objective function is:

$$P = 0.10p + 0.30m \qquad (\text{£/day})$$

If we let P = 100 £ per day, then we can illustrate the objective function graphically. If we then give P another value, the new line will be parallel to the one for P = 100 £ per day. The graph below illustrates the objective function for daily contributions of £50, £100 and £150.

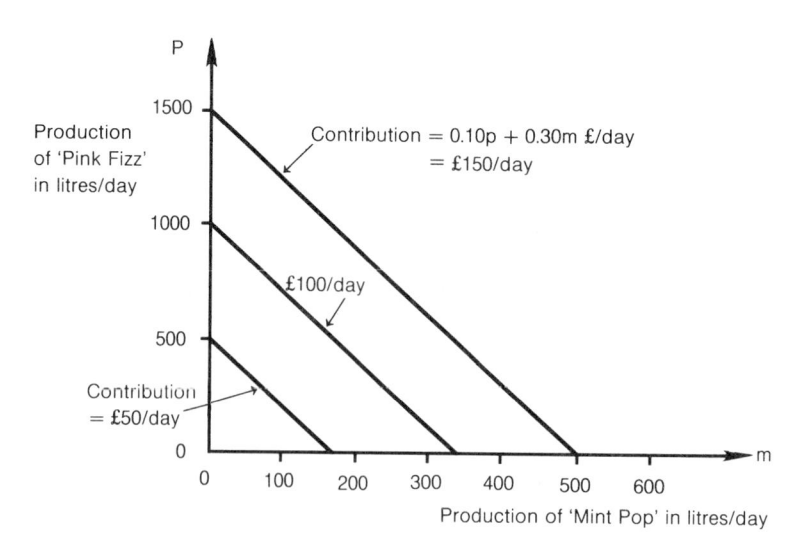

FIGURE 9.7 **Graphical representation of the objective function**

We can generate the entire family of possible contribution lines by drawing one in particular, then moving across the feasible region parallel to it. The further from the origin we move, the larger is the contribution.

If we draw a contribution line on the graph of the linear programme, as in FIGURE 9.8, we can move parallel to this line across the feasible region in the direction of increasing contribution until we reach the last feasible solution(s), before the line moves entirely into the infeasible region:

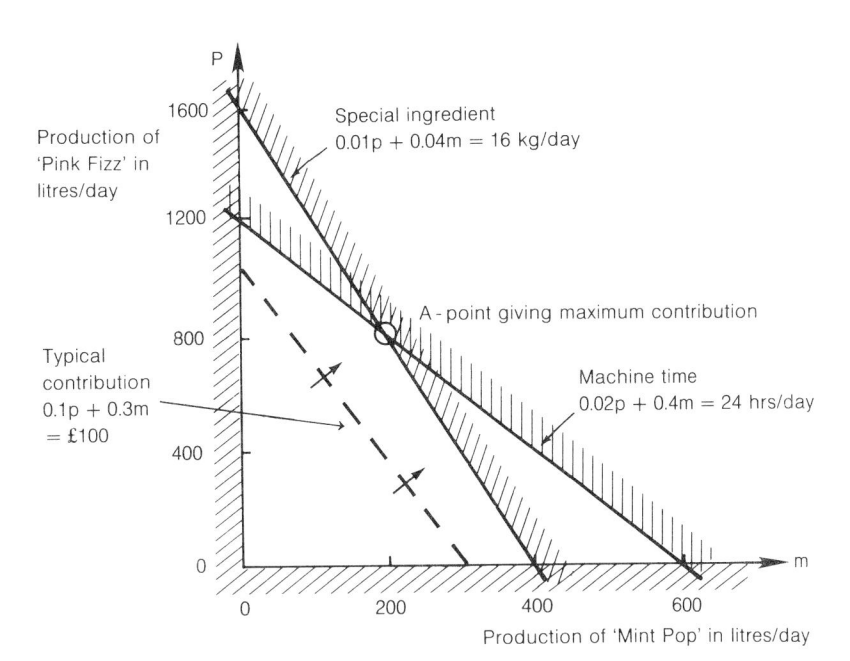

FIGURE 9.8 **Linear programme for Example 9.1**

We can see that point A is the last feasible solution. The co-ordinates of point A give the optimum combination of production for the two drinks. The approximate co-ordinates of point A can be read from the graph, but, for precision, the co-ordinates are calculated by solving simultaneously the equations of the two constraints which form point A.

These two constraints are called the **binding** or **limiting** constraints. They are the resources which are being used fully and therefore prevent the daily contribution from increasing further. The optimum solution is the intersection of:

$$0.02p + 0.04m = 24 \qquad (1)$$

$$0.01p + 0.04m = 16 \qquad (2)$$

Subtract (2) from (1):

$$0.01p = 8$$

Therefore:

$$p = 800 \text{ litres/day}$$

substitute into (2) to find m:

$$0.01 \times 800 + 0.04m = 16$$

Therefore:

$$m = 200 \text{ litres/day}$$

To maximise the daily contribution, the firm should produce 800 litres of 'Pink Fizz' and 200 litres of 'Mint Pop' each day. This will yield a maximum contribution of:

$$0.10 \times 800 + 0.30 \times 200 = £140/\text{day}$$

This combination utilises all of the machine time and special ingredient available each day. There is no spare capacity or slack on either of the constraints.

This method of identifying the optimum corner depends on a suitable profit line being drawn. The following is a practical note which will help to obtain a suitable profit line from which to identify the optimum corner. Choose any convenient point near the middle of the feasible region. Suppose in the above example the point m = 200, p = 200 is chosen. The daily contribution from this product mix is:

$$P = 0.10p + 0.30m = 0.10 \times 200 + 0.30 \times 200 = £80/day$$

All of the other product mixes which give a daily contribution of £80 lie on the line:

$$80 = 0.10p + 0.30m \qquad (£/day)$$

One point on this line is already known, ie m = 200, p = 200. A second point might be m = 0, hence, p = 800. This particular daily contribution line is now drawn on the graph and the procedure described above is followed to identify the optimum solution(s). It is clear from the procedure that the optimum will always be at a corner of the feasible region, or, if the objective function is parallel to one of the constraints, at any point on the line joining two corners.

We have assumed that the variables in the linear programme are continuous, or, if not, then fractions are acceptable. It will often be the case that part units are allowable over the time period of the problem. For example, if two models of car are being produced and the objective of the linear programme is to maximise the machine usage per week, we may find that the optimum solution results in incomplete cars at the end of each week. For such a product 'work in progress' is allowable on a weekly basis.

If, however, we are allocating workers to tasks, part workers are not acceptable. In this case the optimum solution must produce integer values. The feasible solutions are all the points in the feasible region where the variables are integers. The last point within the feasible region which has integer co-ordinates is selected and this may no longer be at a corner of the feasible region.

For two variable linear programmes, it does not make much difference to the solution procedure if the variables must be integer. The feasible region is replaced by the set of feasible points within the constraint boundaries. The typical objective function is moved through these points, rather than through the feasible region as a whole. In the multivariable case, however, the method of integer programming is used. This is not covered by this book.

Example 9.5: A graphical solution for a two variable linear programme

Refer to **Example 9.2** which concerned the production of two machined components. We wish to know the product mix which will achieve maximum total contribution per week.
Solution: The feasible region for each constraint is as follows:

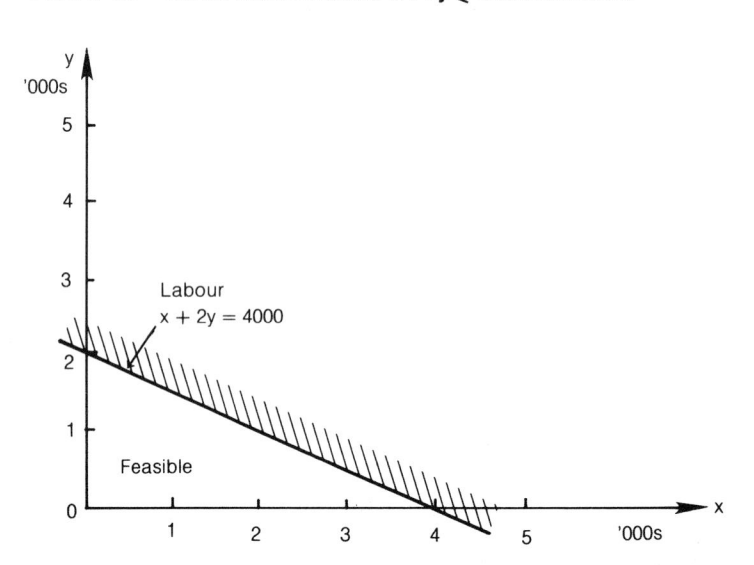

FIGURE 9.9   Labour time constraint: x + 2y ⩽ 4000 hours/week

LINEAR PROGRAMMING: CHAPTER NINE **259**

FIGURE 9.10 **Machine capacity constraint: $x \leqslant 2250$ components/week and $y \leqslant 1750$ components/week**

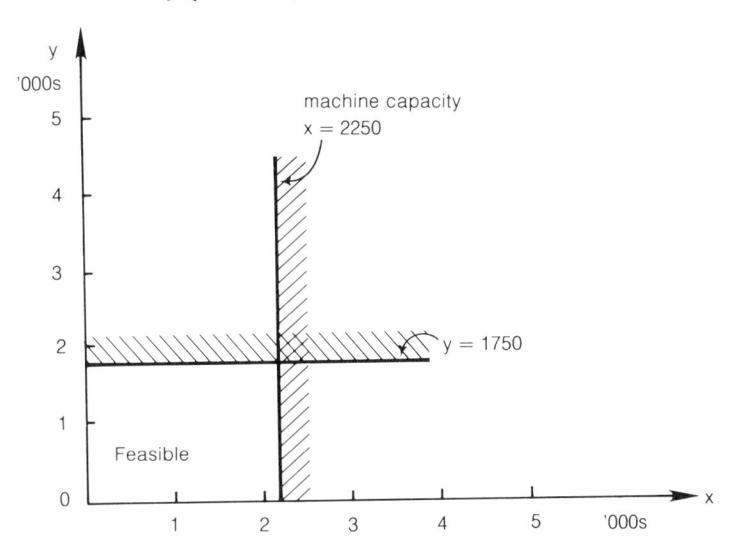

FIGURE 9.11 **Rod constraint: $2x + 5y \leqslant 10000$ kg/week**

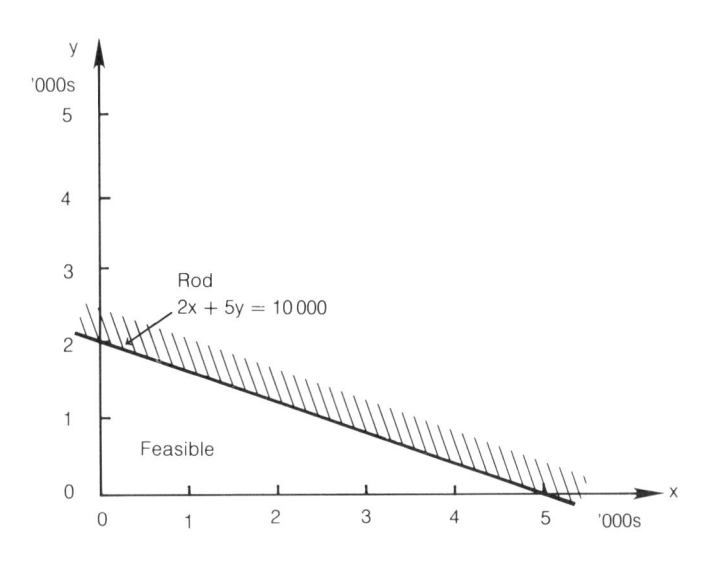

FIGURE 9.12 **Plate constraint: $5x + 2y \leqslant 10000$ kg/week**

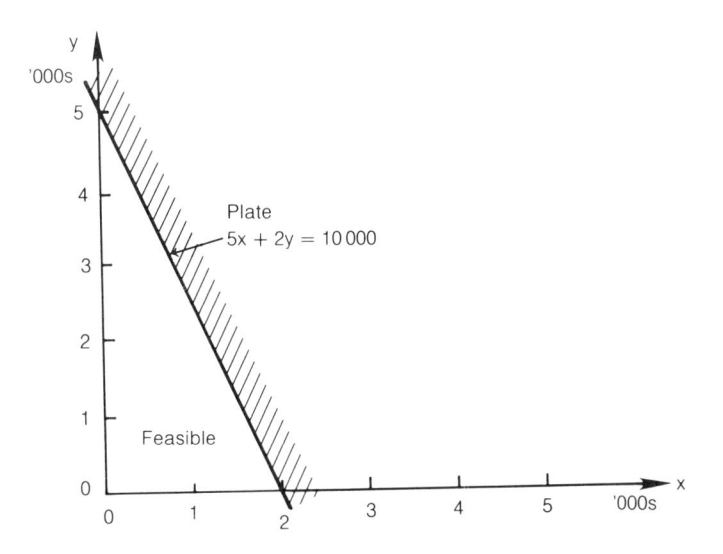

260 QUANTITATIVE ANALYSIS

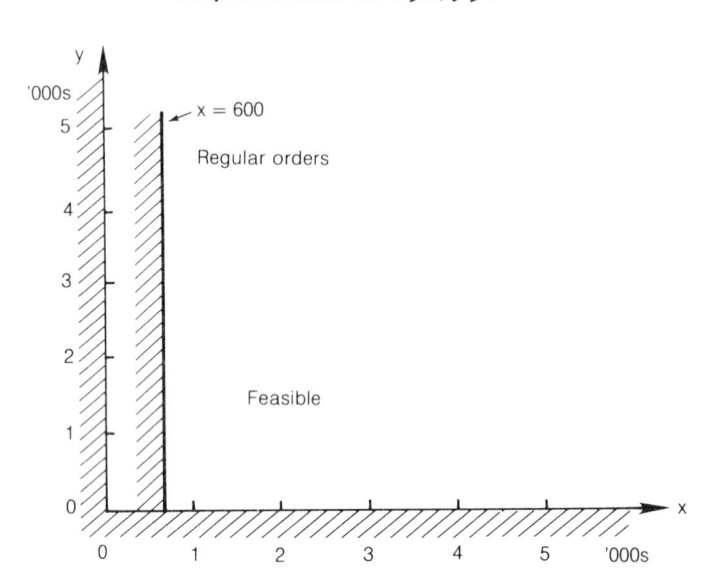

FIGURE 9.13 **Regular orders and non-negativity constraints: $x \geq 600$ components/week and $x \geq 0$, $y \geq 0$**

FIGURE 9.14 **Union agreement constraint: $x + y \geq 1500$ components/week**

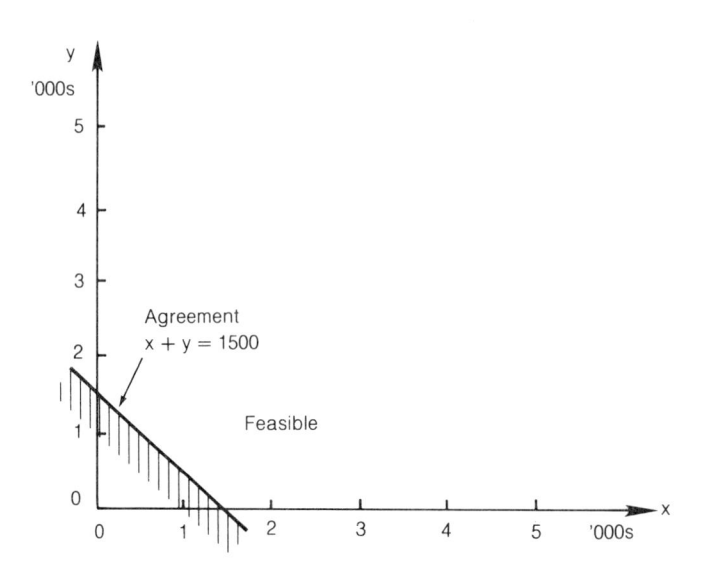

The feasible region, containing all of the possible product mixes for this problem, is shown unshaded.

We wish to identify the optimum product mix which will maximise the weekly contribution. The objective function is:

$$P = 30x + 40y \quad (\text{£/week})$$

To plot this function for a typical value of the weekly contribution, we select the point $x = 1000$, $y = 1000$ which is in the feasible region. The weekly contribution for this product mix is:

$$P = 30 \times 1000 + 40 \times 1000 = £70{,}000/\text{week}$$

We will use the contribution line:

$$70{,}000 = 30x + 40y \quad (\text{£/week})$$

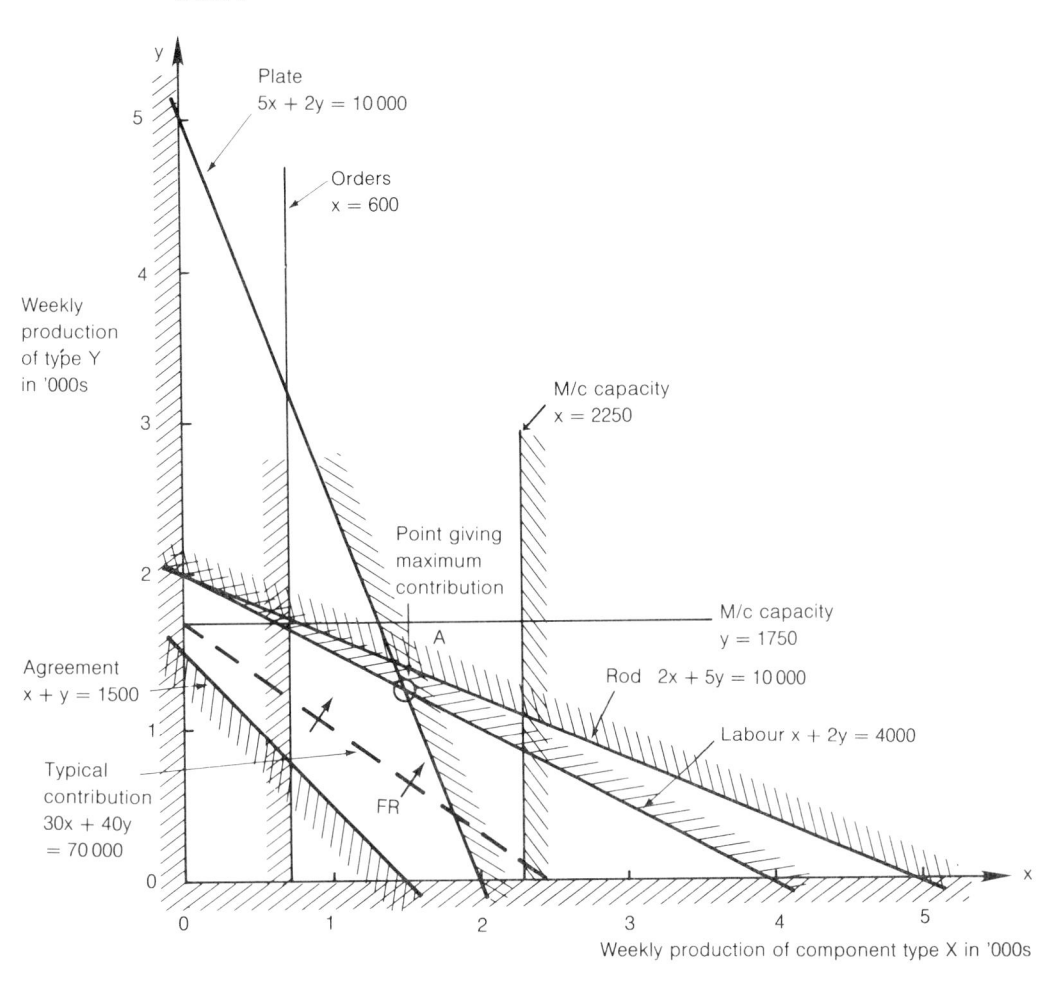

FIGURE 9.15   **A linear programme for the weekly production of machined components of type X and Y**

[Graph showing linear programming constraints with axes: Weekly production of type Y in '000s (y-axis) vs Weekly production of component type X in '000s (x-axis). Constraints shown: Plate $5x + 2y = 10\,000$; Orders $x = 600$; M/c capacity $x = 2250$; M/c capacity $y = 1750$; Rod $2x + 5y = 10\,000$; Labour $x + 2y = 4000$; Agreement $x + y = 1500$; Typical contribution $30x + 40y = 70\,000$. Point A marked as point giving maximum contribution. FR = feasible region.]

as the trial line. This line also passes through the point $x = 0$, $y = 1750$. It is shown by the broken line ---- on FIGURE 9.15. Moving parallel to this line, in the direction of increasing contribution, leads us to point A as the last feasible solution. The binding constraints are therefore:

$$\text{Labour: } x + 2y \leq 4000 \text{ hours/week}$$

$$\text{Plate: } 5x + 2y \leq 10{,}000 \text{ kg/week}$$

Solving the corresponding equations simultaneously gives:

$$x + 2y = 4000 \qquad (1)$$
$$5x + 2y = 10{,}000 \qquad (2)$$
$$(2) - (1) \quad 4x = 6000$$

therefore $x = 1500$, and $y = 1250$ by substitution.

The optimum product mix is 1500 of component X and 1250 of component Y each week. The maximum contribution per week will then be:

$$P_{max} = 30 \times 1500 + 40 \times 1250 = £95{,}000/\text{week}$$

This product mix uses all of the labour hours available and all plate. These are the binding constraints. However, there will be spare capacity on machine time for both components and spare rod capacity. The production will also exceed the minimum required by the regular orders and the minimum required by the union agreement.

We find that the value of the slack variables in the machine time constraints are 750 of component X and 500 of machine tool Y, ie:

$$1500 + s_2 = 2250, \text{ therefore } s_2 = 750 \text{ components/week, and}$$

$$1250 + s_3 = 1750, \text{ therefore } s_3 = 500 \text{ components/week}$$

The slack on the rod constraint is:

$$2 \times 1500 + 5 \times 1250 + s_4 = 10,000$$

therefore:

$$s_4 = 750 \text{ kg/week}$$

The surplus on the regular orders constraint is:

$$1500 - s_6 = 600$$

therefore:

$$s_6 = 900 \text{ components/week}$$

above the minimum needed for the regular orders. The surplus on the union agreement is:

$$1500 + 1250 - s_7 = 1500$$

therefore:

$$s_7 = 1250 \text{ components/week}$$

above the minimum required by the union agreement.

As we have already said, the optimum solution will normally be at a corner of the feasible region. It is possible, therefore, once the graph has been drawn to identify the optimum corner by evaluating the objective function at each corner of the feasible region. A **basic solution** is the name given to the set of variable values at a corner of the feasible region. The **basic variables** are those variables which have non-zero values at a particular corner.

Occasionally problems arise when solving a linear programme. The problem may be **infeasible**. In this case there is no feasible region. No combination of the variables satisfies all of the constraints simultaneously and the linear programme has no solution. If a solution is essential, one or more constraints must be relaxed to open up a feasible region.

Further problems arise if the linear programme is **unbounded**. In this case the solution can be increased indefinitely without violating any constraint. This usually means that the linear programme is formulated incorrectly, with some constraints missing.

The issue of non-unique solutions was mentioned earlier. These arise when the objective function is parallel to a binding constraint. Any point on that constraint between the two optimum corners, will give the optimum value of the objective function. Any one of these points forms an optimum solution to the model. This can be a useful situation since it gives the decision maker some flexibility.

9.4 SENSITIVITY ANALYSIS

In most decision making activities it is prudent to examine the preferred course of action to see what effect changes in the problem will have on the decision. Linear programming is no exception. There are three aspects of the problem which we need to consider:

1 the effect of additional supplies of the limiting resources;
2 the effect of changes in the non-limiting resources;
3 the effect of changes in the coefficients of the objective function.

We will consider each of these situations. In all cases we will assume that only one parameter is being changed at any time.

9.4.1 How do changes in the provision of a limiting resource affect the solution of a linear programme?

The value of the objective function is limited because one or more of the resources is used up. If more of a limiting resource is made available, the optimum solution can be improved. However, a note of caution must be introduced. The change in the optimum solution will result in an improvement in the value of the objective function only if any additional costs, involved in obtaining the additional resource, do not exceed the gains.

If a limiting resource is increased, the corresponding limiting constraint is said to be relaxed. As the limiting constraint is gradually relaxed, it will move parallel to its original position and the optimum corner will move in a direction which improves the objective function. This process continues until some other resource is used up and the constraint in question ceases to be limiting. The **shadow price** of a resource is the amount added to the objective function by the relaxation of a limiting constraint by one unit, ie when the limiting resource is increased by one unit. The shadow price of the resource is the value of one unit of that resource at the optimum solution. If the resource is obtainable at an additional cost which is less than the shadow price, then the additional unit is probably worth obtaining.

Example 9.6: To illustrate a sensitivity analysis of the limiting constraints

Refer to **Example 9.2** and **Example 9.5** which concerned the production of machined components. From **Example 9.5**, we know that the limiting constraints are labour hours and plate. Let us consider plate first. Look at FIGURE 9.16. The plate constraint is relaxed by moving the line parallel to its

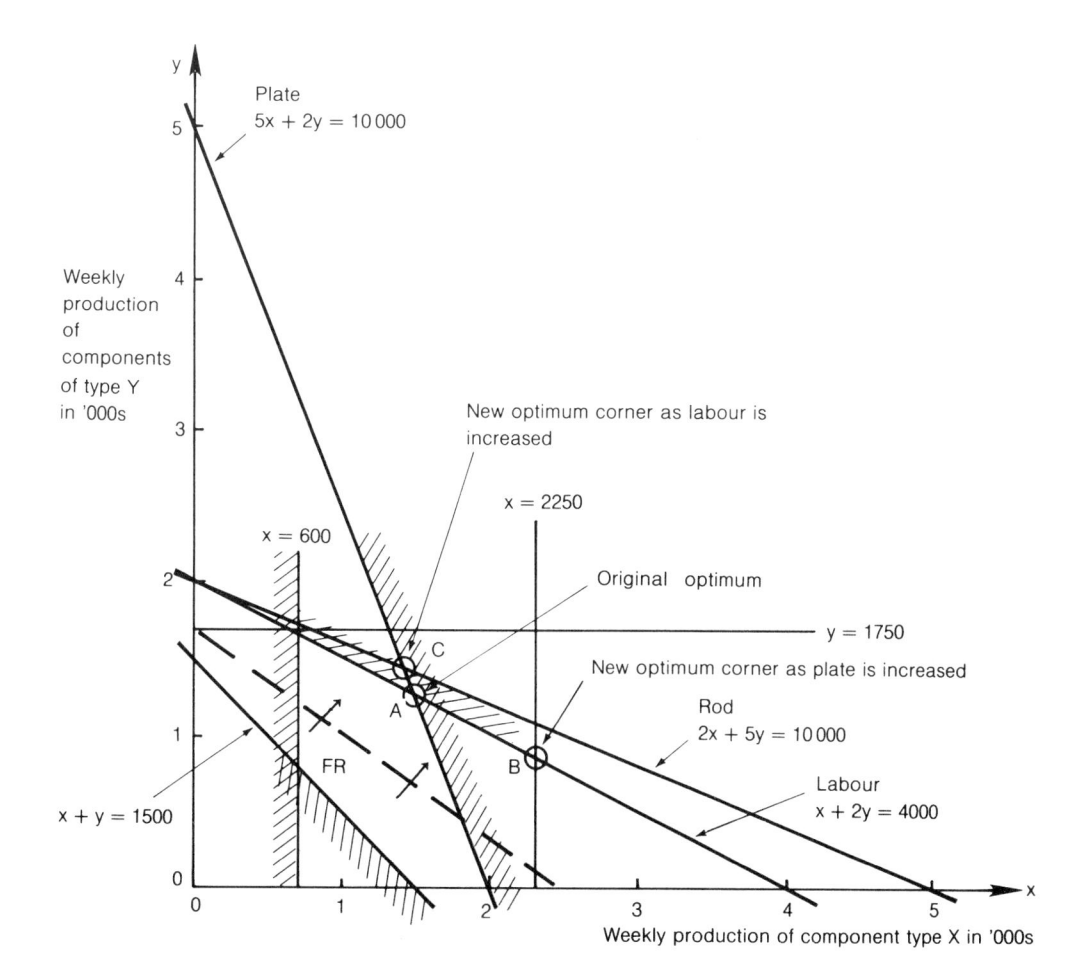

FIGURE 9.16 **A linear programme for the weekly production of components of type X and Y**

original position away from the origin. The feasible region increases and the optimum corner slides down the labour constraint, increasing the value of x and decreasing the value of y. The constraint can be usefully relaxed until it meets the intersection of labour hours and the constraint on the machine capacity for component X, point B. If the plate constraint is relaxed any further, it ceases to be binding and spare plate will be available.

Point B is now the new optimum corner. The co-ordinates of point B may be determined by solving simultaneously the equations for the constraints on labour and the machine capacity for component X.

$$\text{Labour:} \quad x + 2y = 4000 \text{ hours/week}$$

$$\text{Machine capacity for X:} \quad x = 2250 \text{ components/week}$$

Since we know x = 2250 components, we substitute to find y:

$$2250 + 2y = 4000$$

therefore y = 875 components.

The new optimum product mix is to produce 2250 components of type X and 875 components of type Y each week. This mix gives a maximum weekly contribution of $30 \times 2250 + 40 \times 875 = £102,500$ which is an increase of £(102,500 − 95,000) = £7500 per week. At this product mix, the amount of plate being used is:

$$5 \times 2250 + 2 \times 875 = 13,000 \text{ kg/week}$$

This is 3000 kg/week more than the original availability. The labour time and the machine capacity for component X are also fully used at the new optimum.

An additional supply of 3000 kg of plate has generated an additional contribution of £7500 per week, therefore the shadow price for plate is £7500/3000 kg = £2.50 per kg. Each additional kg of plate adds an extra £2.50 to the weekly contribution. It follows that any additional cost of acquiring the extra raw material must be less than £2.50 per kg if the extra material is to be worthwhile.

We will now follow a similar procedure for the second limiting constraint, assuming that the plate constraint is in its original position.

Suppose the labour constraint is relaxed by 1 unit, ie one more hour of manpower becomes available, then the constraint becomes:

$$x + 2y \leqslant 4001$$

This constraint is parallel to the original one but further removed from the origin. You can see in the graph above that the intersection of the plate constraint and the new labour constraint still forms the optimum corner. The optimum solution is now x = 1499.75 and y = 1250.625 which leads to P_{max} = £95,017.50. This is an increase of £17.50 in the value of the objective function. The shadow price of the labour is £17.50 per hour. If one extra hour of labour may be obtained for an extra £17.50 or less, then it is worthwhile to have that extra hour. If the additional cost of the extra hour is in excess of £17.50, then it is not worthwhile.

How many additional hours of labour is it worthwhile buying? As the labour constraint moves away from the origin, parallel to its original position, it moves towards the intersection of the plate constraint and the rod constraint, point C. If the labour constraint is relaxed further, it is no longer binding and further additional labour hours would not be worthwhile. The maximum number of additional man-hours may be determined by solving simultaneously the constraints which intersect at point C:

$$\text{Plate:} \quad 5x + 2y = 10,000$$

$$\text{Rod:} \quad 2x + 5y = 10,000$$

The solution is: x = 10,000/7 and y = 10,000/7 components per week. The number of labour hours used at this point is:

$$x + 2y = 10,000/7 + 2(10,000/7) = 4285.7 \text{ hours/week}$$

This is an increase of 285.7 hours per week on the original 4000 hours. Provided that the additional cost of supplying additional labour hours does not exceed £17.50 per hour, then it is worthwhile securing up to a maximum of 285.7 extra labour hours per week. If the maximum time is provided, then the new maximum value of the weekly contribution is:

$$P_{max} = 30 \times 10{,}000/7 + 40 \times 10{,}000/7 - \text{additional cost}$$

$$= £(100{,}000 - \text{additional cost})/\text{week}$$

It is important to be clear that any costs which are deducted are only those which do not arise in the original problem. For example, in the production of the machined components, suppose the labour normally costs £4.00 per hour. This cost will have been used by the accountant in calculating the unit contributions for the two components. If additional labour is provided as overtime at a cost of £6.00 per hour, then £4.00 per hour is already accounted for in the unit contribution figures. It is only the extra £2.00 per hour which must be charged separately.

9.4.2 How do changes in the non-limiting resources affect the optimum solution?

In **Example 9.6**, we considered the two limiting constraints of labour and plate. The other constraints are not binding at the original optimum solution. These constraints are:

1 Machine time to produce component X.
2 Machine time to produce component Y.
3 Rod.
4 Regular orders.
5 Union agreement.

What happens as each of these constraints is changed? The first three are less-than-or-equal-to constraints. Any increase in their availabilities will not affect the optimum solution. However, any decrease in these three constraints can affect the solution. The tightening of one of the non-limiting constraints will cause it to move towards the origin. At first, the only change will be a reduction in the size of the feasible region. When, however, the particular constraint passes through the original optimum corner, it will itself become limiting and a new optimum solution will emerge.

It is useful to know what the lower limits are on these constraints. The machine capacity for X can be reduced by 750 hours, from 2250 to 1500 hours, before it affects the solution. The machine capacity for Y can be reduced by 500, from 1750 to 1250 hours. The supply of rod can be reduced by 750 kg per week, from 10,000 to 9250 kg. These reductions are the values of the slack variables mentioned earlier. The greater-than-or-equal-to constraints, for the regular orders and the union agreement, act in the opposite way. Any reduction in these requirements will increase the feasible region but will not affect the optimum solution.

Any increase in these constraints will first reduce the feasible region and then affect the optimum solution. If the regular orders for X increase by at least 900 to 1500, the optimum will begin to change. If the union agreement was increased by at least 1250, to more than 2750, there would be no feasible region and no solution. These increases are the surplus variables referred to earlier.

9.4.3 How do changes in the coefficients of the objective function affect the optimum solution?

It is inevitable that the circumstances under which a linear programme is formulated will change. Major changes will probably mean that the work will have to be done again but it may be possible to identify the effect of minor changes from the solution to the original problem. In this section, we consider changes to the objective function. If the objective is to maximise weekly contribution, a change in the cost of a raw material will alter the coefficients in the objective function.

In an investment portfolio problem, if the objective is to maximise the annual return on the investments, a change in the interest rate earned by one of the investments will change that coefficient in the objective function.

QUANTITATIVE ANALYSIS

We will consider situations in which the coefficients change one at a time. Suppose:

$$P = ax + 4y \quad (\text{£/week})$$

represents the objective function for a profit maximising linear programme, where £4 per unit is the profit on product Y and £a per unit is the profit on product X. The profit on product X is liable to change. Suppose this linear programme has been graphed with x and y in the conventional directions. It is helpful to re-arrange the objective function so that y is the subject:

$$y = P/4 - (a/4)x$$

The profit line cuts the y axis at P/4 and the slope of the profit line is $-(a/4)$. The intercept on the y axis is independent of the value of a, but the slope of the line increases as a increases, and decreases as a decreases. In other words, as the value of a changes, the profit lines rotates. Small rotations in either direction will not usually alter the optimum corner. However, larger rotations will result in different corners emerging as the optimum. It is useful to know the range of values which a can take before a particular corner ceases to be the optimum. A similar argument would apply if the coefficient of x was fixed and the coefficient of y was liable to change.

Example 9.7: To illustrate the effect on the optimum solution of changes in one of the objective function coefficients

Refer to **Example 9.2** and **Example 9.5** about the manufacture of machined components. The feasible region is:

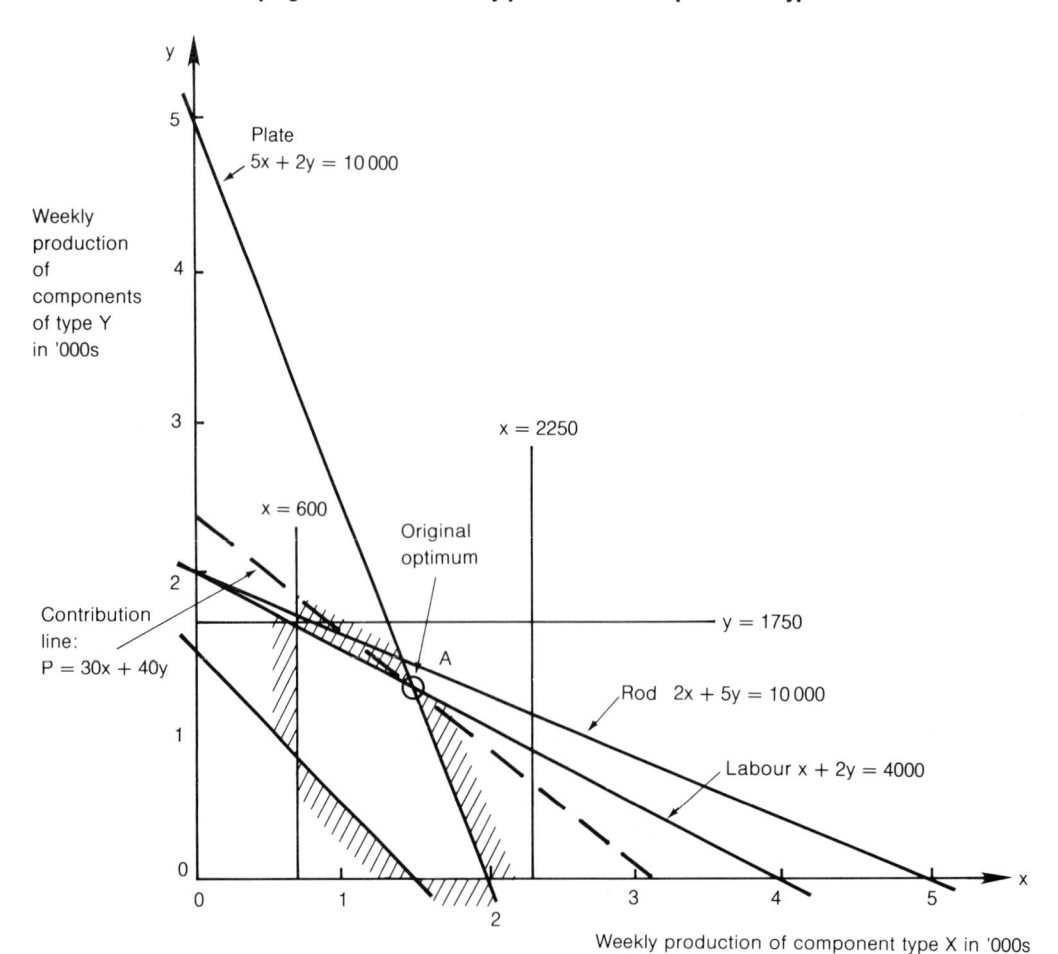

FIGURE 9.17 **A linear programme for the weekly production of components of type X and Y**

The weekly contribution line:

$$P = 30x + 40y \quad (\text{£/week})$$

LINEAR PROGRAMMING: CHAPTER NINE 267

is shown in FIGURE 9.17 drawn through the optimum corner, A. It is now found that the unit contribution for component X is liable to vary. What is the range of values which this unit contribution can take before A ceases to be the optimum corner? The unit contribution for component Y remains constant.

Solution: Re-write the weekly contribution equation as:

$$P = ax + 40y \quad (£/week)$$

where a represents the unit contribution for component X. Re-arranging this equation gives:

$$y = P/40 - (a/40)x$$

The slope of the weekly contribution line is $-(a/40)$. Initially when a = £30/unit, the slope is $-(30/40) = -(3/4)$.

If a decreases below £30 per unit, the slope of the weekly contribution line will become less steep. The line will rotate about point A towards the limiting constraint of labour hours. This reduces the optimum value of P, the weekly contribution. See FIGURE 9.18. If the value of a decreases far enough, the weekly contribution line will became co-incident with the labour constraint. See Figure 9.19.

FIGURE 9.18 **Reducing the contribution for component X**

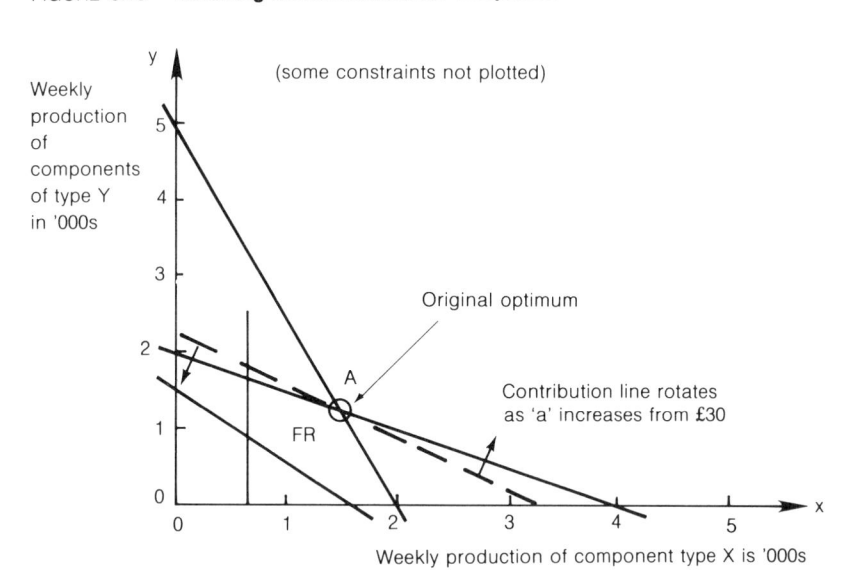

FIGURE 9.19 **The limiting position for reduced contribution**

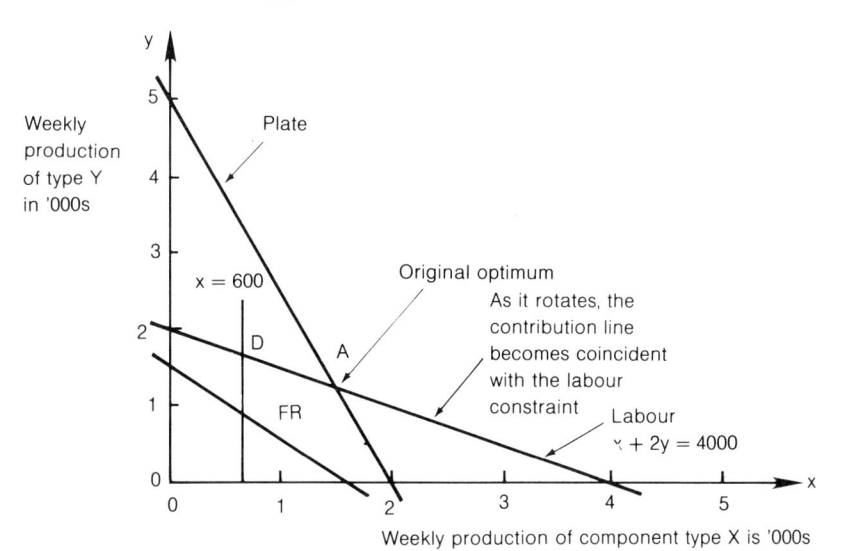

268 QUANTITATIVE ANALYSIS

FIGURE 9.20 **Effect of a further reduction in the contribution for component X**

Weekly production of components of type Y in '000s

- New optimum point as 'a' decreases further
- Original optimum
- D
- A
- FR

Weekly production of component type X in '000s

If the value of a decreases any further, then the optimum corner moves from A to D. See FIGURE 9.20. It follows therefore that the limiting position for the weekly contribution line occurs when it is co-incident with the labour hours limiting constraint. This position fixes the lower limit for the value of a which will keep point A as the optimum corner.

The slope of the labour constraint can be found by re-writing the constraint as

$$y = 4000/2 - (1/2)x.$$

The slope of the limiting constraint is $-(1/2)$. The lower limit for a is found when $-(a/40) = -(1/2)$, and a = £20 per unit. The unit contribution for component X can fall as far as £20 before the optimum corner changes from A to D. The optimum contribution will be reduced, but the optimum product mix will not change, until a falls below £20. The upper limit for the value of a can be found in a similar way. As the value of a increases, the weekly contribution line becomes steeper and eventually becomes parallel to the other limiting constraint, plate. Any further increase in the value of a causes the optimum corner to move from A to E. See FIGURES 9.21 and 9.22:

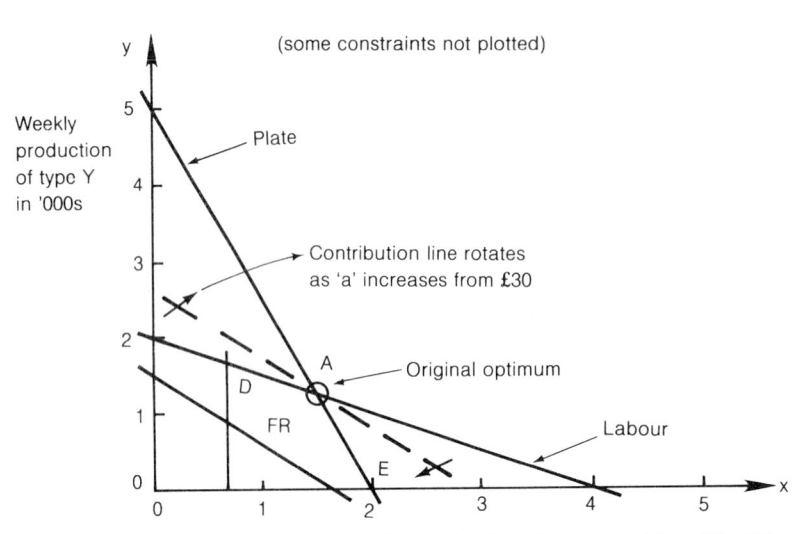

FIGURE 9.21 **Increasing the contribution for component X**

Weekly production of type Y in '000s

- Plate
- Contribution line rotates as 'a' increases from £30
- Original optimum
- A
- D
- FR
- E
- Labour

Weekly production of component type X in '000s

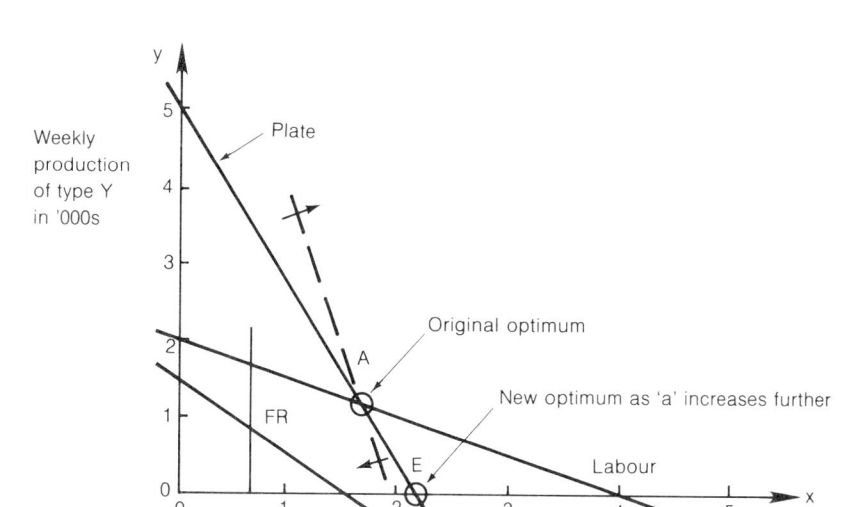

FIGURE 9.22 **Effect of increasing the contribution for component X beyond its limiting position**

The limiting position for the weekly contribution line occurs when it is co-incident with the plate constraint. This position fixes the upper limit for the value of a which will keep point A as the optimum product mix. The slope of the plate constraint can be found by re-writing the equation as:

$$y = 10,000/2 - (5/2)x$$

The slope of the limiting constraint is $-(5/2)$, and the upper limit for a is found when $-(a/40) = -(5/2)$, hence a = £100 per unit. The unit contribution for component X can increase as far as £100 before the optimum product mix changes from A to E.

The corresponding two limits for the value of the unit contribution for component Y can be found in a similar way if the roles of x and y are interchanged. Assume that the value for X is fixed, then:

$$P = 30x + by \quad (£/week).$$

and:

$$x = P/30 - (b/30)y.$$

The same two constraints will form the limiting positions for the weekly contribution line as b increases or decreases. This time we need to write the equations with x as the subject.

$$\text{Labour: } x = 4000 - 2y$$

The slope is -2, therefore at the limit, $-(b/30) = -2$, and b = £60/unit.

$$\text{Plate: } x = 10,000/5 - (2/5)y$$

The slope is $-(2/5)$, and at the limit, $-(b/30) = -(2/5)$, hence b = £12/unit.

The optimum product mix remains at corner A as long as the unit contribution for component Y varies between £12 and £60 per unit only. The optimum contribution will change from £95,000 when the unit contributions of either X or Y change from their original values.

9.5 THE SIMPLEX SOLUTION OF MULTI-VARIABLE LINEAR PROGRAMMES

An algebraic solution method is required if a linear programme contains more than two variables. The basic principle of the solution of a multi-variable model is very simple. It is assumed that the optimum solution is at one of the 'corners' of the feasible region. Therefore, we systematically evaluate the objective function for each corner until we find the one which gives the optimum value of the objective function. We employ the techniques of matrix algebra and an algorithm for moving from corner to corner of the feasible region, in such a way that a move is made only if it improves the value of the objective function. If, at a particular basic solution, no further move is recommended, then we know that the optimum solution has been reached. This algorithm is called the **simplex method**. A detailed knowledge of the simplex method is not necessary, since multi-variable linear programming models are normally solved by using one of the many computer packages which are readily available for this purpose. However, an understanding of the basic principles of the method is helpful for fully interpreting and evaluating the solution to a linear programme which has been obtained by computer package.

The basic simplex method assumes that the linear programming model is a maximising one, subject to a set of $\leq$ constraints. This means that the algorithm can take the origin as the initial corner. The search for the optimum always starts from a zero value for the objective function.

The simplex method can be adapted for minimising problems and for problems with $\geq$ or $=$ constraints. This involves the introduction of artificial, as well as slack and surplus, variables. We have omitted these complications, since the problems are usually solved by a computer package which automatically introduces these variables into the model.

The basic model with which we will work may be formally written as:

$$\text{Maximise } Z = c_1 x_1 + c_2 x_2 + \cdots + c_n x_n$$

The c_i are constants. This is maximised subject to a set of m linear constraints:

$$a_{11}x_1 + a_{12}x_2 + a_{13}x_3 + \cdots + a_{1n}x_n \leq b_1$$
$$a_{21}x_1 + a_{22}x_2 + a_{23}x_3 + \cdots + a_{2n}x_n \leq b_2$$
$$a_{31}x_1 + a_{32}x_2 + a_{33}x_3 + \cdots + a_{3n}x_n \leq b_3$$
$$\vdots \qquad \vdots \qquad \vdots$$
$$a_{m1}x_1 + a_{m2}x_2 + a_{m3}x_3 + \cdots + a_{mn}x_n \leq b_m$$
$$x_i \geq 0$$

There are n variables and m constraints. The double subscripts for the coefficients in the left hand side of the constraints refer first to the constraint, then to the variable. For example, a_{32} is in constraint 3 and is the coefficient of the variable x_2. We will illustrate the use of the simplex method by considering a simple two variable problem which we will first solve graphically. This will enable us to compare the graphical and algebraic solutions.

Example 9.8: To illustrate the use of the simplex method

A firm manufactures two products, X and Y, subject to constraints on three raw materials, RM1, RM2 and RM3. The objective of the firm is to select a product mix which will maximise weekly profit. The linear programme for the problem is:

1   produce x units of product X per week and y units of product Y per week.
2   Maximise the weekly profit, £P, where P = 2x + y   (£/week).
3   Maximise subject to:

$$\text{RM1: } 3x \leq 27 \text{ kg/week}$$
$$\text{RM2: } 2y \leq 30 \text{ kg/week}$$
$$\text{RM3: } x + y \leq 20 \text{ kg/week}$$
$$x, y \geq 0$$

4   Determine the optimum product mix and the maximum value of the weekly profit. State the spare capacity on each resource.

Solution: Graphical method. Add slack variables, s_i, to each constraint. Maximise:

$$P = 2x + y \quad (\text{£/week})$$

subject to:

$$\text{RM1:} \quad 3x \quad\quad + s_1 = 27 \text{ kg/week}$$
$$\text{RM2:} \quad\quad 2y + s_2 = 30 \text{ kg/week}$$
$$\text{RM3:} \quad x + y + s_3 = 20 \text{ kg/week}$$
$$x, y, s_i \geq 0$$

Illustrate the constraints graphically:

FIGURE 9.23 **A linear programme for the weekly production of products X and Y**

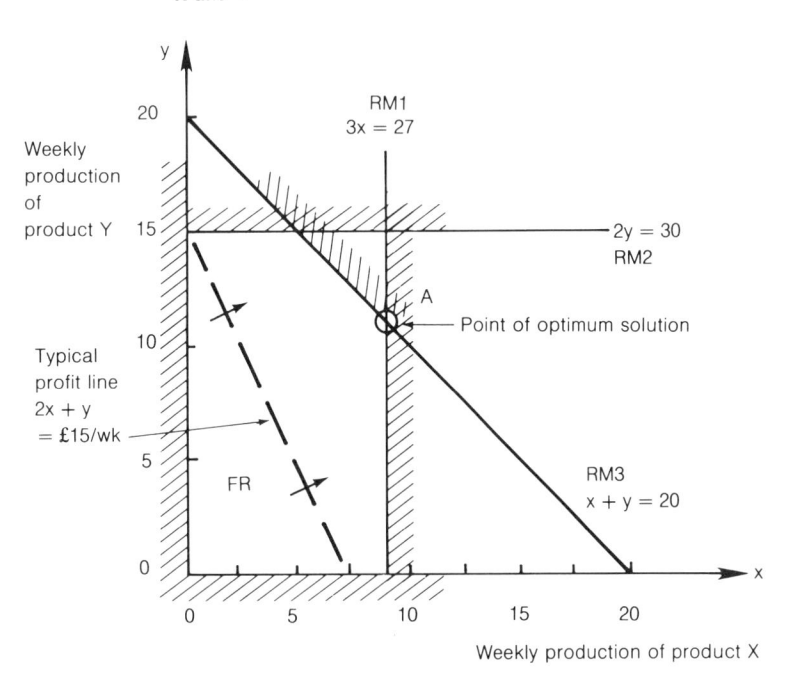

The point $x = 5$, $y = 5$ is in the feasible region. The weekly profit at this point is:

$$P = 2 \times 5 + 5 = \text{£15/week}$$

Take the trial profit line as:

$$15 = 2x + y \quad (\text{£/week})$$

The point $x = 0$, $y = 15$ also lies on this line. The line is shown on the above graph. If the line is moved parallel to this position, in the direction of increasing profit, we can see that the optimum corner is A. A is the intersection of the RM1 and RM3 constraints. Solving these equations gives:

$$3x = 27, \text{ therefore } x = 9$$

substitute in:

$$x + y = 20, \text{ therefore } y = 11$$

The optimum product mix is to produce 9 units of product X and 11 units of product Y each week. The maximum weekly profit is then:

$$P_{max} = 2 \times 9 + 11 = £29/\text{week}$$

Raw materials 1 and 3 are used up but there is spare capacity on raw material 2 of:

$$2 \times 11 + s_2 = 30,$$

therefore $s_2 = 8$ kg per week.

Solution: Simplex method. Arrange the coefficients in the left hand side of the constraint equations in a matrix format. Label the columns with the name of the variables to which they refer. Put the right hand side values of the constraints in a separate column on the right of the matrix. Label the rows with the names of the variables which are basic (have non-zero values) at the initial corner (the origin). Finally, add the objective function as an additional row to the table. There are several slightly different methods of implementing the simplex method. The one explained here requires the coefficients of the objective function to be entered as negative values. The resulting matrix is called the first **simplex tableau**. The above procedure is *Step 1* in the algorithm.

TABLE 9.2 **First simplex tableau**

Basic variables	x	y	s_1	s_2	s_3	Right hand side, b
s_1	3	0	1	0	0	27
s_2	0	2	0	1	0	30
s_3	1	1	0	0	1	20
Objective function, P	−2	−1	0	0	0	0

Step 2: Find the largest negative value in the objective function row (−2). The corresponding column is called the pivotal column, x. Divide the right hand side values (in the 'b' column) by the corresponding numbers in the pivotal column. This produces a set of ratios.

TABLE 9.3 **First simplex tableau with ratios**

Basic variables	x	y	s_1	s_2	s_3	RHS b	Ratios b/pivotal column element
s_1	3*	0	1	0	0	27	27/3 = 9 ← Pivotal row
s_2	0	2	0	1	0	30	30/0 = ∞
s_3	1	1	0	0	1	20	20/1 = 20
Objective function, P	−2 ↑	−1	0	0	0	0	

Pivotal column, x

Step 3: Choose the smallest positive ratio, 9. The corresponding row, s_1, is the pivotal row. The intersection of the pivotal column, x, and the pivotal row, s_1, is the pivotal element, 3, marked by * in TABLE 9.3 above.

Step 4: Divide all of the elements in the pivotal row by the pivotal element, 3. Replace the pivotal row by this new row in TABLE 9.4. Replace the row label, s_1, by the label from the pivotal column, x. The new row labels are the basic variables for the second basic solution.

LINEAR PROGRAMMING: CHAPTER NINE

Step 5: Using arithmetic operations on the rows (row operations in matrix algebra), reduce all of the other elements in the pivotal column, x, to zero. These arithmetic operations must use only the pivotal row as the basis.

R_i denotes the ith row. The notation New R_3 = Old R_3 − New R_1 means that the new row 3 is obtained by subtracting the new pivotal row (row 1) from the old row 3. The operations used are listed at the right hand side of TABLE 9.4.

TABLE 9.4 **Second simplex tableau**

Basic variables	x	y	s_1	s_2	s_3	RHS b	
x	1	0	1/3	0	0	9	New R_1 = Old R_1 ÷ pivotal element (3)
s_2	0	2	0	1	0	30	New R_2 = Old R_2 − 0 × New R_1
s_3	0	1	−1/3	0	1	11	New R_3 = Old R_3 − 1 × New R_1
Objective function, P	0	−1	2/3	0	0	18	New P = Old P − (−2) × New R_1

Step 6: Repeat steps 2 to 5 until all of the elements in the objective function row are positive or zero.

TABLE 9.5 **Second simplex tableau with ratios**

Basic variables	x	y	s_1	s_2	s_3	RHS b	Ratios b/pivotal column element
x	1	0	1/3	0	0	9	9/0 = ∞
s_2	0	2	0	1	0	30	30/2 = 15
s_3	0	1*	−1/3	0	1	11	11/1 = 11 ← pivotal row
Objective function, P	0	−1 ↑	2/3	0	0	18	

pivotal column, y

TABLE 9.6 **Third and final simplex tableau**

Basic variables	x	y	s_1	s_2	s_3	RHS b	
x	1	0	1/3	0	0	9	New R_1 = Old R_1 − 0 × New R_3
s_2	0	0	2/3	1	−2	8	New R_2 = Old R_2 − 2 × New R_3
y	0	1	−1/3	0	1	11	New R_3 = Old R_3 ÷ pivotal element (1)
Objective function, P	0	0	1/3	0	1	29	New P = Old P − (−1) × New R_3

All values in the objective function row are now positive or zero, therefore this tableau represents the optimum solution.

To interpret the final simplex tableau, we will look at the values around the edge of the table first.

TABLE 9.7 **Partial final simplex tableau**

Basic variables	Variables x y s_1 s_2 s_3	RHS b
x		9 = value of x
s_2		8 = value of slack for RM2 = s_2
y		11 = value of y
Objective function, P	0 0 1/3 0 1	29 = value of maximum profit

Cost of forcing 1 unit of a non-basic variable into the solution. → (0 0)

shadow prices → (1/3 0 1)

The basic variables are those which are non-zero at the optimum corner. The values of the basic variables are in the corresponding row of the 'b' column. Therefore:

$$x = 9 \text{ units/week}$$

$$y = 11 \text{ units/week}$$

and the slack for raw material 2 is 8 kg per week.

All other variables are zero, that is, the slack on constraint 1, s_1, and constraint 3, s_3, are zero. This means that these constraints are binding and the available raw material 1 and 3 are fully used.

The optimum value of the objective function is in the 'b' column of the objective function row. The maximum value of weekly profit is £29. This solution corresponds exactly with the graphical solution. The figures in the objective function row and the slack variable columns shown in TABLE 9.7 give the **shadow prices**. The shadow price on constraint 1, RM1, is £1/3 per kg and the shadow price for constraint 3 is £1 per kilogramme. This means that if an extra kilogramme of RM1 becomes available, the weekly profit will increase by 33 pence (less any additional costs above the normal cost of RM1). Similarly, if an extra kilogramme of RM3 becomes available, the weekly profit will increase by £1 (less any additional costs). The figures for the shadow prices may be checked from the graphical solution. We will look at constraint 1 only to illustrate the point.

Constraint 1, for raw material 1, is 3x = 27 kg per week. If this constraint is relaxed by one kilogramme, 3x = 28. The optimum corner will still be the intersection of constraints 1 and 3. Look back at the graph to check this. The new optimum corner has the co-ordinates:

$$x = 28/3 = 9\tfrac{1}{3}$$

and:

$$28/3 + y = 20$$

giving:

$$y = 32/3 = 10\tfrac{2}{3}$$

The new value of the maximum weekly profit is:

$$2 \times (28/3) + (32/3) = 88/3 = £29.33/\text{week}$$

This is a gain of 33 pence for one kilogramme. The shadow price for RM1 is 33 pence per kilogramme.

The remaining values on the edge of the final tableau, are those in the objective function row and the variable columns. In this example, the values in both the x and the y column are zero. These numbers will be non-zero if any of the variables are non-basic in the optimum solution. For example, if the optimum solution had said that we should produce only product X, then y would be non-basic, ie y = 0. In that case, the figure in the y column of the objective function row tells us by how much the maximum value of the objective function would decrease if we insisted on producing one unit of y.

Suppose the following had been the final tableau for the current problem:

TABLE 9.8 **Modified final tableau**

Basic variables	x	y	s_1	s_2	s_3	RHS b
x						9
s_2						8
s_3						11
Objective function, P	0	0.5	1/3	0	0	18 = value of maximum profit

In this solution, the optimum product mix is to produce 9 units of X and none of Y. If we feel that we must produce some units of Y, the value of the objective function will decrease by £0.5 for each unit of Y which is produced.

9.6 SENSITIVITY ANALYSIS AND THE SIMPLEX METHOD

The final tableau of the simplex algorithm can be used to carry out a sensitivity analysis of the solution of a linear programming model. For a limiting constraint, the values in the corresponding slack variable column represent the change in the values of the basic variables if one additional unit of the limiting resource is available.

Example 9.9: To illustrate the use of the final simplex tableau for a sensitivity analysis of the limiting constraints

We will use the final simplex tableau in TABLE 9.8 of **Example 9.8** to determine:

1. the effect on the optimum solution if one additional kilogramme of RM1 becomes available;
2. the effect on the optimum solution if two additional kilogrammes of RM1 become available;
3. the effect on the optimum solution if five additional kilogrammes of RM3 become available;
4. the maximum number of additional kilogrammes of RM3 which can be used without spare capacity being created;
5. the effect on the optimum solution if two fewer kilogrammes of RM1 are available.

Solution: the linear programme and the final simplex tableau are now reproduced.

Maximise the weekly profit, £P, where P = 2x + y   (£/week)

$$\text{subject to RM1: } 3x \leq 27 \text{ kg/week}$$
$$\text{RM2: } 2y \leq 30 \text{ kg/week}$$
$$\text{RM3: } x + y \leq 20 \text{ kg/week}$$
$$x, y \geq 0$$

TABLE 9.9 **Final simplex tableau**

Basic variables	|	x	y	Variables s_1	s_2	s_3	|	RHS b
x	|	1	0	1/3	0	0	|	9
s_2	|	0	0	2/3	1	−2	|	8
y	|	0	1	−1/3	0	1	|	11
Objective function, P	|	0	0	1/3	0	1	|	29

1 If one additional kilogramme of RM1 is available, this limiting constraint is relaxed by one kilogramme. The values in the s_1 column are the changes in the basic variables which result from this relaxation. The final tableau is re-written below with only the relevant values and calculations shown.

TABLE 9.10 **Partial modified final simplex tableau (1 kg extra of RM1)**

Basic variables	|	x	y	Variables s_1	s_2	s_3	|	RHS modified b
x	|			1/3			|	$9 + (1/3) = 9\frac{1}{3}$
s_2	|			2/3			|	$8 + (2/3) = 8\frac{2}{3}$
y	|			−1/3			|	$11 − (1/3) = 10\frac{2}{3}$
Objective function, P	|			1/3			|	$29 + (1/3) = 29\frac{1}{3}$

One additional kilogramme of RM1 causes the value of x to increase by $\frac{1}{3}$ of a unit, the slack for RM2 to increase by $\frac{2}{3}$ of a kilogramme, the value of y to decrease by $\frac{1}{3}$ of a unit and the maximum value of the weekly profit to increase by £$\frac{1}{3}$, ie by the shadow price. The new

FIGURE 9.24 **A linear programme for the weekly production of X and Y 1 extra kg of RM1**

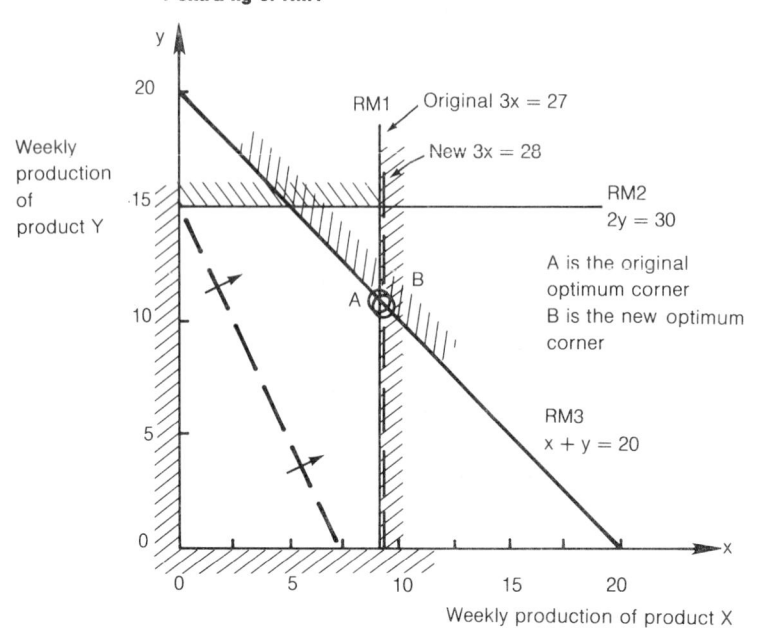

optimum solution requires $9\frac{1}{3}$ of product X and $10\frac{2}{3}$ of product Y to be produced each week. The slack on constraint 2, the amount of raw material 2 not used, is $8\frac{2}{3}$ kg. The other variables are zero. The slack on constraints 1 and 3 is zero, therefore all RM1 and RM3 are used. This means that these constraints are binding. The maximum value of weekly profit is now £29.33. This corresponds with the graphical solution. (See FIGURE 9.24).

2 If two additional kilogrammes of RM1 are available this limiting constraint is relaxed by two kilogrammes. The values in the s_1 column are multiplied by two. The resulting values are then the changes in the values of the basic variables which arise from the additional two kilogrammes. The final tableau is shown in Table 9.11 with only the relevant values and calculations shown.

TABLE 9.11 **Partial modified final simplex tableau (2 kg extra of RM1)**

Basic variables	x	y	s_1	s_2	s_3	RHS modified b
x			1/3 × 2			9 + (2/3) = $9\frac{2}{3}$
s_2			2/3 × 2			8 + (4/3) = $9\frac{1}{3}$
y			−1/3 × 2			11 − (2/3) = $10\frac{1}{3}$
Objective function, P			1/3 × 2			29 + (2/3) = $29\frac{2}{3}$

The new optimum solution requires $9\frac{2}{3}$ of product X and $10\frac{1}{3}$ of product Y to be produced each week. The slack on constraint 2 is $9\frac{1}{3}$ kg. The other variables, s_1 and s_3, are zero. This means that these constraints are binding. The maximum value of weekly profit is £29.67. This solution can be illustrated in the same way as **1** above.

3 If five additional kilogrammes of RM3 are available this limiting constraint is relaxed by five kilogrammes. The values in the s_3 column are multiplied by five. The resulting changes in the basic variables, arising from the additional five kilogrammes, are shown in the modified final tableau below (TABLE 9.12).

TABLE 9.12 **Partial modified final simplex tableau (5 kg extra of RM3)**

Basic variables	x	y	s_1	s_2	s_3	RHS modified b
x					0 × 5	9 + (0) = 9
s_2					−2 × 5	8 + (−10) = −2 ←
y					1 × 5	11 + (5) = 16
Objective function, P					1 × 5	29 + (5) = 34

A problem has now arisen. The value of the slack variable, s_2, for raw material 2, has become negative. This is not allowed, since variables must always be positive or zero. If we consult the graphical solution we can see at once what has happened. The RM3 constraint has been relaxed so far that it is no longer limiting. The tableau gives a point outside the feasible region. We are not able to use all of the extra five kilogrammes of RM3. This problem is discussed further in section **4**.

278 QUANTITATIVE ANALYSIS

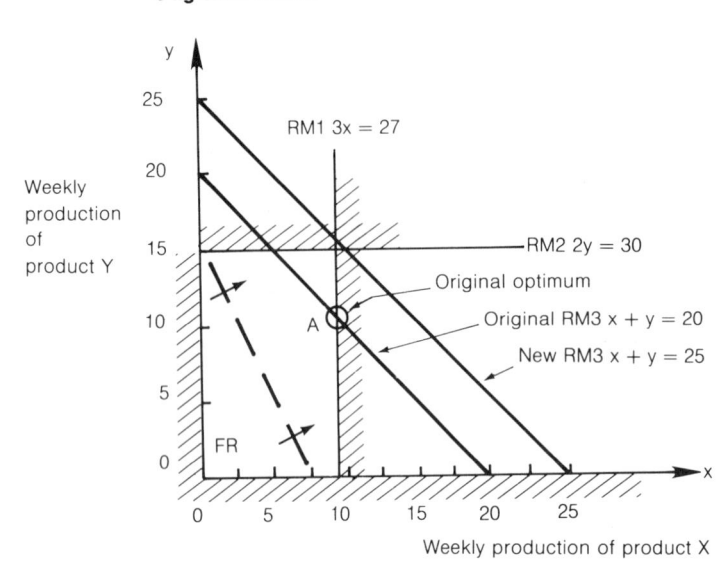

FIGURE 9.25 **A linear programme for the weekly production of X and Y 5 kg extra of RM3**

4 The RM3 constraint is represented by the s_3 column of the final tableau. The only negative value in the s_3 column is the marginal value of s_2, which is -2. As RM3 is relaxed, the value of s_2 decreases by 2, but it cannot be negative. When s_2 reaches zero, the limiting position for the RM3 constraint will occur. Suppose this limiting position is reached when the RM3 constraint has been relaxed by r kg, the value of s_2 will then be zero, therefore:

$$8 + (-2 \times r) = 0$$

which means that r = 4 kg. The RM3 constraint may be relaxed by 4 kg, from 20 to 24 kg, before it ceases to be binding. This takes it to the intersection of the RM1 and RM2 constraints, where x = 9, y = 15 and the RM3 constraint is x + y = 24.

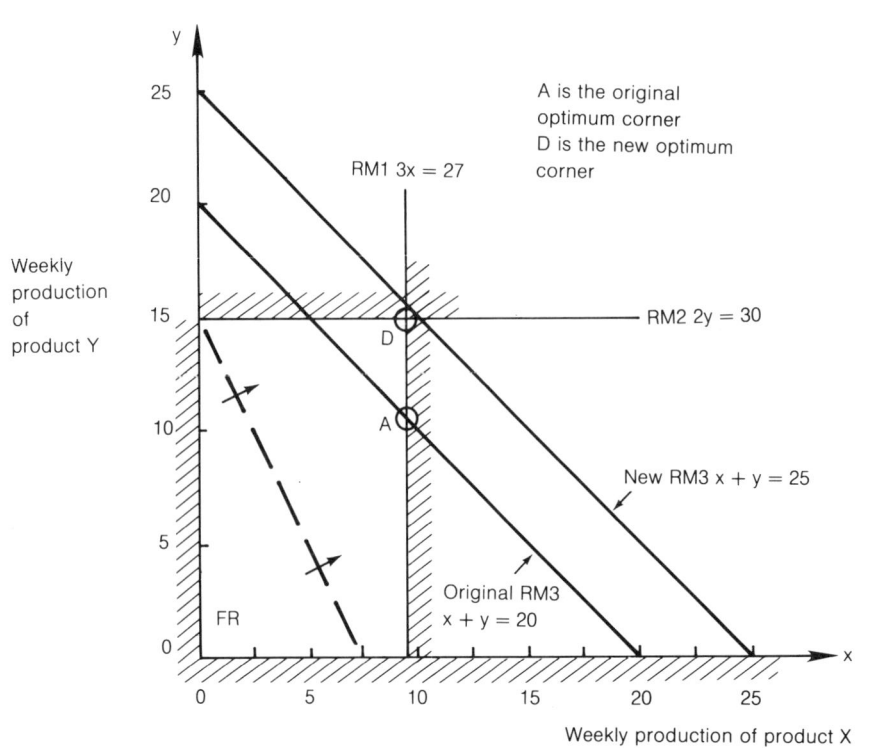

FIGURE 9.26 **A linear programme for the weekly production of X and Y with maximum usable RM3**

5 If two fewer kilogrammes of RM1 are available, this limiting constraint is tightened by two kilogrammes. The values in the s_1 column are multiplied by two. The resulting values are deducted from the values of the basic variables. The final tableau is re-written below.

TABLE 9.13 **Partial modified final simplex tableau (2 kg less of RM1)**

Basic variables	Variables x y s_1 s_2 s_3	RHS modified b
x	1/3 × 2	9 − (2/3) = $8\frac{1}{3}$
s_2	2/3 × 2	8 − (4/3) = $6\frac{2}{3}$
y	−1/3 × 2	11 − (−2/3) = $11\frac{2}{3}$
Objective function, P	1/3 × 2	29 − (2/3) = $28\frac{1}{3}$

The new optimum solution requires $8\frac{1}{3}$ of product X and $11\frac{2}{3}$ of product Y to be produced each week. The slack on constraint 2 is $6\frac{2}{3}$ kg. The slack on constraints 1 and 3 is zero. This means that these constraints are binding. The maximum value of weekly profit is £28.33. This solution is shown in FIGURE 9.27:

FIGURE 9.27 **A linear programme for the weekly production of X and Y (2 kg less of RM1)**

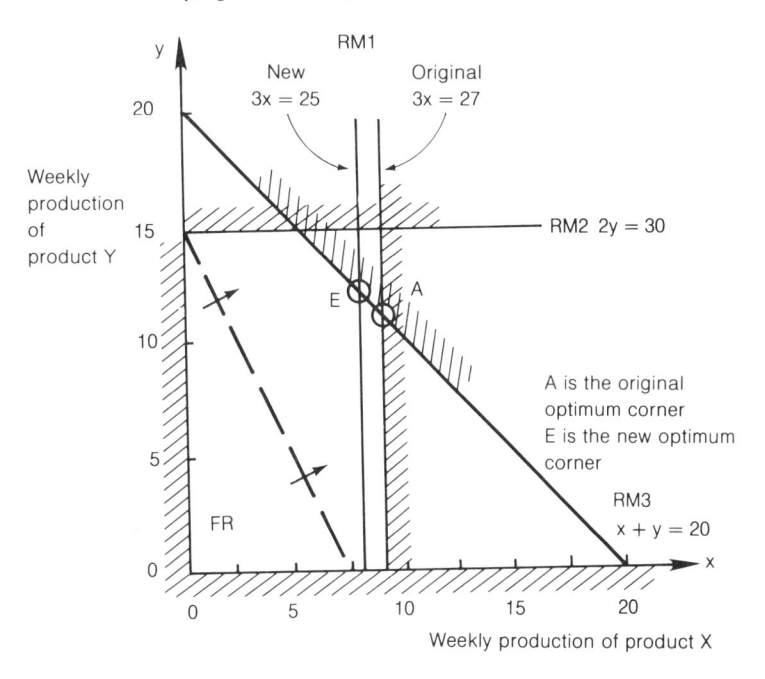

A is the original optimum corner
E is the new optimum corner

This analysis is tedious to do by hand using the simplex method, even with the simplest two variable model. All commercial standard linear programming computer packages will provide the information. This is how all multivariable sensitivity analysis is done in practice. The principles are exactly the same as those for the two variable model we have just completed.

9.7 THE DUAL LINEAR PROGRAMMING MODEL

The dual linear programming model is used to investigate a problem from a different perspective to the one obtained from the usual primal model. The primal and dual models give the same solution and the same sensitivity information. The only reason for using one rather than the other is that computationally one may be easier to solve. With increasingly powerful computer packages the need for the primal/dual switch is becoming less relevant. The variables in the dual model are the shadow prices of the original, or primal, model. The structure of the dual and primal models are similar. If the primal model has been built, the corresponding dual model is derived from it. In general, a linear programming problem can be described by:

$$\text{Maximize } Z = c_1 x_1 + c_2 x_2 + \cdots + c_n x_n$$

Subject to a set of m linear constraints:

$$a_{11} x_1 + a_{12} x_2 + a_{13} x_3 + \cdots + a_{1n} x_n \leq b_1$$
$$a_{21} x_1 + a_{22} x_2 + a_{23} x_3 + \cdots + a_{2n} x_n \leq b_2$$
$$a_{31} x_1 + a_{32} x_2 + a_{33} x_3 + \cdots + a_{3n} x_n \leq b_3$$
$$\vdots \qquad \vdots \qquad \vdots \qquad \vdots$$
$$a_{m1} x_1 + a_{m2} x_2 + a_{m3} x_3 + \cdots + a_{mn} x_n \leq b_m$$
$$x_i \geq 0$$

The above linear programme maximises and has all $\leq$ constraints. Any linear programming model can be put into this form and converted into its dual, as we show below. The dual model is:

$$\text{Minimise } G = b_1 y_1 + b_2 y_2 + \cdots + b_m y_m$$

subject to a set of n linear constraints.

$$a_{11} y_1 + a_{21} y_2 + a_{31} y_3 + \cdots + a_{m1} y_m \geq c_1$$
$$a_{12} y_1 + a_{22} y_2 + a_{32} y_3 + \cdots + a_{m2} y_m \geq c_2$$
$$a_{13} y_1 + a_{23} y_2 + a_{33} y_3 + \cdots + a_{m3} y_m \geq c_3$$
$$\vdots \qquad \vdots \qquad \vdots$$
$$a_{1n} y_1 + a_{2n} y_2 + a_{3n} y_3 + \cdots + a_{mn} y_m \geq c_n$$
$$y_i \geq 0$$

There are m dual variables, y, one for each of the m primal constraints, and n constraints, one for each of the x variables in the primal. The coefficients, c, in the primal objective function and the right hand side values, b, of the constraints in the primal, are interchanged in the dual. The coefficients in the left hand sides of the primal constraints are interchanged row to column in the dual. The dual variables, y, are the shadow prices in the primal problem and vice versa. In this case the dual minimises, the primal maximises. The primal has $\leq$ constraints and the dual $\geq$ constraints.

Example 9.10: To set up and interpret the dual linear programme

A firm makes two products, R and Q, both of which require two raw materials, RM1 and RM2. Each kilogramme of product R requires 2 kg of RM1 and 3.5 kg of RM2. Each kilogramme of product Q requires 3 kg of RM1 and 1.5 kg of RM2. Each week, 10 kg of RM1 and 12 kg of RM2 are available. There is an unlimited supply of labour and machine time and the firm can sell all its production. The unit profit on product R is £5 and on product Q is £8.

1   Set up a profit maximising linear programming model for this problem.
2   Set up the dual linear programming model.
3   Explain the relationship between the two models in **1** and **2**.
4   Find the optimum solution for the two models graphically.

Solution:

1   Produce x_1 kg of product R and x_2 kg of product Q each week. Maximise weekly profit, £P where:

$$P = 5x_1 + 8x_2 \qquad (\text{£/week})$$

subject to:

$$\text{RM1:} \quad 2x_1 + 3x_2 \leq 10 \text{ kg/week}$$
$$\text{RM2:} \quad 3.5x_1 + 1.5x_2 \leq 12 \text{ kg/week}$$
$$x_1, x_2 \geq 0$$

2 Using the primal model, the dual model is:

$$\text{Minimise: } G = 10y_1 + 12y_2 \quad (\text{£/week})$$

subject to:

$$\text{Product R: } 2y_1 + 3.5y_2 \geq 5 \text{ £/unit}$$
$$\text{Product Q: } 3y_1 + 1.5y_2 \geq 8 \text{ £/unit}$$
$$y_1, y_2 \geq 0$$

3 *Primal model:* The variables are the amount of each product to be produced each week. The objective function is the total profit per week from the production of R and Q. Each constraint refers to one raw material. The left hand side of the constraint gives the total requirement for that raw material by both of the products. The right hand side gives the total raw material available each week.

Dual model: the variables are the primal shadow prices, that is, the amount which would be added to the value of the objective function if one more unit of the raw material was available. The shadow prices represent the value of one unit of the raw material. The objective function is the total value per week of the raw materials used in the production of R and Q. Each constraint refers to one product. The left hand side of the constraint gives the total value of both raw materials used to make one kilogramme of that product. The right hand side gives the unit profit generated by that product. Let us look at the dual model again and try to identify the individual components.

Minimise G:

$$\underset{\begin{pmatrix}\text{RM1 availability}\\ \text{per week}\end{pmatrix} \times \begin{pmatrix}\text{value/kg}\\ \text{of RM1}\end{pmatrix}}{10 \quad y_1} + \underset{\begin{pmatrix}\text{RM2 availability}\\ \text{per week}\end{pmatrix} \times \begin{pmatrix}\text{value/kg}\\ \text{of RM2}\end{pmatrix}}{12 \quad y_2} \quad (\text{£/week})$$

Product R:

$$\underset{\begin{pmatrix}\text{kg of RM1/kg}\\ \text{of R}\end{pmatrix} \times \begin{pmatrix}\text{value/kg}\\ \text{of RM1}\end{pmatrix}}{2 \quad y_1} + \underset{\begin{pmatrix}\text{kg of RM2/kg}\\ \text{of R}\end{pmatrix} \times \begin{pmatrix}\text{value/kg}\\ \text{of RM2}\end{pmatrix}}{3.5 \quad y_2} \geq £5 \text{ profit per R}$$

Product Q:

$$\underset{\begin{pmatrix}\text{kg of RM1/kg}\\ \text{of Q}\end{pmatrix} \times \begin{pmatrix}\text{value/kg}\\ \text{of RM1}\end{pmatrix}}{3 \quad y_1} + \underset{\begin{pmatrix}\text{kg of RM2/kg}\\ \text{of Q}\end{pmatrix} \times \begin{pmatrix}\text{value/kg}\\ \text{of RM2}\end{pmatrix}}{1.5 \quad y_2} \geq £8 \text{ profit per Q}$$

Each constraint says that the total value of the raw materials used in that product must be more than or equal to the unit profit on that product. The solution of either the primal or the dual model enables us to solve the other model.

4 The graphical solution for the primal problem is given below.

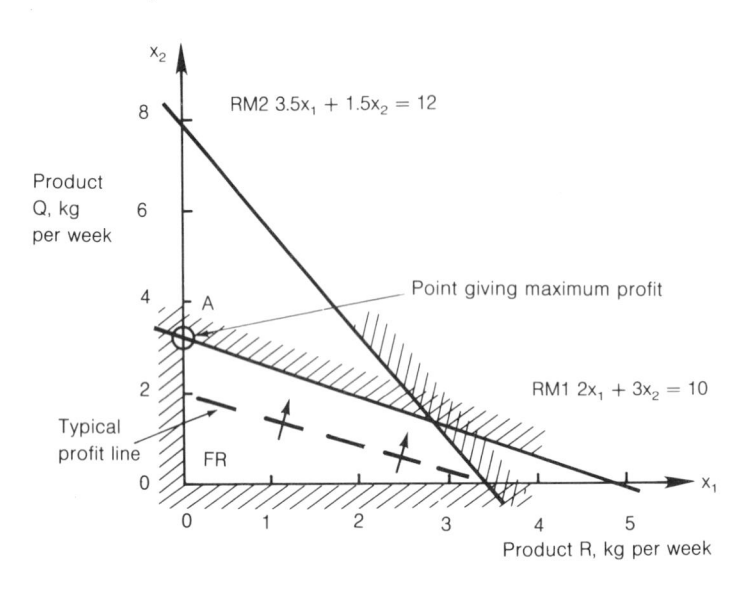

FIGURE 9.28 **Primal model**

The optimum solution is at point A, the intersection of the raw material 1 constraint and the Q axis. To maximise profit we should produce $3\frac{1}{3}$ kg of Q and zero kilogrammes of R. All raw material 1 will be used, but not all of RM2. The maximum profit is $3\frac{1}{3} \times 8 = £26.67$ per week. The graphical solution for the dual is shown below.

FIGURE 9.29 **Dual model**

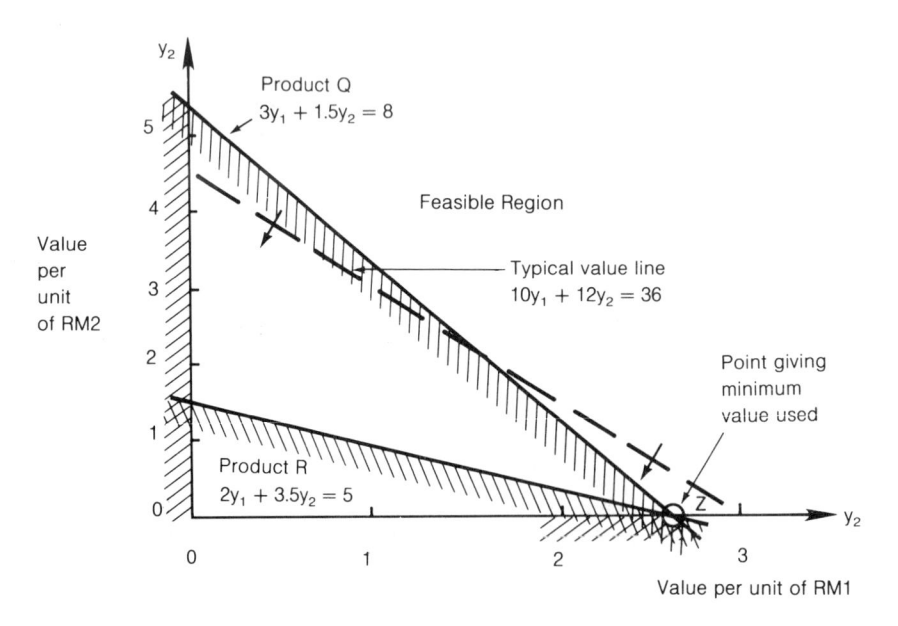

This is a minimisation problem. We wish to make the objective function as small as possible, hence we move towards the origin, parallel to a trial objective function. The last point in the feasible region is given by Z. This is the optimum solution for the dual. Z is the intersection of the product Q constraint and the y_1 axis ie:

$$y_2 = 0$$

and:

$$3y_1 + 1.5y_2 = 8$$

therefore:

$$y_2 = 0 \quad \text{and} \quad y_1 = 2\tfrac{2}{3}$$

The minimum value for the dual problem is:

$$G = 10 \times 2\tfrac{2}{3} + 12 \times 0 = £26.67 \text{ per week}$$

This is the same value for the objective function as the one for the primal model.

The two solutions combined tell us that the maximum profit is £26.67 per week, when we produce $3\tfrac{1}{3}$ kg of Q and none of R. The value of the raw materials, the shadow price, is £2.67 per kg of RM1 and zero for RM2. This is the same as the information which we would derive from a full analysis of the primal problem alone.

SUMMARY

Linear programming models are used to allocate scarce resources in a way which meets a business objective. The objective might be to maximise weekly profit or to minimise daily costs. The steps in formulating a linear programming model are:

Step 1: Identify the decision variables.
Step 2: Identify the linear objective function and the constraints.
Step 3: Express the objective function in terms of the variables.
Step 4: Express the constraints in terms of the variables.

The same method of formulation applies to two variable and multivariable models. The two variable model, however, can be solved graphically. The constraints, which are usually inequalities, either $\leq$ or $\geq$, are represented on the graph by lines and areas. Each constraint divides the graph into a rejected area and an acceptable area. The area in which all of the constraints are satisfied is called the feasible region. This feasible region contains all possible solutions to the problem.

The optimum point, which is always at a corner of the feasible region, is found by plotting a typical objective function on the graph. The objective function is moved, parallel to this trial line, away from the origin if the objective is to maximise, or towards the origin if the objective is to minimise. The last point that this line touches before it completely leaves the feasible region gives the values of the variables which will optimise the objective function.

Sensitivity analysis is very important in linear programming since the values used in the model may be subject to uncertainty. The procedure allows us to consider variation and uncertainty in the objective function coefficients and in the right hand side values of the constraints.

Multivariable linear programming models are solved by computer using the simplex method. The final tableau of the simplex algorithm provides the optimum value of the objective function, the values of the decision variables and the values of the slack or surplus variables. In addition, it gives the shadow prices of the resources. The final tableau can also be used in sensitivity analysis to show the full effect of variation in the scarce resources, on the objective function, and on each of the constraints.

Each primal linear programme has a dual formulation. The solutions to the primal and the dual are identical. The dual may be derived from the primal model by interchanging the role of the coefficients in the model. There are sometimes advantages in solving a simpler dual, rather than a complex primal, formulation.

EXERCISES

Exercise 9.1

GRM plc produce two types of breakfast cereal, 'Crunchy' and 'Chewy' at their Pond Street plant. The two products use basically the same ingredients and there is not usually any shortage of these. The main constraint on production is the availability of labour in each of three processing departments.

Joe Dyson, the Production Manager, plans production on a monthly basis. The labour availability and requirements for the two products are given in the table below:

	Labour requirements hours/ton 'Crunchy'	'Chewy'	Labour availability hours/month
Department A, Roasting	10	4	1000
Department B, Blending	3	2	360
Department C, Packing	2	5	800

The contribution per tonne is £150 for 'Crunchy' and £75 for 'Chewy'. At the moment there is no limit to potential sales. GRM sell all that they produce.

Required:
Formulate a linear programming model for this problem which maximises the total contribution per month.

Answer on page 538.

Exercise 9.2

Oliver A Peters is about to retire and is considering what to do with the lump sum which he will receive from his firm's pension scheme. Mr Peters and his wife intend to make an extended visit to their daughter in Australia in two years' time, hence, any investment made now will be released at that time. Mr Peters' objective is, therefore, to maximise the total return on his investment over this two year period.

He has been advised that a managed fund would be the best option and he is now considering one which consists of three types of investment, A, B and C. The lump sum will amount to £25,000 but Mr Peters does not think it is essential to invest all the money in the managed fund; he is willing to put some into his building society account which will earn 9% per annum interest.

According to the firm's accountant, Mr Peters should try to balance his investment by generating both income and capital growth. Hence, he advises investing at least 40% of the total sum in the building society and investment A. In order to achieve good capital growth at least 25% of the monies invested in the fund should be in investment B, but no more than 35% since B is slightly risky. For safety, at least 50% of the fund monies should be in A and C.

At the moment, investment A earns 10% per annum and has 1% per annum capital growth; investment B, however, has 15% per annum capital growth; investment C earns 4% per annum but has 5% per annum capital growth.

Required:
Given Mr Peters' objective, formulate a linear programming model to show how the lump sum should be distributed between the various investments.

Answers on page 538.

Exercise 9.3

The Caterpillar China Company Ltd produces a range of five similar products, A, B, C, D and E. The following table shows the quantity of each of the required inputs necessary to produce one unit of each product, together with the weekly inputs available and the selling price per unit of each product.

	Product					Weekly inputs
Inputs	A	B	C	D	E	available
Raw materials (kg)	6.0	6.5	6.1	6.1	6.4	35,000
Forming (hours)	1.00	0.75	1.25	1.00	1.00	6000
Firing (hours)	3	4.5	6	6	4.5	30,000
Packing (hours)	0.50	0.50	0.50	0.75	1.00	4000
Selling price (£)	40	42	44	48	52	

The costs of each input are as follows:

> Materials £2.10 per kg
> Forming £3.00 per hour
> Firing £1.30 per hour
> Packing £8.00 per hour

Required:

1. In order to maximise the weekly contribution to profit a linear programming package on a computer is to be used.

   (a) Formulate this problem so that the data can be input, giving both the objective function and the constraints.

   (b) State briefly the assumptions necessary for your model to be suitable.

(ACCA, December 1987)

Answers on page 539.

Exercise 9.4

Refer to Exercise 9.1.

Required:
Use the linear programming model you set up in this question to advise Joe Dyson on the optimum production mix for the next month, if the company policy is to maximise total contribution per month. What will be the maximum contribution?

Answers on page 540.

Exercise 9.5

The PT Oil Company adds chemicals to its diesel fuel oil to improve the starting performance and to reduce the freezing point. Each 1000 litre tank of diesel should contain at least 40 mg of chemical X, at least 14 mg of chemical Y and at least 18 mg of chemical Z. Two chemical companies, A and B, supply PT Oil with these chemicals in a ready-mixed form. The amounts of each chemical in the two products is given in the table below:

	Chemical		
	X mg/litre	Y mg/litre	Z mg/litre
Product A	4	2	3
Product B	5	1	1

Product A costs £1.50 per litre and product B costs £3.00 per litre.

Required:

Find the mix of products A and B which minimises the total cost of adding the chemicals to the diesel.

Answers on page 541.

Exercise 9.6

Refer once again to **Exercises 9.1** and **9.4**. The Managing Director of GRM plc is willing to agree to additional labour being created by overtime work from the existing employees. Joe Dyson has assembled the following information on the costs of overtime in the three departments and the likely availability.

Department	Normal labour cost/hour, £	Overtime labour cost/hour, £	Maximum number of overtime hours available/month
Roasting	4.50	6.50	150
Blending	4.75	6.50	100
Packing	3.50	4.50	80

In order to keep the administrative work and costs to a minimum, Joe Dyson decides that the overtime will be given to one department only, at least in the short term.

Required:

Using the information given above, decide which department should be given the overtime and the maximum number of overtime hours it can use.

Answers on page 542.

Exercise 9.7

Slims plc produce three kinds of malted drink powder. One of these they sell as a health drink because it has less sugar; one they sell to hospitals as an invalid food as it has added vitamins; the third one is a standard product.

The main ingredients, with their costs and normal weekly availabilities, are given in the table below, as are the estimated maximum weekly demands for the three products.

	Required kg per kg of product			Estimated maximum demand per week kg	Sale price per kg £
	Sugar	Malt extract	Skimmed milk powder		
Standard drink	0.30	0.30	0.35	2000	1.00
Health drink	0.15	0.25	0.55	1800	1.20
Invalid drink	0.15	0.30	0.25	1200	1.50
Costs per kg of raw material	20p	60p	50p		
Availability per week of raw materials, kg	1000	1250	2200		

There is an unlimited supply of vitamin additives. Other variable costs are 10p per kg for the standard drink, 9p per kg for the health drink and 12p per kg for the invalid drink.

Required:
1 Formulate the linear programming model for this problem with the objective of maximising total contribution per week.
2 The final simplex tableau for the solution to this problem is given below:

	m	h	i	s_1	s_2	s_3	s_4	s_5	s_6	b
s_1	0	0	0	1	−1	0	0	0.1	0.15	110
m	1	0	0	0	3.333	0	0	−0.833	−1	1466.67
s_3	0	0	0	0	−1.167	1	0	−0.258	0.1	396.67
s_4	0	0	0	0	−3.333	0	1	0.833	1	533.33
h	0	1	0	0	0	0	0	1	0	1800
i	0	0	1	0	0	0	0	0	1	1200
P	0	0	0	0	1.617	0	0	0.251	0.56	3144.33

Determine:
(a) the optimum product mix;
(b) the maximum value of the weekly contribution;
(c) the spare capacity on the constraints.

3 Use the tableau in part **2** to answer the following questions:
(a) Recent market research suggests that the health drink is gaining popularity. A revised maximum demand is for 2500 kg per week. How will this affect the optimum product mix?
(b) Management are thinking of buying in some extra malt extract. However, they would have to use a different supplier and would have to pay 80p per kg for it. Would this allow the total weekly contribution to be improved? If so, what is the maximum amount which should be purchased from this new supplier?

Answers on page 543.

Exercise 9.8

Refer to **Exercise 9.1**.

Required:
1 Set up the dual linear programming model for this problem.
2 Interpret the meaning of the dual variables in the context of the question.

Answers on page 545.

Exercise 9.9

Princetown Paints Ltd manufactures three basic types of paint—emulsion, gloss and undercoat—using the same mixing machines and direct labour for each of the three products. The management accountant of Princetown Paints was faced with the task of arranging the weekly production for his company. Information about the sales price and costs per 100 litres is given in the following table (all figures in £).

	Emulsion	Gloss	Undercoat
Sales price (per 100 litres)	120	126	110
Variable costs (per 100 litres):			
Direct material costs	11	25	20
Direct labour costs	30	36	24
Mixing costs	32	20	36
Other variable costs	12	15	10

The cost of direct labour is £3 per hour and the variable cost of mixing is £4 per hour. In any one week, direct labour hours are restricted to 8000 hours and mixing machine hours are restricted to 5900 hours.

Due to contractual arrangements the company must produce exactly 25,000 litres of undercoat each week. There is a maximum weekly demand for emulsion paint of 35,000 litres and for gloss paint of 29,000 litres.

Required:

1. Formulate the linear programming problem to determine the weekly production levels for emulsion and gloss paint that maximise contribution to profit.
2. Solve the problem graphically. State the optimum weekly production levels and the corresponding contribution to profit.
3. Calculate by how much the sales price of emulsion paint must change from its current level before the optimal solution will change.
4. Suppose that the labour force is prepared to work overtime at a premium of £1 per overtime hour worked. Would it be profitable to work overtime? If so, how many hours would you recommend should be worked, and what would be the extra profit from these hours?

(*ACCA, June 1988*)

Answers on page 545.

Exercise 9.10

As part of a corporate rationalisation programme, the Nemesis Company has decided to merge its two factories at Abbotsfield and Birchwood. The Abbotsfield factory will be closed down and production capacity will be expanded at Birchwood. At the present time, the numbers of skilled and unskilled men employed at the two factories are:

	Abbotsfield	Birchwood
Skilled	200	100
Unskilled	300	200
Total	500	300

whereas after the merger, the enlarged factory at Birchwood will employ 240 skilled men and 320 unskilled men.

After extensive negotiation with the unions involved, the following financial arrangements have been agreed:

1. All workers made redundant will receive redundancy payments as follows:

    Skilled men   £2000

    Unskilled men £1500

2. Workers from the Abbotsfield factory who are retained will receive a relocation payment of £2000.
3. To avoid any bias towards the workers from the Birchwood factory, the proportion of the Abbotsfield men retained should be equal to the proportion of the Birchwood men retained.

Required:

1. Construct a linear programming model to determine how the new labour force will be selected from the two factories in order to minimise total redundancy and relocation costs. You should use the following variables:

    S1   Number of skilled men retained from Abbotsfield
    S2   Number of skilled men retained from Birchwood
    U1   Number of unskilled men retained from Abbotsfield
    U2   Number of unskilled men retained from Birchwood.

2 Using two of the equality constraints, eliminate two of the four variables and solve the problem graphically. What is the minimum total redundancy and relocation cost?

(*ACCA, June 1987*)

Answers on page 547.

Exercise 9.11

A company has £100,000 to invest in projects and there are three projects available for selection. These projects are not mutually exclusive so that any combination can be chosen, and any amount can be invested in each project.

The company's criteria for capital investments are that the payback period must be no more than five years and the accounting rate of return should be at least 20%. For the £100,000 being invested, the projects involved will be treated as one investment using a weighted average to determine the overall payback period and rate of return. Relevant details of the individual projects are as follows:

Project	Payback (years)	Accounting rate of return	Net present value per £1,000 invested
A	3	18%	£200
B	4	20%	£250
C	6	25%	£450

With projects of this nature, the company aims to maximise the net present value of the investment.

Required:
1 Formulate a linear programming model to determine the optimum allocation of funds to the three projects. You should assume that all of the £100,00 will be invested.
2 Using the equality constraint, eliminate one of the three variables from your model and solve the reduced problem graphically. What is the optimum allocation of funds and what will be the net present value of this investment?

(*ACCA, June 1986*)

Answers on page 549.

Exercise 9.12

This is a continuation of **Exercise 9.3**, to which you should refer.

Required:
1 The output from the computer package produces the following final tableau of a simplex solution to this problem:

Basis	A	B	C	D	E	X	S	T	U	Value
A	1	1.18	1.04	0.46	0	0.36	0	0	−2.29	3,357
S	0	−0.34	0.23	0.02	0	−0.18	1	0	0.14	321
T	0	1.37	2.97	2.28	0	−0.27	0	1	−2.79	9,482
E	0	−0.09	−0.02	0.52	1	−0.18	0	0	2.14	2,321
	0	1.26	1.06	0.51	0	2.02	0	0	8.81	105,791

where A, B, C, D and E are the weekly production levels for the five products; X is the amount of raw materials that falls short of the maximum available; S, T, U are the respective number of hours short of the maximum weekly input of forming time, firing time, and packing time.
- (a) Use this tableau to find the optimum weekly production plan for the Caterpillar China Company.
- (b) Describe the implications of using this plan in terms of the unused resources and overall contribution to profit.

2 In the context of this problem explain the meaning of 'the dual or shadow price of a resource'.

3 There is a proposition that the company manufactures an additional product which would sell at £50 per unit. Each unit made would need 6 kg of raw materials, 1 hour of forming time, 5 hours of firing time and 1 hour to pack. Is this a worthwhile proposition?

(ACCA, December 1987)

Answers on page 550.

Exercise 9.13

Python Properties plc have recently acquired a building site on which they have planning permission to build at most 38 houses. The site covers an area of 30,000 square metres but 10% of the area is normally taken up by access roads and services so that the area of land on which houses can be built amounts of only 27,000 square metres. The company's architect has prepared plans for three types of houses which could be built on the site, namely:

- A Blocks of 6 linked terraced houses
- B Blocks of 2 semi-detached houses
- C Single detached houses

The details of each type of house are as follows:

House type	Land required (m^2/house)	Variable cost (£/house)	Selling price (£/house)
A	300	15,000	22,000
B	675	18,000	27,000
C	900	24,000	36,000

Required:
1. Formulate the two constraints and the objective function which make up a linear programming model to determine the mix of house types which maximises the contribution to fixed costs and profits.
2. Write down the dual of the problem which you have formulated in **1**.
3. Solve the dual problem graphically or otherwise.
4. From the form of the dual solution, deduce which variables in the primal problem are basic (ie non-zero). What is the mix of houses which yields a maximum contribution?

(ACCA, December 1985 Alternative)

Answers on page 551.

Transportation and assignment models

10.1 INTRODUCTION

The linear programming methods discussed in Chapter Nine are suitable for a range of allocation problems. The work involved in solving the model can be drastically reduced by use of a computer package. This leaves the decision maker free to concentrate on the interpretation and evaluation of the solution. However, the package still requires the formulation of the linear programming model. This can be a major task for large problems. The variables must be identified and the constraints formulated.

For certain types of allocation problem, the use of specially designed algorithms simplifies the building of the initial model. In this chapter we will look at two related examples of this type of algorithm which are suitable for solving transportation and assignment problems.

In both cases the allocation concerns items which are transferred from a number of origins to a set of destinations according to a particular objective. The objective is often one which minimises the total cost of the transfer. For example, a company has three factories and five regional distribution centres. The management requires the transfer of the finished goods from the factories to the distribution centres to be achieved at minimum cost. This is a situation in which the transportation method would be appropriate.

Assignment is a particular case of the transportation problem. For each combination of origin and destination, the transfer involves one item only. For example, a machine shop has six lathes of varying ages and design. On a given morning, the machine shop manager has six jobs to allocate. The jobs will take different lengths of time on different lathes. The manager wishes to allocate one job to each lathe to minimise the total working time. The assignment algorithm may be used to solve problems such as this.

In this chapter we will describe the application of these two algorithms using small problems. It should be borne in mind that, in practice, the problems will be much larger and are solved using computer packages. In addition, transportation models often involve several stages, for example, factory-to-depot-to-retail outlet. In these cases the basic algorithm has to be modified and more sophisticated methods used.

10.2 TRANSPORTATION PROBLEM AND ALGORITHM

This problem is concerned with the allocation of items between suppliers (called origins) and consumers (called destinations) so that the total cost of the allocation is minimised. The problem can be solved using either linear programming methods or the special transportation algorithm. The linear programming method is illustrated in **Example 10.1.**

10.2.1 The transportation problem

Example 10.1: To illustrate a basic transportation problem

Ace Foods Ltd manufacture soft drinks at two plants, A and B. Bottles for the two plants are supplied by two firms, P and Q. For the month of November, plant A requires 5000 bottles and plant B requires 3500 bottles. Firm P is able to supply a maximum of 7500 bottles and firm Q is able to supply a maximum of 4000 bottles. The cost per bottle of transport between each supplier and each plant is shown in the TABLE 10.1:

TABLE 10.1 **Costs, requirements and availabilities of bottle supply**

		Transport cost, pence/bottle to Plant A	Transport cost, pence/bottle to Plant B	Available bottles
From supplier	P	4	4	7500
	Q	3	2	4000
Bottles required		5000	3500	

How should the bottles be supplied to the plants to minimise the total transport cost?
Solution: It is always useful with transportation to see if there is an obvious solution. Ideally, we would like to use only the cheapest routes. Supplier Q will be preferred by both plants, since it is cheaper than P. Unfortunately Q has only 4000 bottles, compared with the total requirement for 8500. The cheapest solution will probably be to use the 2p per bottle route from Q to plant B, to supply all the requirement at B (3500). The balance from Q (500) should be sent to A at 3p per bottle. The rest of the demand at plant A will come from P at a cost of 4p per bottle. The total cost of this allocation is:

$$0.02 \times 3500 + 0.03 \times 500 + 0.04 \times 4500 = £265/\text{month}$$

We have no proof that this is the most economic allocation. One of the important aspects of the model which we are about to consider, is that it provides a solution, demonstrates that it is the optimum solution and shows the effect on the solution of any changes which arise in the problem.

We will now solve the above problem using a conventional linear programming model with a graphical solution.

Suppose firm P supplies x bottles to plant A and y bottles to plant B. Firm Q must supply the remaining $(5000 - x)$ bottles to A and the remaining $(3500 - y)$ bottles to B. The objective is to minimise the total transport cost, C, pence, where:

$$C = 4x + 4y + 3(5000 - x) + 2(3500 - y)$$

therefore:

$$C = x + 2y + 22000$$

and:

$$Z = C - 22{,}000 = x + 2y$$

Z will take its minimum value, when C takes its minimum value. The values of x and y which minimise $Z = x + 2y$, will also minimise C. The objective function is minimised subject to:

Requirement at A:	x		$\leqslant 5000$ bottles
Requirement at B:		y	$\leqslant 3500$ bottles
Supply from P:	x	+ y	$\leqslant 7500$ bottles
Supply from Q:	$(5000 - x) + (3500 - y)$		$\leqslant 4000$ bottles
ie:	x	+ y	$\geqslant 4500$ bottles
	x,y		$\geqslant 0$

These constraints are illustrated in FIGURE 10.1.

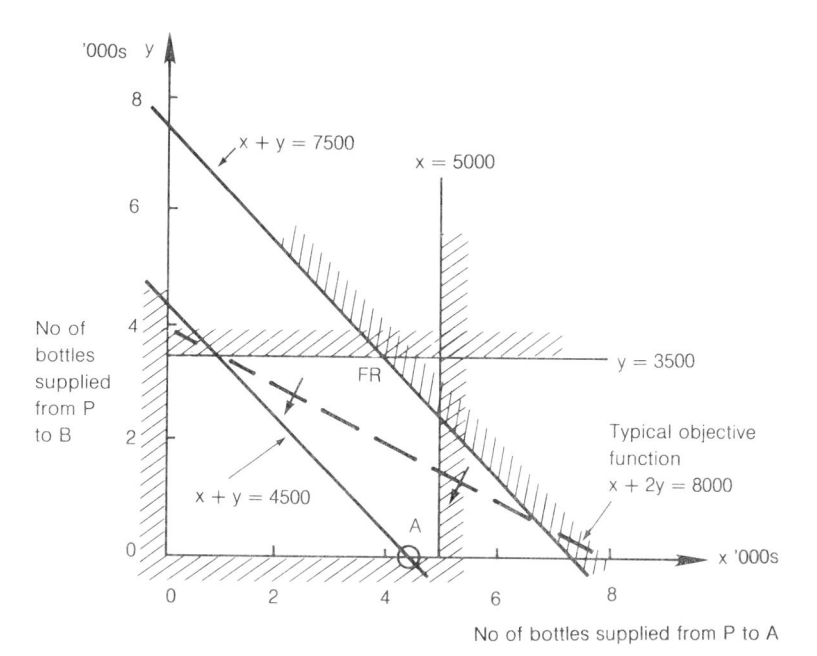

FIGURE 10.1 Linear programme for the supply of bottles to Ace Foods Ltd

The point x = 4000, y = 2000 is a feasible solution. At this point:

$$Z = 4000 + 2 \times 2000 = 8000 \text{ p}$$

The trial objective function is 8000 = x + 2y. This is shown in FIGURE 10.1. Moving in the direction of decreasing values of Z, corner A is the optimum. At this corner, x = 4500 and y = 0. Therefore the optimum solution is for P to supply 4500 bottles to A and none to B, while Q supplies 500 bottles to A and 3500 bottles B. At this solution, the minimum cost is:

$$C_{min} = 4500 + 2 \times 0 + 22{,}000 = 26{,}500 \text{ p} = £265$$

The only spare capacity is at firm P which retains 3000 bottles. This is the solution that we thought would be the minimum. We have now shown that this is the case.

10.2.2 The transportation algorithm

The problem in section **10.2.1** may be solved using the **transportation algorithm**. To use this algorithm, a number of conditions must be satisfied:

1   The cost per item for each combination of origin and destination must be specified.
2   The supply of items at each origin must be known.
3   The requirement of items at each destination must be known.
4   The total supply must equal the total demand.

Example 10.1 satisfies the first three conditions but not the last one. However, a **dummy** plant can be included for which the requirement is the difference between the total available and the total required. In **Example 10.1** the dummy plant would have a requirement for (11,500 − 8500) = 3000 bottles. Any items allocated to a dummy destination represent items which do not leave the supplier. In a similar way, if the total supply is less than the total demand, a dummy supplier is included to supply the shortfall. Any items allocated from this dummy represent items not supplied.

The transportation algorithm has four stages:

Stage 1: Arrange the data in tableau format and find any feasible allocation. A feasible allocation is one in which all demand at the destinations is satisfied and all supply at the origins is allocated.

Stage 2: Test the allocation to see if it is the optimal solution.

Stage 3: If the first allocation is not optimal, re-allocate in order to move to a better, lower cost, solution.
Stage 4: Test again for optimality.
Repeat this iterative process until the optimum allocation is found.

10.2.3 Finding an initial allocation

The initial allocation can be made using any method which will produce a feasible solution. However, a systematic approach tends to produce more useful solutions. We will look at two methods of finding an initial allocation, the **minimum cost method** and **Vogel's method**. The procedure is explained in **Example 10.2**.

Example 10.2 To illustrate methods of finding an initial allocation

Three warehouses P, Q and R can supply 9, 4 and 8 items respectively. Three stores at A, B and C require 3, 5 and 6 items respectively. What is the minimum cost of allocating the items from the warehouses to the stores if the unit transportation costs are as shown in TABLE 10.2?

TABLE 10.2 Costs, requirements and availabilities for Example 10.2

		Transport costs £/item, to stores			Total available
		A	B	C	
From warehouses	P	10	20	5	9
	Q	2	10	8	4
	R	1	20	7	8
Total required		3	5	6	

Solution: The information on costs, availability and requirements are given but total supply is bigger than demand. The warehouses have 21 items available but the stores require only 14. A dummy store is needed to absorb the 7 items which are surplus to requirements. These 7 items will never actually leave the warehouse, and therefore their transportation costs are assumed to be zero. The first transportation tableau is given below:

TABLE 10.3 **Balanced transportation tableau**

		Transport cost £/item, to stores				Total available
		A	B	C	Dummy	
From warehouse	P	10	20	5	0	9
	Q	2	10	8	0	4
	R	1	20	7	0	8
Total required		3	5	6	7	21

To find the first feasible allocation we will use the minimum cost method and Vogel's method in turn. However, it should be remembered that only one method is actually required.

METHOD 1: MINIMUM COST METHOD

1   We allocate as much as possible to the cell with the minimum unit cost.
2   We adjust the remaining availabilities and requirements.

TRANSPORTATION AND ASSIGNMENT MODELS: CHAPTER TEN

3 We choose the next smallest cost and allocate as much as possible to this cell, and so on, until supply and demand are all zero.
4 If more than one cell has the smallest value of unit cost, then we choose at random.

In TABLE 10.4, the transport costs are placed in the separate boxes in the upper right of each cell. The subscripts indicate the order in which the allocations are made and should help you to follow the explanation. The dash in a cell indicates that this cell is no longer available.

TABLE 10.4 **First allocation using minimum cost method**

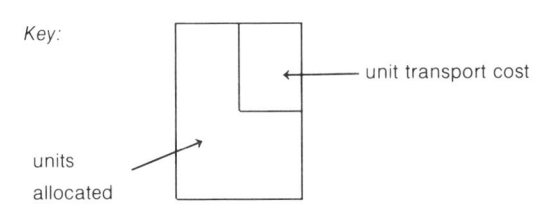

From warehouse	A	B	C	Dummy	Total available
P	10 —	20 —	5 2_3	0 7_1	9 2 0
Q	2 —	10 4_5	8 —	0 —	4 0
R	1 3_2	20 1_6	7 4_4	0 —	8 5 1 0
Total required	3 0	5 1 0	6 4 0	7 0	21

To store across top.

Key: unit transport cost (upper right box); units allocated (main cell).

1 The smallest cost is zero. We choose any one of cells (P, Dummy), (Q, Dummy) or (R, Dummy). Cell (P, Dummy) is chosen and we allocate the maximum amount, 7 units, to the cell. We reduce by 7 the amount available at P and the amount required by the dummy. Cross out the cells which now cannot be used, that is, (Q, Dummy) and (R, Dummy).
2 Neither of the other zero costs are available, so the next smallest cost is 1 in cell (R,A). We allocate as many units as possible, 3, to this cell. Adjust the row and column totals and cross out cells which are no longer available, that is (P,A) and (Q,A).
3 The smallest cost, still available, is 5 in cell (P,C). The remaining 2 items at P are allocated to this cell. The row and column totals are adjusted and we cross out the remaining cell in row P.
4 The final allocations are, in order, (R,C), (Q,B) and (R,B).

296 QUANTITATIVE ANALYSIS

If the allocation is feasible, the total available at each warehouse and the total required at each store should now be zero. The above allocation is feasible.

$$\text{Cost} = £((3 \times 1) + (4 \times 10) + (1 \times 20) + (2 \times 5) + (4 \times 7) + (7 \times 0)) = £101$$

We do not know yet whether this allocation is the cheapest but it should give a reasonable cost.

METHOD 2: VOGEL'S METHOD

This method uses penalty costs. For each row and column the penalty cost is the difference between the cheapest available route and the next cheapest. We try to minimise these penalties.

1. To calculate the penalty cost for each row and column, we look at the least cost cell and next smallest cost cell. For each row and column subtract the smallest cost from the next smallest. This gives the penalty cost of not allocating into the cell with the cheapest cost.
2. We choose the row or column with the largest penalty cost and allocate as much as possible to the cell with the smallest cost in that row or column. In this way, the high penalty costs are avoided as far as possible.
3. As with the previous method, we adjust the row and column totals.
4. We cross out the remaining cells in any row or column for which the supply or demand is now zero, since these cells are no longer available.
5. Return to **1** and re-calculate the penalty costs, ignoring the cells which have been used or crossed out.

We repeat these steps until all demand is satisfied. The subscripts in TABLE 10.5 show the order of choosing the penalty costs and making the allocations.

TABLE 10.5 **First allocation using Vogel's method**

		To store A	B	C	Dummy	Total available	Penalty cost 1 2 3
	P	10 —	20 1	5 6	0 2	9 8 2 0	5 5 5
From warehouse	Q	2 —	10 4_1	8 —	0 —	4 0	2 — —
	R	1 3_2	20 —	7 —	0 5_3	8 5 0	1 1 7_3
Total required		3 0	5 1 0	6 0	7 2 0	21	
1st penalty		1	10_1	2	0		
2nd penalty		9_2	0	2	0		
3rd penalty		—	0	2	0		

Key:

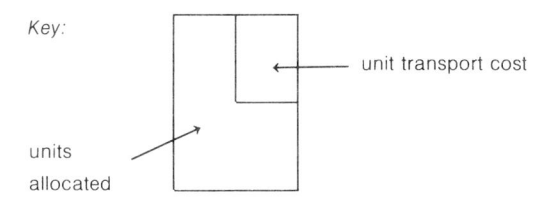

units allocated → [box with "unit transport cost" label]

After the third allocation, there is only one way of completing the solution. The remaining items are allocated as follows—(P,B), (P,C) and (P, Dummy).

$$\text{Cost} = £(1 \times 20 + 6 \times 5 + 2 \times 0 + 4 \times 10 + 3 \times 1 + 5 \times 0) = £93$$

Again, we do not yet know whether this is optimal but we do know that it is a cheaper allocation than the £101 for the minimum cost method.

10.2.4 Testing for optimalilty

To test for optimality, we must first determine whether the initial allocation is basic, that is, whether it is a solution at a corner of the feasible region. The tableau shown in TABLE 10.4 gives a feasible solution, that is, a solution inside or on the edge of the feasible region. If the allocation is basic, there should be one basic variable for every constraint. In a problem with m warehouses and n stores (including the dummy), there are $(m + n - 1)$ independent constraints. A basic solution will therefore have $(m + n - 1)$ allocated cells. These $(m + n - 1)$ variables must be in independent positions. It is not necessary to worry about independence at this stage since any problems will emerge during the test for optimality.

If the allocation has $(m + n - 1)$ independent variables, the methods for testing for optimality may be applied directly. If there are fewer variables, the tests have to be modified, as will be illustrated in Section **10.2.6**. However, if there are more than $(m + n - 1)$ variables, then the allocation procedure has been used incorrectly. It should be possible to modify the allocation to give a cheaper cost with the correct number of variables.

Refer to **Example 10.2**. We will test each of the allocations for basicness. The tableau has 3 rows and 4 columns, therefore a basic solution will have $(3 + 4 - 1) = 6$ allocated cells. We can see that this is the case for both allocation methods. The two methods have given solutions at different corners of the feasible region. The testing procedures may be used without modification.

The initial allocation is tested to determine whether it is the cheapest solution and, if it is not, how it should be changed. We will illustrate two methods of testing for optimality. In the **stepping stone** method, the costs of using the unallocated cells—the shadow costs—are calculated. The procedure is long and rather clumsy but the physical meaning is clear. The **MODI (modified distribution)** method is a mathematical procedure which gives the same shadow costs much more quickly, although the physical meaning is not so obvious. In both methods, if the allocation is not optimal, a stepping stone procedure is used to move to the next basic allocation. Once a basic solution has been found, the algorithm enables us to move from corner to corner of the feasible region, until the optimum solution is found.

Example 10.3: To illustrate the test for optimality using the stepping stone method

We will use the allocation produced by the minimum cost method to illustrate the procedure. The allocation is repeated in TABLE 10.6 overleaf.

298 QUANTITATIVE ANALYSIS

TABLE 10.6 **First allocation using minimum cost method**

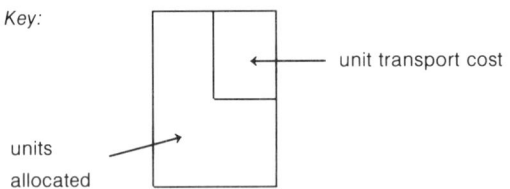

	A	B	C	Dummy	Total available
P	10 / —	20 / —	5 / 2	0 / 7	9
Q	2 / —	10 / 4	8 / —	0 / —	4
R	1 / 3	20 / 1	7 / 4	0 / —	8
Total required	3	5	6	7	21

From warehouse: P, Q, R
To store: A, B, C, Dummy

Key: unit transport cost (top right); units allocated (bottom left)

The stepping stones are the cells which have allocations in them—(P, C), (P, dummy), (Q, B), (R, A), (R, B) and (R, C). We take one of the empty cells and pretend that we move one item into it. This move upsets the totals for the row or column in which the empty cell lies. The amounts in some of the allocated cells are then adjusted to restore the balance. We use these allocated cells, the stepping stones, to calculate the cost of the transfer of this one item into the empty cell. If the cost is positive, using the empty cell will increase total costs and we do not want to do this. If the cost is negative, using the cell will reduce costs. This means that the present allocation is not optimal and we can find a better solution using this cell.

It does not matter which empty cell is chosen as the start. We will choose (P, A). We add 1 item to (P, A). The allocation is no longer correct. Store A is receiving 4 items, when it wants only 3. Warehouse P is supplying 10 when it has only 9 items. We must adjust the A column and the P row. To balance the A column, we must deduct 1 item from the stepping stone (R,A). This corrects column A, but unbalances row R, reducing its supply from 8 to 7.

We can re-balance row P by subtracting 1 item from either (P, C) or (P, Dummy). If we choose (P, Dummy), there is no other allocated cell in the dummy column which could be used to re-adjust the dummy column, therefore we do not make this choice. Adjustments can be made using only those cells which have allocations already. We must use (P, C). We deduct 1 item from (P, C). This corrects row P, but unbalances column C. We now have problems with row R and column C. Both can be adjusted simultaneously by adding 1 item into (R, C). The physical effect of using the empty cell (P, A) and returning to a balanced allocation is shown in TABLE 10.7 opposite.

The net cost effect of moving 1 item into (P, A) is:

$$+1 \times (P, A) \text{ cost} - 1 \times (R, A) \text{ cost} + 1 \times (R, C) \text{ cost} - 1 \times (P, C) \text{ cost}$$
$$= +(1 \times 10) - (1 \times 1) + (1 \times 7) - (1 \times 5) = +£11/\text{item}$$

TABLE 10.7 **Testing empty cell (P,A)**

Physical change—items

	A	C
P	test cell +1	allocated cell −1
R	allocated cell −1	allocated cell +1

Using (P, A) would cost an extra £11 for each item sent from P to A. The shadow price is positive therefore we do not choose to use this empty cell.

We return to the original allocation and repeat the procedure for the other empty cells in turn. Look next at the cell (R, Dummy) and use the stepping stones (P, Dummy), (P, C), and (R, C) to show the physical and cost changes of moving 1 unit into (R, Dummy):

TABLE 10.8 **Testing empty cell (R, Dummy)**

Physical change—items

	C	Dummy
P	allocated cell +1	allocated cell −1
R	allocated cell −1	test cell +1

TABLE 10.9 **Testing empty cell (R, Dummy)**

Cost change, £

	C	Dummy
P	allocated cell +5	allocated cell −0
R	allocated cell −7	test cell +0

The net cost change of adding 1 item to (R, Dummy) is:

$$+0 - 0 + 5 - 7 = -£2 \text{ per item}$$

By allocating into cell (R, Dummy), it is possible to reduce the costs, therefore, the present allocation is not optimal. We can find a cheaper allocation, saving £2 per item, by using (R, Dummy) and this stepping stone route. We must, however, complete the testing of all the empty cells, since there may be a cell which gives an even better saving.

Let us next construct the stepping stone path for the empty cell (Q, Dummy). We must remember that to re-balance rows and columns we can step on allocated cells only. A four step circuit is not possible this time. We must look for a more complex route. We allocate 1 item to cell (Q, Dummy). There is only 1 allocated cell in row Q and only 1 allocated cell in the Dummy column. Suppose we choose to move from (Q, Dummy) to (Q, B). We deduct 1 item from this cell which balances row Q. Column B can be balanced by (R, B) only, therefore we add 1 item to this cell. We can balance row R via (R, A) or (R, C) but (R, A) is the only allocation in column A, so we do not use this cell. If we did, we would not be able to balance column A. We deduct 1 item from (R, C). The route home should now be clear. We balance column C by adding 1 item to (P, C) and balance row P by deducting one item from (P, Dummy). This last move also balances the Dummy column and the

300 QUANTITATIVE ANALYSIS

circuit is complete. We should remember that as long as the initial allocation is basic, it is possible to find a suitable route round the tableau which starts and ends with the chosen empty cell. The physical effects and cost changes are summarised in TABLES 10.10 and 10.11.

TABLE 10.10 **Testing empty cell (Q, Dummy)**

Physical change—items

	B	C	Dummy
P	empty	allocated +1	allocated −1
Q	allocated −1	empty	test +1
R	allocated +1	allocated −1	empty

TABLE 10.11 **Testing empty cell (Q, Dummy)**

Cost change £

	B	C	Dummy
P	empty	allocated +5	allocated −0
Q	allocated −10	empty	test +0
R	allocated +20	allocated −7	empty

The net cost affect of adding 1 item to the empty cell (Q, Dummy) is:

$$+0 - 0 + 5 - 7 + 20 - 10 = +£8/\text{item}$$

The total cost of allocation will increase by £8 per item if we allocate into this empty cell. We do not choose to make this change. The shadow costs for the remaining empty cells are calculated in a similar way to that described above. The full set of values is shown circled in TABLE 10.12:

TABLE 10.12 **Testing the initial allocation—stepping stone method**

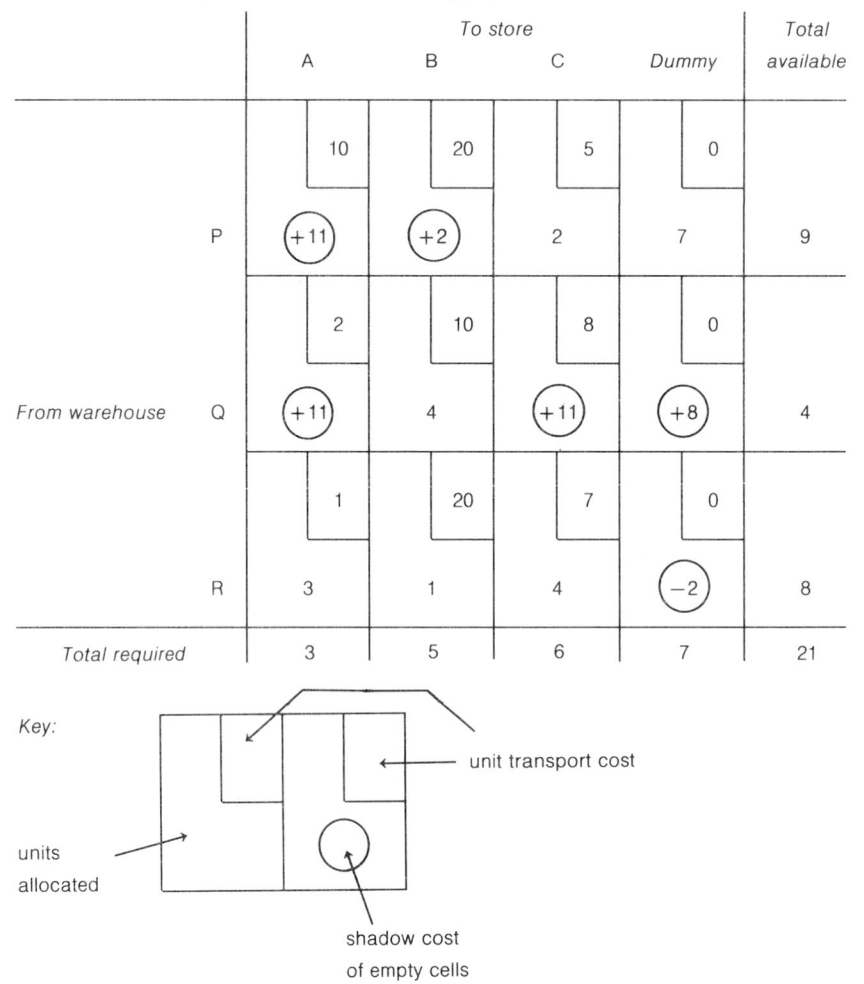

		A	B	C	Dummy	Total available
From warehouse	P	10 (+11)	20 (+2)	5 2	0 7	9
	Q	2 (+11)	10 4	8 (+11)	0 (+8)	4
	R	1 3	20 1	7 4	0 (−2)	8
Total required		3	5	6	7	21

Key:
- unit transport cost
- units allocated
- shadow cost of empty cells

This is not the optimal allocation because cell (R, Dummy) has a negative shadow cost of −£2. The present cost of the allocation, £101, can be reduced by using this cell and the stepping stone circuit which gave the net saving of £2 per item.

We will continue with this example, to find the optimum allocation, in section **10.2.5**, but first we will discuss the MODI method of calculating the shadow costs. The stepping stone procedure is a clumsy method and it is easy to make errors. It is more sensible to use the mathematical approach of the MODI method to test for optimality. The procedure does not give an insight into the physical problem but it does produce the same shadow prices which much less effort.

To begin with, consider only the allocated cells. Each unit cost, c_{ij}, for these cells is split into two components, u_i for the row and v_j for the column. For example, cell (R,B), which is in row 3 and column 2, has a unit cost $c_{32} = £20$. This is split into the row component u_3 and the column component v_2, ie:

$$c_{32} = 20 = u_3 + v_2$$

For each empty (non-basic) cell, we calculate the shadow cost from:

$$s_{ij} = c_{ij} - (u_i + v_j)$$

This shadow cost is the extra cost of transporting one item by the route i to j. If all of the shadow costs are positive or zero, that is, $s_{ij} \geq 0$, then the solution is optimal. In this case, if an allocation is moved into an empty cell for which the shadow cost is positive, the total costs will increase. If the shadow cost is zero, the total costs remains unchanged.

Example 10.4: To test a basic allocation for optimality using the MODI method

Refer again to the initial allocation obtained using the minimum cost method. We will test this allocation for optimality using the MODI method. The initial allocation is repeated below:

TABLE 10.13 **First allocation using minimum cost method**

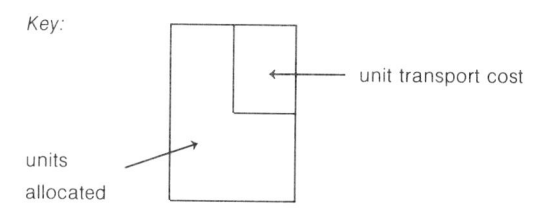

From warehouse	To store A	B	C	Dummy	Total available
P	10 / —	20 / —	5 / 2	0 / 7	9
Q	2 / —	10 / 4	8 / —	0 / —	4
R	1 / 3	20 / 1	7 / 4	0 / —	8
Total required	3	5	6	7	21

Key: unit transport cost (top right), units allocated (bottom left)

The row components, u_i, and the column components, v_j, are calculated using the allocated cells. The allocated cells are (P, C), (P, Dummy), (Q, B), (R, A), (R, B), and (R, C), which give the following six simultaneous equations. These six equations contain seven variables, hence there is no unique solution. The actual values given to the components are not important as long as the set of values is consistent.

$$c_{13} = 5 = u_1 + v_3 \quad \text{for allocated cell (P, C)}$$
$$c_{14} = 0 = u_1 + v_4 \quad \text{for allocated cell (P, Dummy)}$$
$$c_{33} = 7 = u_3 + v_3 \quad \text{for allocated cell (R, C)}$$
$$c_{31} = 1 = u_3 + v_1 \quad \text{for allocated cell (R, A)}$$
$$c_{32} = 20 = u_3 + v_2 \quad \text{for allocated cell (R, B)}$$
$$c_{22} = 10 = u_2 + v_2 \quad \text{for allocated cell (Q, B)}$$

A value is assigned to any one of the components, then the values of the other components are found from the equations. We will choose to set $u_1 = 0$. It follows that $v_3 = 5$, $v_4 = 0$, $u_3 = 2$, $v_1 = -1$, $v_2 = 18$, and $u_2 = -8$. We can now calculate the shadow costs for the unallocated cells, from the equation:

$$s_{ij} = c_{ij} - (u_i + v_j)$$

Substituting gives the following shadow costs:

$$s_{11} = 10 - (0 - (-1)) = +11 \quad \text{for empty cell (P, A)}$$
$$s_{12} = 20 - (0 + 18) \quad\;\; = +2 \quad \text{for empty cell (P, B)}$$
$$s_{21} = 2 - (-8 - 1) \quad\;\; = +11 \quad \text{for empty cell (Q, A)}$$
$$s_{23} = 8 - (-8 + 5) \quad\;\; = +11 \quad \text{for empty cell (Q, C)}$$
$$s_{24} = 0 - (-8 + 0) \quad\;\; = +8 \quad \text{for empty cell (Q, Dummy)}$$
$$s_{34} = 0 - (2 + 0) \quad\quad\;\; = -2 \quad \text{for empty cell (R, Dummy)}$$

These values are entered on the tableau in TABLE 10.14:

TABLE 10.14 **Testing the initial allocation—MODI method**

		A	B	C	Dummy	Total available	
From warehouse	P	10 (+11)	20 (+2)	5 / 2	0 / 7	9	$u_1 = 0$
	Q	2 (+11)	10 / 4	8 (+11)	0 (+8)	4	$u_2 = -8$
	R	1 / 3	20 / 1	7 / 4	0 (−2)	8	$u_3 = 2$
Total required		3	5	6	7	21	
		$v_1 = -1$	$v_2 = 18$	$v_3 = 5$	$v_4 = 0$		

Key:

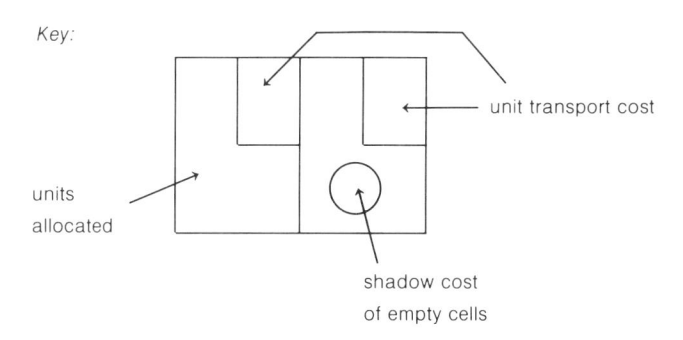

units allocated
unit transport cost
shadow cost of empty cells

The shadow costs are the same as those found using the stepping stone method in TABLE 10.12. Route (R, Dummy) has a negative shadow cost of −£2 per item, therefore the solution is not optimal. Items must be re-allocated using this cell and the associated stepping stone circuit to reduce the costs.

10.2.5 Finding the optimum solution

The iterative procedure for finding the optimum allocation is as follows:

1. If there is more than one empty cell with a negative shadow cost, choose the cell with the largest negative value.
2. Find the stepping stone circuit for this empty cell, as described above.
3. Identify those cells from which items are to be deducted and determine the amount which could be deducted from each cell, without any of the allocations becoming negative. The minimum value of these figures gives the maximum amount that can be allocated to the chosen cell. Re-allocate around the circuit.
4. There is no guarantee that further improvements cannot be made. The new allocation must be checked for optimality using the MODI method. The minimum cost is found when all of the shadow costs are positive or zero.

We will continue with **Example 10.4**.

Example 10.4 continued: To test a basic allocation for optimality using the MODI method

Cell (R, Dummy) is the only one with a negative shadow cost, −£2. We wish to allocate as much as possible into this cell.

The stepping stone circuit for (R, Dummy), which gives the −£2, is shown below with the existing allocations and unit costs.

TABLE 10.15 **Stepping stone circuit for (R, Dummy)**

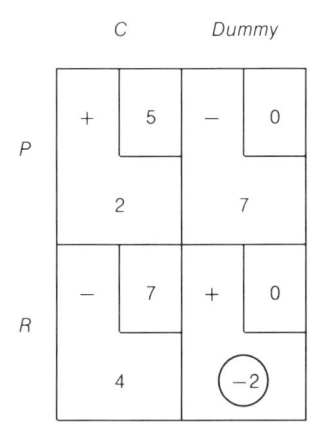

	C	Dummy
P	+ 5	− 0
	2	7
R	− 7	+ 0
	4	(−2)

+ denotes items are to be added to this cell. − denotes that items are to be deducted from this cell.

The − cells are (P, Dummy) and (R, C), which contain allocations of 7 and 4 items. The minimum value in the − cells is 4, which means that 4 items can be moved round the circuit, into the + cells and out of the − cells. The total saving in cost is (2 × 4) = £8. The revised tableau is given in TABLE 10.16.

TABLE 10.16 **Revised allocation**

		To store			Total
From warehouse	A	B	C	Dummy	available
P	10 —	20 —	5 2 + 4	0 7 − 4	9
Q	2 —	10 4	8 —	0 —	4
R	1 3	20 1	7 4 − 4	0 0 + 4	8
Total required	3	5	6	7	21

The solution is still basic since there are 6 allocations. We will re-check for optimality, using the MODI method. Using the allocated cells, (P, C), (P, Dummy), (Q, B), (R, A), (R, B) and (R, Dummy):

$$c_{13} = 5 = u_1 + v_3 \quad \text{Choose } u_1 = 0, \text{ then } v_3 = 5$$
$$c_{14} = 0 = u_1 + v_4 \qquad\qquad\qquad v_4 = 0$$
$$c_{34} = 0 = u_3 + v_4 \qquad\qquad\qquad u_3 = 0$$
$$c_{31} = 1 = u_3 + v_1 \qquad\qquad\qquad v_1 = 1$$
$$c_{32} = 20 = u_3 + v_2 \qquad\qquad\qquad v_2 = 20$$
$$c_{22} = 10 = u_2 + v_2 \qquad\qquad\qquad u_2 = -10$$

The shadow costs for the empty cells are:

$$s_{ij} = c_{ij} - (u_i + v_j)$$
$$s_{11} = 10 - (0 + 1) \quad = +9$$
$$s_{12} = 20 - (0 + 20) \quad = 0$$
$$s_{21} = 2 - (-10 + 1) = +11$$
$$s_{23} = 8 - (-10 + 5) = +13$$
$$s_{24} = 0 - (-10 + 0) = +10$$
$$s_{33} = 7 - (0 + 5) \quad = +2$$

None of the shadow costs is negative, therefore the allocation is optimal.

TABLE 10.17 **Testing the optimum allocation using the MODI method**

	A	B	C	Dummy	Total available	
P	10 (+9)	20 (0)	5 6	0 3	9	$u_1 = 0$
Q	2 (+11)	10 4	8 (+13)	0 (+10)	4	$u_2 = -10$
R	1 3	20 1	7 (+2)	0 4	8	$u_3 = 0$
Total required	3	5	6	7	21	
	$v_1 = 1$	$v_2 = 20$	$v_3 = 5$	$v_4 = 0$		

The minimum cost is:

$$£101 + (4 \times (-2)) = £93$$

Solution

- Warehouse P sends 6 items to store C and retains 3 items.
- Warehouse Q sends 4 items to store B.
- Warehouse R sends 3 items to store A, one to store B and retains four items.

If this second allocation is not optimal, the re-allocation procedure is repeated as many times as is necessary.

It should be noted that the minimum cost was achieved on the first allocation using Vogel's method. This will frequently happen for small scale problems. Vogel's method tends to produce a better first tableau but there is no guarantee that it will give the optimum immediately. It should also be noted that the allocation produced by Vogel's method is different to the one above (see **Example 10.2**). There is an alternative optimum solution:

- Warehouse P sends one item to Store B, 6 items to Store C and retains 2 items.
- Warehouse Q sends four items to Store B.
- Warehouse R sends three items to Store A and retains five items.

The existence of an alternative optimum solution was signalled to us by the zero shadow cost for cell (P, B). These zero shadow costs are associated with alternative allocations which generate the same total cost.

10.2.6. Sensitivity analysis

The final allocation, together with the shadow costs of the empty cells, can be used for sensitivity analysis. The shadow cost tells us by how much the total cost will increase, if we are forced to allocate one item to that empty cell. If we are forced to send one item from warehouse Q to store C, the extra cost will be £13, much more than the unit cost of £8 for the (Q, C) route itself. The extra cost arises because we have to re-balance the allocation using the following stepping stone circuit.

TABLE 10.18 **Stepping stone circuit for (Q, C)**

Physical change—items

	B	C	Dummy
P	empty	allocated −1	allocated +1
Q	allocated −1	test +1	empty
R	allocated +1	empty	allocated −1

TABLE 10.19 **Stepping stone circuit for (Q, C)**

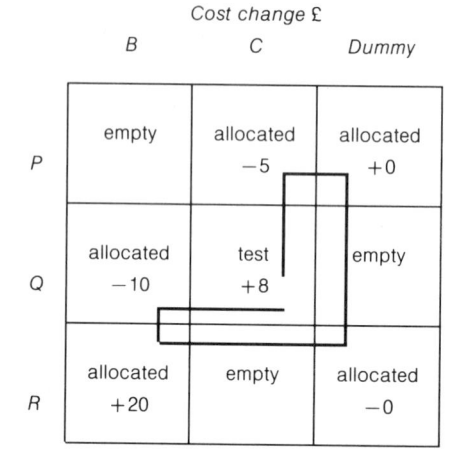

Cost change £

	B	C	Dummy
P	empty	allocated −5	allocated +0
Q	allocated −10	test +8	empty
R	allocated +20	empty	allocated −0

The net change in cost is:

$$+8 - 5 + 0 - 0 + 20 - 10 = +£13/\text{item}$$

The maximum number of items which would be moved round this circuit is the minimum quantity in the − cells, that is:

$$(P,C) = 6, (R, \text{Dummy}) = 4 \text{ or } (Q, B) = 4$$

Four items is the maximum number which could be moved.

The zero shadow cost in cell (P, B) was mentioned in the previous section. The stepping stone circuit for this empty cell is:

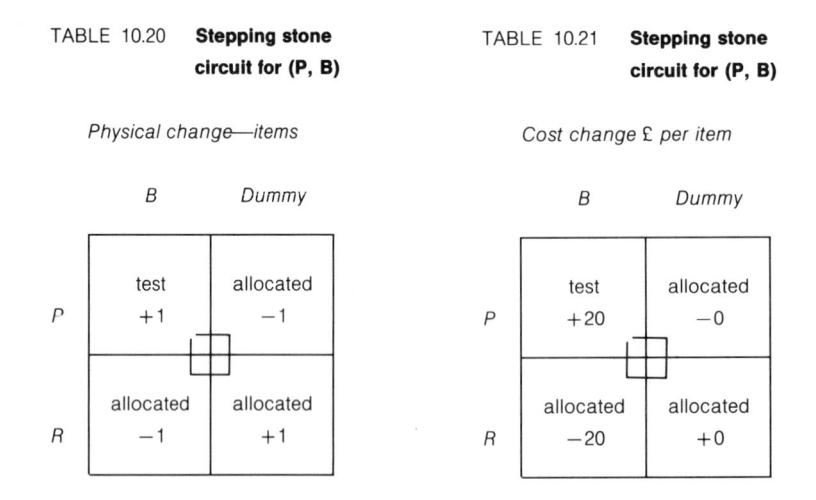

TABLE 10.20 **Stepping stone circuit for (P, B)**

Physical change—items

	B	Dummy
P	test +1	allocated −1
R	allocated −1	allocated +1

TABLE 10.21 **Stepping stone circuit for (P, B)**

Cost change £ per item

	B	Dummy
P	test +20	allocated −0
R	allocated −20	allocated +0

Items can be allocated to cell (P, B) and the net effect on the cost is zero. This means that there is another allocation which will give the same minimum cost of £93. The maximum number of items which can be added to (P, B) is the minimum quantity in the − cells, (R, B) = 1 and (P, Dummy) = 3. Only one item can, therefore, be moved around the circuit into (P, B).

The shadow costs may also be used to indicate how the cost for an empty cell must change before the optimum allocation is affected. The shadow cost for the empty cell (R, C) is +£2, and the actual cost of transfer is £7 per item. The actual cost would have to reduce to at most (7 − 2) = £5 per item, before we would use this cell to reduce the overall costs.

It is more difficult to determine the effect of cost changes in the allocated cells. If the costs reduce, we are encouraged to put more items into that cell. If the costs of an allocated cell increase, at some point we will wish to stop using that cell and transfer to another.

If we look at the allocated cell, (P, C), it has an actual cost of £5 per item. If this cost is reduced, it will not affect the physical allocation since this cell already has the full requirement for Store C allocated to it.

If the cell cost is increased from £5, we must look at the stepping stone circuits which use (P, C). These are the circuits which give the shadow costs of £13 for (Q, C) and £2 for (R, C). In both of these circuits (P, C) is a − cell and any increase in the £5 cost will reduce the shadow cost of these empty cells. The physical allocation will change when the unit transfer cost of (P, C) increases by more than £2, from £5 to more than £7. The shadow cost of (R, C) will then become negative. At this point, it will be advantageous to use this empty cell, changing the (P, C) allocation.

For the current optimum allocation, the cost of (P, C) has an upper limit of £7 and a lower limit of £0. Between these limits the physical allocation is the same but the total costs will change.

10.2.7 Variations in the transportation problem

FORBIDDEN ALLOCATIONS

If an allocation from a particular origin to a particular destination is impossible for some reason, the algorithm can be forced to avoid this allocation by assigning a large cost to the cell. The exact value is unimportant but it must be much larger than the other costs in the tableau. The algorithm will then automatically avoid this cell.

Example 10.5: Forbidden allocations

This example also illustrates how the transportation algorithm can be used to solve problems which are not the straightforward transfer of goods from origins to destinations. In this example, we will deal with the transfer of goods through time. We have a four month production schedule to meet. The demand and production capacities are given below.

TABLE 10.22 **Demand and production capacity**

Month	Production capacity (items)	Demand items
1	300	300
2	350	275
3	325	400
4	375	300

There is an initial stock of 50 items held at the beginning of month 1. Items can be made to meet immediate demand or for stock to meet future demand. If orders are not met during the required month, the sales are lost. The variable costs are £100 per item. The stockholding costs are £2 per item per month. What is the optimum production schedule?

Solution: The situation can be modelled using a transportation tableau with the rows representing the initial stock and monthly production, and the columns representing monthly requirements. The forbidden cells are those which involve meeting past orders from future production. These cells are given an infinite cost in TABLE 10.23.

TABLE 10.23 **Data for production schedule for months 1-4**

		M1	M2	M3	M4	Total available
Stock	M1	2	4	6	8	50
Production	M1	100	102	104	106	300
	M2	∞	100	102	104	350
	M3	∞	∞	100	102	325
	M4	∞	∞	∞	100	375
Total demand		300	275	400	300	

Cost £ per item — Months

The normal procedure is now followed to solve this transportation problem to minimise the cost of meeting the schedule (See Example 10.8).

DEGENERACY

A solution is degenerate when there are fewer than $(m + n - 1)$ allocations in the tableau. This problem can be overcome by allocating a very small amount, essentially zero, to an independent cell. The number of allocations is then increased to $(m + n - 1)$. The procedure in the MODI test for optimality will show us which empty cells to use.

Example 10.6: Degenerate solutions

Three warehouses (X, Y and Z) can supply 6, 3 and 4 items to 3 shops (L, M and N), which require 4, 5 and 1 items respectively. The unit costs of transport are given in the tableau.

TABLE 10.24 **Data for example 10.6**

		L	M	N	Total available
From warehouse	X	6	4	9	6
	Y	5	3	2	3
	Z	2	3	6	4
Total required		4	5	1	

To shop (£/item)

How should the items be allocated in order to minimise the total cost of transport?

Solution: There is a total of 13 items available which is more than the total requirement for 10 items, therefore, we include a dummy shop which absorbs the surplus from the warehouses. We will use Vogel's method to find an initial allocation:

TABLE 10.25 **Initial allocation for Example 10.6 using Vogel's method**

		To shop L	M	N	Dummy	Total available	Penalty cost 1 2 3
From warehouse	X	6 —	4 3	9 —	0 3_1	6 3 0	4_1 2 2
	Y	5 —	3 2	2 1_2	0 —	3 2 0	2 1 2
	Z	2 4_3	3 —	6 —	0 —	4 0	2 1 1
Total required		4 0	5 3 0	1 0	3 0	13	
1st penalty 2nd penalty 3rd penalty		3 3 3_3	0 0 0	4 4_2 —	0 — —		

The cost of the allocation is:

$$4 \times 3 + 0 \times 3 + 3 \times 2 + 2 \times 1 + 2 \times 4 = £28$$

For a basic solution, there should be $(3 + 4 - 1) = 6$ allocations, but here there are only 5. The allocation is degenerate. As the MODI method proceeds, we will have to make one zero allocation, to convert an empty cell into a pretend allocated cell. This gives the required total of 6 allocated cells. It will then be possible to calculate all of the u and v components and hence the shadow costs.

We begin the MODI procedure using the initial 5 allocated cells. The extra zero allocation is made when we can go no further. See TABLE 10.26.

The allocated cells are used to find the row and column components from $c_{ij} = u_i + v_j$ with $u_1 = 0$. We can calculate v_2, v_4, u_2 and v_3 without any problem but we cannot find u_3 or v_1. We need an additional allocated cell. We can put the zero allocation into any empty cell in the v_1 column or the u_3 row. It does not matter which of these cells is chosen. Cell (Z, N) is used. The procedure can be completed and the shadow costs found for the empty cells from $s_{ij} = c_{ij} - (u_i + v_j)$. The figures are given in the table:

TABLE 10.26 **Testing a degenerate allocation—MODI method**

		To shop L	M	N	Dummy	Total available
From warehouse	X	6 (+7)	4 3	9 (+6)	0 3	6 $u_1 = 0$
	Y	5 (+7)	3 2	2 1	0 (+1)	3 $u_2 = -1$
	Z	2 4	3 (−4)	6 [0]	0 (−3)	4 $u_3 = 3$
Total required		4	5	1	3	13
		$v_1 = -1$	$v_2 = 4$	$v_3 = 3$	$v_4 = 0$	

Two of the shadow costs are negative. The allocation is not optimal. It is necessary to re-allocate into cell (Z, M) or (Z, Dummy). We will start with (Z, M), since this has the larger negative shadow cost. The stepping stone circuit for (Z, M), showing the items allocated to each cell is:

TABLE 10.27 **Stepping stone circuit for (Z, M)**

	M	N
Y	allocated − 2	allocated + 1
Z	test +	zero allocation [0] −

To find the number of items to move round the circuit, we look at the − cells, (Y, M) and (Z, N), which contain 2 and zero units. This means we must move the zero allocation round the circuit so that cell (Z, N) becomes empty again, cell (Z, M) takes the zero allocation and becomes the 'pretend' allocated cell. The other allocations remain unchanged. When we test for optimality this time, we find that all of the shadow costs are positive. The allocation is optimal. This means that the initial solution with the 5 allocations is actually the optimal solution. See the tableau in TABLE 10.28:

TABLE 10.28 **Testing the optimal allocation—MODI method**

From warehouse	L	M	N	Dummy	Total available
X	6 / (+3)	4 / 3	9 / (+6)	0 / 3	6, $u_1 = 0$
Y	5 / (+3)	3 / 2	2 / 1	0 / (+1)	3, $u_2 = -1$
Z	2 / 4	3 / [0]	6 / (+4)	0 / (+1)	4, $u_3 = -1$
Total required	4	5	1	3	13
	$v_1 = 3$	$v_2 = 4$	$v_3 = 3$	$v_4 = 0$	

This type of result does not always happen when the initial allocation is degenerate. There will be some problems in which the re-allocation process adds items to the zero allocation cell and it becomes an ordinary allocated cell. The degeneracy of the solution then disappears. The usual procedures will then lead to a better allocation.

MAXIMISATION

The transportation algorithm assumes that the objective function is to be minimised. However, if a suitable problem requires the objective function to be maximised, the algorithm can be modified slightly to deal with this. For example, we may wish to transfer the items in **Example 10.5** in such a way that the total contribution is maximised. In such a case the data required are the unit contributions between each origin and destination. The procedure is to multiply all of the unit contributions by (-1) and then to proceed in the usual way.

10.3 THE ASSIGNMENT PROBLEM

The **assignment problem** is a special case of the transportation problem, in which the number of origins must equal the number of destinations, that is, the tableau is square. Also, at each destination, the 'demand' $= 1$ and at each origin the 'supply' $= 1$. Any assignment problem may be solved using either linear programming or the transportation algorithm. However, the particular structure of this problem has resulted in the development of a specially designed solution procedure called the **Hungarian algorithm**.

10.3.1 The assignment algorithm

The algorithm has three stages to it.
Stage 1:
1. Set out the problem in tableau format, as in the transportation algorithm.
2. For each row in the tableau, find the smallest row element and subtract it from every element in the row.
3. Repeat for the columns.

There is now at least one zero in every row and every column. The assignment problem represented by this 'reduced' tableau is equivalent to the original problem and the optimum allocation will be the same for both. The objective of the Hungarian algorithm is to continue to reduce the matrix until all of the items to be assigned, can be allocated to a cell with a zero value. This means that the total value of the reduced objective function will be zero. Since negative values are not allowed, an objective function value of zero is the optimum.

Stage 2: For a feasible solution, there must be exactly one assignment in every row and every column. If the assignments are made only to cells with zero values, this will give us the minimum value of the objective function.

1 Find a row with only one zero in it, and make an assignment to this zero. If no such row exists, begin with any zero.
2 Cross out all other zeros in the same column.
3 Repeat **1** and **2** until no further progress can be made.

If, at this stage, there are still zeros which are not either assigned or deleted, then:

4 Find a column with only one zero and assign to it.
5 Cross out all other zeros in the same row.
6 Repeat **4** and **5** until no further movement is possible.

If all zeros are still not accounted for, repeat **1** to **6**. If the solution is feasible, that is, all of the allocations have been made to zeros, then the solution must also be optimal. If the solution is not feasible, go on to *Stage 3*.

Stage 3:
1 Draw the minimum number of straight lines through the rows and columns (not diagonals) so that all zeros in the tableau are covered.
2 Find the smallest element without a line through it.
3 Subtract this number from every element without a line through it.
4 Add the chosen number to every element with two lines through it.
5 Leave alone all elements with one line through them.

This procedure has now created at least one new zero. Return to *Stage 2* and repeat the procedure until the optimum solution is reached.

Example 10.7: To illustrate the application of the assignment algorithm

A company has 4 distribution depots and 4 orders to be delivered to separate customers. Each depot has one lorry available which is large enough to carry one of these orders. The distances between each depot and each customer are given in TABLE 10.29:

TABLE 10.29 **Distances from depots to customers**

		\multicolumn{4}{c}{Distance in miles Customer}
		I	II	III	IV
Depot	A	68	72	75	83
	B	56	60	58	63
	C	38	40	35	45
	D	47	42	40	45

How should the orders be assigned to the depots in order to minimise the total distance travelled? *Solution:* It will help you to understand the problem if you try to find a solution to it using a familiar technique, before applying the mechanics of the Hungarian algorithm. Try using Vogel's penalty cost method. See how close you are to the optimum given at the end of the section. The availabilities and requirements are one for each row and column.

Stage 1 of the Hungarian algorithm: Find the smallest row elements.

TABLE 10.30 **To find the smallest row elements**

	Customer				
	I	II	III	IV	Smallest row element
A	68	72	75	83	68
B	56	60	58	63	56
C	38	40	35	45	35
D	47	42	40	45	40

Subtract the smallest element from each element in its row.

TABLE 10.31 **Subtract row element and find smallest column element**

0	4	7	15
0	4	2	7
3	5	0	10
7	2	0	5
0	2	0	5 ← Smallest column element

Subtract the smallest column element from each element in its column.

TABLE 10.32 **Subtract smallest column element**

0	2	7	10
0	2	2	2
3	3	0	5
7	0	0	0

Make assignments as described in *Stage 2* above. $\boxed{0}$ denotes an assignment.

TABLE 10.33 **Make assignments to zeros**

$\boxed{0}$	2	7	10
⌀	2	2	2
3	3	$\boxed{0}$	5
7	$\boxed{0}$	⌀	⌀

We can make only three zero assignments and we require four. This is not a feasible solution. Go on to *stage 3*. We draw the least number of lines to cover all of the zeros.

TABLE 10.34 **Covering zeros with lines**

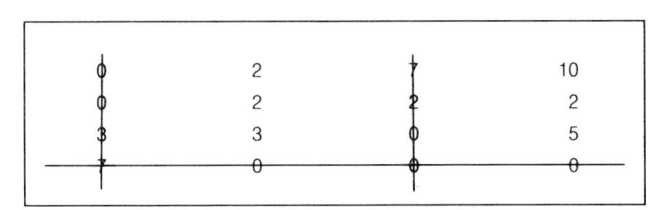

0	2	0	10
0	2	2	2
3	3	0	5
7	0	0	0

The smallest number without a line through it is 2. Adjust the tableau as described in *Stage 3* above, that is, deduct 2 from any number without a line through it, add 2 to any number at the intersection of two lines and leave all the remaining numbers, which are cut by one line. We then re-allocate depots to customers.

TABLE 10.35   **Adjusted tableau with assignments to zeros**

	I	II	III	IV
A	[0]	0	7	8
B	0	[0]	2	0
C	3	1	[0]	3
D	9	0	2	[0]

We have now made the required four allocations to zeros and therefore the solution is optimal. We allocate depot A to customer I, depot B to customer II, depot C to customer III and depot D to customer IV. The solution, though optimal, is not unique. (C, III) must always be allocated because it is the only zero in the C row. There are two other optimum allocations.

TABLE 10.36   **First alternative optimum allocation**

	I	II	III	IV
A	[0]	0	7	8
B	0	0	2	[0]
C	3	1	[0]	3
D	9	[0]	2	0

TABLE 10.37   **Second alternative optimum allocation**

	I	II	III	IV
A	0	[0]	7	8
B	[0]	0	2	0
C	3	1	[0]	3
D	9	0	2	[0]

The minimum mileage for each of these three allocations can be calculated from the original tableau:

$$\text{Allocation 1} = 68 + 60 + 35 + 45 = 208 \text{ miles}$$

$$\text{Allocation 2} = 68 + 63 + 35 + 42 = 208 \text{ miles}$$

$$\text{Allocation 3} = 72 + 56 + 35 + 45 = 208 \text{ miles}$$

All three solutions give the same total mileage.

Note: for larger problems than the one in **Example 10.7**, it may be more difficult to be sure that, at *Stage 3*, step **1**, the minimum number of lines has been drawn to cover all of the zeros. The following 'rule of thumb' may be helpful:

1 Choose any row or column which has a single zero in it.
2 If a row has been chosen, draw a line through the column in which the zero lies.
3 If a column is chosen, draw a line through the row in which the zero lies.
4 Repeat steps **1** to **3** until all of the zeros have been covered.

10.3.2 Special cases of the assignment problem

MAXIMISE THE OBJECTIVE FUNCTION

The assignment algorithm is designed to minimise the objective function. If we have an assignment problem, but we wish to maximise the objective function, we deal with this as we would for the transportation algorithm. We set up the first tableau and multiply all the values in the cells by -1.

Example 10.8: To use the assignment algorithm to maximise an objective

A company has 6 sales areas and 6 salesmen. From past experience it is known that the salesmen perform differently in the different areas. The company's sales director has estimated sales for each person in each area. These are given in TABLE 10.38:

TABLE 10.38 **Sales by area and salesmen**

Sales, £'000

		Area				
Salesman	I	II	III	IV	V	VI
A	68	72	75	83	75	69
B	56	60	58	63	61	59
C	35	38	40	45	25	27
D	40	42	47	45	53	36
E	62	70	68	67	69	70
F	65	63	69	70	72	68

How should the sales director assign the salesmen to the areas to maximise total sales?

Solution: We multiply all of the values in the tableau by (-1):

TABLE 10.39 **Modify data and find smallest row elements**

	Area						
	I	II	III	IV	V	VI	Smallest element
A	−68	−72	−75	−83	−75	−69	−83
B	−56	−60	−58	−63	−61	−59	−63
C	−35	−38	−40	−45	−25	−27	−45
D	−40	−42	−47	−45	−53	−36	−53
E	−62	−70	−68	−67	−69	−70	−70
F	−65	−63	−69	−70	−72	−68	−72

We subtract the smallest (most negative) element from each element in the row.

TABLE 10.40 **Subtract row elements and find smallest column elements**

15	11	8	0	8	14
7	3	5	0	2	4
10	7	5	0	20	18
13	11	6	8	0	17
8	0	2	3	1	0
7	9	3	2	0	4
7	0	2	0	0	0	Smallest element

We subtract the smallest column element from each element in the column.

TABLE 10.41 **Subtract smallest column element**

8	11	6	0	8	14
0	3	3	0	2	4
3	7	3	0	20	18
6	11	4	8	0	17
1	0	0	3	1	0
0	9	1	2	0	4

The solution may now be continued using the normal procedure. See **Exercise 10.9.**

FORBIDDEN ALLOCATIONS

Again this problem is solved in the same way as it was in the transportation algorithm. If a particular assignment is impossible for some reason, we insert a value into the relevant cell which is much bigger than any other value. The algorithm will then automatically avoid this allocation.

UNEQUAL NUMBERS OF ORIGINS AND DESTINATIONS

If the tableau is not square, then additional dummy rows or columns must be included to make it square. The values assigned to these dummy cells will usually be zero.

Destinations which receive allocations from dummy rows (origins), are the ones which, in practice, will not receive an allocation. Allocations which are made to dummy columns represent items which are not allocated.

SUMMARY

The transportation model is a special case of the linear programming model. The basic problem consists of a set of origins, such as a group of warehouses, which supply a set of destinations, such as a group of shops. The objective is to minimise the total transport costs within the constraints of supply and demand. The problem can be solved using a normal linear programming method. The simple structure of the model has enabled special algorithms to be developed which are much easier to use than multivariable linear programming.

The first step is to set up the tableau containing the transport costs. Rows represent the origins and columns represent the destinations.

The second step is to find an initial allocation. We have described two methods for doing this. The minimum cost method allocates to the cheapest route first. In Vogel's method, penalty costs are calculated and allocations made to avoid large penalties. Neither method is guaranteed to give the optimum allocation.

The third step is to test the initial allocation to see if it is the optimum. We have illustrated two ways of testing for optimality. Both methods involve the calculation of the shadow costs of the unallocated cells. If all of the empty cells have positive or zero shadow costs, the allocation is optimal.

In the stepping stone method one item is allocated to an empty cell. We then determine the physical and cost consequences of doing this. The MODI method is more mathematical. Using the costs for each allocated cell, we derive a row and column cost from:

$$c_{ij} = u_i + v_j$$

Using these row and column components, u and v, the shadow costs are calculated for all of the empty cells from:

$$s_{ij} = c_{ij} - (u_i + v_j)$$

The fourth step is needed only if the allocation is not optimal. The stepping stone circuit associated with a negative shadow cost is used to reallocate items. This new allocation is re-tested to see if it is the optimum.

There are various special features of the problem. If demand and supply do not balance, dummy origins or destinations are introduced. An optimum solution must be at a corner of the feasible region, that is, it must be basic. A basic solution is one in which there are (number of rows + number of columns − 1) allocations in the tableau. If we have fewer than this, the solution is degenerate and independent, empty cells are converted to pseudo-allocated cells by a zero allocation.

Unusable routes can be blocked by the introduction of very large costs. The objective function may be maximised instead of minimised.

The assignment model is an even more specialised problem with its own solution algorithm. The number of origins and destinations must be equal and only one allocation can be made to each row and column. The Hungarian algorithm has been developed to solve this special transportation problem.

EXERCISES

Exercise 10.1

Two warehouses supply four stores. The transport costs from the warehouse to the stores, the availabilities at the warehouses and the requirements at the stores are given in the table below.

Transport costs £ per item

		To store G	H	I	J	Items available
From warehouse	1	4	3	5	6	100
	2	8	2	4	7	200
Items required		50	100	75	75	

Required:
How should the goods be allocated to minimise transport costs?

Answers on page 554.

Exercise 10.2

Three factories supply five warehouses with a particular type of steel. The demand for December from each of the warehouses, and the availability at each factory, together with the transport costs per tonne, are given in the table.

Transport costs £ per tonnes

		To warehouse 1	2	3	4	5	Tonnes available
From factory	A	20	27	33	25	34	200
	B	22	36	34	28	26	250
	C	26	29	27	26	28	300
Tonnes required		100	150	200	100	200	

Required:
What is the minimum transport cost for December?

Answers on page 555.

Exercise 10.3

Three bakeries supply four shops with bread each day. The demand, available supply and the transport costs are given below.

		Transport cost pence per loaf				Total available
		To shop				
		I	II	III	IV	
From bakery	X	1·5	2·5	1·0	2·0	700
	Y	2·0	3·0	2·0	1·5	650
	Z	1·0	1·5	2·5	3·0	800
Total required		400	500	350	1000	

Required:
How should the bakeries supply the shops to minimise transport costs?

Answers on page 557.

Exercise 10.4

A retailer has four large stores in different towns, P, Q, R, S. At present the four stores are supplied from two warehouses. A and B, each with a capacity of 40 loads per day.

It is planned to expand the stores so that they will require 27, 25, 30 and 35 loads per day, respectively, from the warehouse system. To meet this and future increases, it is planned to build a third warehouse, capable of supplying 60 loads per day. Two locations are being considered. The transport costs in pounds per load from each warehouse to each store, are given below.

		Transport cost £ per load			
		To store			
		P	Q	R	S
From warehouse	A	70	85	55	120
	B	110	90	75	110
	New 1	115	115	70	90
	New 2	135	95	80	75

Required:
Evaluate the two transportation models and decide which is the better location of the new warehouse. Assume all other costs are the same.

Answers on page 561.

Exercise 10.5

Zeit plc own three farms which grow vegetables for processing in the company's two freezing plants. One of the vegetables is broad beans, which the freezing plants sell for £200 per tonne. The forecast demand for next season is 2750 tonnes from the Craft plant and 3250 tonnes from the Liver plant. The variable costs at each farm and at each freezer plant, and the maximum production tonnages at the farms, are given below.

		Variable cost £/tonne	Maximum tonnage (t)
Farm:	Ascent Hill	90	2000
	Midrow Top	95	3000
	Alum Up	87	1500
Plant:	Craft	20	2750
	Liver	23	3250

The transport costs are:

		To freezer plant £/tonne	
		Craft	Liver
From farm	Ascent Hill	10	15
	Midrow Top	12	12
	Alum Up	18	9

Required:
1. Plan next season's production at the farms and the freezer plants to maximise contribution.
2. Zeit are looking closely at the high cost, £95 per tonne, at Midrow Top, since they want to develop this farm as the centre of excellence for broad bean production. By how much would its variable costs have to be reduced before you would recommend a change in your optimum allocation?

Answers on page 564.

Exercise 10.6

The woodworking firm, Vibra, employs five joiners. Each man has different abilities and skills and takes a different amount of time to do each job. At present, there are five jobs to be allocated. The times are given below:

		Time per job (hours)				
		Job 1	Job 2	Job 3	Job 4	Job 5
Employee	M1	25	16	15	14	13
	M2	25	17	18	23	15
	M3	30	15	20	19	14
	M4	27	20	22	25	12
	M5	29	19	17	32	10

Required:
The jobs have to be assigned one job to one man. How should this be done in order to minimise the total man-time needed to finish all of the jobs?

Answers on page 567.

Exercise 10.7

Vibra, in **Exercise 10.6**, can employ an additional part time joiner who can do the same jobs in the following times.

		Time per job (hours)				
		Job 1	Job 2	Job 3	Job 4	Job 5
Part timer	M6	28	16	19	16	15

Required:
How would this affect the assignment of the five jobs to minimise total time?

Answer on page 569.

Exercise 10.8

Complete the production scheduling problem, **Example 10.5**, given in section **10.2.7**.

Answer on page 570.

Exercise 10.9

Complete the assignment problem, **Example 10.8**, given in section **10.3.2**. Assign the six salesmen to the six sales areas to maximise total sales.

Answer on page 572.

Exercise 10.10

In the Kingdom of the Republic of Idion there are five coal mines which have the following outputs and production costs:

Mine	Output (tonnes/day)	Production cost (£/tonne)
1	120	25
2	150	29
3	80	34
4	160	26
5	140	28

Before the coal can be sold, it must be 'cleaned' and graded at one of three coal preparation plants. The capacities and operating costs of these three plants are as follows:

Plant	Capacity (tonnes/day)	Operating cost (£/tonne)
A	300	2
B	200	3
C	200	3

All coal is transported by rail at a cost of £0.5 per tonne kilometre, and the distances (in kilometres) from each mine to the three preparation plants are:

Preparation plant	Mine 1	2	3	4	5
A	22	44	26	52	24
B	18	16	24	42	48
C	44	32	16	16	22

Required:
1. Using a transportation model, determine how the output of each mine should be allocated to the three preparation plants.
2. Following the installation of new equipment at coal mine number 3, the production cost is expected to fall to £30 per tonne. What effect, if any, will this have on the allocation of coal to the preparation plants?
3. It is planned to increase the output of coal mine number 5 to 180 tonnes per day which can be achieved without any increase in production cost per tonne. How will this affect the allocation of coal to the preparation plants?

(ACCA, June 1986)

Answers on page 573.

Exercise 10.11

1. Briefly describe and contrast two methods of finding an initial feasible solution to a transportation problem.
2. The Braintree Electronics Company produces video cassette tapes for purchase by the general public. The demands (in hundreds) and production capacities (in hundreds) for the three months in the fourth quarter of the year are shown below.

Month	Demand	Normal time capacity	Overtime capacity
October	300	400	150
November	450	400	150
December	800	400	150

Note that capacity exists to produce the video cassette tapes by working normal and overtime hours, where capacity remains constant over time but demand increases for the Christmas sales. The company does not have any initial inventory and does not wish to have any inventory on hand after December.

The costs for production of the video cassette tapes are £150 (per hundred) if produced during normal working hours and £180 (per hundred) if produced during overtime. It has been determined that inventory costs are £20 (per hundred) per month. You should assume that all orders are satisfied on time and that all demands and outputs occur at the midpoint of each month.

Required:
(a) Formulate this production scheduling situation as a transportation problem with six 'sources' and three 'destinations', showing the unit cost associated with each source/destination combination.
(b) Use the transportation algorithm to find the optimum production schedule over this period. State the total production cost of your solution.

(ACCA, June 1988)

Answers on page 575.

Exercise 10.12

The Albion Typewriter Company is currently planning its production schedule for the next three month period. The company works on the basis of a three month supply lead time and orders received during the last three months have produced the following demands for the forthcoming period.

Month	Demand (units)
January	26,000
February	28,000
March	25,000

Under normal conditions, the company carries a finished goods stock of 10,000 units and it is policy to plan for this stock level at the end of each period. In recent months, however, there have been some unexpected production difficulties and it is anticipated that stock at the end of December will amount to only 5000 units. The company would like to build up to their usual stock levels by the end of March. In a typical month, an output of 24,000 units can be maintained with normal working. This figure can, however, be increased by up to 6000 units with overtime working as necessary.

Albion can produce each unit at a cost of £80 under normal working but this increases to £90 if overtime is used. Stockholding costs amount to £4 per unit per month.

Required:
1. Explain how this production scheduling problem can be formulated as a transportation situation with the following 'sources' and 'destinations'.

Sources	Destinations
Stock at start of period	January demand
January—normal working	February demand
January—overtime	March demand
February—normal working	Stock at end of period
February—overtime	
March—normal working	
March—overtime	

Determine the unit cost associated with each cell of the table assuming that all demands and outputs occur at the mid-point of each month, and that all orders will be satisfied on time.

2. By inspection or otherwise, derive a feasible production schedule. Explain how your schedule can be tested to determine whether or not it is the optimum (ie least cost) schedule. Interpret your solution regardless of whether it is, in fact, an optimum solution.

3. Explain briefly how your transportation table would change if it were not considered essential to supply orders on time. You may assume that a month's delay in supplying any order is estimated to cost the company £5 per unit. (It is not necessary to actually solve this amended problem).

(*ACCA, June 1985*)

Answers on page 576.

Exercise 10.13

Amphion Airlines is a small air freight company based in Holland and operating throughtout Western Europe. The company has six aircraft of different types, namely:

3 Type 'A'
2 Type 'B'
1 Type 'C'

whose operating costs and load carrying capacities are as follows:

Aircraft type	Fixed cost (£ per day)	Variable cost (£) per mile	per tonne	Capacity (tonnes)
A	800	0·60	30	10
B	700	0·40	35	8
C	500	1·00	25	4

For any journey, variable costs are assigned on the basis of both the distance flown and the load being carried. All fixed and variable cost components are then added to obtain the relevant total cost. On one particular day, there are five loads to be delivered to various destinations, the size of load and distances being:

Load	Size (tonnes)	Distance (miles)
1	10	200
2	2	550
3	8	320
4	5	280
5	3	450

The distances given above are direct from the company's base in Holland to the location involved and in each case there is no return load so that the aircraft will fly back empty. This means that mileage costs are incurred on both the outward and return journeys, but the tonnage cost is only incurred on the outward journey. All five return journeys can be completed within one day and each aircraft can only fly one load in any one day. You may also assume that loads cannot be divided up and delivered in parts.

Required:
1 For each of the five loads, determine the cost of delivery using each of the feasible aircraft types.
2 Set up the cost matrix for assigning each load to each particular aircraft. Using an appropriate procedure, decide which aircraft should be used for delivering each load so that total cost is minimised. What is the total cost of all five deliveries?

(ACCA, December 1985)

Answers on page 578.

Network analysis and project scheduling

11.1 INTRODUCTION

The linear programming models described in the previous chapters are primarily used to solve repetitive, long-term problems. For example, having set up a model, a production manager would expect to use the optimum weekly product mix over a number of weeks. **Network analysis**, on the other hand, is a technique for planning work of a project nature, that is, the work is not generally repeated. For example, the technique is appropriate if we wish to schedule the activities involved in the installation of a computer system in a company, or the activities required in building an office block. A company may continually be involved in installing computer systems or building office blocks, but it is unlikely that any two projects will be exactly the same. We cannot, therefore, assume that once the scheduling model is built it will be suitable for all projects.

The techniques of network analysis enable us to analyse a project which involves a large number of inter-related activities. We can determine how long the work is likely to take, how much it will cost, where savings might be made in either time or money and which activities cannot be delayed without delaying the entire project. The availability of resources is also an important question. The network analysis may be used to help us determine how the activities should be scheduled in order to satisfy constraints on resources.

The analysis of a project falls into three stages:

1. The breakdown of the project into a set of individual jobs (or activities) which are then arranged into a logical network. An activity is an operation or process which uses time and/or other resources.
2. The estimation of the duration of each activity; the setting up of the project schedule and the identification of the jobs which control the completion of the project.
3. The estimation of the resource requirements of each activity; the re-scheduling of the activities to meet a resource objective, or the re-allocation of money or other resources to improve the schedule.

We will consider each of these stages in turn.

11.2 NETWORK DIAGRAMS

The first step in the analysis of any project is to produce a list of activities which make up the project. The details of the list will depend on the nature of the individual project. In each case, however, the immediately preceding activity or activities must be identified. These immediate predecessors are those activities which must be completed before a given activity can begin. For example, when building a house, the roof cannot be built until the walls are complete.

Once this list has been completed, the logical sequence of the activities may be illustrated by a diagram. Various types of diagram exist. The most commonly used ones are the **arrow diagram** and the **activity-on-node diagram**. There are advantages and disadvantages for each type of diagram. The choice is usually a matter for personal preference, or is dictated by the purpose for which the diagram is to be used.

11.2.1 Arrow diagrams

Each activity is represented by an arrow. The length of the arrow is not significant. The direction of the arrow shows the flow of time and is conventionally from left to right. The beginning and end of each activity is called an event, and is represented by a circle or node in the diagram.

FIGURE 11.1 **An activity as represented in an arrow diagram**

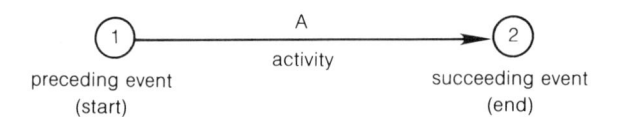

preceding event   activity   succeeding event
(start)   (end)

Activities are identified by a letter or a name; events are identified by a number. Since an activity is defined by a pair of events, it can also be referred to by the numbers of these events. For example, activity A means the same as activity (1,2) in FIGURE 11.1. More than one activity can emerge from or lead into a given node. The event, represented by the node, is not complete until all of the activities leading into it have been completed. An activity, emerging from a particular node, cannot start until its **start event** has been reached, that is, all of the activities leading into the **start node** have been completed.

If activity C cannot start until both jobs A and B have been completed, we represent the logic of the situation by:

FIGURE 11.2 **Logical connections in an arrow diagram**

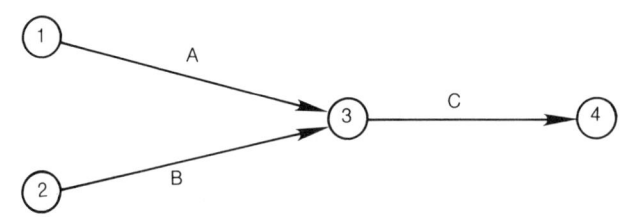

The start event for C is the end event for A and B. It is essential to ensure that the arrow diagram preserves the logical dependence of the activities. In order to do this, it may be necessary to include one or more **logic dummy activities** in the network.

A **logic dummy arrow** is inserted in the diagram when it is necessary to establish that a particular event cannot occur before some other event, and this cannot be shown with the normal activity arrows. The function of the logic dummy is to indicate the sequence of events.

Dummies take zero time and are usually represented by broken arrows. For example, if job C can start when activity A is completed, and job D can start when both job A and B are completed, the arrow network is drawn as follows:

FIGURE 11.3 **The use of a logic dummy in an arrow diagram**

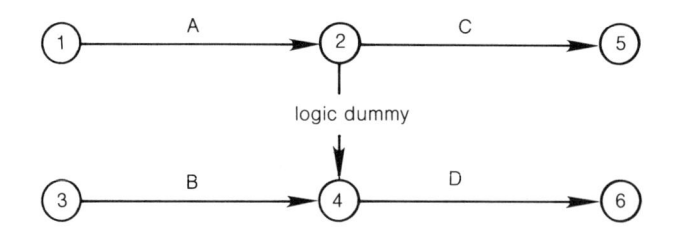

Identification dummies are also used in the arrow diagrams, to avoid ambiguities. Some computer programmes, used in network analysis, define the activities in terms of the event numbers, rather than activity letters or names. If two or more activities take place together, and have the same start and end events, the programme cannot distinguish between them and will reject the input. The inclusion of an identification dummy activity avoids this problem, as we can see in FIGURE 11.4.

FIGURE 11.4 **The use of an identification dummy in an arrow diagram**

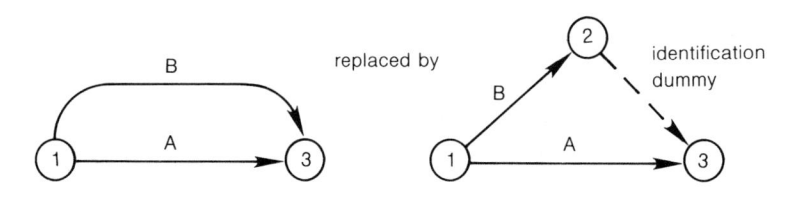

It is normal practice to number the events so that the end event number is always bigger than the start event number.

The first step, after establishing the list of activities which make up the project, is to write out an activity table which shows each activity and its immediate predecessors. This list does not include the logic or identification dummies. The arrow network diagram is constructed from this list, using normal and dummy activities to ensure that the logic relationships are established. Once the initial diagram is complete, unwanted dummies can be identified and removed. The diagram can now be redrawn to improve the layout.

Unnecessary logic dummies can be identified by a simple rule of thumb. If the logic dummy is the only activity emerging from the node, then it is probably not necessary.

Example 11.1: To construct a network arrow diagram

Delco plc is an engineering firm which has a contract to produce a batch of machines to be used by a large shoe manufacturer in the mass production of shoes. The activities required in the design and manufacture of these machines are listed below:

TABLE 11.1 **Activity table for Example 11.1**

Activity		Immediately preceding activity
A	Draw up estimate of costs	—
B	Agree estimate	A
C	Purchase internal machinery	B
D	Prepare design drawings	B
E	Construct main frame	D
F	Assemble machinery	C,E
G	Test machinery	F
H	Determine model type	D
I	Design outer casing	D
J	Construct outer casing	H,I
K	Final assembly	G,J
L	Final check	K

Illustrate the activities by an arrow diagram.

Solution: The network diagram must begin from a single start event, shown by a circle in FIGURE 11.5. It must end with a single end event. We begin drawing the diagram from the first event. The activities with no preceding jobs all start from this first event. It is useful to start with a rough sketch of the network:

FIGURE 11.5 **Rough sketch for Example 11.1**

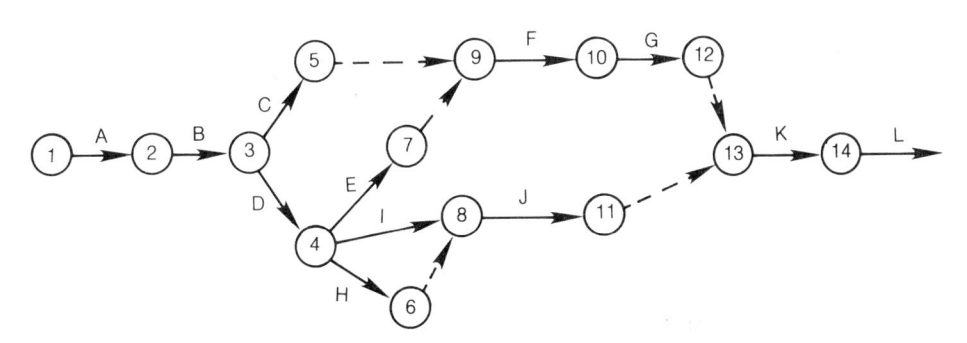

The logic of this first attempt should be checked, activity by activity, against the information given in the activity table. It should be corrected as necessary and redrawn to improve the layout. All of the logic dummies can be removed. One identification dummy remains.

328 QUANTITATIVE ANALYSIS

FIGURE 11.6   **Redrawn arrow diagram for Example 11.1**

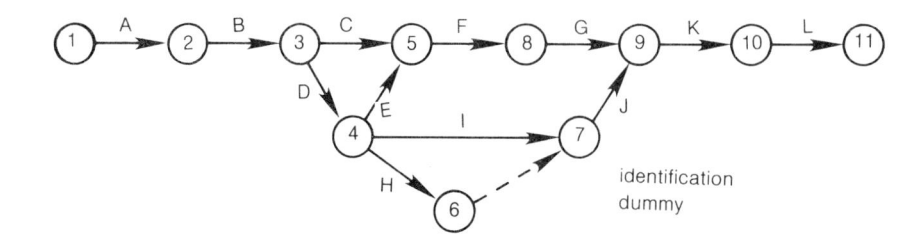

Example 11.2: To construct an arrow diagram

Delco plc are also involved in a second project, the details of which are given below:

TABLE 11.2   **Activity table for Example 11.2**

Activity	Immediately preceding activities
A	—
B	—
C	—
D	A,B
E	B,C
F	C
G	D,E
H	F,G

Illustrate the project by an arrow diagram.

Solution: We begin with one start event, the first circle. From the activity table there are three activities, A, B and C, which have no preceding activities. All three arrows emerge from this start event. The activity table looks deceptively simple, but it is quite difficult to represent the logic with an arrow diagram. We must use three logic dummies.

FIGURE 11.7   **Arrow diagram for Example 11.2**

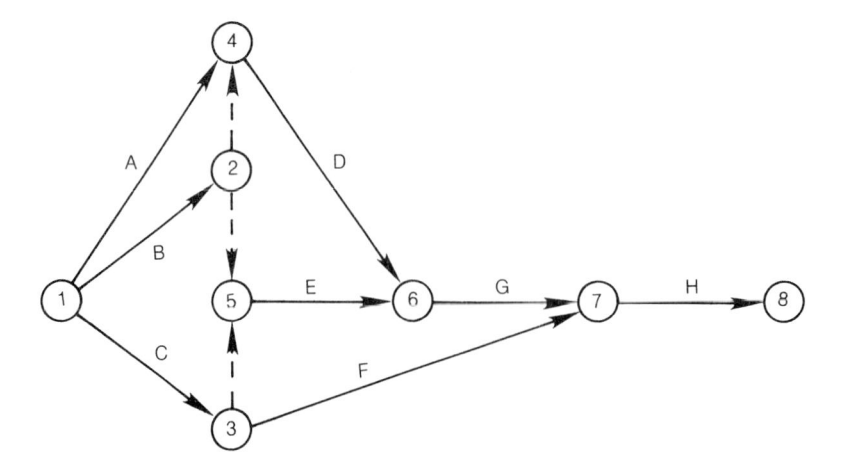

11.2.2 Activity-on-node diagrams

In this type of diagram, the activities are represented by the nodes of the network, and the arrows show the inter-dependencies. There is no need for dummy activities. It is still important to show the flow of time from left to right.

Example 11.3: To construct an activity-on-node network diagram

Refer to **Example 11.2**. We will re-draw the network for this example using the activity-on-node system.

Solution: In this example, the network is much easier to illustrate using the activity-on-node method, but it is less easy to gain a picture of the overall flow from one activity to the next. The activity-on-node diagram begins from the start node, which leads to the first three activities, A, B, C. The diagram is relatively easy to construct.

FIGURE 11.8   **Activity-on-node diagram—Example 11.3**

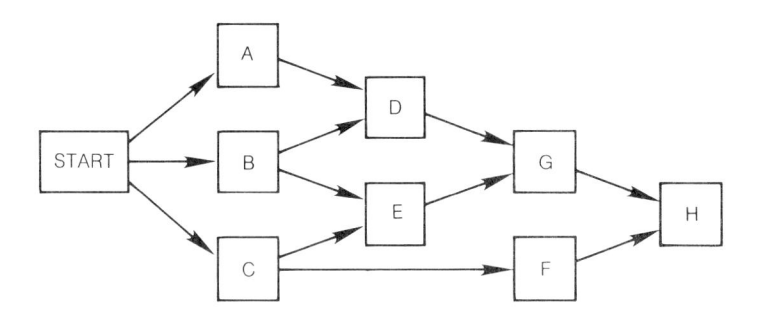

Each type of diagram has advantages and disadvantages. It does not usually matter which system is used. If the arrow diagram involves the use of a large number of dummy activities, then the activity-on-node diagram may be the preferred choice. A comparison of the main features of the two systems is shown below.

FIGURE 11.9   **Comparison of Arrow and Activity-on-node diagrams**

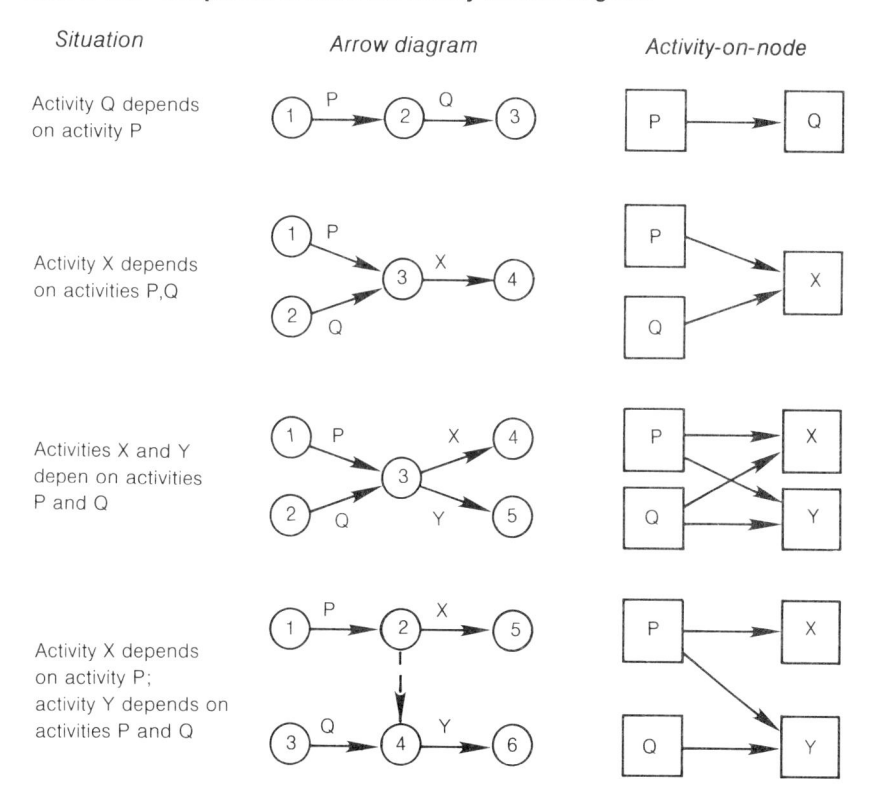

11.3   CRITICAL PATH ANALYSIS

Once the activities have been identified, it is possible to assess their durations. The overall project duration can be found from the activity durations and the network logic. At this stage, these activity durations are treated as fixed and not subject to any uncertainty. We will see in a later section, how to

330 QUANTITATIVE ANALYSIS

amend the analysis to deal with uncertainty in the activity times. There will be several possible paths through any network. The overall time taken to complete any one path is the sum of its individual activity times. The duration of the entire project will be determined by the path which takes the longest time. The activities on this longest path are known as **critical activities**. Any delay in the start or completion of these jobs will delay the entire project. Critical activities form a continuous chain running through the network. This chain is called the **critical path**. Each network will have at least one critical path.

To determine the overall duration of the project, we must find the duration of the critical path. For most networks, it is too laborious to identify all of the paths through the network to find the one which takes the longest time. There are two possible methods of tracing the passage of time through the network:

1. determine the earliest time at which each activity can start and the earliest time at which it can finish; or
2. determine the earliest time at which each event can be completed.

The second method is suitable for use with arrow diagrams only.

11.3.1 Critical path analysis using activity-on-node diagrams

Example 11.4: To calculate the overall duration of a project using an activity-on-node diagram

The table below gives the duration of each activity for the project used in **Examples 11.2** and **11.3**. Determine the overall duration of the project.

TABLE 11.3 **Activity table and activity durations for Example 11.4**

Activity	Immediately preceding activities	Time in days
A	—	8
B	—	10
C	—	6
D	A,B	8
E	B,C	9
F	C	14
G	D,E	14
H	F,G	6

The activity-on-node diagram, illustrating the project, was drawn in **Example 11.3**.
Solution: We assume that the initial activities, A, B and C, each begin at time zero. This is the **earliest start time, ES**, for these activities. The earliest time at which they can finish is given by:

earliest finish time, EF = ES + activity duration

These times are usually written directly onto the diagram, but we will tabulate them first to illustrate the method of calculation.

TABLE 11.4 **Calculation of earliest activity start and finish times for Example 11.4**

Activity	Duration, days	Earliest start time	Earliest finish time	Comment
A	8	0	0 + 8 = 8	
B	10	0	0 + 10 = 10	
C	6	0	0 + 6 = 6	
D	8	10	10 + 8 = 18	Cannot begin until both A and B are complete.
E	9	10	10 + 9 = 19	Cannot begin until both B and C are complete.
F	14	6	6 + 14 = 20	Cannot begin until C is complete.
G	14	19	19 + 14 = 33	Cannot begin until both D and E are complete.
H	6	33	33 + 6 = 39	Cannot begin until both F and G are complete.

The earliest start and earliest finish times have been entered on the activity-on-node in FIGURE 11.10. We can see that activity H is finished by day 39, therefore this is the overall duration of the project.

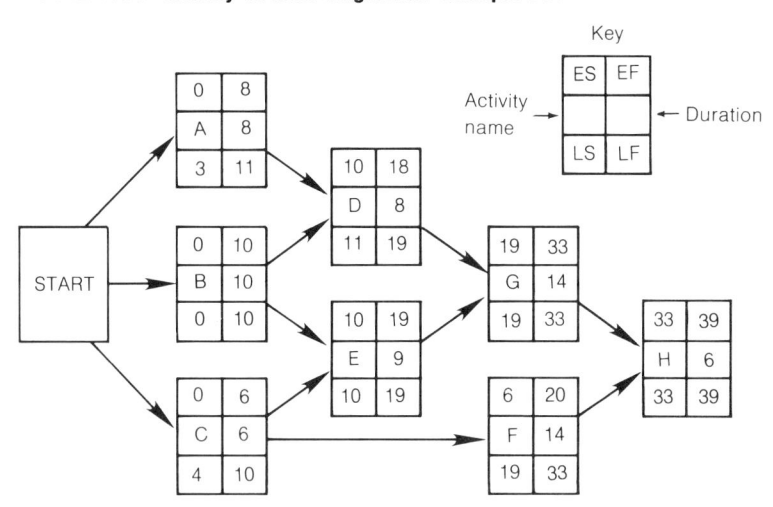

FIGURE 11.10 **Activity-on-node diagram for Example 11.4**

We cannot at this stage identify the critical activities. To do this we must calculate two more times for each activity. These are the **latest start time, LS**, and the **latest finish time, LF**. We now begin with the last activity in the network and assume that the latest finish time and the earliest finish time are equal. The latest start time is then found by subtracting the activity duration. See TABLE 11.5:

TABLE 11.5 **Calculation of latest activity start and finish times for Example 11.4**

Activity	Duration, days	Latest finish time	Latest start time	Comment
H	6	39	39 − 6 = 33	
G	14	33	33 − 14 = 19	G must finish in time for the latest start of H.
F	14	33	33 − 14 = 19	F must finish in time for the latest start of H.
E	9	19	19 − 9 = 10	E must finish in time for the latest start of G.
D	8	19	19 − 8 = 11	D must finish in time for the latest start of G.
C	6	10	10 − 6 = 4	C must finish in time for the latest starts of E and F. The earliest of these, day 10, must be used.
B	10	10	10 − 10 = 0	B must finish in time for the latest starts of D and E. The earliest of these, day 10, must be used.
A	8	11	11 − 8 = 3	A must finish in time for the latest start of D.

A critical activity is one for which:

$$ES = LS \text{ and } EF = LF$$

that is, an activity for which there is no spare time between the earliest start time and the latest finish time. We can see that in this example, the critical activities are B, E, G and H. The path through the activity-on-node network, linking these activities is called the critical path. The critical path is B-E-G-H.

11.3.2 Critical path analysis using an arrow diagram

The above analysis can be carried out in the same way using an arrow diagram. The ES, EF, LS and LF times are written along the activities:

FIGURE 11.11 **Recording activity times on an arrow diagram**

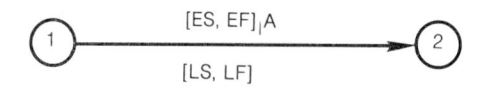

Alternatively, the analysis can be done in terms of the times for each event. Calculate the earliest time by which each event can be completed. This is called the **earliest event time**, **EET**. The EET of the final node determines the overall duration of the project. The initial event has an EET of zero.

To identify the critical activities, we again work back through the network, calculating the **latest event times**, the LET, by which each event must be complete. Events for which:

$$LET_{start} - EET_{end} - duration = 0$$

or:

$$EET_{start} - LET_{end} - duration = 0$$

are critical.

Example 11.5: To calculate the overall duration of a project, using the arrow diagram

We will repeat **Example 11.4** using EET and LET, remembering that dummies have a duration of zero.
Solution: We first calculate the earliest event times. When more than one activity enters an event, we have a choice of values. The largest value must be chosen, since the event is not complete until all of the activities which comprise it have been completed.

TABLE 11.6 **Calculations of EETs for Example 11.5**

Node	EET, days	Comment
1	0	Start event.
2	0 + 10 = 10	EET node 1 + duration of activity B.
3	0 + 6 = 6	EET node 1 + duration of activity C
4	0 + 8 = 8	EET node 1 + duration of activity A
	or 10 + 0 = 10*	or EET node 2 + duration of dummy.
		Choose the longer time, ie 10 days.
5	10 + 0 = 10*	EET node 2 + duration of dummy
	or 6 + 0 = 6	or EET node 3 + duration of dummy.
		Choose the longer time, ie 10 days.
6	10 + 8 = 18	EET node 4 + duration of activity D
	or 10 + 9 = 19*	or EET node 5 + duration of activity E.
		Choose the longer time, ie 19 days.
7	19 + 14 = 33*	EET node 6 + duration of activity G
	or 6 + 14 = 20	or EET node 3 + duration of activity F.
		Choose the longer time, ie 33 days.
8	33 + 6 = 39	EET node 7 + duration of activity H.

Note: * denotes the EET when there is a choice.

The times are entered on the arrow diagram constructed in **Example 11.2** and repeated in FIGURE 11.12.

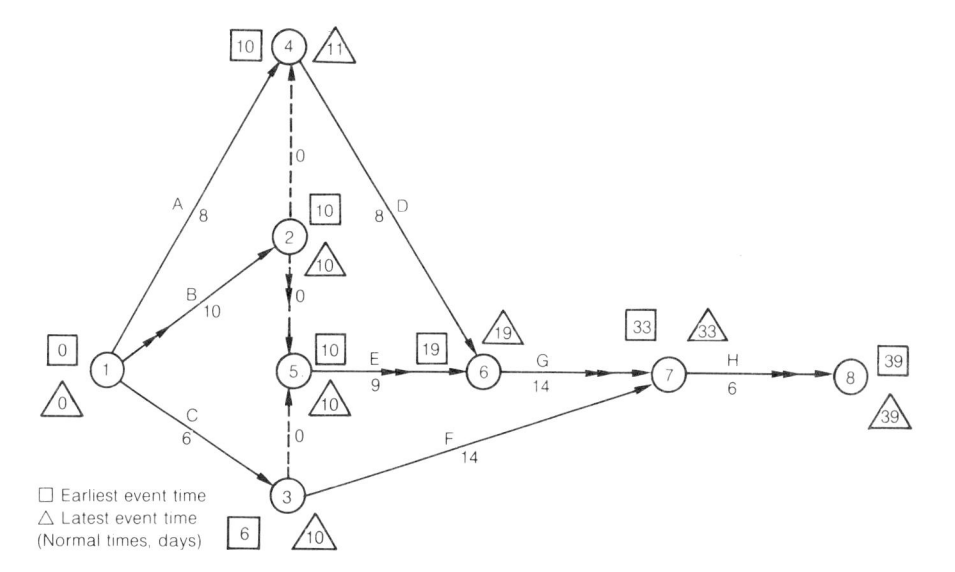

FIGURE 11.12 **Arrow diagram with even EETs and LETs for Example 11.5**

☐ Earliest event time
△ Latest event time
(Normal times, days)

The EET of the final node is 39 days, which is the overall project completion time.

To determine the critical activities, we work back through the network calculating the LET for each node. We assume that the EET = LET for the final node. When more than one arrow leads back into an event, we have a choice of values for the LET. The smallest value must be chosen, since the event must be completed in time to satisfy all of the latest start times for the events leading out of the node.

TABLE 11.7 **Calculation of LETs for Example 11.5**

Node	LET, days	Comment
8	39	Final node LET = EET.
7	39 − 6 = 33	LET node 8 − duration of activity H.
6	33 − 14 = 19	LET node 7 − duration of activity G.
5	19 − 9 = 10	LET node 6 − duration of activity E.
4	19 − 8 = 11	LET node 6 − duration of activity D.
3	10 − 0 = 10*	LET node 5 − duration of dummy
	or 33 − 14 = 19	or LET node 7 − duration of activity F.
		Choose the shortest time, ie 10 days.
2	10 − 0 = 10*	LET node 5 − duration of dummy.
	or 11 − 0 = 11	or LET node 4 − duration of dummy.
		Choose the shorter time, ie 10 days.
1	11 − 8 = 3	LET node 4 − duration of activity A
	or 10 − 10 = 0*	or LET node 2 − duration of activity B
	or 10 − 6 = 4	or LET node 3 − duration of activity C.
		Choose the shorter time, ie 0 days.

Note: * denotes the LET, when there is a choice.

The times are entered on the arrow diagram in FIGURE 11.12.

The activities are critical if:

$$EET_{start} = LET_{start}$$

and:

$$EET_{end} = LET_{end}$$

and:

$$LET_{end} - EET_{start} - duration = 0$$

We can see from FIGURE 11.12, that the critical activities are B, E, G and H as before.

Delay of any kind on the critical path will delay the whole project. However, there is scope for some delay or re-scheduling on the non-critical paths. The spare time available in a network is called **float**. A number of different kinds of float are defined, which arise from the different effects the spare time has in the network. **Total float** is the amount of time by which an activity can be lengthened or re-scheduled without affecting the overall duration of the project. **Free float** is the amount of time by which an activity can be lengthened or re-scheduled without affecting the earliest start time of any following activity. A third type, **independent float**, is sometimes used. It has no effect on earlier or later activities. For any activity:

$$\text{total float} = \text{LET}_{end} - \text{EET}_{start} - \text{duration}$$

and:

$$\text{free float} = \text{EET}_{end} - \text{EET}_{start} - \text{duration}$$

and:

$$\text{independent float} = \text{EET}_{end} - \text{LET}_{start} - \text{duration}$$

It is sometimes useful to illustrate the float which is available in a network, particularly if it is necessary to re-schedule activities. One possible way of doing this is the **Gantt Chart**.

Example 11.6: To construct a Gantt Chart

Refer to **Example 11.5**. Calculate the total float for each activity.

TABLE 11.8 **Calculation of activity total float for Example 11.5**

Activity	LET$_{end}$ days	EET$_{start}$ days	Duration, days	Total float days
A	11	0	8	3
B	10	0	10	0
C	10	0	6	4
D	19	10	8	1
E	19	10	9	0
F	33	6	14	13
G	33	19	14	0
H	39	33	6	0

The activities with zero total float are the critical activities. The Gantt Chart is shown in FIGURE 11.13.

FIGURE 11.13 **Gantt Chart for Example 11.5**

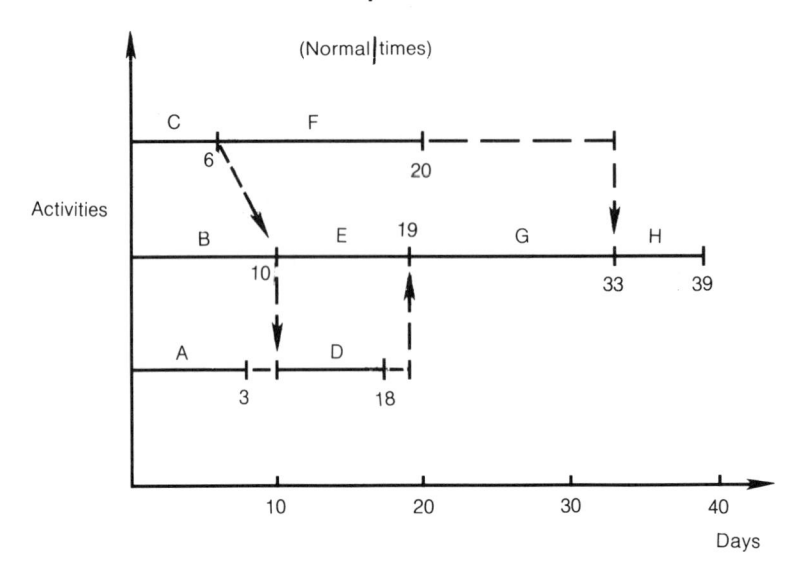

11.4 THE COSTS OF A PROJECT

The total cost of a project will depend on the cost of the individual activities plus any additional variable or fixed costs. Since all activities must be completed, whether they are critical or not, the total cost of the activities is simply the sum of the individual values.

It may be possible to reduce the duration of a specific activity by employing additional resources. The implication of this is that the cost of that activity will increase. However, if the activity is critical, a saving in its duration may result in an overall saving in time on the project, and a consequent reduction in the total project costs.

The shortest possible time in which an activity can be completed is referred to as the **crash time**. In some cases, it will be possible to complete activities in either their normal times or their crash times, but not in between. In other projects it will be possible gradually to reduce the time taken for the activities until the crash time is reached. We will consider the effect of reductions in activity time, on the network and on the cost of the project, using two separate objectives:

1   to minimise the overall time for the project;
2   to minimise the overall cost of the project.

11.4.1 To minimise the overall project duration with minimum additional cost.

We must now provide information on the costs of each activity, any possible reductions in time and the costs incurred if the times are reduced.

Example 11.7: To minimise the overall project duration

Refer to **Example 11.2**. Additional information on activity costs and possible reductions are given below:

TABLE 11.9   **Normal and crash times and costs for Example 11.7**

Activity	Immediately preceding activities	Normal time, days	cost, £	Cash time, days	cost, £
A	—	8	7500	4	9000
B	—	10	8500	8	11000
C	—	6	6000	5	7000
D	A,B	8	13000	5	16000
E	B,C	9	14000	6	16500
F	C	14	14500	11	18000
G	D,E	14	13500	10	18750
H	F,G	6	5500	4	6500
	Total activity costs		£82500		£102750

The crash information represents the minimum time in which the activity could be completed and the total cost of completing the activity within this shorter time. The choice is between normal time and cost, or crash time and cost. It is not possible to save one day on a particular activity for a proportionate increase in cost. In addition to the costs for each activity, there is a site cost of £1000 per day.

1   What is the minimum time in which the project can be completed?
2   What is the associated minimum additional cost?

336 QUANTITATIVE ANALYSIS

Solution: The minimum time can be found by crashing all of the activities in the network, both critical and non-critical. The arrow diagram, produced in **Example 11.2** is reproduced below. The crash times have been added and the EET and LET determined:

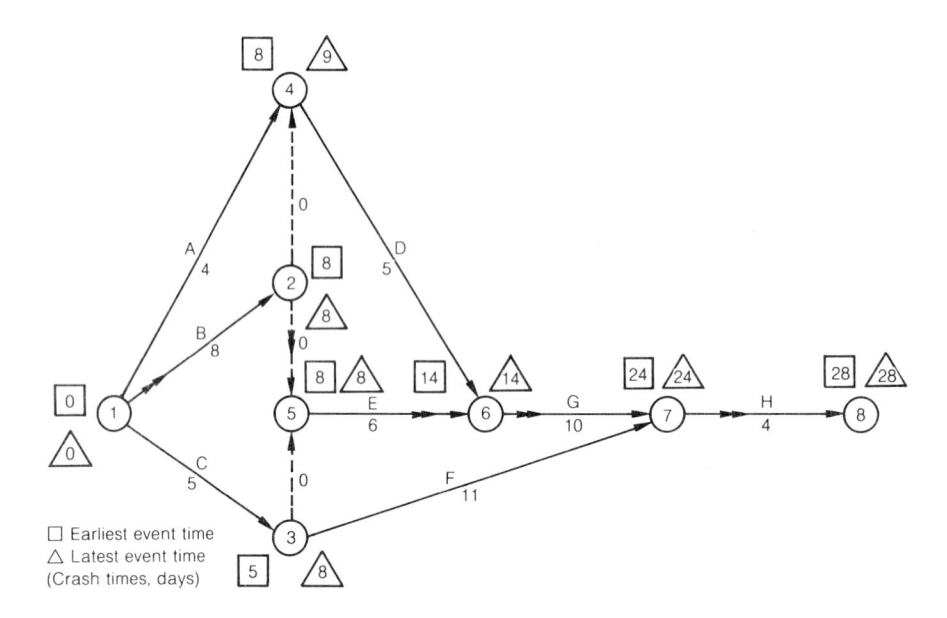

FIGURE 11.14 **Arrow diagram for Example 11.7 using crash times**

☐ Earliest event time
△ Latest event time
(Crash times, days)

We can see that the EET of node 8 is 28 days, therefore the minimum time for the project is 28 days. The critical path remains B-E-G-H.

The total cost is found by using the following equation:

$$\text{total cost} = (\text{crash cost of activities}) + 28 \times \text{site cost/day}$$

$$= £102{,}750 + 28 \times £1000 = £130{,}750.$$

This is not, however, the minimum cost for this completion time, since it may not be necessary to crash the non-critical activities. The non-critical activities are A, C, D and F. We must look at the effect of uncrashing these activities. If it is possible to restore them to their normal durations without increasing the overall project time, this should be done and the cost of crashing saved.

It is not necessary to crash activity A, since restoring A to 8 days does not change the EET of node 4, therefore there is no follow-on effect in the rest of the network. If we uncrash activity A, we save £1500.

If activity C is uncrashed from 5 to 6 days, it will increase the EET at node 3 to 6, but will have no effect on the EET at nodes 5 or 7. This will have no effect on the project duration. C should be uncrashed saving £1000.

If activity D is not crashed, the EET of node 6 becomes 16 days. Node 6 is on the critical path, therefore, activity D must be crashed to achieve the minimum overall time of 28 days.

If activity F is not crashed then the EET of node 7 is unchanged. The overall duration of the project will not be increased. F should be uncrashed, saving £3500.

The minimum cost of completing the project in 28 days is:

$$£130{,}750 - £1500(A) - £1000(C) - £3500(F) = £124{,}750$$

The cost of completion at normal time is:

$$£82{,}500 \text{ (activities)} + £39{,}000 \text{ (site costs)} = £121{,}500$$

The additional cost of completing 11 days earlier is therefore:

$$£(124{,}750 - 121{,}500) = £3250$$

Example 11.8: To minimise the overall project duration

Refer to **Example 11.1** Additional information on activity costs and possible reductions are given in TABLE 11.10.

TABLE 11.10 **Normal and crash times and cost for Example 11.8**

Activity	Normal time, weeks	Normal cost, £	Possible reduction, weeks	Crash time, weeks	Extra cost for reduction of one week, £
A	2	400	1	1	400
B	1	0	0	1	0
C	4	200	2	2	125
D	6	450	4	2	175
E	3	700	2	1	250
F	3	200	2	1	200
G	4	600	3	1	125
H	2	0	0	2	0
I	3	250	1	2	200
J	8	600	4	4	100
K	2	450	1	1	250
L	2	200	1	1	150
Activity costs		£4050			

Variable overheads are £300 per week for the project duration.

1. Determine the normal overall completion time and total cost of the project, and
2. the minimum time in which the project can be completed and the associated minimum cost.

Solution: The arrow diagram which was drawn in **Example 11.1** is repeated in FIGURE 11.15. The EET and LET have been added. With normal times the project can be completed in 24 weeks and the critical path is:

$$A - B - D - I - J - K - L$$

The total cost of the project is:

$$£4050 \text{ (activities)} + 24 \times £300 \text{ (variable overheads)} = £11,250$$

FIGURE 11.15 **Arrow diagram for Example 11.8 using normal times**

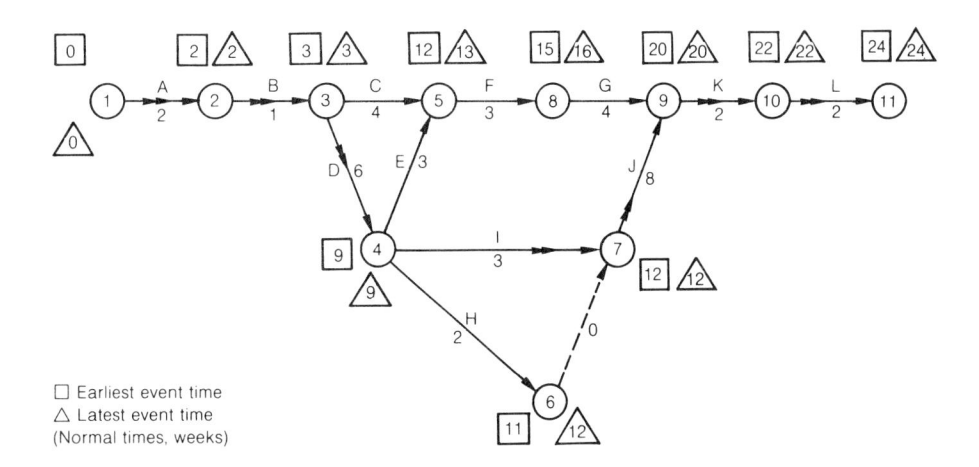

☐ Earliest event time
△ Latest event time
(Normal times, weeks)

In order to determine the minimum completion time, we will complete all activities at their minimum times. These times and the resulting EET and LET are shown in FIGURE 11.16:

FIGURE 11.16 **Arrow diagram for Example 11.8 using crash times**

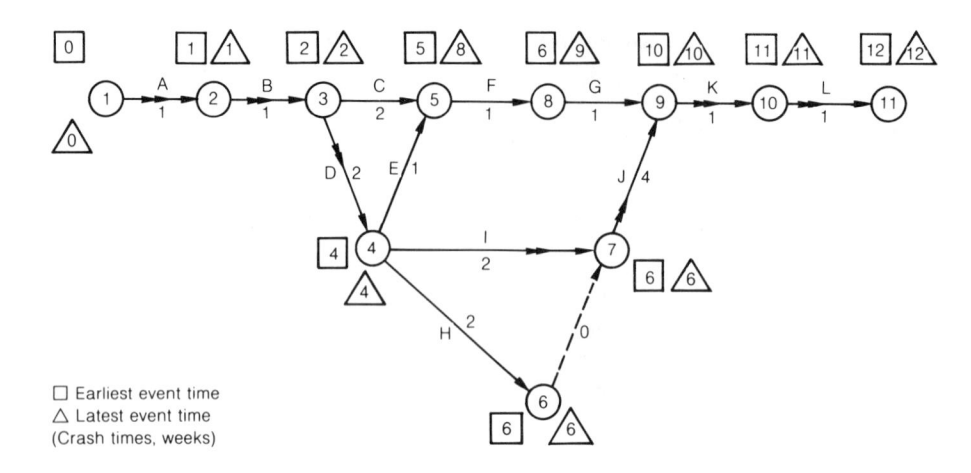

□ Earliest event time
△ Latest event time
(Crash times, weeks)

The minimum duration of the project is 12 weeks. The critical paths are now:

$$A - B - D - I - J - K - L$$

and

$$A - B - D - H - J - K - L$$

Let us find out whether money can be saved by uncrashing some of the non-critical activities.

The non-critical activities are C, E, F, and G. In this project we can change the activity durations on a week by week basis. We will deal first with the activities which are the most expensive to crash. The order is E (£250), F (£200) and C (£125) or G (£125):

TABLE 11.11 **Uncrashing the non-critical activities in Example 11.8**

Activity	Change in duration	Effect
E	increase by 2 weeks	EET at node 5 becomes 7 weeks; EET at node 8 becomes 8 weeks, but EET at node 9, on the critical path, is not affected; no overall effect. E is now at normal time.
F	increase by 1 week	EET at node 8 becomes 9 weeks. Activities E, F and G are now critical.
G	no increase now possible	
C	increase by 2 weeks	There is no effect at node 5. C is now at normal time.

The minimum cost of completing the project in 12 weeks is:

£4050 (activities normal cost) + (1 × 400(A) + 4 × 175(D) + 1 × 200(F) + 3 × 125(G)
+ 1 × 200(I) + 4 × 100(J) + 1 × 250(K) + 1 × 150(L))(crash costs)
+ 12 × £300 (variable overheads) = 4050 + 2675 + 3600 = £10,325

11.4.2 Completing a project at minimum cost

For projects which incur variable costs, such as site costs, it may be beneficial to reduce the duration of the project. Since such reductions will themselves incur costs, we must strike a balance. Savings in

Example 11.9: To determine the minimum cost of a project

We will refer again to **Examples 11.7**, to determine the minimum cost at which the project can be completed, together with the associated project time. Activities can be completed in their normal or crash times, but not in between.

Solution: Using the network with normal times in **Example 11.5**, we begin by listing all of the critical activities, together with the maximum time which could be saved and the net savings in cost.

TABLE 11.12   Calculation of the minimum cost of the project in Example 11.9

Activity	Number of days saved at crash time	Cost at crash time, £	Saving, £	Net saving, £	Comment
B	2	2500	2 × 1000	−500	Do not crash
Either E	1*	2500	1 × 1000	−1500	Do not crash
or E and D	3	5500	3 × 1000	−2500	Do not crash
G	4	5250	4 × 1000	−1250	Do not crash
H	2	1000	2 × 1000	+1000	Crash. Reduce project time from 39 to 37 days.

* The full 3 days cannot be saved, because A-D-G-H takes over as the critical path. This reduces the overall duration by only one day. If both E and D are crashed, all 3 days can be saved. It now costs (£2500 + £3000) and is still not worth doing.

The minimum cost of the project is £(121,500 − 1000) = £120,500
The associated time is 37 days.

11.5 UNCERTAINTY IN ACTIVITY TIMES

In the analysis so far, it has been assumed that the activity times are known with certainty. In practice, these times are usually uncertain. The project manager will have some idea how long each job will take, but he cannot predict problems or delays. This uncertainty in the activity times means that the overall project duration is also subject to uncertainty.

The approach which is adopted to deal with this uncertainty depends on the type of project and the nature of the uncertainty. If maximum and minimum durations can be specified for each activity, it may be possible to work in terms of the expected (average) durations and to determine an expected project completion time. The most commonly used procedure is the **Project Evaluation and Review Technique, (PERT)**. PERT uses the expected activity times to calculate the expected project completion time. The procedure is the same as that described above for fixed activity times.

Non-critical paths are of more importance when the activity times are uncertain. If all times are subject to variation it is possible, in practice, for a path which is non-critical on the basis of expected times to take over from the designated critical path.

PERT is based on an assumption about the behaviour of the activity durations. It is assumed that each individual activity time follows a beta distribution. If this is true the completion time for the project as a whole is normally distributed. PERT is a suitable technique only if this assumption is reasonable for the specific project. The beta distribution is shown in FIGURE 11.17. The shortest possible activity time is called the **optimistic time** (a) and the longest possible is called the **pessimistic time** (b). The peak in the distribution gives the most likely activity time (m). These three times must be estimated for each activity in the network.

FIGURE 11.17 **Typical beta distribution for an activity time**

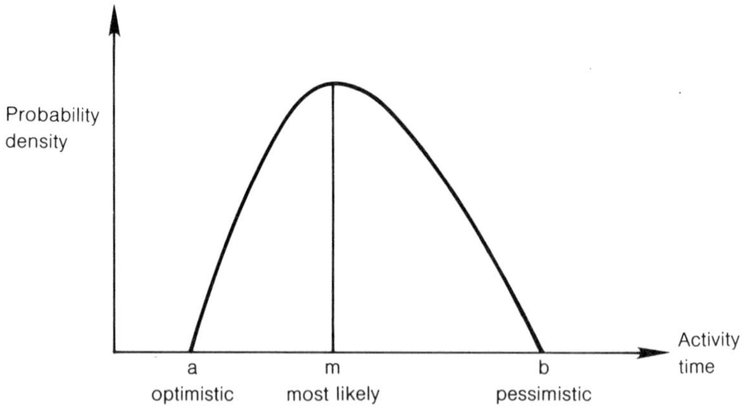

From the three specified times, the expected activity duration and its variance may be calculated. The expected activity time, t, is given by:

$$t = \frac{a + 4m + b}{6}$$

The associated variance of the expected duration is given by:

$$\sigma_t^2 = \left(\frac{b - a}{6}\right)^2$$

The project completion time is found from the network, using the expected activity durations. This overall completion time is assumed to be normally distributed.

On the assumption that the activity times are independent of each other, the mean of the normal distribution is given by the sum of the expected durations of the critical activities. The variance is given by the sum of the variances of the critical activities. This normal distribution may be used to estimate the probability of meeting a specified completion date.

The procedure for using PERT is similar to the analysis of a network with fixed durations.

1. Develop a list of the activities for the project, with the immediate predecessors and the optimistic, most likely and pessimistic times.
2. Draw the network.
3. Estimate the expected activity time and variance for each activity, assuming a beta distribution in each case.
4. Using the expected activity times, find the overall expected project completion time.
5. Identify the critical activities and the critical path.
6. Use the variance of the critical activities to estimate the variance of the expected project duration.

Example 11.10: Application of PERT

The following activities have been identified as those necessary for the development and launch of a new product by ABC plc:

NETWORK ANALYSIS AND PROJECT SCHEDULING: CHAPTER ELEVEN

TABLE 11.13 **Activity table and times for Example 11.10**

Activity	Immediate predecessors	Optimistic a	Most likely m	Pessimistic b
A	—	1.5	2	2.5
B	A	2	2.5	6
C	—	1	2	3
D	C	1.5	2	2.5
E	B,D	0.5	1	1.5
F	E	1	2	3
G	B,D	3	3.5	7
H	G	3	4	5
I	F,H	1.5	2	2.5

Time, weeks

1 Determine the expected number of weeks until the project is complete. Which are the critical activities?

2 What is the probability that the project will take more than 16 weeks?

Solution: The expected activity times and variances are:

TABLE 11.14 **Calculation of expected times and variances for Example 11.10**

Activity	Expected time, weeks	Variance, weeks^2
A	$\dfrac{1.5 + 4 \times 2 + 2.5}{6} = 2$	$\left(\dfrac{2.5 - 1.5}{6}\right)^2 = 1/36$
B	$\dfrac{2 + 4 \times 2.5 + 6}{6} = 3$	$\left(\dfrac{6 - 2}{6}\right)^2 = 16/36$
C	$\dfrac{1 + 4 \times 2 + 3}{6} = 2$	$\left(\dfrac{3 - 1}{6}\right)^2 = 4/36$
D	2	= 1/36
E	1	= 1/36
F	2	= 4/36
G	4	= 16/36
H	4	= 4/36
I	2	= 1/36

The network diagram with the expected activity durations is given below:

FIGURE 11.18 **Arrow diagram with expected times for Example 11.10**

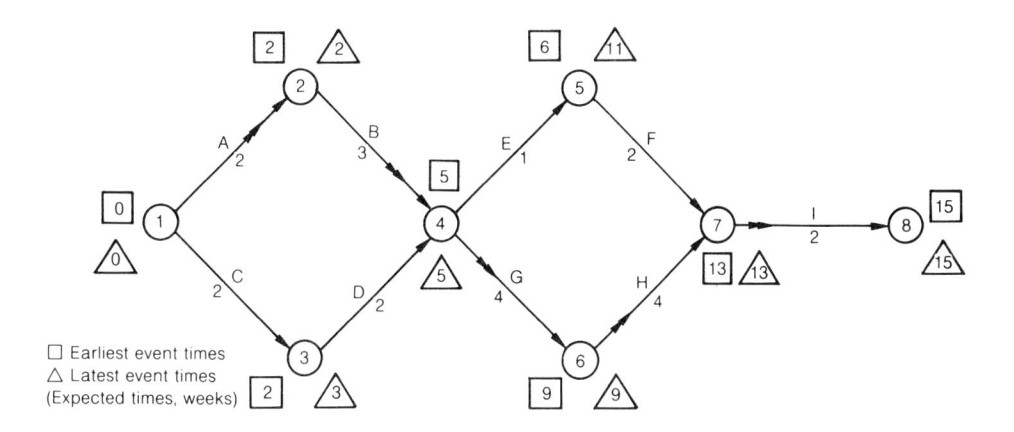

□ Earliest event times
△ Latest event times
(Expected times, weeks)

The expected overall completion time is calculated in the usual way. From FIGURE 11.18, we can see that the project is expected to be finished in 15 weeks. The critical activities are A, B, G. H, I. For comparison purposes, the other paths through the network are:

$$A, B, E, F, I \text{ which takes } 10 \text{ weeks}$$
$$C, D, E, F, I \text{ which takes } 9 \text{ weeks}$$
$$C, D, G, H, I \text{ which takes } 14 \text{ weeks}$$

The last path, C-D-G-H-I, takes only one week less than the critical path. A small amount of variation in some of the activity times could easily change the critical path.

The variance of the expected project completion time is given by the sum of the variances of the critical activities:

$$\sigma^2 = \sigma_A^2 + \sigma_B^2 + \sigma_G^2 + \sigma_H^2 + \sigma_I^2$$

therefore:

$$\sigma^2 = 1/36 + 16/36 + 16/36 + 4/36 + 1/36 = 38/36 = 1.11 \quad \text{weeks}^2$$

The standard deviation of the project completion time is:

$$\sqrt{1.11} = 1.03 \text{ weeks}$$

The probability that the project take more than 16 weeks to complete is found as follows:

FIGURE 11.19 **Distribution of project completion time for Example 11.10**

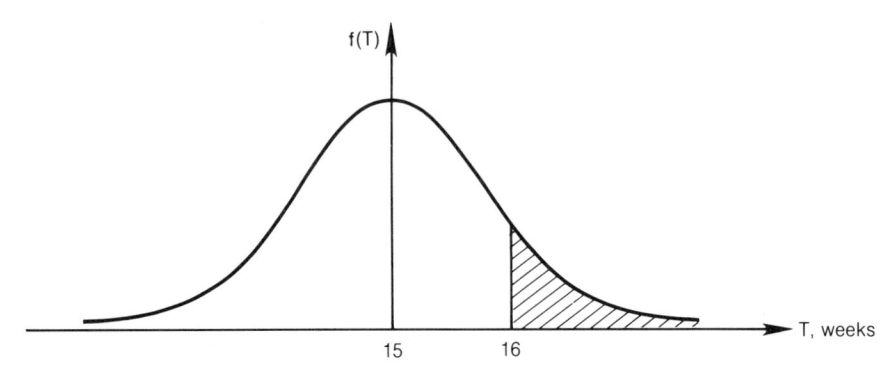

Overall project completion time

Sixteen weeks is z standard deviations above the mean where:

$$z = \frac{16 - 15}{1.03} = 0.97$$

From the standard normal tables:

$$P(z \geqslant 0.97) = 0.166$$

Therefore, there is a 16.6% probability that the project will take more than 16 weeks to complete.

11.6 RESOURCE ALLOCATION

The network diagram gives the logical sequence of activities in a project. The analysis we have discussed so far takes no account of any limitations imposed by availability of resources. The initial activity schedule was drawn up on the implied assumption that all of the resources which are called for will be available. This is not necessarily so, and even if it is, using the resources as required by this first schedule may be uneconomic.

The way in which activities are scheduled when resource considerations are taken into account, depends on the objectives of the project controllers. It may be of greatest importance that the project is finished in a certain time irrespective of the resource implications—a time-constrained network. Alternatively, if money is limited, then the project must not utilise more than a certain level of resources, irrespective of the time taken—a resource-constrained network. In this context resources include manpower, machinery, materials, money, space, etc.

At the outset of the project, the Project Manager should determine the objective for the allocation of resources. The objective may be:

1   To maximise the utilisation of the resources. The utilisation can be assessed in terms of a utilisation factor:

$$\text{utilisation factor} = \frac{\text{total resources employed}}{\text{total resources provided}}$$

2   To minimise the peak resource requirements.
3   To minimise the maximum change in resource requirements.

There are many other possibilities.

There are also many possible methods of solving the resource allocation problem, for example, heuristic methods, linear programming or other forms of mathematical programming. We will look at one simple method which uses resource profiles and 'trial and error'.

11.6.1 Resource profiles

If the total requirement for a particular resource is determined at regular intervals, for example on a daily or weekly basis, then a **resource profile** may be assembled. The resources specified for each job are totalled for all concurrent jobs, assuming that each job is begun at its earliest start time. A separate profile is required for each resource. FIGURE 11.20 illustrates a possible manpower resource profile. At times the required manpower exceeds the number of men available, but the total number of man-hours needed does not exceed the total number available.

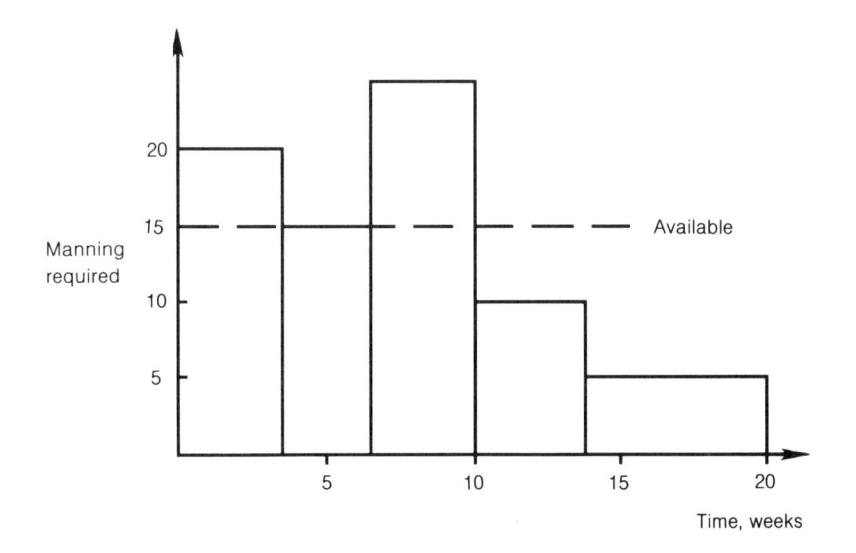

FIGURE 11.20   **Manpower resource profile.**

Whenever a resource limit is exceeded, either more resources must be put into the project, or activities must be re-scheduled in some way. It may even be necessary to delay the project completion. Some activities, although they do not have a logical sequential relationship in the project, cannot go on at the same time, because of resource limitations. This restriction can be allowed for in the resource profile by limiting the resource available. This prevents these activities being scheduled for the same time periods.

Example 11.11: Resource allocation

XYZ Ltd have been awarded a contract to re-surface a car park. The Project Manager has identified eight essential activities in this work. The details are listed below.

TABLE 11.15 **Activity table, times and manpower needs for Example 11.11**

Activity	Preceding activities	Time, days	Number of men required for the activity
A	—	3	1
B	—	6	1
C	—	7	2
D	A	8	2
E	C	4	1
F	B,E	3	2
G	C	10	2
H	F,G	3	1

Unfortunately, due to the pressure of work on other jobs, XYZ can spare only 4 men for the car park job. We will determine how long the job will take and how the men should be allocated. Assume that all of the men can do all of the activities.

Solution: We will build up a resource profile, starting all activities at the earliest times. We can then re-schedule the activities in order to meet the limitations on the number of men available. We first draw the network and identify the critical path.

FIGURE 11.21 **Arrow diagram for Example 11.11**

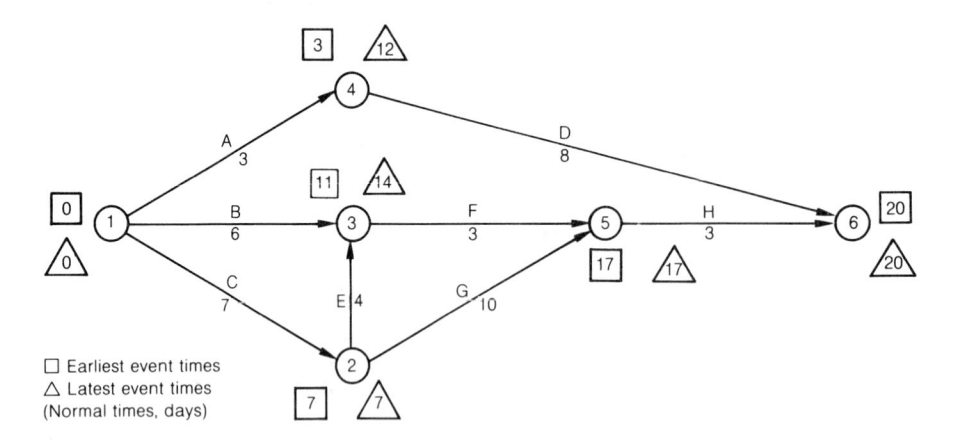

□ Earliest event times
△ Latest event times
(Normal times, days)

The overall completion time, without resource considerations, is 20 days. The critical path is C-G-H.

We will construct a Gantt chart (see section **11.3.2**) and build up a resource profile, assuming that all jobs are started as soon as possible. The chart shows the float allocated at the end of each activity. This diagram helps us to identify which activities are concurrent and which activities it is possible to re-schedule, without causing an overall delay in the project.

FIGURE 11.22 **Gantt Chart for Example 11.11.**

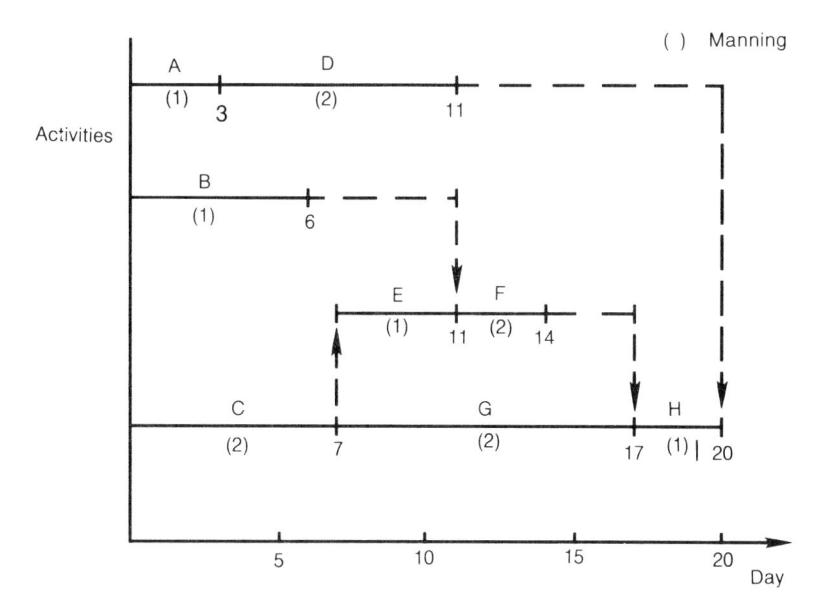

FIGURE 11.23 **Resource profile for earliest start times for Example 11.11**

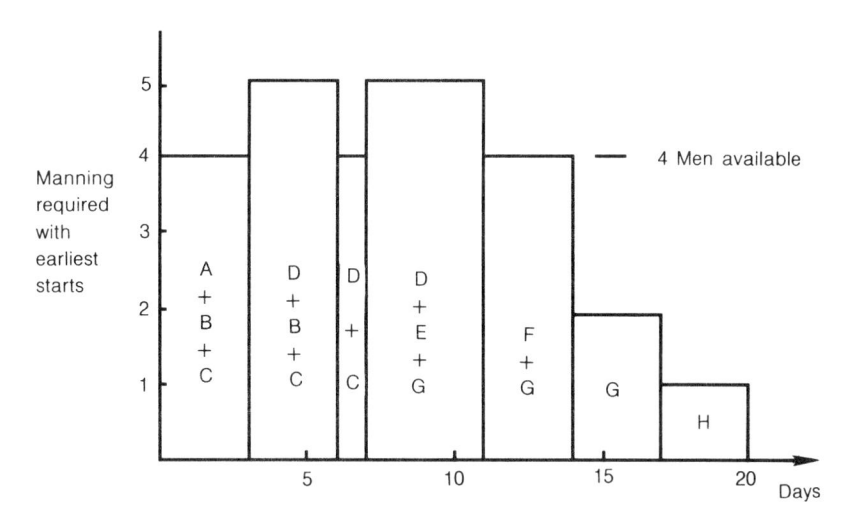

We can see from the resource profile that the limit of 4 men available is exceeded when activity D is in progress between days 3 and 11. It is not possible to re-schedule activities in order to accommodate D entirely. The critical activities C and G both require 2 men, so that D cannot be started during days 0 — 17, unless no other non-critical jobs are being carried out. If D is moved to its latest start time of 12 days, there will still be days 12 to 14, when we require more than 4 men. G (2 men), F (1 man) and D (2 men) would be in progress during these 2 days. We either hire one more man for this period, or we must start activity D when F has finished in day 14. This will delay the project by 2 days, extending the duration from 20 to 22 days.

SUMMARY

Network analysis is used in project planning and scheduling. A project is broken down into individual jobs or activities. The network shows the logical connections between activities. Activity times are used to calculate the total project duration, the possible start and finish times for each job and the critical path activities.

Networks can be illustrated using either arrows or nodes to represent activities. The Gantt Chart is an alternative way of presenting a network, using a time scale.

In most projects, jobs may be done more quickly than normal, but at extra cost. These are referred to as crash times and costs. They can be used to calculate the 'minimum time' or 'minimum cost' project schedule.

There may be uncertainty in the activity times. A PERT analysis can be used on the assumption that the activity times follow a beta distribution, with minimum value (a), most likely value (m) and maximum value (b). The expected activity time is then:

$$\frac{(a + 4m + b)}{6}$$

and its variance is:

$$\frac{(b - a)^2}{6^2}$$

The project duration is normally distributed with a mean which is the sum of the expected activity times on the critical path. The variance of this normal distribution is the sum of the variances of the critical activities. The distribution of the overall project time is used to calculate the probability of completing the project within a given period.

Resource considerations can also be incorporated into the analysis. Gantt Charts and resource profiles are used to determine how the activities should be scheduled to meet the given objective.

EXERCISES

Exercise 11.1

MR plc are undertaking a small building project. The essential activities, immediate predecessors and activity durations are given in the table below:

Activity	Immediate predecessor	Duration, weeks
A	—	4
B	—	6
C	A,B	7
D	B	3
E	C	4
F	D	5
G	E,F	3

Required:
1 Illustrate the project using an arrow diagram.
2 Determine the critical activities and the overall project duration.

Answers on page 580.

Exercise 11.2

Using the project data in **Exercise 11.1**:
1 Illustrate the project by an activity-on-node diagram.
2 On the basis of the diagram in **1**, determine the effect of a delay of 4 weeks in activity D.

Answers on page 581.

Exercise 11.3

The Chartered Institute of Quantitative Decision Makers (CIQDM) runs an annual programme of meetings for its members. The staff at the Institute's Head Office begin to prepare the programme for the year in the previous autumn. The programme includes details of the speakers and their talks as well as a list of current members. The activities necessary in the preparation of this programme are given below, together with the immediately preceding activities:

	Activity	Immediate preceding activities	Normal time, days	Crash time, days	Additional cost, £
A	Select dates for meetings.	—	5	5	—
B	Arrange speakers and agree titles for their talks.	A	20	10	100
C	Obtain advertising material for the programme.	—	15	10	150
D	Mail membership renewal notices.	—	15	5	200
E	Prepare list of paid-up members.	D	30	25	50
F	Send programme and membership list to printers.	B,C,E	10	5	100
G	Proof read programme and list.	F	10	5	50
H	Print and assemble programme.	G	15	10	75
I	Obtain computer printed address labels of members.	E	5	2	50
J	Send out programmes.	H,I	5	2	50

If the programme is prepared using the normal headquarters staff, then each activity is estimated to take the normal time given above. It is assumed that the staff work a 5 day week. However, it would be possible to employ additional temporary staff to help out at this busy time. The duration of each activity under these circumstances is given above under crash times, together with the associated additional cost of doing the activity in the crash time. For simplicity, assume that the activities take the normal or the crash times only.

Required:
1. Illustrate the project by a network diagram.
2. Determine the overall time required to produce and distribute this programme if no additional staff are taken on. Which activities are critical?
3. What would be the effect on the overall duration of the project if the time taken to obtain the advertising material had been underestimated and it took 30 days to assemble?
4. What is the shortest possible time in which the programme can be produced and distributed? What is the minimum additional cost of this time?

Answers on page 582.

Exercise 11.4

Jubilee Computer Systems Ltd are preparing a customer order. The relevant data are given below:

Activity	Immediately preceding activity	Time, days Optimistic	Most likely	Pessimistic	Cost, £, at expected duration
A	—	3	4	5	1000
B	—	4	7	10	1400
C	—	4	5	6	2000
D	A	5	6	7	1200
E	B	2	2.5	6	900
F	C	10	10.5	14	2500
G	D,E	3	4	5	800
H	G,F	1	2	9	300

The project's indirect costs are £300 per day. The contract with the customer specifies a penalty of £100 per day if the project is not finished by the end of day 15.

Required:
1. Draw the network. What is the overall expected project completion time? What is the associated cost?
2. What is the critical path? Comment on the durations of the non-critical paths.
3. What is the probability that the project could be completed without incurring any penalty charge?

Answers on page 582.

Exercise 11.5

Rogers plc is about to launch a small publishing subsidiary. The necessary activities, together with their dependancies and durations, are given in the table below.

Activity	Immediate predecessor	Duration, weeks
A	—	3
B	A	4
C	A	2
D	A	6
E	B	3
F	D	2
G	D	4
H	G	7
I	C,E,F	5
J	G,I	3

Required:
1. What is the expected overall completion time for the project?
2. Assuming that each activity requires one person to complete it in the specified time, determine the revised expected project duration if only two people are available to undertake the work and only one person can work on any one activity.

Answers on page 584.

Exercise 11.6

The Chief Accountant of Mercury Manufacturing Ltd has identified the major activities involved in the annual preparation of year end accounts. These activities along with their durations and the immediately preceding activities, are as follows:

	Activity	Duration (weeks)	Immediately preceding activities
A	Check outstanding purchase invoices.	2	—
B	Close cash account and post to general ledger.	2	—
C	Complete sales invoices.	3	—
D	Check accrued salaries and wages.	1	—
E	Count and verify stock levels.	4	—
F	Calculate stock valuation.	2	E
G	Close purchase ledger and post to general ledger.	1	A
H	Close sales ledger and post to general ledger.	2	C
I	Post salaries and wages to general ledger.	1	D
J	Close general ledger.	1	B,G,H,I
K	Prepare trial balance.	3	J
L	Apply closing adjustments to trial balance.	2	F,K
M	Re-open general ledger for new period.	1	N
N	Prepare final accounts.	6	L
O	Submit final accounts to Board for approval.	1	N

Required:
(a) Draw a network to represent the inter-relationships between the activities indicated, and insert earliest and latest event times throughout.
(b) If the preparation of the accounts starts at the beginning of the first week in January, how many weeks will elapse before the final accounts can be presented to the Board?
(c) The end of week 13 is the end of the financial year and the Board has indicated that the final accounts must have been approved by that time. The chief accountant has suggested that, to achieve this, activities A, C and D can all be completed by the end of December and also that the verification of stock levels, E, could be done in only two weeks if additional labour was available. Comment on these suggestions.
(d) Discuss briefly the relative advantages and disadvantages of 'activity-on-node' and 'activity-on-arrow' networks.

(ACCA, June 1983)

Answers on page 585.

Exercise 11.7

On 1 September each year, Salemis Ltd begins the task of preparing budgets for the coming year. The various stages in the budgeting process have been identified as follows:

	Stage	Preceding stage(s)	Time, (weeks)
A	Estimate wage rates.	—	2
B	Develop a market forecast.	—	4
C	Fix selling prices.	—	3
D	Prepare sales quantities budget.	B	3
E	Prepare sales revenue budget.	C,D	1
F	Prepare selling expenses budget.	A,D	3
G	Prepare production quantities budget.	D	6
H	Prepare overheads budget.	A	4
I	Prepare manpower budget.	A,G	2
J	Prepare materials budget.	G	3
K	Prepare plant and equipment budget.	G	5
L	Produce overall profit forecast.	E,F,H,I,J,K	1

The whole process must be completed by the end of December which gives a period of 17 working weeks.

Required:
(a) Draw a network to represent the sequence of stages involved in the preparation of budgets. Can the whole process be completed within the required period of 17 weeks?
(b) If it is necessary to reduce the time taken to complete the budgeting process, which stages should be investigated and why?
(c) Explain the difference between total float, free float and independent float. Show that stage I has a free float of three weeks, of which two weeks is independent float.

(ACCA, June 1986)

Answers on page 586.

QUANTITATIVE ANALYSIS

Exercise 11.8

A particular project comprises 10 activities which have the following durations and precedences:

Activity	Duration, days	Immediately preceding activities
A	6	—
B	1	A
C	2	A
D	1	B
E	1	D
F		B
G	1	C
H		F,G
I	4	E,H
J	5	I

Activities F and H have uncertain durations which at this stage are difficult to estimate.

Required:
- (a) Draw a suitable network to represent the inter-relationships between the 10 activities.
- (b) What is the minimum time that the project could take, ignoring the effects of activities F and H?
- (c) If the project must be completed in 19 days, what restrictions does this place on the durations of activities F and H?
- (d) After further investigation it is estimated that the expected times for activities F and H are two days and one day respectively. Furthermore, it may be assumed that the uncertainty in these two activity durations may be represented by a Poisson distribution. On the basis of this, what is the probability that the project will be completed in no more than 19 days? A selection of Poisson probabilities is given in the following table:

Mean (μ)	\multicolumn{5}{c}{Probability of}				
	0	1	2	3	4 or more
1	0.368	0.368	0.184	0.061	0.019
2	0.135	0.271	0.271	0.180	0.143

(ACCA, June 1987)

Answers on page 587.

Exercise 11.9

The Hydra Company manufactures a range of hair care and shaving products, including disposable razors. A competitor has recently introduced a new style of disposable razor which, in the last six months, has taken a significant share of the market, with adverse effects on Hydra's sales. The management at Hydra has decided that a competitive product must be introduced as quickly as possible and has asked Jim Sharp, the management accountant, to draw up a plan for developing and marketing this new product.

As the first step in planning the project, Jim Sharp has identified the following major tasks which will be involved in the new product launch. He has also estimated how long each task will take and what other tasks must precede each one.

	Task	Time, weeks	Preceding tasks
A	Design new product.	8	—
B	Design packaging.	4	—
C	Organise production facilities.	4	A
D	Obtain production materials.	2	A
E	Manufacture trial batch.	3	C,D
F	Obtain packaging.	2	B
G	Decide on test market area.	1	—
H	Package trial batch.	2	E,F
I	Distribute product in test area.	3	H,G
J	Conduct test market.	4	I
K	Assess test market results.	3	J
L	Plan national launch.	4	K

Required:
(a) Draw a network to represent the logical sequence of tasks and determine how long it will be before the new product can be launched. (You may assume that the national launch can take place immediately after it has been planned.)
(b) Calculate the float which is available for each of the non-critical activities.
(c) The time taken to complete tasks, A, B, D, K, and L is somewhat uncertain and so the following optimistic and pessimistic estimates have also been made to supplement the most likely figure given above. These additional estimates are:

Task	Optimistic time, weeks	Pessimistic time, weeks
A	5	13
B	2	6
D	1	4
K	2	6
L	2	8

What now is the expected time until the product can be launched and what is the probability of this time exceeding 35 weeks? (You should assume that the overall project duration follows a normal distribution.)

(ACCA, December 1985)

Answers on page 588.

Inventory planning and control

12.1 INTRODUCTION

One of the consequences of changes in the economic climate is that manufacturing companies have had to review their policies towards the holding and control of stocks, both raw materials and finished goods.

When a company holds stocks of goods, capital is tied up in these goods. This unusable capital represents a cost to the company in the form of lost interest payments, or investment opportunities. In addition, the stock will incur costs since accommodation must be provided to house it; personnel must be employed to manage it; it must be insured, and so on. It seems reasonable to suppose, therefore, that a company will wish to hold as little stock as possible. However, there are other considerations. Demand for the item will probably be uncertain. The smaller the stock level, the more likely it is that shortages will arise. These shortages will themselves represent a cost to the company, either in lost production or lost custom.

If a company manufactures a stock item on a batch production basis, there are likely to be economic advantages of producing large batches but then this also generates large initial stocks. There are clearly many issues to consider in this field. The effective analysis and appraisal of stock control policy is an important activity in most companies today. The implications are likely to be far-ranging within the company. If modern procedures, such as 'just-in-time' are adopted, then it is particularly important that the company understands how its stock system works. Techniques for modelling stock systems have been available for some years. These models range from basic ones for simple deterministic systems, to more complex models which can accommodate uncertainty in demand or delivery times. If the system is more complex, then simulation techniques can be employed (see Chapter Fourteen). In all cases, the modelling process is designed to help the decision-maker determine how much to order and when to order, in the light of some decision criterion.

The objective of the decision is almost always to minimise the total costs associated with the holding of stocks. As we have seen previously, it is also important to examine the consequences of not using the optimum scheme, that is, a sensitivity analysis is required.

In this chapter, we will look at the basic stock model and see how it can be adapted to deal with different stock problems, including uncertainty. There is criticism of these basic models on the grounds that they lack realism. While this criticism is justified, the models serve as a useful introduction to the subject area and the issues which must be considered.

12.2 THE BASIC STOCK MODEL

Firstly, we will consider stock problems which involve either the ordering of goods in batches from an outside supplier or the batch production of goods. The ordering or producing policy is to be designed so that the total costs involved are minimised.

In any stock system, the stock level will vary over time in a cyclical pattern. The pattern of the demand will dictate how the stock level drops. At some point, an order will be placed to replenish the stocks. After an elapse of time, known as the **lead time**, the order will arrive and the stock level will instantly increase. A new stock cycle begins. See FIGURE 12.1.

The objective of the analysis and the model is the minimisation of total cost.

12.2.1 Assumptions for the basic model

In order to simplify the modelling process, a number of assumptions are made:

1 The demand for the items is constant, or approximately constant. If the rate of usage is constant, the stock level will fall at a constant rate.

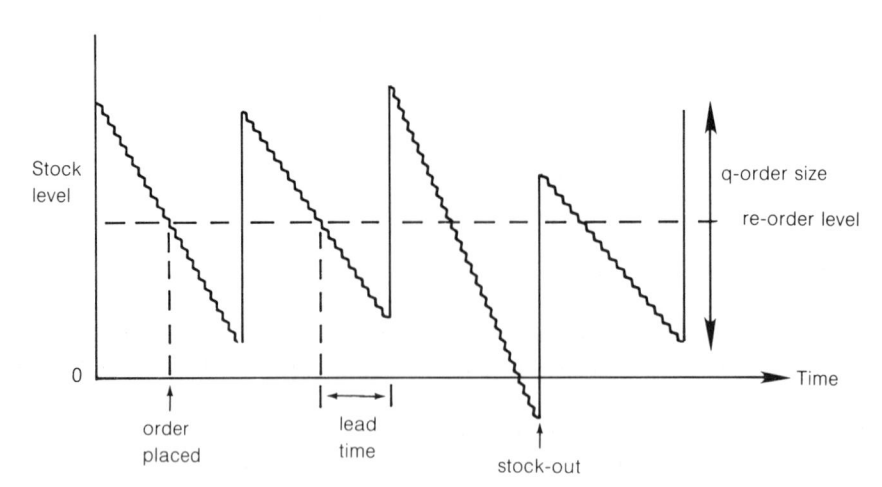

FIGURE 12.1 **Typical stock holding pattern**

2   The lead time is assumed to be known and constant. This means that the orders can be placed at a time or stock quantity (the **re-order level**) which will ensure that delivery takes place as the stock level reaches zero.
3   Stock-outs are not permitted.
4   The same amount (q) is ordered each stock cycle.

The resulting stock pattern is:

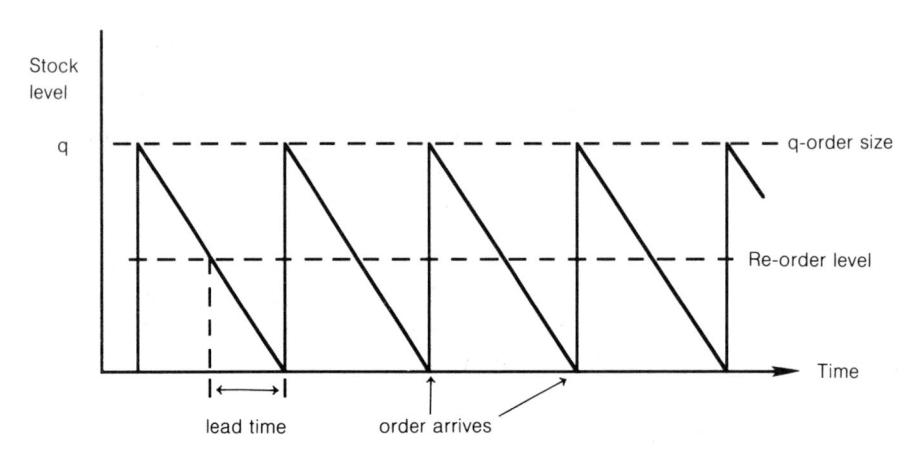

FIGURE 12.2 **Stock pattern for the basic model**

All stock cycles are identical. The maximum amount ever held in stock is equal to the order quantity, q.

12.2.2 The costs of stocking goods

If the goods are ordered in batches from an outside supplier, then costs will be incurred in the ordering and delivery process. Provision must be made to store and administer the stock. Further costs will arise here. For particular cases, there will be other costs involved such as the cost of shortages or the costs of holding a buffer stock. We will introduce these costs as they arise in the discussion. For the time being, we will consider only the costs of ordering and the costs of holding, or storing, the stock, together with the cost of purchase.

If the goods are produced by a batch process, rather than purchased from outside, similar costs arise. The cost of ordering is replaced by the cost of setting up the batch process and the purchase cost is replaced by the cost of manufacture. The analysis is the same. Each type of cost will comprise a fixed and a variable element. For the purposes of the basic stock model, we are interested in the variable costs only. At this stage, a further assumption is made: the variable costs of placing an order, or setting up a batch run, and the unit variable costs of holding stock are known, fixed and independent of the size of the order.

12.2.3 The total cost equation

We must now build a model which describes the total stock costs incurred over a period of time. The length of the period used does not matter: it can be a day, a month, a year, etc. We will choose to work on an annual basis for now. We will define the following symbols:

D = annual demand for the stock item
C_0 = variable cost of placing one order, £/order
C_h = variable cost of holding one item in stock for one year, £/item/year
C = the purchase price of one item of stock, £/item
q = order quantity, items/order
Total stock cost per year = total cost of ordering/year + total cost of holding the stock/year

Let us take each part of this equation in turn.

ANNUAL COST OF ORDERING

If D items are required per year and they are ordered in batches of q, then (D/q) orders will be needed each year.

Annual cost of ordering = cost of placing one order × number of orders placed in a year

$$= C_0 \times (D/q) \quad (£/year)$$

ANNUAL COST OF HOLDING THE STOCK

This cost is usually based on the average amount held in stock over a single stock cycle. In the simple situation with which we are dealing, the stock level varies linearly from q to zero, therefore, the average stock level is (q/2). In more complex situations, it is necessary to use more sophisticated mathematical techniques to calculate the average stock level.

The unit holding cost, C_h, is specified as either a fixed amount per year or as a percentage of the value of the item per year. Companies will have differing ways of allocating costs in this area, but C_h will generally reflect the cost of borrowing the money which is tied up in stock, the cost of stock deterioration or maintenance, plus a contribution towards the cost of the stockholding system.

Annual cost of holding the stock
= cost of holding one unit for a year × average amount held in stock
$$= C_h \times (q/2) \quad (£/year)$$

It follows that the total cost per year of stocking an item is given by:

$$TC = C_0(D/q) + C_h(q/2) \quad (£/year)$$

This equation is the total cost equation for the basic stock model. We must now determine the value of q which will give the total cost its smallest value.

12.2.4 The optimum order quantity, q_o

Differentiation is used to determine the optimum value for q as follows:

$$TC = C_o(D/q) + C_h(q/2)$$

TC takes its minimum value when:

$$\frac{dTC}{dq} = 0 \quad \text{and} \quad \frac{d^2TC}{dq^2} > 0$$

$$\frac{dTC}{dq} = -C_o\frac{D}{q^2} + C_h\frac{1}{2}$$

and:

$$\frac{d^2TC}{dq^2} = 2C_o\frac{D}{q^3} + 0 > 0 \qquad \text{if } q > 0$$

Set:

$$\frac{dTC}{dq} = 0, \text{ then } -C_o\frac{D}{q^2} + C_h\frac{1}{2} = 0$$

therefore:

$$C_o\frac{D}{q^2} = C_h\frac{1}{2}$$

$$q^2 = \frac{2C_oD}{C_h}$$

$$q_o = \pm\sqrt{\frac{2C_oD}{C_h}}$$

When $q_o = +\sqrt{2C_oD/C_h}$, TC takes its minimum value. This optimum order quantity is known as the **economic order quantity (EOQ)**. If this quantity is ordered at regular intervals throughout the year, then the total cost of holding the stock will take its minimum value. It is normal practice simply to quote the EOQ model formula rather than to derive it each time from the total cost equation.

It is useful to look at a graphical representation of the total cost equation and the component costs. The holding cost is proportional to the order quantity and is, therefore, a straight line through the origin. The ordering costs are proportional to $1/q$. These costs and their sum are plotted below.

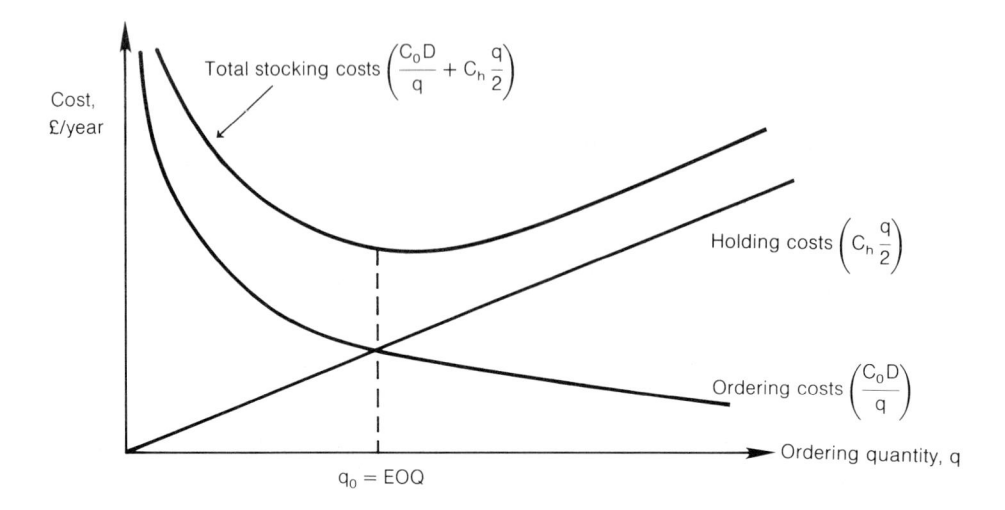

FIGURE 12.3 **Graph of ordering, holding and total stocking assets**

We can see that, for small order quantities, the cost of ordering is dominant—frequent orders but small quantities to store. For large order sizes, the cost of holding is the major cost—few orders but large quantities to store. The turning point in the total cost equation occurs when the two costs are equal. This is a useful check of EOQ calculations. We can also see that the total cost curve is fairly flat at the turning point. This means that the total cost is not very sensitive to changes in the order quantity in this region. Once the EOQ has been calculated, there is usually a reasonable amount of flexibility about the quantity actually chosen. A convenient order size can be selected without increasing the total costs significantly.

12.2.5 The re-order level and re-order interval

We now know how much to order but we still do not know when to order.

If the lead time from the supplier is L weeks, then during the lead time $L \times (D/52)$ units of stock will be used, assuming a 52 week year. Since the demand is constant, the usage during the lead time is also the re-order level. When the stock level falls to $L \times (D/52)$, the order should be placed. The new order will arrive just as the stock level reaches zero.

We require (D/q) orders per year at regular intervals, therefore, a new stock cycle begins every:

$$\frac{1 \text{ year}}{(D/q) \text{ orders}} = q/D \text{ years}$$

Since the stock cycles are identical, the re-order interval is also (q/D) years.

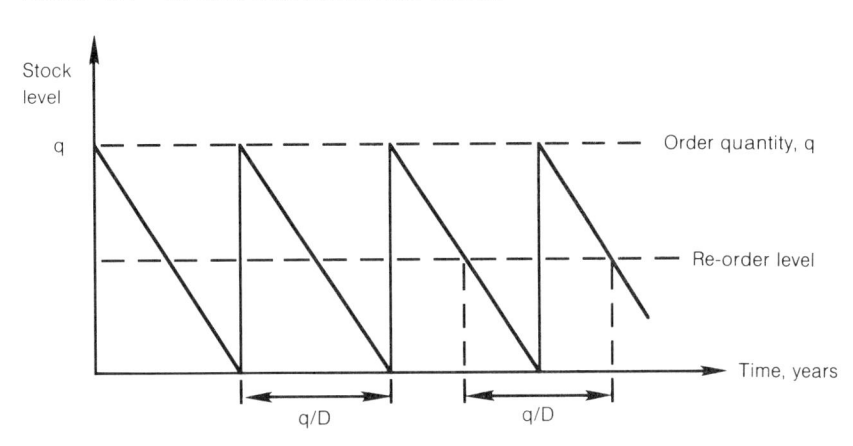

FIGURE 12.4 **Re-order level and re-order interval**

Example 12.1: To calculate an economic order quantity, re-order level and re-order interval

A shop sells 500 packets of soap powder per year. The demand is spread evenly over the year. The packets are purchased for £2 each. It costs the proprietor £10 to place an order. The supplier's lead time is 12 working days (assume a 6-day week). The holding costs are estimated to be 20% per annum of the average stock value. How many packets should the proprietor order at a time if he wishes to minimise his total stocking costs? Assuming that the shop is open for 300 days a year, determine how often the orders should be placed and the re-order level.
Solution: The economic order quantity is:

$$q_0 = \sqrt{2C_0 D/C_h}$$

D = 500 packets per year
C_0 = £10 per order
C_h = 20% per annum of the stock value per item
　　 = 0.2 × £2 per packet per year

Therefore:

$$q_0 = \sqrt{\frac{2 \times 10 \times 500}{0.2 \times 2}} = 158.11$$

We must order a whole number of packets, therefore, we will take the EOQ as 158 packets. Later, we may wish to round further to produce a more convenient order quantity. The minimum value of the total annual stocking cost is given by:

$$TC = C_o D/q_0 + C_h q_0/2 \qquad (\text{£/year})$$

Therefore:

$$TC = 10 \times 500/158 + 0.2 \times 2 \times 158/2 = 31.6 + 31.6 = \text{£}63.2/\text{year}$$

The total cost to the shopkeeper of buying 500 packets of soap powder per year is:

$$\text{Stocking cost} + \text{Purchase cost} = \text{£}63.2 + \text{£}2 \times 500 = \text{£}1063.2/\text{year}$$

The stocking costs are 6% of the annual costs.

If the shopkeeper ordered in batches of 150 packets, the total stocking cost per year would be:

$$TC_{150} = 10 \times 500/150 + 0.2 \times 2 \times 150/2 = 33.33 + 30.0 = \text{£}63.33/\text{year}$$

This is a very small increase of 13 pence per year compared to the cost at the calculated EOQ. The shopkeeper must place his order every $q/D = 158/500$ years. Since there are 300 working days, the re-order interval is:

$$\frac{158 \times 300}{500} = 94.8 \simeq 95 \text{ working days}$$

The sales of the soap powder during the 12 day lead time will be:

$$(\text{demand/day}) \times \text{lead time} = (500/300) \times 12 = 20 \text{ packets}$$

The re-order level is therefore 20 packets. A new order is placed when the stock level reaches 20 packets.

12.2.6 The economic batch quantity model

Companies which produce a number of different lines may organise their production on a batch basis rather than a continuous one. For example, a bakery may decide to make a batch of large wholemeal loaves and then a batch of small bread rolls, followed by a batch of scones. When batch production is

FIGURE 12.5 **Economic batch quantity model**

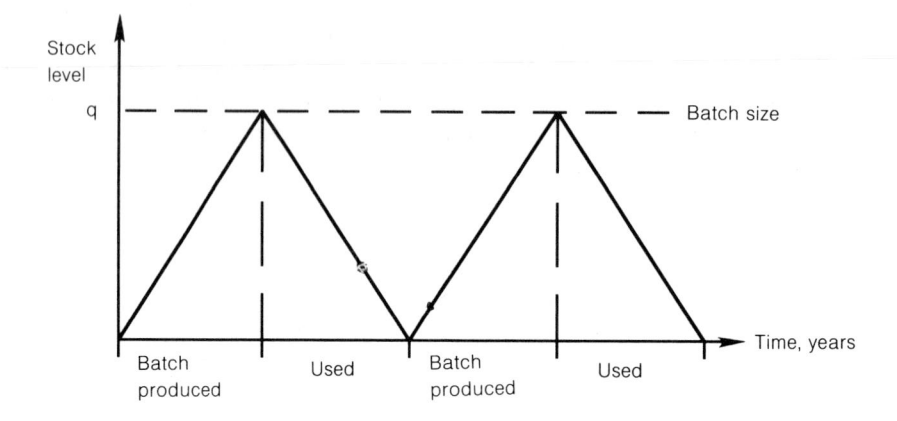

used, the company has to decide how large a batch to make at a time and how often to make a batch of a particular product. The problem is similar to the economic order quantity problem. The ordering of a fixed quantity from an outside supplier is replaced by the production of a fixed amount. The cost per order in the previous model is, therefore, replaced by the cost of set up.

Total annual cost of production = annual cost of set up + annual holding cost

If £C_s is the cost of setting up each production run, then:

$$TC = C_s \times (D/q) + C_h \times (q/2) \quad (£/year)$$

where q is the batch size. By comparison with the previous problem, we can see that TC will take its minimum value when:

$$q_0 = \sqrt{2C_sD/C_h}$$

This optimum batch quantity is known as the **economic batch quantity (EBQ).**

Example 12.2: To calculate an economic batch quantity and production interval

A pottery company produces several different designs of coffee mug. These mugs are made on a batch production basis at a rate of 500 per week. The demand for the most popular design, X, is 2500 per year, spread evenly over the year. Whenever a batch of X is to be made, set up costs of £200 are incurred. The company estimates that the annual cost of holding these mugs is £1.50 per mug.

How many mugs should the company produce in a batch if they wish to minimise the total annual cost of production and holding? How often should there be a production run and how long will it last? Assume a 50 week year.
Solution:

$$D = 2500 \text{ mugs/year}$$
$$C_s = £200/\text{run}$$
$$C_h = £1.50/\text{mug/year}$$

The economic batch quantity is:

$$q_0 = \sqrt{2C_sD/C_h} = \sqrt{2 \times 200 \times 2500/1.50} = 816.5$$

Since the total cost curve is not very sensitive to small changes in the value of q, we can probably take the EBQ as 820 mugs without increasing the cost significantly. This can be checked easily. When q = 816.5 mugs per batch:

$$TC = 200 \times 2500/816.5 + 1.5 \times 816.5/2 = 612.37 + 612.37 = £1224.74/\text{year}$$

When q = 820 mugs per batch:

$$TC = 200 \times 2500/820 + 1.5 \times 820/2 = 609.76 + 615 = £1224.76/\text{year}$$

When q = 800 mugs per batch:

$$TC = 200 \times 2500/800 + 1.5 \times 800/2 = 625 + 600 = £1225/\text{year}$$

A convenient batch quantity of 800 mugs increases the total cost of producing and holding the mugs by 26 pence compared with the optimum quantity.

Let us take the EBQ as 800 mugs. The number of production runs per year is 2500/800 = 3.125 (25 runs every 8 years!), therefore, the interval between production runs is 800 × 50/2500 = 16 weeks. If the production rate is 500 mugs per week, then it takes 800/500 = 1.6 weeks to make a batch.

12.3 QUANTITY DISCOUNTS

When ordering from an outside supplier, the price charged for the goods may depend on the quantity purchased. These discounts are usually offered for large orders. We must examine how the acceptance of the discount will affect the total costs. The larger batches will increase the costs of stocking the item (ordering plus holding costs) but this increase will be off-set to some extent by a reduction in the purchase price.

If the cost of purchase is also included, the total cost equation becomes:

$$\text{Total cost of purchase and stocking} = \frac{C_o D}{q} + \frac{C_h q}{2} + CD \qquad (\text{£/year})$$

where C is the unit purchase cost. If the purchase cost is a constant, independent of q, its addition to the cost equation has the effect of moving the graph of the cost equation parallel to the q axis, without changing its shape (see FIGURE 12.6).

FIGURE 12.6 **Annual purchasing and stocking costs**

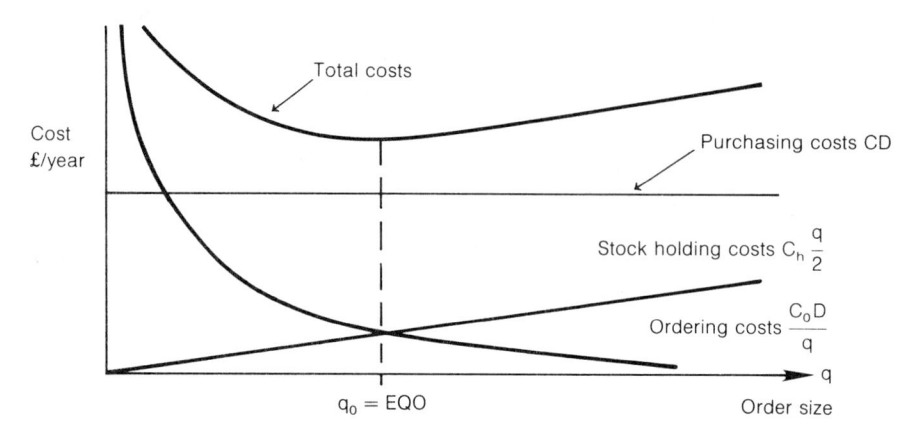

The purchase cost is likely to be much larger than the combined stocking costs.

If the item is normally sold at a unit price C, but for order sizes in excess of q_1, a discount is offered which reduces the unit price to C_1, then the variation in the total cost will be as shown in FIGURE 12.7.

FIGURE 12.7 **Effect on annual costs of one quantity discount**

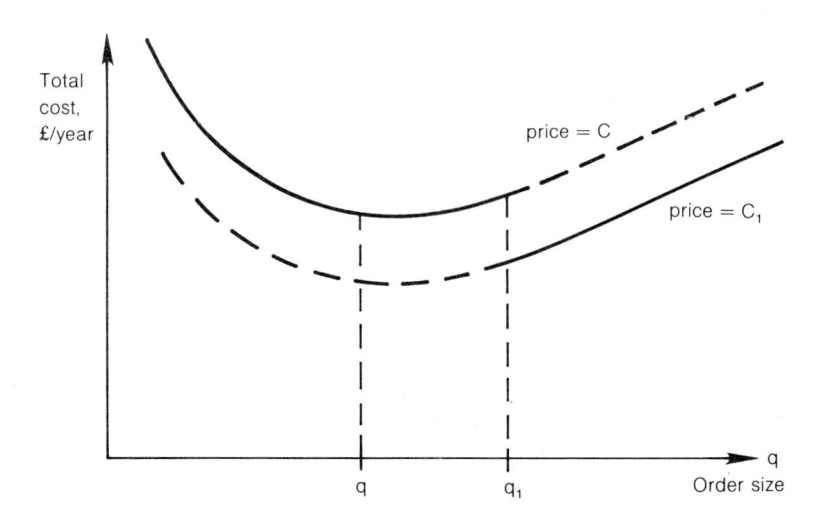

If a further discount to a unit price of C_2 is offered for order sizes in excess of q_2, then we might have:

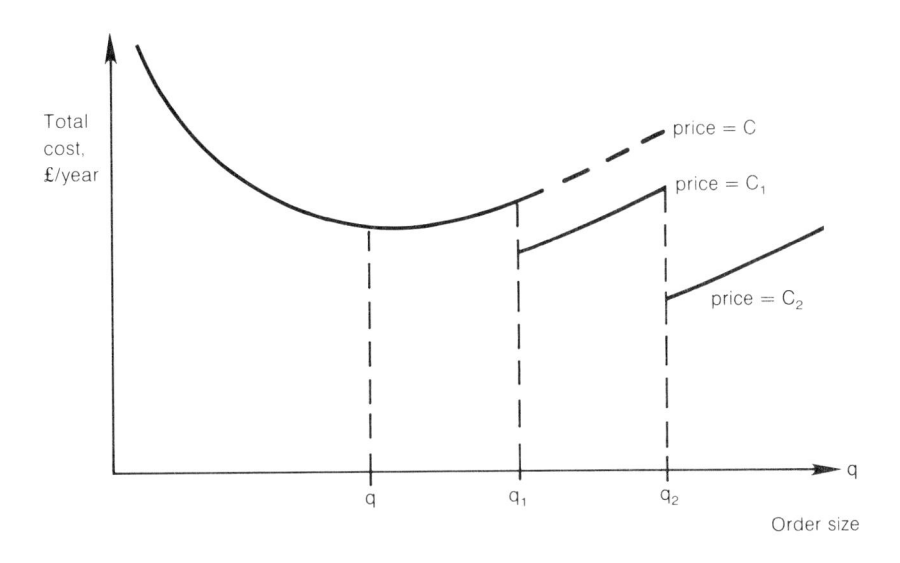

FIGURE 12.8 **Effect on annual costs of two quantity discounts**

Clearly, the discount is advantageous only for certain ranges of order size. The order quantity at which the discount is first offered is known as the **price break**.

In effect, FIGURE 12.8 shows three cost curves, one for each of the unit prices £C, £C_1 and £C_2, but only part of each curve can be used. If the value of q at the turning point of the curve is not included in the discount range, then the turning point no longer gives the optimum value of the order size. In order to find the optimum value for q, we first of all ignore the restrictions on its value and determine the order size which gives the minimum value of cost at each price level. If this value of q is within the discount range, then it is also the optimum value of order size. If the value of q at the turning point is below the discount range, then we re-calculate the total costs using the smallest possible value of q which carries the discount price.

Example 12.3: To illustrate the effects of purchase price discounts

Refer again to **Example 12.1** which concerns the purchase of soap powder by a shopkeeper. The soap powder was bought in batches of 158 packets for £2 each. The supplier now offers the following discounts:

TABLE 12.1 **Quantity discounts offered by supplier**

Order quantity	Discount	Cost per packet
0–199	0	£2.00
200–499	2%	£1.96
500 or more	4%	£1.92

Should the shopkeeper accept one of the discounts?

Solution: We know that the stock held will increase if the shopkeeper decides to take up the discount because he must order at least 200, rather than the present 158 packets. Will the increase in stockholding costs be balanced by the reduction in purchasing costs and ordering costs?

As the order size increases the variation in the total cost of the soap powder is given by combining the relevant parts of the three cost curves, one for each unit price of £2, £1.96 and £1.92 (see FIGURE 12.9 overleaf).

362 QUANTITATIVE ANALYSIS

FIGURE 12.9   **Effect of discounts on annual cost of stocking soap powder**

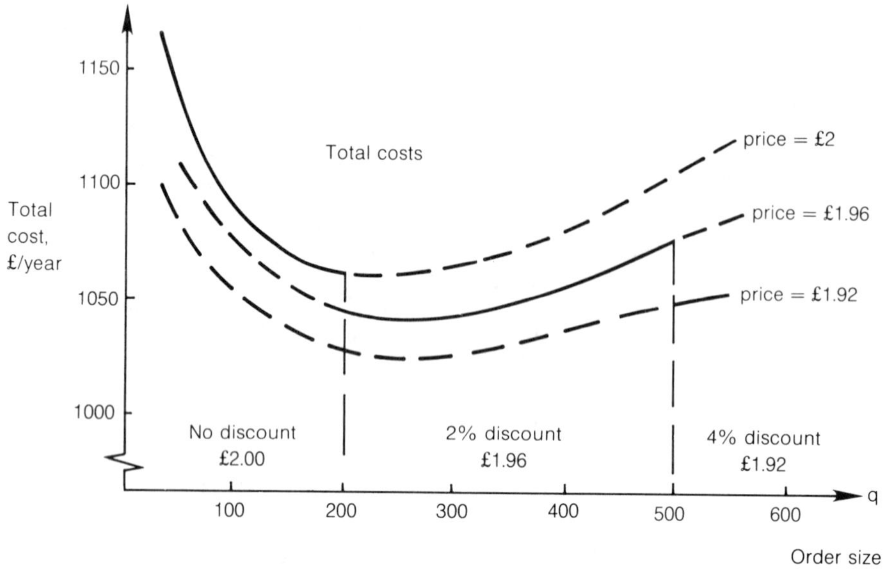

When the price is £2.00 per packet, we know from **Example 12.1** that the order quantity at the turning point in the total cost curve is 158 packets and that the minimum annual total cost of purchasing 500 packets is £1063.2.

Let us now consider a price of £1.96 per packet. The holding cost is:

$$C_h = 20\% \text{ of } £1.96 = £0.392/\text{packet/year}.$$

The optimum order quantity is:

$$EOQ = \sqrt{2C_oD/C_h} = \sqrt{2 \times 10 \times 500/0.392} = 159.72$$

This is below the range for the first discount, 200–499, therefore, the turning point in the £1.96 per packet curve is not a permissible order size. The minimum achievable cost will occur when q = 200 packets.

$$\text{The minimum total annual cost (at £1.96 per packet)} = \frac{10 \times 500}{200} + 0.392 \times \frac{200}{2} + 1.96 \times 500$$

$$= 25 + 39.20 + 980 = £1044.20/\text{year}$$

For a price of £1.92 per packet, the holding cost is:

$$C_h = 20\% \text{ of } £1.92 = £0.384/\text{packet/year}.$$

The optimum order quantity is:

$$EOQ = \sqrt{2C_oD/C_h} = \sqrt{2 \times 10 \times 500/0.384} = 161.37$$

This is below the range for the second discount, 500 and above, therefore, the turning point in the £1.92 per packet curve is not a permissible order size. The minimum achievable cost will occur when q = 500 packets.

$$\text{The minimum total annual cost (at £1.92 per packet)} = \frac{10 \times 500}{500} + 0.384 \times \frac{500}{2} + 1.92 \times 500$$

$$= 10 + 96 + 960 = £1066/\text{year}$$

Comparing the three solutions:

TABLE 12.2 **Comparison of minimum total costs for three price levels**

Cost/packet, £	Order quantity	Min total cost, £/yr
2.00	158	1063.20
1.96	200	1044.20
1.92	500	1066.00

The shopkeeper should take advantage of the first discount offered and place orders for 200 packets. This will reduce costs by £19.00 per year.

12.4 FURTHER STOCK MODELS

The basic model which we have developed for the purchasing of goods from an outside supplier can be adapted to suit other situations. We have already used the model in a batch production problem. In this case, it was assumed that the goods produced in the batch all went into stock. The maximum stock level was the same as the batch size. However, in many batch production processes, the goods will be used from stock while the production is in progress, so that the maximum stock level will be less than the batch size.

In the situations examined so far, it has been assumed that stock-outs are not permitted. There may be occasions in which it is cheaper to go out of stock than to carry the size of stock needed to avoid stock-outs. The next two sections examine the ways in which the basic model can be adapted to include these variations in conditions.

12.4.1 The batch production model

Suppose a machine produces a batch of components, some of which pass directly to a slower machine for immediate use. The rest are stocked until needed by the second machine. Instead of the stock arriving all together, and the stock held jumping from zero to q, the stock increases steadily during the time for which the first machine is producing and then decreases as the second machine uses up the stock. The rate of production is P and the rate of usage from stock is D, where $P \geqslant D$. The stock level varies over time as shown in FIGURE 12.10:

FIGURE 12.10 **Variation in stock levels for the batch production model**

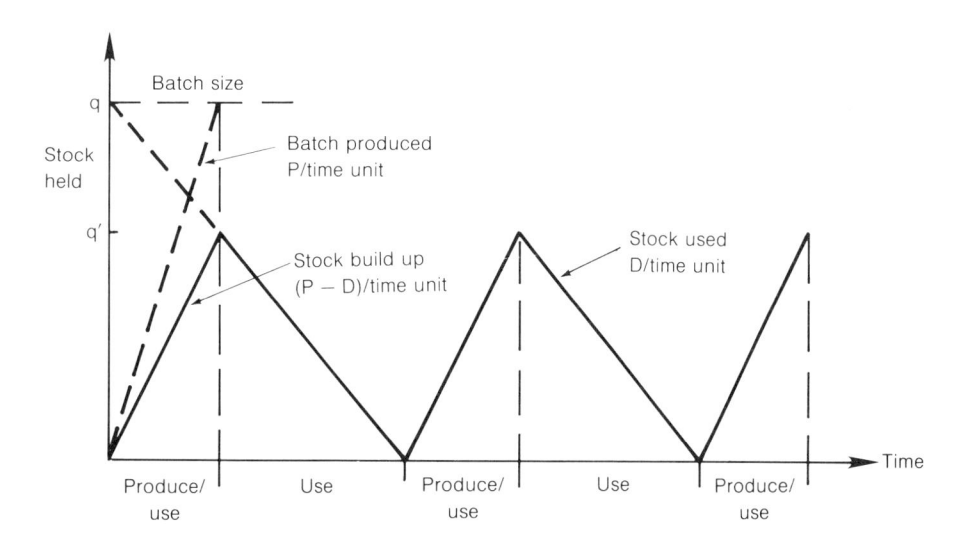

What is the optimum batch size, q, for the first machine? How often should the batch be made? The total annual variable costs of batch production, TC, comprise the set up costs and the holding costs. Therefore.

$$TC = C_s \times \text{number batch runs/year} + C_h \times \text{average stock held}$$

$$\text{Number batch runs/year} = \text{annual demand/batch size} = D/q$$

To calculate the average quantity in stock, we will consider one stock cycle.

FIGURE 12.11 Average stock level for the batch production model

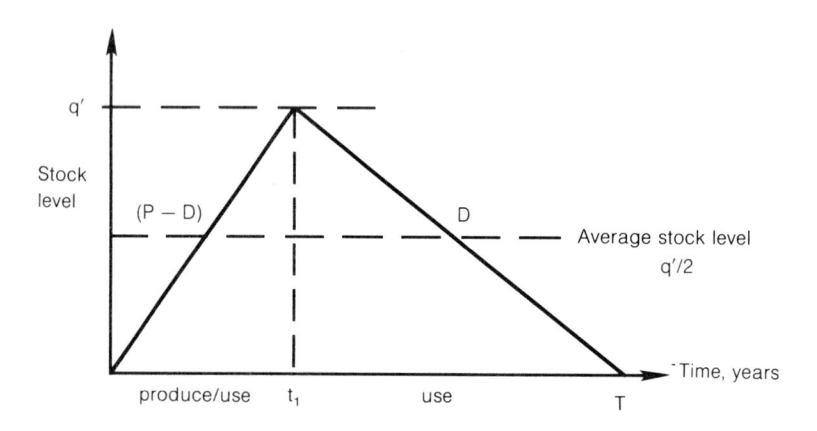

The batch size is q but, since items are used as they are produced, the maximum stock level, q', is less than q. If the items are produced at a rate of P per year and used at a rate of D per year $(P \geqslant D)$, then the stock builds up at a rate of $(P - D)$. The average stock level is still half the maximum level, as it was for the EOQ model.

If the production cycle lasts for t_1 years, then the total quantity produced during a cycle is:

$$q = Pt_1$$

therefore:

$$t_1 = q/P \text{ years}$$

The maximum stock level is $(P - D) \times t_1$ items. Substituting for t_1, the maximum stock level is $(P - D) \times (q/P)$ items. The average stock level is therefore:

$$\frac{(P - D)q}{2P} \text{ items}$$

We can now write down the total variable cost equation:

$$TC = C_s \frac{D}{q} + C_h \frac{(P - D)q}{2P} \qquad (\text{£/year})$$

TC takes its minimum value when:

$$\frac{dTC}{dq} = 0 \quad \text{and} \quad \frac{d^2TC}{dq^2} > 0$$

$$\frac{dTC}{dq} = \frac{-C_s D}{q^2} + \frac{C_h(P - D)}{2P} \quad \text{and} \quad \frac{d^2TC}{dq^2} = \frac{2C_s D}{q^3} > 0 \quad \text{if} \quad q > 0$$

When

$$\frac{dTC}{dq} = 0, \quad \text{then} \quad \frac{C_s D}{q^2} = \frac{C_h(P-D)}{2P}$$

therefore:

$$q^2 = \frac{2C_s DP}{C_h(P-D)}$$

The economic batch quantity which minimises the total variable cost of production is:

$$q = \sqrt{\frac{2C_s D}{C_h} \frac{P}{(P-D)}} = \sqrt{\frac{P}{(P-D)}} \times \text{EBQ} \text{ (see section } \mathbf{12.2.6})$$

Example 12.4: To illustrate the use of the economic batch quantity stock model, with simultaneous usage

A machine manufactures parts at the rate of 2000 per month. A second machine uses these parts at the rate of 500 per month; the remainder being put into stock. It costs £1000 to set up the first machine. The company estimate their stock holding costs as 20% per annum of the average stock value. Each part costs £2.50 to make.

1 What batch size should be produced on the first machine and at what frequency? What is the total annual variable cost of production?
2 If the set up costs could be reduced to £500, how would the change affect the answers in **1**?
3 If a further reduction in set up costs to £250 is possible, how would the answers in **1** be changed?

Solution:
1 C_s = £1000/batch run
D = 500 items/month = 6000 items/year
P = 2000 items/month = 24,000 items/year
C_h = 0.2 × £2.50 = £0.50 per item per year

The economic batch quantity is given by:

$$q = \sqrt{\frac{2C_s D}{C_h} \frac{P}{(P-D)}}$$

therefore:

$$q = \sqrt{\frac{2 \times 1000 \times 6000}{0.5} \times \frac{24000}{(24000-6000)}} = 5656.85$$

The optimum batch size is 5657 components. In practice, this value would be rounded further for convenience. The number of batches required per year is:

$$D/q = 6000/5657 = 1.06$$

Therefore, the frequency of the batch production is q/D years:

$$q/D = 5657/6000 = 0.94 \text{ years or } 11.24 \text{ months.}$$

The total variable cost of production is given by:

$$TC = C_s D/q + C_h(P-D) \times q/(2P)$$

$$= 1000 \times 6000/5657 + 0.5 \times 18000 \times 5657/(2 \times 24000)$$

$$= 1060.63 + 1060.69 = £2121.32/\text{year}$$

If one batch of 6000 items was made each year, then the total variable cost would be:

$$TC = 1000 \times 6000/6000 + 0.5 \times 18000 \times 6000/(2 \times 24000)$$

$$= 1000 + 1125 = £2125/\text{year}$$

Clearly, the company would choose to produce just one batch each year for the full 6000 items required.

2 If the set up cost is reduced to £500 per run, then the economic batch quantity is given by:

$$q = \sqrt{\frac{2C_s D}{C_h} \frac{P}{(P-D)}}$$

therefore:

$$q = \sqrt{\frac{2 \times 500 \times 6000}{0.5} \times \frac{24000}{(24000-6000)}} = 4000$$

The optimum batch size is 4000 components. The number of batches required per year is $D/q = 6000/4000 = 1.5$. Therefore, the frequency of the batch production is q/D years:

$$q/D = 4000/6000 = 2/3 \text{ years or 8 months}$$

The total variable cost of production is given by:

$$TC = C_s D/q + C_h(P-D)q/(2P)$$

$$= 500 \times 6000/4000 + 0.5 \times 18000 \times 4000/(2 \times 24000)$$

$$= 750 + 750 = £1500/\text{year}$$

If the set up cost could be halved, the saving in the total variable costs would be £625 per year. The part would be made in batches of 4000 every 8 months.

3 If the set up cost is reduced to £250 per run, then:

$$q = \sqrt{\frac{2C_s D}{C_h} \frac{P}{(P-D)}}$$

therefore:

$$q = \sqrt{\frac{2 \times 250 \times 6000}{0.5} \times \frac{24000}{(24000-6000)}} = 2828.43$$

The optimum batch size is 2828 components. The number of batches required per year is $D/q = 6000/2828 = 2.12$. Therefore, the frequency of the batch production is q/D years:

$$q/D = 2828/6000 = 0.47 \text{ years or 5.64 months}$$

The total variable cost of production is given by:

$$TC = C_s D/q + C_h(P-D)q/(2P)$$

$$= 250 \times 6000/2828 + 0.5 \times 18000 \times 2828/(2 \times 24000)$$

$$= 530.41 + 530.25 = £1060.66/\text{year}$$

If the set up cost could be halved again, a further saving in the total variable costs of £439 per year could be achieved.

12.4.2 The planned shortages model

We will now return to the example in which goods are purchased from an outside supplier. In some cases, the cost of holding the items will be greater than any costs incurred by being out of stock for a short while. It is possible to build a stock model which includes regular stock-out periods.

Two possible situations arise. The first occurs when the demand which arises during the stock-out period remains unsatisfied. For example, a supermarket may decide to reduce the stocks it holds of bulk items such as soap powder or breakfast cereal. The result of this is that, for a few days in each stock cycle, these items are out of stock. The resulting costs will arise due to lost sales and to some loss of customer goodwill. The supermarket will have to off-set these costs against the savings made in stocking the goods. A further example arises when an electrical goods retailer decides to reduce the stocks of a particular type of washing machine because of the cost of the capital tied up in the stock. In this case, however, if a customer demands this type of washing machine during the stock-out period, the shopkeeper is likely to offer to take an order and to supply the item as soon as the next delivery arrives. The shopkeeper will incur some cost in operating the ordering system but again this must be balanced against the savings in the stocking costs.

The basic difference between these two situations is that, in the first, orders are not filled when the new delivery arrives, therefore, the maximum stock level is the same as the order quantity. In the second case, the orders are filled from the next delivery, so the maximum stock level is the order quantity less the maximum demand during the stock-out period (see FIGURES 12.12 and 12.13).

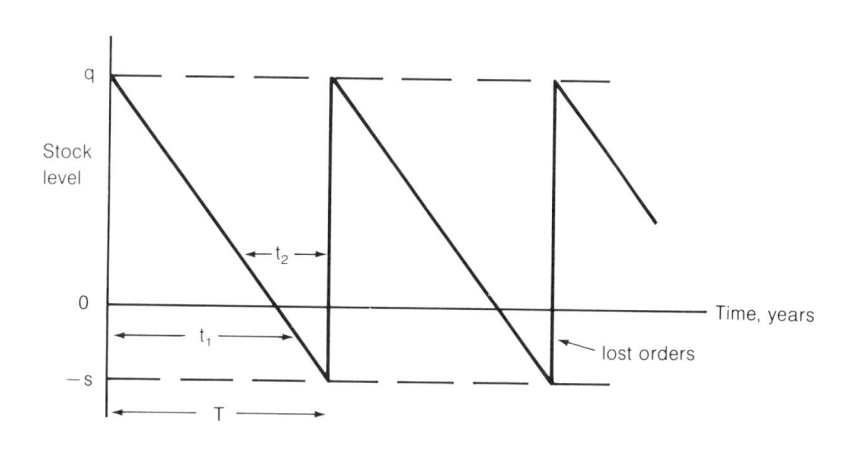

FIGURE 12.12 **Planned shortages—orders are not filled from the next delivery**

FIGURE 12.13 **Planned shortages—orders are filled from the next delivery**

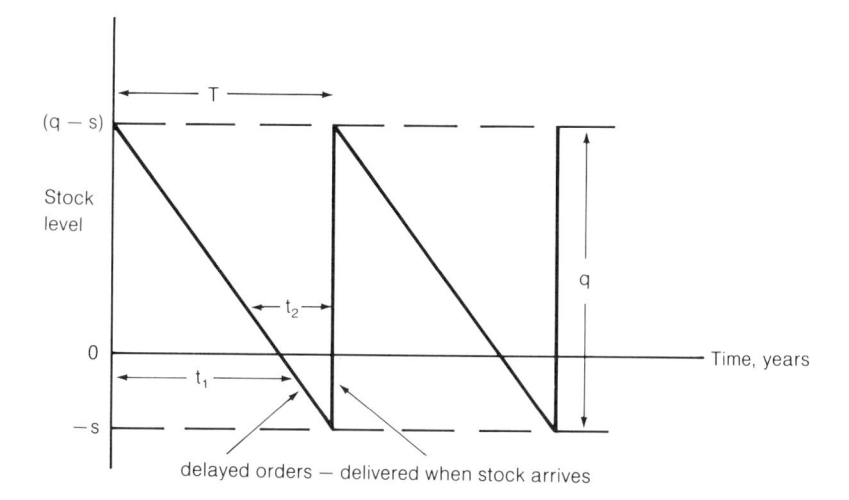

We will consider first the case in which orders are filled. The maximum stock level is the order size, q, less the maximum demand during the stock-out period, S. The maximum stock level is, therefore, $(q - S)$.

The total annual variable cost of stocking, TC, has three components:

1. annual cost of ordering = cost of one order × number of orders/year
2. annual cost of holding = unit cost of holding/year × average stock level
3. annual cost of stock-out = unit cost of stock-out × average shortage

As before:

C_o = cost of placing one order (£/order)

C_h = cost of holding one item for one year (£/year)

D = annual demand

C_b = cost of being out of stock of one item for one year (£/year)

q = order size

TC = C_o × D/q + C_h × average amount held + C_b × average amount short (£/year)

To calculate the average amount held, we consider one stock cycle of duration T years. We use stock for time t_1 years and are out of stock for time t_2 years: $t_1 + t_2 = T$.

For the period for which we have stock, t_1, the average stock level is $(q - S)/2$. We are, therefore, holding an average of $(q - S)/2$ items of stock for time t_1. This is a total of $(q - S)t_1/2$ items. We then hold zero items for the rest of the cycle, time t_2; a total of $0 \times t_2$ items. We require the average number of items held over the duration of the whole cycle T. Hence, the average number of items held over the stock cycle is:

$$\frac{((q - S)t_1/2 + 0t_2)}{T} = \frac{(q - S)t_1}{2T}$$

Now the rate of usage, D items/year, can be expressed as:

$$D = (q - S)/t_1 \quad \text{or} \quad D = q/T$$

Therefore:

$$t_1 = (q - S)/D \quad \text{and} \quad T = q/D$$

Substituting for t_1 and T, the average stock held over the stock cycle is:

$$\frac{(q - S) \times (q - S)/D}{2q/D} = \frac{(q - S)^2}{2q}$$

To calculate the average shortage, we can use the method described above. We are short, on average, of S/2 items for time t_2 and zero items for time t_1. Hence, the average number short over the whole stock cycle is:

$$\frac{(0 \times t_1 + S \times t_2/2)}{T} = \frac{St_2}{2T}$$

$D = S/t_2$, therefore, $t_2 = S/D$. The average number short is therefore:

$$\frac{S \times (S/D)}{2q/D} = \frac{S^2}{2q}$$

We can now complete the total cost equation:

$$TC = C_0 \frac{D}{q} + C_h \frac{(q-S)^2}{2q} + C_b \frac{S^2}{2q} \quad (\text{£/year})$$

This equation is different from the others because it has two independent variables, q and S. The minimum value of TC is found by using the mathematical process of partial differentiation. In this text, we will simply quote the results of this process.

The optimum order size is:

$$q = \sqrt{\frac{2C_0 D}{C_h} \frac{(C_h + C_b)}{C_b}} = EOQ \sqrt{\frac{(C_h + C_b)}{C_b}}$$

and the maximum number short is:

$$S = \sqrt{\frac{2C_0 D}{C_b} \frac{C_h}{(C_h + C_b)}}$$

If we now consider the case in which orders are not filled, then the analysis follows the same pattern except that the maximum amount in stock is q. We can simply replace $(q - S)$ by q, and q by $(q + S)$, in the above expression for average stock held and average shortage. The total variable cost equation becomes:

$$TC = \frac{C_0 D}{(q+S)} + \frac{C_h q^2}{2(q+S)} + \frac{C_b S^2}{2(q+S)} \quad (\text{£/year})$$

Again, using partial differentiation it can be shown that the optimum order quantity is given by:

$$q = \sqrt{\frac{2C_0 D}{C_h} \frac{C_b}{(C_h + C_b)}} = EOQ \sqrt{\frac{C_b}{(C_h + C_b)}}$$

and the maximum number short is:

$$S = \sqrt{\frac{2C_0 D}{C_b} \frac{C_h}{(C_h + C_b)}}$$

Example 12.5: To illustrate the use of the planned shortages stock model

Greens Ltd is a large, independent retailer of electronic and audio equipment. One of their more popular lines is a stereo radio cassette player. The demand is for 2000 a year, spread evenly over the year. This item costs Greens £50 to buy direct from the manufacturer. The cost of placing an order is estimated to be £50 and the cost of holding the radios in stock is charged at 15% per annum of the average stock value.

The manager of Greens is considering reducing the stocks held of this item in order to help improve the company's cashflow. He has estimated that the cost of administering an out-of-stock ordering system, together with a charge for any lost sales and loss of goodwill, amounts to £5 per radio per year.

1 Determine the minimum value of the total variable costs of stocking the radios if stock-outs are not permitted. What is the optimum order size?
2 How much could be saved if a system of planned shortages was introduced? Assume stock-outs are supplied from the next order. What is the new optimum order quantity?

Solution:

1   C_0 = £50/order
                          D = 2000 radios/year
                          C = £50/radio
                          C_h = 0.15 × £50 = £7.50/item/year

The economic order quantity is:

$$q = \sqrt{2C_oD/C_h} = \sqrt{2 \times 50 \times 2000/7.5} = 163.3$$

Greens should order 163 radio cassette players for each stock cycle.
The total annual variable cost of stocking is given by:

$$TC = C_oD/q + C_hq/2 \quad (\pounds/year)$$

$$TC = 50 \times 2000/163 + 7.5 \times 163/2 = 613.5 + 611.25 = \pounds1224.75/year$$

2 Planned shortages:

$$C_b = \pounds5/radio/year$$

The optimum order quantity is:

$$q = \sqrt{\frac{2C_oD}{C_h} \frac{(C_h + C_b)}{C_b}}$$

$$q = \sqrt{\frac{2 \times 50 \times 2000}{7.5} \frac{(7.5 + 5)}{5}} = 258.2$$

Greens should now order in batches of 258 radios. The maximum number out of stock is:

$$S = \sqrt{\frac{2C_oD}{C_b} \frac{C_h}{(C_h + C_b)}}$$

$$S = \sqrt{\frac{2 \times 50 \times 2000 \times 7.5}{5(7.5 + 5)}} = 154.9$$

The maximum shortfall in the stock is 155 radios. The total annual variable costs are:

$$TC = \frac{C_oD}{q} + \frac{C_h(q - S)^2}{2q} + \frac{C_bS^2}{2q}$$

$$TC = \frac{50 \times 2000}{258} + \frac{7.5(258 - 155)^2}{2 \times 258} + \frac{5 \times 155^2}{2 \times 258}$$

$$= 387.6 + 154.2 + 232.8 = \pounds774.6/year$$

This is a saving of £(1224.75 − 774.6) = £450.15 per year, compared to the basic model. If Greens use a planned shortages model, they will save £450 per year on the total variable stocking costs.

12.5 UNCERTAINTY AND THE BASIC STOCK MODEL

The models which we have considered so far have all been based on the assumption that the demand and the lead time are constant. However, in practice, many stock systems will have to contend with uncertainty in both lead time and demand. We may also find that demand changes over time, that is, the average demand per year is changing. Problems in which both the lead time and the demand are subject to uncertainty and the demand is varying over time, are extremely complex. Mathematical models of the type we have been using so far are unlikely to be suitable. Other techniques such as simulation (see Chapter Fourteen), must be employed. However, if we limit the increase in complexity to uncertainty in the lead time or the demand, we can build a suitable mathematical model. We must still make assumptions about the behaviour of the system. If the demand is uncertain, we assume a pattern for the demand. This pattern may be derived from empirical records of actual demand or it may be assumed to be a standard statistical pattern such as a Poisson or a normal distribution. When the demand or the lead time is variable, there is a chance that we will run out of stock. If a re-order level is set which simply meets the average demand in the average lead time, then we will run out of stock during a large proportion of the stock cycles in the year.

The importance of the proportion of stock-outs varies depending on the ordering frequency. For example, we estimate that there is a 0.2 probability of running out of stock on any order cycle. If the item concerned is ordered only once a year, there is a small chance of a stock-out each year. The expected number of stock-outs per year is given by:

$$E(\text{number stock-outs/year}) = \text{number of stock cycles/year} \times \text{probability of stock-out on each cycle}$$

$$= 1 \times 0.2 = 0.2$$

If, however, the stock item is ordered 50 times a year, then:

$$E(\text{number stock-outs/year}) = 50 \times 0.2 = 10$$

The same probability of a stock-out per stock cycle may be acceptable in one system but not in the other. We must decide whether a given probability of stock-out is acceptable. To do this, we must decide on the **level of service** which we wish to provide. If the probability of a stock-out per stock cycle is 0.2, that is 20%, then the level of service is 80%. If this is not adequate, then the probability of a stock-out must be reduced. This can be done by changing the **re-order level**. The re-order level is increased by adding a **buffer** or **safety stock** to the average demand during the average lead time.

$$\text{Re-order level} = \text{average demand during lead time} + \text{buffer stock}$$

The higher the buffer stock, the lower is the probability of a stock-out, but the higher is the cost of holding the stock. The decrease in the cost of stock-outs must be balanced against the increase in the cost of holding the stock.

The choice of a suitable size for the buffer stock will depend on the objective to be achieved. We may wish to achieve a minimum level of service, regardless of the extra cost involved. On the other hand, a stock-out may disrupt vital production; it may lead to extra manufacturing costs, to higher purchasing costs from another source, to increased ordering costs for an emergency order, or to lower customer satisfaction and, hence, to lower demand. It may be possible to establish a cost of a stock-out. The buffer stock may then be selected on the basis of minimising total variable stocking costs. Traditionally, there are two types of model which deal with uncertainty:

1 **The re-order level model**—a fixed quantity is ordered at variable time intervals, that is, whenever the stock level falls to a pre-determined value.
2 **The re-order cycle model**—a variable quantity is ordered at a fixed time interval.

You should note that the basic model, discussed in section **12.2,** combined both of these features—a fixed amount was ordered at fixed time intervals. It is the variability in the lead time and in demand which forces us to choose one or the other system. In the following sections, we will look at examples of each type of model.

12.5.1 A re-order level system

MODEL I: A MINIMUM LEVEL OF SERVICE IS REQUIRED

We have two decisions to make:

1 What should be the fixed order quantity, q?
2 At what stock level should the order be placed? This is the re-order level, R.

372 QUANTITATIVE ANALYSIS

The procedure is to fix the re-order quantity, using the EOQ model, and then to select the re-order level. This process will not necessarily give the 'best' solution but it will give a good solution. In order to fix the re-order level, we must know how the demand is varying during the lead time and the level of service which is expected. The general procedure is illustrated in the following example.

Example 12.6: To determine a re-order quantity and re-order level which maintains a minimum level of service

James plc use a certain component, X, in one of their manufacturing processes. X is purchased from an outside supplier. James' demand for X is variable but may be approximately described by a normal distribution with a mean of 80 components per day. The standard deviation of the demand is 10 components per day. Each component costs £0.50. When an order is placed with the supplier, it is estimated that a cost of £25 is incurred. The supplier's lead time is fixed at 8 days. James assess their stockholding costs as 20% per annum of the average stock value. James plc work a 5 day week for 50 weeks per year.

How many components should James order at a time and what should be the re-order level if James does not wish a stock-out to occur more than once in 20 stock cycles on average? How big is the safety stock with this re-order level?

Solution: To determine a suitable order size, assume that the demand is constant at the average demand.

$$C_o = £25/\text{order}$$

$$D = 80 \times 5 \times 50 = 20{,}000 \text{ components/year (average)}$$

$$C_h = 20\% \text{ of } £0.50 = £0.10 \text{ per component/year}$$

If the demand is assumed constant, then the economic order quantity is:

$$q_0 = \sqrt{2C_o D/C_h} = \sqrt{2 \times 25 \times 20000/0.1} = 3162.3$$

Take the order quantity as 3162 components. The maximum permitted level of stock-outs is set at 1 in 20 cycles, that is, on average only 5% of stock cycles should include stock-outs. The level of service is 95%. Since the daily demand is normally distributed, then the demand during the lead time will be normally distributed, as long as we can assume that the demand for each day is independent of the demand for the other days.

The average demand during the 8 day lead time is $80 \times 8 = 640$ components. The variance of the demand during the lead time is 8 × variance of demand per day = 8×10^2, therefore, the standard deviation of the demand during the lead time is $\sqrt{8 \times 10^2} = 28.28$ components. The distribution of the demand during the lead time is illustrated below:

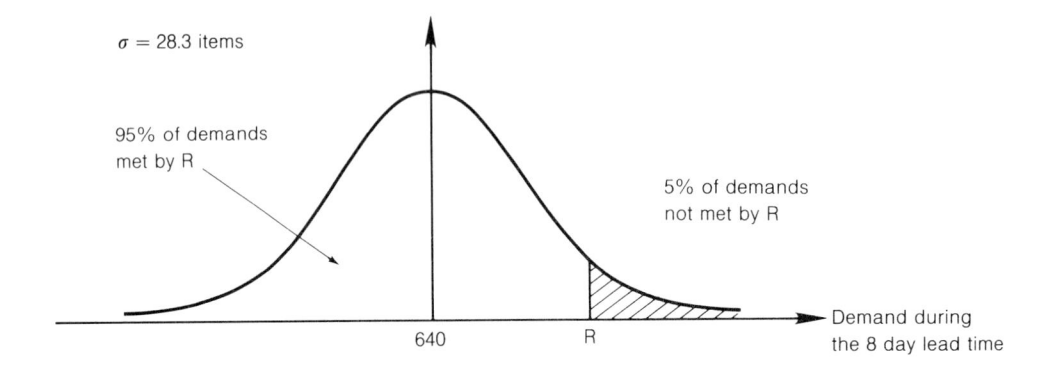

FIGURE 12.14   **Distribution of demand during the lead time**

The re-order level, R, is chosen so that the probability that the demand during the lead time is less than the re-order level, is at least 0.95. That is, P(demand during the lead time < R) > 0.95. R is z standard deviations above the mean, where:

$$z = \frac{R - 640}{28.3}$$

From the standard normal tables, if $P(z > (R - 640)/28.3) = 0.05$, then $z = 1.645$. Therefore:

$$1.645 = \frac{R - 640}{28.3}$$

and:
$$R = 686.55$$

Take the re-order level as 687 components. The safety stock is therefore 47 components. This stock is needed to allow for the variability of the demand and the level of service which is required. The 47 components are assumed to be held in stock all of the time, therefore, the average stock level per year is now (q/2 + 47) components.

If the costs of the stock-outs are ignored, then the total annual variable cost is given by:

$$TC = C_o D/q + C_h(q/2 + \text{safety stock})$$
$$= 25 \times 20000/3162 + 0.10 \times (3162/2 + 47) = 158.13 + 162.8 = £320.93/\text{yr}$$

The safety stock costs £(0.1 × 47) = £4.70 per year

MODEL II: COST MINIMISATION IS REQUIRED

The necessary decisions and procedure are the same as those for Model I. We will re-examine **Example 12.6** using the objective of minimising the total annual variable costs.

Example 12.7: To determine a re-order quantity and re-order level which minimises the total annual variable costs

Refer to **Example 12.6**. The production at James plc is disrupted if there is a stock-out, therefore, when this is imminent, the stores van is sent out to a local supplier to purchase additional stock. The firm estimates that this activity costs an extra £1 per component.

How many components should James order at a time and what should be the re-order level if James wish to minimise the total annual variable costs? How big is the safety stock with this re-order level?

Solution:

Total annual variable costs = annual ordering cost
+ annual holding costs for ordinary stock
+ annual holding cost for buffer stock
+ annual stock-out costs

The fixed order quantity is the same as that in **Example 12.6**, that is, 3162 components per order, therefore:

$$TC = C_o D/q + C_h q/2 + C_h \times (\text{buffer stock})$$
$$+ C_b \times (\text{expected number of items out of stock/year})$$
$$= 25 \times 20000/3162 + 0.1 \times 3162/2 + 0.1 \times (\text{buffer stock})$$
$$+ 1 \times (\text{expected number out of stock/year})$$

We must choose a value for the buffer stock which minimises the total value of the last two terms. As the buffer stock increases, the cost of holding it also increases but the expected number of items out of stock decreases and, therefore, the cost of stock-outs decreases and vice versa. We are

QUANTITATIVE ANALYSIS

looking for a buffer stock which produces the best balance of these two costs. The method we will use is based on 'trial and error'.

In this example, the demand during the lead time has been approximated by a continuous distribution, therefore, we must select specific points in the distribution at which we will examine the costs of stock-out and of holding the buffer stock. We choose to examine the costs at intervals of 10 components. This choice is made on the basis of convenience, and, since the components are relatively cheap, steps of 10 components is a reasonable cost interval.

If the demand during the lead time is average or less than average, then no stock-out will occur. The problems arise only if the demand during the lead time is greater than average.

TABLE 12.3 **Calculation of the probability of various lead time demands, using a normal distribution with $\mu = 640$, $\sigma = 28.3$ items per 8 days**

Approximate demand during lead time	Probability of this demand arising*	Buffer stock required to meet demand
640	0.135	0
650	0.134	10
660	0.109	20
670	0.082	30
680	0.052	40
690	0.030	50
700	0.016	60
710	0.007	70
720	0.003	80

* To estimate the probability, 640 is taken to represent the range from 635 up to 645. We find $P(635 \leqslant \text{demand} \leqslant 645) = 0.135$ using the standard normal tables. The other values are calculated similarly.

For each of the chosen values of buffer stock, we calculate the expected number of stock-outs per cycle. This number is then multiplied by the number of stock cycles per year to give the expected number of stock-outs per year. The cost of holding the extra stock (£0.10 per unit) and the expected cost of stock-out (£1 per unit) are added to give a total annual expected cost for that level of buffer stock. The costs will usually fall steadily to the minimum and then rise again. Once the costs begin to rise, then no further calculations are needed. The number of stock cycles per year is 20,000/3162 = 6.3. (*Note:* In the following calculations, it is assumed that the probability that the lead time demand will exceed 720 components, is zero.)

TABLE 12.4 **Costs for different buffer stocks**

Buffer stock	Demand met	Expected number of stock-outs /cycle	/year	Stock-out	Costs, £/year buffer stock	Total
80	720	0	0	0	80 × 0.10 = 8	8.00
70	710	10 × 0.003 = 0.03	0.03 × 6.3 = 0.19	0.19 × 1	70 × 0.10 = 7	7.19
60	700	20 × 0.003 + 10 × 0.007 = 0.13	0.13 × 6.3 = 0.82	0.82 × 1	60 × 0.10 = 6	6.82
50	690	30 × 0.003 + 20 × 0.007 + 10 × 0.016 = 0.39	0.39 × 6.3 = 2.46	2.46 × 1	50 × 0.10 = 5	7.46

The total expected cost per year is rising, therefore, we can assume that it takes its minimum value when the buffer stock is 60 components. The total annual variable cost is:

$$TC = 25 \times 20000/3162 + 0.1 \times 3162/2 + 0.1 \times \text{(buffer stock)}$$
$$+ 1 \times \text{(expected number out of stock/year)}$$
$$= 158.1 + 158.1 + 0.1 \times 60 + 1 \times 0.82$$
$$= £323.02 \text{ per year}$$

This value is arrived at after substantial approximation but it is probably the best that can be done relatively simply. In this case, the variable costs of stocking are small compared to the purchase cost (£0.50 × 20,000 = £10,000 per year).

Example 12.8: To determine a re-order quantity and re-order level which minimises the total annual variable costs

P&R is a large cash and carry warehouse which sells electrical goods. The most popular model of television set is purchased directly from the manufacturer at a cost of £250 per set. Average sales are 475 sets during a 300 day year. Whenever an order is placed, P&R incur a cost of £50. The stockholding costs are estimated at 15% per annum of the stock value.

The lead time is three days. During the last 50 stock cycles, the demand during the lead time has generated the following frequency distribution:

TABLE 12.5 **Lead time demand for television sets**

Lead time demand (TV sets)	0	1	2	3	4	5	6	7	8
Number of stock cycles	1	2	6	8	10	8	8	5	2

Each time the warehouse runs out of stock, an emergency order is placed. The extra cost of this, together with the costs of taking customer orders, is estimated at £20 per set.

How many television sets should P&R order at a time and what should be the re-order level if management wish to minimise the total annual variable costs? How big is the safety stock with this re-order level?

Solution:

Total annual variable costs = annual ordering cost
+ annual holding costs for ordinary stock
+ annual holding cost for buffer stock
+ annual stock-out costs

C_o = £50 per order
D = 475 TV sets per year on average
C_h = 0.15 × £250 = £37.50 per set per year
C = £250 per set
C_b = £20 per set

The fixed order quantity is the taken to be the EOQ using the average demand, therefore:

$$q = \sqrt{2 \times 50 \times 475/37.5} = 35.6$$

We will take the fixed order quantity as 36 sets.

$$TC = C_o D/q + C_h q/2 + C_h \times \text{(buffer stock)}$$
$$+ C_b \times \text{(expected number of items out of stock/year)}$$
$$= 50 \times 475/36 + 37.5 \times 36/2 + 37.5 \times \text{(buffer stock)}$$
$$+ 20 \times \text{(expected number out of stock/year)}$$
$$= 1334.72 + 37.5 \times \text{(buffer)} + 20 \times \text{(expected number out of stock/year)} \quad \text{(£/year)}$$

We must choose a value for the buffer stock which minimises the total value of the last two terms. If the demand during the lead time is average or less than average, then no stock-out will occur. The problems arise only if the demand during the lead time is greater than average.

The average demand per day is 475/300 = 1.58 sets. The average demand during the lead time = 1.58 × 3 = 4.75. We will err on the side of caution and call this average lead time demand 4 sets. The probability distribution of lead time demand is found from the frequency distribution:

TABLE 12.6 **Probability distribution of lead time demand**

Lead time demand (sets)	0	1	2	3	4	5	6	7	8
Number of stock cycles	1	2	6	8	10	8	8	5	2
Probability	0.02	0.04	0.12	0.16	0.20	0.16	0.16	0.10	0.04

TABLE 12.7 **Buffer stock required for a given lead time demand**

Demand during lead time	Probability of this demand arising	Buffer stock required to meet demand
4	0.20	0
5	0.16	1
6	0.16	2
7	0.10	3
8	0.04	4

For each of the chosen values of buffer stock, we calculate the expected number of stock-outs per cycle. This number is then multiplied by the number of stock cycles per year to give the expected number of stock-outs per year. The cost of holding the extra stock (£37.5 per unit) and the expected cost of stock-out (£20 per unit) are added to give a total annual expected cost for that level of buffer stock. The number of stock cycles per year is 475/36 = 13.2.

TABLE 12.8 **Costs for different buffer stocks**

Buffer stock	Demand met	Expected Number of stock-outs /cycle	/year	Stock-out	Costs, £/year buffer stock	Total
4	8	0	0	0	4 × 37.5 = 150	150.0
3	7	1 × 0.04	0.04 × 13.2 = 0.528	0.528 × 20 = 10.56	3 × 37.5 = 112.5	123.1
2	6	2 × 0.04 + 1 × 0.10 = 0.18	0.18 × 13.2 = 2.376	2.376 × 20 = 47.52	2 × 37.5 = 75	122.5
1	5	3 × 0.04 + 2 × 0.10 + 1 × 0.16 = 0.48	0.48 × 13.2 = 6.336	6.336 × 20 = 126.72	1 × 37.5 = 37.5	164.2

The total expected cost per year is rising, therefore, we can assume that it takes its minimum value when the buffer stock is 2 television sets. If the average lead time demand is assumed to be 4 sets, then the re-order level is (4 + 2) = 6 sets. The total annual variable cost is:

TC = 1334.72 + 37.5 × (buffer stock) + 20 × (expected number out of stock/yr)
   = 1334.72 + 75.0 + 47.52 = £1457.24/year

In order to minimise the total annual variable costs of stocking, P&R should order the sets in batches of 36 whenever the stock level falls to 6.

12.5.2 A re-order cycle system

In the basic re-order cycle model, two decisions have to be made:

1 What is the fixed interval at which the orders should be placed?
2 What is the quantity to be ordered?

Again, we approach this problem in two stages in order to obtain a reasonable, but not necessarily optimum, solution. We will fix the cycle time, T, ignoring the variability in the demand or lead time. The value of T will be rounded to a convenient figure. The stock system must be easy to operate, therefore, we do not wish the store keeper to have to inspect the stock at awkward intervals. The objective of the stock system will be used to choose the order quantity. Again, we will look at two objectives—minimum service level and minimum cost.

MODEL I: A MINIMUM LEVEL OF SERVICE IS REQUIRED

To determine the fixed re-order interval, we ignore any variability in the demand or lead time and find the re-order interval which will minimise the total variable costs of ordering and holding.

Total annual variable costs = annual ordering cost + annual holding costs

If T years is the re-order interval, the number of orders placed per year is 1/T. The order size is q where D = q/T, therefore, q = DT. The average stock level is q/2 = DT/2, ignoring any buffer stock held. The total annual variable cost is now given by:

$$TC = C_0 \times (1/T) + C_h \times (DT/2) \quad (\text{£/year})$$

TC takes its minimum value when:

$$\frac{dTC}{dT} = 0 \quad \text{and} \quad \frac{d^2TC}{dT^2} > 0$$

$$\frac{dTC}{dT} = \frac{-C_0}{T^2} + \frac{C_h D}{2} \quad \text{and} \quad \frac{d^2TC}{dT^2} = \frac{2C_0}{T^3} > 0 \text{ if } T > 0$$

When

$$\frac{dTC}{dT} = 0, \quad \frac{-C_0}{T^2} + \frac{C_h D}{2} = 0$$

therefore

$$T = \sqrt{\frac{2C_0}{C_h D}} \quad \text{compared to the EOQ} = \sqrt{\frac{2C_0 D}{C_h}}$$

Having calculated this value for T, the value is adjusted to a suitable inspection interval. For example, if the value of T is calculated to be 4.2 days, then we would adjust this to a weekly inspection interval.

We now have to choose the stock level which will control the order size. For example, we may decide that, at the time of placing the order, we will choose an order quantity which would make the stock level up to 100 items if the order were delivered immediately. Hence, if the stock level is 35, the order size will be 65 but if the stock level is 43, the order size will be 57.

Example 12.9: To determine a re-order quantity for a given re-order cycle, to meet a minimum level of service

Let us assume a service level which is equivalent to one stock-out per year for an item with a 4 working week re-order cycle. Assume 50 working weeks per year. The lead time is fixed at 2 weeks.

The weekly demand for the item is approximately normally distributed with a mean of 300 items per week and a standard deviation of 50 items per week.

Solution: The number of stock cycles per year is 50/4 = 12.5. The probability of a stock-out per cycle is 1/12.5 = 0.08. The required service level is 0.92.

The variable demand which must be covered is the demand arising from the time of making the decision, to the arrival of the next re-order quantity, that is, demand during this re-order cycle, plus the next lead time, not just the lead time demand as in the re-order level model. The distribution of the demand over a 6 week period (4 weeks cycle time + 2 weeks lead time) can be assumed to be normal with a mean of (6 × 300) = 1800 items and a standard deviation of ($\sqrt{6 \times 50^2}$) = 122.5 items. The distribution of demand is shown below:

FIGURE 12.15 **Demand during re-order cycle and lead time**

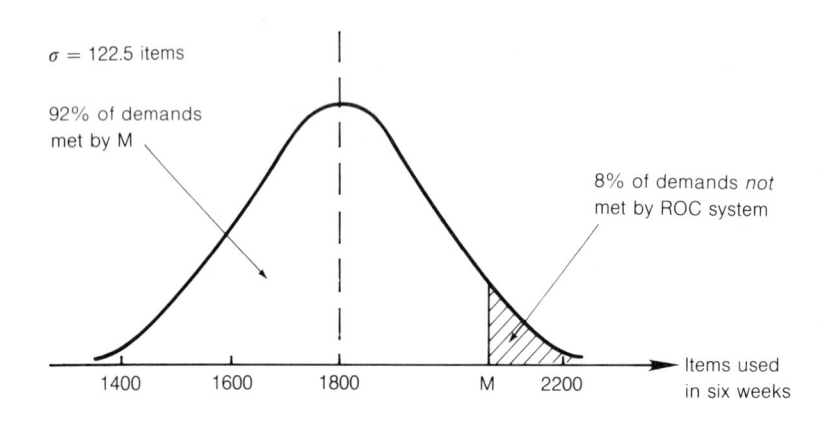

The order size is chosen so that the stock level is increased to M, where M is chosen so that the probability of meeting demand during the stock cycle is 92%. M is z standard deviations above the mean where:

$$z = \frac{M - 1800}{122.5}$$

P(z > (M − 1800)/122.5) = 0.08, therefore, from the standard normal table, z = 1.405. Hence:

$$1.405 = \frac{M - 1800}{122.5}$$

Therefore:

$$M = 1800 + (1.405 \times 122.5) = 1972.1$$

At each 4 weekly inspection, an order will be placed which would bring the stock level upto 1972 items if the order were received immediately. This will give a service level of 92%, or one stock-out per year on average.

MODEL II: A MINIMISATION OF COST IS REQUIRED

The procedure, described for Model I, is used to fix a suitable re-order cycle time. The stock level, M, which will minimise the total annual variable cost, is found in a similar way to the method described in section **12.5.1**, Model II, in which the required buffer stock was determined. Using the data of **Example 12.8** we can determine a fixed re-order interval.

$$C_o = £50/\text{order}$$
$$D = 475 \text{ sets/year on average}$$
$$C_h = 0.15 \times £250 = £37.5/\text{set/year}$$
$$C = £250/\text{set}$$
$$C_b = 20/\text{set}$$
$$L = 3 \text{ days}$$
$$\text{Working year} = 300 \text{ days}$$

The optimum re-order interval is given by:

$$T = \sqrt{\frac{2C_0}{C_h D}} = \sqrt{\frac{2 \times 50}{475 \times 37.5}} = 0.07 \text{ years}$$

The optimum re-order interval is $0.07 \times 300 = 21$ working days. Assuming that the 300 day year is based on a 6 working day week, then the most convenient re-order interval is 4 weeks. The order quantity must be chosen on each occasion so that the stock level would be made up to the level, N, if the order arrived at once, where M will minimise the cost of holding buffer stock and the cost of stock-outs per year. The buffer stock is:

$$B = (M - \text{mean demand in lead time plus re-order cycle})$$

The re-order cycle is (4×6) working days, and the lead time is 3 working days. We must cover these 27 working days. We are told that the annual demand is 475 sets in 300 working days, so the mean demand in 27 days $= (475/300) \times 27 = 42.75$ sets. The buffer stock equals $(M - 42.75)$, and the cost per year will be $(M - 42.75) \times 37.5$ £ per year. The expected stock-out costs per year depend on the variability of demand during these 27 days. Unfortunately, we have not been provided with sufficient data to do the calculations. In practice, we would have to guess a distribution and check it by collecting additional data. When we have done this, we would have to go through the calculations, using the same method as in section **12.5.1**, except that they would be much longer.

12.6 OTHER ISSUES IN THE CONTROL OF STOCK

Most practical inventory systems deal with hundreds or thousands of items. A supermarket or a large manufacturer is a good example. Not all of the stock items should, or need to, be dealt with in the same way. It is sensible to concentrate efforts on the items which have a high annual value, rather than on those which have a small annual value. One way of doing this is to make a list of all of the stock items in descending order of sales value per year. It is likely that this list will display the **Pareto effect**, that is, about 20% of the items will cover 80% of the value. It is this 20% which should be looked at first, because we would expect these items to give the highest return on the effort supplied to model the stock systems. Even amongst these high annual value items, there may be sub-divisions. An item may be in this group because the quantity used is very large, or because the value of individual items is very large.

An airline, for example, uses a large amount of fuel, costing pence per litre but with a high annual value due to the volume. The airline will also have a small number of spare engines which will also give a high annual value by virtue of their high unit cost. Both of these items are likely to be in the list of the top 20% high annual value items, but the stock problems would be modelled in different ways.

Multi-item problems can also be complicated by limitations of space or capital invested in the stock. Any shop can be regarded as a large inventory. The shelf or floor space is a limiting factor, which is fixed for a particular shop layout. The shop must decide how much space to allocate to each product. The large stores have had many years to refine their systems and to understand their customers' demands. Bar codes, automated checkouts and hand-held computers, for example, are all part of the shops' efforts to ensure that the right goods are in the right place at the right time. We have been looking at models which minimise the total annual variable cost, whereas retailers will generally organise their shops in order to maximise profits.

Many stock systems contain a series of stores—a central store feeding subsidiary stores, for example. Decisions then have to be made about which items should be held in the central store only, which in the subsidiary stores only and which items should be held in both stores. We must also decide how much of each item to order and how often. The holding costs at different levels of stock must be balanced against the administrative and transport costs of moving items frequently from the central store to the subsidiary stores. For this type of problem, a mathematical model is possible only if numerous simplifying assumptions are made. Once this level of complexity in the stock system is reached, it is likely that a simulation technique will be more useful than the mathematical models of this chapter.

SUMMARY

The simple stock model is based on the equation for stocking costs per time period:

$$TC = \text{ordering costs} + \text{stockholding costs} \quad (\text{£/time unit})$$

$$= \frac{C_o D}{q} + \frac{C_h q}{2}$$

The model assumes complete control over all aspects of the stocking system. The economic order quantity that minimises TC is:

$$EOQ = \sqrt{\frac{2 C_o D}{C_h}}$$

This basic model is modified in a variety of complex ways to deal with different situations. Quantity discounts can be evaluated by including the total purchase cost in the equation. Planned shortages can be incorporated using the cost, C_b, the cost of one shortage per time unit. The shortages may be treated as either lost orders or as being supplied from the next order.

The model can be applied to batch production of stock within a company, rather than ordering from outside. The EOQ becomes the economic batch quantity, EBQ, with C_s, the cost of the set up, replacing C_o, the cost of placing an order. Sometimes, as the batch is produced, part of it is stocked and part of it is used. The model can be adapted for this situation.

The introduction of uncertainty (reality) into the model makes it more difficult to handle. Different models and different objectives can be developed to deal with uncertain demand and uncertain lead times.

The re-order level system sets a stock level R, at which a fixed quantity, the EOQ is ordered. The re-order cycle system sets a fixed time interval, T, at which the stock level will be raised to a level, M, by ordering a variable amount. For either of these systems, we can use one of two criteria, achieving a given service level or minimising the total costs of ordering, stockholding and stock-outs.

EXERCISES

Exercise 12.1

Dekkers Ltd is an electrical goods retailers. One of its lines is a calculator. The demand for the calculators is 25 per week, evenly spread over the week. Dekkers buy the calculators for £9 each. The ordering costs are £15 per order and the stockholding costs are 50p per item of average stock per year plus 15% per annum of the average stock value. Assume 50 weeks per year.

Required:
1. What is the optimum order quantity?
2. Currently, Dekkers are ordering in batches of 300 calculators. How much money will they save each year if they change to the order quantity found in part **1**?
3. If the order costs could be reduced to £5 per order, how should Dekkers change the decision made in part **1**?

Answers on page 590.

Exercise 12.2

A firm employs 1000 engineers who leave at a steady rate of 150 per year. The engineers are recruited in groups and are put through a company training course. Each course costs £25,000 to run. If the engineers are not needed immediately on the job, this costs the company £500 per month per man.

Required:

1. How many engineers should be taken onto each training course?
2. How often should the courses be run? What is the total annual variable cost of training the engineers?
3. If the courses are restricted to 25 engineers, how would this affect the solution to part **2**?

Answers on page 591.

Exercise 12.3

Systems plc is a large firm of consultants for business computer systems. The firm requires a supply of floppy disks for the system programmes. The disks are purchased from an outside supplier and it is estimated that the annual usage will be 20,000 over the foreseeable future. The cost of placing each order for the disks is £32. For any disk in stock it is estimated that the annual holding cost is equal to 1% of its cost. The disks cost £0.80. No stock-outs are permitted and the rate of usage may be assumed constant.

Required:

1. What is the optimal order size and how many orders should be placed in a year?
2. What is the total relevant inventory cost per annum?
3. If the demand has been underestimated and the true demand is 24,200 disks per annum, what would be the effect of keeping to the order quantity calculated in **1** and still meeting demand, rather than using the new optimal order level?
4. What does your answer to **3** tell you about the sensitivity of your model to changes in demand?

Answers on page 591.

Exercise 12.4

A car showroom sells 200 cars per year. The ordering costs are £500 per order and the stockholding costs are 30% per annum of the average stock value. The cars cost £6000 each for orders of less than 50 cars. There is a discount of 1.5% for orders from 50 to 99 cars, and 3% for orders of 100 or more.

Required:
1. What should be the order size?
2. If the supplier increases the 3% discount to 5%, would this affect the answer to part **1**?

Answers on page 592.

Exercise 12.5

A small tools retailer has an average weekly demand of 3 for a particular tool. The weekly demand can be assumed to follow a Poisson distribution. There is a fixed lead time of 2 weeks from the supplier. Each tool costs the retailer £40. The cost of ordering is £50 per order. Stockholding costs are 30% per annum of the average stock value and the cost of running out of stock is £100 per unit. Assume a 50 week year.

Required:
What should the retailer do to minimise (approximately) the total annual variable cost of stocking this item?

Answers on page 593.

Exercise 12.6

A company carries an item with an annual demand of approximately 4800 units. Stock holding costs are estimated at £20 per unit per year and it costs £30 to place an order, independent of size. The demand during the fixed lead-time is variable but may be approximated by a normal distribution with a mean of 100 units and a standard deviation of 10 units. It is estimated that the cost of a stock-out over the lead-time is £10 per item.

Required:
1. What is the economic order quantity?
2. Determine the re-order level and safety stock if the company desires a 3% probability of a stock-out on any given order cycle.
3. If the manager sets the re-order level at 115 units, what is the probability of a stock-out on any given order cycle? If this re-order level is used, how many times would you expect a stock-out during the year?
4. Determine the re-order level which minimises expected costs.

Answers on page 595.

Exercise 12.7

The Oxygon Office Supplies Company Ltd is a well established firm of paper merchants and stationers, which is open for 50 weeks each year and specialises in the retailing of general office supplies. Its many customers include financial institutions, legal establishments and insurance companies. However, steadily increasing operating costs have diminished their financial reserves which has prompted the chief accountant to recommend a reduction in overall stock levels. Whereas in previous times it was common for the company to hold over 12 months' stock for many stock items in order to guarantee availability, pressures on liquidity seemed to demand a reduction in inventory levels.

The company's main selling item was a high quality typing paper which tended to have erratic demand but can be assumed to have a normal distribution with a mean of 800 boxes each week and a standard deviation of 250 boxes per week. This paper is supplied by the Tiara Paper Company at a cost of £2.50 per box. It was found that the lead time of supply of this paper recently had been very consistent at 3 weeks.

The annual cost of stockholding was estimated at 15% of the stock item value and is based on the cost of storage and the company's cost of capital. In order to estimate the cost of a delivery of paper from Tiara the cost of making and receiving the order together with the associated accounting and stock control tasks requires a total effort of approximately 12 man hours, where the average wage rate is £160 per week for a 40 hour week.

Required:
1. Outline the basic principles of inventory control policy and explain why a good inventory policy is of value to Oxygon.
2. Calculate the economic order quantity for this stock item, together with the average length of time between replenishments.
3. Determine the recommended re-order level if there is to be no more than a 1% chance that a stock out will occur in any one replenishment period.
4. Determine the total stockholding cost (storage and delivery costs) per annum using the calculated values of the economic order quantity and re-order level.

(ACCA, December 1987)

Answers on page 597.

Exercise 12.8

The Plutonic Pharmaceutical Co uses a particular chemical in many of its products, the chemical being stored in special refrigerated units which are provided by the supplier at a nominal rental of £40 per month. The demand for the chemical is reasonably constant from month to month and averages about 1000 litres per month. The company currently rents one storage unit which has a capacity of 1000 litres, so that replenishment takes place every month when stock falls to zero. The process of stock replenishment involves cleaning and sterilising the unit each time at a cost of £50.

As a result of an expansion of the company's product range, the demand for this chemical is expected to increase to 2500 litres per month and the company accountant has been asked to recommend an appropriate purchasing and storage policy. Additional storage units could be obtained but this would involve a further rental cost of £40 per month for each additional unit. However, there would be some economies in the cleaning and sterilising costs as these would only increase by £25 for each additional unit involved.

The cost of the chemical is £5 per litre and the company's cost of capital is 24% per annum.

Required:
1. Show that the present policy of ordering 1000 litres every month is the most economical ordering policy in the current demand situation with just the one storage unit. What is the total annual cost associated with the storing of this chemical?
2. Given the projected increase in demand, advise the company on whether an additional storage unit should be rented if the objective is to minimise the storage costs involved.
3. Show that demand has to increase to 7200 litres per month before it becomes economical to rent a second storage unit.

(ACCA, June 1987)

Answers on page 598.

Exercise 12.9

A certain item is produced in batches of size Q to meet a constant demand of D per year. The manufacturing cost is £C per item and the stockholding cost is £H per item per year. Each production run commences when stock falls to zero and the cost of setting up the production facilities is £S per run. You may assume that production takes place relatively quickly such that stock replenishment is effectively instantaneous.

Required:
1. Write down an expression for the total annual production cost which comprises manufacturing, set-up and stockholding costs. Show that, at the economic batch quantity (EBQ), the total annual set-up cost is equal to the total annual stockholding cost. (It is not necessary to derive the expression for the economic batch quantity.)
2. Experience has shown that the manufacturing cost accounts for about 75% of the total production cost when economic batch quantities are used.
   (a) What percentage increase in total production cost will result if the batch size is increased by 50% from the EBQ?
   (b) What percentage increase in total production cost will result if the batch size is decreased by 50% from the EBQ?
   (c) If the batch size is doubled, there will be a saving of 5% in the manufacturing cost due to more efficient working.
   Show that this saving of 5% will not lead to a lower total annual production cost. How large a saving in manufacturing cost is necessary for it to be economical to double the batch size?

(ACCA, June 1986)

Answers on page 599.

Exercise 12.10

Proteus Products is a distributor of spare parts to the motor trade. One of the main problems faced by the purchasing manager of Proteus Products is the complicated discount structure operated by most of its suppliers. It is not uncommon for an item to have as many as five alternative prices according to the quantity which is ordered, as in the case of item 53/X2 whose price structure is as follows:

Quantity Ordered	Unit Price (£)
1– 99	10.20
100–249	9.95
250–499	9.65
500–999	9.30
1000 +	8.80

One of the reasons for the variations in price is that all delivery costs are borne by the supplier. As a result, the total ordering cost incurred by Proteus for most items is small enough to be regarded as negligible. On the other hand, the company's cost of capital makes it impossible to disregard the cost of holding stock. When additional allowance has been made for the variable costs of storage, the stockholding cost amounts to 20% per annum of the value of the stock held. You may assume that the demand for item 53/X2 is reasonably constant at about 10 per week for a 50 week year.

Required:
1. Write down an expression for the total annual cost of purchasing and holding stocks of item 53/X2 assuming that replenishment is instantaneous. You should define carefully the meaning of any symbols that you use.
2. Draw a graph of the total annual cost for a range of order quantities between 1 and 1000 and determine the most economical order quantity.
3. Show that the stockholding cost must be down to about 12% before it becomes economical to purchase item 53/X2 in quantities of 1000.

(ACCA, December 1985)

Answers on page 602.

Exercise 12.11

The management of Leander Products Ltd is becoming increasingly concerned about the level of the company's investment in stocks, particularly in the form of manufactured components. All components are produced in the machine shop on a batch basis and, on examining the stock records, it is evident that stocks of one particular component (XT/24) are abnormally high. This component is produced on a specific machine, which is manned by one operator, who is paid a basic wage of £96 for a 40 hour week. In addition to the cost of the operator's time, which may be regarded as fixed, the machine incurs running costs of £1.40 per hour.

The expected demand for component XT/24 is 50 per week, which is not likely to show any marked variation from week to week. Raw material costs amount to £7.40 per unit and the manufacturing time is 30 minutes per unit, with a machine set-up time of 6 hours prior to the production of each batch. Taking into account the specialist labour required, machine set-up has been estimated to cost the company an average of £10 per hour.

After manufacture, all components go into stock in a special storage area which is the responsibility of one storekeeper who is paid a fixed weekly wage of £80. Insurance for this storage area currently costs £230 a year and other overheads (all of which are fixed) amount to £2600 a year. You may assume that the company works a 5 day week, for 50 weeks a year and that the cost of capital is 18%.

Required:
1. Indicate, with reasons, which of the above costs are relevant to the determination of optimum batch sizes and stock levels for component XT/24.

INVENTORY PLANNING AND CONTROL: CHAPTER TWELVE   385

2   An appropriate expression for the 'economic batch size' of any component which is being produced at a rate of R per week and simultaneously used at a rate of D per week is:

$$Q\sqrt{\frac{R}{R-D}}$$

where Q is the usual 'economic order quantity' in a situation of instantaneous stock replenishment. Assuming that production of all component commences when stock falls to zero, calculate the optimum batch size for component XT/24 and determine the maximum stock level of this component.

3   Assuming that all components are withdrawn from stock on a 'first-in-first-out' basis, what are the maximum and minimum times that any item will remain in stock?

(ACCA, June 1984)

Answers on page 603.

Queueing models

13.1 INTRODUCTION

We are all familiar with queues in our everyday lives. A queue is an ordered group of people or items which are waiting for some event or some activity to happen. This activity is referred to as the 'service'. Whenever a group of items waits for service, we have a queue, for example, letters waiting in an in-tray, cars at traffic lights, goods in stock waiting to be used, customers at a supermarket checkout, machines waiting to be repaired, lorries waiting to be unloaded, travellers waiting to buy tickets, planes waiting to land, patients waiting for a doctor, telephone callers waiting for a line, partly made components waiting for the next process, and so on. The variety and complexity of queueing situations are very large.

Since queues are so common, the study and modelling of queueing problems were important in the early development of operational research. The modelling proceeded on two parallel courses. The first approach, which we will consider in this chapter, is probabilistic and mathematical. The second approach uses simulation and the computer. We will consider this technique in the following chapter. There are strong similarities between the approach to the modelling of stock problems and the approach to the modelling of queueing problems. In both cases, as long as we make simplifying assumptions, the mathematics can help us to understand the underlying process. As soon as we relax these assumptions, and introduce practical complexities, the mathematics becomes difficult or impossible to handle.

In this chapter, after a general discussion, we will look at simple queueing problems which are based on the Poisson process. In the Poisson process it is assumed that the rate of arrival at the queue follows a Poisson distribution, and the time for service follows a negative exponential distribution. The models are extended to deal with more complex situations, in which several servers are used.

13.2 COMPONENTS OF THE QUEUEING SYSTEM

Before a model can be built, we must think systematically about the queueing problem, to identify its major components. In this section, we will look at the features of the different components which make up the queueing system, and at some of the criteria used to make decisions about the system.

FIGURE 13.1 **A typical queueing system**

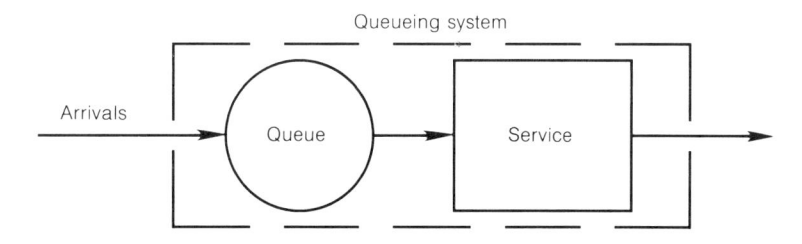

Items are generated from some source and arrive at a queue. We must describe this arrival pattern. The items wait in the queue. There are one or more servers, which select the next item from the queue, and service it. We need to know the queue discipline, and the service pattern.

13.2.1 Arrivals and arrival patterns

We will consider an example of patients arriving at a doctor's surgery. Let us suppose there are appointments every 10 minutes, and the patients arrive exactly on time. The doctor takes exactly 10 minutes to see each patient. No one waits. No one is delayed. No one is idle. Exactly every 10 minutes, someone arrives and someone leaves. There are no interruptions, no emergencies and no problems in this ideal situation. This is an example of a perfect queueing system.

The arrival rate in this example is 6 patients per hour, and there is no variability. The inter-arrival time is constant at 10 minutes. The arrivals can be described either in terms of the rate of arrival, that is the number of patients per hour, 6, or in terms of the time between patients, 1/6 hours. These are different ways of giving the same information.

This fixed arrival pattern is unusual. Even with an appointments system, some patients will arrive early and some late. What happens to the system if the doctor always takes 10 minutes to see each patient, but the patients are a little irregular in their arrival times? If a patient is early, a queue will form, but the doctor is not affected. If however a patient is late, the doctor and all subsequent patients are delayed. The doctor must wait, and, since it always takes him exactly 10 minutes to deal with each patient, he will be late for the next patient. If a patient is 2 minutes late, the doctor will be 2 minutes late for all subsequent appointments. As soon as any variation, however small, arises in the arrival pattern, a queue will develop.

The above arrival pattern is controlled. Imagine building up an arrival pattern for the number of people in a large airport. There are three main flows of people arriving at the airport; those who are boarding planes, those who are landing at the airport, and those who are meeting passengers or seeing them off. The three flows are connected, since one group of passengers will board the planes the others have left, and both are connected to the third group of non-travellers. This complex pattern will change throughout the year, throughout the week and throughout the day. The departing passengers may arrive at the airport by bus, car or train, and with a spread of arrival times dependent on how early they wish to be for their flight. When a train arrives, a large group of passengers enters the airport together. In the same way, when each plane arrives, the arrivals system has to switch from doing nothing, to dealing with large numbers of people and baggage. It is a highly irregular arrival pattern, which changes through time.

The arrival pattern of the planes at a busy airport may be regular. If you watch aircraft coming into a large airport, you will see that they wait in a queue and land at regular intervals, for safety reasons. However, this does not lead to a regular arrival pattern for passengers, which is what governs the size and services at the terminal.

13.2.2 Queue Discipline

There are many ways in which a server can deal with a queue. At the surgery, the doctor takes the patient with the next appointment. Even if a patient arrived 20 minutes early, with an appointment system, he would not jump the queue. At traffic lights, the first car to arrive at the red stop light is the first to go through the green go light. This is an example of the **FIFO (first-in-first-out)** queue discipline, which is a common queueing system. If the doctor does not have an appointment system, then the FIFO system is the one he would normally operate.

Another system is the **LIFO (last-in-first-out)** system. A cable car is an example of this. The last person in is nearest to the door, and is probably the first person out. When tins are stacked on supermarket shelves, the top ones are taken first by the customers. There are many other systems. How do we deal with letters in an in-tray? The earlier letters are at the bottom. What we might do is look at each letter. If it is easy to deal with, we do so immediately and put it in the out-tray. If it is difficult, it goes into the pending-tray, on top of all the other awkward jobs. This system is some kind of priority queue discipline, with a LIFO system for the pending-tray.

The queue discipline is important when it is people who are in the queue. An understanding of the psychology of queues is essential for anyone providing a service. If the discipline is clearly stated, and is felt to be fair by the queuers, the service will appear to be more satisfactory. For example, the change in the arrangement of banks from individual queues at each service point to one single queue for all customers gives the queuers a feeling of greater fairness, or at least of less frustration. The tedium of waiting for a lift may be reduced by providing mirrors, so that those waiting have something to look at. The indicators showing the position of the lifts, are a similar psychological device, giving the queuer some information about the likely arrival of the next lift.

13.2.3 Servers, service time and idleness

The number of servers or service channels will affect the queue length and waiting time. Many queueing problems are concerned with the number of servers which should be provided. How many checkouts should a supermarket have, and how many should be manned at any particular time? It is not sensible to man all 25 checkouts when the store is almost empty. Most would be idle, and the employees should be doing something else. Equally, it would be wrong to have only one checkout manned when the store is busy, and have a vast queue of irate shoppers. A balance must be found between the costs of providing more servers, and the customers' waiting time and feelings about the service.

The queueing models in this chapter cannot deal with this kind of complexity. Simulation models would be more appropriate. As with the arrivals, service can be expressed either as a rate, that is, the number of arrivals that can be serviced per hour, or as a service time. The two are interchangeable. A service time of 2 minutes per customer, 3 hours per car, or 2 days per machine can be converted into a service rate of 30 customers per hour, 1/3 cars per hour, or 1/2 machines per day. It is more useful to use service times, rather than rates, because the effective service rate depends on the arrival pattern. If there is no one in the queue, the server is idle, with nothing to do. The number of people served per hour depends on both the service rate and the arrival rate. It is the interaction of these two which makes mathematical queueing models and the simulation of queueing problems interesting and important.

13.2.4 Criteria for judging a 'good' queue

The adequacy of a queueing system changes depending on who you are—the customer, the server or the owner. The customer requires a fast service, ideally with no queue. Most customers know that it is unlikely that this will happen. They will be satisfied with a reasonable wait, and an organised queueing system. The customers are willing to compromise. The server requires a reasonable loading. If the loading is too light or too heavy, service time and service quality may suffer. The skill of the servers and the system that they operate will determine what is too light or too heavy. The server is usually, however, excluded from the decision making process. The owner of the service is concerned with the costs of his system and its quality. He can reduce costs by removing a server, or lengthening the service time. This will increase the customer's waiting time and reduce the quality of the system. The owner has to consider the short and long term effects on his service. As with most business decisions, the simple cost criterion is inadequate. The measures which are useful for comparing different queueing systems are:

1. the distribution and mean value of the queue length;
2. the distribution and mean value of the queueing times;
3. service times;
4. costs.

If two systems have the same costs, we can use queue length and queueing times to decide between them. More usually, however, the better system will cost more, and we have to consider the cost-service balance.

13.3 THE POISSON PROCESS AND THE M/M/1 QUEUE

In the previous sections, we considered some of the general complexities of queueing systems—the arrival pattern, the queue discipline, the number of servers and the service time. In this section, we will look at a particular, simple queueing system with variable arrival and service times.

The first models of queueing systems to be built, described the arrival of telephone calls at an exchange with a fixed number of lines. It was found that a Poisson distribution could be used to describe both the arrival rate and the service rate. In this situation, the system can be described using just two values, λ, the mean arrival rate, and μ, the mean service rate.

13.3.1 The Poisson process

The simple model assumes that both the arrival pattern and the service pattern are Poisson processes. This process can be described by either:

1. The rate of occurrence, that is, the number of arrivals per unit time, using the discrete Poisson distribution:

$$P(r \text{ arrivals/unit time}) = \frac{\lambda^r e^{-\lambda}}{r!} \qquad r = 0, 1, 2, 3, \ldots$$

where λ is the mean arrival rate and is equivalent to m, used in Chapter Two.

2. The inter-arrival times, using the continuous negative exponential distribution:

$$P(\text{time, t, between arrivals}) = \lambda e^{-\lambda t}$$

The Poisson process assumes that:

1. events occur randomly through time;
2. the number of arrivals in one time period is independent of the number in the previous time period, which means that the Poisson process has no memory;
3. the mean arrival rate remains constant;
4. one event only can occur in each small time interval.

Many of the situations which we described in the previous section do not satisfy these conditions. The 10 minute appointment system at the doctor's surgery is designed to avoid random arrivals. A constant 10 minute service time cannot be random. At the airport, the mean arrival rate changes as the passengers arrive in planeloads. Our simple Poisson process model will not help in these situations. It has, however, been found to fit many other cases; telephone systems, demand for stock, equipment failures and machine breakdowns, for example. The Poisson process model has also been applied to events spread through space rather than time. It has been fitted to the occurrence of misprints on a page, faults along a length of wire, bombs falling over an area, and refractory inclusions in a volume of steel.

The Poisson distribution, which we considered in Chapter Two, section **2.4**, describes the distribution of arrivals. For example, if we know that the mean number of arrivals of cars at a petrol station is $\lambda = 2$ cars per minute, then the probability distribution for the number of arrivals in any one minute interval is:

$$P(r) = \frac{\lambda^r e^{-\lambda}}{r!} = \frac{2^r e^{-2}}{r!} \qquad r = 0, 1, 2, \ldots$$

$$P(0 \text{ cars in a 1 minute interval}) = P(0) = \frac{2^0 e^{-2}}{0!} = 0.135$$

$$P(1 \text{ car in a 1 minute interval}) = P(1) = \frac{2^1 e^{-2}}{1!} = 0.271$$

$$P(2 \text{ cars in a 1 minute interval}) = P(2) = \frac{2^2 e^{-2}}{2!} = 0.271$$

$$P(3 \text{ cars in a 1 minute interval}) = P(3) = \frac{2^3 e^{-2}}{3!} = 0.180$$

We can use the same information to calculate the distribution of the number of cars arriving in a 5 minute interval. If the mean rate is 2 per minute, then the mean rate per 5 minutes is $(2 \times 5) = 10$ cars per 5 minutes. The probabilities are:

$$P(0 \text{ cars in a 5 minute interval}) = P(0) = \frac{10^0 e^{-10}}{0!} = 0.00005$$

$$P(1 \text{ car in a 5 minute interval}) = P(1) = \frac{10^1 e^{-10}}{1!} = 0.0005$$

This discrete distribution of the number of arrivals can be converted into a continuous distribution of the time between arrivals. The distribution of the inter-arrival times for a Poisson process is a negative exponential distribution. The probability density function is:

$$P(t) = \lambda e^{-\lambda t}$$

This has a mean inter-arrival time of $(1/\lambda)$, with a standard deviation of $(1/\lambda)$. For example, if $\lambda = 2$ cars per minute, the mean inter-arrival time is $1/2$ minute.

The probability that the inter-arrival time is less than T, is:

$$P(t < T) = 1 - e^{-\lambda T}$$

and the probability that an inter-arrival time lies between T_1 and T_2 is:

$$P(T_1 < t < T_2) = e^{-\lambda T_1} - e^{-\lambda T_2}$$

For example, using $\lambda = 2$ per minute, the probability that an inter-arrival time will be between 1 and 2 minutes is:

$$e^{-2 \times 1} - e^{-2 \times 2} = 0.135 - 0.018 = 0.117$$

The two distributions are shown in FIGURES 13.2 and 13.3.

FIGURE 13.2 **Poisson distribution of number of cars in a minute interval, $\lambda = 2$ cars per minute**

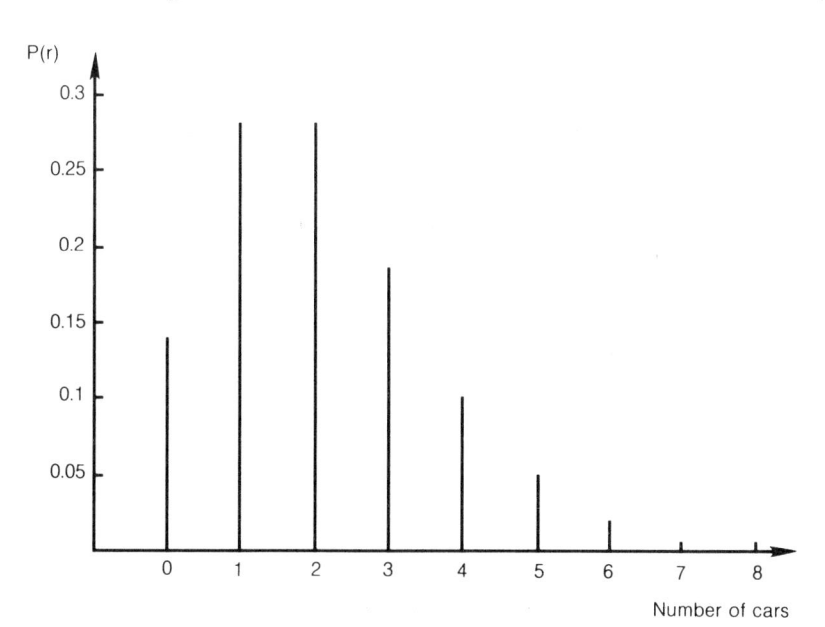

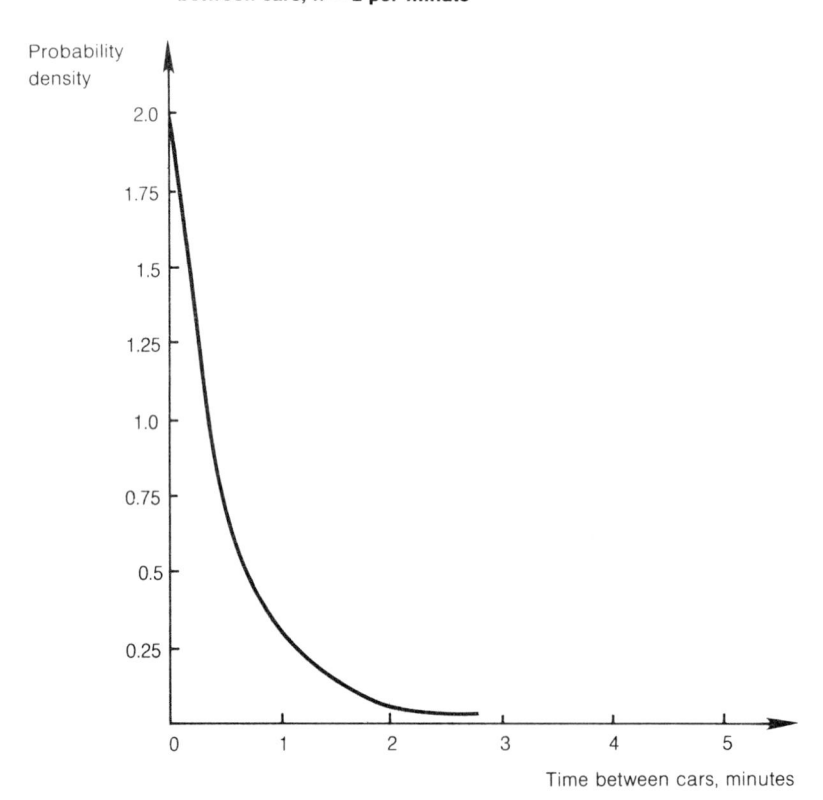

FIGURE 13.3 **Equivalent exponential distribution of inter-arrival times—minutes between cars, $\lambda = 2$ per minute**

13.3.2 The M/M/1 queueing system

Mathematicians have developed a code to describe queueing models. The three parts of the code represent the arrival distribution, the service distribution and the number of servers.

M/M/1 represents a model with a Poisson process arrival pattern, M, a Poisson process service pattern, M, and a single server, 1. The mean arrival rate of the Poisson process is λ items per time unit, and the mean service rate is μ items per time unit. There are three further assumptions made in this model:

1 there is an infinite number of items which can join the queue;
2 there is a single queue with a queueing discipline of first-in-first-out;
3 the system has settled down into a steady state.

When a queueing system first starts up, there will be fluctuations in the arrival and service patterns. These erratic conditions cannot be modelled easily. However, in time the system will settle down so that the average arrival rate and average service rates are constant over time. It is this situation which the M/M/1 model is representing. As long as the arrival rate, λ, is less than the service rate, μ, the system will reach this steady state. This means that it does not matter how many people are in the queue at time 0. This is a further limitation of these models. If the practical situation does not have time to stabilise before the arrival rate changes, we cannot use these M/M/1 models.

Notice that we distinguish between the queue and the system, which is the queue plus the server. The M/M/1 queueing model provides us with the means of calculating the average length of the queue and the average time spent queueing, both for the queue alone and for the whole system.

The ratio of the mean arrival rate to the mean service rate is referred to as the **traffic intensity**, ρ:

$$\rho = \frac{\lambda}{\mu}$$

The traffic intensity is the proportion of time for which the server is busy.

The steady state results for the M/M/1 queueing model are given below:

Queue Lengths:

$$P(\text{server idle}) = P(\text{no queue}) = P(0) = \left(1 - \frac{\lambda}{\mu}\right)$$

$P(\text{server busy and } (n-1) \text{ items in queue})$

$$= P(n \text{ items in system}) = P(n) = \left(\frac{\lambda}{\mu}\right)^n P(0)$$

$$\text{Mean number of items waiting in queue} = \frac{\lambda^2}{\mu(\mu - \lambda)}$$

Mean number of items waiting and being served

$$= \text{Mean number in the system} = \frac{\lambda^2}{\mu(\mu - \lambda)} + \frac{\lambda}{\mu} = \frac{\lambda}{(\mu - \lambda)}$$

Queue times:

$$\text{Mean time waiting in the queue} = \frac{\lambda}{\mu(\mu - \lambda)}$$

Mean time waiting and being served

$$= \text{Mean time in the system} = \frac{\lambda}{\mu(\mu - \lambda)} + \frac{1}{\mu} = \frac{1}{(\mu - \lambda)}$$

The traffic intensity, λ/μ, is sometimes known as the **utilisation factor**, because it is equal to the probability that an arrival has to wait for service.

An important point to note is what happens to the queueing system as the arrival rate, λ, approaches μ, the service rate. When the arrival rate equals the service rate, the queue explodes. It slowly gets bigger and bigger, and the system never settles down.

$$P(0) = (1 - \lambda/\mu) = 0$$

which means that the server is never idle. This does not happen to the doctor with his 10 minute appointment system, because he has a certain number of appointments only each day. The arrival rate drops to zero at the end of the surgery, and he clears the backlog. Our model, however, assumes that the appointment system goes on indefinitely. For the M/M/1 system to be appropriate, the arrival rate, λ, must be less than the service rate, μ.

Example 13.1: To illustrate the M/M/1 queueing system

A small shop has one assistant. Customers arrive at random at the rate of 20 per hour. It takes 2 minutes on average for the assistant to serve each customer. The distribution of service times is negative exponential.

1. If we go into the shop, how long will we be waiting in the queue on average?
2. What is the probability that there is no one in the shop when we arrive?
3. For what proportion of the time will the assistant be idle?
4. What is the mean number of people in the system?
5. What is the probability that there are at least 4 customers in the shop?
6. If the arrival rate rose to 25 per hour, how would this change **1** to **5**?

Solution:

We are assuming that:

(a) the arrivals are random, and follow a Poisson distribution with $\lambda = 20$ customers per hour;
(b) the service distribution is Poisson with a service rate $\mu = 60/2 = 30$ customers per hour;
(c) neither rate changes during the day;
(d) the queueing discipline is first-in-first-out;
(e) the system has reached a steady state.

An M/M/1 queueing model is appropriate.

1 The mean time spent waiting in the queue is:

$$\frac{\lambda}{\mu(\mu - \lambda)} = \frac{20}{30 \times 10} = \frac{1}{15} \text{ hours} = 4 \text{ minutes}$$

2 The probability that no one is in the shop is:

$$P(0) = (1 - \lambda/\mu) = (1 - 20/30) = 1/3$$

3 The probability that no one is in the shop is also the proportion of time that the assistant is not serving customers.

4 The mean number of people in the system is:

$$\frac{\lambda}{\mu - \lambda} = \frac{20}{30 - 20} = 2 \text{ customers}$$

5 The probability that there are 4 or more customers in the system is:

$$1 - (P(0) + P(1) + P(2) + P(3))$$
$$= 1 - (0.333 + 0.222 + 0.148 + 0.099) = 0.197$$

6 If the arrival rate rose to 25, the time spent waiting increases from 4 minutes to:

$$25/(30 \times 5) = 1/6 \text{ hours} = 10 \text{ minutes.}$$

$P(0)$ becomes 1/6 instead of 1/3

The mean number of people in the system increases from 2 to 5 customers, and the probability that 4 or more customers are in the system increases from 0.197 to 0.482.

13.4 EXTENSIONS TO THE M/M/1 MODEL

In this section we will indicate how the basic model can be extended to include a second server and how we can check that the process involved is Poisson. We will then consider how the costs of the system can be included in the model.

13.4.1 The M/M/2 model

The M/M/2 model assumes a Poisson process for the arrivals, with mean λ, which join one queue, which then feeds two identical servers, using the FIFO queue discipline. The two servers are Poisson processes with the same mean service rate, μ. The steady state equations for the two server model are given below:

$$P(\text{both servers idle}) = P(0) = \frac{1}{1 + \frac{\lambda}{\mu} + \left(\frac{\lambda}{\mu}\right)^2 \frac{\mu}{(2\mu - \lambda)}}$$

$$P(n \text{ customers in system}) = P(n) = \frac{(\lambda/\mu)^n}{n!} P(0) \text{ for } 0 < n \leqslant 2$$

$$P(n) = \frac{(\lambda/\mu)^n}{2! \, 2^{n-2}} P(0) \text{ for } n \geqslant 2$$

$$\text{Mean number waiting in queue} = \frac{\lambda^3}{\mu(2\mu - \lambda)^2} P(0)$$

$$\text{Mean number waiting in the system} = (\text{mean number in queue}) + (\lambda/\mu)$$

$$\text{Mean time waiting in the queue} = (\text{mean number in queue})/\lambda$$

$$\text{Mean time waiting and being served} = \text{Mean time in the system}$$
$$= (\text{mean time waiting}) + 1/\mu$$

We can use these equations to examine how best to improve a system which has long queues and bad service.

13.4.2 Checking the basic Poisson assumptions

The procedure used to check the assumption that the arrival pattern is a Poisson distribution is the χ^2 Goodness of Fit test which we discussed at the end of Chapter Six. Data are collected for the number of arrivals per time unit, and a frequency table is established. From this observed distribution, the observed mean, λ, is calculated.

The null hypothesis is that the distribution of the number of arrivals per unit time follows a Poisson distribution with mean λ. The alternative hypothesis is that the arrival pattern does not follow the distribution.

The observed mean is used to calculate the probabilities of 0, 1, 2, ... arrivals per unit time, assuming a Poisson distribution. These probabilities are multiplied by the total number of observations, to give the expected frequencies, based on the null hypothesis. If any expected frequencies are less than 5, categories are amalgamated in the usual way.

$$\text{Calculated } \chi^2 = \sum \frac{(f_0 - f_E)^2}{f_E}$$

This test statistic is compared with the appropriate value at the 5% decision level, $\chi^2_{0.05, (n-2)}$, where n is the number of categories. If the calculated χ^2 is less than $\chi^2_{0.05, (n-2)}$, we accept the null hypothesis. The process is assumed to be Poisson with mean λ. If we reject the null hypothesis, we assume that the process is not Poisson with mean λ. In this latter case, we must use some other modelling process which will probably involve simulation.

The service system is dealt with in a similar way, except that the service times are recorded, rather than the number of service completions. The mean service time, $1/\mu$, is calculated from the observations. We use this mean and:

$$P(t < T) = 1 - e^{-\mu T}$$

to construct the expected exponential distribution. The distribution is subdivided into time intervals which match those of the observations. The expected frequencies for service times are calculated for each time interval.

Both of these applications of the χ^2 test follow closely **Examples 6.20** and **6.21** in sections **6.11.2** and **6.11.3** of Chapter Six.

13.4.3 The costs of a queueing system

Example 13.2: To illustrate the inclusion of costs in the queueing model

We will look again at the shop described in **Example 13.1**. The arrival rate, λ, is assumed to be 25 customers per hour, and the service rate, μ, 30 customers per hour. To improve the service to customers, we can either reduce the mean time taken to serve each customer, or we can have two independent servers working at the old rate. We will assume that there are two options:

1 Spend £100 per week on a packer to reduce the mean service time from 2 minutes to 1.5 minutes per customer.
2 Spend £150 per week to employ a second assistant, who will work at the same rate of 2 minutes per customer.

These are the costs to the owner of the shop. We must evaluate the benefits to the customer. In practice, this is not easy. The shop is open for 60 hours a week. We will cost the average time waiting for service at £2 per hour per customer.

Solution: The total number of customers per week = 25 × 60 = 1500. With the present system of one assistant taking an average of 2 minutes per customer, the mean waiting time in the queue is:

$$\frac{\lambda}{\mu(\mu - \lambda)} = \frac{25}{30 \times 5} = 1/6 \text{ hours} = 10 \text{ minutes}$$

Therefore, for 1500 customers, the weekly waiting costs are:

$$1500 \times 1/6 \times £2 \text{ per hour} = £500/\text{week}$$

With the first change, we employ a packer. The system is still assumed to be M/M/1, but the service rate, μ, is increased to 60/1.5 = 40 customers per hour. The mean waiting time in the queue is now:

$$\frac{\lambda}{\mu(\mu - \lambda)} = \frac{25}{40 \times 15} \text{ hours} = 2.5 \text{ minutes}$$

Therefore, for 1500 customers per week, the waiting costs are:

$$1500 \times (2.5/60) \times 2 = £125/\text{week}$$

The owner has spent £100 per week on the packer, and reduced the customers' notional waiting costs by £375 per week. If the waiting costs are realistic, then this is well worth doing.

What happens if we now look at the two server model, M/M/2, with λ = 25 customers per hour, and μ back to the original 30 customers per hour? The calculations are more lengthy. To find the mean waiting time, we must first calculate P(0):

$$P(0) = \frac{1}{1 + \frac{\lambda}{\mu} + \left(\frac{\lambda}{\mu}\right)^2 \frac{\mu}{(2\mu - \lambda)}} = \frac{1}{1 + \frac{25}{30} + \left(\frac{25}{30}\right)^2 \frac{30}{(60 - 25)}} = \frac{1}{2.429} = 0.412$$

The mean waiting time is:

$$\frac{\lambda^2}{\mu(2\mu - \lambda)^2} P(0) = \frac{25^2}{30(60 - 25)^2} \times 0.412 \text{ hours} = 0.42 \text{ minutes}$$

Therefore, for 1500 customers per week, the waiting costs are:

$$1500 \times (0.42/60) \times 2 = £21.00 \text{ per week}$$

By spending £150 on a second assistant, we have reduced the waiting cost to £21 per week. The extra £50 per week paid for the assistant compared with the packer has reduced customer notional waiting costs by about £120 per week. The owner will have to judge whether this last step is worth taking.

SUMMARY

The model of a queue is made up of the following components:

1. the arrival pattern
2. the queue discipline
3. the number of servers
4. the service pattern

In practice there are many different types of queue and model. We consider only the simplest situation. The arrival and service patterns are represented by Poisson processes, either in the form of the Poisson distribution or the negative exponential distribution. The queue discipline is first-in-first-out (FIFO). With one server, this is the M/M/1 model. If the arrival pattern has a mean rate of λ items per time unit and the service pattern μ items per time unit, the steady state results are:

$$\text{Mean number of items waiting in the queue} = \frac{\lambda^2}{\mu(\mu - \lambda)}$$

$$\text{Mean number waiting and being served} = \frac{\lambda}{(\mu - \lambda)}$$

$$\text{Mean time waiting in the queue} = \frac{\lambda}{\mu(\mu - \lambda)}$$

$$\text{Mean time waiting and being served} = \frac{1}{(\mu - \lambda)}$$

Results are also given for the two server, M/M/2, model. These results can be used to compare the effects of different changes in the queueing system. More complicated situations are dealt with by other methods, often using simulation.

EXERCISES

Exercise 13.1

The number of cars passing a point on a road follows a Poisson distribution with a mean of three cars per minute. What is the probability that:

1. no cars pass in a one minute interval?
2. two cars pass in a one minute interval?
3. four or more cars pass in a one minute interval?

Answers on page 606.

Exercise 13.2

The telephone calls into an office follow a Poisson process with a mean of three per 15 minutes. What is the probability that:

1. there are no calls in a 15 minute interval?
2. there are no calls in a 5 minute interval?
3. there are at least two calls in a 10 minute interval?

Answers on page 606.

Exercise 13.3

Lorries enter a cold store unloading area randomly throughout the week, 24 hours, 7 days a week, at a mean rate of 2.5 per hour. If it takes exactly 15 minutes to unload each lorry, for what number of hours in a week will the unloading bay be unable to cope?

Answer on page 606.

Exercise 13.4

The demand for a stores item is Poisson with a mean number of 5 per 5 working days. What is the probability that the time between requests is:

1   greater than 2 days?
2   between 3 and 4 days?
3   less than 1 day?

Answers on page 607.

Exercise 13.5

Vehicles using the northern bound lanes of a motorway pass a point at the rate of 1200 per hour. What is the probability that:

1   no vehicles pass in a 30 second interval?
2   the inter arrival time is between 5 and 15 seconds?

Answers on page 607.

Exercise 13.6

New clients visit a solicitor's office at random throughout the week. One solicitor is designated to deal with the initial interviews. The office is open from 9 am to 5 pm. The mean arrival rate is 2 per hour, and it takes an average of 20 minutes to deal with each client. Assume Poisson processes for both arrival and service times.

Required:
1   How many clients on average will be waiting to see the solicitor?
2   How long will they have to wait on average?
3   What is the probability that no one will be waiting when a client arrives?

Answers on page 607.

Exercise 13.7

A library has an enquiries section, staffed by 1 librarian. Enquiries are randomly spaced throughout the day at 5 minute intervals on average. It takes the librarian 3.5 minutes on average to deal with the enquiries. Assuming that both are Poisson processes, what can you say about the queue?
   If 2 librarians were employed, each taking 4 minutes on average to deal with queries, how would this affect queueing times and lengths?
   The library is open for 50 hours per week. A librarian is paid £6 per hour, and it is estimated that waiting costs each customer £2 per hour. Comment on the costs for each system.

Answers on page 607.

Exercise 13.8

Carnucopia is a large motor dealer with 12 showrooms in the South of England. In recent months problems have arisen with the supply of new cars to customers as unanticipated delays have been occurring in the pre-delivery inspection stage. The system operated by Carnucopia is that all stocks of new cars are held in a central compound and, when a sale is made at any one of the showrooms, the order is telephoned through to the compound and the car in question is taken from stock and submitted for pre-delivery inspection at the company's main workshop. After the car has been inspected and any faults repaired, the car is delivered to the appropriate showroom for collection by the customer.

In recent months, the company has been selling new cars at a rate of about 38 per week with no significant seasonal variation. Furthermore it seems reasonable to assume that sales are occurring randomly throughout the week as no regular patterns are discernible. Pre-delivery inspection is carried out on a 'first-come-first-served' basis and takes on average about 2 hours for all models. The workshop has 2 inspection bays and the engineers who carry out the pre-delivery inspections work a 40 hour week. When the inspection is completed, delivery to the showroom usually takes 1 working day. You may assume that inspection engineers work a 5 day week and you should ignore the effects of weekends.

Required:

(a) State the two main conditions which must be satisfied by the pre-delivery inspection stage if it is to be regarded as a multiple server (M/M/c) queueing situation.

(b) Show that P_0 (the proportion of time that the workshop is idle) is 0.0256 and determine the average time which elapses between an order being placed and the car being delivered to the showroom.

(c) In view of the delays which have been occurring, the company accountant has been asked to advise whether it would be worthwhile employing 2 extra engineers in the workshop at a total cost of £220 per week. It has been estimated that this would reduce the average inspection time to $1\frac{1}{2}$ hours and the resulting earlier delivery would produce a saving of £2 per day for an average car as a result of earlier payment by the customer. Would you recommend that the 2 extra engineers should be employed?

You are reminded that, with an arrival rate of λ and a service rate of μ in *each* of c channels:

$$\text{Average time in queue} = \frac{(\rho c)^c P_0}{c!\,(1-\rho)^2 c\mu}$$

$$\text{Average time in system} = \frac{(\rho c)^c P_0}{c!\,(1-\rho)^2 c\mu} + \frac{1}{\mu}$$

where:

$$P_0 = \left\{ \sum_{i=0}^{c-1} \frac{(\rho c)^i}{i!} + \frac{(\rho c)^c}{c!\,(1-\rho)} \right\}^{-1}$$

and

$$\rho = \frac{\lambda}{c\mu}$$

(ACCA, December 1985)

Answers on page 608.

Exercise 13.9

In a large machine shop there are 50 identical machines, each manned by 1 operative and producing a certain component at a rate of 20 an hour. The demand for this component is such that all output can be sold at the current price of £1 per unit. The estimated production costs amount to £0.85 per unit and are broken down as follows:

	£/unit
Direct Materials	0.15
Direct Labour	0.20
Variable Overhead	0.15
Fixed Overhead	0.35

(The fixed overhead absorption rate is based on a production capacity of 40,000 units per week.)

When the machines break down, they are attended by a maintenance engineer who is responsible for all repairs to the 50 machines. Service records show that there have been, on average, 3 breakdowns a week during the past year and that these breakdowns have occurred randomly. Furthermore, it has been estimated that it takes, on average, 10 hours to repair a machine and that this time follows a negative exponential distribution. The cost of employing a maintenance engineer is £160 a week and the cost of employing a machine operator is £140 a week for a 40 hour week. Overtime is not permitted.

Required:

(a) According to the work's accountant, all direct labour costs must be regarded as fixed, and 60% of the variable overheads are still incurred even when machines are idle. On this basis, estimate the average weekly cost of idle time (ie the opportunity cost of lost output) on the 50 machines.

(b) A suggestion has been made by the production manager that savings could be introduced if a second maintenance engineer were employed to work with the existing engineer as a two man team. If this were done, the average repair time could be cut to 8 hours. What would be your advice? Explain briefly how the situation would be different if the second maintenance engineer worked independently of the first.

(c) Assuming that the second maintenance engineer is *not* employed, the production manager has further suggested that the number of operatives should be reduced, as there usually seems to be at least one machine idle. Determine how many operatives should be employed to achieve a maximum profit.

(ACCA June 1983)

Answers on page 609.

Exercise 13.10

The manager of a large supermarket is concerned about the long delays which are occurring to the lorries waiting to deliver goods to the supermarket. Occasionally there are as many as 100 deliveries a week, and in some cases the lorry drivers have had to wait several hours before they can unload at the supermarket's one unloading bay. This has resulted in congestion in the streets surrounding the supermarket and frequent complaints from the lorry drivers. You have been asked to make recommendations for improving the situation and, as a first step, have collected the following data relating to the deliveries which occurred last week:

Number of lorries arriving per hour	Number of hours	Unloading time (minutes)	Number of lorries
0	7	0– 20	38
1	10	20– 40	26
2	8	40– 60	10
3	8	60– 80	3
4	5	80–100	2
5	2	100–120	1

Deliveries of goods are permitted between 9 am and 5 pm, Monday to Friday. Any lorry that has arrived by 5 pm can join the queue of lorries that are waiting to be unloaded. If necessary, the staff of the unloading bay work overtime each evening to clear the queue of lorries which arrived before 5 pm.

Required:
- (a) Explain carefully the conditions which must be satisfied in order to apply the basic single server queueing model (M/M/1) to this situation. Explain also how you would use the data which have been collected to test the appropriateness of the required assumptions. (It is not necessary to actually perform the tests you have described.)
- (b) Assuming that an M/M/1 model is appropriate, estimate how many lorries, on average, are waiting to be unloaded and also the time that a lorry would expect to spend at the supermarket.
- (c) The unloading bay is currently staffed by 2 employees who are each paid £100 for a 40 hour week, with any overtime being paid at time and a third. A suggestion has been made that a third person should be employed in the unloading bay which, it has been estimated, would result in a saving of 7 minutes in the average time to unload a lorry. This, it has been claimed, would not only reduce the lorry waiting time but would also produce a saving in cost to the supermarket. Analyse this suggestion and make a recommendation.

(ACCA, June 1984)

Answers on page 610.

Exercise 13.11

A large firm of accountants and financial advisors uses a mini-computer to handle routine tasks such as calculations and data searches. Requests to perform these tasks are made on several office terminals linked to the mini-computer. The time to process any particular task is, on average, 7.5 minutes while the number of requests arriving at the computer each hour is given by the following frequency table.

Number of tasks arriving in one hour	Frequency
0– 2	8
3– 5	34
6– 8	44
9–11	12
12–14	1
15	1

Required:
- (a) Without carrying out the test, explain how you would perform a statistical test to find out whether or not tasks arrive at the mini-computer according to the Poisson distribution.
- (b) It is assumed that the queueing process satisfies an M/M/1 model. Explain the conditions that are necessary for this assumption.
- (c) Calculate the following operating characteristics of the system:
  - (i) The proportion of the time the mini-computer is busy.
  - (ii) The average time a task is in the system.
- (d) Due to employee waiting time it has been estimated that for each hour a task is in the system there is a 'task system cost' to the firm of approximately £5. An additional processor could be hired for £8,000 per year and would result in a reduction in time to process a task of, on average, 1.5 minutes. Assuming a 40 hour week and a 50 week working year decide whether or not it is worthwhile for the firm to purchase this processor. For what values of the 'task system cost' would it be profitable for the company to purchase the processor?

(ACCA, December 1987)

Answers on page 612.

Simulation

14.1 INTRODUCTION

In the previous two chapters we have used mathematical models to represent stock and queueing situations. In both cases, the problems must be straightforward and must fit all of the basic assumptions implicit in the models. However, it was also mentioned that once the problems become more complex, the solutions provided by these mathematical models are inadequate. We must look to some other kind of technique which can cope with situations which do not conform to the assumptions built into the simple models.

Simulation takes over when the mathematical, analytical models fail or become too complex. It is a very flexible and powerful, if inelegant, technique. Simulation replicates the operation of a system, step by step. The system may contain a number of stochastic variables. For example, in a stock system, the demand per year and the lead time may both be subject to uncertainty. In a queueing system, the time between the arrival of units in the system and the time taken to serve such units may not be Poisson processes. The times may follow other distributions or simply be described by their observed frequency distribution.

Using sample data, we can simulate the behaviour of the system. If the simulation is carried on for long enough, recurring patterns may emerge, or we can calculate the expected values of certain parameters. The simulation gives us a clear view of how the systems can be expected to behave.

We will concentrate on the method which is known as **Monte Carlo simulation**. All variables are assigned discrete values, even though they may, in fact, be continuous. Time, for example, is split into minute intervals, or hours or days, depending on the simulation. The probability of each value is found and random numbers are used to select values of the variable from the probability distribution. This procedure generates a series of values from which the simulation is constructed.

In simulation, as in most of the operational research techniques which we have discussed, a computer would normally be used to build and run the model. This is particularly important in this area, since meaningful information can be extracted from the simulation only after a number of runs with different random numbers. If we are interested in the steady state of the model, the simulation must be allowed to proceed for a long period of simulated time so that average values of the relevant statistics may be calculated. If the period is too short, the initial start-up fluctuations can affect the mean values.

It is not possible to show realistic simulations in this book but the basic principles of the method remain the same.

14.2 THE PRINCIPLES OF THE DISCRETE SIMULATION MODEL

A market research organisation is putting together a proposal for a client who wants to assess public opinion on a matter of topical interest. The client wishes to know how long the work will take and how much it will cost. The first stage is to design the sample and the questionnaire. The second stage is to collect the data. Let us assume that a suitable questionnaire has been prepared and a sampling plan chosen. It has been decided that the data will be collected by an interviewer questioning pedestrians in a town centre. The time and costs will depend on how long it takes the interviewer to collect the data. How can the organisation assess how long this will be?

We must analyse the situation. The interviewer will stop passers-by, ask them if they are willing to be interviewed and, if they agree, the interview will take place. The variables are:

1   The interviewer has to wait for someone to pass by. We need to know the **inter-arrival times (IAT)** of the passers-by;
2   the willingness of the passer-by to be interviewed;
3   the duration of the interview.

If we can generate some data which represents passers-by stopping and perhaps being interviewed, we can simulate the problem and derive the time it would take to conduct a given number of interviews. The data must reflect the normal pattern of the variables we have identified. Each one of these variables is stochastic, that is, it is subject to uncertainty. The simplest way is to collect some data in a trial run.

If we take as a trial 100 passers-by, we can record the time intervals between their arrivals, whether they are willing to be interviewed and, if they are, how long the interview takes. The precision of recording will depend on the problem. In this case, it is not necessary for the times to be recorded with very great precision. It is at this stage that we decide on the discrete values of time to be used. For example, the time between successive passers-by is required to the nearest minute, and the time each interview takes to the nearest two minutes.

Once the data have been collected for the trial of 100 people, a frequency distribution can be drawn up for each variable and the corresponding probability distributions calculated. Let us suppose the following data are recorded:

TABLE 14.1 **Arrival pattern—inter-arrival times**

Time between arrivals at the interviewer, minutes	0	1	2	3	4	5
Number of occasions, f	25	35	18	10	8	4
Probability (f ÷ 100)	0.25	0.35	0.18	0.10	0.08	0.04

Of the people who were asked, 75 were willing to be interviewed. Therefore, the probability that any passer-by will agree to be interviewed may be taken as 0.75.

TABLE 14.2 **Interview times**

Interview time, minutes	2	4	6
Number of interviews, f	40	45	15
Probability (f ÷ 100)	0.40	0.45	0.15

How do we use these data to generate imaginary passers-by? One method is to use random number tables (see Appendix Two). The random number tables comprise the digits 0 to 9 selected in random order. The groupings in the tables are only for convenience of reading. The tables are used by selecting a place to start—anywhere will do. The digits are chosen singly, or in pairs or in threes as required, moving across or down the table. The random numbers are used by assigning a range of random numbers (eg 0–9, 00–99) to a range of values of the variable. The random numbers are allocated to the variable values in proportion to the probabilities.

A random number is selected from the tables and the corresponding value is assigned to the variable. Since the probabilities in this problem are to two decimal places, we will use two-digit random numbers. The random number range 00–99 is divided up as shown below:

TABLE 14.3 **Allocation of random number ranges to inter-arrival times**

Time between arrivals, minutes (IAT)	Probability	Cumulative probability	Random number
0	0.25	0.25	00–24
1	0.35	0.60	25–59
2	0.18	0.78	60–77
3	0.10	0.88	78–87
4	0.08	0.96	88–95
5	0.04	1.00	96–99

This means that if we select the random number 03, it is in the range (00-24) and represents an inter-arrival time (IAT) of zero minutes. The random number 47 is in the group (25–59) and represents an IAT of one minute. By taking the random numbers systematically along a row or down a column of the table, we use the above data to give each person an inter-arrival time. These values of IAT are accumulated, starting at time zero, to give an arrival time for each passer-by.

TABLE 14.4 **Allocation of random number ranges to willingness to be interviewed**

Passer-by willing to be interviewed	Probability	Cumulative probability	Random number
Yes	0.75	0.75	00–74
No	0.25	1.00	75–99

To find if a simulated passer-by is willing to be interviewed, we select a random number, from another column or row of the table. Let us assume that it is 35. This is in the range (00–74), so this person is willing to be interviewed. If the next number is 64, this falls in the same range so this person is also willing to be interviewed.

TABLE 14.5 **Allocation of random number ranges to interview times**

Interview time, minutes	Probability	Cumulative Probability	Random number
2	0.40	0.40	00–39
4	0.45	0.85	40–84
6	0.15	1.00	85–99

The interview times are found in the same way but using a different set of random numbers.

We are now ready to begin the simulation. We will continue until 10 interviews have been conducted. Random numbers are selected for each variable and values are generated for the variable needed to progress the simulation (the arrival time) and for the variables needed to describe the behaviour of the system (willingness to be interviewed and the interview time).

For convenience, an extract from the random number tables is given below. This will help you to follow the simulation.

```
03 47 43 73 86 97 74 24 67 62 16 76 62 27 66 12 56 85 99 26 55 59 56
35 64 16 22 77 94 39 84 42 17 53 31 63 01 63 78 59 33 21 12 34 29 57
```

For the inter-arrival times, we will choose the random numbers from the beginning and work across the row. The series begins 03 47 43. The random numbers for people's willingness to be interviewed are taken from the second row, beginning with 35 64 16. The random numbers for the interview times are also taken from the second row but starting at the end and working backwards, 57 29 34. We will assume that the simulation clock starts at time zero. The first passer-by arrives at time (0 + the first inter-arrival time). We will also assume that an interview can end and the next one begin immediately.

The tenth interview will finish 36 minutes after the start of the session. A different set of random numbers will produce a different result. If we require an estimate for the time taken to complete 10 interviews, we must let the simulation run for a large number of interviews—100 or 200, for example. An average time for 10 interviews can then be calculated.

One of the problems with a simulation is knowing what information to record as the simulation progresses. In this small hand simulation we can note each step and go back over it if we change our minds about what is required. With a large computer simulation, it is important to decide at the outset what is needed and how the data should be collected and presented.

406 QUANTITATIVE ANALYSIS

TABLE 14.6 **Simulation of 10 interviews using 1 interviewer**

Passer-by number	Arrival pattern Random number	IAT minutes	Arrival time, minutes	Willingness Random number	Stop Yes/No	Interview pattern Random number	Duration minutes	Time Start	Time End
1	03	0	0	35	yes	57	4	0	4
2	47	1	1	interviewer busy					
3	43	1	2	interviewer busy					
4	73	2	4	64	yes	29	2	4	6
5	86	3	7	16	yes	34	2	7	9
6	97	5	12	22	yes	12	2	12	14
7	74	2	14	77	no		refusal		
8	24	0	14	94	no		refusal		
9	67	2	16	39	yes	21	2	16	18
10	62	2	18	84	no		refusal		
11	16	0	18	42	yes	33	2	18	20
12	76	2	20	17	yes	59	4	20	24
13	62	2	22	interviewer busy					
14	27	1	23	interviewer busy					
15	66	2	25	53	yes	78	4	25	29
16	12	0	25	interviewer busy					
17	56	1	26	interviewer busy					
18	85	3	29	31	yes	63	4	29	33
19	99	5	34	63	yes	01	2	34	36

10 interviews completed

There are two main purposes in collecting the data. We can use it to check that the model is behaving as we would expect. This is part of the model validation. For example, from the input distribution, the expected interview time is:

$$E(\text{interview time}) = 2 \times 0.4 + 4 \times 0.45 + 6 \times 0.15 = 3.5 \text{ minutes.}$$

From our short simulation, the interviewer spent 28 minutes carrying out the 10 interviews, giving a mean of 2.8 minutes, which is shorter than we might expect. This variation is not surprising for such a small sample. If we obtain the same result from the first 100 interviews, however, it would suggest that the model was somehow incorrect and required careful checking.

The second use of the data is to obtain information from the model. For example, how long does it take to complete 10 interviews?—36 minutes. For what proportion of the time is the interviewer idle?—8 minutes out of 36. How many people passed before 10 interviews were obtained?—19: 6 when the interviewer was busy, 3 refused and 10 accepted.

We can extend the investigation by introducing, for example, a second interviewer. The cost of the second interviewer can be balanced against the reduction in the time taken to do 10 interviews.

The introduction of the second interviewer forces us to make additional rules for operating the simulation. What happens if both interviewers are free? Who approaches the next passer-by? We will assume that Interviewer 1 always selects a passer-by first. It is usually necessary to make rules such as this at the start of any simulation. In this case, it does not matter which interviewer is chosen but in all simulations it is important to be consistent and to formalise the rules for operating the system.

We now find that 10 interviews require 27 minutes. Again, in practice, an average value is calculated from a much longer simulation. In order to make a decision about whether to use a second interviewer, other data, such as the idle time of the interviewers, will be required.

Once the simulation model has been constructed, it is important to validate it. We must be sure that the model accurately reflects the situation it is representing. The simplest way to validate is to use historical data and compare the results of the simulation with the way the system actually behaved at

TABLE 14.7 **Simulation of 10 interviews using 2 inviewers**

Passer-by number	Arrival pattern Random number	IAT minutes	Time minutes	Willingness Random number	Stop Yes/No	Interview pattern Random number	Duration minutes	Number 1 Start	End	Number 2 Start	End
1	03	0	0	35	yes	57	4	0	4		
2	47	1	1	64	yes	29	2			1	3
3	43	1	2	both interviewers busy							
4	73	2	4	16	yes	34	2	4	6		
5	86	3	7	22	yes	12	2	7	9		
6	97	5	12	77	no	refusal					
7	74	2	14	94	no	refusal					
8	24	0	14	39	yes	21	2	14	16		
9	67	2	16	84	no	refusal					
10	62	2	18	42	yes	33	2	18	20		
11	16	0	18	17	yes	59	4			18	22
12	76	2	20	53	yes	78	4	20	24		
13	62	2	22	31	yes	63	4			22	26
14	27	1	23	both interviewers busy							
15	66	2	25	63	yes	01	2	25	27		

10 interviews complete

the time. In the example above, past data may not be available and the validation must hinge on careful checking and evaluation of the probability distributions used. This is a very important stage in simulation modelling which it is tempting to neglect.

14.3 THE APPLICATION OF A SIMULATION MODEL TO A QUEUEING PROBLEM

The precise nature of the simulation will depend on the details of the problem. We will look at two different types of queueing problem to illustrate the general procedure.

Example 14.1: To illustrate the use of simulation in a queueing problem

Doctor Abbott and Doctor Booth jointly hold a morning surgery which begins at 9.00am. The door to the waiting room is opened at 8.30am and closed again at 10.00am. The receptionist has been keeping records of the arrival pattern of patients over the past few weeks and the doctors have also recorded their consultation times. The arrival pattern of patients is:

TABLE 14.8 **Arrival pattern for patients**

Inter-arrival time, minutes	1	2	3	4	5	6	7	8
Probability	0.05	0.05	0.10	0.20	0.40	0.10	0.05	0.05

Half the patients are registered with Dr Abbott and half with Dr Booth and they form two separate queues which operate on a first-in-first-out (FIFO) principle. If, however, the other doctor is free when it is their turn, 90% of patients are willing to see the other doctor. Each doctor has the following distribution of consultation times.

TABLE 14.9 **Consultation times—service pattern**

Consultation time, minutes	6	8	10	12	14
Probability	0.10	0.20	0.50	0.10	0.10

Each patient receives the same consultation time whichever doctor he sees. If we thought it more appropriate, we could have two different consultation distributions, one for each doctor.

Use simulation to examine the flow of patients during the morning surgery and to estimate the following:

1 What is the number of patients waiting at 9.00am?
2 What is the average time patients have to wait?
3 What is the time the last patient leaves each doctor?

Solution: The stochastic variables in the problem are:

(a) the inter-arrival times, from which we calculate the arrival time of each patient;
(b) the doctor the patient belongs to;
(c) the patient's willingness to see the other doctor, if he is free;
(d) the consultation time, which we are assuming depends on the patient, not on the doctor seen.

We allocate random numbers to each value of the variables.

TABLE 14.10 **Inter-arrival time, minutes**

Value minutes	Probability	Cumulative probability	Random number
1	0.05	0.05	00–04
2	0.05	0.10	05–09
3	0.10	0.20	10–19
4	0.20	0.40	20–39
5	0.40	0.80	40–79
6	0.10	0.90	80–89
7	0.05	0.95	90–94
8	0.05	1.00	95–99

TABLE 14.11 **Consultation time, minutes**

Value minutes	Probability	Cumulative probability	Random number
6	0.10	0.10	00–09
8	0.20	0.30	10–29
10	0.50	0.80	30–79
12	0.10	0.90	80–89
14	0.10	1.00	90–99

TABLE 14.12 **Patient's doctor**

Doctor	Probability	Random number
A	0.5	0–4
B	0.5	5–9

TABLE 14.13 **Willingness to see other doctor**

	Probability	Random number
Yes	0.9	0–8
No	0.1	9

We are now ready to begin the simulation. We will start the simulation clock at 8.30am. The first patient arrives at 8.30am + the first inter-arrival time (IAT).

1 Five people were waiting at 9.00am.
2 Average patient waiting time is:

    Doctor Abbott—38.9 minutes for 11 patients
    Doctor Booth—24.9 minutes for 11 patients
        Overall—31.9 minutes, with a minimum of 4 and a maximum of 54 minutes

3 Last patient left Dr Abbott at 10.56am; last patient left Dr Booth at 10.46am

Before making use of this information, the doctors should simulate several morning surgeries and calculate average values for each of the statistics. They should begin to ask questions about how to provide a better service. Why do they open the surgery half an hour early? Could they operate an appointment system? Could they devise a fairer queueing discipline for the patients who

TABLE 14.14 **Simulation of doctors' morning surgery**

	Arrival IAT		Patient's doctor		See other doctor?		Time	Consultation Doctor Abbott		Doctor Booth		Patient waiting
RN	minutes	Time	RN	Type	RN	Y/N	RN	minutes	Start	End	Start	End	time, minutes
63	5	8:35	5	B	6	yes	69	10			9:00	9:10	25
27	4	8:39	4	A	2	yes	39	10	9:00	9:10			21
15	3	8:42	2	A	0	yes	39	10	9:10	9:20			28
99	8	8:50	2	A	4	yes	27	8			9:10	9:18	20
86	6	8:56	3	A	8	yes	85	12	9:20	9:32			24
71	5	9:01	1	A	3	yes	49	10	9:32	9:42			31
74	5	9:06	3	A	3	yes	90	14	9:42	9:56			36
45	5	9:11	3	A	7	yes	25	8	9:56	10:04			45
11	3	9:14	5	B	1	yes	84	12			9:18	9:30	4
02	1	9:15	7	B	1	yes	47	10			9:30	9:40	15
15	3	9:18	5	B	2	yes	42	10			9:40	9:50	22
14	3	9:21	5	B	5	yes	04	6			9:50	9:56	29
18	3	9:24	3	A	2	yes	83	12	10:04	10:16			40
07	2	9:26	1	A	9	no	03	6	10:16	10:22			50
14	3	9:29	7	B	0	yes	78	10			9:56	10:06	27
58	5	9:34	2	A	2	yes	87	12	10:22	10:34			48
68	5	9:39	7	B	1	yes	61	10			10:06	10:16	27
39	4	9:43	3	A	5	yes	82	12	10:34	10:46			51
31	4	9:47	8	B	2	yes	69	10			10:16	10:26	29
08	2	9:49	2	A	6	yes	33	10			10:36	10:46	47
13	3	9:52	4	A	0	yes	40	10	10:46	10:56			54
55	5	9:57	7	B	2	yes	64	10			10:26	10:36	29

waiting room closed

are willing to see either doctor? Each of these alternatives could be simulated to see the theoretical effect before trying them out on the patients.

Example 14.2: The simulation of a queueing problem

AMC Tyres Ltd sells and fits car tyres at a workshop based in the centre of a town. Customers arrive casually and no appointments are made. Customers who ring in advance are told to arrive when convenient. The time interval between the arrival of successive customers has been observed and the following data recorded.

TABLE 14.15 **Arrival patern of cars at AMC Tyres Ltd**

Time interval between arrivals, minutes	0	5	10	15	20	25	30	35
Probability	0.04	0.08	0.15	0.30	0.20	0.13	0.08	0.02

The time taken to examine and replace tyres has been recorded to the nearest minute. The times range between 21 minutes and 40 minutes; all values being equally likely.

At the moment, inside their workshop, AMC have one service bay with all the necessary equipment and parking for one additional car. There is further parking outside which will accommodate only one car. Parking is not allowed in the access road, so that any customer who arrives when the service bay is in use and there are two cars parked, cannot stop and is lost to the company. A lost customer represents an average cost to AMC of £50. With a minor re-organisation, a second service bay could be equipped inside the workshop but the parking there would be lost. This is not thought to be a problem since the queue length is likely to be a deterrent to a passing customer anyway. The operating cost of the second service bay would be £35 per hour. Simulate this problem for 25 potential customers. Should AMC install a second service bay?

Solution: The variables in the problem are:

1. The inter-arrival time of the customers. This is difficult to collect because we need the inter-arrival pattern of potential customers, many of whom are forced to drive past when the parking spaces are full.
2. The service time for the cars. There are 20 one minute intervals from 21 to 40, inclusive. If they are all equally likely, then each has a probability of 1/20 = 0.05 of occurring.

TABLE 14.16 **Allocate random number ranges to inter-arrival times**

IAT minutes	Probability	Cumulative probability	Random number
0	0.04	0.04	00–03
5	0.08	0.12	04–11
10	0.15	0.27	12–26
15	0.30	0.57	27–56
20	0.20	0.77	57–76
25	0.13	0.90	77–89
30	0.08	0.98	90–97
35	0.02	1.00	98–99

TABLE 14.17 **Allocate random number ranges to service times**

Time minutes	Probability	Cumulative probability	Random number
21	0.05	0.05	00–04
22	0.05	0.10	05–09
23	0.05	0.15	10–14
24	0.05	0.20	15–19
25	0.05	0.25	20–24
26	0.05	0.30	25–29
27	0.05	0.35	30–34
28	0.05	0.40	35–39
29	0.05	0.45	40–44
30	0.05	0.50	45–49
31	0.05	0.55	50–54
32	0.05	0.60	55–59
33	0.05	0.65	60–64
34	0.05	0.70	65–69
35	0.05	0.75	70–74
36	0.05	0.80	75–79
37	0.05	0.85	80–84
38	0.05	0.90	85–89
39	0.05	0.95	90–94
40	0.05	1.00	95–99

We will run the simulation with the single service bay and maximum queue length of three cars, starting the simulation clock at time zero. The first customer arrives at time (0 + the first IAT).

TABLE 14.18 **Simulation of the existing situation at AMC Tyres**

Car	Arrival pattern RN	IAT mins	Minutes after 0	Service bay	Park 1	Park 2	Service RN	Time mins	Start	End	Car waiting, minutes
A	44	15	15	A	—	—	95	40	15	55	0
B	22	10	25	A	B	—	23	25	55	80	30
C	78	25	50	A	B	C	10	23	80	103	30
D	84	25	75	B	C	D	76	36	103	139	28
E	26	10	85	C	D	E	30	27	139	166	54
F	04	5	90	C	D	E	F cannot join queue				
G	33	15	105	D	E	G	76	36	166	202	61
H	46	15	120	D	E	G	H cannot join queue				
I	09	5	125	D	E	G	I cannot join queue				
J	52	15	140	E	G	J	28	26	202	228	62
K	68	20	160	E	G	J	K cannot join queue				
L	07	5	165	E	G	J	L cannot join queue				
M	97	30	195	G	J	M	75	36	228	264	33
N	06	5	200	G	J	M	N cannot join queue				
O	57	20	220	J	M	O	39	28	264	292	44
P	16	10	230	M	O	P	30	27	292	319	62
Q	90	30	260	M	O	P	Q cannot join queue				
R	82	25	285	O	P	R	90	39	319	358	34
S	66	20	305	P	R	S	14	23	358	381	53
T	59	20	325	R	S	T	80	37	381	418	56
U	83	25	350	R	S	T	U cannot join queue				
V	62	20	370	S	T	V	52	31	418	449	48
W	64	20	390	T	V	W	91	39	449	488	59
X	11	5	395	T	V	W	X cannot join queue				
Y	12	10	405	T	V	W	Y cannot join queue				

During the simulation of 25 arrivals, 10 are turned away due to lack of space. This represents a cost of 10 × £50 = £500. They are the lucky ones. The waiting time averages about 50 minutes, designed to drive customers away.

We will now run the simulation again with 2 service bays and 1 parking space. The arrival times are taken from TABLE 14.18. The same service distribution is used but with different random numbers. To simplify the table the random numbers are not given.

We can see that the pattern is quite different this time. Only 2 arrivals are turned away. The customer waiting time is greatly reduced and the parking bay is often empty. The general pattern will lead to much greater customer satisfaction.

As far as the cost is concerned, we must combine the cost of the lost customers with the additional cost of the second service bay. It took 461 minutes, 7.68 hours, to deal with 25 arrivals. The total cost for the 25 arrivals is:

$$7.68 \times £35 + 2 \times £50 = £368.80$$

This is a saving of £131.20 compared to the single bay problem. However, we must remember that the simulation should be continued for much longer, and an average saving per hour calculated, before any decision is made.

In addition to providing information about potential savings, the simulation gives us a picture of the pattern of service. Average customer waiting times and idle times for the service bays can be determined.

TABLE 14.19 Simulation of the proposed system at AMC Tyres

Car	Arival minutes after 0	Car location Bay 1	Bay 2	Park	Time mins	Servicing Bay 1 Start	End	Bay 2 Start	End	Car waiting minutes
A	15	A	—	—	40	15	55			0
B	25	A	B	—	38			25	63	0
C	50	A	B	C	23	55	78			5
D	75	C	D	—	27			75	102	0
E	85	E	D	—	40	85	125			0
F	90	E	D	F	23			102	125	12
G	105	E	F	G	36	125	161			20
H	120	E	F	G		H cannot join queue				
I	125	G	I	—	37			125	162	0
J	140	G	I	J	26	161	187			21
K	160	G	I	J		K cannot join queue				
L	165	J	L	—	29			165	194	0
M	195	M	—	—	36	195	231			0
N	200	M	N	—	24			200	224	0
O	220	M	N	O	37			224	261	4
P	230	M	O	P	27	231	258			1
Q	260	Q	O	—	37	260	297			0
R	285	Q	R	—	39			285	324	0
S	305	S	R	—	23	305	328			0
T	325	S	T	—	36			325	361	0
U	350	U	T	—	32	350	382			0
V	370	U	V	—	35			370	405	0
W	390	W	V	—	39	390	429			0
X	395	W	V	X	21			405	426	10
Y	405	W	X	Y	35			426	461	21

14.4: THE APPLICATION OF A SIMULATION MODEL TO A STOCK CONTROL PROBLEM

A similar procedure to the one described above can be used to simulate complex stock problems. Uncertainties in both demand and lead time can be accommodated. Data must be collected which enables probability distributions to be constructed for the variables.

Example 14.3: To simulate a stock control problem

ELA plc manufacture cars. The batteries for their Lunar model are bought from an outside supplier. From past experience, ELA find that the weekly demand for the batteries can be approximated by a normal distribution with a mean of 500 and a standard deviation of 10, over the range 470 to 530.

There is an initial stock of 2000 batteries and the company has decided to order in batches of 2500 whenever the stock level falls below 1500 batteries. Again, past experience indicates that the time between the order being placed and the delivery, varies as follows:

TABLE 14.20 Lead time distribution, ELA plc

Lead time, weeks	1	2	3	4
Probability	0.20	0.50	0.25	0.05

The unit cost of holding is £0.50 per week, applied to the total stock held at each week ending. The cost associated with placing an order is £50 and the unit cost of being out of stock is put at £20 per week.

Use a simulation over a period of 20 weeks to estimate the average cost per week of the above policy. Assume that all accounting is done at the end of the week and that all ordering and delivery occurs at the beginning of the week.

Solution: The variables in the problem are the demand and the lead time. Since the demand is approximated by the continuous normal distribution, we will consider demand in steps of 5 batteries. For example, the probability of a demand for 510 batteries will be estimated by P(507.5 < demand < 512.5).

TABLE 14.21 **Allocate random number ranges to the lead time**

Lead time, weeks	Probability	Cumulative probability	Random number
1	0.20	0.20	00–19
2	0.50	0.70	20–69
3	0.25	0.95	70–94
4	0.05	1.00	95–99

TABLE 14.22 **Allocate random number ranges to weekly demand**

Demand /week	Probability	Cumulative probability	Random number
470	0.003	0.003	000–002
475	0.009	0.012	003–011
480	0.028	0.040	012–039
485	0.066	0.106	040–105
490	0.121	0.227	106–226
495	0.175	0.402	227–401
500	0.197	0.599	402–598
505	0.175	0.774	599–773
510	0.121	0.895	774–894
515	0.066	0.961	895–960
520	0.028	0.989	961–988
525	0.009	0.998	989–997
530	*0.003	1.000	998–999*

* slight discrepancy due to rounding

We can now carry out the simulation.

TABLE 14.23 **Stock control simulation—ELA plc**

Week number	Opening stock	Demand RN	Demand Amount	Closing stock	Re-order? Yes/No	Lead time RN	Lead time Weeks	Shortage
1	2000	034	480	1520				
2	1520	743	505	1015				
3	1015	738	505	510	Yes	95	4	
4	510	636	505	5				
5	5	964	520	0				515
6	0	736	505	0				505
7	2500	614	505	1995				
8	1995	698	505	1490				
9	1490	637	505	985	Yes	73	3	
10	985	162	490	495				
11	495	332	495	0				
12	2500	616	505	1995				
13	1995	804	510	1485				
14	1485	560	500	985	Yes	10	1	
15	3485	111	490	2995				
16	2995	410	500	2495				
17	2495	959	515	1980				
18	1980	774	510	1470				
19	1470	246	495	975	Yes	76	3	
20	975	762	505	470				
		Total	10050	22865				1020

Mean demand = 10,050/20 = 502.5 batteries/week
Mean closing stock = 22,865/20 = 1143.25 batteries/week
Mean shortage = 1020/20 = 51.0 batteries/week

Number of orders placed during the 20 week period = 4, therefore, mean number of orders/week = 4/20 = 0.2.

$$\text{The expected weekly cost} = 1143.25 \times £0.50 + 51 \times £20 + 0.2 \times £50$$
$$= £1602$$

As in the previous examples, the simulation should be continued for much longer to ensure that the steady state conditions have been reached.

This simulation can be used to investigate the behaviour of the stock system under a variety of ordering policies. The management can then select the policy which best satisfies its objectives.

SUMMARY

Simulation is one of the techniques which modellers use when mathematical models become too complex or their assumptions do not match reality. It can be used in complex situations without making assumptions about the input data.

We consider Monte Carlo simulation, in which all variables are converted into sets of discrete values. Data are collected and probability distributions constructed for these variables. Random numbers are used to sample from the distributions to give the values used in the simulation. For each model we define the variables involved and the rules which connect them. In small hand simulations, the results are presented in tables which can be analysed easily.

The model may be modified, re-run and the new results compared with the original ones. Simulation does not give an optimum solution in the way that linear programming does but it can guide us to a better solution. Before any output can be used, the model must be validated and the runs must be long enough to give representative results. Simulations are normally carried out using computer packages.

EXERCISES

Exercise 14.1

Butterby City Council operates a mini-bus service to take shoppers and tourists from the bus and railway stations to various locations in the city. The following data have been collected for the arrival of passengers at the bus stop outside the railway station:

Time between successive arrivals, minutes	0	1	2	3	4	5	6
Probability	0.04	0.16	0.24	0.28	0.16	0.10	0.02

The mini-buses are scheduled to run every 10 minutes but variation in traffic conditions results in the following distribution:

Time between successive buses, minutes	8	10	12	14	16
Probability	0.10	0.38	0.28	0.15	0.09

The number of empty seats on the bus is found to follow the distribution:

Number of empty seats	0	1	2	3	4	5	6
Probability	0.06	0.18	0.27	0.34	0.11	0.03	0.01

Required:
1. Simulate the arrival of 30 passengers at the bus stop. Assume that the simulation clock begins at time zero.
2. Estimate the average time a passenger must wait for a bus and the average length of the queue.

Answers on page 614.

Exercise 14.2

Repeat **Example 14.3** in section **14.4** but placing an order for new stock when the stock level falls below 1000 batteries. Compare the results with those in **Example 14.3**.

Answers on page 615.

Exercise 14.3

1 Describe the advantages and disadvantages of using simulation to investigate queueing situations compared with the use of queueing theory formulae.
2 The time between arrivals at a complaints counter in a large department store has been observed to follow the distribution shown below:

Time between arrivals (minutes)	Probability
0–4	0.25
4–8	0.45
8–12	0.20
12–16	0.10

Customers' complaints are handled by a single complaints officer but all customers who consider their complaints to be 'serious' or who have to wait 5 minutes or more before being seen by the complaints officer demand to see the store manager, who deals with them separately. The time to deal with complaints by the complaints officer has a normal distribution with a mean of 7 minutes and a standard deviation of 2 minutes. It is estimated that 20% of customers with complaints consider that their complaint is 'serious'.

Required:
(a) Using the above information, a table of random digits and the following table derived from the standard normal cumulative distribution table, describe how you would simulate the arrival and service flows of this system.

Random number	Number of deviations from mean	Random number	Number of deviations from mean
00–01	−2.5	61–77	+0.5
02–04	−2.0	78–88	+1.0
05–10	−1.5	89–94	+1.5
11–21	−1.0	95–97	+2.0
22–38	−0.5	98–99	+2.5
39–60	0		

(b) Use the following random digits to simulate the handling of 10 complaining customers, some of whom may have 'serious' complaints.

Inter-arrival time	09	06	51	62	83	61	59	20	82	68
Serious complaint	5	0	7	3	8	2	9	8	1	6
Service time	39	60	50	31	02	02	83	90	71	16

(c) Use your simulation to estimate the proportion of customers who eventually see the store manager. Hence estimate the total amount of time that the store manager spends dealing with complaints, assuming an 8 hour day and that the time he spends per complaint has the same distribution as that of the complaints officer.

(d) Explain briefly how to use simulation to decide if it would be worth while employing an additional complaints officer.

(ACCA, June 1988)

Answers on page 616.

QUANTITATIVE ANALYSIS

Exercise 14.4

Romulus Products plc operates 30 day terms for all its customers. Experience has shown that 80% of all accounts are settled within one month and 70% of the remainder are settled during the second month after the customer has been sent a standard 'overdue account' letter. Of those accounts still unpaid after 2 months, 50% are settled during the third month after a 'final demand' has been sent.

Any accounts still not paid after 3 months are dealt with in one of two ways. If the amount owing exceeds £1000, the company institutes legal proceedings to recover the money. Taking into account the legal costs involved, the proportion of the original sum owing which is ultimately recovered varies as follows:

Proportion recovered	Probability
0%– 40%	0.1
40%– 60%	0.3
60%– 80%	0.4
80%–100%	0.2

This process takes a further 3 months before payment is finally received.

If the amount owing is less than £1000, the debt is sold to a debt collecting company in return for 50% of the sum involved, which is obtained after a further month, ie at the end of month 4.

In recent months, the size of the accounts issued by Romulus is shown by the following distributions:

Size of account	Probability
£0–£ 200	0.1
£200–£ 500	0.2
£500–£1,000	0.3
£1,000–£2,000	0.3
£2,000–£5,000	0.1

You may assume that there is no relationship between the size of the account when it is settled and the proportion recovered, and that all accounts are settled on the last day of the month. The company's cost of capital is the equivalent of 1.5% per month.

Required:
1. What is the probability that, for any particular account, payment is received at the end of:
   (a) the second month
   (b) the third month
   (c) the fourth month
   (d) the sixth month?
2. What is the expected present value of a new account which has £2000 outstanding?

   Monthly discount factors are:

Month	1	2	3	4	5	6
Factor	0.9852	0.9707	0.9563	0.9422	0.9283	0.9145

3. Show how the system as a whole may be simulated by using the following random digits to determine the present value of two simulated accounts.

Account 1	8	8	7	5	7
Account 2	9	9	8	2	9

(ACCA, December 1986)

Answers on page 618.

Exercise 14.5

Mentor Products plc are considering the purchase of a new computer controlled packing machine to replace the 2 machines which are currently used to pack product X. The new machine would result in reduced labour costs because of the more automated nature of the process and, in addition, would permit production levels to be increased by creating greater capacity at the packing stage. With an anticipated rise in the demand for product X, it has been estimated that the new machine will lead to increased profits in each of the next 3 years. Due to uncertainty in demand however, the annual cash flows (including savings) resulting from purchase of the new machine cannot be fixed with certainty and have therefore been estimated probabilistically as follows:

Annual cash flows (£'000)

Year 1	Prob	Year 2	Prob	Year 3	Prob
10	0.3	10	0.1	10	0.3
15	0.4	20	0.2	20	0.5
20	0.3	30	0.4	30	0.2
		40	0.3		

Because of the overall uncertainty in the sales of product X, it has been decided that only 3 years' cash flows will be considered in deciding whether to purchase the new machine. After allowing for the scrap value of the existing machines, the net cost of the new machine will be £42,000. The effects of taxation should be ignored.

Required:
1. Ignoring the time value of money, identify which combinations of annual cash flows will lead to an overall negative net cash flow, and determine the total probability of this occurring.
2. On the basis of the average cash flow for each year, calculate the net present value of the new machine given that the company's cost of capital is 15%. Relevant discount factors are as follows:

Year	Discount factor
1	0.8696
2	0.7561
3	0.6575

3. Analyse the risk inherent in this situation by simulating the net present value calculation. You should use the random numbers given at the end of the question to simulate 5 sets of cash flows. On the basis of your simulation results, what is the expected net present value and what is the probability of the new machine yielding a negative net present value?

	Set 1	Set 2	Set 3	Set 4	Set 5
Year 1	4	7	6	5	0
Year 2	2	4	8	0	1
Year 3	7	9	4	0	3

(ACCA, December 1985)

Answers on page 620.

Exercise 14.6

The personnel manager of the Orpheus Life Assurance Company has recently been analysing the turnover of secretarial and clerical staff at the company's head office. Currently there are three grades of secretarial and clerical staff (SC1, SC2 and SC3) and it is the company's policy to recruit new staff only into the lowest grade SC1. To maintain appropriate numbers of staff in the two higher grades, promotions are made on a basis of strict seniority (ie length of service) from the grade immediately below. In the past, a reasonably high turnover of staff has meant that most employees have been able to progress to the highest grade (SC3) within a few years. More recently, however, the turnover of staff has fallen and as a consequence, there have been fewer vacancies in the two higher grades which has slowed down the rate of promotion.

The number of employees currently required in each of the three grades is:

$$SC1:54$$
$$SC2:48$$
$$SC3:32$$

and, on the basis of present figures, the distributions of 'wastage' due to employees leaving the company are as follows:

Number of employees leaving per year

Grade SC1	Probability	Grade SC2	Probability	Grade SC3	Probability
0– 4	0.1	0– 4	0.2	0– 4	0.4
5– 9	0.3	5– 9	0.5	5– 9	0.4
10–14	0.4	10–14	0.3	10–14	0.2
15–19	0.2				

Required:

1. Using the random numbers given at the end of the question, simulate wastage and recruitment over a 5 year period assuming that there is no change in the number of employees required in each grade.
2. From your simulation, what is the average number of new employees recruited per year and the average number of promotions from SC1 and SC2? How do these figures compare with the expected values based on the given probability distributions?
3. Explain in detail how the procedure you have used in **1** would have to be changed in order to simulate the length of time for an employee to reach grade SC3.

Random Numbers SC1:32882
SC2:54163
SC3:95221

(ACCA, December 1984)

Answers on page 621.

Solutions to exercises

Solutions to Chapter One exercises

ANSWERS TO EXERCISES

Exercise 1.1

The outcomes may be illustrated by a tree diagram.

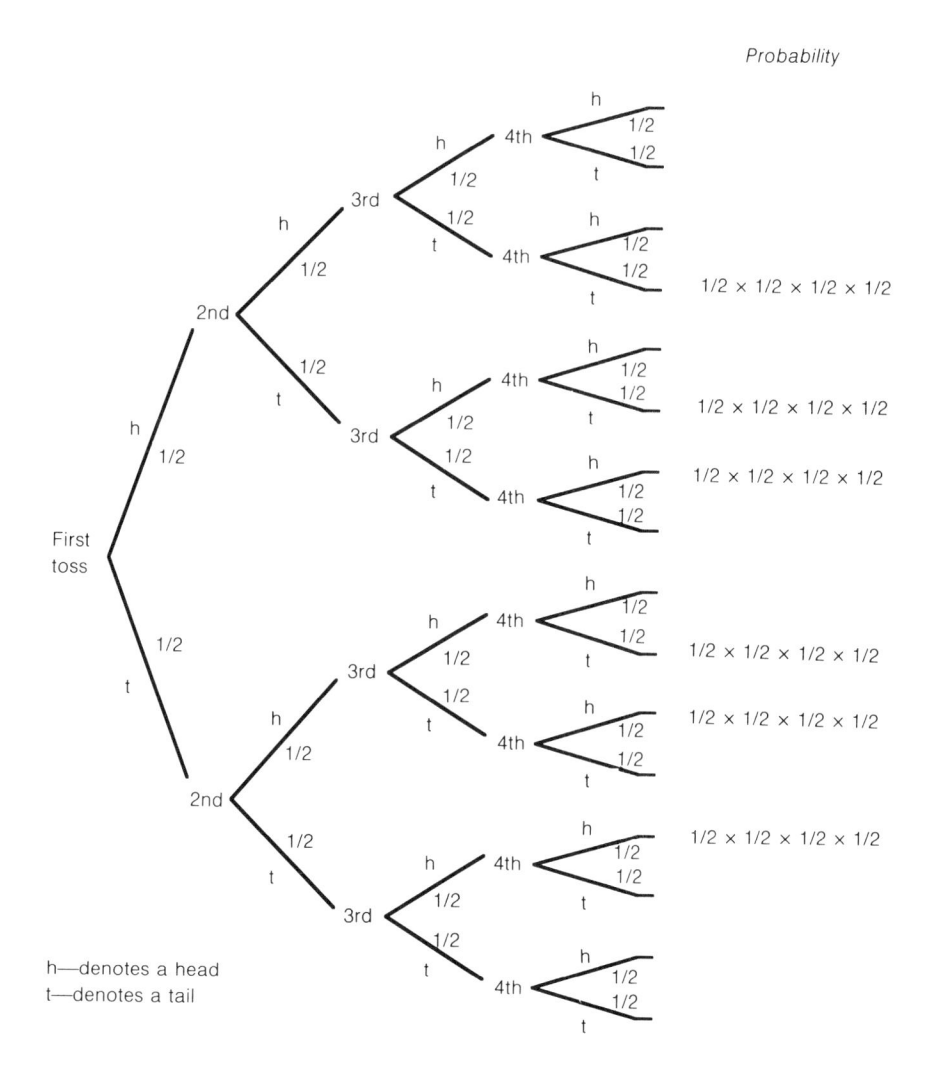

The probabilities are the same at the end of all the routes through the tree.

(i)   P(hhtt) = 1/2 × 1/2 × 1/2 × 1/2 = 1/16.

(ii) If the order of occurrence does not matter, we find that there are 6 routes through the tree which give 2 heads and 2 tails. Hence, P(2 heads and 2 tails in any order) = 6 ways × 1/16 = 3/8.

Exercise 1.2

The possible outcomes may be tabulated as follows:

Score on first toss

		1	2	3	4	5	6
	1	1	2	3	4	5	6
Score	2	2	4	6	8	10	12
on	3	3	6	9	12	15	18
second	4	4	8	12	16	20	24
toss	5	5	10	15	20	25	30
	6	6	12	18	24	30	36

(i) There are no outcomes with the value 23, therefore $P(23) = 0$.

(ii) There are 4 outcomes with the value 12, therefore since all outcomes are equally likely:

$$P(12) = \frac{\text{number of successful outcomes}}{\text{total number of outcomes}} = \frac{4}{36}$$

Exercise 1.3

Department	Female	Male	Total
Production	6	20	26
Maintenance	3	10	13
Stores	5	5	10
Transport	2	8	10
Sales	5	10	15
Total	21	53	74

For each selection, the probability is:

$$\frac{\text{number of successful choices}}{\text{total number of possible choices}}$$

(i) $P(\text{female}) = 21/74$

(ii) $P(\text{maintenance}) = 13/74$

(iii) $P(\text{male from stores}) = 5/74$

(iv) $P(\text{female from stores or transport}) = P(\text{female from stores}) + P(\text{female from transport})$
$= 5/74 + 2/74 = 7/74$

(v) $P(\text{person from production or sales}) = P(\text{person from production}) + P(\text{person from sales})$
$= 26/74 + 15/74 = 41/74$

(vi) $P(\text{female and female}) = 21/74 \times 20/73 = 0.0777$

(vii) $P(\text{production and production}) = 26/74 \times 25/73 = 0.1203$

(viii) $P(\text{sales and transport}) = 2 \times 15/74 \times 10/73 = 0.0555$

(ix) P(female from maintenance and male from stores) = 2 × 3/74 × 5/73 = 0.0056

(x)   P(both females from stores) + P(person from production and a male from sales)
      = (5/74 × 4/73) + (2 × 26/74 × 10/73) = 0.09996

Exercise 1.4

Select two invoices:

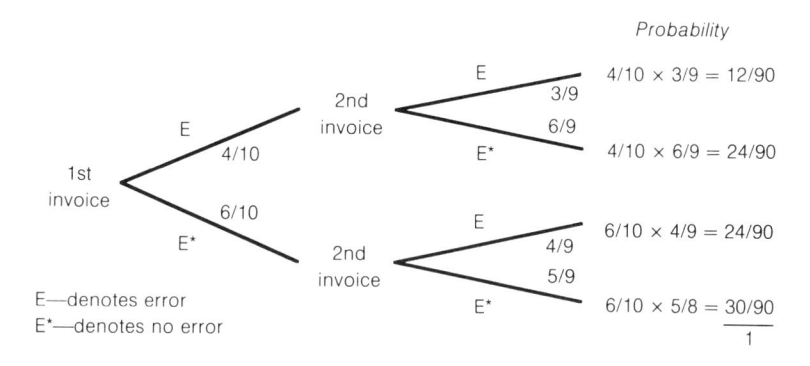

Probability

1st invoice → E (4/10) → 2nd invoice → E (3/9): 4/10 × 3/9 = 12/90
                                     → E* (6/9): 4/10 × 6/9 = 24/90
            → E* (6/10) → 2nd invoice → E (4/9): 6/10 × 4/9 = 24/90
                                      → E* (5/9): 6/10 × 5/8 = 30/90

E—denotes error
E*—denotes no error

(i)   P(both in error) = 4/10 × 3/9 = 2/15
(ii) P(no error in one and error in the other) = 4/10 × 6/9 + 6/10 × 4/9 = 8/15

Two invoices are examined:

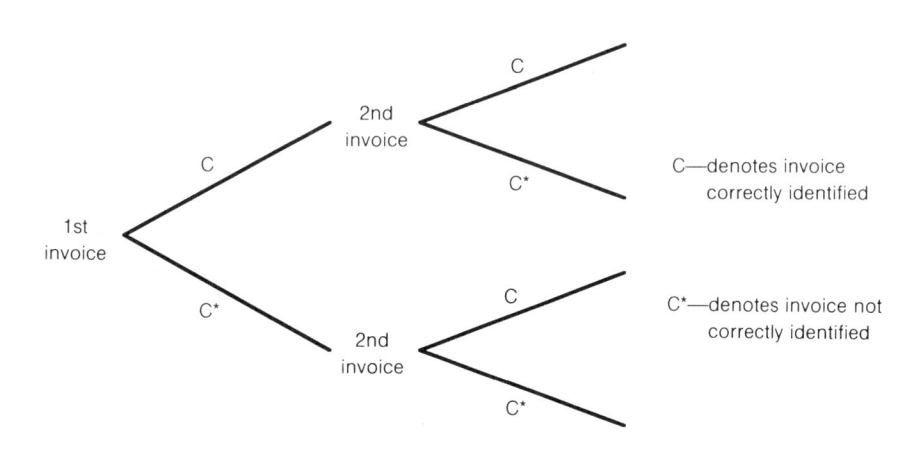

1st invoice → C → 2nd invoice → C
                              → C*
            → C* → 2nd invoice → C
                               → C*

C—denotes invoice correctly identified
C*—denotes invoice not correctly identified

(iii) Both invoices are in error, therefore the probability of the trainee correctly identifying each one is 0.8

   P(both correctly identified) = 0.8 × 0.8 = 0.64

(iv) One invoice is correct, therefore the probability of the trainee correctly identifying it is 0.9.
      One invoice is in error, therefore the probability of the trainee correctly identifying it is 0.8.

   P(both correctly identified) = 0.9 × 0.8 = 0.72

424 QUANTITATIVE ANALYSIS

Exercise 1.5

Accept the batch if all 5 items are satisfactory. Part of the probability tree is shown below:

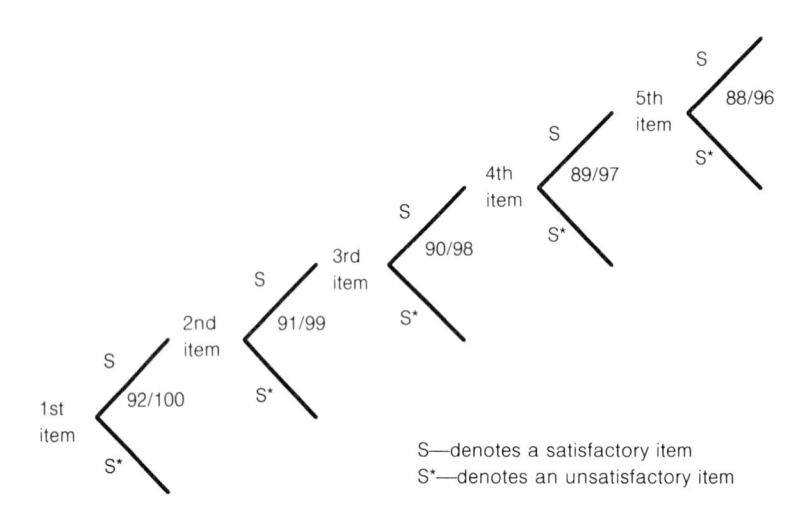

S—denotes a satisfactory item
S*—denotes an unsatisfactory item

$$P(\text{batch accepted}) = P(\text{all 5 items are satisfactory})$$
$$= P(\text{1st is satisfactory}) \times P(\text{2nd is satisfactory}|\text{1st is satisfactory}) \times \cdots$$
$$\times P(\text{5th is satisfactory}|1,2,3,4 \text{ are satisfactory})$$
$$= 92/100 \times 91/99 \times 90/98 \times 89/97 \times 88/96$$
$$= 0.6532$$

Exercise 1.6

In this example, each stage is independent of the others, therefore at every stage, $P(\text{error}) = 0.002$ and $P(\text{no error}) = 0.998$. Again, part of the probability tree is shown below:

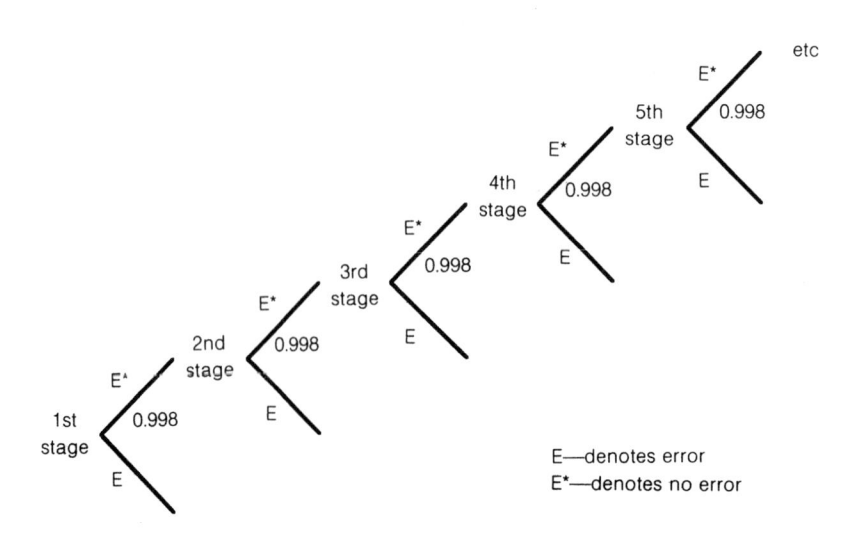

E—denotes error
E*—denotes no error

(i) $P(\text{completely accurate}) = P(\text{all 10 stages are error free})$.
Using the multiplication law with independent events:

$$P(\text{all 10 stages are error free}) = 0.998 \times 0.998 \times 0.998 \times \cdots \times 0.998$$
$$= 0.998^{10} = 0.9802$$

(ii) P(one error) = P(9 stages are error free and one stage is in error).
There are 10 possible ways of achieving this, since any one of the 10 stages could be the one in error.
Hence: P(one error) = $10 \times 0.998^9 \times 0.002 = 0.0196$

(iii) Reduce to 5 stages: P(error at each stage) = 0.003
Repeating the calculations in (i) and (ii) above:

$$P(\text{completely accurate}) = 0.997^5 = 0.9851$$
$$P(\text{one error only}) = 5 \times 0.997^4 \times 0.003 = 0.0148$$

The reorganisation has improved the probability that the documentation will be completely accurate and has reduced the probability of one error occurring.

Exercise 1.7

The probability tree for this problem is:

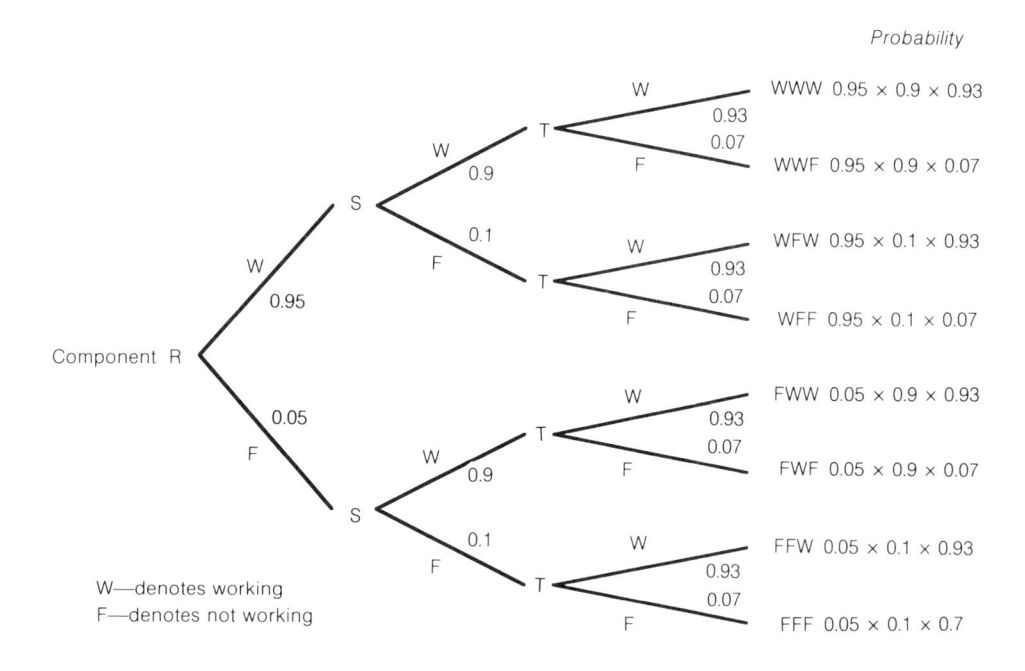

	Probability
WWW	$0.95 \times 0.9 \times 0.93$
WWF	$0.95 \times 0.9 \times 0.07$
WFW	$0.95 \times 0.1 \times 0.93$
WFF	$0.95 \times 0.1 \times 0.07$
FWW	$0.05 \times 0.9 \times 0.93$
FWF	$0.05 \times 0.9 \times 0.07$
FFW	$0.05 \times 0.1 \times 0.93$
FFF	$0.05 \times 0.1 \times 0.7$

W—denotes working
F—denotes not working

(i) P(system will work all year) = P(R works) × P(S works) × P(T works)
= $0.95 \times 0.9 \times 0.93 = 0.7952$

(ii) P(system will work all year) = P(2 or 3 components work)
= P(all 3 work) + P(2 only work)
From the tree, P(2 only work) = $0.95 \times 0.9 \times 0.07 + 0.95 \times 0.1 \times 0.93$
$+ 0.05 \times 0.9 \times 0.93$
= 0.19005
hence, P(system will work all year) = $0.7952 + 0.19005 = 0.9853$

(iii) P(system will work all year) = P(1, 2 or 3 components work)
P(system does not work all year) = P(no components work)
P(no components work ie all 3 fail) = $0.05 \times 0.1 \times 0.07$
= 0.00035
Hence, P(system works all year) = 1 − P(system does not work)
= 1 − 0.00035 = 0.99965

426 QUANTITATIVE ANALYSIS

Exercise 1.8

The probability tree for this problem is:

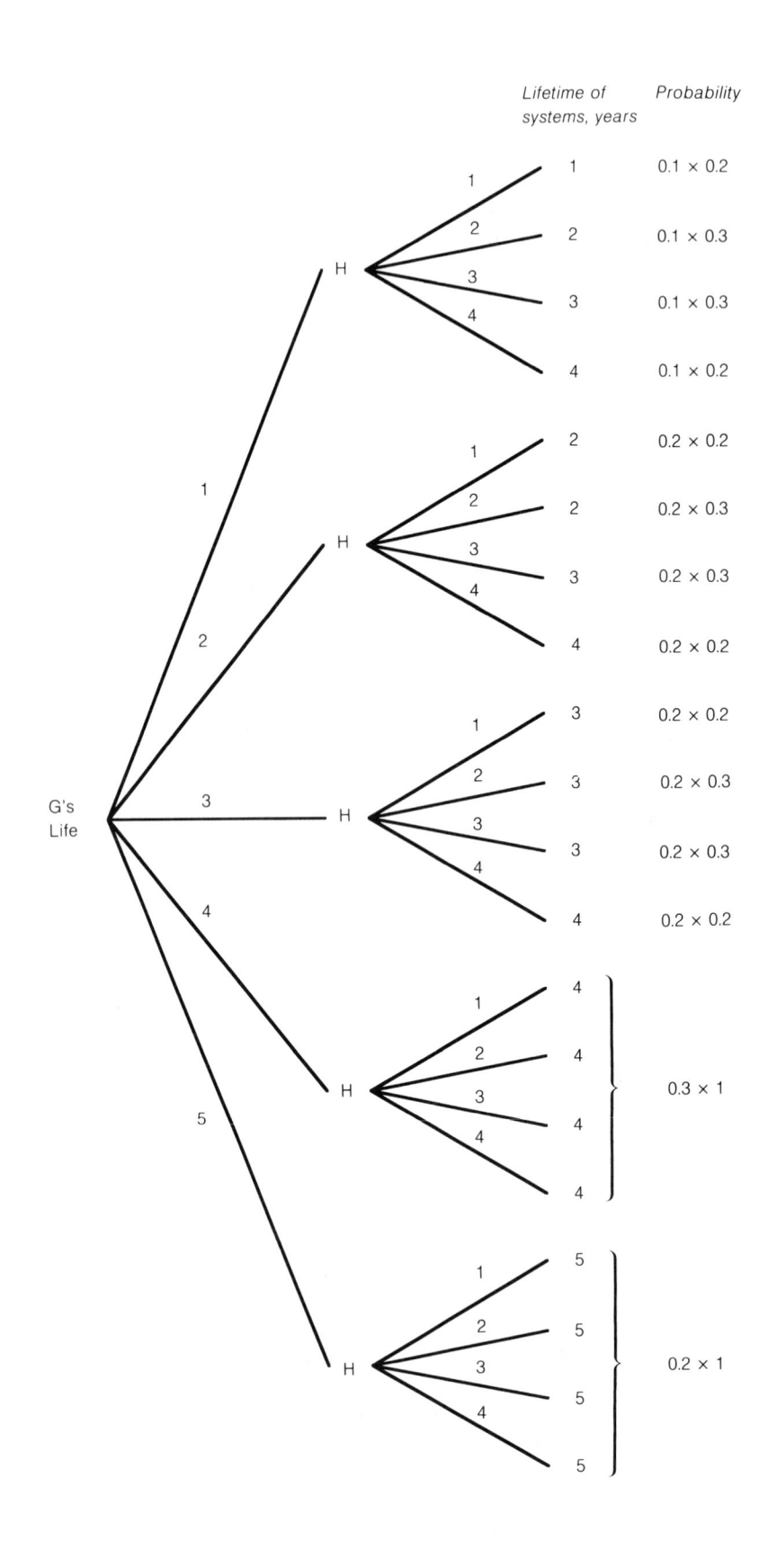

(i)

Lifetime of system, years	Probability	
1	0.02	0.02
2	0.03 + 0.04 + 0.06	= 0.13
3	0.03 + 0.06 + 0.16	= 0.25
4	0.02 + 0.04 + 0.04 + 0.30	= 0.40
5	0.20	0.20
Total		1.0

Hence, the probability distribution for the lifetime of the system is:

Lifetime, years	1	2	3	4	5
Probability	0.02	0.13	0.25	0.40	0.20

(ii) Expected lifetime = $\sum$ life × probability
G: E(life)   = 1 × 0.1 + 2 × 0.2 + 3 × 0.2 + 4 × 0.3 + 5 × 0.2 = 3.3 years
H: E(life)   = 1 × 0.2 + 2 × 0.3 + 3 × 0.3 + 4 × 0.2   = 2.5 years
System: E(life) = 1 × 0.02 + 2 × 0.13 + 3 × 0.25 + 4 × 0.4 + 5 × 0.2 = 3.63 years

Exercise 1.9

(i) The probability tree is:

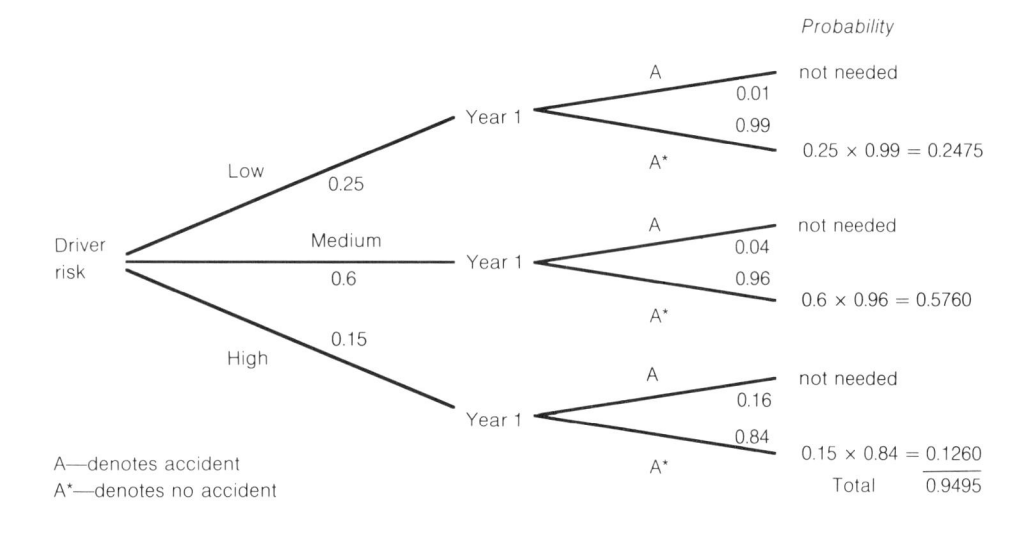

[Probability tree diagram:
- Driver risk → Low (0.25) → Year 1 → A (0.01) not needed; A* (0.99) → 0.25 × 0.99 = 0.2475
- Driver risk → Medium (0.6) → Year 1 → A (0.04) not needed; A* (0.96) → 0.6 × 0.96 = 0.5760
- Driver risk → High (0.15) → Year 1 → A (0.16) not needed; A* (0.84) → 0.15 × 0.84 = 0.1260
- Total = 0.9495

A—denotes accident
A*—denotes no accident]

We require P(High risk driver|no accident in the year)
Using Bayes' Rule:

$$P(\text{High risk}|\text{no accident}) = \frac{P(\text{High risk and no accident})}{P(\text{no accident})}$$

$$= \frac{0.1260}{0.9495} = 0.1327$$

Given the additional information that there has been no accident, the probability that the driver is high risk has been revised downwards from 0.15 to 0.133

428 QUANTITATIVE ANALYSIS

(ii) The probability tree in part (i) must now be extended to include the outcomes for years 2 to 4. A portion only of the extended tree is shown below.

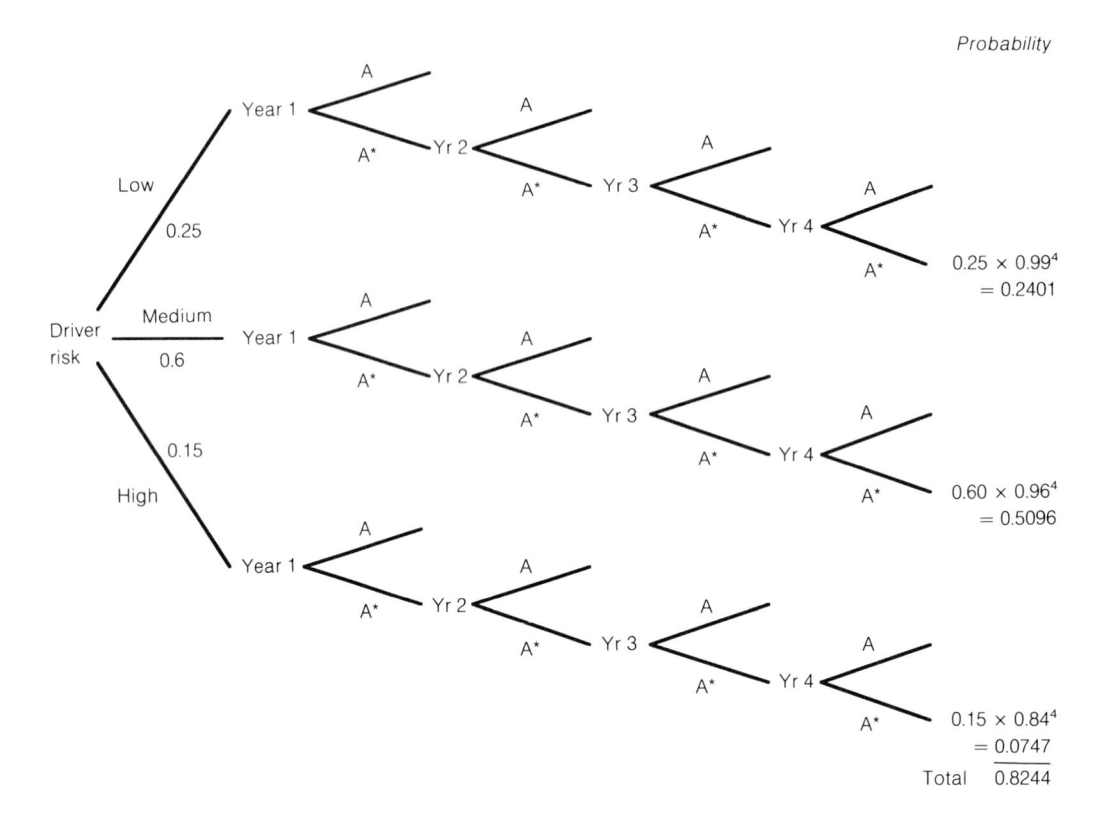

Probability

$0.25 \times 0.99^4 = 0.2401$

$0.60 \times 0.96^4 = 0.5096$

$0.15 \times 0.84^4 = 0.0747$

Total $\overline{0.8244}$

We require P(Low risk driver | no accidents for 4 years)
Using Bayes' Rule:

$$P(\text{Low risk} | \text{no accident for 4 years}) = \frac{P(\text{Low risk } and \text{ no accident for 4 years})}{P(\text{no accidents for 4 years})}$$

$$= \frac{0.2401}{0.8244} = 0.2912$$

Given the additional information that there has been no accident for 4 years, the probability that the driver is low risk has been revised upwards from 0.25 to 0.29.

Exercise 1.10

The probability tree for this problem is:

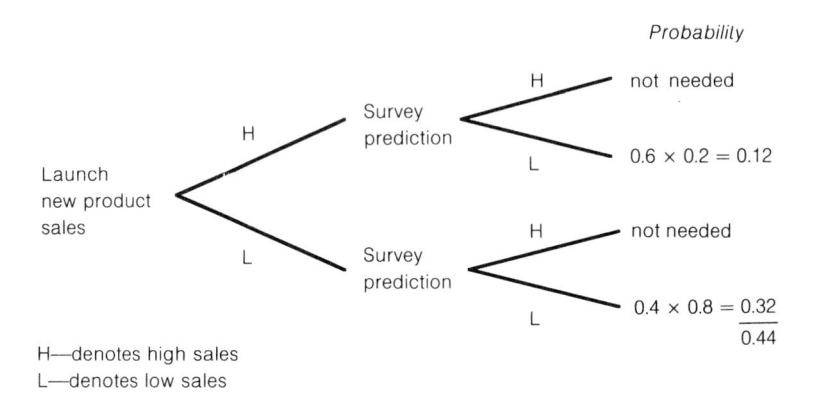

Probability

Launch new product sales
- H — Survey prediction
  - H — not needed
  - L — $0.6 \times 0.2 = 0.12$
- L — Survey prediction
  - H — not needed
  - L — $0.4 \times 0.8 = 0.32$

$\overline{0.44}$

H—denotes high sales
L—denotes low sales

$$P(\text{high sales}|\text{prediction of low sales}) = \frac{P(\text{high sales } and \text{ low pred.})}{P(\text{low prediction})}$$

$$= \frac{0.12}{0.44} = 0.273$$

$$P(\text{low sales}|\text{prediction of low sales}) = \frac{P(\text{low sales } and \text{ low pred.})}{P(\text{low prediction})}$$

$$= \frac{0.32}{0.44} = 0.727$$

Hence, the additional information that the survey predicts low sales, revises downwards the probability that the sales will actually be high and revises upwards the probability that the sales will actually be low.

Exercise 1.11

The problem may be illustrated as follows:

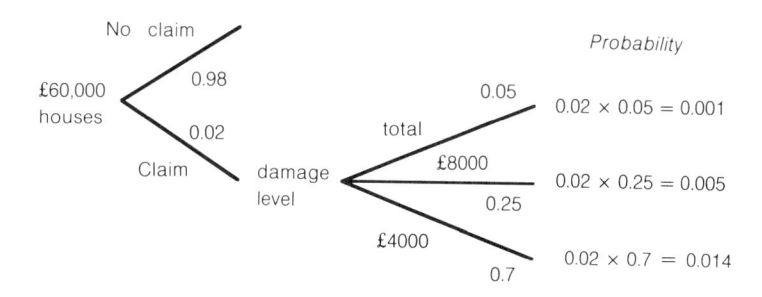

The minimum value of the premium = expected payout on the policies.

$$E(\text{payout}) = \sum (\text{payout}) \times (\text{probability})$$
$$= £60{,}000 \times 0.001 + £8000 \times 0.005 + £4000 \times 0.014 = £156$$

Hence, the company must charge a premium of £156 per house per year.

Exercise 1.12

(i) Select Chairman, Secretary and Treasurer:
Any one of the 25 members could be chosen as Chairman, leaving 24 members; any one of the 24 members could be chosen as Secretary, leaving 23 members; any one of the 23 members could be chosen as Treasurer. The total number of possible selections is $25 \times 24 \times 23 = 13{,}800$. In this case the order of selection is important.

(ii) Select the 4 committee members from 22. The order of selection is no longer important, hence we require the number of combinations of 4 people from 22.

$$^{22}C_4 = \frac{22!}{4!\,(22-4)!} = \frac{22!}{4! \times 18!} = \frac{22 \times 21 \times 20 \times 19}{4 \times 3 \times 2 \times 1} = 7315$$

There are 7315 different ways of choosing the other 4 committee members.

(iii) The total number of committees which could be chosen is:

$$13{,}800 \times 7315 = 100{,}947{,}000$$

Exercise 1.13

(i) Suppose the factories are A, B, C, D, and E.
There are 10 choices of driver to go to A, leaving 9 drivers.
There are 9 choices of driver to go to B, leaving 8 drivers.
There are 8 choices of driver to go to C, leaving 7 drivers.
There are 7 choices of driver to go to D, leaving 6 drivers.
There are 6 choices of driver to go to E.
Total numbers of ways of selecting the drivers is:

$$10 \times 9 \times 8 \times 7 \times 6 = 30,240$$

(ii) The first driver chosen to go to A can be selected in 10 ways and the second driver in 9 ways. Since the order in which they are chosen does not matter, the number of ways in which 2 drivers can be chosen to go to A is $(10 \times 9)/2$. This leaves 8 drivers. Similarly for the other factories:

The drivers to go to B can be chosen in $(8 \times 7)/2$ ways, leaving 6.
The drivers to go to C can be chosen in $(6 \times 5)/2$ ways, leaving 4.
The drivers to go to D can be chosen in $(4 \times 3)/2$ ways, leaving 2.
The drivers to go to E can be chosen in $(2 \times 1)/2$ ways.

$$\text{The total number of ways} = \frac{10!}{(2!)^5} = \frac{10 \times 9 \times 8 \times 7 \times 6 \times 5 \times 4 \times 3 \times 2 \times 1}{2^5}$$

$$= 3,628,800/32 = 113,400$$

Exercise 1.14

(i) The problem may be illustrated by a probability tree:

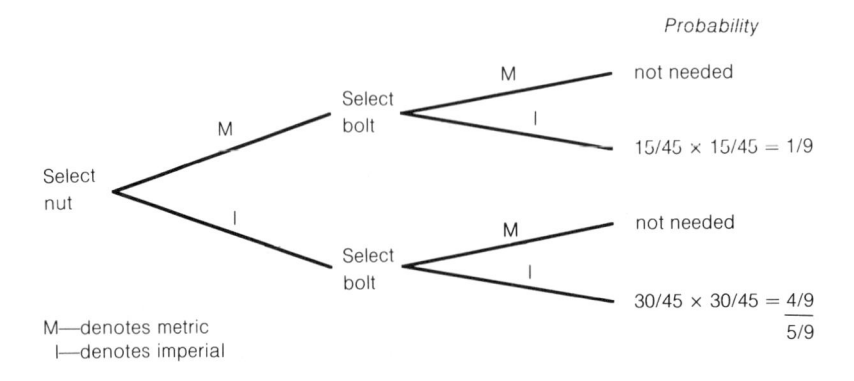

```
                                              Probability
                                  M ── not needed
                      Select bolt
              M ──┤
                      I ── 15/45 × 15/45 = 1/9
Select nut ──┤
                      M ── not needed
              I ──┤
                    Select bolt
                      I ── 30/45 × 30/45 = 4/9
                                              ─────
                                               5/9
```

M—denotes metric
I—denotes imperial

P(selecting a nut and bolt which fit) = 5/9

(ii) We must extend the tree for the new problem:

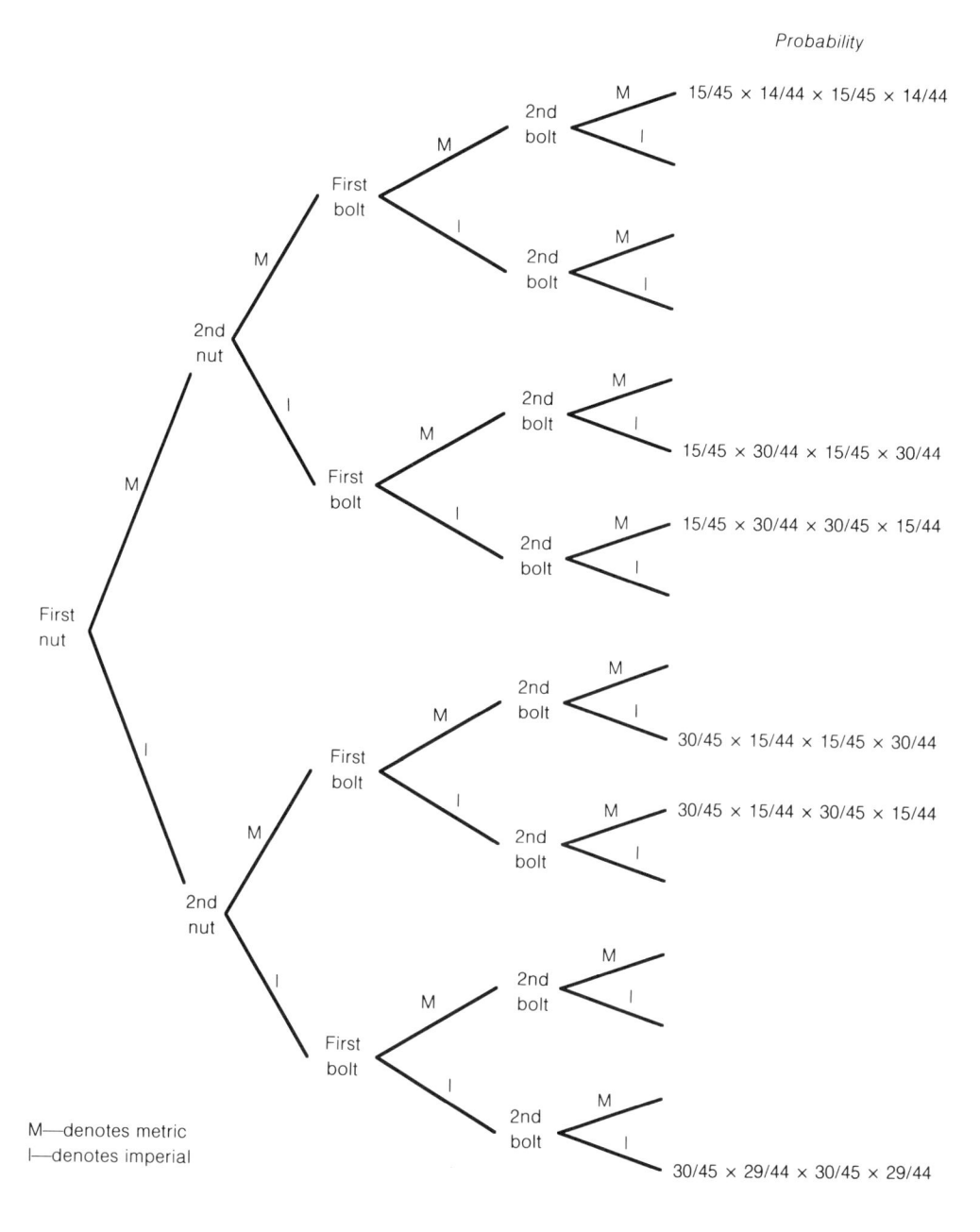

$$P(2 \text{ matching pairs}) = \frac{(15^2 \times 14^2) + 4 \times (15^2 \times 30^2) + (30^2 \times 29^2)}{(45^2 \times 44^2)}$$

$$= \frac{1611000}{3920400} = 0.4109$$

Exercise 1.15

Calculate the expected annual returns $= \sum \text{return} \times P(\text{return})$

(i) Chemicals:
$$E(\text{return}) = (-2) \times 0.05 + (-1) \times 0.1 + 0 \times 0.2 + 1 \times 0.2$$
$$+ 2 \times 0.2 + 3 \times 0.2 + 4 \times 0.05$$
$$= 1.2 \quad (\text{£'}000)$$

432 QUANTITATIVE ANALYSIS

(ii) Brewers:

$$\begin{aligned}E(\text{return}) &= (-2) \times 0.05 + (-1) \times 0.2 + 0 \times 0.3 + 1 \times 0.2 \\ &\quad + 2 \times 0.1 + 3 \times 0.1 + 4 \times 0.05 \\ &= 0.6 \quad (\text{£'000})\end{aligned}$$

The expected annual returns are £1200 for Chemicals and £600 for Brewers

Exercise 1.16

1 (a) Using the table given in the exercise, on average 748 people out of 1000 survive to the age of 60. Therefore, on average, $(1000 - 748) = 252$ people out of 1000 do not survive to the age of 60. Hence:

$$P(\text{person will die before age 60}) = \frac{252}{1000} = 0.252$$

(b) On average, out of 1000 people, 944 survive to the age of 30. The average number who die between 30 and 60 is $(944 - 748) = 196$. Hence:

$$P(\text{person now aged 30 will die before 60}) = \frac{196}{944} = 0.208$$

(c) Using the same process as in (b),

$$P(\text{person now aged 50 will die before 60}) = \frac{(880 - 748)}{880} = 0.150$$

The 3 probabilities steadily decrease in size. This implies that the older a person is now, the more likely he/she is to survive to the age of 60.

2 If the company is to break even in the long run, then they should charge a premium (P) so that:

$$P = \text{expected value of the money paid out on the policy}$$
$$\begin{aligned}E(\text{payout}) &= \text{sum insured} \times \text{probability (insured dies before 60 given that age is now 50)} \\ &= £10{,}000 \times 0.150 \text{ (from } \mathbf{1}(c)) \\ &= £1{,}500\end{aligned}$$

Therefore, the premium should be £1,500.

3 The company will make a loss if the amount paid out $> 12 \times £2000 = £24{,}000$. Since each pay out is £10,000, the company will make a loss if more than 2 people from the group of 12 die before they are 60. Since we may regard the 12 people as 12 identical, independent trials with 2 possible outcomes—they die before the age of 60 or they do not—and we can assume that the probability that each one will die before 60 is the same (this assumption is the major weakness in this type of analysis), then we can describe the probability of r deaths before the age of 60 by the binomial distribution with $n = 12$, $p = 0.15$. Hence:

$$P(r \text{ deaths before 60}) = {}^{12}C_r \times (0.15)^r \times (0.85)^{12-r} \qquad r = 0, 1, 2, \ldots, 12$$

Therefore:

$$\begin{aligned}P(r > 2) &= P(r = 3) + P(r = 4) + \ldots P(r = 12) \\ &= 1 - (P(r = 0) + P(r = 1) + P(r = 2)) \\ &= 1 - \left(0.85^{12} + (12 \times 0.15 \times 0.85^{11}) + \left(\frac{12 \times 11}{2} \times 0.15^2 \times 0.85^{10}\right)\right) \\ &= 1 - 0.7358 = 0.2642\end{aligned}$$

so that, the probability that the company makes a loss is 0.2642.

SOLUTIONS TO EXERCISES: CHAPTER ONE

4 Anyone aged 50 in 1986 who would take out the proposed insurance policy will not be 60 until 1996. Hence the probabilities derived from the data may be out of date. Some investigation of trends in the mortality data is needed so that the required probabilities may be revised. We know that with improved health care, new medicines, etc people are in general living longer.

5 (a) The expected age of death = $\sum$(age of death × probability of this age). In the new table, the ages increase in steps of two years, hence we will take the age of death as the mid-point of each interval, ie 51, 53, 55, 57, 59 years. For each 1000 of population, $(880 - 748) = 132$ people aged 50 die before they reach 60. Therefore, expected age of death is given by:

$$51 \times \left(\frac{880 - 866}{132}\right) + 53\left(\frac{866 - 846}{132}\right)$$

$$+ 55 \times \left(\frac{846 - 822}{132}\right) + 57\left(\frac{822 - 788}{132}\right)$$

$$+ 59 \times \left(\frac{788 - 748}{132}\right)$$

$$= \left(51 \times \frac{14}{132}\right) + \left(53 \times \frac{20}{132}\right) + \left(55 \times \frac{24}{132}\right)$$

$$+ \left(57 \times \frac{34}{132}\right) + \left(59 \times \frac{40}{132}\right) = 56 \text{ years}$$

On average, a person who is 50 now and who does not survive until 60, will die aged 56 years.

(b) We require:

Income from premium = Expected present value of the claims paid + expenses

$$\text{Present value of payout on claims} = \frac{10000}{1.08} \times \frac{14}{132} + \frac{10000}{1.08^3} \times \frac{20}{132}$$

$$+ \frac{10000}{1.08^5} \times \frac{24}{132} + \frac{10000}{1.08^7} \times \frac{34}{132} + \frac{10000}{1.08^9} \times \frac{40}{132}$$

$$= £6441.08$$

Therefore:

$$\text{Premium} = 6441.08 \times \frac{(880 - 748)}{880} + 100 + 200$$

$$= 6441.08 \times 0.15 + 300 = £1266.16$$

The premium which should be charged under this scheme is £1266.

Exercise 1.17

1 (a) The first firm visited can be chosen in 5 ways, the second one visited can be chosen in 4 ways, the third firm visited can be chosen in 3 ways, etc. Therefore, the total number of possible routes $= 5! = 5 \times 4 \times 3 \times 2 \times 1 = 120$. Therefore:

$$P(\text{choosing one particular route}) = 1/120$$

(b) In week 1, the driver has free choice of any of the 120 routes. In week 2, the driver has a choice of 119 of the routes, ie the route chosen in week 1 is excluded. In week 3, the driver has a choice of 118 routes, ie the routes used in weeks 1 and 2 are excluded. Therefore, in week 4, the driver has a choice of 117 routes. Hence:

$$P(4 \text{ different routes}) = \frac{119}{120} \times \frac{118}{120} \times \frac{117}{120} = 0.951$$

(c) Consider AB as a single visit. There are now 4 places to visit. Therefore, the number of possible routes $= 4! = 24$. However, firms A and B could be visited in either order, A to B or B to A, hence the total number of possible routes $= 24 \times 2 = 48$. Therefore:

$$P(A \text{ and } B \text{ visited consecutively}) = 48/120 = 0.4$$

2 (a) Total number of possible routes $= 3! \times 2 = 12$
Therefore:

$$P(\text{one particular route}) = 1/12 = 0.083$$

(b) $P(4 \text{ different routes}) = 11/12 \times 10/12 \times 9/12$
$= 0.573$

(c) Within the first group A and B can be visited consecutively in 4 ways. There are two arrangements for D and E. Therefore:

$$P(A \text{ and } B \text{ visited consecutively}) = \frac{4 \times 2}{12} = \frac{2}{3}$$

3 In week 1, any one of the 12 routes can be chosen. In week 2, any one of the remaining 11 routes can be chosen. In week 3, the route must be different from that in week 2, so there are 11 from which to choose, but we also want this route to differ from that used in week 1, hence only 10 of these 11 are acceptable. Therefore:

$$P(\text{routes in weeks 1,2,3, are all different}) = \overset{Week1}{\frac{12}{12}} \times \overset{Week2}{\frac{11}{11}} \times \overset{Week3}{\frac{10}{11}}$$

$$= \frac{10}{11}$$

Similarly in week 4, the route must be different from that in week 3, and we also want it to differ from both weeks 1 and 2, so that only 9 of the 11 available routes are suitable. Hence:

$$P(\text{routes are all different in weeks 1,2,3 and 4}) = \frac{10}{11} \times \frac{9}{11} = \frac{90}{121} = 0.744$$

Solutions to Chapter Two exercises

ANSWERS TO EXERCISES

Exercise 2.1

1 This situation is suitable for modelling using a binomial distribution since the 10 cars may be regarded as 10 identical and independent trials. There are 2 possible outcomes for each trial; a claim is filed against the warranty or it is not. We can assume that there is a constant 5% probability that any one customer will file such a claim.

2 This situation is not suitable for modelling using a binomial distribution since the variable, petrol consumption, is a continuous variable.

Exercise 2.2

This situation is suitable for modelling using a binomial distribution since the 20 invoices may be regarded as 20 identical and independent trials. There are 2 possible outcomes for each trial; an invoice is faulty or it is not. We can assume that there is a constant 7% probability that any one invoice will be faulty.

$$P(r \text{ faulty invoices in a sample of } 20) = {}^{20}C_r \times (0.07)^r \times (0.93)^{20-r} \quad r = 0, 1, 2, \ldots, 20$$

1
$$P(r = 3) = {}^{20}C_3 \times (0.07)^3 \times (0.93)^{17}$$

$$= \frac{20!}{3!\,17!} \times (0.07)^3 \times (0.93)^{17} = 0.1139$$

2 $P(r \geqslant 3) = 1 - P(r < 3) = 1 - \{P(r = 0) + P(r = 1) + P(r = 2)\}$
$P(r = 0) = {}^{20}C_0 \times (0.07)^0 \times (0.93)^{20} = 1 \times 1 \times (0.93)^{20} \qquad = 0.2342$
$P(r = 1) = {}^{20}C_1 \times (0.07)^1 \times (0.93)^{19} = 20 \times (0.07) \times (0.93)^{19} \quad = 0.3526$
$P(r = 2) = {}^{20}C_2 \times (0.07)^2 \times (0.93)^{18} = 190 \times (0.07)^2 \times (0.93)^{18} = 0.2521$
$\phantom{P(r = 2) = {}^{20}C_2 \times (0.07)^2 \times (0.93)^{18} = 190 \times (0.07)^2 \times (0.93)^{18}}$ Total 0.8389

Therefore, P(at least 3 faulty invoices) $= 1 - 0.8389 = 0.1611$

Exercise 2.3

We may regard the 6 stockbrokers as 6 identical and independent trials. There are 2 possible outcomes for each trial; the issue is recommended or it is not. We can assume that there is a constant

80% probability that any one stockbroker will recommend the issue. Hence, the number of recommendations from the 6 stockbrokers will follow a binomial distribution with n = 6 and p = 0.8.

$$P(r \text{ recommendations}) = {}^6C_r \times (0.8)^r \times (0.2)^{6-r} \quad r = 0, 1, 2, \ldots, 6$$

$$P(\text{at least 4 recommendations}) = P(r \geq 4) = P(r = 4) + P(r = 5) + P(r = 6)$$

$$P(r = 4) = {}^6C_4 \times (0.8)^4 \times (0.2)^2 = \frac{6!}{4!\,2!} (0.8)^4 \times (0.2)^2 = 0.24576$$

$$P(r = 5) = {}^6C_5 \times (0.8)^5 \times (0.2)^1 = \frac{6!}{5!\,1!} (0.8)^5 \times (0.2) = 0.39322$$

$$P(r = 6) = {}^6C_6 \times (0.8)^6 \times (0.2)^0 = \frac{6!}{6!\,0!} (0.8)^6 \times (0.2)^0 = 0.26214$$

Hence, P(at least 4 rcommendations) = 0.24576 + 0.39322 + 0.26214
$$= 0.90112$$

Exercise 2.4

A probability tree for this problem is:

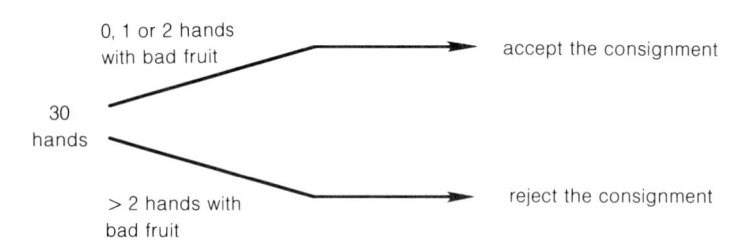

Regard the sample of 30 hands of bananas as 30 identical, independent trials; 2 possible outcomes. The probability of any one hand containing bad bananas is constant at 5%. Hence, the number of hands containing bad bananas in the sample of 30, follows a binomial distribution with n = 30 and p = 0.05.

$$P(r \text{ hands with bad bananas}) = {}^{30}C_r \times (0.05)^r \times (0.95)^{30-r} \quad r = 0, 1, \ldots, 30$$

$$P(\text{consignment rejected}) = P(r > 2) = 1 - P(r \leq 2)$$
$$= 1 - \{P(r = 0) + P(r = 1) + P(r = 2)\}$$

$$P(r = 0) = \frac{30!}{0!\,30!} \times (0.05)^0 \times (0.95)^{30} = 0.21464$$

$$P(r = 1) = \frac{30!}{1!\,29!} \times (0.05)^1 \times (0.95)^{29} = 0.33890$$

$$P(r = 2) = \frac{30!}{2!\,28!} \times (0.05)^2 \times (0.95)^{28} = 0.25864$$

$$P(\text{consignment rejected}) = 1 - \{0.21464 + 0.33890 + 0.25864\}$$
$$= 1 - 0.81218 = 0.18782$$

There is an 18.8% chance that the wholesaler will reject a consignment which actually satisfies his condition of up to 10% unusable bananas.

Exercise 2.5

The binomial distribution with n = 10 and p = 0.01. We may regard the 10 transistors as ten identical and independent trials. There are 2 possible outcomes for each trial; the transistor is defective or it is not. We can assume that there is a constant 1% probability that any one transistor is defective.

Exercise 2.6

1. The number of lorries arriving in an hour will follow a Poisson distribution approximately. The hour may be regarded as a large number of small time intervals for each of which there is a small probability of an arrival. These intervals are trials and a success in a trial is the arrival of a lorry. We have a large number of trials with a small probability of a success in each. This leads to the Poisson distribution for the number of successes in an hour.

2. $P(r \text{ lorries arrive in an hour}) = \dfrac{3^r e^{-3}}{r!} \quad r = 0, 1, 2, \ldots.$

   $P(r > 4) = 1 - P(r \leq 4) = 1 - \{P(r=0) + P(r=1) + P(r=2) + P(r=3) + P(r=4)\}$

r	P(r)	
0	$\dfrac{3^0 e^{-3}}{0!} = \dfrac{1 \times 0.04979}{1}$	= 0.04979
1	$\dfrac{3^1 e^{-3}}{1!} = \dfrac{3 \times 0.04979}{1}$	= 0.14936
2	$\dfrac{3^2 e^{-3}}{2!} = \dfrac{9 \times 0.04979}{2}$	= 0.22404
3	$\dfrac{3^3 e^{-3}}{3!} = \dfrac{27 \times 0.04979}{6}$	= 0.22404
4	$\dfrac{3^4 e^{-3}}{4!} = \dfrac{81 \times 0.04979}{24}$	= 0.16803
Total		0.81526

   P(more than 4 lorries arrive in an hour) = 1 − 0.81526 = 0.1847

Exercise 2.7

This problem may be described by a Poisson distribution with a mean of 2. The hour may be regarded as a large number of small time intervals for each of which there is a small probability of a machine breakdown. We have a large number of trials each with a small probability of a success, ie a machine breakdown. This leads to a Poisson distribution for the number of successes in an hour.

$$P(r \text{ breakdowns in an hour}) = \dfrac{2^r e^{-2}}{r!} \quad r = 0, 1, 2, \ldots.$$

$P(\text{assistance needed}) = P(r > 2) = 1 - P(r \leq 2)$
$\qquad\qquad\qquad\qquad\quad = 1 - \{P(r=0) + P(r=1) + P(r=2)\}$

r	$P(r)$	
0	$\dfrac{2^0 e^{-2}}{0!} = \dfrac{1 \times 0.13534}{1}$	$= 0.13534$
1	$\dfrac{2^1 e^{-2}}{1!} = \dfrac{2 \times 0.13534}{1}$	$= 0.27067$
2	$\dfrac{2^2 e^{-2}}{2!} = \dfrac{4 \times 0.13534}{2}$	$= 0.27067$
Total		0.67668

$$P(\text{assistance needed}) = 1 - 0.67668 = 0.32332$$

Hence, expected number of calls for assistance in 120 hours = $120 \times 0.32332 = 38.8$
Approximately, 39 times per week.

Exercise 2.8

Call the number of old files accessed per day, N. Call the number of days in which files are accessed, f. This is the frequency of occurrence.

1 The mean number of old files accessed per day $= \dfrac{\sum f \times N}{\sum f}$

The standard deviation number of old files accessed per day $= \sqrt{\dfrac{\sum fN^2}{\sum f} - \left(\dfrac{\sum fN}{\sum f}\right)^2}$

Number of old files accessed/day, N	Number of days f	fN	fN^2
0	10	0	0
1	30	30	30
2	27	54	108
3	17	51	153
4	6	24	96
5	5	25	125
6	4	24	144
7	1	7	49
Total	100	215	705

Mean number of old files accessed per day $= 215/100 = 2.15$

Standard deviation number of old files accessed/day $= \sqrt{(705/100) - 2.15^2}$

$$= \sqrt{2.4275} = 1.558$$

2 There is a very large number of files and a small probability that any particular one will be required in a given day. We may consider these files as identical and independent trials with a very small probability that any particular file will be required on a given day.

The observed distribution has a mean of 2.15 old files accessed in a day and a variance of 2.4275. These 2 values are approximately equal. All of the above factors are necessary conditions for the Poisson distribution to be the appropriate distribution to describe the number of old files accessed in a day.

3 Assume that:

$$P(r \text{ old files accessed in a day}) = \frac{2.15^r e^{-2.15}}{r!} \quad r = 0, 1, 2, \ldots$$

Expected frequency $= 100 \times P(r)$

Number of old files accessed per day, r	P(r old files accessed in a day)	Expected frequency
0	$\frac{2.15^0 e^{-2.15}}{0!} = 0.1165$	$100 \times 0.1165 = 11.65$
1	$\frac{2.15^1 e^{-2.15}}{1!} = 0.2504$	$100 \times 0.2504 = 25.04$
2	$\frac{2.15^2 e^{-2.15}}{2!} = 0.2692$	$100 \times 0.2692 = 26.92$
3	$\frac{2.15^3 e^{-2.15}}{3!} = 0.1929$	$100 \times 0.1929 = 19.29$
4	$\frac{2.15^4 e^{-2.15}}{4!} = 0.1037$	$100 \times 0.1037 = 10.37$
5	$\frac{2.15^5 e^{-2.15}}{5!} = 0.0446$	$100 \times 0.0446 = 4.46$
6	$\frac{2.15^6 e^{-2.15}}{6!} = 0.0160$	$100 \times 0.0160 = 1.60$
7	$\frac{2.15^7 e^{-2.15}}{7!} = 0.0049$	$100 \times 0.0049 = 0.49$
>7	(1 − total of above) = 0.0018	$100 \times 0.0018 = 0.18$
		Total 100.00

A statistical test could now be used to compare the observed and expected frequencies to see whether the assumption of a Poisson distribution is supported or not. See Chapter Six.

Exercise 2.9

It may be assumed that the demand for floppy disks follows a Poisson distribution with a mean of 4 boxes per month, therefore:

$$P(r \text{ boxes demanded in a month}) = \frac{4^r e^{-4}}{r!} \quad r = 0, 1, 2, \ldots$$

If the department stocks q boxes at the beginning of each month, we require:

$$P(\text{demand in a month} > q) < 0.04$$

or:

$$P(\text{demand in a month} \leq q) > 0.96$$

These 2 probabilities are equivalent but the second one is slightly quicker to evaluate.

Stock at the beginning of the month, q	P(demand in a month = q)	P(demand in a month ≤ q)
0	$\dfrac{4^0 e^{-4}}{0!} = 0.0183$	0.0183
1	$\dfrac{4^1 e^{-4}}{1!} = 0.0733$	$0.0183 + 0.0733 = 0.0916$
2	$\dfrac{4^2 e^{-4}}{2!} = 0.1465$	$0.0916 + 0.1465 = 0.2381$
3	$\dfrac{4^3 e^{-4}}{3!} = 0.1954$	$0.2381 + 0.1954 = 0.4335$
4	$\dfrac{4^4 e^{-4}}{4!} = 0.1954$	$0.4335 + 0.1954 = 0.6289$
5	$\dfrac{4^5 e^{-4}}{5!} = 0.1563$	$0.6289 + 0.1563 = 0.7852$
6	$\dfrac{4^6 e^{-4}}{6!} = 0.1042$	$0.7852 + 0.1042 = 0.8894$
7	$\dfrac{4^7 e^{-4}}{7!} = 0.0595$	$0.8894 + 0.0595 = 0.9489$
8	$\dfrac{4^8 e^{-4}}{8!} = 0.0298$	$0.9489 + 0.0298 = 0.9787$

If 8 boxes of disks are stocked at the beginning of each month:

$$P(\text{demand in a month} \leqslant 8) > 0.96$$

hence:

$$P(\text{running out}) < 0.04$$

Exercise 2.10

The number of bad oranges in a crate will follow a binomial distribution with $n = 250$ and $p = 0.006$, however, since n is large, p is small and $np = 1.5 < 5$, the Poisson distribution may be used as an approximation. Approximately:

$$P(r \text{ bad oranges in a crate}) = \frac{1.5^r e^{-1.5}}{r!} \quad r = 0, 1, 2, \ldots$$

$$P(r \leqslant 2) = P(r = 0) + P(r = 1) + P(r = 2)$$
$$= \frac{1.5^0 e^{-1.5}}{0!} + \frac{1.5^1 e^{-1.5}}{1!} + \frac{1.5^2 e^{-1.5}}{2!}$$
$$= 0.2231 + 0.3347 + 0.2510 = 0.8088$$

The probability that a given crate contains no more than 2 bad oranges is 80.88%.

Exercise 2.11

The weight dispensed to the packets is normally distributed with $\mu = 930$ g and $\sigma = 20$ g:

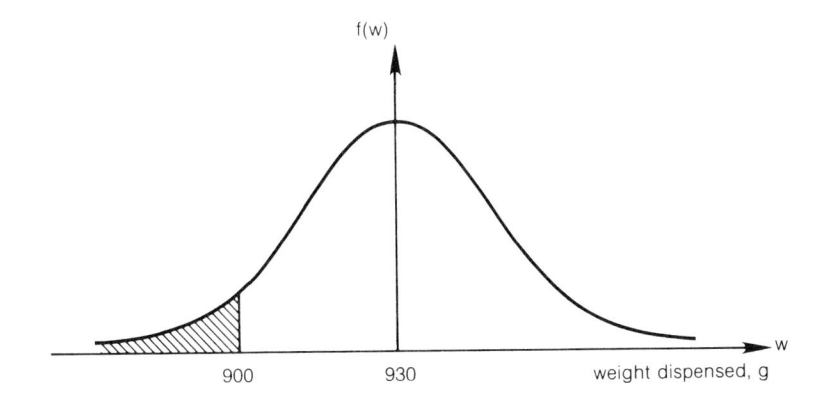

1 We require P(weight dispensed < 900 g) Nine hundred grammes is z standard deviations below the mean where:

$$z = \frac{900 - 930}{20} = -1.5$$

From the standard normal tables, $P(z > +1.5) = 0.06681$
From the symmetry of the distribution, $P(z < -1.5) = 0.06681$
Therefore, 6.68% of the packets will contain less than 900 g.

2 In order to reduce the proportion of packets containing less than 900 g, the average fill must be increased. We assume that the standard deviation does not change. Using the standard normal tables, if $P(z > z') = 0.025$, then $z' = 1.96$. Hence, with the new mean, 900 g is 1.96 standard deviations below the mean. Therefore:

$$-1.96 = \frac{900 - \mu}{20}$$

hence:

$$\mu = 900 + 1.96 \times 20$$
$$= 939.2 \text{ g}$$

The machine should be reset to a mean of 940 g.

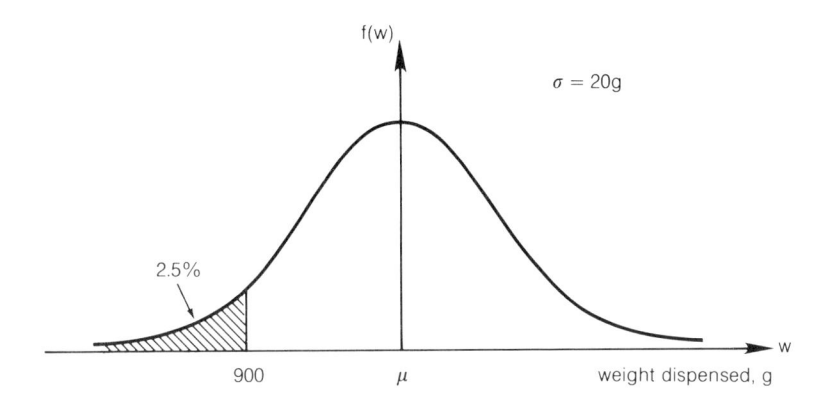

Exercise 2.12

The distribution of lifetimes is:

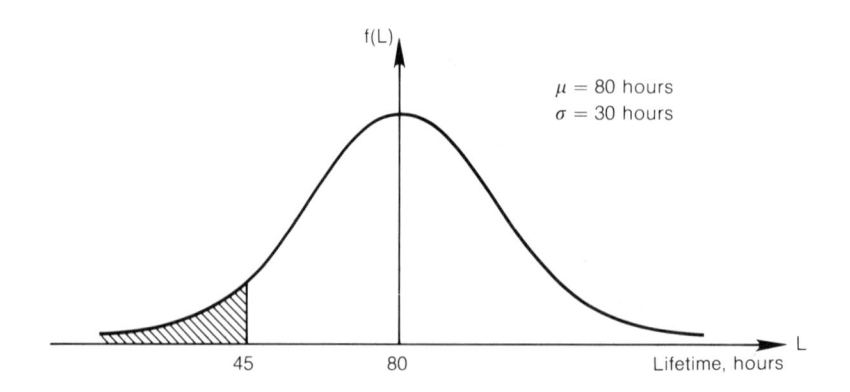

$\mu = 80$ hours
$\sigma = 30$ hours

Forty five hours is z standard deviations below the mean where:

$$z = \frac{45 - 80}{30} = -1.167$$

From the standard normal tables, $P(z > +1.167) = 0.1210$.
From the symmetry of the distribution, $P(z < -1.167) = 0.1210$.
Therefore, 12.1% of the components will have to be replaced.
If the manufacturer wishes to reduce this proportion to 10%, then the guaranteed lifetime must be reduced. We assume that the standard deviation does not change. Using the standard normal tables, if $P(z > z') = 0.10$, then $z' = 1.28$. Hence, with the new guaranteed life it is 1.28 standard deviations below the mean. Therefore:

$$-1.28 = \frac{L_g - 80}{30}$$

hence, $L_g = 80 - (1.28 \times 30) = 41.6$ hours.

The guarantee should be set at 40 hours for convenience.

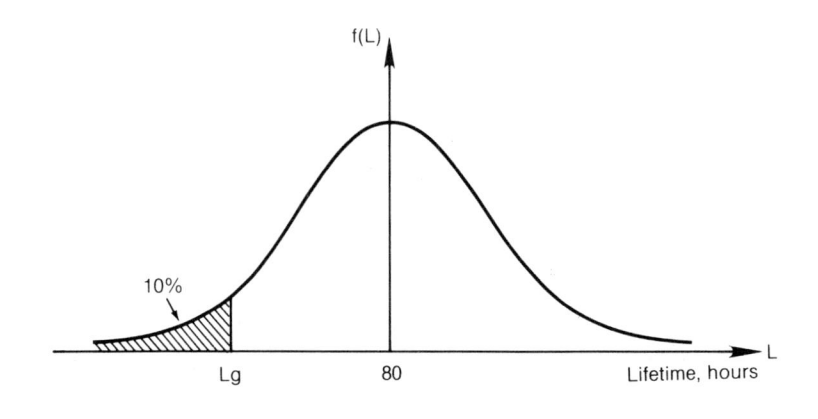

Exercise 2.13

The distribution of the rod diameters is:

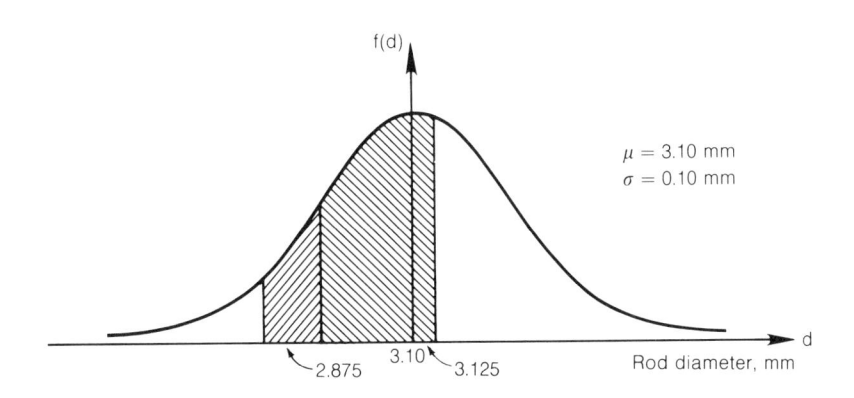

[Figure: Normal distribution curve with $\mu = 3.10$ mm, $\sigma = 0.10$ mm, showing shaded region between 2.875 and 3.125 mm]

Number 11 needles may have diameters between $(3.00 - 0.125)$ mm and $(3.00 + 0.125)$ mm, that is, 2.875 mm and 3.125 mm.

Upper Limit: 3.125 mm is z standard deviations above the mean, where:

$$z = \frac{3.125 - 3.10}{0.10} = 0.25$$

From the standard normal tables, $P(z > 0.25) = 0.40129$.

Lower Limit: 2.875 mm is z standard deviations below the mean, where:

$$z = \frac{2.875 - 3.10}{0.10} = -2.25$$

From the standard normal tables, $P(z > 2.25) = 0.01222$. Hence, $P(z < -2.25) = 0.01222$. It follows that $P(-2.25 < z < 0.25) = 1 - (0.40129 + 0.01222) = 0.58649$. Therefore the proportion of rods which is suitable for making into number 11 needles is 58.6%.

Exercise 2.14

The distribution of packet weights is:

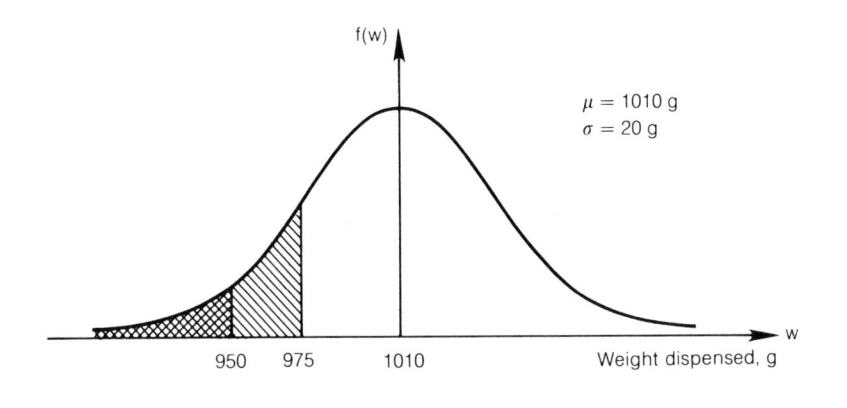

[Figure: Normal distribution curve with $\mu = 1010$ g, $\sigma = 20$ g, showing shaded regions at 950 and 975 g]

444 QUANTITATIVE ANALYSIS

1 The proportion of packets containing less than 975 g is given by P(packet contains < 975 g). 975 g is z standard deviations below the mean where:

$$z = \frac{975 - 1010}{20} = -1.75$$

From the standard normal tables, $P(z < -1.75) = 0.04006$.
Hence, 4.01% of packets contain less than 975 g. This does not satisfy the second condition.

2 950 g is z standard deviation below the mean number:

$$z = \frac{950 - 1010}{20} = -3.00$$

From the standard normal tables, $P(z < -3.00) = 0.00135$. Hence, 0.135% of packets contain less than 950 g. This does not satisfy the third condition (1 in 10,000 = 0.01%).

3 If the third condition is just satisfied, then:

$$P(\text{weight} < 950 \text{ g}) = 0.0001$$

If 950 g is z standard deviations below the mean, then:

$$P(\text{standard normal variable} < z) = 0.0001$$

Using the standard normal tables, we find that $z = -3.73$. Therefore:

$$-3.73 = \frac{950 - \mu}{20}$$

Hence:

$$\mu = 950 + (20 \times 3.73) = 1024.6$$

This is the lowest setting for the mean fill which will satisfy condition **3**. This value also satisfies condition **1**. We will check that it also satisfies condition **2**. If 975 g is z standard deviations below the mean, where:

$$z = \frac{975 - 1024.6}{20} = -2.48$$

From the standard normal tables, $P(z < -2.48) = 0.00657$. Since this value is less than the maximum value of 0.025, condition 2 is also satisfied. The minimum new setting for the filling machine is 1024.6 g or 1025 g to the nearest 1 g. This calculation assumes that the standard deviation of the filling machine is not altered.

Exercise 2.15

Consider first the cheaper machine. The distribution of the weights of the bags is:

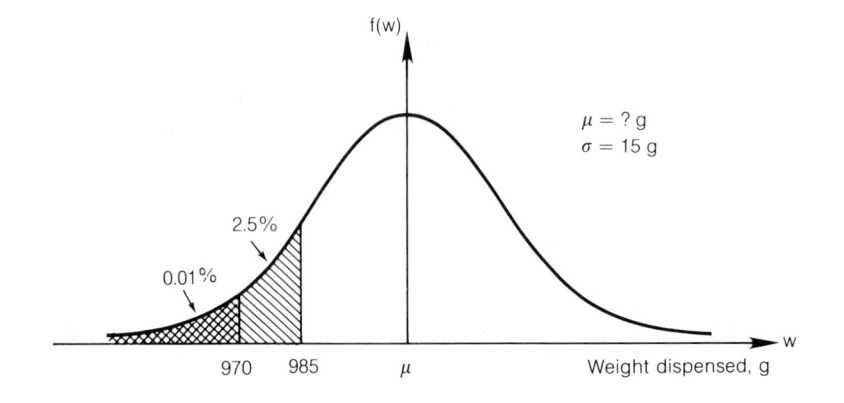

SOLUTIONS TO EXERCISES: CHAPTER TWO

We must determine the minimum value of the average fill which will satisfy all of the packing regulations. We require:

$$P(\text{bag contains} < 985 \text{ g}) \leqslant 0.025 \text{ and}$$
$$P(\text{bag contains} < 970 \text{ g}) \leqslant 0.0001$$

970 g is z standard deviation below the mean where:

$$z = \frac{970 - \mu}{15}$$

From the standard normal tables, if:

$$P(\text{standard normal variable} < z) = 0.0001, \text{ then } z = -3.73$$

Therefore:

$$-3.73 = \frac{970 - \mu}{15}$$

Hence:

$$\mu = 1025.95 \text{ g}$$

This value for the mean setting of the machine satisfies regulations **1** and **3**, and we will check that it also satisfies regulation **2**.

985 g is z standard deviations below the mean where:

$$z = \frac{985 - 1025.95}{15} = -2.73$$

From the standard normal tables $P(z < -2.73) = 0.00317$. Hence, 0.32% of packets contain less than 985 g. This is well within the requirement to satisfy the second condition. Therefore, the cheaper machine must be set up to give an average fill of 1025.95 g. The annual variable cost will be:

$$C_1 = \text{average fill/bag} \times \text{unit variable cost} \times \text{number of bags/year}$$
$$= 1025.95 \times (0.15/1000) \times 4,000,000 \text{ £/year}$$
$$= £615,570$$

We must repeat this calculation for the more expensive machine. The only difference is that the standard deviation of the weight delivered is now 5 g instead of 15 g.

970 g is z standard deviation below the mean where:

$$z = \frac{970 - \mu}{5}$$

From the standard normal tables, if:

$$P(\text{standard normal variable} < z) = 0.0001, \text{ then } z = -3.73$$

therefore:

$$-3.73 = \frac{970 - \mu}{5}$$

hence:

$$\mu = 988.65 \text{ g}$$

This value for the mean setting of the machine does not satisfy regulation **1**. The mean setting must be at least 1000g. We will check that a mean of 1000 g also satisfies regulation **2**.

985 g is z standard deviations below the mean where:

$$z = \frac{985 - 1000}{5} = -3.00$$

From the standard normal tables, $P(z < -3.00) = 0.00135$.

Hence, 0.14% of packets contain less than 985 g. This is well within the requirement to satisfy the second condition. Therefore the more expensive machine must be set up to give an average fill of 1000 g. The annual variable cost will be:

$$C_2 = \text{average fill/bag} \times \text{unit variable cost} \times \text{number of bags/year}$$

$$= 1000 \times (0.15/1000) \times 4,000,000 \text{ £/year}$$

$$= £600,000$$

The annual saving in the variable cost of production if the more expensive machine were purchased is £(615,570–600,000) = £15,570. The more expensive machine is worth purchasing if its lifetime is sufficiently long for the additional expenditure to be more than off-set by this saving.

Exercise 2.16

The number of votes received by the particular political party may be described by a binomial distribution with n = 10,800 and p = 0.45. However, since n is large, p is neither small nor large and np > 5 and nq > 5, the binomial distribution may be approximated by a normal distribution with mean = np = 4860 and standard deviation = $\sqrt{npq} = \sqrt{2673} = 51.701$.

The approximate distribution of the number of votes cast for the particular party is:

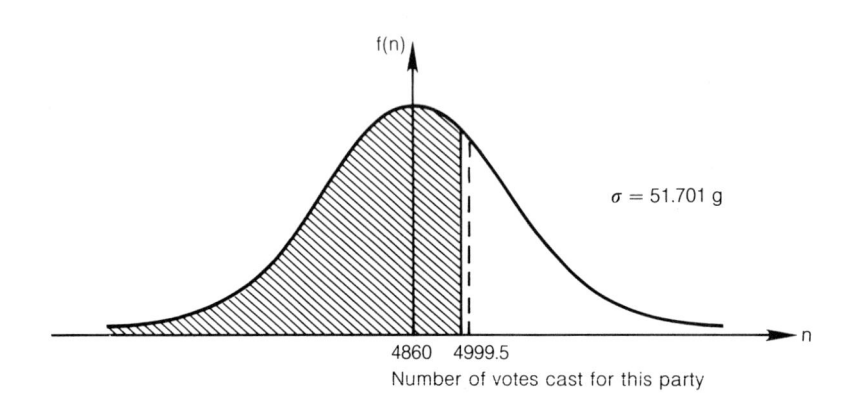

We require $P(n < 5000)$. Since we are using the normal distribution to approximate the binomial distribution we should incorporate the 'continuity correction', although with such a large sample it is unlikely to make a significant difference. We take the variable value as 4999.5.

4999.5 is z standard deviations above the mean where:

$$z = \frac{4999.5 - 4860}{51.701} = 2.70$$

From the standard normal tables $P(z > 2.70) = 0.00347$. Hence:

$$P(\text{party receives} < 5000 \text{ votes}) = 1 - 0.00347 = 0.99653$$

There is a 99.65% probability that the party in question will receive less than 5000 votes.

SOLUTIONS TO EXERCISES: CHAPTER TWO 447

Exercise 2.17

The number of questions answered correctly may be described by a binomial distribution with n = 100 and p = 1/3. However, since n is large, p is neither small nor large and np ⩾ 5 and nq ⩾ 5, the binomial distribution may be approximated by a normal distribution with mean = np = 100/3 and standard deviation = $\sqrt{npq} = \sqrt{200/9} = 4.714$. The approximate distribution of the number of questions answered correctly is:

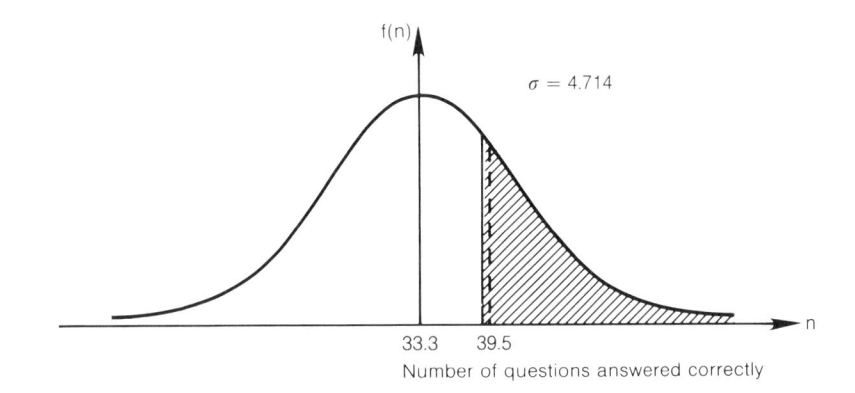

f(n)

σ = 4.714

33.3   39.5
Number of questions answered correctly

We require P(n ⩾ 40). Since we are using the normal distribution to approximate the binomial distribution we should incorporate the 'continuity correction', hence we take the variable value as 39.5.

39.5 is z standard deviations above the mean where:

$$z = \frac{39.5 - 100/3}{4.714} = 1.31$$

From the standard normal tables P(z > 1.31) = 0.0951. Hence:

$$P(\text{number of correct answers} \geqslant 40) = 0.0951$$

There is a 9.5% probability that a candidate will pass the examination by choosing his answers at random.

Exercise 2.18

1

Number of defectives per box, r	Number of boxes f	Number of defectives f × r
0	392	0
1	73	73
2	30	60
3	3	9
4	2	8
5	0	0
Total	500	150

(a)   Proportion of defectives = $\dfrac{\text{total number of defectives}}{\text{total number of pens}}$

$$= 150/(500 \times 5) = 0.06$$

448 QUANTITATIVE ANALYSIS

(b) Assume the number of defectives per box is binomially distributed, with n = 5 and p = 0.06.

$$P(r \text{ defectives per box}) = {}^5C_r \times 0.06^r \times 0.94^{5-r} \quad \text{for} \quad r = 0, 1, 2, 3, 4, 5$$

r	P(r defectives per box)	Expected frequency = 500 × P(r)
0	0.7339	366.95 (367)
1	0.2342	117.1 (117)
2	0.0299	14.95 (15)

We expect 367, 117 and 15 boxes to contain 0, 1, 2 defectives respectively.

Exercise 2.19

1 Expected number accepting early retirement

$$= \sum (\text{Number of employees} \times \text{probability of retiring})$$
$$= 5 \times 0.6 + 3 \times 0.7 + 2 \times 0.8 = 6.7$$

For each age group, the number accepting early retirement follows a binomial distribution. For the 62 years old age group:

$$P(r \text{ accept early retirement}) = {}^5C_r \times 0.6^r \times 0.4^{5-r} \quad r = 0,1,2,3,4,5.$$

Therefore:
$$P(r = 5) = 0.6^5 = 0.07776$$

Similarly for the 63 years old age group, $P(r = 3) = 0.7^3 = 0.343$, and for the 64 years old age group, $P(r = 2) = 0.8^2 = 0.64$. Therefore:

$$P(\text{all 10 accept early retirement}) = 0.6^5 \times 0.7^3 \times 0.8^2$$
$$= 0.0171$$

2 Age group 62 years: P(1 does not accept early retirement) = $5 \times 0.6^4 \times 0.4 = 0.2592$

Age group 63 years: P(1 does not accept early retirement) = $3 \times 0.7^2 \times 0.3 = 0.4410$

Age group 64 years: P(1 does not accept early retirement) = $2 \times 0.8 \times 0.2 = 0.32$

Now consider all 3 age groups. The one person who does not accept could come from any one of the 3 groups. The alternatives are:

$$(\overline{62})(63)(64) \text{ or } (62)(\overline{63})(64) \text{ or } (62)(63)(\overline{64})$$

Where $(\overline{62})$ denotes that one person in that age group does not accept and (62) denotes that all in that age group accept. Therefore:

P(exactly 9 employees accept early retirement)
$= 0.2592 \times 0.343 \times 0.64 + 0.07776 \times 0.4410 \times 0.64 + 0.07776 \times 0.343 \times 0.32$
$= 0.0569 + 0.02195 + 0.00853$
$= 0.0874$

3 Variance of total number accepting early retirement

$= $ Variance of number of 62 age group accepting + Variance of number of 63 age group accepting + Variance of number of 64 age group accepting
$= 5 \times 0.6 \times 0.4 + 3 \times 0.7 \times 0.3 + 2 \times 0.8 \times 0.2 = 2.15$

Approximate the binomial distribution of the total number of employees accepting early retirement by a normal distribution with mean = 6.7 and variance = 2.15

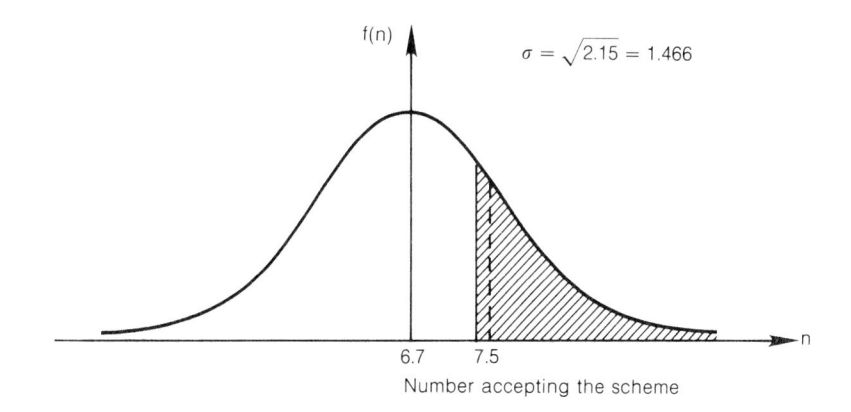

$$\sigma = \sqrt{2.15} = 1.466$$

Using the continuity correction, we require P(number accept > 7.5). 7.5 is z standard deviations above the mean where:

$$z = \frac{7.5 - 6.7}{1.466} = 0.5457$$

Using the standard normal tables, P(number accept > 7.5) = 0.2925. Therefore the probability of at least 8 employees accepting the scheme is approximately 29%.

Exercise 2.20

1   Let w denote the number of £1 notes issued.
    Let f denote the number of £5 notes issued.
    Let t denote the number of £10 notes issued.
    The average amount of money issued:

$$\begin{aligned} \text{in £1 notes} &= 1 \times \bar{w} = £1200 \\ \text{in £5 notes} &= 5 \times \bar{f} = £3000 \\ \text{in £10 notes} &= 10 \times \bar{t} = £500 \end{aligned}$$

The standard deviations of the amount of money issued:

$$\begin{aligned} \text{in £1 notes} &= \sqrt{1^2 \times (sd_w)^2} = £250 \\ \text{in £5 notes} &= \sqrt{5^2 \times (sd_f)^2} = £500 \\ \text{in £10 notes} &= \sqrt{10^2 \times (sd_t)^2} = £50 \end{aligned}$$

The total amount of money issued = w + 5f + 10t = A
If w, f, t, are all normally distributed, then A is also normally distributed with mean and standard deviations as follows:

$$\bar{A} = \bar{1} + 5\bar{f} + 10\bar{t} = 1200 + 3000 + 500 = £4700$$

$$\begin{aligned} sd_A &= \sqrt{1^2 \times (sd_w)^2 + 5^2 \times (sd_f)^2 + 10^2 \times (sd_t)^2} \\ &= \sqrt{250^2 + 500^2 + 50^2} = \sqrt{315000} = £561.25 \end{aligned}$$

450 QUANTITATIVE ANALYSIS

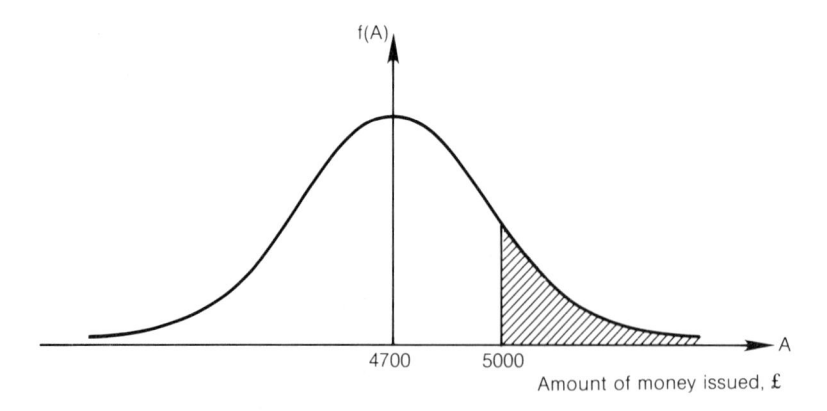

We will take 'amount of money issued' as being approximately a continuous variable, hence the continuity correction is not applied. £5000 is z standard deviations above the mean, where:

$$z = \frac{5000 - 4700}{561.25} = 0.5345$$

Using the standard normal tables, $P(z > 0.53) = 0.2963$.
Therefore, P(amount of money issued $>$ £5000) $= 0.2963$.

2 £1 notes: require P(demand for £w $<$ £1600)

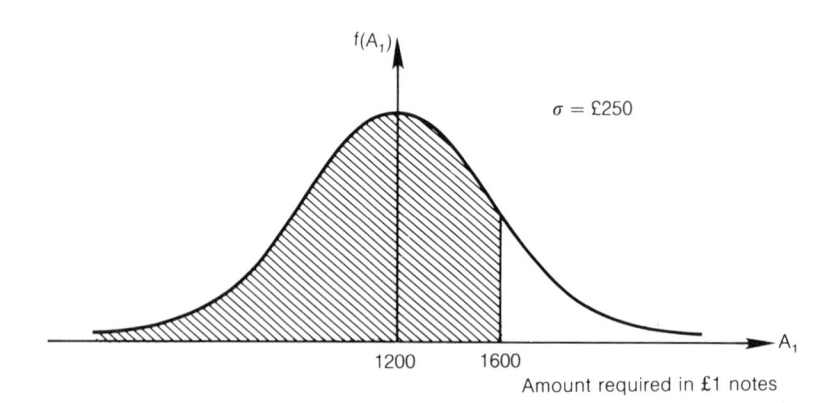

£1600 is z standard deviations above the mean where:

$$z = \frac{1600 - 1200}{250} = 1.6$$

From the normal tables P(demand for £1 $<$ £1600) $= 1 - 0.0548$
$= 0.9452$.

£5 notes: require P(demand for £5 $<$ £3500)

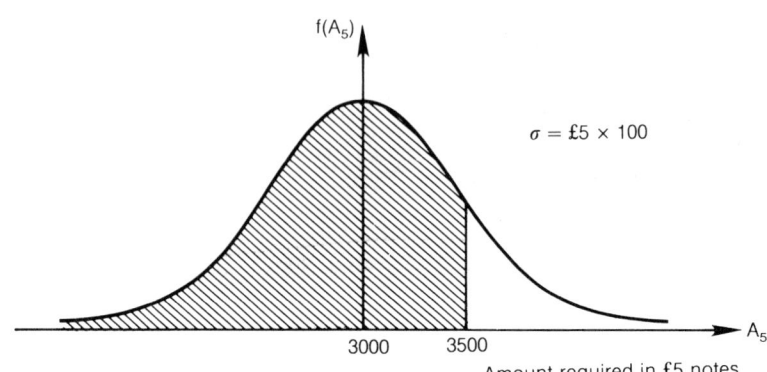

As before:
$$z = \frac{3500 - 3000}{500} = 1.0$$

From tables $P(z > 1.0) = 0.1587$, therefore:
$$P(\text{demand for £5} < £3500) = 1 - 0.1587 = 0.8413$$

£10 notes: require P(demand for £10 notes < 600)

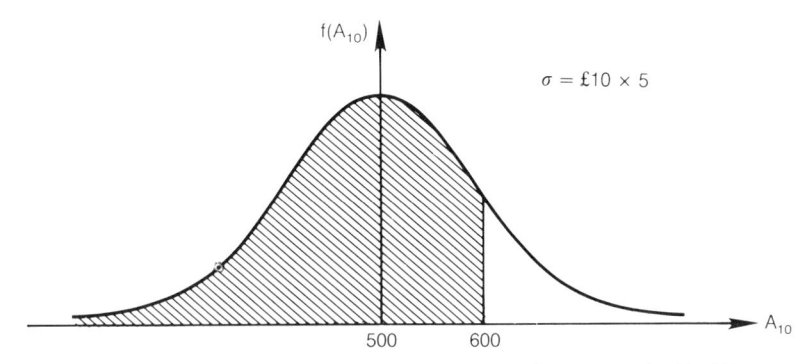

Amount required in £10 notes

As before:
$$z = \frac{600 - 500}{50} = 2.0$$

From tables, $P(z > 2.0) = 0.02275$, therefore:
$$P(\text{demand for £10 notes} < £600) = 1 - 0.02275 = 0.97725$$

Therefore:
$$P(\text{all demands will be met}) = 0.9452 \times 0.8413 \times 0.97725$$
$$= 0.7771$$

The £5 note is the most critical denomination, since this has the lowest probability that the demand will be met.

3 The distributions are now no longer independent and therefore the calculation of the relevant probabilities will no longer be straightforward. We can, however, write down the relationships between the three variables. We still require:

(i) $w \leqslant £1600$, but if $w < £1600$, the surplus can be used instead of £5 notes, hence:

$$5f \leqslant 3500 + (1600 - w) \quad \text{if} \quad w \leqslant £1600$$
(surplus of £1 notes)

Therefore:

(ii) $5f + w \leqslant £5100$ if $w \leqslant £1600$. In addition, surplus £5 and £1 notes could be used instead of £10 notes, hence:

$$10t \leqslant 600 + \text{surplus of £5 and £1 notes.}$$

Therefore:

$$10t \leqslant 600 + (3500 + (1600 - w) - 5f)$$

Hence:

(iii) $10t + 5f + w \leqslant 5700$

As remarked above, the probabilities of event (ii) and (iii) are not simple to evaluate.

Exercise 2.21

1 r represents the demand per day for a hire car, in excess of 10, therefore if 10 cars are rented r = 0, if 11 cars are rented r = 1, etc. If r follows a Poisson distribution with a mean of 4, then:

$$P(r) = \frac{4^r e^{-4}}{r!} \quad r = 0, 1, 2, \ldots$$

$$P(0) = \frac{4^0 e^{-4}}{0!} = 0.0183$$

$$P(1) = \frac{4 e^{-4}}{1!} = 0.0733$$

$$P(2) = \frac{4^2 e^{-4}}{2!} = 0.1465$$

$$P(3) = \frac{4^3 e^{-4}}{3!} = 0.1954$$

$$P(4) = \frac{4^4 e^{-4}}{4!} = 0.1954$$

$$\text{Total} = 0.6289$$

Fifteen cars will be used if the demand is for 15 or more cars. Therefore:

$$P(r \geqslant 5) = 1 - 0.6289 = 0.3711$$

$$P(10 \text{ cars used}) = 0.0183$$
$$P(11 \text{ cars used}) = 0.0733$$
$$P(12 \text{ cars used}) = 0.1465$$
$$P(13 \text{ cars used}) = 0.1954$$
$$P(14 \text{ cars used}) = 0.1954$$
$$P(15 \text{ cars used}) = 0.3711$$

Expected number of cars rented each day = $\sum$ (no rented) × probability

$$= 10 \times 0.0183 + 11 \times 0.0733 + \cdots + 15 \times 0.3711$$
$$= 13.59$$

Therefore the expected number of cars rented per day is 13.6.

2 Net income per car hired per day = £(20 − 2) = £18 per day
Fixed holding cost = 15 × £6 = £90 per day
Therefore expected profit per day

$$= (\text{net income/day} \times \text{average number of cars hired day}) - \text{holding cost}$$
$$= £18 \times 13.59 - £90 \text{ per day}$$
$$= £154.62/\text{day}$$

3 P(demand $\geqslant$ 14) = 0.1954 + 0.3711 = 0.5665
P(demand = 15) = 0.1563
Therefore:
$$P(\text{demand} \geqslant 16) = 0.3711 - 0.1563 = 0.2148$$

If the firm has 14 cars, expected number of cars hired per day

$$= 10 \times 0.0183 + 11 \times 0.0733 + 12 \times 0.1465 + 13 \times 0.1954 + 14 \times 0.5665$$
$$= 13.2185$$

Therefore:

$$\text{expected profit per day} = £18 \times 13.2185 - 14 \times £6$$
$$= £153.93 \text{ per day}$$

If the firm has 16 cars, expected number of cars hired per day

$$= 10 \times 0.0183 + \cdots + 15 \times 0.1563 + 16 \times 0.2148$$
$$= 13.8044$$

Therefore expected profit per day $= £18 \times 13.8044 - 16 \times £6$
$$= £152.48 \text{ per day}$$

The largest profit per day is obtained with 15 cars, hence the firm is not advised to change.

Let the holding cost be £h per day.
The expected profit per day on 15 cars is greater than that on 14 cars if

$$18 \times 13.59 - 15h > 18 \times 13.2185 - 14h.$$

Therefore:

$$6.69 > h.$$

Hence, 15 cars are preferred to 14 if the daily holding cost is less than £6.69. The expected profit per day on 15 cars is greater than that on 16 cars if:

$$18 \times 13.59 - 15h > 18 \times 13.8044 - 16h.$$

Therefore:

$$h > 3.86$$

Hence 15 cars are preferred to 16 if the daily holding cost is more than £3.86. Therefore 15 cars is the optimum number if the daily holding cost is between £3.86 and £6.69.

Exercise 2.22

1

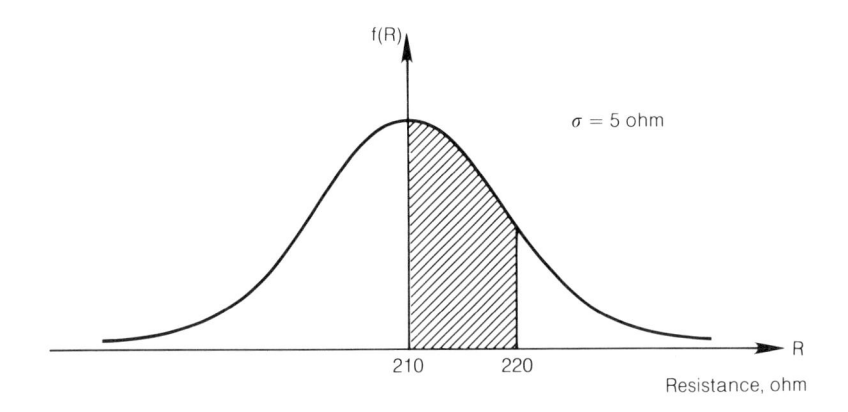

First of all, we require the probability that a solenoid will have a resistance between 210 and 220 ohms. From the symmetry of the normal distribution, we know that $P(R < 210 \text{ ohm}) = 0.5$. 220 ohms is z standard deviations above the mean where:

$$z = \frac{220 - 210}{5} = 2.0$$

From the standard normal tables, $P(z \leqslant 2.0) = 1 - 0.02275$
$ = 0.97725$

Therefore:

$$P(210 < R < 220 \text{ ohm}) = 0.97725 - 0.5$$
$$\phantom{P(210 < R < 220 \text{ ohm})} = 0.47725$$

Hence, the expected number of solenoids with a resistance in the useful range is $0.47725 \times 2000 = 954.5 \approx 955$. Therefore, the expected number to be scrapped and replaced is $2000 - 955 = 1045$.

Costs:
(i) If the whole stock is tested:

Test: 2000 @ £1.40 = £2800
Net Replacement: 1045 @ £3.50 = £3657.5
Total net cost: £6457

(ii) If we scrap the whole stock and replace:

Purchase: 2000 @ £4.00 = £8000
Scrap: 2000 @ £0.50 = £1000
Total net cost: £7000

Hence, there is a saving in cost of £7000 − £6457 = £542 if all the solenoids are tested, and only the unsatisfactory ones replaced.

2

Mean	P(210 < R < 220 ohm)	Exp number usable	Exp number to replace/scrap
205	0.15731	315	1685
210	0.47725	955	1045
215	0.68268	1365	635
220	0.47725	955	1045
225	0.15731	315	1685

Mean	Cost if test all stock, C_1, £	Cost if replace all stock, C_2, £
205	2800 + 5897.5 = 8697.5	7000
210	2800 + 3657.5 = 6457.5	7000
215	2800 + 2222.5 = 5022.5	7000
220	2800 + 3657.5 = 6457.5	7000
225	2800 + 5897.5 = 8697.5	7000

Graph of costs from the two available courses against possible mean values of resistance

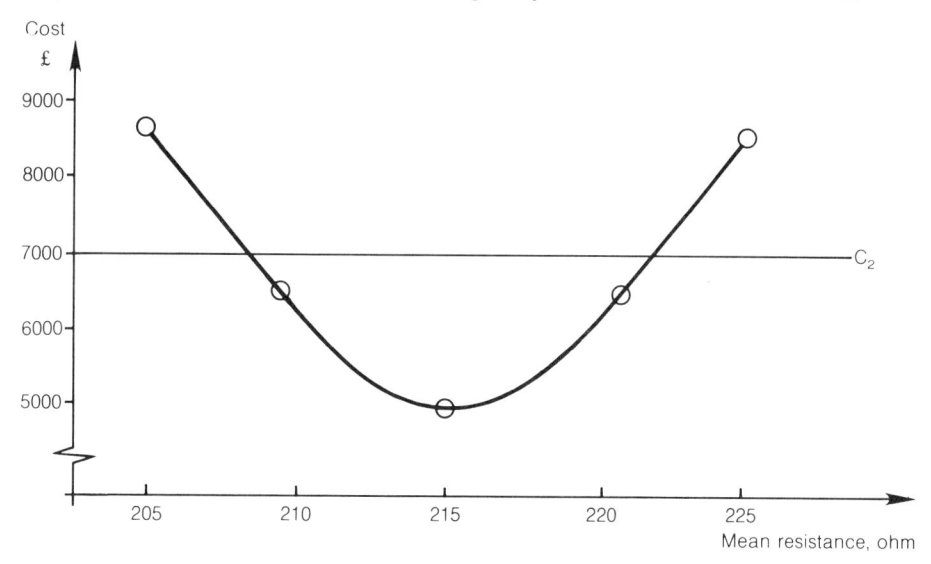

The mean resistance must lie between R_1 and R_2 ohms for the testing method to be more economic.

From the graph, $R_1 = 209$ ohms to nearest ohm
$R_2 = 221$ ohms to nearest ohm

3 If all values are equally likely, the probability of each one $= 0.2$.
Therefore the expected cost of testing $= \sum$ cost $\times$ probability

$$= 8697.5 \times 0.2 + 6457.5 \times 0.2 + 5022.5 \times 0.2 + 6457.5 \times 0.2 + 8697.5 \times 0.2$$
$$= £7066.50$$

This is greater by £66.50 than the cost of replacing the whole stock, hence you would decide to replace the whole stock.

If $P(\mu = 215 \text{ ohms})$ is twice as likely as each of the others, then:

$$P(\mu = 205) = P(\mu = 210) = P(\mu = 220) = P(\mu = 225) = 1/6$$

and $P(\mu = 215) = 1/3$. Therefore expected cost of testing

$$= 1/6(8697.5 + 6457.5 + 6457.5 + 8697.5) + 1/3 \times 5022.5$$
$$= £6725.8$$

Since this is now less than the cost of total replacement, the previous decision would be reversed.

Solutions to Chapter Three exercises

ANSWERS TO EXERCISES

Exercise 3.1

First, calculate the payoff table:

		\multicolumn{5}{c}{Payoff—Profit £/day}
		\multicolumn{5}{c}{Number of loaves made/day}
		\multicolumn{5}{c}{(possible decisions)}
		10,000	12,000	14,000	16,000	18,000
Possible	10,000	1,000	600	200	−200	−600
outcomes:	12,000	1,000	1,200	800	400	0
Number of	14,000	1,000	1,200	1,400	1,000	600
loaves/day	16,000	1,000	1,200	1,400	1,600	1,200
demanded	18,000	1,000	1,200	1,400	1,600	1,800

Using rule (a)—maximax payoff—maximise the maximum payoff:

Loaves made/day	10,000	12,000	14,000	16,000	18,000
Maximum payoff £	1,000	1,200	1,400	1,600	1,800
					↑
					MAX

Therefore, we produce 18,000 loaves per day with this rule.
Using rule (b)—maximin—maximise the minimum payoff.

Loaves made/day	10,000	12,000	14,000	16,000	18,000
Minimum payoff, £/day	1,000	600	200	−200	−600
	↑				
	MAX				

Therefore, we produce 10,000 loaves per day with this rule.
Using rule (d)—maximum expected payoff—the probability of each demand actually occurring can be determined from the frequency distribution given:

Demand/day, '000s	10	12	14	16	18	Total
Number of days	5	10	15	15	5	50
Probability	5/50 = 0.1	10/50 = 0.2	0.3	0.3	0.1	1.0

Use these probabilities with the payoff table to evaluate the expected payoff for each course of action. Each row of the payoff table is multiplied by the probability of that outcome arising. The expected payoff table is:

SOLUTIONS TO EXERCISES: CHAPTER THREE

		Expected payoff—expected profit £/day				
		Number of loaves made/day (possible decisions)				
		10,000	12,000	14,000	16,000	18,000
Possible	10,000	100	60	20	−20	−60
outcomes:	12,000	200	240	160	80	0
number of	14,000	300	360	420	300	180
loaves	16,000	300	360	420	480	360
demanded/day	18,000	100	120	140	160	180
Expected profit per day, £		1,000	1,140	1,160	1,000	660
				↑		
				MAX		

Therefore, we can produce 14,000 loaves per day to maximise the expected payoff £1160 per day.

For rules (c) and (e) we must construct the opportunity loss table. This is done by taking each *row* of the payoff table, ie each outcome. The best decision gives an opportunity loss of zero for each outcome.

		Opportunity loss £/day					
		Number of loaves made/day					
		(possible decision)					
		10,000	12,000	14,000	16,000	18,000	Probability
Possible	10,000	0	400	800	1,200	1,600	0.1
outcomes:	12,000	200	0	400	800	1,200	0.2
number of	14,000	400	200	0	400	800	0.3
loaves/day	16,000	600	400	200	0	400	0.3
demanded	18,000	800	600	400	200	0	0.1

Using rule (c)—the minimax rule—to minimise the maximum opportunity loss:

Loaves made/day	10,000	12,000	14,000	16,000	18,000
Maximum opportunity loss £	800	600	800	1,200	1,600
		↑			
		MIN			

Therefore, we produce 12,000 loaves per day with the minimax rule.

Rule (e)—minimum expected opportunity loss—uses the same method as rule (c). Multiply each demand by the probability of its occurrence to create the expected opportunity loss table.

		Expected opportunity loss, £/day				
		Number of loaves made/day (possible decisions)				
		10,000	12,000	14,000	16,000	18,000
Possible	10,000	0	40	80	120	160
outcomes:	12,000	40	0	80	160	240
number of	14,000	120	60	0	120	240
loaves/day	16,000	180	120	60	0	120
demanded	18,000	80	60	40	20	0
Expected opportunity loss, £/day		420	280	260	420	760
				↑		
				MIN		

Therefore, produce 14,000 loaves per day to minimise the expected opportunity loss, at £260 per day. (*Note*: this decision is always the same as the one which is made when maximising the expected payoff.)

Calculate the value of perfect information. This can be done in two ways: either use the payoff table or the expected opportunity loss.

(a) Using the payoff table. With perfect information the decision would be matched exactly to the outcome each day:

Outcome: Loaves demanded/day	10,000	12,000	14,000	16,000	18,000
Payoff £/day	1,000	1,200	1,400	1,600	1,800
Probability of outcome	0.1	0.2	0.3	0.3	0.1

Therefore, expected payoff/day = $\sum$ (Probability × Payoff) = £1420/day with perfect information. Therefore, the value of perfect information

= expected payoff with perfect information − without perfect information
= 1420 − 1160 = £260/day.

(b) The minimum expected opportunity loss = the value of perfect information.

Exercise 3.2

Calculate the theatre payoff table:

$$\text{Costs} = £200 + £0.3 \times \text{programmes produced/run}$$

$$\text{Income} = £300 + £0.6 \times \text{programmes sold/run}$$

Where programmes sold = 0.4 × size of audience.

		\multicolumn{5}{c}{Payoff, £/run — Number of programmes produced/run}
		1,600	1,800	2,000	2,200	2,400
Possible	1,600	580	520	460	400	340
outcomes:	1,800	580	640	580	520	460
programmes	2,000	580	640	700	640	580
demanded	2,200	580	640	700	760	700
/run	2,400	580	640	700	760	820

1 Using the rule of maximax payoff.

Decision: Programmes/run	1,600	1,800	2,000	2,200	2,400
Maximum payoff £/run	580	640	700	760	820 ↑ **MAX**

Therefore, we produce 2400 programmes per run to maximise the maximum payoff.

Using the maximin rule—to maximise the minimum payoff:

Decision: programmes/run	1,600	1,800	2,000	2,200	2,400
Minimum payoff £/run	580 ↑ **MAX**	520	460	400	340

Therefore, we produce 1600 programmes per run to maximise the minimum payoff.

Using the minimax rule, to minimise the maximum opportunity loss—the opportunity loss table is constructed from the rows of the payoff table:

		Opportunity loss £/run Number of programmes produced/run (possible decisions)					Probability
		1,600	1,800	2,000	2,200	2,400	
Possible	1,600	0	60	120	180	240	0.1
outcomes:	1,800	60	0	60	120	180	0.3
programmes	2,000	120	60	0	60	120	0.3
demanded/run	2,200	180	120	60	0	60	0.2
	2,400	240	180	120	60	0	0.1
Maximum opportunity loss		240	180	120	180	240	

↑
MIN

Therefore, we produce 2,000 programmes per run to minimise the maximum opportunity loss.

To maximise expected payoff per run—multiply each column of the payoff table by the probabilities given in the exercise. Use the method illustrated in **Exercise 1.1** to construct the expected payoff table:

Decision: programmes	1,600	1,800	2,000	2,200	2,400
Expected payoff £/run	580	628	640	616	568

↑
MAX

Therefore, we produce 2,000 programmes per run to maximise expected payoff.

To minimise expected opportunity loss—multiply each column of the opportunity loss table by the probabilities to calculate the expected loss for each decision:

Decision: programmes	1,600	1,800	2,000	2,200	2,400
Expected opportunity loss £	114	66	54	78	126

↑
MIN

Therefore, produce 2,000 programmes per run to minimise the expected opportunity loss. (Notice again that this is always the same decision as the one to maximise the expected payoff. In addition, notice that the differences between the maximum expected payoff (£640 per run) and the other expected payoffs, are exactly the same as the differences between the minimum expected opportunity loss (£54 per run) and the other expected opportunity losses.)

The value of perfect information. This equals the minimum opportunity loss of £54 per run, first we will check this using the payoff table:

Outcome-audience/run	4,000	4,500	5,000	5,500	6,000
Programmes	1,600	1,800	2,000	2,200	2,400
Payoff £/run	580	640	700	760	820
Probability of outcome	0.1	0.3	0.3	0.2	0.1

Expected payoff with perfect information = £694 per run. Therefore value of perfect information

= expected payoff with perfect information − expected payoff without perfect information
= £694 − £640 = £54/run

which confirms the original figure from the opportunity loss table.

2 If the audience is over 5250 and the demand for programmes is met, the advertisers will increase their payment by £100 per run. How will this affect the decisions in **1**? The payoff table will change but only the last two rows (audiences over 5250) and only then when demand is met. We have added £100 to the * figures. The new payoff table will be:

		Number of programmes produced/run (possible decisions)					
		1600	1800	2000	2200	2400	Prob
Possible	1600	580	520	460	400	340	0.1
outcomes:	1800	580	640	580	520	460	0.3
programmes	2000	580	640	700	640	580	0.3
demanded	2200	580	640	700	860*	800*	0.2
per run	2400	580	640	700	760	920*	0.1

Repeat this for each of the 5 rules:

Maximax rule: the maximum maximum has changed from £820 to £920 but the decision is unchanged at 2400.

Maximum rule: this is not affected by the change.

Minimax opportunity loss rule: the new table is given below. The last 2 rows have changed:

		Number of programmes produced/run (possible decisions)					
		1600	1800	2000	2200	2400	Prob
Possible	1600	0	60	120	180	240	0.1
outcomes:	1800	60	0	60	120	180	0.3
programmes	2000	120	60	0	60	120	0.3
demanded	2200	280	220	160	0	60	0.2
per run	2400	340	280	220	160	0	0.1
Maximum opportunity loss		340	280	220	**180**	240	

↑
MIN

Therefore the decision has changed from 2000 to 2200 programmes per run.

To maximise expected payoff: these values have changed for the last 2 decisions, giving:

Decision: programmes	1600	1800	2000	2200	2400
Expected payoffs £/run	580	628	**640**	636	598

↑
MAX

The maximum expected payoff is still given by the decision to produce 2000 programmes.

To minimise expected opportunity loss: this should give the same solution as rule **4**:

Decision: programmes	1600	1800	2000	2200	2400
Expected opportunity loss	144	96	**84**	88	126

↑
MIN

As expected, this gives the same solution as maximising expected payoff. The decision is still to produce 2000 programmes.

3 Sensitivity—the probabilities change:

From	0.1	0.3	0.3	0.2	0.1
To	0.2	0.2	0.2	0.2	0.2

Would this change the decision to order 2000 programmes using the rule of maximising expected payoff? Using the original payoff table in part **1**, the expected payoffs with the new probabilities are:

Decision: programmes	1600	1800	2000	2200	2400
Expected payoffs (new)	580	616	628	616	580
Expected payoffs (old)	580	628	640	616	568

↑
MAX

Therefore, the decision to produce 2000 programmes is not changed by the change in probabilities.

Exercise 3.3

Calculate Kilroy's beverage payoff table:
Each batch contains 40 pints.
Variable cost per batch = 0.7 × 40 = £28 per batch.
Selling price per batch = 1.5 × 40 = £60 per batch.
Failure to meet demand cost per batch = 0.3 × 40 = £12 per batch.
Probabilities are: 0.1 0.2 0.3 0.2 0.2
(The probabilities are calculated from the frequency distribution as illustrated in **Exercise 3.1**.)

		Payoff £/week Number of batches produced/week (Possible decisions)				
		3	4	5	6	7	Prob.
Possible	3	96	68	40	12	−16	0.1
outcomes:	4	84	128	100	72	44	0.2
number of	5	72	116	160	132	104	0.3
batches	6	60	104	148	192	164	0.2
demanded	7	48	92	136	180	224	0.2

1 Using the rule to maximise maximum payoff:

Decision: batches/week	3	4	5	6	7
Maximum payoff £/week	96	128	160	192	224

↑
MAX

Therefore, we can produce 7 batches per week to maximise maximum payoff.
Using the maximin rule—to maximise minimum payoff.

Decision: batches/week	3	4	5	6	7
Minimum payoff £/week	48	68	40	12	−16

↑
MAX

Therefore, we can produce 4 batches per week to maximise minimum payoff.

Using the rule to minimise the maximum opportunity loss—construct the opportunity loss table from the rows of the payoff table:

		Opportunity loss £/week Number of batches produced/week (possible decisions)					
		3	4	5	6	7	Prob
Possible	3	0	28	56	84	112	0.1
outcomes:	4	44	0	28	56	84	0.2
number of	5	88	44	0	28	56	0.3
batches	6	132	88	44	0	28	0.2
demanded	7	176	132	88	44	0	0.2
Maximum opportunity loss		176	132	88	84	112	

↑
MIN

Therefore, we produce 6 batches per week to minimise the maximum opportunity loss.

Using the rule to maximise expected payoff—multiply each column of the payoff table by the probabilities:

Decision: batches/week	3	4	5	6	7
Expected payoff £/week	69.6	106.4	128.8	129.6	116

↑
MAX

Therefore, we produce 6 batches per week to maximise expected payoff.

Using the rule to minimise expected opportunity loss:

Decision: batches/week	3	4	5	6	7
Expected opportunity loss	96.8	60	37.6	36.8	50.4

↑
MIN

Therefore, we produce 6 batches per week to minimise expected opportunity loss.

Calculate the value of perfect information. From the minimum expected opportunity loss, we know this should be £36.8 per week—again we will check using the payoff table:

Outcome: batches demanded/week	3	4	5	6	7
Payoff £/week	96	128	160	192	224
Probability of outcome	0.1	0.2	0.3	0.2	0.2

expected payoff with perfect information = £166.4. Therefore, value of perfect information = expected payoff with perfect information − expected payoff without perfect information = 166.4 − 129.6 = £36.8 per week, which equals the minimum expected opportunity loss.

2 Sensitivity: if Kilroys increase the price from £1.50 to £1.75 per pint would this change any of the decisions? Assuming that the demand is not affected by the price change, we have to reconstruct the payoff table and repeat all of the above calculations:

		Number of batches produced/week (possible decisions)					
		3	4	5	6	7	Prob
Possible	3	126	98	70	42	14	0.1
outcomes:	4	114	168	140	112	84	0.2
number of	5	102	156	210	182	154	0.3
batches	6	90	144	198	252	224	0.2
demanded/wk	7	78	132	186	240	294	0.2

SOLUTIONS TO EXERCISES: CHAPTER THREE

Using the rule to maximise maximum payoff. There is no change. We would still produce 7 batches, giving a maximum possible payoff of £294 per week.

Using the rule to maximise minimum payoff. There is no change. We would still produce 4 batches, giving a maximum possible payoff of £98 per week.

Using the rule to minimise the maximum opportunity loss—construct the opportunity loss table:

		Number of batches produced/week (possible decisions)					Prob
		3	4	5	6	7	
Possible	3	0	28	56	84	112	0.1
outcomes:	4	54	0	28	56	84	0.2
number of	5	108	54	0	28	56	0.3
batches	6	162	108	54	0	28	0.2
demanded/week	7	216	162	108	54	0	0.2

Using the minimax rule there is no change. We would still produce 6 batches per week giving the same maximum opportunity loss of £84 per week.

Using the rule of maximising expected payoff:

Decision: batches/week	3	4	5	6	7
Expected payoff £/week	99.6	145.4	174.8	179.6	168
				↑	
				MAX	

Again there is no change in the decision. We would still produce 6 batches per week, giving an expected payoff of £179.6 per week.

Using the rule to minimise expected opportunity loss:

Decision: batches/week	3	4	5	6	7
Expected opportunity loss £/week	118.8	73	43.6	38.8	50.4
				↑	
				MIN	

Again there is no change in the decision. We would still produce 6 batches per week.

Exercise 3.4

Calculate the book publisher's payoff table. Every book sold makes a contribution of £9. Every book not sold loses £4 per book. Every demand not met loses £1 per book.

		Payoff table £/3 years Number of books produced in 3 years. (possible Decisions)				Prob
		2000	3000	4000	5000	
Possible	2000	18,000	14,000	10,000	6,000	0.1
outcomes:	3000	17,000	27,000	23,000	19,000	0.5
books demanded	4000	16,000	26,000	36,000	32,000	0.2
per 3 years	5000	15,000	25,000	35,000	45,000	0.2

Using the maximax payoff rule:

Decision: books printed	2000	3000	4000	5000
Maximum payoff £/3years	18,000	27,000	36,000	45,000
				↑
				MAX

Therefore, we print 5000 books for the 3 year period to maximise maximum payoff.

Using the maximin rule—to maximise minimum payoff:

Decision: books printed	2000	3000	4000	5000
Minimum payoff £/3 years	15,000	14,000	10,000	6000

↑
MAX

Therefore, we print 2000 books in the 3 year period to maximise minimum payoff.

Using the rule to minimise the maximum opportunity loss—construct the opportunity loss table from the rows of the payoff table:

		Opportunity loss £/3 years. Number of books printed for 3 years (possible decisions)				
		2000	3000	4000	5000	Prob.
Possible outcome:	2000	0	4000	8000	12,000	0.1
books demanded	3000	10,000	0	4000	8000	0.5
per 3 years	4000	20,000	10,000	0	4000	0.2
	5000	30,000	20,000	10,000	0	0.2
Decision: books printed		2000	3000	4000	5000	
Maximum opportunity loss		30,000	20,000	10,000	12,000	

↑
MIN

Therefore, we print 4000 books for the 3 year period to minimise maximum opportunity loss.

Using the rule to maximise expected payoff—multiply each column of the payoff table by the probabilities.

Decision: books printed	2000	3000	4000	5000
Expected payoff £/3 years	16,500	25,100	26,700	25,500

↑
MAX

Therefore, we print 4000 books for the 3 year period to maximise expected payoff.

Using the rule to minimise expected opportunity loss:

Decision: books printed	2000	3000	4000	5000
Expected opportunity loss £/3 years	15,000	6400	4800	6000

↑
MIN

Therefore, we print 4000 books for the 3 year period to minimise expected opportunity loss.

Calculate the value of perfect information. The minimum expected opportunity loss is £4800 over 3 years. This is the value of perfect information. We will check using the payoff table:

Outcome: books demanded/3 years	2000	3000	4000	5000
Payoffs, £/3 years	18,000	27,000	36,000	45,000
Probability of outcome	0.1	0.5	0.2	0.2

Expected payoff with perfect information = £31,500 per 3 years. Therefore, value of perfect information = expected payoff with perfect information − expected payoff without perfect information = 31,500 − 26,700 = £4800 over 3 years, which is the same as the minimum expected opportunity loss.

Exercise 3.5

1   Plot the utility curves for the marketing and finance directors:

SOLUTIONS TO EXERCISES: CHAPTER THREE

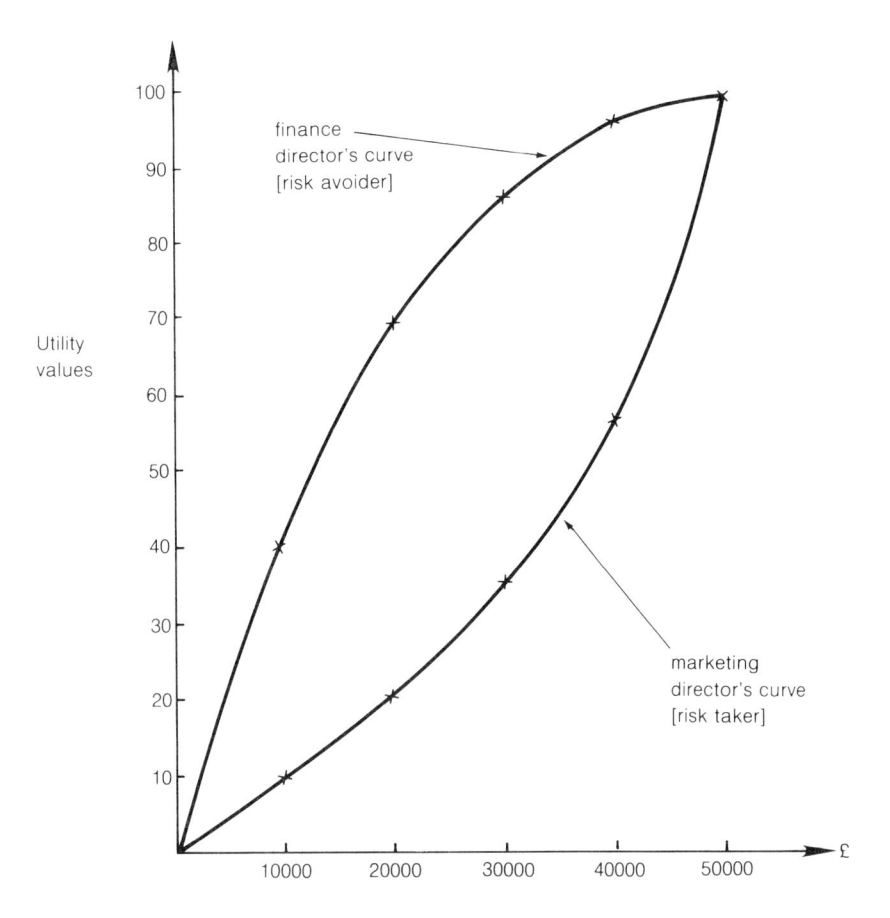

[Graph showing utility curves: finance director's curve (risk avoider) as a concave curve and marketing director's curve (risk taker) as a convex curve. X-axis: £ from 0 to 50000; Y-axis: Utility values from 0 to 100.

Finance director's curve points approximately: (10000, 40), (20000, 69), (30000, 86), (40000, 96), (50000, 100).
Marketing director's curve points approximately: (10000, 10), (20000, 20), (30000, 35), (40000, 57), (50000, 100).]

2   From the payoff table in **Exercise 3.4** we take each payoff and estimate its utility value, first using the finance director's curve, then the marketing director's.

		\multicolumn{4}{c}{Numbers of books produced in 3 years (possible decisions)}			
		2000	3000	4000	5000	Prob.
Possible	2000	65	55	40	26	0.1
outcomes:	3000	63	82	76	67	0.5
books	4000	60	80	92	87	0.2
demanded/3 years	5000	57	78	91	98	0.2
Expected utility: Finance director:		61.4	78.1	78.6 ↑ MAX	73.1	

		2000	3000	4000	5000	Prob.
Possible	2000	18	14	10	6	0.1
outcomes:	3000	17	30	24	19	0.5
books	4000	16	28	44	38	0.2
demanded/3 years	5000	15	27	42	72	0.2
Expected utility: Marketing director:		16.5	27.4	30.2	32.1 ↑ MAX	

466 QUANTITATIVE ANALYSIS

The original decision in **Exercise 3.4** using the criterion of maximising expected payoff was to print 4000 books.

Using the two directors' utilities to reflect their attitudes to the money involved, we obtain two solutions. The (risk avoiding) finance director would still make the decision to print 4000 books, but his expected utility for printing 3000 is almost as large (78.1 compared to 78.6). The marketing director, however, would decide to print 5000 books, showing his risk taking attitude.

Exercise 3.6

Which routes should be taken to minimise the total travel time for a journey from Stafford to Nottingham to Cambridge? The decision tree is given below:

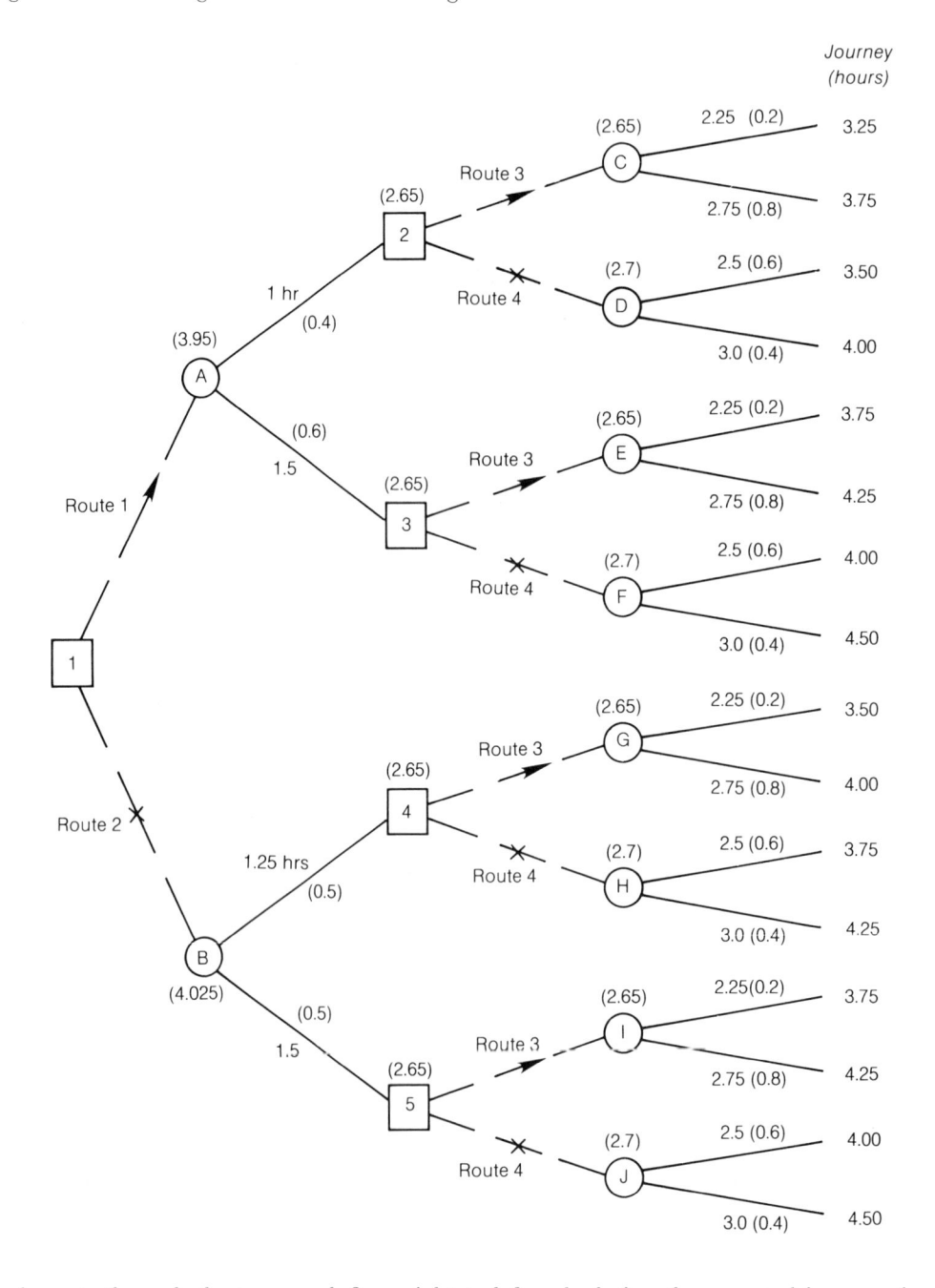

For each route through the tree, work from right to left, calculating the expected journey times at each node. For example, at node C:

Expected time = $2.25 \times 0.2 + 2.75 \times 0.8 = 2.65$ hours

At node D:

$$\text{Expected time} = 2.5 \times 0.6 + 3.0 \times 0.4 = 2.7 \text{ hours}$$

Hence, at decision node 2, choose route 3 since this has the shorter expected time. Transfer the expected time of 2.65 hours to node 2. Repeat this procedure throughout the tree. The final decision is: take route 1 from Stafford to Nottingham, then route 3 from Nottingham to Cambridge.

Exercise 3.7

1   The decision tree is:

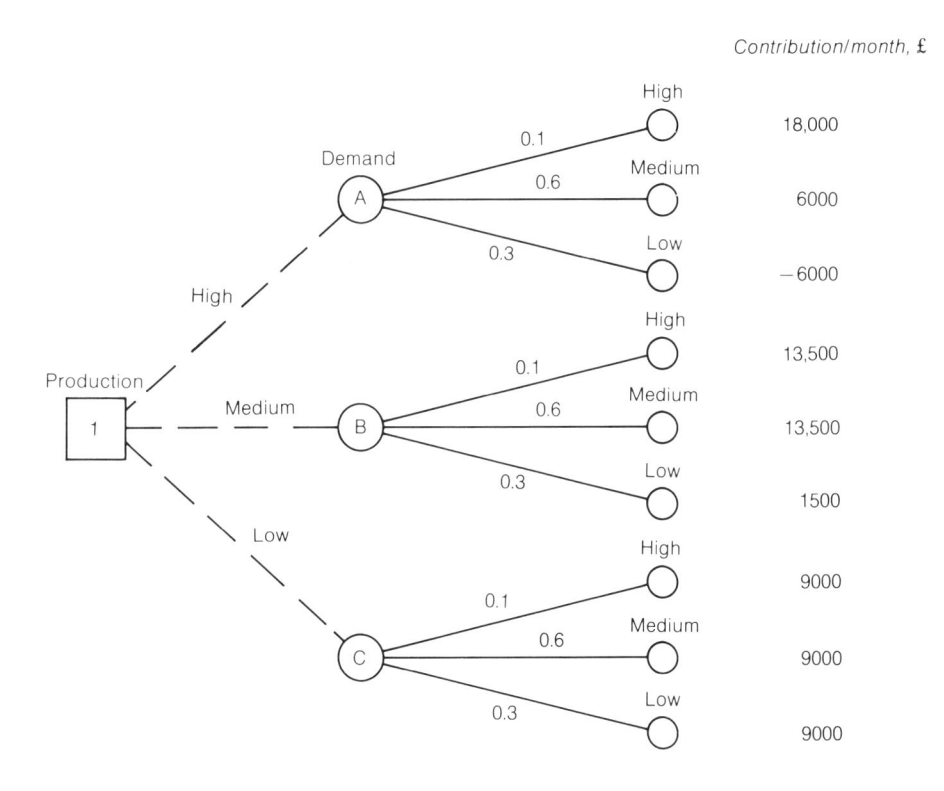

2   The advice is based on the decision which leads to the largest expected contribution per month.

Node A:

$$\text{Expected contribution/month} = \sum (\text{contribution} \times \text{probability})$$
$$= 18{,}000 \times 0.1 + 6000 \times 0.6 + (-6000) \times 0.3$$
$$= £3600 \text{ per month}$$

Node B:

$$\text{Expected contribution/month} = 13{,}500 \times 0.1 + 13{,}500 \times 0.6 + 1500 \times 0.3$$
$$= £9900 \text{ per month}$$

Node C: Expected contribution/month = £9000 per month

Hence, the decision should be made to produce 15 tons per month since this leads to the highest expected contribution per month.

3 We require:

$$P(\text{demand} > Q) = \frac{v}{p} = \frac{1,500}{2,400} = 0.625$$

when it is assumed that the demand is normally distributed.

Demand, tons, D	Probability p	D × P	D² × p
10	0.3	3	30
15	0.6	9	135
20	0.1	2	40
Total	1.0	14	205

The estimate of the mean of the normal distribution of the demand is:

$$\mu = \sum D \times p = 14 \text{ tons.}$$

The estimate of the standard deviation is:

$$\sigma = \sqrt{\sum D^2 p - (\sum Dp)^2} = \sqrt{205 - 14^2} = 3 \text{ tons}$$

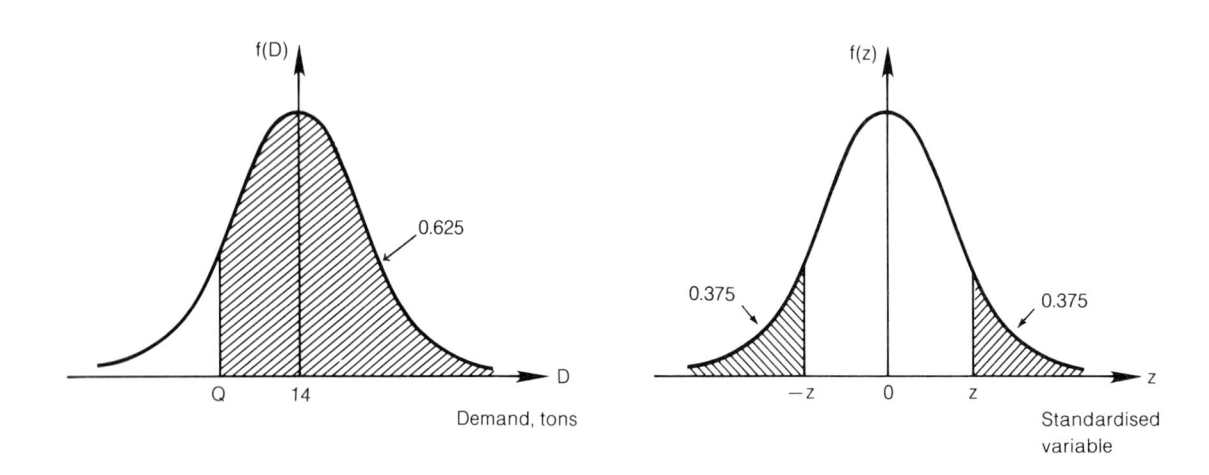

P(demand > Q) = 0.625, hence P(demand < Q) = 0.375. If z standard deviations above the mean cuts off the top 0.375 of the distribution, then from the standard normal tables, z = 0.319. Therefore, by symmetry, P(z < −0.319) = 0.375. Hence:

$$-0.319 = \frac{Q - 14}{3}$$

therefore:

$$Q = 14 - (3 \times 0.319) = 13.04 \text{ tons}$$

The optimum production level is 13 tons per month.

SOLUTIONS TO EXERCISES: CHAPTER THREE

Exercise 3.8

1 (a) The decision tree is:

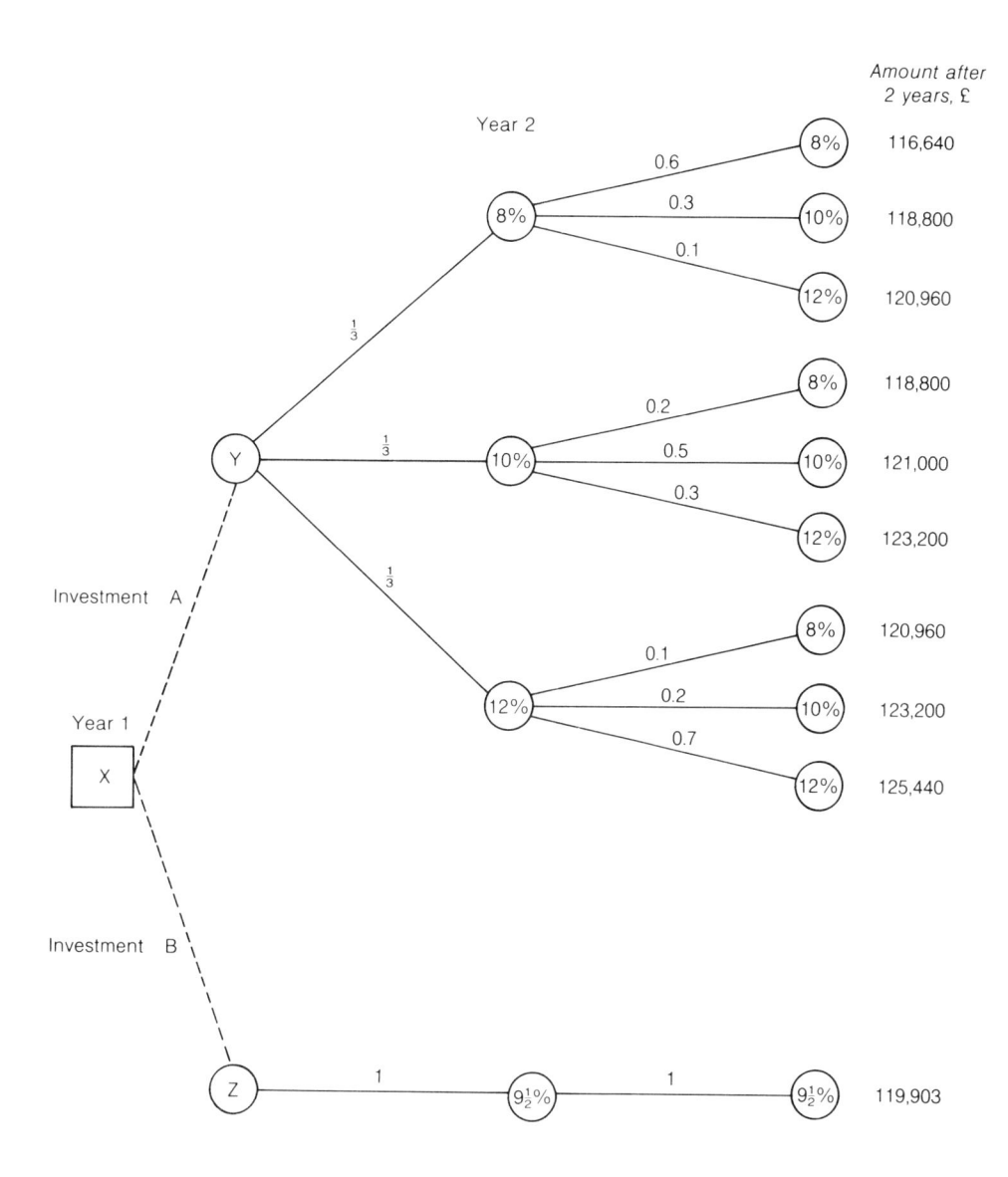

(b) Work through the tree from right to left, calculating the expected amount accumulated at each node. At node Y, the expected accumulated amount is:

$$116{,}640 \times 0.6 \times (1/3) + 118{,}800 \times 0.3 \times (1/3) + 120{,}960 \times 0.1 \times (1/3)$$
$$+ 118{,}800 \times 0.2 \times (1/3) + 121{,}000 \times 0.5 \times (1/3) + 123{,}200 \times 0.3 \times (1/3)$$
$$+ 120{,}960 \times 0.1 \times (1/3) + 123{,}200 \times 0.2 \times (1/3) + 125{,}440 \times 0.7 \times (1/3)$$
$$= £121{,}161$$

At node Z, the expected accumulated amount is £119,903. Hence, investment A is preferred since it yields a larger expected amount.

(c) The 3 outcomes marked * are the only occasions on which investment B yields a larger return than A.

$$P(\text{B yields a larger return than A}) = (1/3) \times 0.6 + (1/3) \times 0.3 + (1/3) \times 0.2$$
$$= 0.367$$

470 QUANTITATIVE ANALYSIS

2 The new tree is:

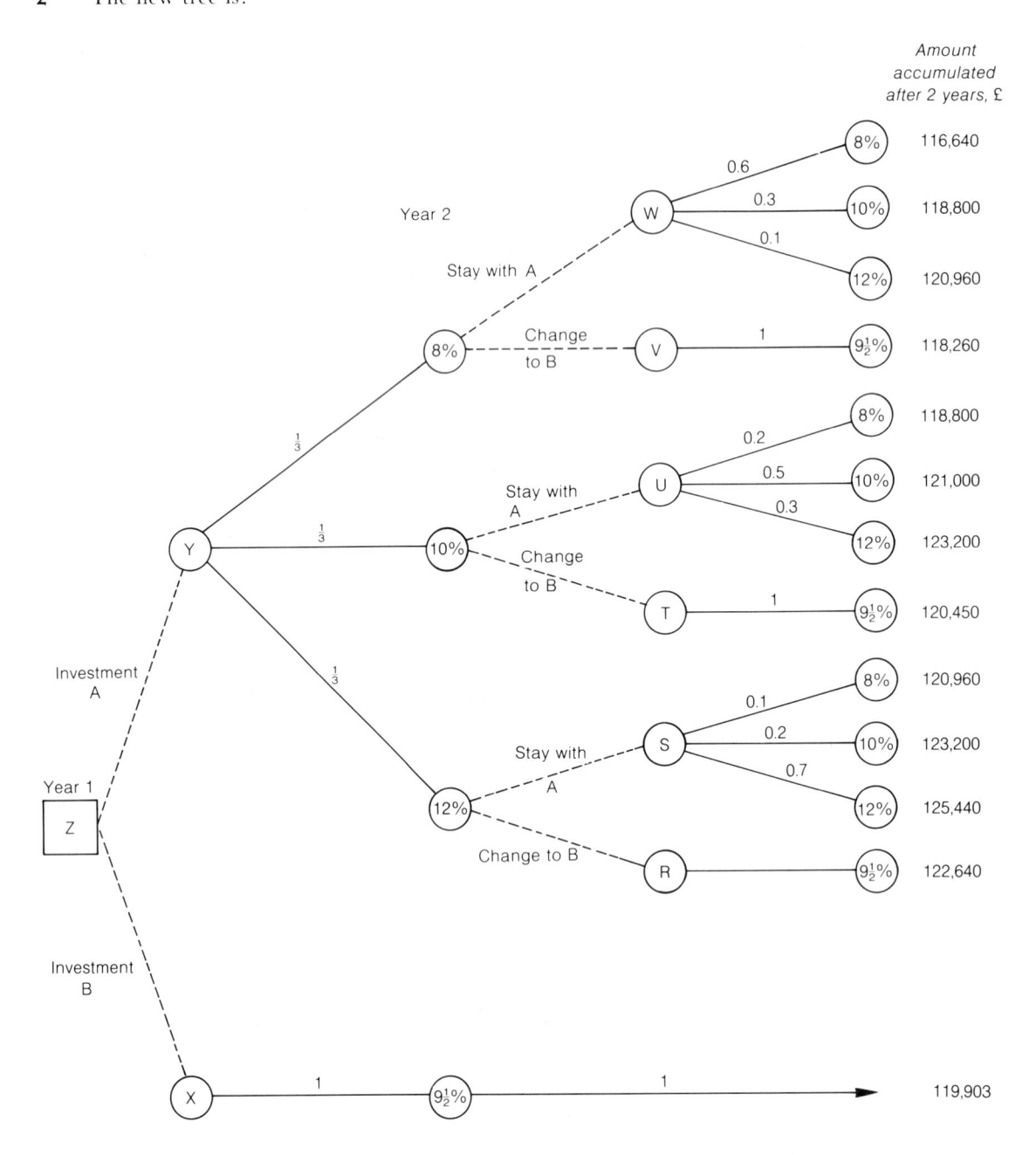

Amount accumulated after 2 years, £

```
                                                    ┌─ 8%   116,640
                                              W ────┼─ 10%  118,800
                                     Stay with A    └─ 12%  120,960
                              8% ────┤
                                     Change to B
                                         V ──── 9½%  118,260

                                                    ┌─ 8%   118,800
                                              U ────┼─ 10%  121,000
                                     Stay with A    └─ 12%  123,200
                       Y ──── 10% ───┤
                                     Change to B
                                         T ──── 9½%  120,450

                                                    ┌─ 8%   120,960
                                              S ────┼─ 10%  123,200
                                     Stay with A    └─ 12%  125,440
                              12% ───┤
                                     Change to B
                                         R ──── 9½%  122,640

   Z ─── Investment B ─── X ──── 9½% ──── 119,903
```

Probabilities at W: 0.6, 0.3, 0.1
Probabilities at U: 0.2, 0.5, 0.3
Probabilities at S: 0.1, 0.2, 0.7
Y branches: each 1/3

Work through the tree from right to left, calculating the expected amount accumulated at each node.

$$\text{Expected value at node W} = £117{,}720$$
$$\text{Expected value at node V} = £118{,}260$$

Hence, choose to change to investment B if the return is 8% in Year 1 with investment A.

$$\text{Expected value at node U} = £121{,}220$$
$$\text{Expected value at node T} = £120{,}450$$

Hence, choose to stay with investment A.

$$\text{Expected value at node S} = £124{,}544$$
$$\text{Expected value at node R} = £122{,}640$$

Hence, choose to stay with investment A.

$$\text{Expected value at node Y} = (118{,}260 + 121{,}220 + 124{,}544) \times (1/3)$$
$$= £121{,}341$$
$$\text{Expected value at node X} = £119{,}903$$

Hence, the preferred strategy is: invest in A, if the return is 8% in Year 1 change to investment B at the end of the year, otherwise stay with investment A.

Solutions to Chapter Four exercises

ANSWERS TO EXERCISES

Exercise 4.1

1. Finite
2. Infinite
3. Finite
4. Infinite
5. Infinite.

Exercise 4.2

1. The public's attitude to smoking: define the area, eg UK, the age of population, eg all people over 16 years of age, a time period.
2. The basic wages of hourly paid workers in the shoe manufacturing industry: define the area, the shoe manufacturing industry, an hourly paid worker, a time period. The population is all people who fit these definitions.
3. A company's accounts for year ending April 19X9 audit. The company and time period are already defined. The population is all the accounts that fit this definition.
4. Average lifetime of a particular electronic component. Define the component, the locations of manufacture to be considered and the time period. The population is all the components that fit the definition.

Exercise 4.3

1. All the households in postal district 7. The electoral register will give a reasonably complete list. Recent movements will give some errors.
2. All the students at the university. The enrolment forms will give a good frame. There are problems of inaccuracy because of students who drop out during the year.
3. All solicitors in the town. There are lists of solicitors available. The Law Society or the Citizen's Advice Bureau are possible sources.
4. A town's retail sales outlets for electrical goods. A clearer definition of retail and wholesale, electrical goods, level of sales in a particular shop is required. For example, supermarkets sell batteries and plugs. Would they be included in the frame? When the definition is refined, we could use trade associations, business directories and other information available at public libraries.

Exercise 4.4

1. Views on vivisection through a tear-off questionnaire in a wildlife magazine. People who take wildlife magazines have a special interest in animals and are not a cross-section of the public. The tear-out method usually means that only those who are strongly interested will reply.
2. Reaction to taxation through a house to house survey during the day. The timing is too limited: the majority of people who pay tax will be at work, not at home, and their opinions will be under-represented.

Exercise 4.5

3 Problems of living in the inner city via the telephone. People without a telephone are excluded from the sampling. This is most likely to exclude a majority of people who live in the inner city.

Exercise 4.5

The manager and the office vending machine—a delicate situation. If the staff know that a survey is to be done, they could change the way they use the machine during the survey. If they do not know, but find out afterwards, how will this improve staff/manager relations? It is probably best to think of other ideas—move the machine, motivate the staff, sack the manager.

Exercise 4.6

1 Interview 500 at a football match. This is biased since they are obviously interested in football and the survey misses out everyone else.
2 Town Hall exhibition and book. Who goes to the Town Hall? Is it a cross section of the people who might use the leisure centre? The book will provide no structure for the comments. They are unlikely to give a clear picture of what people want from a leisure centre, how often they may use it, or what they may be willing to pay for it. Only those who see the book and are strongly interested are likely to put their comments in it.
3 Article in the local paper asking people to write in. A good way to put out information, but not to get responses back. Only the strongly interested, for or against, are likely to write in.
4 Ring up 500 names from the telephone book. If almost all households in the area have telephones, this will be a good method.
5 Send batches of questionnaires to sports club secretaries and ask them to circulate to members. This assumes that the leisure centre is only for fairly committed sports people, which is unlikely. It also relies on the secretaries, who may wish to bias the results in their sport's favour by encouraging their members to respond in a particular way.

Exercise 4.7

1 Traffic flow: direct observation is the obvious choice.
2 Sales of cat food: almost any of the methods could be appropriate. The purpose of the survey needs to be clarified. If last year's sales volume is required, published statistics may be available. If details are needed about the people who buy the brand, and for what kind of cat, then a questionnaire or interview would be appropriate. If the reactions of the cats to this and other brands were needed, direct observation would be the only way.
3 Reaction to a new television series: the questionnaire or interviews would be most appropriate. The population would depend on the type of series and who it is aimed at.
4 Reaction of catalogue agents to a new payment method: a postal questionnaire would be convenient as the agents are already in touch with the company in this way. If, however, the company wants more open ended reactions to the payment system, then interviews would be helpful. More than one method can be used, for example, a postal questionnaire with a follow up interview.
5 Response of consumers to a new food product: there are various stages to the investigation. If a company is interested in a type of new product without having developed a specific product, then personal interviews would be useful in guiding the development. If they had already produced one or two specific products, the company would want the direct reactions of consumers to these through direct observation.

Exercise 4.8

1 The population mean $= \dfrac{295}{24} = 12.29$ g

474 QUANTITATIVE ANALYSIS

2 (a) Three simple random samples of 6 items.

Sample 1 (column 1)	RN	g		Sample 2 (column 2)	RN	g		Sample 3 (column 3)	RN	g
	01	4			15	33			04	11
	17	4			22	12			15	33
	07	31			24	5			21	30
	20	6			07	31			14	5
	21	30			10	13			17	4
	23	7			14	5			01	4
		$\sum x = 82$				$\sum x = 99$				$\sum x = 87$
		$\bar{x}_6 = 13.67$ g				$\bar{x}_6 = 16.5$ g				$\bar{x}_6 = 14.5$ g

(b) Three systematic random samples of 6 items. Therefore, we must pick every fourth number. Use the random numbers to pick the first number to start with from 1, 2, 3, 4.

Sample 1 (column 4)	RN	g		Sample 2 (column 5)	RN	g		Sample 3 (column 6)
	02	6			04	11		The first random number, in the range 1–4, is 04—Sample 3 will be the same as Sample 2.
		35				7		
		13				7		
		5				14		
		5				6		
		12				5		
		$\sum x = 76$				$\sum x = 50$		
		$\bar{x}_6 = 12.67$ g				$\bar{x}_6 = 8.33$ g		

(c) Strata sampling: rearrange the population into its 3 strata.

Stratum 1 Light	Number	1	2	8	9	12	14	17	18	19	20	23	24
	weight, g	4	6	7	3	7	5	4	5	6	6	7	5

Stratum 2 Medium	Number	3	4	5	10	11	13	16	22
	weight, g	12	11	12	13	13	14	14	12

Stratum 3 Heavy	Number	6	7	15	21
	weight, g	35	31	33	30

Each sample will have 3 light, 2 medium, 1 heavy in proportion to the population:

Sample 1 (column 7)		Random number	g
3 light		02	6
		19	6
		23	7
2 medium		16	14
		11	13
1 heavy		21	30
		$\sum x = 76$	
		$\bar{x}_6 = 12.67$ g	

SOLUTIONS TO EXERCISES: CHAPTER FOUR

Sample 2 (column 8)	Random number	g
3 light	09	3
	12	7
	20	6
2 medium	03	12
	04	11
1 heavy	15	33

$$\sum x = 72$$
$$\bar{x}_6 = 12.0 \text{ g}$$

Sample 3 (column 9)	Random number	g
3 light	18	5
	24	5
	14	5
2 medium	13	14
	05	12
1 heavy	15	33

$$\sum x = 74$$
$$\bar{x}_6 = 12.33 \text{ g}$$

(d) Cluster sampling. Rearrange the population into 4 clusters, each of 3 light, 2 medium and 1 heavy. Normally you would use this method if the population was already in clusters. You would not rearrange it like this, into clusters. If there were no clusters, you would use strata sampling.

Cluster 1	Number	g
3 light	1	4
	2	6
	8	7
2 medium	3	12
	4	11
1 heavy	6	35

$$\sum x = 75$$
$$\bar{x}_6 = 12.5 \text{ g}$$

Cluster 2	Number	g
3 light	9	3
	12	7
	14	5
2 medium	5	12
	10	13
1 heavy	7	31

$$\sum x = 71$$
$$\bar{x}_6 = 11.83 \text{ g}$$

Cluster 3	Number	g
3 light	17	4
	18	5
	19	6
2 medium	11	13
	13	14
1 heavy	15	33

$$\sum x = 75$$
$$\bar{x}_6 = 12.5 \text{ g}$$

Cluster 4	Number	g
3 light	20	6
	23	7
	24	5
2 medium	16	14
	22	12
1 heavy	21	30

$$\sum x = 74$$
$$\bar{x}_6 = 12.33 \text{ g}$$

Select 3 clusters using random numbers in the range 01–04 inclusive.
The 3 clusters chosen are from Column 10, Cluster x from Column 1, Cluster 1 and from Column 2 over to Column 3, Cluster 4. (*Note*: all 3 clusters could have been selected from column 10 above.)

476 QUANTITATIVE ANALYSIS

3 Comments and comparisons. The population mean is 12.29 g. Means from the different sampling methods:

Simple random sampling	Systematic random sampling	Strata sampling	Cluster sampling
$\bar{x}_6 = 13.67$ g	12.67 g	12.67 g	12.50 g
$\bar{x}_6 = 16.50$ g	8.33 g	12.00 g	12.50 g
$\bar{x}_6 = 14.50$ g	8.33 g	12.33 g	12.33 g

Simple and systematic sampling give the most erratic results. Strata and cluster sampling are the most consistent. However, they assume you know a lot about the population and that it can be neatly divided into clear strata or clusters.

Exercise 4.9

1 Take all samples of 4 from the population—4, 8, 12, 16, 20, 24.

2 Calculate $\bar{x}_4$ for each sample and the sample standard deviation $= \sqrt{\left(\dfrac{\sum x^2}{4} - \bar{x}^2\right)}$

Sample number	Sample				$\bar{x}_4$	s
1	4	8	12	16	10	4.47
2	4	8	12	20	11	5.92
3	4	8	12	24	12	7.48
4	4	8	16	20	12	6.32
5	4	8	16	24	13	7.68
6	4	8	20	24	14	8.25
7	4	12	16	20	13	5.92
8	4	12	16	24	14	7.21
9	4	12	20	24	15	7.68
10	4	16	20	24	16	7.48
11	8	12	16	20	14	4.47
12	8	12	16	24	15	5.92
13	8	12	20	24	16	6.32
14	8	16	20	24	17	5.92
15	12	16	20	24	18	4.47

$\sum (\bar{x}_4) = 210$
$\div 15 = 14$

Sampling distribution of sample means

$\bar{x}_4$	Frequency
10	1
11	1
12	2
13	2
14	3
15	2
16	2
17	1
18	1
Total	15

Sampling distribution of sample standard deviation

s	Frequency
4.47	3
5.92	4
6.32	2
7.21	1
7.48	2
7.68	2
8.25	1
Total	15

The frequencies for samples of 4 are compared with the frequencies for samples of 2 listed in **3**.

3 Plots comparing the sample means and sample standard deviations of all the samples of size 2 and 4 taken from the population 4, 8, 12, 16, 20, 24:

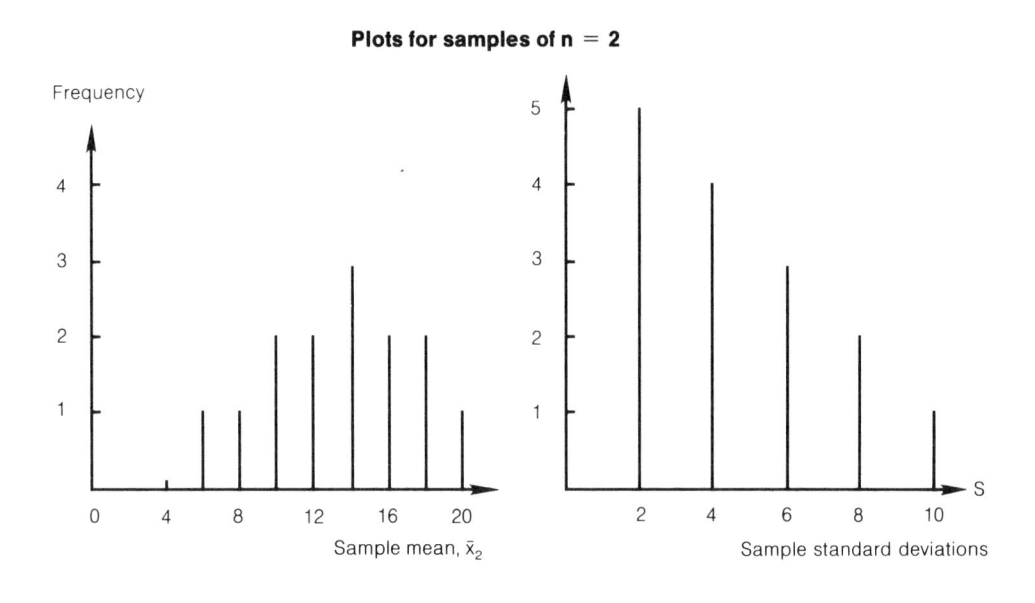

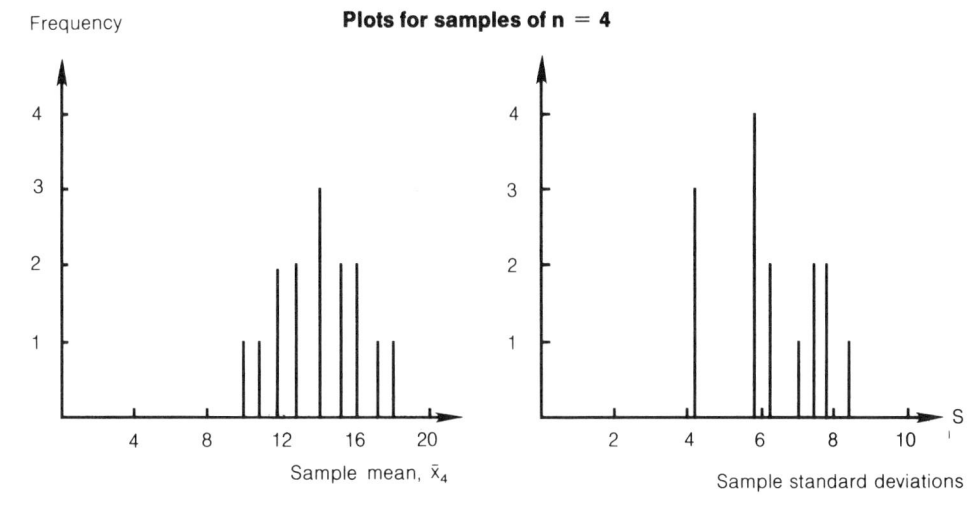

Plots for samples of n = 2

Plots for samples of n = 4

4 Comments: sample means—the bigger the sample size the narrower the spread of the sample means. Both plots are symmetrical about the population mean $\mu = 14$. Sample standard deviation—the population sd $\sigma = 6.83$. Both sample distributions are skewed, particularly n = 2. The distribution for n = 4 is much more closely packed about the population standard deviation.

Solutions to Chapter Five exercises

ANSWERS TO EXERCISES

Exercise 5.1

The 95% confidence interval for the population mean weight of chocolate dispensed is:

$$\bar{x} \pm z_{0.025} \frac{\sigma}{\sqrt{n}}$$

where $z_{0.025}$ is the number of standard errors above and below the mean which enclose 95% of the sampling distribution of sample means.

From the standard normal tables, $z_{0.025} = \pm 1.96$, therefore the 95% confidence interval for the population mean is:

$$99.5 \pm 1.96 \times \frac{2.5}{\sqrt{15}} = (99.5 \pm 1.265)\text{g}$$

The required confidence interval is 98.24g to 100.77g.

Exercise 5.2

Since the population distribution of lifetimes is approximately normal, the sampling distribution of sample mean lifetimes will be even closer to a normal distribution (Central Limit Theorem). Therefore, the 99% confidence interval for the population mean lifetime of the components is approximately:

$$\bar{x} \pm t_{0.005, n-1} \frac{\hat{\sigma}}{\sqrt{n}}$$

where $t_{0.005, n-1}$ is the value of the standard t variable with $n - 1$ degrees of freedom, above which 0.5% of the distribution lies.

The t variable is used in this case because the population standard deviation is unknown. From the t tables, $t_{0.005, 49} = \pm 2.68$ approximately. Since:

$$\hat{\sigma} = \sqrt{\frac{n}{n-1}}\, s, \qquad \frac{\hat{\sigma}}{\sqrt{n}} = \frac{s}{\sqrt{n-1}}$$

therefore the 99% confidence interval for the population mean is:

$$840 \pm 2.68 \times \frac{22}{\sqrt{49}} = (840 \pm 8.42) \text{ hours}$$

The required confidence interval is 831.6 hours to 848.4 hours.

SOLUTIONS TO EXERCISES: CHAPTER FIVE

Exercise 5.3

1 We can judge how good an estimate the sample mean is of the population mean by the relative size of a confidence interval. Let us consider the 95% confidence interval. The 95% confidence interval for the population mean lifetime of the light bulbs is:

$$\bar{x} \pm t_{0.025, n-1} \frac{\hat{\sigma}}{\sqrt{n}}$$

where $t_{0.005, n-1}$ is the value of the standard t variable with $n-1$ degrees of freedom, above which 2.5% of the distribution lies.

The t variable is used in this case because the population standard deviation is unknown. From the t tables, $t_{0.025, 74} = \pm 1.993$ approximately. Since:

$$\hat{\sigma} = \sqrt{\frac{n}{n-1}}\, s, \qquad \frac{\hat{\sigma}}{\sqrt{n}} = \frac{s}{\sqrt{n-1}}$$

therefore the 95% confidence interval for the population mean is:

$$2920 \pm 1.993 \times \frac{400}{\sqrt{74}} = (2920 \pm 92.67)\,\text{hours}$$

There is a 95% probability that the population mean lies within 93 hours of the sample mean of 2920 hours. This is a variation of $\pm 3.2\%$. Hence, the sample mean appears to be a reasonably good estimate of the population mean.

2 If the manager wishes to reduce the range of the confidence interval to ± 50 hours, then he must increase his sample size. We now require $t_{0.025, n-1} \dfrac{\hat{\sigma}}{\sqrt{n}} \leqslant 50$ hours. Since the value of t depends on the size of the sample, we will approximate $t_{0.025, n-1}$ by $z_{0.025}$. The sample size will be large, therefore the approximation is reasonable.

$$z_{0.025} = \pm 1.96$$

As before:

$$\hat{\sigma}/\sqrt{n} = s/\sqrt{n-1}$$

and s, the sample standard deviation will be approximated by the standard deviation of the existing sample, 400 hours. Hence:

$$1.96 \times \frac{400}{\sqrt{n-1}} \leqslant 50$$

therefore:

$$\frac{(1.96 \times 400)^2}{50^2} \leqslant n - 1$$

$$n - 1 \geqslant 245.86 \quad \text{therefore} \quad n \geqslant 247$$

The manager must increase the sample size to 250 (for convenience) in order to reduce the range of the confidence interval to ± 50 hours.

Exercise 5.4

The number and the proportion of housewives who would like the town centre to be traffic-free follows a binomial distribution. However, since the sample size is large, the binomial distribution may be approximated by a normal distribution. The 90% confidence interval for the population proportion of housewives in favour of a traffice-free shopping centre is approximately:

$$\hat{p} \pm z_{0.05} \times SE_{\hat{p}}$$

where $z_{0.05}$ is the number of standard errors above and below the mean which enclose 90% of the sampling distribution of sample means.

The sample proportion in favour, $\hat{p} = 480/800 = 0.6$, and sample proportion not in favour, $\hat{q} = 320/800 = 0.4$. Since the population proportion, p, is unknown, the best estimate of $SE_{\hat{p}}$ is:

$$SE_{\hat{p}} = \sqrt{\frac{\hat{p} \times \hat{q}}{n}} = \sqrt{\frac{0.6 \times 0.4}{800}} = 0.01732$$

From the standard normal tables, $z_{0.05} = \pm 1.645$, therefore the 90% confidence interval for the population proportion is:

$$0.6 \pm 1.645 \times 0.01732 = (0.6 \pm 0.028)$$

The required confidence interval is 0.57 to 0.63.

Exercise 5.5

1. The number and the proportion of outlets willing to stock the new product follows a binomial distribution. However, since the sample size is large, the binomial distribution may be approximated by a normal distribution. The 95% confidence interval for the population proportion of willing outlets is approximately:

$$\hat{p} \pm z_{0.025} \times SE_{\hat{p}}$$

where $z_{0.025}$ is the number of standard errors above and below the mean which enclose 95% of the sampling distribution of sample means.

The sample proportion in favour, $\hat{p} = 50/200 = 0.25$ and sample proportion not in favour, $\hat{q} = 150/200 = 0.75$. Since the population proportion, p, is unknown, the best estimate of $SE_{\hat{p}}$ is:

$$SE_{\hat{p}} = \sqrt{\frac{\hat{p} \times \hat{q}}{n}} = \sqrt{\frac{0.25 \times 0.75}{200}} = 0.03062$$

From the standard normal tables, $z_{0.025} = +1.96$, therefore the 95% confidence interval for the population proportion is:

$$0.25 \pm 1.96 \times 0.03062 = (0.25 \pm 0.060)$$

The required confidence interval is 0.19 to 0.31.

2. We require $z_{0.025} \times SE_{\hat{p}} \leqslant 0.04$, therefore:

$$1.96 \times \sqrt{\frac{\hat{p} \times \hat{q}}{n}} \leqslant 0.04$$

Since the proportion of outlets in Scotland willing to take the new product is not known, we must approximate it by using the value for Wales. Hence:

$$1.96 \times \sqrt{\frac{0.25 \times 0.75}{n}} \leqslant 0.04$$

Therefore:

$$\frac{(1.96^2 \times 0.25 \times 0.75)}{0.04^2} \leqslant n$$

$$450.2 \leqslant n$$

In Scotland, Ms Briggs should use a sample of at least 451 outlets if she wishes to reduce the range of the 95% confidence interval to $\pm 4\%$.

Exercise 5.6

1 The number, and the proportion, of customers interested in applying for a credit card follows a binomial distribution. However, since the sample size is large, the binomial distribution may be approximated by a normal distribution. The 95% confidence interval for the proportion of all customers who would apply for a credit card is:

$$\hat{p} \pm z_{0.025} \times SE_{\hat{p}}$$

where $z_{0.025}$ is the number of standard errors above and below the mean which enclose 95% of the sampling distribution of sample means.

The sample proportion in favour of a credit card, $\hat{p} = 32/80 = 0.4$. Since the population proportion, p, is unknown, the best estimate of $SE_{\hat{p}}$ is:

$$\widehat{SE}_{\hat{p}} = \sqrt{\frac{\hat{p} \times \hat{q}}{n}} = \sqrt{\frac{0.4 \times 0.6}{80}} = 0.05477$$

From the standard normal tables, $z_{0.025} = \pm 1.96$, therefore the 95% confidence interval for the population proportion is:

$$0.4 \pm 1.96 \times 0.05477 = (0.4 \pm 0.107)$$

The required confidence interval is 29.3% to 50.7%.

2 The 95% confidence interval for the population average amount borrowed on the credit card is:

$$\bar{x} \pm z_{0.025} \times SE_{\bar{x}} \qquad (£)$$

Since the population standard deviation is unknown, the best estimate of $SE_{\bar{x}}$ is:

$$\frac{\hat{\sigma}}{\sqrt{n}} = \sqrt{\frac{n}{n-1}} \frac{s}{\sqrt{n}} = \frac{s}{\sqrt{n-1}}$$

Hence the confidence interval is:

$$450 \pm 1.96 \times \frac{150}{\sqrt{31}} = £(450 \pm 52.80)$$

For all customers with a credit card, the 95% confidence interval for the average amount borrowed on the credit card is £397 to £503.

3 We require $z_{0.025} \times SE_{\hat{p}} \leqslant 0.05$ and $z_{0.025} \times SE_{\bar{x}} \leqslant 20$. We use the sample statistics from the pilot survey to estimate the statistics in the forthcoming survey, therefore:

$$1.96 \times \frac{\sqrt{0.4 \times 0.6}}{\sqrt{n}} \leqslant 0.05 \quad \text{and} \quad 1.96 \times \frac{150}{\sqrt{n-1}} \leqslant 20$$

$$\frac{(1.96^2 \times 0.4 \times 0.6)}{0.05^2} \leqslant n \quad \text{and} \quad \left(\frac{1.96 \times 150}{20}\right)^2 \leqslant n-1$$

$$368.8 \leqslant n \quad \text{and} \quad 216.09 \leqslant n-1$$

Hence, in the first case, $n \geqslant 369$ and in the second case, $n \geqslant 218$. To satisfy both requirements, the new sample size must be at least 369 customers.

4 The results of a nationwide survey may differ from the results of the Oxford Street survey since a store in London is unlikely to be typical of the country as a whole.

Exercise 5.7

Using the data given, the total number of invoices which contain at least one error is:

$$(2 + 8 + 8) + (2 + 8 + 6) = 34.$$

There are 200 invoices altogether, therefore, the proportion of invoices with at least one error is $34/200 = 0.17$. Since the sample size is large, the binomial distribution, which describes the proportion of invoices in the sample with at least one error, may be approximated by the normal distribution, hence the approximate 95% confidence interval for the population proportion is:

$$\hat{p} \pm z_{0.025} SE_{\hat{p}}$$

From the standard normal tables $z_{0.025} = 1.96$. The best estimate of $SE_{\hat{p}}$ is:

$$\sqrt{\frac{\hat{p} \times \hat{q}}{n}} = \sqrt{\frac{0.17 \times 0.83}{200}} = 0.02656$$

Therefore the confidence interval is:

$$0.17 \pm 1.96 \times 0.02656$$

$$0.17 \pm 0.052$$

The 95% confidence interval for the proportion of all invoices with at least one error is $17\% \pm 5.2\%$.

Exercise 5.8

1 Total value of the 6 stock items $= \sum (\text{unit cost}) \times (\text{no in stock})$:

Item number	Unit cost, £	Number in stock	Total value of item, £x	x^2
1	0.6	74	44.4	1971.36
2	2.6	10	26.0	676.0
3	1.2	27	32.4	1049.76
4	0.8	49	39.2	1536.64
5	0.4	32	12.8	163.84
6	1.6	6	9.6	92.16
Total			164.4	5489.76

Hence, the total value of the 6 stock items = £164.40
The average value of the stock held per item = £164.4/6
= £27.40 per item
Therefore, an estimate of the total value of all 55 stock items is 55 × £27.40 = £1507.

2 In order to set up a confidence interval for the total value of the 55 stock items, we must first establish a confidence interval for the average value of the stock items. Since the population standard deviation is unknown, the 95% confidence interval for the population average value of the stock items is:

$$\bar{x} \pm t_{0.025, n-1} \times \frac{\hat{\sigma}}{\sqrt{n}}$$

where $t_{0.025, n-1}$ is the standard t variable above which 2.5% of the t distribution lies for $(n-1)$ degrees of freedom. From the tables, $t_{0.025, 5} = 2.57$. Since σ is unknown, it must be estimated from the sample standard deviation. In addition, we note that the population is very small, hence the best estimate of σ is given by:

$$\hat{\sigma} = \sqrt{\frac{N-n}{N-1} \cdot \frac{n}{n-1}} \cdot s$$

where N is the population size and n is the sample size. s is the sample standard deviation. Hence, the best estimate of $SE_{\bar{x}}$ is:

$$\widehat{SE}_{\bar{x}} = \frac{\hat{\sigma}}{\sqrt{n}} = \sqrt{\frac{N-n}{N-1}} \cdot \frac{s}{\sqrt{n-1}}$$

where:

$$s = \sqrt{\frac{\sum (x-\bar{x})^2}{n}} = \sqrt{\frac{\sum x^2}{n} - (\bar{x})^2}$$

From the table drawn up in **1**, $\sum x^2 = 5489.76$, therefore:

$$s = \sqrt{\frac{5489.76}{6} - 27.4^2} = 12.814$$

therefore:

$$\widehat{SE}_{\bar{x}} = \sqrt{\frac{55-6}{55-1}} \times \frac{12.814}{\sqrt{5}} = 5.46$$

The 95% confidence interval for the population average value of the stock items is:

$$27.4 \pm 2.57 \times 5.46 = 27.4 \pm 14.03$$

Therefore the 95% confidence interval for the population total value of the stock items is:

$$55(27.4 \pm 14.03) = £(1507 \pm 771.6)$$

The confidence interval, £(1507 ± 772), means that there is a 95% probability that the true total value of the 55 stock items lies within this range. For 95% of all samples of 6 items which could be randomly selected from the population, the true population total value will lies within £772 of the total estimated from the sample.

We assume that the original sample of 6 items was randomly selected from a normal (or approximately normal) distribution, ie the distribution of the values of the population of stock items is approximately normal.

3 We require the range of the 95% confidence interval to be less than £150, therefore:

$$55 \times t_{0.025, n-1} \times \widehat{SE}_{\bar{x}} < 150$$

Hence:

$$55 \times t_{0.025, n-1} \times \sqrt{\frac{55-n}{55-1}} \times \frac{s}{\sqrt{n-1}} < 150$$

The value of the t statistic depends on the size of the sample, however, it is likely that the new sample size will be quite large and, therefore, we could approximate the t statistic by the standard normal statistic, $z_{0.025}$. From the standard normal tables, $z_{0.025} = 1.96$. In addition, we will estimate the standard deviation of the future larger sample by the value for the current sample of 6 items. It follows that:

$$55 \times 1.96 \times \sqrt{\frac{55-n}{54}} \times \frac{12.814}{\sqrt{n-1}} < 150$$

$$\frac{1381.349}{150 \times \sqrt{54}} \leqslant \sqrt{\frac{n-1}{55-n}}$$

$$1.25319 \leqslant \sqrt{\frac{n-1}{55-n}}$$

Squaring:

$$1.57047(55-n) \leqslant n-1$$

$$87.376 \leqslant 2.5705n$$

$$33.99 \leqslant n$$

At least 34 items should be sampled in future years.

Exercise 5.9

1 The average value per item in the sample $= \dfrac{\sum x}{18} = \dfrac{1800}{18}$

$$= £100 \text{ per item}$$

Therefore an estimate of the total value of all 1200 stock items is:

$$1200 \times £100 = £120,000.$$

The reliability of the estimate depends on the sample being selected randomly from the population to avoid bias. The sample should also reflect any particular characteristics of the population.

2 In order to set up a confidence interval for the total value of the 1200 stock items, we must first establish a confidence interval for the average value of the stock items. Since the population standard deviation is unknown, the 95% confidence interval for the population average value of the stock items is:

$$\bar{x} \pm t_{0.025, n-1} \times \frac{\hat{\sigma}}{\sqrt{n}}$$

where $t_{0.025, n-1}$ is the standard t variable above which 2.5% of the t distribution lies for $(n-1)$ degrees of freedom, provided that it may be assumed that the distribution of the values per item in the population is approximately normal.

From the tables, $t_{0.025, 17} = 2.11$. Since σ is unknown, it must be estimated from the sample standard deviation. Although the population is finite, 1200 items is not a particularly small number, hence the small population correction factor is not necessary. The best estimate of σ is therefore given by:

$$\hat{\sigma} = \sqrt{\frac{n}{n-1}} \cdot s$$

where s is the sample standard deviation. Hence, the best estimate of $SE_{\bar{x}}$ is:

$$\widehat{SE}_{\bar{x}} = \frac{\hat{\sigma}}{\sqrt{n}} = \frac{s}{\sqrt{n-1}}$$

where:

$$s = \sqrt{\frac{\sum (x - \bar{x})^2}{n}} = \sqrt{\frac{\sum x^2}{n} - (\bar{x})^2}$$

Given $\sum x^2 = 207{,}200$, therefore:

$$s = \sqrt{\frac{207{,}200}{18} - 100^2} = £38.87$$

The 95% confidence interval for the population average value of the sales items is:

$$100 \pm 2.11 \times \frac{38.87}{\sqrt{17}} = 100 \pm 19.892$$

Therefore, the 95% confidence interval for the population total value of the sales items is:

$$1200(100 \pm 19.892) = £(120{,}000 \pm 23{,}870)$$

We require the range of the 95% confidence interval to be less than $\pm £5000$ therefore:

$$1200 \times t_{0.025, n-1} \times \widehat{SE}_{\bar{x}} \leqslant 5000$$

Hence:

$$1200 \times t_{0.025, n-1} \times \frac{s}{\sqrt{n-1}} \leqslant 5000$$

The value of the t statistic depends on the size of the sample. However, it is likely that the new sample size will be quite large and, therefore, we could approximate the t statistic by the standard normal statistic, $z_{0.025}$. From the standard normal tables, $z_{0.025} = 1.96$.

In addition, we will estimate the standard deviation of the future larger sample by the value for the current sample of 18 items. It follows that:

$$1200 \times 1.96 \times \frac{38.87}{\sqrt{n-1}} \leqslant 5000$$

$$18.28 \leqslant \sqrt{n-1}$$

Squaring:

$$334.2 \leqslant n - 1$$

The auditor should therefore sample at least 336 items.

Solutions to Chapter Six exercises

ANSWERS TO EXERCISES

Exercise 6.1

First we must write down the null hypothesis and the alternative hypothesis.

H_0: The sample mean is consistent with the sample having been randomly selected from a normal population with mean 5 cm, ie $\mu = 5$ cm.

H_1: The sample mean is not consistent with the sample having been randomly selected from a normal population with mean 5 cm, ie $\mu \neq 5$ cm.

We will choose to test the evidence at the 5% level of significance using a two sided normal test.
Using the standard normal tables, $z_{0.025} = 1.96$, where $\pm z_{0.025}$ is the number of standard deviations from the mean which cut off the top and bottom 2.5% of the normal distribution. The test statistic is:

$$z = \frac{\bar{x} - \mu}{SE_{\bar{x}}} \quad \text{where} \quad SE_{\bar{x}} = \frac{\sigma}{\sqrt{n}}$$

therefore:

$$z = \frac{5.025 - 5}{0.05/\sqrt{25}} = 2.5$$

$2.5 > z_{0.025}$, therefore the result is significant at the 5% level of significance. At this decision level, the sample evidence does not support the null hypothesis. We therefore choose to accept the alternative hypothesis that the average length of the bolts in this production run is not 5 cm.

Exercise 6.2

H_0: The sample mean is consistent with the sample having been randomly selected from a normal population with a mean fill of 500 ml, ie $\mu = 500$ ml.

H_1: The sample mean is less than 500ml, ie $\mu < 500$ml.

We will choose to test the evidence at the 5% level of significance. Since the population standard deviation is unknown, a one sided t test is used with $n - 1$ degrees of freedom:

$$n - 1 = 30 - 1 = 29 \text{ degrees of freedom}$$

Using the standard t tables, $t_{0.05, 29} = 1.70$, where $t_{0.05, 29}$ is the value of the t statistic which cuts off the top or bottom 5% of the t distribution with 29 degrees of freedom. The test statistic is:

$$t = \frac{\bar{x} - \mu}{\widehat{SE}_{\bar{x}}} \quad \text{where} \quad \widehat{SE}_{\bar{x}} = \frac{\hat{\sigma}}{\sqrt{n}} = \sqrt{\frac{n}{n-1}} \frac{s}{\sqrt{n}} = \frac{s}{\sqrt{n-1}}$$

therefore:

$$t = \frac{495.6 - 500}{8.3/\sqrt{29}} = -2.85$$

$-2.85 < -t_{0.05,29}$, therefore the result is significant at the 5% level of significance. At this decision level, the sample evidence does not support the null hypothesis. We therefore choose to accept the alternative hypothesis that the machine is underfilling on average and must be reset.

Exercise 6.3

1   H_0: The sample mean is consistent with the sample having been randomly drawn from a normal population with a mean of 500 g, ie $\mu = 500$ g.

   H_1: The sample mean is consistent with the sample having been drawn from a population with a mean of less than 500 g, ie $\mu < 500$ g.

   Since the population standard deviation, σ, is unknown, test at the 1% decision level using a one tail t test with $10 - 1 = 9$ degrees of freedom.

2   From the standard t tables, $t_{0.01,9} = 2.82$. Test statistic:

$$t = \frac{\bar{x} - \mu}{\widehat{SE}_{\bar{x}}} = \frac{\bar{x} - \mu}{s/\sqrt{n-1}}$$

We must now calculate the values of the sample mean and standard deviation.

Packet Number	1	2	3	4	5	6	7	8	9	10
Weight, x g	497	498	502	503	495	496	497	500	501	496

$$\sum x = 4{,}985 \quad \text{and} \quad \sum x^2 = 2{,}485{,}093$$

therefore:

$$\bar{x} = \frac{\sum x}{n} = \frac{4{,}985}{10} = 498.5 \text{ g}$$

and:

$$s = \sqrt{\frac{\sum x^2}{n} - (\bar{x})^2} = \sqrt{\frac{2{,}485{,}093}{10} - (498.5)^2} = 2.655$$

Hence:

$$t = \frac{498.5 - 500}{2.655/\sqrt{9}} = -1.695$$

$-1.695 > -t_{0.01,9}$, therefore the result is not significant at the 1% level. The evidence is consistent with the null hypothesis at this decision level, hence accept H_0 at this level. Although $\bar{x}$ is below the expected mean of 500 g, the difference is not significant and we must assume that the data are consistent with the claim.

Exercise 6.4

H_0: The sample is drawn from a population with a proportion of errors of 10%, ie $p = 0.1$.

H_1: There has been an improvement in the error rate, ie $p < 0.1$.

The proportion of errors in the sample follows a binomial distribution but, since the sample size is large and the probability of an error is also large, the binomial distribution may be approximated by a normal distribution.

Test at the 1% level using a one sided normal test. Using the standard normal tables, $z_{0.01} = 2.33$, where $z_{0.01}$ is the number of standard deviations from the mean which cut off the top or bottom 1% of the normal distribution.

$$\hat{p} = 7/100$$

The test statistic is:

$$z = \frac{(\hat{p} - p)}{SE_{\hat{p}}}$$

$$SE_{\hat{p}} = \sqrt{\frac{pq}{n}} = \sqrt{\frac{0.1 \times 0.9}{100}} = 0.03$$

Therefore:

$$z = \frac{0.07 - 0.1}{0.03} = -1.0$$

$-1.0 > -z_{0.01}$, therefore the result is not significant at the 1% decision level. The evidence supports the null hypothesis at this level. Hence, we choose to accept H_0 and assume that there has been no improvement in the error rate.

Exercise 6.5

H_0: The sample has been drawn from a normal population with a variance of 0.42 g², ie $\sigma^2 = 0.42$ g².

H_1: The machine is out of control so that the variance is greater than 0.42 g², ie $\sigma^2 > 0.42$ g².

Test at the 5% level of significance using a 1 sided F test.
The test statistic is:

$$F = \frac{s^2 \times (n/n - 1)}{\sigma^2} = \frac{0.516 \times (10/9)}{0.42} = 1.365$$

(*Note*: In order to use the Standard F Tables, arrange the statistic so that $F \geqslant 1$).

Since σ^2 is known, there is an infinite number of degrees of freedom in the denominator. The sample size is 10, therefore there are $10 - 1 = 9$ degrees of freedom in the numerator.

From the standard F tables, $F_{0.05, 9, \infty} = 1.89$.

$1.365 < F_{0.05, 9, \infty}$, therefore the result is not significant at the 5% decision level. At this level the evidence supports the null hypothesis. We choose to accept that the variability of the machine has not changed.

Exercise 6.6

H_0: The 2 samples are drawn from normal populations with the same mean weight, ie $\mu_1 = \mu_2$.

H_1: $\mu_1 > \mu_2$.

The appropriate test to choose depends on whether or not we may assume that the population variances are equal. An F test is used to investigate the population variances.

F Test:

$H_0: \sigma_1^2 = \sigma_2^2$

$H_1: \sigma_1^2 \neq \sigma_2^2$

Test at the 5% level using a two sided F test:

$$\hat{\sigma}_1^2 = \frac{n_1}{n_1 - 1} s_1^2 = \frac{25}{24} \times 5.03^2 = 26.355$$

$$\hat{\sigma}_2^2 = \frac{n_2}{n_2 - 1} s_2^2 = \frac{30}{29} \times 4.39^2 = 19.937$$

The test statistic is:

$$F = \frac{\hat{\sigma}_1^2}{\hat{\sigma}_2^2} = \frac{26.355}{19.937} = 1.32$$

From the standard F distribution tables, $F_{0.025, 24, 29} = 2.16$.

$1.32 < F_{0.025, 24, 29}$, therefore the result is not significant at the 5% decision level. The evidence supports H_0 at this level, hence we assume that the two population variances are equal. We will now continue with the test of the sample means.

Since the population variances are unknown but may be assumed to be equal, we will test at the 1% level of significance using a one sided t test with $n_1 + n_2 - 2 = 25 + 30 - 2 = 53$ degrees of freedom.

From the standard t distribution tables, $t_{0.01, 53} = 2.40$ approximately. The test statistic is:

$$t = \frac{(\bar{x}_1 - \bar{x}_2) - 0}{\widehat{SE}_{\bar{x}_1 - \bar{x}_2}}$$

$$\widehat{SE}_{\bar{x}_1 - \bar{x}_2} = \hat{\sigma} \sqrt{\frac{1}{n_1} + \frac{1}{n_2}} \quad \text{and} \quad \hat{\sigma} = \sqrt{\frac{n_1 s_1^2 + n_2 s_2^2}{n_1 + n_2 - 2}}$$

Hence:

$$\hat{\sigma} = \sqrt{\frac{25 \times 5.03^2 + 30 \times 4.39^2}{53}} = 4.7794$$

therefore:

$$\widehat{SE}_{\bar{x}_1 - \bar{x}_2} = 4.7794 \times \sqrt{(1/25) + (1/30)} = 1.2943$$

The test statistic is:

$$t = \frac{503.6 - 498.2}{1.2943} = 4.17$$

$4.17 > t_{0.01, 53}$, therefore the result is significant at the 1% decision level. The evidence does not support the null hypothesis at this level, therefore we choose to reject H_0 and to accept H_1. In the light of the evidence, we choose to assume that the mean weight in the own brand packets is less than the mean weight in the normal brand packets.

Exercise 6.7

The initial sample, taken before the latest adjustments were made, has a mean % yield of:

$$\bar{x}_1 = \frac{\sum x_1}{n_1} = \frac{725.8}{8} = 90.725$$

and a standard deviation of % yield of:

$$s_1 = \sqrt{\frac{\sum x_1^2}{n_1} - (\bar{x}_1)^2} = \sqrt{\frac{65851.16}{8} - (90.725)^2} = 0.6078$$

The second sample, taken after the latest adjustments were made, has a mean % yield of:

$$\bar{x}_2 = \frac{\sum x_2}{n_2} = \frac{547.2}{6} = 91.2$$

and a standard deviation of % yield of:

$$s_2 = \sqrt{\frac{\sum x_2^2}{n_2} - (\bar{x}_2)^2} = \sqrt{\frac{49905.34}{6} - (91.2)^2} = 0.3416$$

H_0: The 2 samples are drawn from normal populations with the same mean % yield, ie $\mu_1 = \mu_2$.
H_1: $\mu_2 > \mu_1$.

The appropriate test to choose depends on whether or not we may assume that the population variances are equal. An F test is used to investigate the population variances.

F Test:

H_0: $\sigma_1^2 = \sigma_2^2$.
H_1: $\sigma_1^2 \neq \sigma_2^2$.

Test at the 5% level using a two sided F test:

$$\hat{\sigma}_1^2 = \frac{n_1}{n_1 - 1} s_1^2 = \frac{8}{7} \times 0.6078^2 = 0.4222$$

$$\hat{\sigma}_2^2 = \frac{n_2}{n_2 - 1} s_2^2 = \frac{6}{5} \times 0.3416^2 = 0.140$$

The test statistic is:

$$F = \frac{\hat{\sigma}_1^2}{\sigma_2^2} = \frac{0.4222}{0.140} = 3.016$$

From the standard F distribution tables, $F_{0.025,7,5} = 6.85$.

$3.02 < F_{0.025,7,5}$, therefore the result is not significant at the 5% decision level. The evidence supports H_0 at this level, hence we assume that the two population variances are equal. We will now continue with the test of the sample means.

Since the population variances are unknown but may be assumed to be equal, we will test at the 5% level of significance using a one sided t test with $n_1 + n_2 - 2 = 8 + 6 - 2 = 12$ degrees of freedom. From the standard t distribution tables, $t_{0.05,12} = 1.78$.

The test statistic is:

$$t = \frac{(\bar{x}_1 - \bar{x}_2) - 0}{\widehat{SE}_{\bar{x}_1 - \bar{x}_2}}$$

$$\widehat{SE}_{\bar{x}_1 - \bar{x}_2} = \hat{\sigma}\sqrt{\frac{1}{n_1} + \frac{1}{n_2}} \quad \text{and} \quad \hat{\sigma} = \sqrt{\frac{n_1 s_1^2 + n_2 s_2^2}{n_1 + n_2 - 2}}$$

Hence:

$$\hat{\sigma} = \sqrt{\frac{8 \times 0.6078^2 + 6 \times 0.3416^2}{12}} = 0.5519$$

therefore:

$$\widehat{SE}_{\bar{x}_1 - \bar{x}_2} = 0.5519 \times \sqrt{(1/8) + (1/6)} = 0.2981$$

The test statistic is:

$$t = \frac{90.725 - 91.2}{0.2981} = -1.59$$

$-1.59 > -t_{0.05,12}$, therefore the result is not significant at the 5% decision level. The evidence does support the null hypothesis at this level, therefore we choose to accept H_0. In the light of the evidence, we choose to assume that the mean % yield has not changed as a result of the adjustments which were made.

Exercise 6.8

A comparison is required of the mean yield from the two processes. If a test is carried out directly on the two mean values and a significant result is found, it will not be clear whether the result arises because there is a difference between the two processes or whether it arises because of the variable purity of the raw material. Allowance can be made for the variability in the purity by taking the yields from the two processes in pairs. This will then remove purity as a variable in the hypothesis test.

H_0: The average difference between the sample values is consistent with the samples being drawn from normal populations with equal means so that $\mu_d = 0$.

H_1: $\mu_d \neq 0$

Batch Number	1	2	3	4	5	6	7	8
Yield from Process 1	70.1	67.0	68.6	68.8	70.2	68.0	69.5	68.4
Yield from Process 2	73.9	70.3	68.5	71.3	69.7	71.0	69.8	68.2
Difference in yield, d	−3.8	−3.3	0.1	−2.5	0.5	−3.0	−0.3	0.2

$$\text{Mean difference, } \bar{d} = \frac{\sum d}{n} = \frac{-12.1}{8} = -1.5125$$

Standard deviation of the difference is:

$$s_d = \sqrt{\frac{\sum d^2}{n} - (\bar{d})^2} = \sqrt{\frac{40.97}{8} - (-1.5125)^2} = 1.6833$$

Test at the 5% decision level using a two sided t test with $n - 1 = 7$ degrees of freedom.
From the standard t distribution tables, $t_{0.025,7} = 2.36$.

The test statistic is:

$$t = \frac{\bar{x}_d - \mu_d}{\widehat{SE}_{\bar{d}}} \quad \text{where} \quad \widehat{SE}_{\bar{d}} = \frac{s_d}{\sqrt{n-1}}$$

therefore:

$$t = \frac{-1.5125 - 0}{1.6833/\sqrt{7}} = -2.38$$

$-2.38 < -t_{0.025,7}$, therefore the result is significant at the 5% decision level. However, the result is highly marginal and it would be very foolish to make an important decision on the basis of a result which is as close as this is to the boundary value. Hence, although it would appear that the evidence does not support H_0 at the 5% decision level, the best course of action would be to gather additional data and to repeat the test.

Exercise 6.9

H_0: The two samples are drawn from populations with the same proportion of errors, ie $p_1 = p_2 = p$.
H_1: There has been an improvement in the error rate, ie $p_2 < p_1$.

The proportion of errors in the sample follows a binomial distribution but, since the sample sizes are large and the probability of an error is also large, the binomial distribution may be approximated by a normal distribution.

Test at the 1% level using a one sided normal test. Using the standard normal tables, $z_{0.01} = 2.33$, where $z_{0.01}$ is the number of standard deviations from the mean which cut off the top or bottom 1% of the normal distribution.

$$\hat{p}_1 = 56/100 = 0.56 \qquad \hat{p}_2 = 28/75 = 0.373$$

The test statistic is:

$$z = \frac{(\hat{p}_1 - \hat{p}_2) - (p_1 - p_2)}{\widehat{SE}_{\hat{p}_1 - \hat{p}_2}} = \frac{\hat{p}_1 - \hat{p}_2}{\widehat{SE}_{\hat{p}_1 - \hat{p}_2}}$$

$$\widehat{SE}_{\hat{p}_1 - \hat{p}_2} = \sqrt{\frac{p_1 q_1}{n_1} + \frac{p_2 q_2}{n_2}} = \sqrt{pq\left\{\frac{1}{n_1} + \frac{1}{n_2}\right\}} \quad \text{where} \quad p_1 = p_2 = p$$

Since the population proportions are unknown, they must be estimated from the sample values. The best estimate is achieved by pooling the two sample values. Therefore:

$$\hat{p} = \frac{56 + 28}{100 + 75} = \frac{84}{175} = 0.48 \quad \text{and} \quad \hat{q} = 0.52$$

Hence:

$$\widehat{SE}_{\hat{p}_1 - \hat{p}_2} = \sqrt{0.48 \times 0.52\{(1/100) + (1/75)\}} = 0.076315$$

Therefore:

$$z = \frac{(56/100) - (28/75)}{0.076315} = 2.45$$

$2.45 > z_{0.01}$, therefore the result is significant at the 1% decision level. The evidence does not support the null hypothesis at this level. Hence, we choose to reject H_0 and to accept H_1. We assume that there has been an improvement in the error rate.

SOLUTIONS TO EXERCISES: CHAPTER SIX

Exercise 6.10

H_0: There is no preference amongst consumers for any of the brands A to E, therefore $p_A = p_B = p_C = p_D = p_E = 0.2$.

H_1: There is a preference amongst consumers, ie $p_A \neq p_B \neq p_C \neq p_D \neq p_E$.

We can now calculate the expected numbers of people who would choose each of the brands if the null hypothesis was true. These expected frequencies can then be compared with the observed frequencies using a χ^2 test.

Note: For a χ^2 test, the actual frequencies must be used not percentages.

Brand	A	B	C	D	E
Observed frequency, f_o	38	40	30	50	42
Expected frequency, f_e*	40	40	40	40	40

* All brands equally likely.

Test at the 5% decision level using a χ^2 test with $n - 1 = 4$ degrees of freedom. From the standard χ^2 distribution tables, $\chi^2_{0.05,4} = 9.49$.

The test statistic is:

$$\chi^2 = \sum \left\{ \frac{(f_o - f_E)^2}{f_E} \right\} = \frac{(38-40)^2}{40} + \frac{(40-40)^2}{40} + \frac{(30-40)^2}{40} + \frac{(50-40)^2}{40} + \frac{(42-40)^2}{40}$$
$$= 5.2$$

$5.2 < \chi^2_{0.05,4}$, therefore the result is not significant at the 5% decision level. The evidence is consistent with H_0 at this level. We have no reason to reject the null hypothesis, hence, we assume that the consumers showed no preference for any of the brands.

Exercise 6.11

1   Repeat the procedure for **Exercise 6.10** using the new observed frequencies.

H_0: There is no preference amongst consumers for any of the brands A to E, therefore $p_A = p_B = p_C = p_D = p_E = 0.2$.

H_1: The consumers do exhibit a preference, ie $p_A \neq p_B \neq p_C \neq p_D \neq p_E$.

Brand	A	B	C	D	E
Observed frequency, f_o	20	19	14	25	22
Expected frequency, f_e	20	20	20	20	20

Test at the 5% decision level using a χ^2 test with $n - 1 = 4$ degrees of freedom. From the standard χ^2 distribution tables, $\chi^2_{0.05,4} = 9.49$.

The test statistic is:

$$\chi^2 = \sum \left\{ \frac{(0-E)^2}{E} \right\} = \frac{(20-20)^2}{20} + \frac{(19-20)^2}{20} + \frac{(14-20)^2}{20} + \frac{(25-20)^2}{20} + \frac{(22-20)^2}{20}$$
$$= 3.3$$

$3.3 < \chi^2_{0.05,4}$, therefore the result is not significant at the 5% decision level. The evidence is consistent with H_0 at this level. We have no reason to reject the null hypothesis, hence we assume that the consumers still showed no preference for any of the brands.

2   H_0: The market share of Brand E has not changed as a result of the advertising campaign.

H_1: The market share of Brand E has changed as a result of the advertising campaign.

Set up a contingency table for the frequency of choice of Brand E and the other brands, before and after the campaign:

	Brand E	Other Brands	Total
Before	42	158	200
After	22	78	100
Total	64	236	300

If the null hypothesis is true, the number of people whom we expect to choose Brand E before the advertising campaign is:

$$\frac{200}{300} \times 64 = 42.67$$

The rest of the expected frequencies follow, since the row and column totals are fixed. The expected frequency table is:

	Brand E	Other Brands	Total
Before	42.67	157.33	200
After	21.33	78.67	100
Total	64	236	300

Test at the 5% decision level using a χ^2 test with $(2-1)(2-1) = 1$ degree of freedom. (Yates' correction is not used, because the sample is large.) From the standard χ^2 distribution tables, $\chi^2_{0.05,1} = 3.84$.

The test statistic is:

$$\chi^2 = \frac{(42-42.67)^2}{42.67} + \frac{(158-157.33)^2}{157.33} + \frac{(22-21.33)^2}{21.33} + \frac{(78-78.67)^2}{78.67}$$

$$= 0.0401$$

$0.0401 < \chi^2_{0.05,1}$, therefore the result is not significant at the 5% decision level. The evidence is consistent with the null hypothesis at this level. We have no reason to reject H_0, however, the calculated value of χ^2 is very small—perhaps too small to arise by chance?

Let us consider the value of χ^2 which cuts off the lower 5% of the standard distribution. From the tables, $\chi^2_{0.95,1} = 0.00393$. Since $0.0401 > \chi^2_{0.95,1}$, the result is not significant at this side of the distribution either, hence we have no reason to suppose that the calculated value of χ^2 is suspiciously small.

3 H_0: The pattern of preferences has not changed as a result of the advertising campaign.

H_1: The pattern of preferences has changed as a result of the advertising campaign.

Set up a contingency table for the frequency of choice of each brand, before and after the campaign:

	A	B	C	D	E	Total
Before	38	40	30	50	42	200
After	20	19	14	25	22	100
Total	58	59	44	75	64	300

If the null hypothesis is true, the number of people whom we expect to choose Brand A before the advertising campaign is:

$$\frac{200}{300} \times 58 = 38.67$$

The number of people whom we expect to choose Brand B before the advertising campaign is:

$$\frac{200}{300} \times 59 = 39.33$$

The rest of the expected frequencies follow the same pattern. The expected frequency table is:

	\multicolumn{5}{c}{Brand}	
	A	B	C	D	E	Total
Before	38.67	39.33	29.33	50.0	42.67	200
After	19.33	19.67	14.67	25.0	21.33	100
Total	58	59	44	75	64	300

Test at the 5% decision level using a χ^2 test with $(2-1)(5-1) = 4$ degrees of freedom.
From the standard χ^2 distribution tables, $\chi^2_{0.05,4} = 9.49$.
The test statistic is:

$$\chi^2 = \frac{(38 - 38.67)^2}{38.67} + \frac{(40 - 39.33)^2}{39.33} + \cdots + \frac{(25 - 25.0)^2}{25.0} + \frac{(22 - 21.33)^2}{21.33}$$

$$= 0.1522$$

$0.1522 < \chi^2_{0.05,4}$, therefore the result is not significant at the 5% decision level. The evidence is consistent with the null hypothesis at this level. We have no reason to reject H_0, however, as in part **2**, the calculated value of χ^2 is very small—perhaps too small to arise by chance?

Let us consider the value of χ^2 which cuts off the lower 5% of the standard distribution. From the tables, $\chi^2_{0.95,4} = 0.711$. Since $0.1522 < \chi^2_{0.95,4}$, the result is significant at the 95% decision level, hence we have reason to suppose that the calculated value of χ^2 is suspiciously small. It is unlikely that a value of χ^2 as small as this would arise by chance. The implication is that there is some problem with the data. We cannot tell what that problem is, without further investigation.

Exercise 6.12

H_0: There is no relationship between the source and the size of the home loan.

H_1: There is a relationship between the source and the size of the home loan.

Set up a contingency table for the frequency of occurrence in each category.

		\multicolumn{2}{c}{Size}	
	Observed frequencies	Less than £10,000	£10,000 and over	Total
Source	Bank	25	25	50
	Building society	125	75	200
	Total	150	100	250

If the null hypothesis is true, the number of people whom we expect to have mortgages of less than £10,000 from a bank is:

$$\frac{50}{250} \times 150 = 30$$

The rest of the expected frequencies follows since the row and column totals are fixed. The expected frequency table is:

	Expected frequencies	Less than £10,000	£10,000 and over	Total
Source	Bank	30	20	50
	Building Society	120	80	200
	Total	150	100	250

Test at the 5% decision level using a χ^2 test with $(2-1)(2-1) = 1$ degree of freedom. From the standard χ^2 distribution tables, $\chi^2_{0.05,1} = 3.84$.

The test statistic is:

$$\chi^2 = \frac{(25-30)^2}{30} + \frac{(25-20)^2}{20} + \frac{(125-120)^2}{120} + \frac{(75-80)^2}{80}$$

$$= 2.60$$

$2.60 < \chi^2_{0.05,1}$, therefore the result is not significant at the 5% decision level. The evidence is consistent with the null hypothesis at this level. We have no reason to reject H_0, hence we assume that there is no relationship between the size and the source of the home loans.

Exercise 6.13

Estimate the average number of telephone answering systems sold per week.

$$\text{Average number sold/week} = \frac{\sum fx}{\sum f} = \frac{206}{50} = 4.12$$

$$\text{The variance of the number sold/week} = \frac{\sum fx^2}{\sum f} - (\bar{x})^2 = \frac{1044}{50} - (4.12)^2$$

$$= 3.91$$

The mean and variance are similar. This is a necessary condition for the Poisson distribution.

H_0: The number of units sold in a week follows a Poisson distribution with a mean of 4.12 units per week.

H_1: The number of units sold in a week does not follow a Poisson distribution with a mean of 4.12 units per week.

On the basis of the null hypothesis:

$$P(r \text{ units sold in a week}) = \frac{4.12^r e^{-4.12}}{r!} \quad r = 0, 1, 2, \ldots$$

The expected frequencies may be found from $50 \times P(r)$.

r	P(r)	Expected frequency	Observed frequency
0	$e^{-4.12} = 0.01624$	0.81 ⎫ Combine	1 ⎫
1	$4.12 \times e^{-4.12} = 0.06693$	3.35 ⎬ to exceed	3 ⎬
2	$= 0.13787$	6.89 ⎭ 5	6 ⎭
3	$= 0.18934$	9.47	11
4	$= 0.19502$	9.75	10
5	$= 0.16070$	8.03	7
6	$= 0.11035$	5.52	5
7	$= 0.06495$	3.25 ⎫	3 ⎫
8	$= 0.03345$	1.67 ⎬ Combine to exceed 5	4 ⎬
>8	$= 1 - 0.97485$ $= 0.02515$	1.26 ⎭	0 ⎭
	Total	50.00	50

For a χ^2 test the expected frequencies must exceed 5. Since we have some small values in the above table, some of the categories must be combined.

Number sold in a week	0,1,2	3	4	5	6	>6
Observed frequency	10	11	10	7	5	7
Expected frequency	11.05	9.47	9.75	8.03	5.52	6.18

Test at the 5% level using a χ^2 test with $(6 - 1 - 1) = 4$ degrees of freedom. *Note*: An extra degree of freedom is lost because the population average number sold in a week was estimated from the sample data. From the standard χ^2 distribution tables, $\chi^2_{0.05,4} = 9.49$.

The test statistic is:

$$\chi^2 = \frac{(10 - 11.05)^2}{11.05} + \frac{(11 - 9.47)^2}{9.47} + \cdots + \frac{(7 - 6.18)^2}{6.18}$$

$$= 0.6433$$

$0.6433 < \chi^2_{0.05,4}$, therefore the result is not significant at the 5% decision level. The evidence is consistent with H_0 at this level, hence we assume that the distribution of the number of units sold in a week is Poisson with a mean of 4.12 units per week.

Exercise 6.14

Refer to **Exercise 2.18**, Chapter Two, where it was calculated that the sample proportion of defectives is 0.06.

H_0: The sample proportion is consistent with the sample having been randomly selected from a population with a proportion of defectives of 0.06, ie p = 0.06.

H_1: The sample proportion is not consistent with the sample having been randomly selected from a population with a proportion of defectives of 0.06.

If the null hypothesis is true, the number of defectives in a box will follow a binomial distribution with p = 0.06, hence the expected frequencies may be calculated.

P(r defectives in a box) = $^5C_r \times 0.06^r \times 0.94^{5-r}$ for r = 0, 1, 2, 3, 4, 5. We will now complete the table of expected frequencies, begun in **Exercise 2.18** in answer to part **1** of this question:

r	P(r defectives per box)	Expected Frequency = 500 × P(r)
0	0.7339	366.95
1	0.2342	117.10
2	0.0299	14.95 ⎫
3	0.0019	0.95 ⎬ Combine to exceed 5
4	0.0001	0.05 ⎪
5	0.00000	0.00 ⎭

We will compare the expected frequencies with the observed frequencies using a χ^2 test. This requires the expected frequencies to exceed 5, hence the last 4 values must be combined:

r	Expected frequency	Observed frequency
0	366.95	392
1	117.10	73
2 to 5	15.95	35
Total	500.00	500

Test at the 5% decision level using a χ^2 test with n − 1 − 1 = 1 degree of freedom. The extra degree of freedom is lost because the population proportion of defectives was estimated from the sample. From the standard χ^2 distribution tables, $\chi^2_{0.05,1}$ = 3.84.

The test statistic is:

$$\chi^2 = \frac{(392 - 366.95)^2}{366.95} + \frac{(73 - 117.1)^2}{117.1} + \frac{(35 - 15.95)^2}{15.95} = 41.07$$

41.07 > $\chi^2_{0.05,1}$, therefore the result is significant at the 5% decision level. The evidence is not consistent with H_0 at this level. We choose to reject H_0 in favour of H_1—the observed distribution of the number of defectives in a box does not follow a binomial distribution with p = 0.06.

There are now two possible situations:

1 The number of defectives in a box follows a binomial distribution but the proportion of defectives is not 0.06. This means that the defectives are being produced at random and infrequently but for some reason the sample does not reflect the population proportion of defectives—the sample itself has not been randomly selected.

2 The number of defectives in a box does not follow a binomial distribution at all. This means that there is something wrong with the process and the defectives are not being produced infrequently and at random.

Exercise 6.15

1

Number of regular statements/year, n	Number of customers f	fn	fn²
0	11	0	0
2	13	26	52
4	61	244	976
12	65	780	9360
	150	1050	10388

$$\text{Mean number of regular statements/year, } \bar{n} = \frac{\sum fn}{\sum f} = \frac{1050}{150} = 7$$

Standard deviation of the number of regular statements per year is:

$$s = \sqrt{\frac{\sum fn^2}{\sum f} - (\bar{n})^2} = \sqrt{\frac{10388}{150} - 7^2} = 4.50$$

2 The 95% confidence interval for the population mean number of regular statements per year is approximately:

$$\bar{n} \pm z_{0.025} SE_{\bar{n}}$$

This assumes that the mean number of regular statements is approximately normally distributed. Since the sample size is large, this assumption is likely to be reasonable.

From the standard normal tables, $z_{0.025} = 1.96$.

$$\widehat{SE}_{\bar{n}} = \frac{\hat{\sigma}}{\sqrt{n}} = \sqrt{\frac{n}{n-1}} \frac{s}{\sqrt{n}} = \frac{s}{\sqrt{n-1}} = \frac{4.50}{\sqrt{149}}$$

Therefore the 95% confidence interval for the population mean is:

$$7 \pm 1.96 \times \frac{4.50}{\sqrt{149}} = 7 \pm 0.723$$

Since the sample size is large, the standard normal value is approximately equal to the t value.

The confidence interval tells us that for 95% of all samples, the population average number of regular statements required per year lies within 0.72 of the sample mean. Hence, in this case, there is a 95% probability that the population mean lies between 6.28 and 7.72 statements per year.

3 Assume that the demand for the statements can be smoothed over the year so that the same number can be despatched each day. This assumption includes the despatch of irregular statement requests. On the basis of the sample mean, the estimated total number of statements required each year is:

$$7 \times 1{,}200{,}000 + 600{,}000 = 9{,}000{,}000, \text{ ie 9 million}$$

This means that the bank must despatch:

$$9 \text{ million}/300 = 30{,}000 \text{ statements/day on average}$$

The bank is in difficulties if the population mean number of statements required for despatch per day is greater than 33,000.

H_o: The sample is drawn from a normal population for which the mean number of statements required for despatch each day is 33,000, ie $\mu = 33{,}000$.

H_1: The population mean is $<33{,}000$, ie $\mu < 33{,}000$.

Test at the 5% decision level using a one sided test. Since the population variance is unknown, the appropriate test is a t test with 149 degrees of freedom. Since the sample size is so large, a normal test could readily be used as an approximation to the t test.

From the standard t distribution tables, $t_{0.05,149} = 1.655$. The test statistic is:

$$t = \frac{\bar{x} - \mu}{\widehat{SE}_{\bar{x}}}$$

where x̄ is the average number of statements despatched per day.

$$\widehat{SE}_{\bar{x}} = \frac{1{,}200{,}000}{300} \times \widehat{SE}_{\hat{n}} \quad \text{statements per day}$$

$$= \frac{1{,}200{,}000}{300} \times \frac{4.50}{\sqrt{149}} = 1474.6 = 1475 \text{ approximately.}$$

Therefore:

$$t = \frac{30{,}000 - 33{,}000}{1475} = -2.03$$

$-2.03 < -t_{0.05, 149}$, therefore the result is significant at the 5% decision level. The evidence does not support the null hypothesis at this level, hence H_0 is probably untrue. We therefore choose to adopt the alternative hypothesis that the population mean is less than 33,000 statements per day.

It should be noted that the result is not significant at the 1% decision level ($t_{0.01, 149} = 2.35$ approximately), hence the best course of action would be to collect more data and to carry out further statistical tests.

Exercise 6.16

1 H_0: The sample proportion of errors is consistent with the sample having been drawn from a population in which the proportion of errors is binomially distributed with $p = 0.02$.

H_1: The sample is drawn from a population in which $p > 0.02$.

Since the sample is large, we can assume a normal approximation to the binomial.

We will test the null hypothesis at the 5% level using a one sided normal test. From the standard normal tables, $z_{0.05} = 1.645$. The sample proportion is not a continuous variable and, hence, we should consider the use of a continuity correction when we use the normal approximation. However, since the sample size is large, we could assume that the proportion of errors is approximately continuous. For the purpose of illustration, the problem will be worked using both methods.

Assume that the sample is large enough for the values of p to be approximately continuous—no continuity correction is used. The sample proportion of errors, $\hat{p}, = 9/300 = 0.03$, and:

$$SE_{\hat{p}} = \sqrt{\frac{pq}{n}} = \sqrt{\frac{0.02 \times 0.98}{300}} = 0.00808$$

The test statistic is:

$$z = \frac{\hat{p} - p}{SE_{\hat{p}}} = \frac{0.03 - 0.02}{0.00808} = 1.24$$

Alternatively, apply a continuity correction. In the binomial distribution, we require:

$$P \text{ (number of errors} \geqslant 9)$$

This is equivalent in the normal distribution to:

$$P(\text{number of errors} > 8.5).$$

Therefore, the sample proportion, $\hat{p}, = 8.5/300 = 0.02833$.

The test statistic is:

$$z = \frac{0.0283 - 0.02}{0.00808} = 1.03$$

In both cases, the test statistic $< z_{0.05}$, therefore the result is not significant at the 5% decision level. The evidence is consistent with H_0 at this level, hence we have no reason to reject the null hypothesis. We assume that the stipulated error rate has not been violated.

2 If the result is to be significant at the 5% level, then:

$$\frac{\hat{p} - 0.02}{0.00808} \geqslant 1.645$$

therefore:

$$\hat{p} \geqslant 0.0333$$

hence:

$$n/300 \geqslant 0.0333 \qquad \text{where n is the number required.}$$

therefore:

$$n \geqslant 9.99$$

If the continuity correction is not being used, then 10 items would need to contain a minor irregularity before the result of the test would be significant. However, if we feel that the continuity correction should be used, then 11 (9.9 + 0.5 = 10.49) items are required before the test gives a significant result. In either case we can see that the result is sensitive to small changes in the number of errors found in the sample.

Exercise 6.17

1 *Point 1*: the data consist of the annual earnings for each share over a period of time. The earnings for each share will vary from year to year for at least two reasons:

(i) the nature of the share;
(ii) the passage of time, ie specific conditions which arose in a given year.

If the two average earnings are compared, we will not know how much of the difference between the averages is due to the nature of the shares and how much is due to the variation in time. Hence, it is necessary to eliminate the effect of time. This can be done by pairing the data. The difference in the earnings of the two shares is calculated for each year. The average difference is then compared with zero.

Point 2: no null hypothesis was specified before the test was carried out. It is undesirable to examine the test statistic on both a one and two sided basis. The objective of the test should be clearly thought out and stated at the beginning.

Point 3: The test statistic used:

$$t = \frac{\bar{x}_1 - \bar{x}_2}{\sqrt{\frac{s_1^2}{n_1} + \frac{s_2^2}{n_2}}}$$

which is appropriate when the population variances cannot be assumed to be equal but the sample sizes are large. Then the test statistic is approximately normal. The sample sizes in this case are too small for the approximation to be valid.

For such small samples, the t test for the comparison of means is valid only if the two population variances can be assumed to be equal. Then the best estimate of the standard error of the difference between the sample means is given by:

$$\widehat{SE}_{\bar{x}_1 - \bar{x}_2} = \sqrt{\frac{n_1 s_1^2 + n_2 s_2^2}{n_1 + n_2 - 2}\left[\frac{1}{n_1} + \frac{1}{n_2}\right]}$$

where s_1^2 and s_2^2 are the variances of the two samples, calculated by:

$$s^2 = \frac{\sum (x - \bar{x})^2}{n}$$

and not by:

$$\widehat{SE}_{\bar{x}_1 - \bar{x}_2} = \sqrt{\frac{s_1^2}{n_1} + \frac{s_2^2}{n_2}}$$

as in the exercise.

Point 4: the phrasing of the conclusion is far too definite for a test of significance. If the null hypothesis has been defined as:

H_0: The two sample means are consistent with the samples having been drawn from normal populations with the same mean.

then a non-significant result means that the evidence is consistent with the null hypothesis. We have no reason to reject the hypothesis that the average earnings per share are the same. This is not the same as saying that the average earnings are the same.

2 A paired data test will be used.

	Earnings/share		Difference in the	
	A	B	earnings/share, d	d^2
	14.3	13.8	0.5	0.25
	15.6	14.6	1.0	1.00
	17.2	16.4	0.8	0.64
	16.4	16.8	−0.4	0.16
	14.9	15.0	−0.1	0.01
	17.6	16.4	1.2	1.44
Total			3.0	3.5

$$\bar{d} = \frac{\sum d}{n} = \frac{3.0}{6} = 0.5 \quad \text{and} \quad s_d = \sqrt{\frac{\sum d^2}{n} - (\bar{d})^2} = \sqrt{\frac{3.5}{6} - 0.5^2} = 0.577$$

H_0: The samples are drawn from normal populations with equal means, therefore $\mu_1 - \mu_2 = 0$.

H_1: $\mu_1 - \mu_2 \neq 0$.

Since the population variances are unknown, test at the 5% decision level using a two sided t test with $6 - 1 = 5$ degrees of freedom. From the standard t distribution tables, $t_{0.025,5} = 2.571$.

The test statistic is:

$$t = \frac{\bar{d} - 0}{\widehat{SE}_{\bar{d}}}$$

$$\widehat{SE}_{\bar{d}} = \frac{s_d}{\sqrt{n-1}} = \frac{0.577}{\sqrt{5}}$$

therefore:

$$t = \frac{0.5}{0.577/\sqrt{5}} = 1.94$$

$1.94 < t_{0.025,5}$ therefore the result is not significant at the 5% decision level. The evidence is consistent with H_0 at this level. We have no reason to reject the null hypothesis, therefore we assume that the average earnings per share are not different.

Exercise 6.18

1   Average cost/square metre, $\bar{x} = \dfrac{\sum x}{n}$

Standard deviation of cost/m², $s = \sqrt{\dfrac{\sum x^2}{n} - (\bar{x})^2}$

For schools: $\bar{x}_s = \dfrac{210}{7} = 30$

$$s_s = \sqrt{\frac{6492}{7} - 30^2} = 5.237$$

Note: No units are given.

For offices: $\bar{x}_0 = \dfrac{185}{5} = 37$

$$s_0 = \sqrt{\frac{6883}{5} - 37^2} = 2.757$$

Since the population variances are unknown, we cannot choose the appropriate test until we know whether or not we may assume that these unknown variances are equal.

F test:

H_0: The samples are drawn from normal populations with equal variances, ie $\sigma_s^2 = \sigma_0^2$.

H_1: $\sigma_s^2 \neq \sigma_0^2$.

Test at the 5% decision level using a two sided F test:

$$\hat{\sigma}^2 = \frac{n}{n-1} s^2$$

therefore:

$$\hat{\sigma}_s^2 = \frac{7}{6} \times 5.237^2 = 31.9972$$

and:

$$\hat{\sigma}_0^2 = \frac{5}{4} \times 2.757^2 = 9.5013$$

Since $\hat{\sigma}_s^2 > \hat{\sigma}_0^2$, $F = \hat{\sigma}_s^2/\hat{\sigma}_0^2$ and has 6 and 4 degrees of freedom. From the standard F distribution tables, $F_{0.025,6,4} = 9.20$. The test statistic is:

$$F = \frac{31.9972}{9.5013} = 3.368$$

$3.368 < F_{0.025,6,4}$, therefore the result is not significant at the 5% decision level. The evidence is consistent with H_0 at this level, hence we choose to accept the null hypothesis that the two population variances are equal.

It now follows that the two sample means may be compared using a t test with the common population variance being estimated by pooling the two sample variances.

t test:

H_0: The sample means are consistent with the samples having been drawn from normal populations with equal means, ie $\mu_s = \mu_0$.

H_1: $\mu_s < \mu_0$

Test at the 5% decision level using a one sided t test with $(7 + 5 - 2) = 10$ degrees of freedom. From the standard t distribution tables, $t_{0.05,10} = 1.812$.

The test statistic is:

$$t = \frac{\bar{x}_s - \bar{x}_0}{\widehat{SE}_{\bar{x}_s - \bar{x}_0}}$$

$$\widehat{SE}_{\bar{x}_s - \bar{x}_0} = \sqrt{\frac{n_s s_s^2 + n_0 s_0^2}{n_s + n_0 - 2}\left(\frac{1}{n_s} + \frac{1}{n_0}\right)}$$

$$= \sqrt{\frac{7 \times 5.237^2 + 5 \times 2.757^2}{7 + 5 - 2}\left(\frac{1}{7} + \frac{1}{5}\right)} = 2.8081$$

therefore:

$$t = \frac{30 - 37}{2.8081} = -2.493$$

$-2.493 < -t_{0.05,10}$, therefore the result is significant at the 5% decision level. The evidence does not support H_0 at this level, hence we choose to reject the null hypothesis and to accept the alternative hypothesis that the average cost per square metre is lower for schools than for office blocks.

2 The procedure adopted in **1** was based on the assumption that the data represented two samples drawn randomly from two large normal populations. We now have information which implies that the populations are small. This fact must be taken into account when estimating the standard error of the difference between the means. When the populations are small:

$$SE_{\bar{x}_s - \bar{x}_0} = \sqrt{\frac{(N_s - n_s)}{(N_s - 1)}\frac{\sigma_s^2}{n_s} + \frac{(N_0 - n_0)}{(N_0 - 1)}\frac{\sigma_0^2}{n_0}}$$

Since, as in **1**, we may assume that $\sigma_s^2 = \sigma_0^2 = \sigma^2$, and the best estimate of σ^2 is:

$$\hat{\sigma}^2 = \frac{n_s s_s^2 + n_0 s_0^2}{n_s + n_0 - 2}$$

then the best estimate of $SE_{\bar{x}_s - \bar{x}_0}$ is:

$$\widehat{SE}_{\bar{x}_s - \bar{x}_0} = \hat{\sigma}\sqrt{\frac{(N_s - n_s)}{(N_s - 1)} \times \frac{1}{n_s} + \frac{(N_0 - n_0)}{(N_0 - 1)} \times \frac{1}{n_0}}$$

$$= 4.7957 \times \sqrt{\frac{(24 - 7)}{(24 - 1)} \times \frac{1}{7} + \frac{(27 - 5)}{(27 - 1)} \times \frac{1}{5}} = 2.5141$$

The null and alternative hypotheses are the same as in part **1**. The test statistic is:

$$t = \frac{\bar{x}_s - \bar{x}_0}{\widehat{SE}_{\bar{x}_s - \bar{x}_0}} = \frac{30 - 37}{2.5141} = -2.784$$

Again, the result is significant at the 5% decision level. Hence, we continue to reject the null hypothesis and to accept the alternative hypothesis that the average cost per square metre is less for schools than for office blocks.

3 The t test is based on the assumption that the samples are randomly drawn from normal populations with equal variances. We do not know whether or not these assumptions are valid with the rather small populations.

There may be many factors which affect the unit cost of buildings and ignoring these factors may invalidate the test since the samples may not be independent of each other. These factors include the size of the buildings, the geographical location, variation in architectural styles, fittings and furnishings included in the price, etc.

Exercise 6.19

1 The term 'significantly different' means that the difference between the value of a sample statistic (or a sample observation) and the value of the corresponding population parameter (or expected value) is too large to be likely to have arisen due to chance sampling fluctuation.

2 (a) We will test the given null hypothesis at the 5% decision level using a χ^2 test with $(2 - 1)(4 - 1) = 3$ degrees of freedom. From the standard χ^2 distribution tables, $\chi^2_{0.05,3} = 7.81$. The value of the test statistic, $7.34 < \chi^2_{0.05,3}$, therefore the result is not significant at the 5% decision level. The evidence is consistent with the null hypothesis at this level, hence we have no reason to reject H_0.

In view of the closeness of 7.34 to $\chi^2_{0.05,3}$, the decision is not really as clear cut as we might wish. An increase in the size of the sample would probably produce a more clear cut result.

(b) Under 40's: use the new figures with the original ones to calculate the number of people at each plant who are against the wage deal.

	A	B	Plant C	D	Total
In favour	60	28	30	45	163
Against	15	20	24	12	71
Total	75	48	54	57	234

H_0: There is no relationship between the plants and the views on the wage deal.

H_1: There is a relationship between the plants and the views on the wage deal.

Test at the 5% decision level using a χ^2 test with $(2-1)(4-1) = 3$ degrees of freedom, hence, as in part (a) $\chi^2_{0.05,3} = 7.81$.

We must now calculate the frequencies which we would expect if the null hypothesis were true. If there is no relationship between the plant and the views expressed, then we would expect the employees to be distributed between the cells of the contingency table according to the total proportion in each category. Hence, for the top left hand cell $(1,1)$, the expected frequency is:

$$\frac{163}{234} \times 75 = 52.24$$

For the next cell $(1,2)$, $f_E = \frac{163}{234} \times 48 = 33.44$

and for the cell $(1,3)$, $f_E = \frac{163}{234} \times 54 = 37.62$

The remaining expected frequencies are now determined since the row and column totals are fixed.

	Plant A	B	C	D	Total
In favour	52.24	33.44	37.62	39.71	163
Against	22.76	14.56	16.38	17.29	71
Total	75	48	54	57	234

The test statistic is:

$$\chi^2 = \Sigma\left\{\frac{(O-E)^2}{E}\right\} = \frac{(60-52.24)^2}{52.24} + \cdots + \frac{(12-17.29)^2}{17.29}$$

$$= 14.13$$

$14.13 > \chi^2_{0.05,3}$, therefore the result is significant at the 5% decision level. The evidence is not consistent with H_0 at this level, hence we choose to reject H_0 and to accept H_1—that there is a relationship between the plant and the views about the wage deal.

If we examine the contingency table of observed frequencies, we can see that there is a higher percentage of people in favour of the deal at plants A and D compared to B and C.

Over 40's: using the data given in the question and the data for the under 40's, the observed frequencies can be derived for a contingency table for the over 40's:

	Plant A	B	C	D	Total
In favour	20	12	20	15	67
Against	20	10	16	13	59
Total	40	22	36	28	126

The null hypothesis, alternative hypothesis and test conditions are the same as those used for the under 40's test.

The expected frequencies are calculated on the same basis as those for the under 40's.

	A	B	Plant C	D	Total
In favour	21.27	11.70	19.14	14.89	67
Against	18.73	10.30	16.86	13.11	59
Total	40	22	36	28	126

The test statistic is:

$$\chi^2 = \sum \left\{ \frac{(O-E)^2}{E} \right\} = \frac{(20-21.27)^2}{21.27} + \cdots + \frac{(13-13.11)^2}{13.11}$$
$$= 0.2626$$

$0.2626 < \chi^2_{0.05,3}$, therefore the result is not significant at the 5% decision level. The evidence is consistent with H_0 at this level, hence we choose to accept H_0 at this level, hence we assume that there is no relationship between the plant and the views about the wage deal for the over 40 age group.

(c) Under 40's; percentage in favour of the wage deal $= \dfrac{163}{234} \times 100$

$$= 69.66\,\%$$

Over 40's; percentage in favour of the wage deal $= \dfrac{67}{126} \times 100$

$$= 53.17\,\%$$

To test whether these percentages are significantly different, we will work with proportions and, since the sample sizes are large, we can use the normal test as an approximation, ie we will approximate the binomial distribution of the proportion in favour of the wage deal by a normal distribution.

H_0: The sample proportions are consistent with the samples having been drawn from approximately normal populations with the same proportion in favour of the wage deal, ie $p_1 = p_2 = p$.

H_1: $p_1 \neq p_2$.

Test at the 5% level using a two sided normal test. From the standard normal tables, $z_{0.025} = 1.96$.

The test statistic is:

$$z = \frac{(\hat{p}_1 - \hat{p}_2) - 0}{SE_{\hat{p}_1 - \hat{p}_2}}$$

$$SE_{\hat{p}_1 - \hat{p}_2} = \sqrt{\frac{pq}{n_1} + \frac{pq}{n_2}} = \sqrt{pq\left(\frac{1}{n_1} + \frac{1}{n_2}\right)}$$

Since p is unknown, the best estimate is obtained by pooling $\hat{p}_1$ and $\hat{p}_2$. Therefore:

$$\hat{p} = \frac{163 + 67}{234 + 126} = \frac{230}{360} \quad \text{hence} \quad \hat{q} = \frac{130}{360}$$

and:

$$\widehat{SE}_{\hat{p}_1 - \hat{p}_2} = \sqrt{\frac{230}{360} \times \frac{130}{360} \left(\frac{1}{234} + \frac{1}{126} \right)} = 0.05308$$

Therefore:

$$z = \frac{(163/234) - (67/126)}{0.05308} = 3.11$$

$3.11 > z_{0.025}$, therefore the result is significant at the 5% decision level. The evidence is not consistent with H_0 at this level, that is, there is no evidence that the two population proportions are the same. Hence, we choose to reject H_0 and to accept the alternative hypothesis that the proportions in favour of the wage deal are different in the two age groups.

Solutions to Chapter Seven exercises

ANSWERS TO EXERCISES

Exercise 7.1

1. What factors could affect the quantity of lettuce sold? Price, quantity available, yesterday's price, demand, the time of year, the weather. The quantity price relationship could be complex.
2. (a) Plot the data—to see if a linear model is appropriate. Quantity is the variable we wish to explain. It is the dependent variable, y. Price is the independent variable, x. In this complicated case, the two variables interact and you could argue that price depends on quantity. However, we are asked to explain the variation in quantity, not the variation in price.

   (b) **Plot of quantity of lettuce sold per day against price per lettuce:**

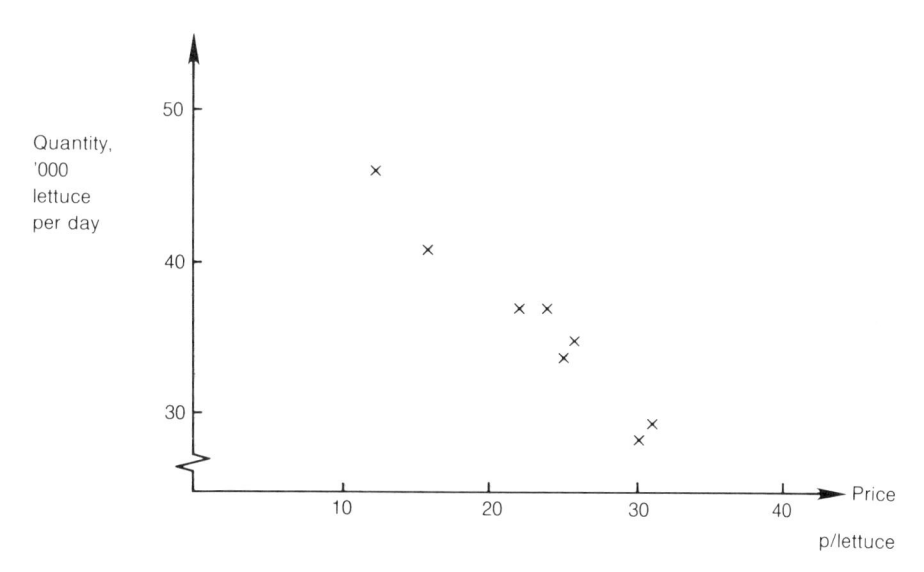

   The plot shows a good linear relationship:

   $$\sum x = 186, \sum y = 287, \sum xy = 6409, \sum x^2 = 4622$$

   $$\sum y^2 = 10541, n = 8$$

   The correlation coefficient $r = -0.977$, using the standard calculation, and $r^2 = 0.955$, ie 95.5% of the variation in quantity is explained by the variation in price.

   Test the correlation coefficient using a two sided t test with $(n - 2) = 6$ degrees of freedom. We choose to test at the 1% decision level. The hypotheses for the test are:

   $H_0: \rho = 0$: there is no linear relationship between quantity and price.
   $H_1: \rho \neq 0$: there is a linear relationship.

   The test statistic is:

   $$t = \sqrt{\frac{r^2(n-2)}{(1-r^2)}} = \sqrt{\frac{0.977^2(8-2)}{(1-0.977^2)}} = 11.2$$

510 QUANTITATIVE ANALYSIS

This is compared with the boundary value of:

$$t_{0.005, 6} = 3.71$$

The test statistic is a long way outside this boundary. The result is highly significant. We choose to reject H_0 and accept H_1. Assume there is a strong linear relationship between quantity and price.

(b) The regression equation, using the standard calculations, is:

$$\text{Quantity} = 56.5 - 0.887 \times \text{price} \quad (\text{'000/day})$$

The meaning of the coefficients. The equation is calculated as a whole to minimise the sums of squares of the errors between predicted and observed quantities. Splitting the equation into its separate parts, and interpreting these separately, must be done with care, if at all. (This is particularly true for multiple regression.) The regression coefficient, or slope, (-0.887) may be interpreted, within the range of the data, (from 12p to 31p per lettuce), as: for every 1p increase in price, the demand for lettuce will decrease by approximately 877 lettuces per day. The interpretation of the intercept term (56.5) is more difficult. We cannot think of it as the quantity demanded, if the price were zero. Since our lowest price is 12p, we have no evidence to suggest that this linear model is still valid when the price is zero. It is safest to regard the intercept as a base quantity from which we deduct $(0.887 \times \text{price per lettuce})$ to determine the quantity demanded per day when the price is in the range 12p to 31p per lettuce.

(c) If the price were 45p, the quantity sold would be:

$$Q = 56.5 - 0.887 \times 45 = 16.6, \text{ ie } 16,600/\text{day}$$

The assumption is that the linear relationship will hold a long way outside the range of the price data. Thirty one pence is the highest price in the data. Forty five pence is too far outside for us to predict, with any kind of certainty, what will happen.

Exercise 7.2

1 Savings and income. Plot of annual savings against annual income:

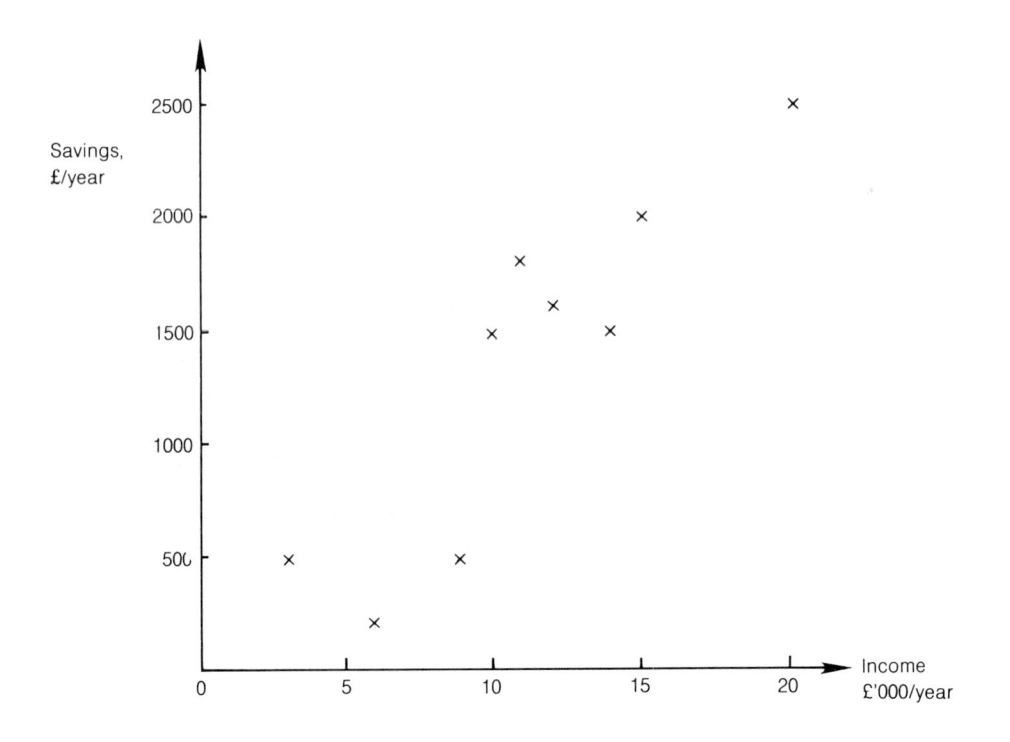

SOLUTIONS TO EXERCISES: CHAPTER SEVEN

The relationship looks approximately linear:

$$\sum x = 100, \sum y = 12{,}100, \sum xy = 162{,}200, \sum x^2 = 1312, \sum y^2 = 21{,}090{,}000$$

Calculate the correlation coefficient:

$$r = \frac{n\sum xy - \sum x \sum y}{\sqrt{(n\sum x^2 - (\sum x)^2)(n\sum y^2 - (\sum y)^2)}} = +0.892$$

$H_0: \rho = 0$: there is no linear relationship between savings and income.

$H_1: \rho \neq 0$: there is a relationship.

Test using a two sided t test with $(n - 2) = 7$ degrees of freedom:

$$t_{0.005, 7} = 3.50$$

The test statistic is:

$$t = \sqrt{\frac{r^2(n-2)}{1-r^2}} = \sqrt{\frac{0.796 \times 7}{0.204}} = 5.23$$

This is outside the boundary. There is very strong evidence to reject H_0 and to accept H_1. Assume there is a good linear relationship between savings and income.

2 The regression equation, using the standard calculations, is:

$$\text{Savings} = -191 + 0.138 \text{ income} \qquad (\pounds/\text{year})$$

See the comments on interpretation in **Exercise 7.1**. The $-\pounds 191$ has no practical meaning. The 0.138 coefficient could be interpreted as the estimated proportion of income that is saved. The people in this sample are saving about 14% of their marginal income.

3 Other factors to consider are as follows. What are the circumstances of each individual? Do they live on their own or have a family to support from their income? Where do they live and how may this affect their expenditure? Someone in expensive London will probably save less than someone in the North with the same income.

Exercise 7.3

1 Establish a model to explain variations in the number of guests. The following is a plot of numbers of guests visiting a hotel against that hotel's advertising expenditure.

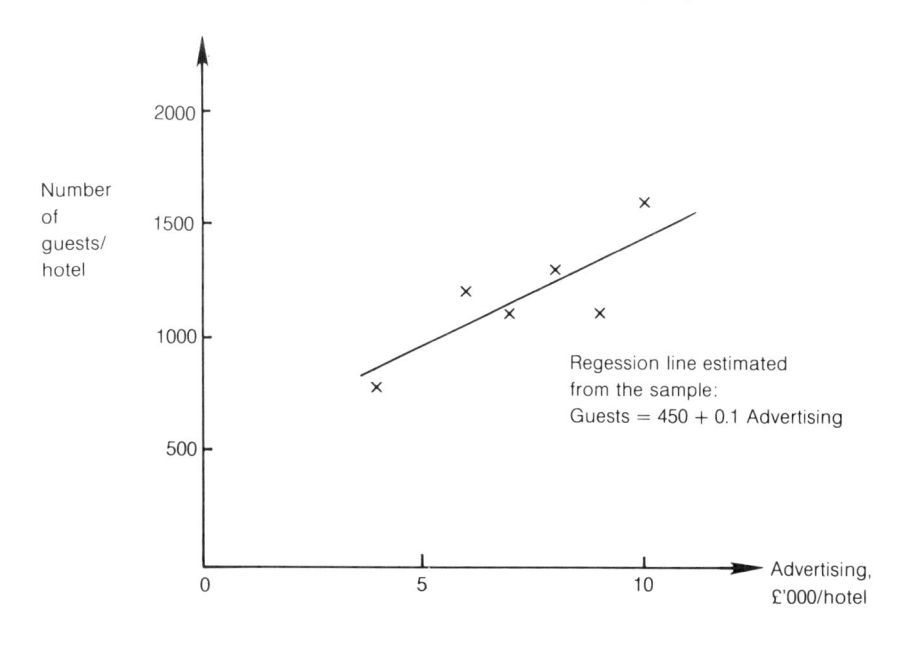

512 QUANTITATIVE ANALYSIS

The plot indicates that a linear relationship may exist.

$$\sum x = 44,000, \sum y = 7100, \sum xy = 54,400,000, \sum x^2 = 346,000,000, \sum y^2 = 8,750,000$$

We will calculate r and test its significance. By calculation, $r = 0.82$.

$H_0: \rho = 0$: there is no linear relationship between guests and advertising at individual hotels.
$H_1: \rho \neq 0$: there is a relationship.

Test using a two sided t test with $(n - 2) = 4$ degrees of freedom. Test against the boundary values:

$$t_{0.025,4} = 2.78 \text{ and } t_{0.005,4} = 4.60$$

The test statistic is:

$$t = \sqrt{\frac{r^2(n-2)}{(1-r^2)}} = \sqrt{\frac{0.6724(4)}{0.3276}} = 2.87$$

This is just outside the 5% boundary. There is reasonable evidence to reject H_0 and accept H_1. There probably is a linear relationship between guests and advertising expenditure. The model, using the standard calculations is:

$$\text{Number of guests} = 450 + 0.10 \times \text{Advertising}$$

2   Comment on the likely accuracy of forecasts. $r = 0.82$, therefore $r^2 \times 100$, the coefficient of determination = 67%. The model explains 67% of the variation in number of guests and does not explain 33%. There are likely to be some large errors and uncertainties.

Plotting the regression line on the diagram gives a clear indication of the errors for this sample. The errors range from 250 guests ($y - \hat{y} = 1100 - 1350$), when advertising is £9000 to 50 guests, when advertising is £4000, £7000 and £8000.

We should be very cautious about using a model which is only just significant, from such a small sample.

Exercise 7.4

1   Set preliminary standards for a production department. The following is a plot of departmental cost per day against output per day.

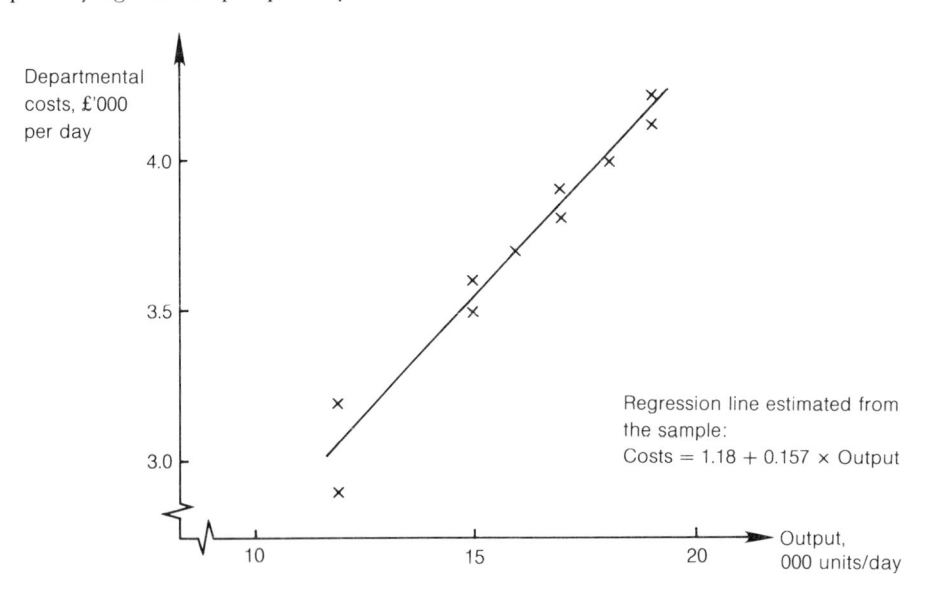

$$\sum x = 160, \sum y = 36, \sum xy = 599.5, \sum x^2 = 2618, \sum y^2 = 137.65$$

The plot shows that a linear relationship probably exists. Using the formula $r = +0.98$:

$H_0: \rho = 0$
$H_1: \rho \neq 0$, two sided test.

The boundary values are $t_{0.025,8} = \pm 2.31$, $t_{0.005,8} = \pm 3.36$. The test statistic is:

$$t = \sqrt{\frac{0.98^2 \times 8}{1 - 0.98^2}} = 13.9$$

which is very significant. We reject H_0 and accept H_1. Assume there is a very strong linear relationship between departmental costs and daily output. Using the standard formulae, the relationship is:

$$\text{Costs} = 1.18 + 0.157 \times \text{output} \qquad (£'000/\text{day})$$

This is plotted on the diagram.

The coefficients can be thought of as: 1.18 is £1,180 per day fixed costs. This is a doubtful figure, because we have no data lower than 12,000 units per day and the model fits less well for these points. $+0.157$ is £0.157 per unit—the variable cost of producing 1 unit.

2 Comment on accuracy of the standards. r is very large and, from the diagram, the model fits well between 15,000 and 20,000 units per day with errors of about £50 per day. The two low output figures have larger errors and it is unlikely that these standards would be adequate at lower output rates. This small sample suggests that more data is required at the lower output.

Other factors which should be considered are as follows. Is the output always the same product? Did this sample of days represent the full range of products? How are maintenance/breakdown costs represented? How do the costs vary at lower output rates? What do the costs used in the model include? Is it all fixed and variable costs?

Exercise 7.5

1 Garden Groceries—store amalgamation. The following is a plot of annual profit against turnover for the 12 shops in Garden Groceries.

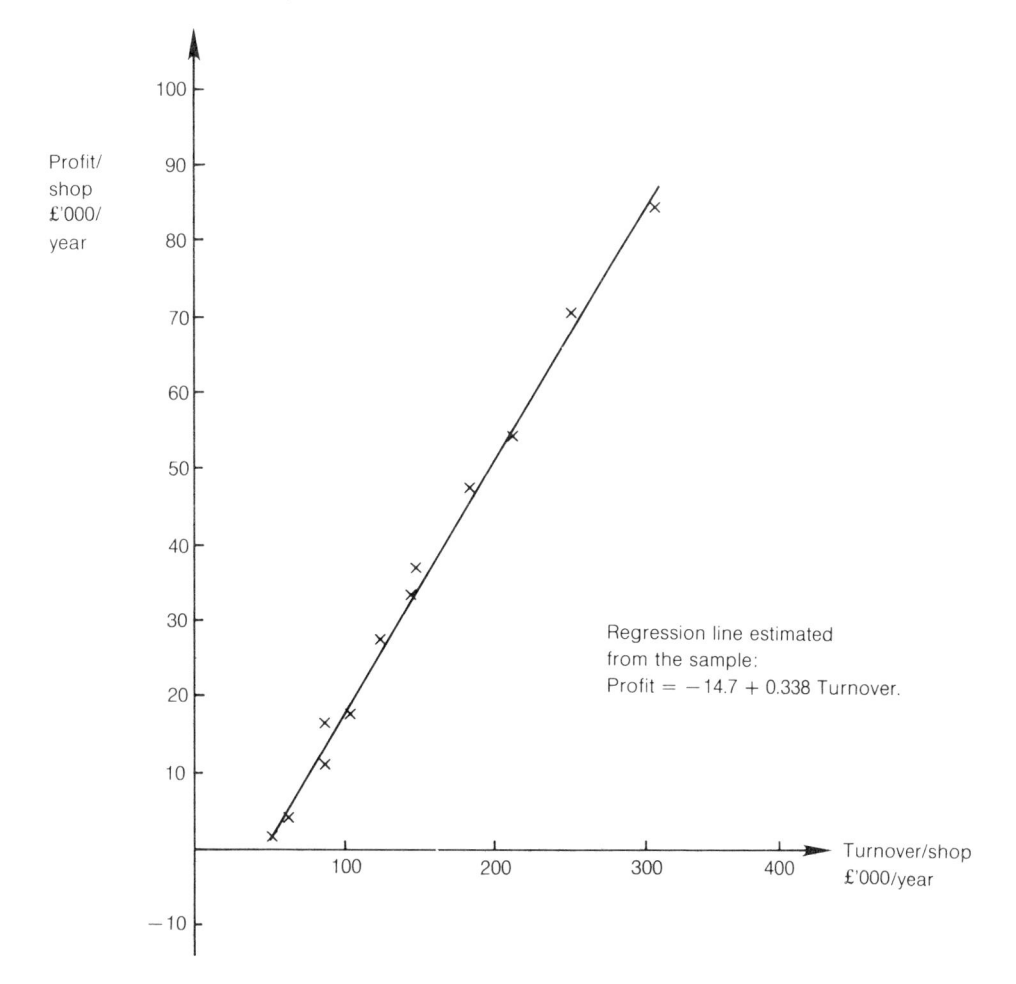

Regression line estimated from the sample:
Profit = −14.7 + 0.338 Turnover.

There is a remarkably good linear relationship between profit and turnover:

$$\sum x = 1735, \sum y = 409, \sum xy = 81{,}660, \sum x^2 = 317{,}575, \sum y^2 = 21{,}585$$

r = +0.997, almost perfect correlation, which is clearly significant. Using the formulae, the relationship is:

$$\text{Profit} = -14.7 + 0.338 \times \text{turnover} \quad (\pounds'000)$$

Bearing in mind the dangers of interpreting the separate coefficients in the equation, −£14,700 could be thought of as a kind of fixed cost. The regression coefficient, +0.338, suggests that a very good 33.8% of turnover is profit. Profit is not defined, so this may not be as good as it looks, in practice.

2 Amalgamation. The finance director's assumption is that the company will lose no turnover if shops are amalgamated. If this is true, each shop he amalgamates with another will increase profits by £14,700 per year. For example, let us use the model for Shops 1 and 2.
As separate shops:

$$\text{Profit} = -14.7 + 0.338 \times 50 + (-14.7 + 0.338 \times 60)$$
$$= -29.4 + 0.338 \times 110 \quad (\pounds'000)$$

As one amalgamated shop:

$$\text{Profit} = -14.7 + 0.338 \times 110$$

ie an increase of £14,700.

Exercise 7.6

Calculate r_s, Spearman's Rank Correlation coefficient.

Student	Ranks		d	d²
1	10	13	−3	9
2	5	4	1	1
3	12	10	2	4
4	1	1	0	0
5	6	11	−5	25
6	2	2	0	0
7	7	8	−1	1
8	11	9	2	4
9	15	14	1	1
10	3	5	−2	4
11	9	7	2	4
12	14	12	2	4
13	13	15	−2	4
14	4	3	1	1
15	8	6	2	4

Therefore: $\sum d^2 = 66$, so that:

$$r_s = 1 - \frac{6 \sum d^2}{n(n^2 - 1)} = 1 - \frac{6 \times 66}{15 \times 224} = 1 - 0.118 = 0.882$$

The hypothesis test is:

$H_0: \rho_s = 0$: there is no association between the ranks.

$H_1: \rho_s \neq 0$: there is an association between the ranks.

The association may be positive or negative, therefore the test is two sided.

The test statistic for $n \geq 10$ is approximately:

$$z = \frac{r_s}{1/\sqrt{(n-1)}} = \frac{0.882}{1/\sqrt{14}} = 3.30$$

At the 1% level of significance the boundary values are:

$$z_{0.005} = \pm 2.575$$

The result is significant at the 1% level. We reject H_0 and accept H_1. We assume there is an association between the two rankings and by inspection, we can see that it is a positive association.

Exercise 7.7

Atlas Stores plc has collected data on turnover per week and customers per week.

1 and **2**: the following is a plot of a shop's customers per week against turnover per week (Note: This could have been plotted with 2 false zeros).

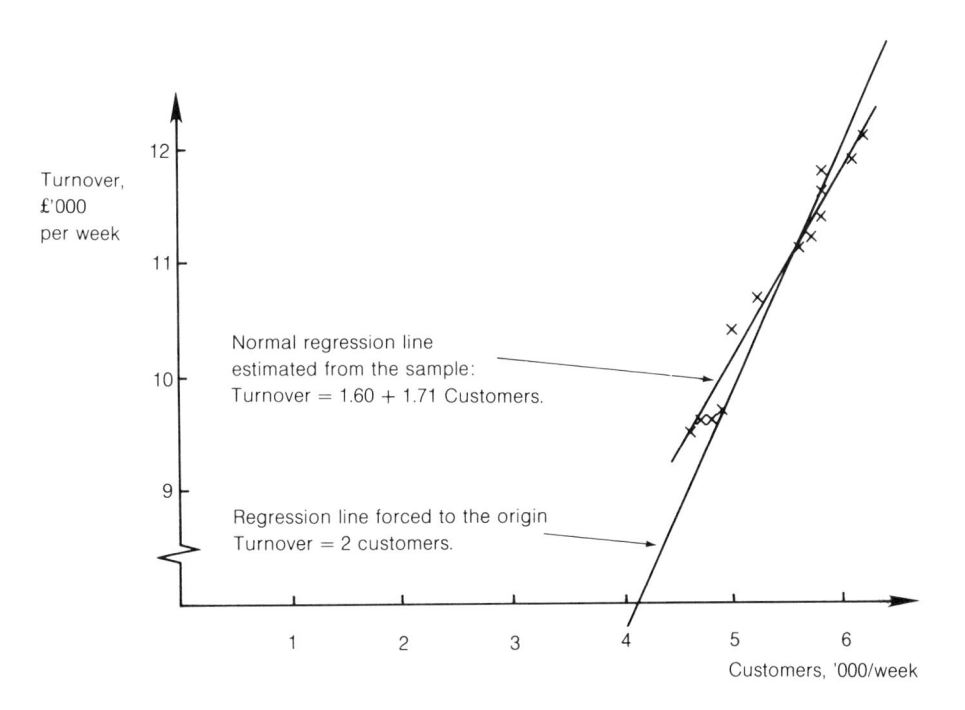

The diagram shows a good linear relationship. The question gives the regression line as:

$$\text{Turnover} = 1.60 + 1.71 \text{ customers} \quad (\text{\pounds}'000/\text{week})$$

and comments that according to this equation the turnover is £1600 per week with zero customers; note that the customer range, is from 4600 to 6200 per week. The £1.71 is the amount spent by each customer (a very low figure for most shops).

516 QUANTITATIVE ANALYSIS

The question suggests the model should be fitted with a constant of zero, ie:

$$y = bx \quad (a = 0)$$

We are told that:

$$b = \frac{\sum xy}{\sum x^2}$$

$$b = \frac{765.52}{382.76} = 2.0$$

The model is now:

$$\text{Turnover} = 2 \times \text{Customer} \quad (\text{£'000/week})$$

The turnover per customer is now £2 per person. Plot these two models on the diagram.
Comment: The two lines coincide at 5520 cm. The normal regression line fits the data better, but the zero-intercept line is reasonable. Below 5520 it underestimates turnover, above this it overestimates turnover.

3 Two reasons why a non-zero intercept may arise:
   (i) Even if the linear model through the origin is correct for the population, the estimation of the intercept in the sample is unlikely to give exactly zero.
   (ii) We have no evidence to say that the relation is linear for all customer values. We have evidence only for the range 4600 to 6200 customers per week. There is no reason why the same model should apply outside this range. It could be a curve.

   The normal regression line should be used by the management accountant since he has only the sample to work from. He should be careful to use the model within the range of the data only.

Exercise 7.8

Pan Products' accountant has analysed the manufacturing times for various batch sizes of a particular product. He has established a linear relationship, with $r = 0.99$. It is:

$$\text{Manufacturing Time/Batch} = 3.5 + 0.6 \text{ Batch Size}$$

1 The meaning of the two regression coefficients: 3.5 hours is the time to set up the machinery to produce a batch of any size. 0.6 hours is the time to produce each item.

2 Calculate the deviations of times from the line.

Batch number	Observed hours y	Estimated from Line, ŷ, hours	$(y - \hat{y}) = d$	d^2	Batch size
1	21.4	22.7	−1.3	1.69	32
2	17.0	17.9	−0.9	0.81	24
3	20.4	21.5	−1.1	1.21	30
4	29.6	30.5	−0.9	0.81	45
5	12.6	12.5	+0.1	0.01	15
6	19.1	19.1	+0.0	0	26
7	34.2	33.5	+0.7	0.49	50
8	15.2	14.3	+0.9	0.81	18
9	16.3	15.5	+0.8	0.64	20
10	29.2	27.5	+1.7	2.89	40
			$\sum d = 0$	$9.36 = \sum d^2$	

Show that the derivations sum to 0.

$$\text{Each } d_i = y_{obs} - y_{est} = y_i - (a + bx_i) = y_i - bx_i - a$$

Therefore:
$$\sum d_i = \sum y_i - b \sum x_i - na$$

Assuming that the usual least squares method has been used to obtain the regression line, the intercept is given by:
$$a = \frac{\sum y_i}{n} - \frac{b \sum x_i}{n}$$

Therefore:
$$\sum d_i = \sum y_i - b \sum x_i - \sum y_i + b \sum x_i = 0 \quad \text{QED}$$

3 Calculate the $\sum d^2$ and explain its importance. The values for d^2 have been entered in the table at **2**.

$$\sum d^2 = 9.36$$

This is the quantity that is minimised when we calculate a and b for the regression line. It is also used to calculate the estimated standard deviation of the residuals:

$$\hat{\sigma}_e^2 = \frac{9.36}{(10 - 2)} = 1.17 \quad \text{therefore} \quad \sigma_e = 1.08 \text{ hours}$$

4 Plot the deviations against batch size and batch number:

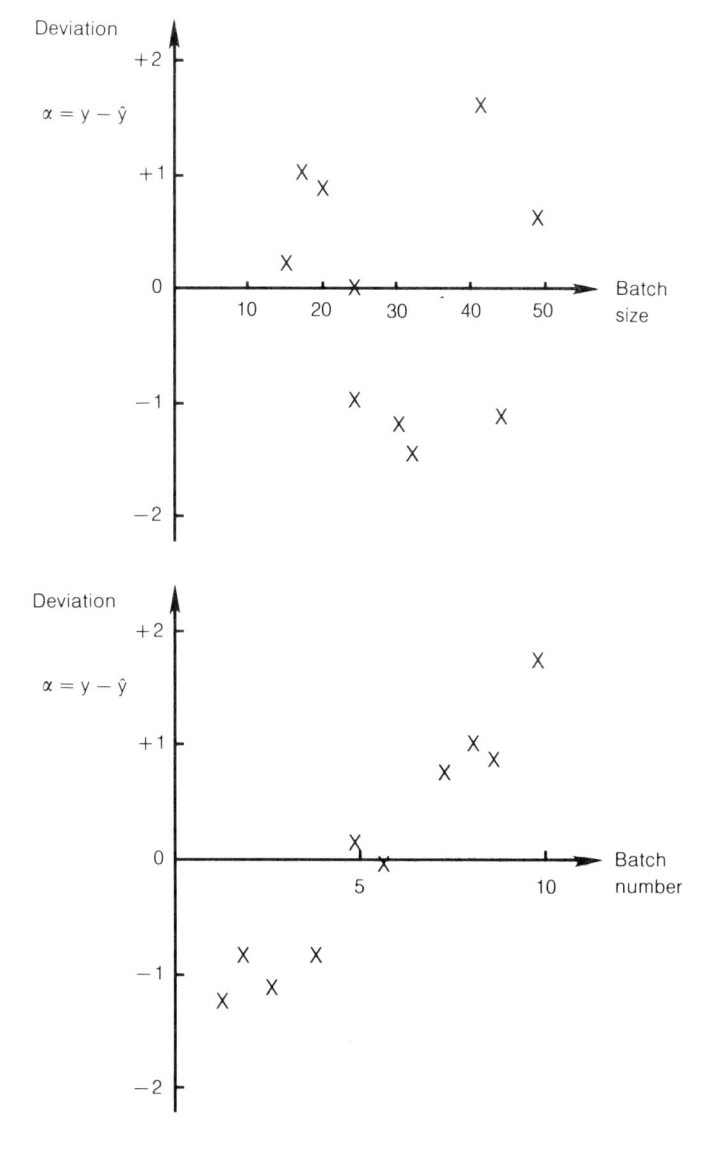

518　QUANTITATIVE ANALYSIS

Comment: The deviations are randomly spread about zero for batch size, but not for batch number. All the early batches (1–4) have over-estimated times, all the last 4 have under-estimated times.

The relationship should not be used until this variation with time is explained and hopefully included in the model.

Exercise 7.9

Pandora's distribution system has 7 depots. A small sample of deliveries have been taken and the time calculated with the number of customers visited and the total cartons delivered.

1　Plot. We are trying to explain time taken, y, by number of customers visited and total cartons delivered. We plot hours against cartons, hours against customers and customers against cartons.

Plot of hours/journey against load carried

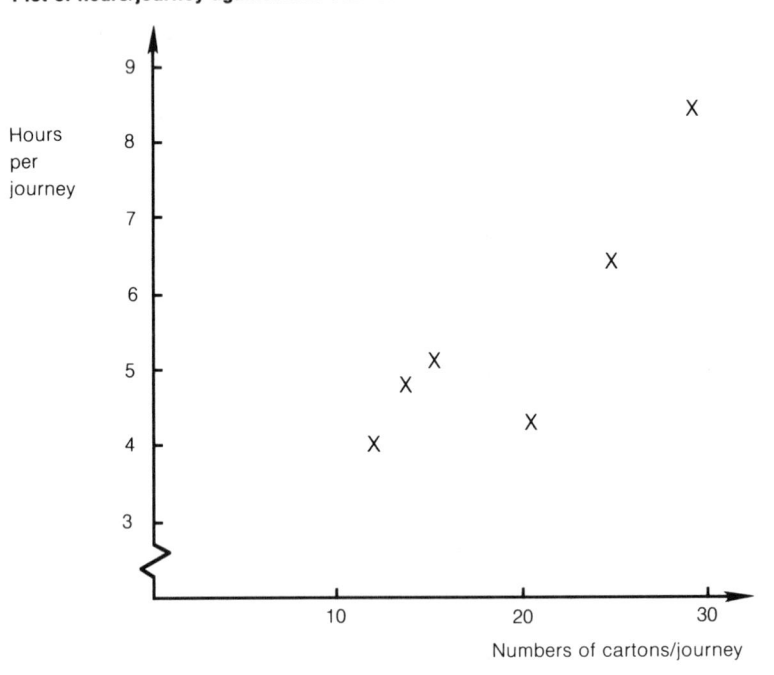

Hours per journey vs *Numbers of cartons/journey*

Plot of hours/journey against customers

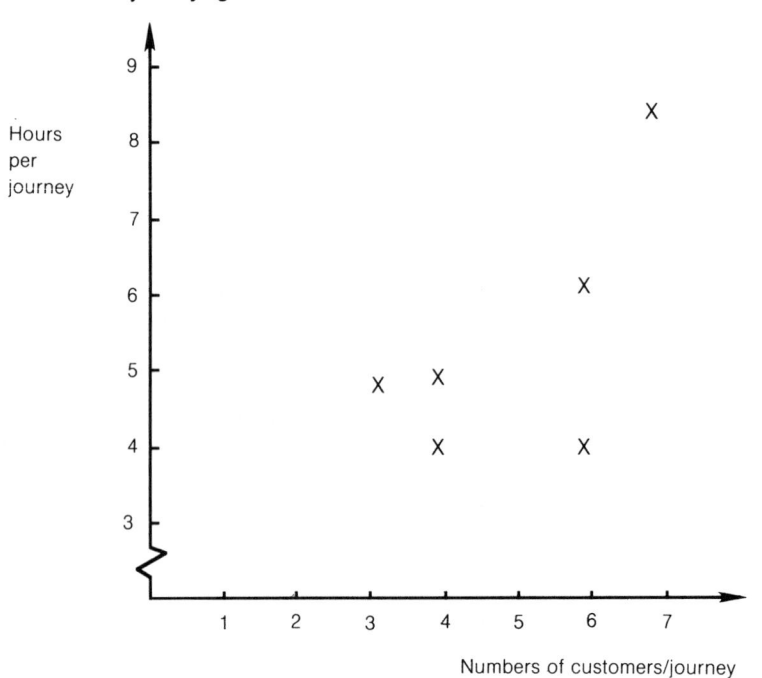

Hours per journey vs *Numbers of customers/journey*

Plot of customers visited against cartons carried

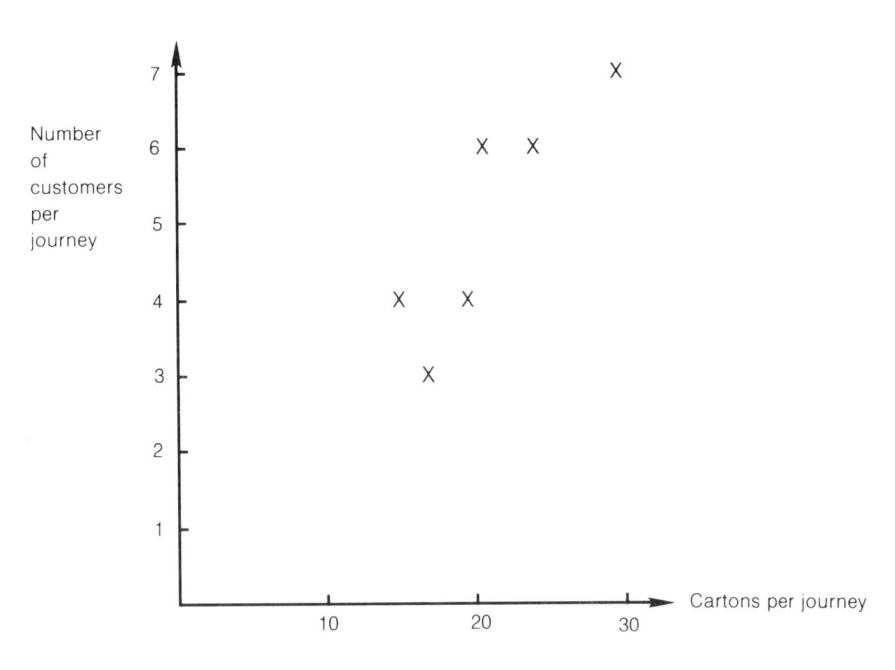

[Scatter plot: Number of customers per journey (y-axis, 0–7) vs Cartons per journey (x-axis, 0–30). Points approximately at (15,3), (15,4), (20,4), (20,6), (25,6), (28,7).]

All three plots show some linearity. Customers against load and time against load indicate reasonable linear relationships. Time against customer indicates a weaker linear relationship.

2 Using the formula provided, the correlation coefficients, r:

Time against load: r = 0.916
Time against customers: r = 0.679.

We have only six sets of data, so we must be careful with the interpretation. From this, time against load will be the better predictor. Load size explains $0.916^2 \times 100 = 84\%$ of the variability in time taken.

3 Multiple regression has been carried out. The model is:

$$\text{Hours/journey} = a + b_1 \text{ cartons} + b_2 \text{ customers}$$

and we are told that the multiple correlation coefficient is r = 0.99, ie the model explains $0.99^2 \times 100 = 98\%$ of variability. The introduction of the second variable, customers, has increased the explained variation in time from 84% to 98%. We will use a partial F test to see if this is a significant improvement:

$H_0: \beta_2 = 0$: Customer does not contribute and should not be included in the model, use the 1 variable model.

$H_1: \beta_2 \neq 0$: Customers should be included in the model and we should use the 2 variable model.

The test statistic is:

$$F = \frac{0.98 - 0.84}{(1 - 0.98)/(6 - 2 - 1)} = 21.0$$

This is tested against:

$$F_{0.05,1,3} = 10.13 \quad \text{and} \quad F_{0.01,1,3} = 34.1$$

The result is significant at the 5% level, but not the 1%. There is reasonably strong evidence to reject H_0 and accept H_1. The sample size is very small. Ideally we would like more data, but at the present time, choose to use the 2 variable model. (Note: In this examination students were not given F tables and were expected to explain the purpose of the test, only.)

Exercise 7.10

Themis Processing Ltd recovers iron from slag. They want to estimate average processing cost (£/tonne) for different plant capacities. The data are:

1

Average cost £/tonne (y)	Capacity t/month (x)	1/x
51.95	900	0.0011
57.18	500	0.0020
46.90	1750	0.000571
45.37	2000	0.000500
46.03	1400	0.000714
48.15	1500	0.000667
44.22	3000	0.000333
48.72	1100	0.000909
45.40	2600	0.000385
44.69	1900	0.000526

Plot Cost(y) against Capacity(x) and against (1/x)

Plot of average cost against capacity

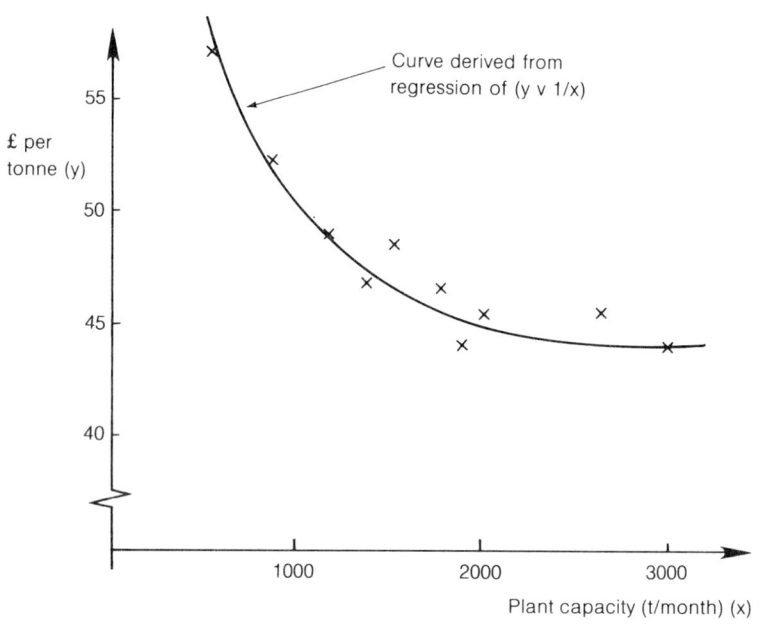

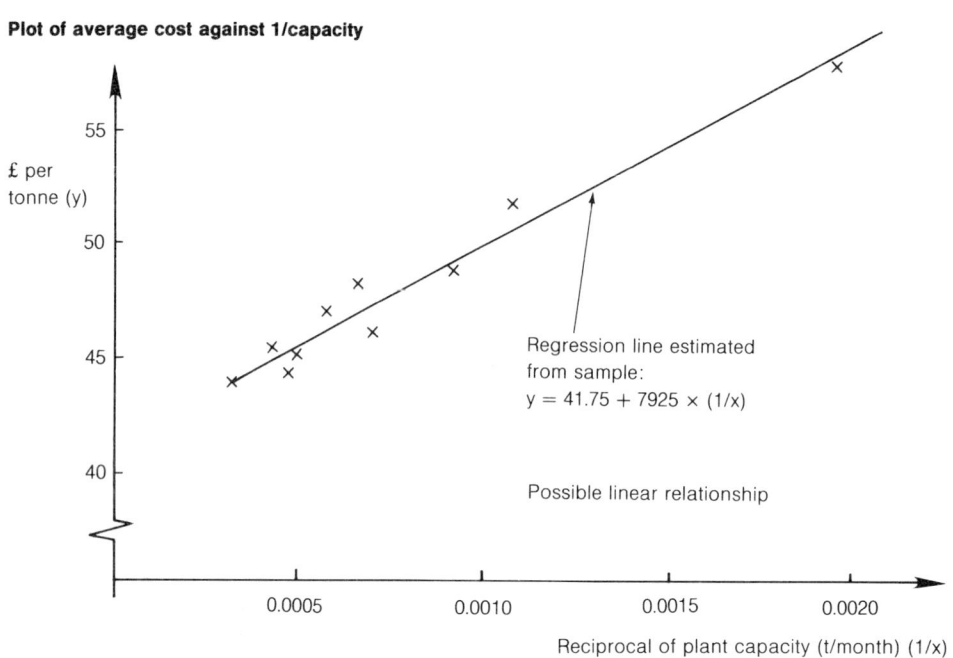

Curve derived from regression of (y v 1/x)

£ per tonne (y)

Plant capacity (t/month) (x)

Plot of average cost against 1/capacity

£ per tonne (y)

Regression line estimated from sample:
$y = 41.75 + 7925 \times (1/x)$

Possible linear relationship

Reciprocal of plant capacity (t/month) (1/x)

2 The plot of cost and capacity gives a curve. The plot $(1/x)$ gives a reasonable linear relationship. The results of regression analysis of cost against 1/capacity are:

$$\text{Cost} = 41.75 + 7925 \times (1/\text{Capacity})$$

Plot onto both diagrams:

For Graph 1:

Capacity, tonne	500	1000	1500	2000	2500	3000
Cost (£/tonne)	57.60	49.7	47.0	45.7	44.9	44.4

For Graph 2:

When $(1/\text{Capacity}) = 0$ Cost = 41.75
When $(1/\text{Capacity}) = 0.002$ Cost = $41.75 + 7.925 \times 2 = 57.6$

Explain the meaning of the various pieces of information:

(i) The standard errors of the intercept and gradient. If repeated samples of data were taken, different values would be obtained for the intercept and slope of the regression model. The intercept and slope each have a sampling distribution of values.

The standard error is a measure of the variability of these sampling distributions. The significance of the individual values of the intercept and slope may be tested using the standard errors to create the appropriate t statistics.

Test the intercept, a:

$H_0: \alpha = 0$.

$H_1: \alpha \neq 0$, ie a two tail test.

The test statistic is:

$$t = \frac{a}{se_a} = \frac{41.75}{0.602} = 69.4$$

Compare this with $t_{0.005,(10-2)} = 3.36$.
A very significant result; reject H_0, accept H_1, $\alpha \neq 0$.
Test the slope, b.

$H_0: \beta = 0$.

$H_1: \beta \neq 0$, ie a two tail test.

The test statistic is:

$$t = \frac{b}{se_b} = \frac{7925}{667.2} = 11.9$$

which is also very significant. $\beta \neq 0$.

(ii) The residual standard error (s—we have used $\hat{\sigma}_e$) = 0.984
This estimates the standard deviation of individual cost values about the true population regression line.

(iii) $r^2 = 0.946$. The model has explained 94.6% of the variability in costs and therefore not explained 5.4%.

3 Do not plot—as the question suggests, they do not mean anything. Our estimate of the 95% limits (if our line were the population line) would be:

$$\pm t_{0.025,8} \times 0.984 = \pm 2.31 \times 0.984 = \pm 2.27, \text{£/tonne}$$

The proper 95% limits are a combination of this and the uncertainty due to our estimation of the line itself. We do not know the population line, therefore the proper 95% limits are bigger than ±2.27 and are curves.

Exercise   7.11

Odin Chemicals want to explain variations in power costs (y) during the last six months.

1   Use the method of least squares to calculate the regression line for costs and output. Using the formulae:

$$\text{Power costs} = 2.30 + 0.342 \times \text{Output} \quad (\pounds'000/\text{month})$$

2   If a multiple regression is done, the question says:

$$\text{Power Costs} = 4.42 + 0.82 \text{ Output} + 0.10 \text{ Month}$$

Unfortunately, if you do the multiple regression this is incorrect, as are the t values in the exercise. The coefficients 0.82 and 0.10 have been reversed and the t values are those for the constant a, and the regression coefficient for output. Fortunately these errors do not matter in answering the exercise.

The coefficient of multiple correlation $r = 0.976$, therefore:

$$r^2 \times 100 = 95.2\%$$

This is the percentage of the variation in costs explained by the two variables. However, we must test the β coefficients against $t_{(n-k-1)} = t_{6-2-1}$. From the standard tables:

$$t_{0.025,3} = 3.18$$

The hypotheses are:

$H_0: \beta_1 = 0.$

$H_1: \beta_1 \neq 0.$

The test statistic is given as 2.64, which is not significant at the 5% level. We accept H_0, $\beta_1 = 0$. The hypothesis and the test is exactly the same for β_2. We find $\beta_2 = 0$ also. The two variable model is not valid and should not be used.

We must now calculate the r for the one variable model in **1**. Using the standard calculation $r = 0.956$, therefore, the percentage of variation in costs explained by output on its own is $0.956^2 \times 100 = 91.4\%$.

Test this:
$H_0: \rho = 0;$
$H_1: \rho \neq 0.$
There is a linear relationship between the output and costs. Test against:

$$t_{0.025,(6-2)} = 2.78 \text{ and } t_{0.005,4} = 4.60$$

The test statistic is:

$$t = \sqrt{\frac{r^2(n-2)}{1-r^2}} = \sqrt{\frac{0.914 \times 4}{0.086}} = 6.5$$

The result is very significant. There is a very strong relationship between output and costs. This is the better model of the two suggested in the question. You could also develop the model of cost against time, though in an exam you would not be expected to. This has a marginally higher r of 0.973, explaining 94.7% of the variability in costs.

What about month 7? It is more likely that you would try the model that relates output and costs, treating time as the intervening variable, to which both of the others are related. To forecast costs, you must forecast output and the data will probably be of no use.

Exercise 7.12

1 The total tableware production for Blueland for the last 8 years is:

Date	Year Number	Production '000t	$\log_{10}$ (Production)
19X0	0	744	2.872
19X1	1	773	2.888
19X2	2	828	2.918
19X3	3	900	2.954
19X4	4	936	2.971
19X5	5	977	2.990
19X6	6	1007	3.003
19X7	7	1066	3.028

Plot of $\log_{10}$ of Tableware production against years.

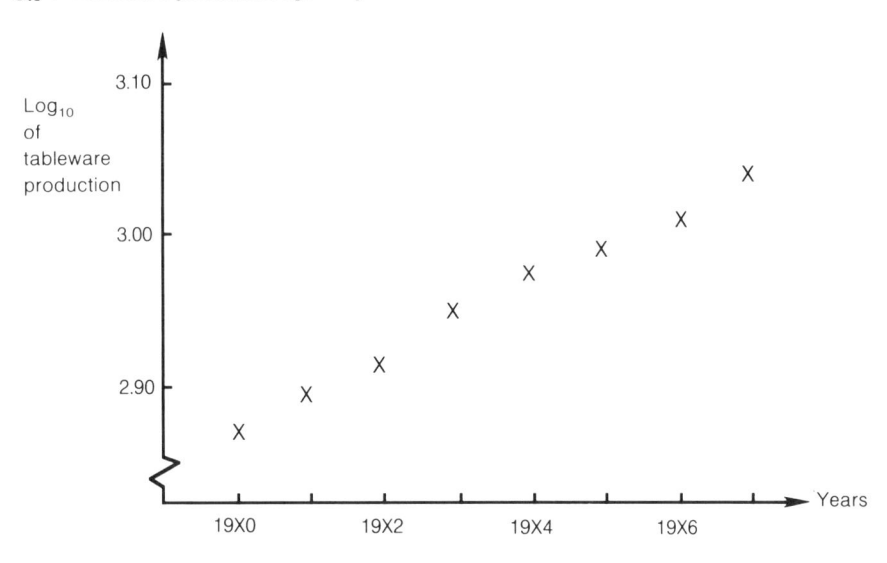

There is a good linear relationship between log of production and year, ie:

$$\log(\text{production}) = a + b\,(\text{year})$$

2 We are given:

Production (000t) = a b$^{(X)}$ where X = year − 19X0

To find a and b, take logs:

$$\log(\text{Prod}) = \log a + X \log b$$
$$\log(\text{Prod}) = a' + b'X$$

From the graph above, we know this is a linear relationship and we can find a', b' from the standard formulae which give:

$$\log(\text{Prod}) = 2.874 + 0.0226\,X$$

Therefore, log a = 2.874, taking antilogs, a = 748 and log b = 0.0226, taking antilogs, b = 1.053, so that:

$$\text{Production} = 748 \times 1.053^X \quad (000t)$$

The model therefore estimates the production in 19X0 when X = 0 to be 748,000t and to be growing at a compound growth rate of 5.3% per annum.

3 Predictions:

Year	X	Production, 000t/year
19X8	8	$748 \times 1.053^8 = 1131$
19X9	9	$748 \times 1.053^9 = 1191$
19X0	10	$748 \times 1.053^{10} = 1254$

4 Optimistic and Pessimistic forecasts for Venus.

(a) Pessimistic—market share is 16.4% of above figures.
(b) Optimistic—16.6% for 19X8; 16.8% for 19X9; 17% for 19X0.

Therefore:

	19X8	19X9	19X0
Pessimistic	186,000t	195,000t	206,000t
Optimistic	188,000t	200,000t	213,000t

5 Shortcomings of regression-based forecasting. One of the golden rules of regression analysis is that the model is valid only within the range of the data. This kind of forecasting inevitably means we move outside the range and who knows whether the model will still be valid?

Solutions to Chapter Eight exercises

ANSWERS TO EXERCISES

Exercise 8.1

The following is a plot of turnover against time for Amada plc.

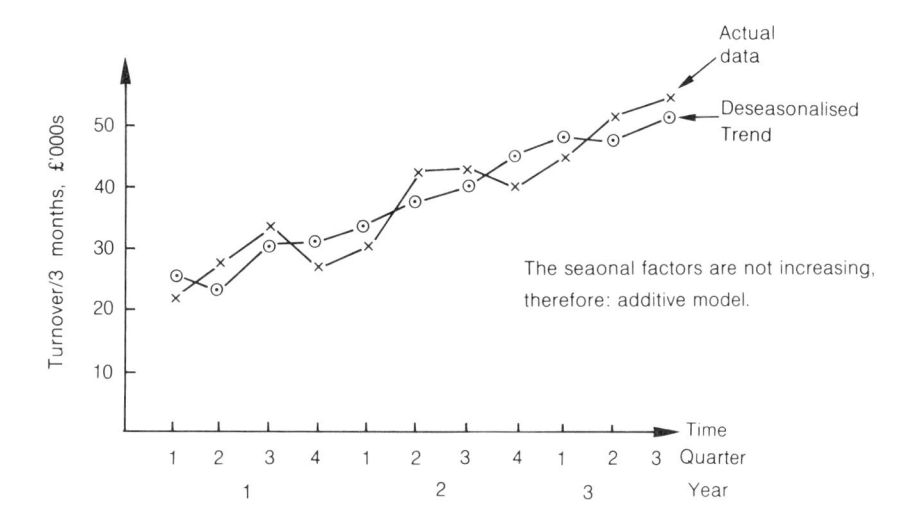

The seaonal factors are not increasing, therefore: additive model.

1 Fit an additive component model A = T + S + E.

Date Quarter	Year	Quarter Number	Actual A	4pt ma	Centred est T	A − T = S + E	AdjS (see below)	A − AdjS = T + E (Deseasonalised)
1	1	1	22	—	—		−3.1	25.1
2		2	28		—		+4.8	23.2
				27.75				
3		3	34		28.9	+5.1	+2.9	31.1
				30.0				
4		4	27		31.9	−4.9	−4.6	31.6
				33.75				
1	2	5	31		34.9	−3.9	−3.1	34.1
				36.0				
2		6	43		37.7	+5.3	+4.8	38.2
				39.5				
3		7	43		41.4	+1.6	+2.9	40.1
				43.25				
4		8	41		44.5	−3.5	−4.6	45.6
				45.75				
1	3	9	46		47.4	−1.4	−3.1	49.1
				49.0				
2		10	53		—		+4.8	48.2
3		11	56		—		+2.9	53.1

Calculation of the seasonal factors.

Year	Quarter 1	Quarter 2	Quarter 3	Quarter 4	
1	—	—	+5.1	−4.9	
2	−3.9	+5.3	+1.6	−3.5	
3	−1.4				
Total	−5.3	+5.3	+6.7	−8.4	
	÷2	÷1	÷2	÷2	
Estimated seasonal	−2.7	+5.3	+3.4	−4.2	Sum = +1.8
Adjusted seasonal, AdjS	−3.1	+4.8	+2.9	−4.6	Sum = 0 ✓

The adjusted seasonal components are used to deseasonalise the time series. The deseasonalised data is used with linear regression to calculate the linear trend, which appears on the graph.

$$\text{Trend} = a + b \times \text{Quarter Number}$$

This gives:

$$\text{Trend Turnovers (£000s/3 months)} = 20.2 + 3.0 \times \text{Quarter Number}$$

Date Quarter	Year	Quarter Number	Actual A	Adjusted S	Trend (20.2 + 3 quarter number)	Forecast T + S = F	Errors (A − F)
1	1	1	22	−3.1	23.2	20.1	+1.9
2		2	28	+4.8	26.2	31.0	−3.0
3		3	34	+2.9	29.2	32.1	+1.9
4		4	27	−4.6	32.2	27.6	−0.6
1	2	5	31	−3.1	35.2	32.1	−1.1
2		6	43	+4.8	38.2	43.0	0
3		7	43	+2.9	41.2	44.1	−1.1
4		8	41	−4.6	44.2	39.6	+1.4
1	3	9	46	−3.1	47.2	44.1	+1.9
2		10	53	+4.8	50.2	55.0	−2.0
3		11	56	+2.9	53.2	56.1	−0.1
Forecasts							
4		12	—	−4.6	+56.2	=51.6	—
1	4	13	—	−3.1	+59.2	=56.1	—
2		14	—	+4.8	+62.2	=67.0	—

2

From the errors in the above table, the likely accuracy is ±2 (£'000 per quarter), increasing the further ahead the forecast.

SOLUTIONS TO EXERCISES: CHAPTER EIGHT

Exercise 8.2

Peace Retailers.

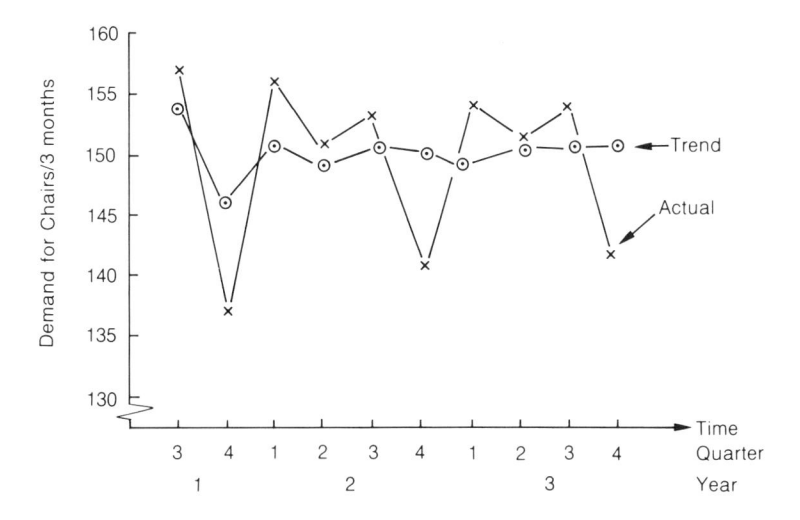

1 Fit a suitable additive component model. The underlying trend demand looks constant, therefore:

$$A = \text{Constant} + S + E$$

Deseasonalise the data using moving averages to find the constant.

Date Quarter	Year	Quarter Number	Actual A	4pt m.a.	Centred est T	A − T = S + E	AdjS (see below)	A − S = T + E
3	1	1	157				+2.8	154.2
4		2	137				−9.0	146.0
				150.25				
1	2	3	156		149.75	+6.25	+4.9	151.1
				149.25				
2		4	151		149.75	+1.25	+1.3	149.7
				150.25				
3		5	153		150.0	+3.0	+2.8	150.2
				149.75				
4		6	141		149.9	−8.9	−9.0	150.0
				150.0				
1	3	7	154		150.1	+3.9	+4.9	149.1
				150.25				
2		8	152		150.4	+1.6	+1.3	150.7
				150.5				
3		9	154				+2.8	151.2
4		10	142				−9.0	151.0

528 QUANTITATIVE ANALYSIS

Calculate the seasonal components, taking the figures from the table above:

	Year	Quarter 1	Quarter 2	Quarter 3	Quarter 4	
	2	+6.25	+1.25	+3.0	−8.9	
	3	+3.9	+1.6	—	—	
Total		+10.15	+2.85	+3.0	−8.9	
		÷2	÷2	÷1	÷1	
Unadjusted Seasonal		+5.07	+1.42	+3.0	−8.9	Sum = +0.59
Adjusted Seasonal		+4.9	+1.3	+2.8	−9.0	Sum = 0 ✓

The adjusted seasonal components are used in the table above to deseasonalise the data. From the plot the trend is a constant, ie:

$$T = 150 \text{ (Chairs per 3 months)}$$

2 Actuals, forecasts and errors are given below:

Date Quarter	Year	Quarter Number	Actual A	T +	S	= Forecast F	Errors A − F
3	1	1	157	150 +	2.8	= 152.8	+4.2
4		2	137	150 −	9.0	141.0	−4.0
1	2	3	156	150 +	4.9	154.9	+1.1
2		4	151	150 +	1.3	151.3	−0.3
3		5	153	150 +	2.8	152.8	+0.2
4		6	141	150 −	9.0	141.0	0
1	3	7	154	150 +	4.9	154.9	−0.9
2		8	152	150 +	1.3	151.3	+0.7
3		9	154	150 −	2.8	152.8	+1.2
4		10	142	150 +	9.0	141.0	+1.0
Forecasts							
1	4	11	—	150 +	4.9	154.9	—
2		12	—	150 +	1.3	151.3	—

SOLUTIONS TO EXERCISES: CHAPTER EIGHT

The forecasts for the first two quarters of year 4 are for 155 and 151 chairs. If the historical pattern holds, the errors should be ±1, ignoring the first two estimates.

Exercise 8.3

Cobournes plc—output.

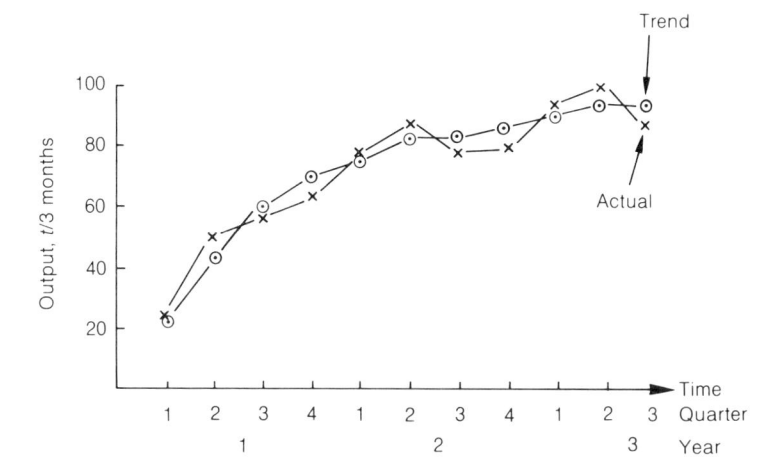

[Graph: Output, t/3 months vs Time/Quarter/Year, showing Trend and Actual curves rising from about 25 to about 95 over quarters 1-3 across years 1-3]

1 Analyse the time series—take 4 point moving averages and see what happens—A = T + S + E

Date Quarter	Year	Quarter Number	Actual A	4 point m.a.	Centred est T	A − T = S + E	Adjusted S (see below)	A − S = T + E
1	1	1	24	—	—	—	+2.5	21.5
2		2	50		—	—	+7.8	42.2
				48.25				
3		3	56		55.1	+0.9	−3.4	59.4
				62.0				
4		4	63		66.9	−3.9	−6.9	69.9
				71.75				
1	2	5	79		74.6	+4.4	+2.5	76.5
				77.5				
2		6	89		79.6	+9.4	+7.8	81.2
				81.75				
3		7	79		83.5	−4.5	−3.4	82.4
				85.25				
4		8	80		86.6	−6.6	−6.9	86.9
				88.0				
1	3	9	93		89.1	+3.9	+2.5	90.5
				90.25				
2		10	100		—	—	+7.8	92.2
3		11	88		—	—	−3.4	91.4

Calculation of the seasonal components:

Year	Quarters			
	1	2	3	4
1	—	—	+0.9	−3.9
2	+4.4	+9.4	−4.5	−6.6
3	+3.9	—	—	—
Total	+8.3	+9.4	−3.6	−10.5
	÷2	÷1	÷2	÷2
Unadjusted Seasonal	+4.15	+9.4	−1.8	−5.25	Sum = +6.5
Adjusted Seasonal	+2.5	+7.8	−3.4	−6.9	Sum = 0 ✓

2 Comment on the trend. The trend is not linear. It is clearly some kind of curve. We are, therefore, not able to continue the analysis to find the equation for the trend curve.

Exercise 8.4

Doble-Flood plc.

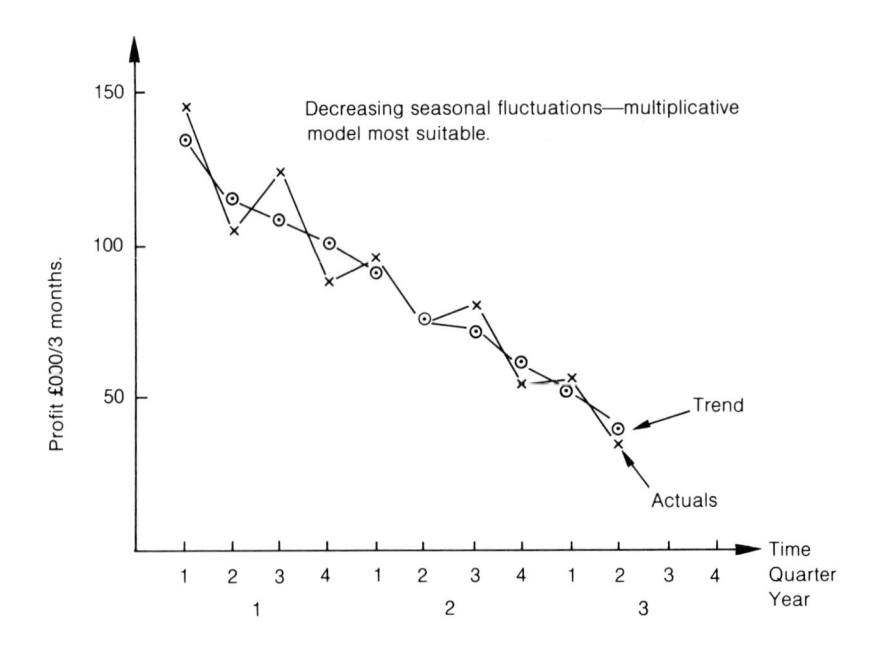

Decreasing seasonal fluctuations—multiplicative model most suitable.

1 Fit a multiplicative model: $A = T \times S \times E$.

SOLUTIONS TO EXERCISES: CHAPTER EIGHT

Date Quarter	Year	Quarter Number	Actual A	4 point m.a.	Centred est T	A/T = S × E	A/S = T × E
1	1	1	146	—	—		136.0
2		2	106		—	—	115.2
				116			
3		3	123		109.9	1.119	109.4
				103.75			
4		4	89		99.8	0.892	100.9
				95.75			
1	2	5	97		90.4	1.073	90.3
				85			
2		6	74		80.5	0.919	80.4
				76			
3		7	80		70.9	1.129	71.2
				65.75			
4		8	53		60.9	0.871	60.1
				56			
1	3	9	56		—	—	52.2
2		10	35		—	—	38.0

Calculations of the seasonal components—Doble-Flood.

Year	Quarter 1	2	3	4	
1	—	—	1.119	0.892	
2	1.073	0.919	1.129	0.871	
3	—	—			
Total	1.073	0.919	2.248	1.763	
	÷1	÷1	÷2	÷2	
Estimated Seasonal	1.073	0.919	1.124	0.881	Sum = 3.997
Adjusted Seasonal	1.074	0.920	1.124	0.882	Sum = 4 ✓

Use the adjusted seasonals to calculate: A/S = T × E in the first table. Use these figures to calculate the linear trend equation using linear regression, because the trend is linear from the plot:

$$\text{Trend} = a + b \times \text{Quarter Number}$$

$$\text{Trend Profit } (\pounds 000/3 \text{ months}) = 141.0 - 10.1 \times \text{Quarter Number}$$

532 QUANTITATIVE ANALYSIS

Date Quarter	Year	Quarter Number	Actual A	Adjusted S	Trend T = (141 − 10.1 Q No)	Forecast T × S	Error A/(T × S)
1	1	1	146	1.074	130.9	140.6	1.04
2		2	106	0.920	120.8	111.1	0.95
3		3	123	1.124	110.7	124.4	0.99
4		4	89	0.882	100.6	88.7	1.00
1	2	5	97	1.074	90.5	97.2	1.00
2		6	74	0.920	80.4	74.0	1.00
3		7	80	1.124	70.3	79.0	1.01
4		8	53	0.882	60.2	53.1	1.00
1	3	9	56	1.074	50.1	53.8	1.04
2		10	35	0.920	40.0	36.8	0.95
Forecasts 3		11	—	1.124	29.9	33.6	
4		12	—	0.882	19.8	17.5	

2

The forecast profits for the next two quarters are £33,600 and £17,500. The largest error is likely to be ±5% from the above table.

Exercise 8.5

Banham and Barsey plc.

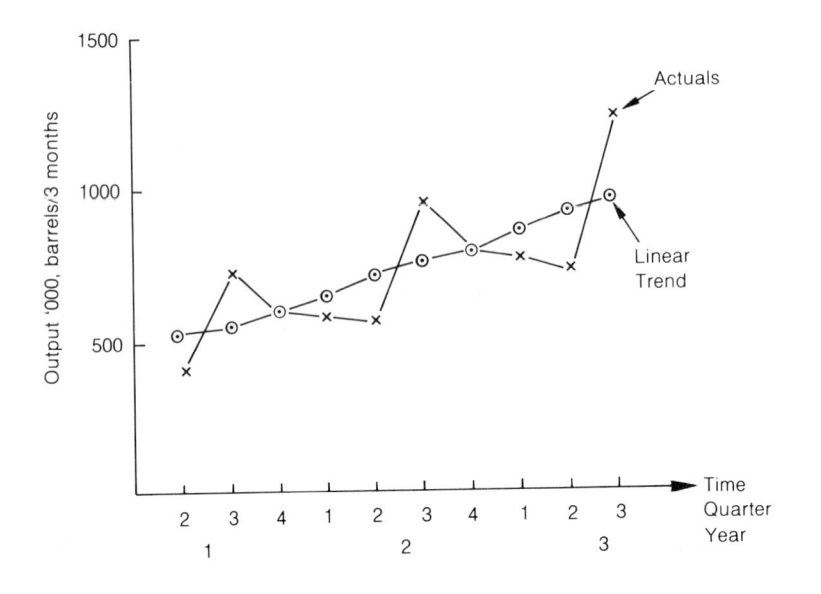

1 Fit a suitable component model—calculate the deseasonalised values using 4 point moving averages then decide on the model.

Date Quarter	Year	Quarter Number	A	4 pt m.a.	Centred est T	A/T = S × E	A − T = S + E	A/S = T × E
2	1	1	400	—	—			503.8
3		2	715		—	—		551.7
				575				
4		3	600		595	1.008	+5	595.8
				615				
1	2	4	585		647.5	0.903	−62.5	647.8
				680				
2		5	560		705.0	0.794	−145	705.3
				730				
3		6	975		752.5	1.296	+222.5	752.3
				775				
4		7	800		795	1.006	+5	794.4
				815				
1	3	8	765		847.5	0.903	−82.5	847.2
				880				
2		9	720		—	—		906.8
3		10	1235		—	—		952.9

It is not clear from the two calculations of the seasonal estimates which model would be better. There is not enough data with only six values to see if the seasonals are increasing as the trend increases. From the diagram, it looks as though the seasonals are probably increasing and so we will calculate the model $A = T \times S \times E$.

Year	Quarter 1	2	3	4
1	—	—	—	1.008
2	0.903	0.794	1.296	1.006
3	0.903	—	—	—
Total	1.806	0.794	1.296	2.014
	÷2	÷1	÷1	÷2
Unadjusted Seasonal	0.903	0.794	1.296	1.007 Sum = 4 ✓
Adjusted Seasonal				

From the plot a linear trend is appropriate. Linear regression, using the deseasonalised (A/S) figures gives:

Trend Output (000 Barrels/3 months) = 450.1 + 50.1 × Quarter Number

534 QUANTITATIVE ANALYSIS

Banham and Barsey forecasts and errors:

Date Quarter	Year	Quarter Number	Actual A	Adjusted S	T (450.1 + 50.1 Q)	Forecast F = T × S	Error A/(T × S)
2	1	1	400	0.794	500.2	397.2	1.01
3		2	715	1.296	550.3	713.2	1.00
4		3	600	1.007	600.4	604.6	0.99
1	2	4	585	0.903	650.5	587.4	1.00
2		5	560	0.794	700.6	556.3	1.01
3		6	975	1.296	750.7	972.9	1.00
4		7	800	1.007	800.8	806.4	0.99
1	3	8	765	0.903	850.9	768.4	0.99
2		9	720	0.794	901.0	715.4	1.01
3		10	1235	1.296	951.1	1232.6	1.00
2 Forecasts							
4	3	11	—	1.007	1051.2	1058.6	—
1	4	12	—	0.903	1101.3	994.5	—

The forecasts for the next two quarters are 1,059,000 and 995,000 barrels per three months. If the very consistent historical pattern continues, the errors will be small at ±1%.

Exercise 8.6

Peace Retailers—refer to **Exercise 8.2** for diagram.

1 Use the model developed in **Exercise 8.2**:

$$F = T + S = 150 + \left\{ \begin{array}{c} Q1 \\ +4.9 \end{array} \text{ or } \begin{array}{c} Q2 \\ +1.3 \end{array} \text{ or } \begin{array}{c} Q3 \\ +2.8 \end{array} \text{ or } \begin{array}{c} Q4 \\ -9.0 \end{array} \right\}$$

2 and the simple exponential smoothing model:

$$F_t = a A_t + (1 - a) F_{t-1}, \quad \text{where} \quad a = 0.2$$

ie

$$F_t = 0.2 A_t + 0.8 F_{t-1}, \quad \text{where} \quad F_0 = 150$$

SOLUTIONS TO EXERCISES: CHAPTER EIGHT

Date Quarter	Year	Quarter Number	A	F = T + S	E = A − F	F = 0.2A$_t$ + 0.8F$_{t-1}$	E = A − F
3	1	1	157	152.8	+4.2	150	+7.0
4		2	137	141.0	−4.0	151.4	−14.4
1	2	3	156	154.9	+1.1	148.5	+7.5
2		4	151	151.3	−0.3	150.0	+1.0
3		5	153	152.8	+0.2	150.2	+2.8
4		6	141	141.0	0	150.8	−9.8
1	3	7	154	154.9	−0.9	148.8	+5.2
2		8	152	151.3	+0.7	149.9	+2.1
3		9	154	152.8	+1.2	150.3	+3.7
4		10	142	141.0	+1.0	151.0	−9.0

$$\text{MAD} = \frac{\sum |E_t|}{10} \qquad 1.36 \qquad 6.25$$

$$\text{MSE} = \frac{\sum (E_t)^2}{10} \qquad 3.87 \qquad 54.3$$

3 The time series model is the much better model, because of the seasonality of the data. The simple exponential smoothing model cannot adapt to the seasons.

Exercise 8.7

Purchase plc.

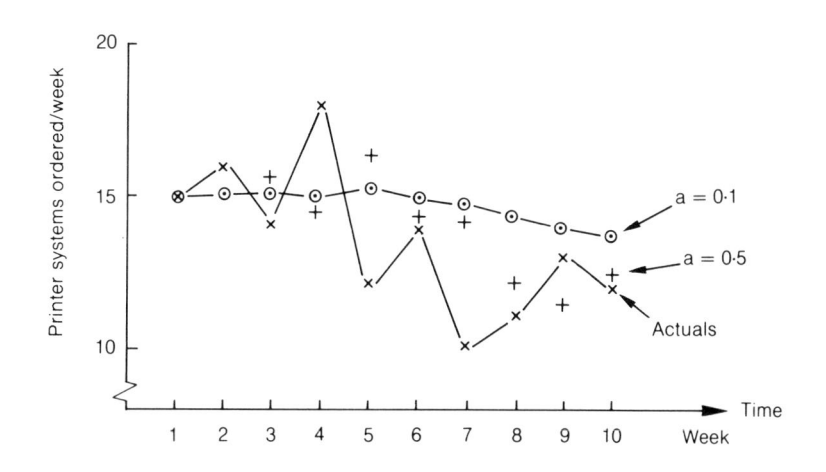

1

Date Week	A	(a = 0.1) $F_t = 0.1 A_t + 0.9 F_{t-1}$	A − F = E	(a = 0.5) $F_t = 0.5 A_t + 0.5 F_{t-1}$	A − F = E
1	15	15.0	0	15.0	0
2	16	15.0	+1.0	15.0	+1.0
3	14	15.1	−1.1	15.5	−1.5
4	18	15.0	+3.0	14.7	+3.3
5	12	15.3	−3.3	16.4	−4.4
6	14	15.0	−1.0	14.2	−0.2
7	10	14.9	−4.9	14.1	−4.1
8	11	14.4	−3.4	12.1	+1.1
9	13	14.0	−1.0	11.5	+1.5
10	12	13.9	−1.9	12.3	−0.3

MSE = $\sum (E_t)^2 / 10$ 6.3   5.4

2 Similar results; a = 0.5 responds better to the possible decline towards the end of the series, but is more erratic.

Exercise 8.8

Droco plc—central stores demand for spare part A.

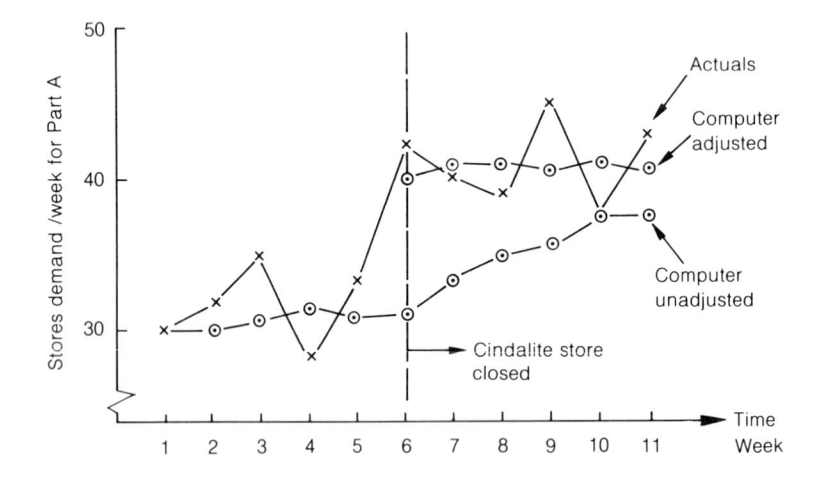

SOLUTIONS TO EXERCISES: CHAPTER EIGHT

1 Exponential smoothing forecast with $F_0 = 30$ and $a = 0.2$.

Date Week	Actual A	F_t – Forecast $= 0.2 A_t + 0.8 F_{t-1}$	Error E	
1	30	30.0	0	
2	32	30.0	+2.0	
3	35	30.4	+4.6	
4	28	31.3	−3.3	
5	33	30.7	+2.3	
		closed.		
6	42	31.1	+10.9	All forecasts
7	40	33.3	+6.7	below actuals, as
8	39	34.6	+4.4	we expected.
9	45	35.5	+9.5	
10	38	37.4	+0.6	
11	43	37.5	+5.5	

MSE/MAD   30.5/4.5

2 Suggest how the computer system should have been changed—a new forecast should have been started in week 6 with $F_0 = 40$, because this is the approximate total demand for spare part A from Droco + Cindalite. The new forecasts would be:

Date Week	Actual A	Forecast F_t (a = 0.2)	Error E	
1	30	30	0	
2	32	30.0	+2.0	
3	35	30.4	+4.1	As before.
4	28	31.3	−3.3	
5	33	30.7	+2.3	
Closed—reprogramme with $F_0 = 40$				
6	42	40	+2.0	
7	40	40.4	−0.4	Much better
8	39	40.3	−1.3	forecasts,
9	45	40.1	+4.9	again as
10	38	41.0	−3.0	we expected.
11	43	40.4	+2.6	

MSE/MAD   7.9/2.4

A sensible adjustment to the stores computer has significantly improved the forecasts.

Solutions to Chapter Nine exercises

ANSWERS TO EXERCISES

Exercise 9.1

Produce r tonnes of Crunchy per month and h tonnes of Chewy per month. Maximise the total monthly contribution, £P, where:

$$P = 150r + 75h \quad \text{£/month}$$

Maximise P subject to:

$$\begin{aligned}
\text{Roasting Dept. Hours:} & \quad 10r + 4h \leqslant 1000 \text{ hours/month} \\
\text{Blending Dept. Hours:} & \quad 3r + 2h \leqslant 360 \text{ hours/month} \\
\text{Packing Dept. Hours:} & \quad 2r + 5h \leqslant 800 \text{ hours/month} \\
& \quad r, h \geqslant 0
\end{aligned}$$

Exercise 9.2

Invest £s in the building society, £a in investment A, £b in investment B and £c in investment C. Maximise total return over two years, £R. Monies invested in the building society earn 9% per annum, therefore, each £1 invested here will accumulate to £$(1 + 0.09)^2$ after two years. Investment A earns 10% per annum interest and 1% per annum capital growth, giving 11% per annum increase in value. After two years, each £1 invested in A will accumulate to £$(1 + 0.11)^2$. A similar situation exists for the other two investments. The return after two years:

$$R = \text{total accumulated sum}$$

therefore:

$$R = (1 + 0.09)^2 s + (1 + 0.11)^2 a + (1 + 0.15)^2 b + (1 + 0.09)^2 c \quad £$$

that is:

$$R = 1.1881s + 1.2321a + 1.3225b + 1.1881c \quad £$$

Maximise R subject to:

$$\begin{aligned}
\textit{Total sum}: & \quad s + a + b + c = 25{,}000 \quad £ \\
\textit{Balance}: & \quad s + a \geqslant 0.4 \times 25{,}000 \quad £ \\
\textit{Capital growth}: & \quad b \geqslant 0.25(a + b + c) \quad £ \\
\textit{Risk}: & \quad b \leqslant 0.35(a + b + c) \quad £ \\
\textit{Safety}: & \quad a + c \geqslant 0.5(a + b + c) \quad £ \\
& \quad s, a, b, c \geqslant 0
\end{aligned}$$

Hence, the final linear programming model is:

$$\text{Maximise: } R = 1.1881s + 1.2321a + 1.3225b + 1.1881c \quad £$$

subject to:

$$
\begin{array}{lrl}
\textit{Total sum:} & s + a + b + c = 25{,}000 & £ \\
\textit{Balance:} & s + a \geqslant 10{,}000 & £ \\
\textit{Capital growth:} & -0.25a + 0.75b - 0.25c \geqslant 0 & £ \\
\textit{Risk:} & -0.35a + 0.65b - 0.35c \leqslant 0 & £ \\
\textit{Safety:} & +0.50a - 0.50b + 0.50c \geqslant 0 & £ \\
& s, a, b, c \geqslant 0 &
\end{array}
$$

Exercise 9.3

1 (a) Produce a units of product A, b units of product B, c units of product C, d units of product D and e units of product E, each week. Calculate the unit contribution for each product:

A: unit contribution is $40 - (2.10 \times 6 + 3.0 \times 1 + 1.3 \times 3 + 8.0 \times 0.5)$ £
$\qquad = £16.50$ per unit
B: unit contribution is $42 - (2.10 \times 6.5 + 3.0 \times 0.75 + 1.3 \times 4.5 + 8.0 \times 0.5)$ £
$\qquad = £16.25$ per unit
C: unit contribution is $44 - (2.10 \times 6.1 + 3.0 \times 1.25 + 1.3 \times 6 + 8.0 \times 0.5)$ £
$\qquad = £15.64$ per unit
D: unit contribution is $48 - (2.10 \times 6.1 + 3.0 \times 1 + 1.3 \times 6 + 8.0 \times 0.75)$ £
$\qquad = £18.39$ per unit
E: unit contribution is $52 - (2.10 \times 6.4 + 3.0 \times 1 + 1.3 \times 4.5 + 8.0 \times 1)$ £
$\qquad = £21.71$ per unit

Maximise total weekly contribution, £P, where:

$$P = 16.5a + 16.25b + 15.64c + 18.39d + 21.71e \quad £/\text{week}$$

subject to:

$$
\begin{array}{lrl}
\textit{Materials:} & 6.0a + 6.5b + 6.1c + 6.1d + 6.4e \leqslant 35000 & \text{kg/week} \\
\textit{Forming:} & 1.0a + 0.75b + 1.25c + 1.0d + 1.0e \leqslant 6000 & \text{hr/week} \\
\textit{Firing:} & 3.0a + 4.5b + 6.0c + 6.0d + 4.5e \leqslant 30000 & \text{hr/week} \\
\textit{Packing:} & 0.5a + 0.5b + 0.5c + 0.75d + 1.0e \leqslant 4000 & \text{kg/week} \\
& a, b, c, d, e \geqslant 0 &
\end{array}
$$

(b) The model is suitable if:
  (i) the selling prices, costs, and usage of the resources are fixed and independent of the quantities produced;
  (ii) all production can be sold;
  (iii) the weekly production hours given do not include time to change over between products.

Exercise 9.4

Plot the constraints from **Exercise 9.1** on a graph as shown below:

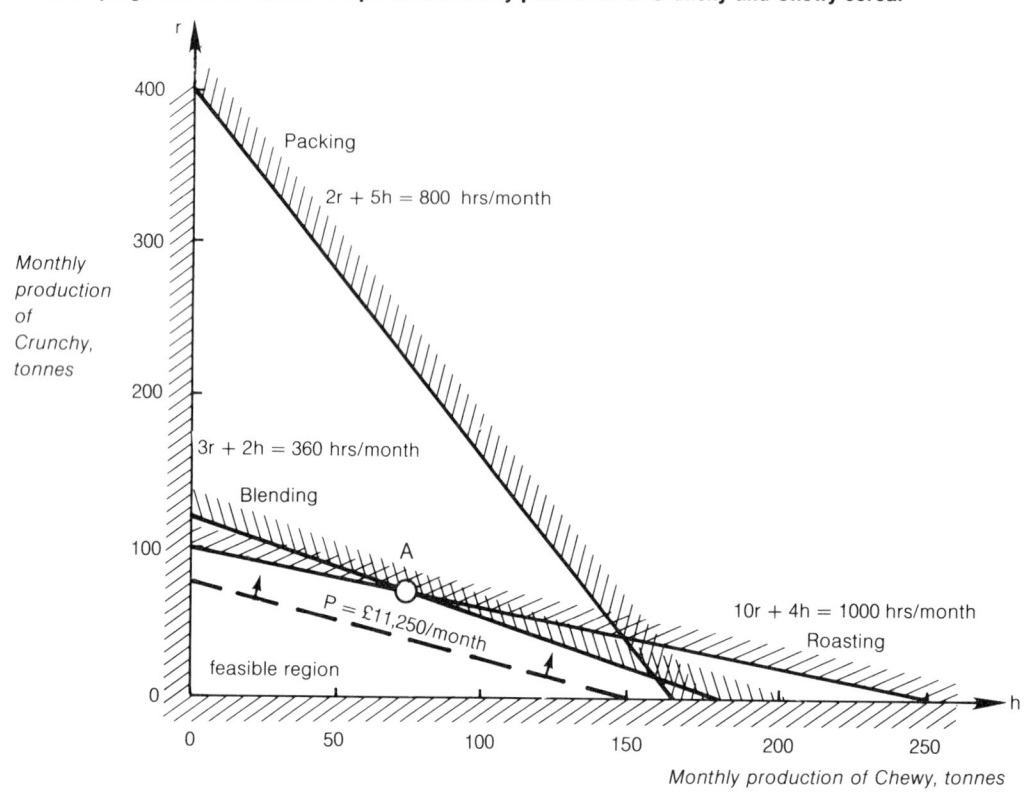

Linear programme to illustrate the possible monthly production of Crunchy and Chewy cereal

Monthly production of Crunchy, tonnes

Packing 2r + 5h = 800 hrs/month
3r + 2h = 360 hrs/month
Blending
A
P = £11,250/month
10r + 4h = 1000 hrs/month
Roasting
feasible region

Monthly production of Chewy, tonnes

The feasible region is left unshaded. The next step is to add a trial monthly contribution line. The point h = 50 tonnes, r = 50 tonnes is a feasible solution. At this point, the contribution is P = 150 × 50 + 75 × 50 = £11,250 per month. We will take as the trial monthly contribution line:

$$11{,}250 = 150r + 75h \qquad £/month$$

This line also passes through the point r = 0, h = 150. The line is shown dotted on the graph above. Moving parallel to this line in the direction of increasing contribution, we can see that the optimum corner is A. Corner A is the intersection of the constraints on Roasting and Blending time. These are the limiting constraints. Solving the equations of these constraints simultaneously:

$$10r + 4h = 1000 \text{ hours/month} \quad (1)$$
$$3r + 2h = 360 \text{ hours/month} \quad (2)$$

Multiply equation (2) by 2 and subtract from equation (1):

$$4r = 280, \text{ therefore } r = 70 \text{ tonnes/month}$$

Substituting in (1), 700 + 4h = 1000 therefore h = 75 tonnes per month. The optimum product mix is 70 tonnes of Crunchy per month and 75 tonnes of Chewy per month. At this point the contribution is:

$$P_{max} = 150 \times 70 + 75 \times 75 = £16{,}125 \text{ per month}$$

At this mix, the roasting and blending departments are fully occupied but there are 285 hours per month spare in the packing department.

SOLUTIONS TO EXERCISES: CHAPTER NINE

Exercise 9.5

Use a litres of product A per tank and b litres of product B per tank. Minimise cost per tank, £C, where:

$$C = 1.5a + 3.0b \quad \text{£/tank}$$

Minimise C subject to:

$$\begin{aligned}
\textit{Chemical X}: 4a + 5b &\geq 40 \text{ mg}/1{,}000 \text{ litres} \\
\textit{Chemical Y}: 2a + 1b &\geq 14 \text{ mg}/1{,}000 \text{ litres} \\
\textit{Chemical Z}: 3a + 1b &\geq 18 \text{ mg}/1{,}000 \text{ litres} \\
a, b &\geq 0
\end{aligned}$$

Illustrate the constraints graphically.

Linear programme to illustrate the possible combinations of products A and B

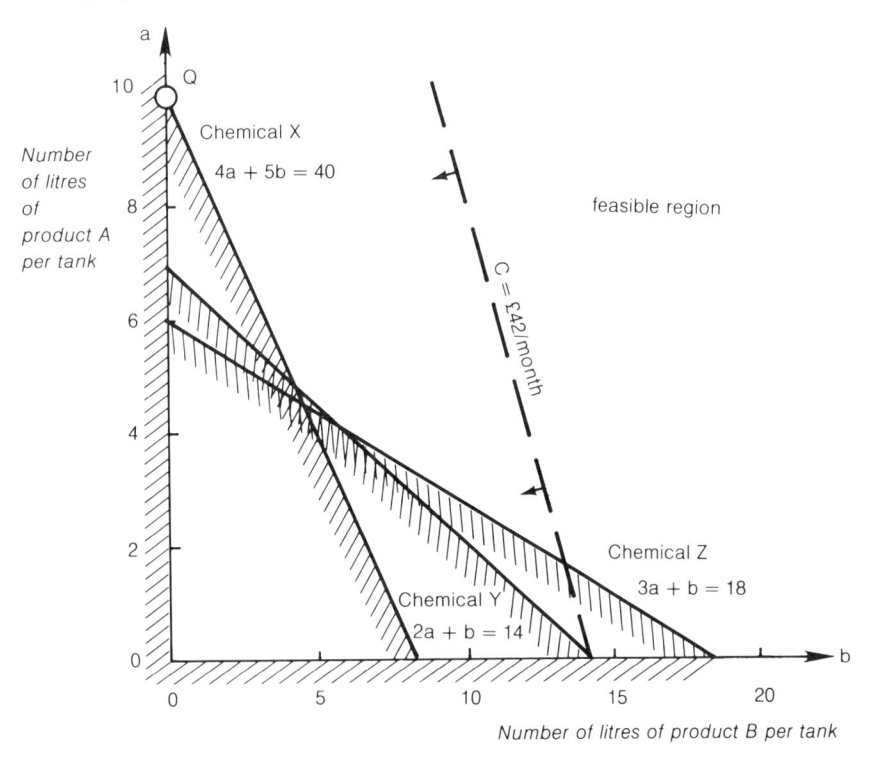

The feasible region is left unshaded. The next step is to add a trial cost line. The point b = 10 litres, a = 8 litres is a feasible solution. At this point the cost is C = 1.5 × 8 + 3.0 × 10 = £42.00 per tank. We will take as the trial cost line:

$$42 = 1.5a + 3.0b \quad \text{£/tank}$$

This line also passes through the point b = 14, a = 0. The line is shown dotted on the graph above. Moving parallel to this line in the direction of decreasing cost, we can see that the optimum corner is Q. Corner Q is the intersection of the constraint on Chemical X and the a axis. These are the limiting constraints. The optimum product mix is 10 litres of Product A per tank and none of Product B. At this point the cost is:

$$C_{min} = 1.5 \times 10 + 3.0 \times 0 = £15.00 \text{ per tank}$$

At this mix, the provision of chemical X is at the minimum, but the provision of the other two chemicals is above the minimum.

Exercise 9.6

Let us look again at the solution to this problem.

Linear programme to illustrate the possible monthly production of Crunchy and Chewy cereal

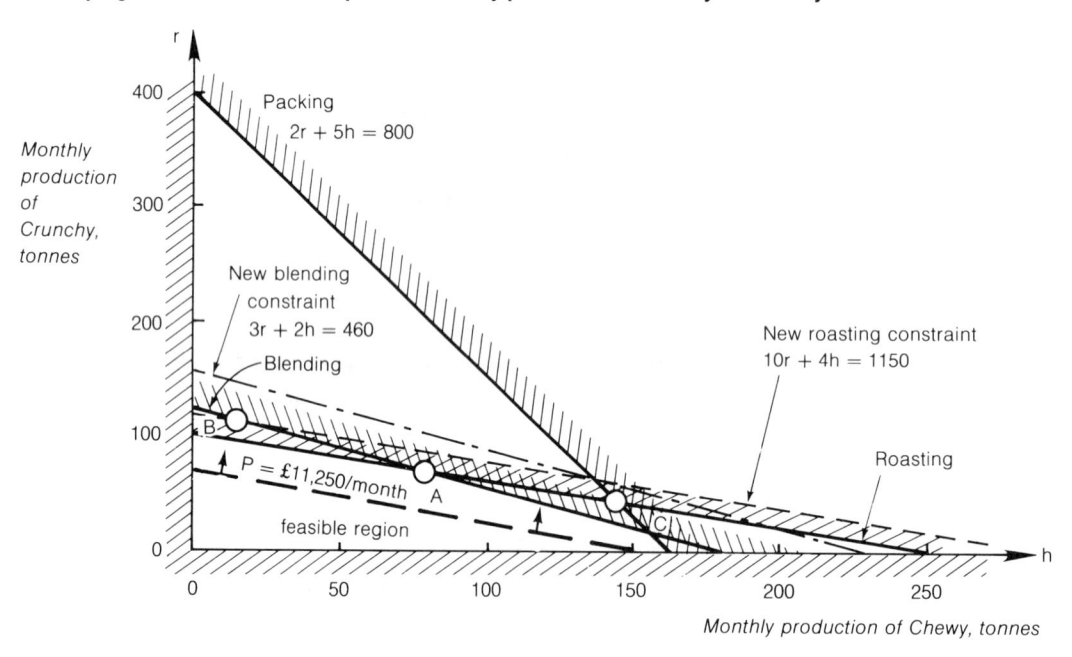

[Graph showing feasible region with axes: r (Monthly production of Crunchy, tonnes) vertical, h (Monthly production of Chewy, tonnes) horizontal. Constraints shown: Packing $2r + 5h = 800$, New blending constraint $3r + 2h = 460$, Blending, New roasting constraint $10r + 4h = 1150$, Roasting. Objective line $P = £11,250/month$. Corners labelled A, B, C.]

The optimum product mix is 70 tonnes per month of Crunchy and 75 tonnes per month of Chewy (corner A) with a contribution of £16,125 per month.

Let us consider the constraints one at a time:

(a) The packing constraint is not limiting, therefore, additional hours need not be considered for this department.

(b) If the roasting constraint is relaxed by 150 hours per month, then:

$$10r + 4h \leqslant 1150 \text{ hours/month}$$

This new constraint is shown — — on the graph above. We can see that the optimum corner is still formed by the intersection of the roasting and blending constraints (corner B). Solving the equations again:

$$10r + 4h = 1150 \text{ hours/month} \quad (1)$$
$$3r + 2h = 360 \text{ hours/month} \quad (2)$$

Using the same procedure as in **Exercise 9.4**, we find $r = 107.5$ tonnes per month and $h = 18.75$ tonnes per month. The monthly contribution is now:

$$P_B = 150 \times 107.5 + 75 \times 18.75 = £17,531$$

This is an increase of £1406 per month compared to the original optimum solution of £16,125. The shadow price for the roasting department is £1406 per 150 hours = £9.37 per hour. The additional cost of the overtime is £(6.50 − 4.50) = £2.00 per hour. This is less than the shadow price therefore a net gain is made from the use of the overtime.

Hence, if the overtime is given to the roasting department, the optimum solution will be to produce 107.5 tonnes per month of Crunchy and 18.75 tonnes per month of Chewy. This yields a net maximum monthly contribution of (£17,531 − 150 × £2.00) = £17,231.

(c) If the blending constraint is relaxed by 100 hours per month, then:

$$3r + 2h \leq 460 \text{ hours/month}$$

This new constraint is shown ·—·—· on the graph opposite. We can see that this constraint is no longer binding. The maximum relaxation for this constraint is to point C, the intersection of the roasting and packing constraints. If we solve the equations for roasting and packing simultaneously, we find that at point C, r = 42.857 and h = 142.857 tonnes per month. The number of blending hours used at this point is:

$$3 \times 42.857 + 2 \times 142.857 = 414.285 \text{ hours/month}$$

We will call this 414 hours per month, an increase of 54 hours per month. The monthly contribution at point C is:

$$P_c = 150 \times 42.857 + 75 \times 142.857 = £1,7143$$

This is an increase of £1018 per month compared to the original optimum solution. The shadow price for the blending department is £1018 per 54 hours = £18.85 per hour. The additional cost of the overtime is £(6.50 − 4.75) = £1.75 per hour. This is less than the shadow price therefore a net gain is made from the use of the overtime.

Hence, if the overtime is given to the blending department, the optimum solution will be to produce 42.857 tonnes per month of Crunchy and 142.857 tonnes per month of Chewy. This yields a net maximum monthly contribution of:

$$(£17,143 − 54 \times £1.75) = £17,049.$$

This is slightly less than the maximum possible if the overtime is given to the roasting department, hence the roasting department should be chosen if there are no other issues to consider.

Note: The use of overtime changes the quantities in the optimum product mix by quite large amounts. This may not be desirable.

Exercise 9.7

1. Produce m kg of the Standard drink, h kg of the Health drink and i kg of the Invalid drink, each week. Maximise total contribution per week, P, £ per week, where:

$$P = 0.485m + 0.655h + 1.045i \quad £/week$$

subject to:

Sugar: $\quad 0.30m + 0.15h + 0.15i \leq 1000$ kg/week
Malt: $\quad 0.30m + 0.25h + 0.30i \leq 1250$ kg/week
Skim milk powder: $0.35m + 0.55h + 0.25i \leq 2200$ kg/week
Maximum demand: $\quad m \leq 2000$ kg/week
Maximum demand: $\quad h \leq 1800$ kg/week
Maximum demand: $\quad i \leq 1200$ kg/week
$\quad m, h, i \geq 0$

2. From the tableau:

   (a) m = 1466.67 kg per week; h = 1800 kg per week; i = 1200 kg per week. Hence, the optimum product mix is to produce 1466⅔ kg of the Standard drink, 1800 kg of the Health drink and 1200 kg of the Invalid drink each week.

   (b) The contribution for the product mix in (a) is £3144.33 per week.

(c) There is spare capacity on constraints 1 (sugar), 3 (milk) and 4 (demand for the Standard drink). All other slack variables are zero.

> *Sugar:* there are 110 kg spare each week.
> *Milk:* there are $396\frac{2}{3}$ kg spare each week.
> *Demand:* the actual production of the Standard drink is below the maximum demand by $533\frac{1}{3}$ kg per week.

3 (a) The demand for the Health drink is a limiting constraint, therefore, the revised figure for the maximum demand allows this constraint to be relaxed by 700 kg per week. Refer to the s_5 column in the final tableau. Multiply each value in this column by the additional demand (700 kg) and add the result to the existing values of the basic variables.

	s_5		b		
s_1	0.1 × 700	110	+ 70	=	180
m	−0.833 × 700	1466.67	− 583.1	=	883.57
s_3	−0.258 × 700	396.67	− 180.6	=	216.07
s_4	0.833 × 700	533.33	+ 583.1	=	1116.43
h	1 × 700	1800	+ 700	=	2500
i	0 × 700	1200	+ 0	=	1200
P	0.251 × 700	3144.33	+ 175.7	=	3320.03

The new optimum solution is as follows: the optimum product mix is to produce 883.57 kg of the Standard drink, 2500 kg of the Health drink and 1200 kg of the Invalid drink each week. The contribution for the product mix in (a) is £3320.03 per week. There is spare capacity on constraints 1 (sugar), 3 (milk) and 4 (demand for the Standard drink). All other slack variables are zero.

> *Sugar:* there are 180 kg spare each week.
> *Milk:* there are 216.07 kg spare each week.
> *Demand:* the actual production of the Standard drink is below the maximum demand by 1116.43 kg per week.

(b) From the tableau, the shadow price for malt extract is £1.62 per kg (1.617). The current cost of the malt extract is 60p per kg, therefore, the additional supply will cost an extra 20p per kg. This is less than the shadow price for malt and, therefore, a net increase in the weekly contribution of £1.42 per kg could be achieved, if additional malt was available. To determine the maximum amount which should be purchased, refer to the s_2 column in the final tableau.

	s_2	b
s_1	−1	110
m	3.333	1466.67
s_3	−1.167	396.67
s_4	−3.333	533.33
h	0	1800
i	0	1200
P	1.617	3144.33

We consider only those rows with negative values in the s_2 column. If r kg per week is the maximum additional amount of malt extract which can be used, then the largest value which r can take is the one which reduces one of the basic variables to zero.

s_1 *row*: if $110 + (-1 \times r) = 0$, then $r = 110$
s_3 *row*: if $396.67 + (-1.167 \times r) = 0$, then $r = 339.91$
s_4 *row*: if $533.33 + (-3.333 \times r) = 0$, then $r = 160.02$

The smallest of these three values for r is r = 110 kg per week, hence this is the maximum value which r can take. The maximum additional amount of malt extract which should be purchased is 110 kg per week. At this optimum solution all of the sugar is used up (ie $s_1 = 0$).

The contribution per week will increase by:

$$110 \times 1.417 = £155.87$$

from £3144.33 to £3300.20.

Exercise 9.8

1. Since there are three constraints in the primal linear programme, we require three dual variables. Call these x, y and z. The dual linear programme is: minimise the total value per month of the resource used, £V per month, where:

$$V = 1000x + 360y + 800z \quad £/month$$

Minimise V subject to:

$$10x + 3y + 2z \geqslant 150 \quad £/tonne$$
$$4x + 2y + 5z \geqslant 75 \quad £/tonne$$
$$x,y,z \geqslant 0$$

2. The dual variables are the shadow prices of the constraints in the primal model. £x per hour is the amount which would be added to the value of the objective function if one additional hour was available in the roasting department, that is, £x per hour is the value of one extra hour in that department. y and z may be similarly explained for the blending and packing departments, respectively.

Exercise 9.9

1. Let the weekly production of emulsion be e 100 litres, and the weekly production of gloss be g 100 litres. The objective is to maximise the weekly contribution to profit, £P per week, where:

$$P = \text{Revenue} - \text{Variable costs}$$
$$= 120e + 126g - (11 + 30 + 32 + 12)e - (25 + 36 + 20 + 15)g \quad £/week$$
$$= 35e + 30g \quad £/week$$

The direct labour costs for 100 litres of emulsion is £30. The cost of direct labour is £3.00 per hour, therefore, we may assume that 100 litres of emulsion requires 30/3 = 10 hours of labour. A similar calculation can be done for the gloss paint and for the mixing process.

Also, the company is producing exactly 25,000 litres per week of undercoat. This quantity will generate a contribution of:

$$(110 - (20 + 24 + 36 + 10)) \times 250 = £5000/week$$

This production will use $(24/3) \times 250 = 2000$ hours per week of labour, leaving 6000 hours per week for the other two paints. The production of undercoat will use $(36/4) \times 250 = 2250$ hours per week of labour, leaving 3650 hours per week for the other two paints. Maximise contribution subject to:

$$\begin{aligned}
\textit{Labour}: \quad & 10e + 12g \leqslant 6000 \text{ hours/week} \\
\textit{Mixing}: \quad & 8e + 5g \leqslant 3650 \text{ hours/week} \\
\textit{Demand}: \quad & e \leqslant 350 \text{ 100 litres/week} \\
\textit{Demand}: \quad & g \leqslant 290 \text{ 100 litres/week} \\
& e,g \geqslant 0
\end{aligned}$$

2 The linear programme is illustrated below:

Linear programme to illustrate the feasible production of emulsion and gloss paint at Princetown Paints Ltd

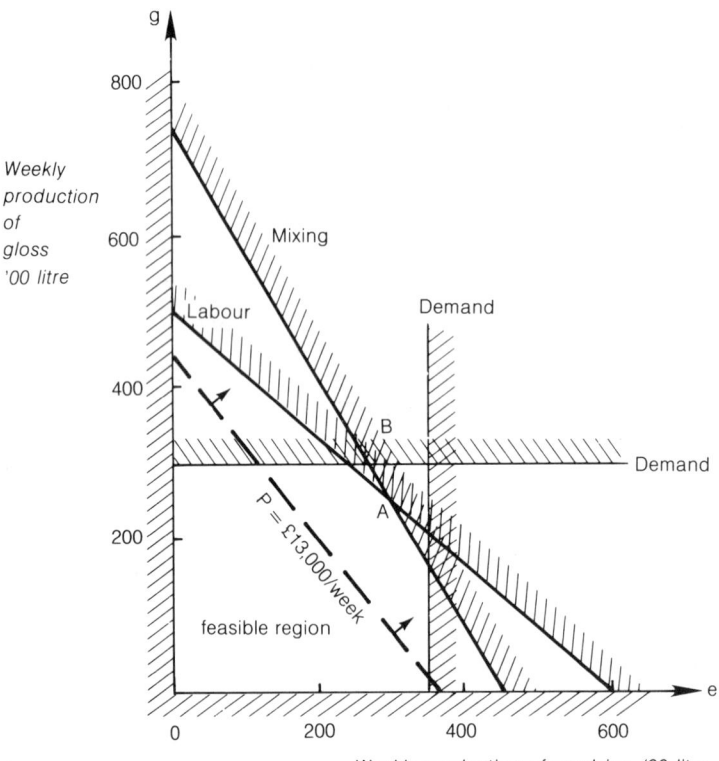

We will choose a point in the feasible region, X, say. The co-ordinates of X are $e = g = 200$. At X the weekly contribution is $P_x = 35 \times 200 + 30 \times 200 = £13,000$ per week. We will take $13,000 = 35e + 30g$ as a trial weekly contribution line. This line also passes through the point $e = 0$, $g = 433.3$. The line is shown --- on the graph above. If we move parallel to this line in the direction of increasing contribution, we see that corner A is the optimum corner. A is the intersection of the labour and the mixing constraints. Solving these equations simultaneously gives:

$$10e + 12g = 6000 \quad (1)$$
$$8e + 5g = 3650 \quad (2)$$
$$(1) \times 8 - (2) \times 10 \text{ gives} \quad 46g = 11,500$$
$$\text{therefore } g = 250$$

Substituting in (1), $e = 300$. The optimum weekly production levels are 25,000 litres of gloss, 30,000 litres of emulsion and the 25,000 litres of undercoat. The maximum weekly contribution is:

$$P_{max} = 35 \times 300 + 30 \times 250 + 5000 \quad £/\text{week}$$
$$= £23,000 \text{ per week}$$

3 The total weekly contribution for gloss and emulsion is:

$$P = ae + 30g \quad £/\text{week}$$

therefore:

$$g = \frac{P}{30} - \frac{a}{30}e$$

The slope of the contribution line is $-(a/30)$. The value of a can decrease until the contribution line is parallel to the labour constraint, without the optimum corner changing. The value of a can increase until the contribution line is parallel to the mixing constraint without the optimum corner changing. The slope of the labour constraint is $-(10/12)$ and the slope of the mixing constraint is $-(8/5)$. The optimum corner will remain unchanged if $(a/30) > (10/12)$, that is, $a > 25$. The optimum corner will remain unchanged if $(a/30) < (8/5)$, that is, $a < 48$.

$$a = \text{selling price} - \text{variable costs} = \text{selling price} - 85 \quad £/100 \text{ litre}$$

Therefore, if $25 < a < 48$, $110 < \text{selling price} < 133$ £/100 litre. The selling price may vary between £110 and £133 per 100 litre.

4 Since labour is a binding constraint, additional labour hours are potentially profitable. The labour constraint may be relaxed upto point B on the graph above, before it ceases to be binding. The co-ordinates of point B are $g = 290$, $8e = 3650 - (5 \times 290) = 2200$, therefore $e = 275$.

At this level of production, the number of labour hours used is:

$$10 \times 275 + 12 \times 290 = 6230 \text{ hours/week}$$

This is an additional 230 hours per week. The weekly contribution at this point is:

$$P_{max} = 35 \times 275 + 30 \times 290 + 5000 - 1 \times 230 \text{ £/week}$$
$$= £23,095 \text{ per week}$$

This is an extra £95 per week compared to the original problem. An additional 230 hours per week of labour will generate an additional £95 of contribution.

Exercise 9.10

1 Minimise the total redundancy and relocation costs, £C, where:

$$\text{Relocation costs} = 2000S1 + 2000U1$$
$$\text{Redundancy costs} = (300 - S1 - S2) \times 2000 + (500 - U1 - U2) \times 1500$$
$$= 1,350,000 - 2000S1 - 2000S2 - 1500U1 - 1500U2$$

Therefore:

$$C = 500U1 - 2000S2 - 1500U2 + 1,350,000 \quad £$$

Minimise C subject to:

Skilled men: $S1 + S2 = 240$
Unskilled men: $U1 + U2 = 320$
Bias: $(S1 + U1)/500 = (S2 + U2)/300$
Abbotsfield: $S1 \leq 200; U1 \leq 300$
Birchwood: $S2 \leq 100; U2 \leq 200$
                 $S1, S2, U1, U2 \geq 0$

The linear programme is:

Minimise:

$$C = 500U1 - 2000S2 - 1500U2 + 1,350,000 \quad £$$

subject to:

$$\begin{aligned}
\text{Skilled Men:} \quad & S1 + S2 & = 240 \quad (1) \\
\text{Unskilled men:} \quad & U1 + U2 & = 320 \quad (2) \\
\text{Bias:} \quad & 3S1 + 3U1 - 5S2 - 5U2 & = 0 \quad (3) \\
\text{Abbotsfield:} \quad & S1 & \leq 200 \quad (4) \\
& U1 & \leq 300 \quad (5) \\
\text{Birchwood:} \quad & S2 & \leq 100 \quad (6) \\
& U2 & \leq 200 \quad (7) \\
& S1, S2, U1, U2 & \geq 0
\end{aligned}$$

From constraint (1), $S2 = 240 - S1$; from constraint (2), $U2 = 320 - U1$. Therefore, constraint (3) becomes:

$$\begin{aligned}
3S1 + 3U1 - 5(240 - S1) - 5(320 - U1) &= 0 \\
8S1 + 8U1 &= 2800 \quad (3) \\
S1 &\leq 200 \quad (4) \\
U1 &\leq 300 \quad (5) \\
240 - S1 &\leq 100
\end{aligned}$$

hence:

$$\begin{aligned}
S1 &\geq 140 \quad (6) \\
320 - U1 &\leq 200
\end{aligned}$$

hence

$$U1 \geq 120 \quad (7)$$

Minimise:

$$\begin{aligned}
C &= 500U1 - 2000(240 - S1) - 1500(320 - U1) + 1{,}350{,}000 \\
C &= 2000S1 + 2000U1 + 390{,}000 \quad £
\end{aligned}$$

The linear programme is illustrated below:

Linear programme for the man power rationalisation at the Nemesis Company

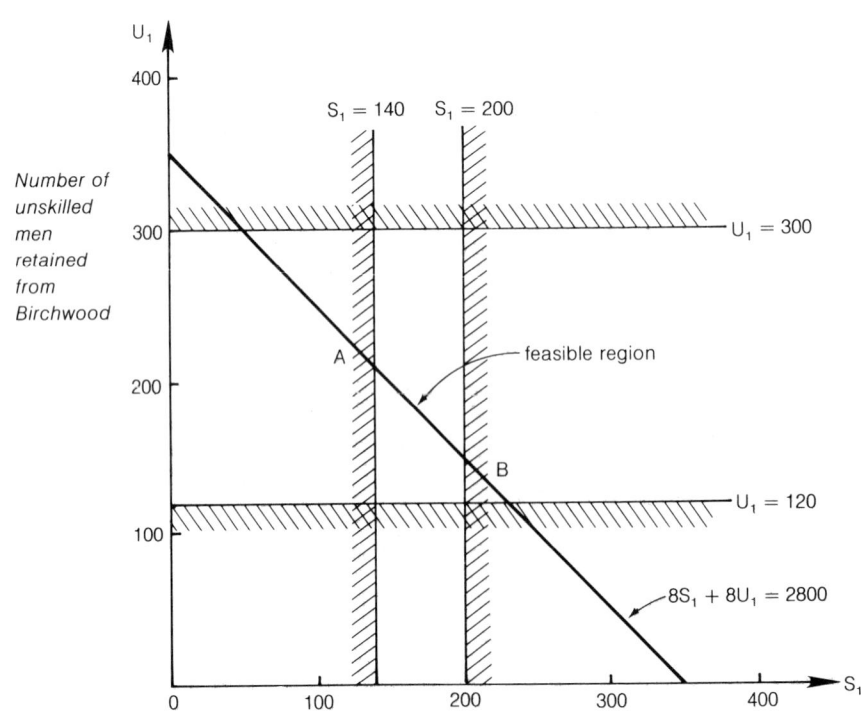

SOLUTIONS TO EXERCISES: CHAPTER NINE

Since one of the constraints is an equality, the feasible region in this problem is the set of points on the equality constraint between the points marked A and B on the graph. The objective function is C = 2000(S1 + U1) + £390,000. We can see that the cost, C, will take its minimum value when (S1 + U1) takes its minimum value. We will, therefore, minimise Z = S1 + U1.

We should now note that this line is parallel to constraint (3), 8(S1 + U1) = 2800, therefore, the portion of this constraint, between the points A and B marked on the graph above, represents the optimum solution. This corresponds to the set of feasible values. Any point on this section of line will give the optimum value of the objective function. The minimum value of Z is given by Z = 2800/8 = £350, therefore, the minimum value of the cost is given by:

$$C = (2000 \times 350) + 390{,}000 = £1{,}090{,}000$$

The minimum total redundancy and relocation cost is £1,090,000.

Exercise 9.11

1 Invest £a in Project A, £b in Project B and £c in Project C.
Maximise the net present value of the investment, NPV, £, where:

$$\text{NPV} = \frac{200a}{1000} + \frac{250b}{1000} + \frac{450c}{1000} \quad £$$
$$\text{NPV} = 0.20a + 0.25b + 0.45c \quad £$$

Maximise subject to:

Total sum: a + b + c = 100000 £
Average payback period: 3(a/100000) + 4(b/100000) + 6(c/100000) ⩽ 5 years

where (a/100000) is the proportion of the total money invested which is put into Project A. Similarly, for the other two terms. The left hand side of this constraint is a weighted average of the actual payback periods.

Rate of return: 0.18a + 0.20b + 0.25c ⩾ 0.20 × 100,000 £
a,b,c ⩾ 0

The linear programme is:

Maximise: NPV = 0.20a + 0.25b + 0.45c £

subject to:

Total sum:	a + b + c	= 100000 £	(1)
Average payback period:	3a + 4b + 6c	⩽ 500000	years	(2)
Rate of return:	0.18a + 0.20b + 0.25c	⩾ 20000 £	(3)
	a,b,c ⩾ 0		

2 From constraint (1), we can write c = 100,000 − a − b. Substituting for c in the linear programme, gives:

Maximise NPV = 0.20a + 0.25b + 0.45(100000 − a − b) £
NPV = 45000 − 0.25a − 0.20b £
NPV = 45000 − (0.25a + 0.20b) £

The NPV will take its maximum value, when (0.25a + 0.20b) takes its minimum value, therefore minimise Z = 0.25a + 0.20b £, subject to:

$$\text{Average payback period: } 3a + 4b + 6(100000 - a - b) \leqslant 500000 \quad \text{years}$$
$$3a + 2b \geqslant 100000 \quad \text{years}$$
$$\text{Rate of Return: } 0.18a + 0.20b + 0.25(100000 - a - b) \geqslant 20000 \quad £$$
$$0.07a + 0.05b \leqslant 5000 \quad £$$

Also, since c ⩾ 0, 100,000 − a − b ⩾ 0, therefore:

$$a + b \leqslant 100,000 \quad £$$
$$a, b \geqslant 0$$

These constraints and a trial objective function are illustrated below:

Linear programme for the investment in three projects

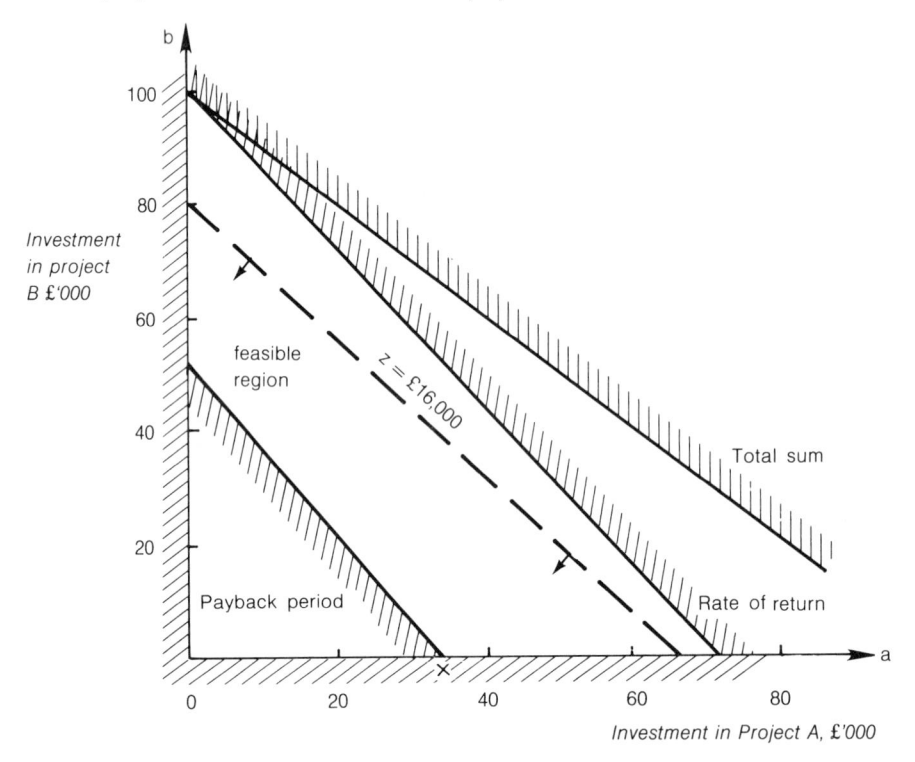

Moving the objective function towards the origin, we find that point X is the optimum corner. At point X, a = £33,333 and b = 0. Therefore, c = 100,000 − 33,333 − 0 = £66,667 and the maximum value of the NPV = 45,000 − 0.25 × 33,333 = £36,667.

Exercise 9.12

1. (a) The optimum weekly production plan is to produce 3357 units of product A, 2321 units of product E and none of B, C or D. The resulting maximum weekly contribution is £105,791.
   (b) There is spare capacity of 321 hours per week on the forming process and 9482 hours per week on the firing process. All raw materials and all packing time are used up. Raw materials and packing time are the limiting constraints in the problem.
2. The shadow price is the amount which would be added to the value of the total weekly contribution if one extra unit of a limiting resource were made available, provided that:
   (a) no additional costs were incurred;
   (b) the resource remains limiting.

   Alternatively, the shadow price is the amount by which the total weekly contribution would fall if the provision of a limiting resource was reduced by 1 unit.

From the tableau, we can see that the shadow price for raw materials is £2.02 per kg and for packing time it is £8.81 per hour. One additional kg of raw material will generate an extra £2.02 of contribution, subject to the conditions above. One extra hour of packing time will, similarly, generate an additional £8.81 of contribution.

3   The additional product would also have to be made at the expense of one or both of the other products, since all raw material and packing time are currently used.

$$\text{Unit contribution of the new product} = 50 - (2.1 \times 6 + 3.0 \times 1 + 1.3 \times 5 + 8.0 \times 1)$$
$$= £19.9$$

If one unit of this new product was made, the provision of raw materials for the other two products would effectively be reduced by 6kg. This would reduce the current total contribution by $6 \times £2.02 = £12.12$. Similarly, the available packing time would be reduced by 1 hour, this reduces the total contribution by £8.81. The total reduction in the weekly contribution would be:

$$£12.12 + £8.81 = £20.93$$

The gain from one unit of the new product is £19.90, therefore, if one unit of the new product is made, there will be a net loss of £19.90 − £20.93 = £1.03. The proposition is not worthwhile.

Exercise   9.13

1   Build a blocks of 6 linked terraced houses, b blocks of semi-detached houses and c detached houses. Maximise contribution, P, £'000, where:

$$P = 6(22 - 15)a + 2(27 - 18)b + (36 - 24)c \quad £'000$$
$$P = 42a + 18b + 12c \quad £'000$$

Maximise subject to:

$$\begin{aligned} \textit{Total number}: \quad & 6a + 2b + c \leqslant 38 \\ \textit{Area}: 6 \times 300a &+ 2 \times 675b + 900c \leqslant 27000 \text{ m}^2 \\ & a,b,c \geqslant 0 \end{aligned}$$

The linear programme is:

$$\text{Maximise: } P = 42a + 18b + 12c \quad £'000$$

subject to:

$$\begin{aligned} \textit{Total number}: \quad & 6a + 2b + c \leqslant 38 \\ \textit{Area}: \quad & 1800a + 1350b + 900c \leqslant 27000 \text{ m}^2 \\ & a,b,c \geqslant 0 \end{aligned}$$

2   We require two dual variables, call these x and y. The dual LP is:

$$\text{Minimise: } Z = 38x + 27000y$$

subject to:

$$\begin{aligned} \textit{Terraced block}: \quad & 6x + 1800y \geqslant 42 \quad £/\text{unit} \\ \textit{Semi-detached block}: \quad & 2x + 1350y \geqslant 18 \quad £/\text{unit} \\ \textit{Detached house}: \quad & x + 900y \geqslant 12 \quad £/\text{unit} \\ & x,y \geqslant 0 \end{aligned}$$

3 The dual is illustrated graphically below:

Dual linear programme for house building project

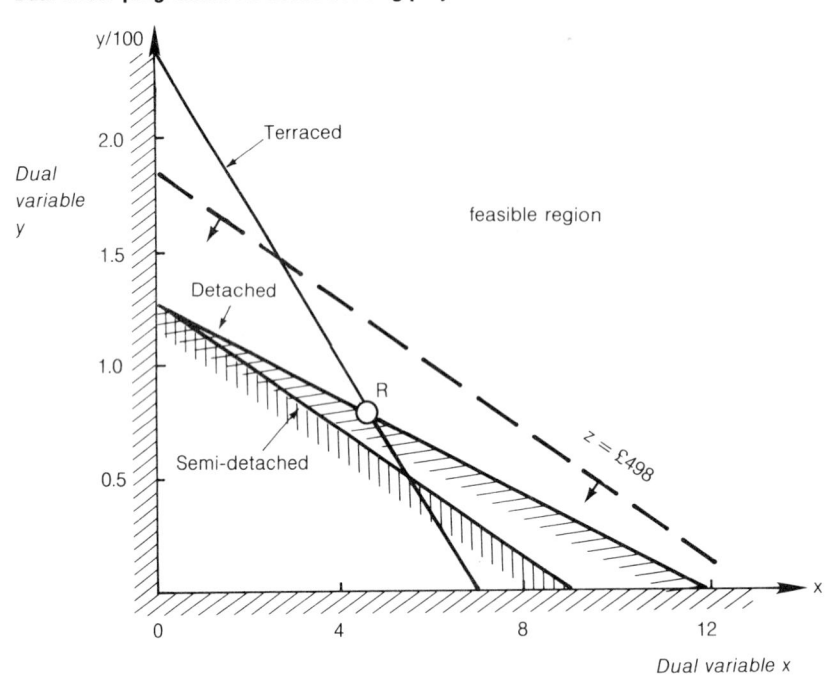

[Graph showing dual variable y (y/100) on vertical axis from 0 to 2.0+, and dual variable x on horizontal axis from 0 to 12. Lines labelled Terraced, Detached, Semi-detached bound a feasible region. Point R marked at intersection. Dashed line shows z = £498.]

A trial objective function has been added to the graph. When this is moved towards the origin, point R is seen to be the optimum corner. This is the intersection of the constraints for the terraced blocks and for the detached houses:

$$6x + 1800y = 42$$
$$x + 900y = 12$$

Solving these equations simultaneously gives $x = 4.5$ and $y = 7.5/900 = 0.0083$. Therefore:

$$Z_{min} = 38 \times 4.5 + 27000 \times (7.5/900) = £396.$$

4 In the dual, constraints 1 and 3 are binding and constraint 2 has spare capacity. Since the primal variables are equal to the dual shadow prices, then the primal variable $b = 0$ (dual shadow price on constraint 2 is zero), but a and c are not zero.

The dual variables, x and y, are the primal shadow prices. Since neither of these is zero, then both of the primal constraints must be binding. Hence:

Total number of houses: $6a + 2 \times 0 + c = 38$
Area: $1800a + 1350 \times 0 + 900c = 27000$ sq m

Therefore:

$$6a + c = 38$$

and:

$$18a + 9c = 270$$

Solve the two equations simultaneously for a and c:

$$a = 2 \quad \text{and} \quad c = 26$$

Therefore, Python Properties should build 2 blocks of terraced houses, 26 detached houses and no semi-detached houses. This will yield a maximum contribution of:

$$P_{max} = 42 \times 2 + 12 \times 26 = £396,000$$

(the same as the optimum value in the dual).

Solutions to Chapter Ten exercises

ANSWERS TO EXERCISES

Exercise 10.1

The problem satisfies the requirements of a transportation problem. The total number of items available = total number of items required, therefore, no dummy origin or destination is required. The minimum cost method will be used to find the first allocation. The subscripts show the order in which the allocations are made.

		| G	| H	| I	| J	| Items available
From warehouse	1	4 / 50$_2$	3 / —	5 / —	6 / 50$_4$	100 50 0
	2	8 / —	2 / 100$_1$	4 / 75$_3$	7 / 25$_5$	200 100 25 0
Items required		50 / 0	100 / 0	75 / 0	75 / 25	0

This initial allocation is basic if there are $(m + n - 1) = (2 + 4 - 1) = 5$ allocations. We can see from the tableau that this is the case. We have a basic solution.

To test for optimality using the MODI method:
For the allocated cells:

$$c_{11} = 4 = u_1 + v_1 \quad \text{let } u_1 = 0, \text{ then } v_1 = 4$$
$$c_{14} = 6 = u_1 + v_4 \quad\quad\quad\quad\quad v_4 = 6$$
$$c_{22} = 2 = u_2 + v_2 \quad\quad\quad\quad\quad v_2 = 1$$
$$c_{23} = 4 = u_2 + v_3 \quad\quad\quad\quad\quad v_3 = 3$$
$$c_{24} = 7 = u_2 + v_4 \quad\quad\quad\quad\quad u_2 = 1$$

For the empty cells, the shadow costs are:

$$s_{ij} = c_{ij} - (u_i + v_j)$$

Therefore:

$$s_{12} = 3 - (0 + 1) = 2$$
$$s_{13} = 5 - (0 + 3) = 2$$
$$s_{21} = 8 - (1 + 4) = 3$$

All of the shadow costs are positive, therefore, the allocation is optimal.

Warehouse 1 should send 50 items to store G and 50 items to store J. Warehouse 2 should send 100 items to store H, 75 items to store I and 25 items to store J. The minimum cost of transfer is:

$$£(50 \times 4 + 50 \times 6 + 100 \times 2 + 75 \times 4 + 25 \times 7) = £1175$$

Exercise 10.2

The problem satisfies the requirements of a transportation problem. The total number of items available = total number of items required, therefore, no dummy origin or destination is required.

Vogel's method will be used to find the first allocation. The subscripts show the order in which the penalties are used and the allocations are made.

		Transport costs £ per tonne To warehouse					Tonnes	
		1	2	3	4	5	available	Penalty cost
From factory	A	20 100_2	27 100	33 —	25 —	34 —	200 100 0	5 5_2 2 2 2
	B	22 —	36 —	34 —	28 50_4	26 200_3	250 50 0	4 4 2_3 8_4
	C	26 —	29 50	27 200_1	26 50_5	28 —	300 100 50 0	0 0 2 3 3_5
Tonnes required		100 0	150 0	200 0	100 50 0	200 0		
Penalty cost		2 2	2 2 2 2 2	6_1	1 1 1 1 1	2 2 2 1		

This initial allocation is basic if there are $(m + n - 1) = (3 + 5 - 1) = 7$ allocations. We can see from the tableau that this is the case. We have a basic solution.

Test for optimality using the MODI method:
For the allocated cells:

$$c_{11} = 20 = u_1 + v_1 \quad \text{let } u_1 = 0, \text{ then } v_1 = 20$$
$$c_{12} = 27 = u_1 + v_2 \quad\quad\quad\quad\quad v_2 = 27$$
$$c_{24} = 28 = u_2 + v_4 \quad\quad\quad\quad\quad u_2 = 4$$
$$c_{25} = 26 = u_2 + v_5 \quad\quad\quad\quad\quad v_5 = 22$$
$$c_{32} = 29 = u_3 + v_2 \quad\quad\quad\quad\quad u_3 = 2$$
$$c_{33} = 27 = u_3 + v_3 \quad\quad\quad\quad\quad v_3 = 25$$
$$c_{34} = 26 = u_3 + v_4 \quad\quad\quad\quad\quad v_4 = 24$$

For the empty cells, the shadow costs are:

$$s_{ij} = c_{ij} - (u_i + v_j)$$

Therefore:

$$s_{13} = 33 - (0 + 25) = 8$$
$$s_{14} = 25 - (0 + 24) = 1$$
$$s_{15} = 34 - (0 + 22) = 12$$
$$s_{21} = 22 - (4 + 20) = -2$$
$$s_{22} = 36 - (4 + 27) = 5$$
$$s_{23} = 34 - (4 + 25) = 5$$
$$s_{31} = 26 - (2 + 20) = 4$$
$$s_{35} = 28 - (2 + 22) = 4$$

There is one negative shadow cost, therefore, the solution is not optimal. We must re-allocate into cell (B,1). The stepping stone circuit is:

	1	2	3	4
A	20 100 −	27 100 +	33	25
B	22 test +	36	34	28 50 −
C	26	29 50 −	27 200	26 50 +

Look at the − cells, ... the minimum is 50 tonnes, therefore, move 50 tonnes around the stepping stone circuit. The new allocation is:

		To warehouse				
		1	2	3	4	5
From factory	A	20 50	27 150	33 —	25 —	34 —
	B	22 50	36 —	34 —	28 —	26 200
	C	26 —	29 —	27 200	26 100	28 —

There are now only 6 allocations. We have a degenerate solution. Test for optimality as before. For the allocated cells:

$$c_{11} = 20 = u_1 + v_1 \quad \text{let } u_1 = 0, \text{ then } v_1 = 20$$
$$c_{12} = 27 = u_1 + v_2 \qquad\qquad v_2 = 27$$
$$c_{21} = 22 = u_2 + v_1 \qquad\qquad u_2 = 2$$
$$c_{25} = 26 = u_2 + v_5 \qquad\qquad v_5 = 24$$

At this stage we must make a zero allocation to an empty cell, in order to complete the process of finding the row and column costs. We will choose to allocate to cell (C,2).

$$c_{32} = 29 = u_3 + v_2 \qquad\qquad u_3 = 2$$
$$c_{33} = 27 = u_3 + v_3 \qquad\qquad v_3 = 25$$
$$c_{34} = 26 = u_3 + v_4 \qquad\qquad v_4 = 24$$

For the empty cells, the shadow costs are:

$$s_{ij} = c_{ij} - (u_i + v_j)$$

Therefore:

$$s_{13} = 33 - (0 + 25) = 8$$
$$s_{14} = 25 - (0 + 24) = 1$$
$$s_{15} = 34 - (0 + 24) = 10$$
$$s_{22} = 36 - (2 + 27) = 7$$
$$s_{23} = 34 - (2 + 25) = 7$$
$$s_{24} = 28 - (2 + 24) = 2$$
$$s_{31} = 26 - (2 + 20) = 4$$
$$s_{35} = 28 - (2 + 24) = 2$$

All of the shadow costs are now positive, therefore, we have found the optimum allocation.

Factory A sends 50 tonnes to warehouse 1 and 150 tonnes to warehouse 2. Factory B sends 50 tonnes to warehouse 1 and 200 tonnes to warehouse 5. Factory C sends 200 tonnes to warehouse 3 and 100 tonnes to warehouse 4.

The minimum cost of the transfer is:

$$£(50 \times 20 + 150 \times 27 + 50 \times 22 + 200 \times 26 + 200 \times 27 + 100 \times 26) = £19,350$$

Exercise 10.3

The problem satisfies the requirements of a transportation problem, but the total number of items available < total number of items required, therefore, a dummy origin is required.

Total number of loaves available = 2150
Total number of loaves required = 2250

The dummy bakery supplies 100 loaves.

558 QUANTITATIVE ANALYSIS

Vogel's method will be used to find the first allocation. The subscripts show the order in which the penalties are used and the allocations are made.

		\multicolumn{4}{c}{Transport cost pence per loaf To shop}	Total available	Penalty cost
		I	II	III	IV		
From bakery	X	1.5 —	2.5 —	1.0 350$_4$	2.0 350	~~700~~ 350 0	0.5 0.5 0.5 1.0$_4$
	Y	2.0 —	3.0 —	2.0 —	1.5 650	~~650~~ 0	0.5 0.5 0.5 0.5
	Z	1.0 400$_3$	1.5 400$_2$	2.5 —	3.0 —	~~800~~ ~~400~~ 0	0.5 0.5 1.5$_3$
	Dummy	0 —	0 100$_1$	0 —	0 —	~~100~~ 0	0
Total required		~~400~~ 0	~~500~~ ~~400~~ 0	~~350~~ 0	~~1000~~ ~~650~~ 0		
Penalty cost		1.0 0.5 0.5	1.5$_1$ 1.0$_2$	1.0 1.0 1.0 1.0	1.5 0.5 0.5 0.5		

This initial allocation is basic if there are (m + n − 1) = (4 + 4 − 1) = 7 allocations. We can see from the tableau that this is not the case. We have only 6 allocations, therefore, the solution is degenerate.

Test for optimality using the MODI method:
For the allocated cells:

$$c_{13} = 1.0 = u_1 + v_3 \quad \text{let } u_1 = 0, \text{ then } v_3 = 1.0$$
$$c_{14} = 2.0 = u_1 + v_4 \quad\quad\quad\quad\quad\quad v_4 = 2.0$$
$$c_{24} = 1.5 = u_2 + v_4 \quad\quad\quad\quad\quad\quad u_2 = -0.5$$

At this stage, it is necessary to make a zero allocation in row Z in order to find u_3. We will choose to make the allocation to cell (Z,III), therefore, this cell is now treated as an allocated cell.

$$c_{33} = 2.5 = u_3 + v_3 \quad\quad u_3 = 1.5$$
$$c_{31} = 1.0 = u_3 + v_1 \quad\quad v_1 = -0.5$$
$$c_{32} = 1.5 = u_3 + v_2 \quad\quad v_2 = 0$$
$$c_{42} = 0 \ \ = u_4 + v_2 \quad\quad u_4 = 0$$

For the empty cells, the shadow costs are:

$$s_{ij} = c_{ij} - (u_i + v_j)$$

Therefore:

$$\begin{aligned}
s_{11} &= 1.5 - (0 - 0.5) &= 2.0 \\
s_{12} &= 2.5 - (0 + 0) &= 2.5 \\
s_{21} &= 2.0 - (-0.5 - 0.5) &= 3.0 \\
s_{22} &= 3.0 - (-0.5 + 0) &= 3.5 \\
s_{23} &= 2.0 - (-0.5 + 1.0) &= 1.5 \\
s_{34} &= 3.0 - (1.5 + 2.0) &= -0.5 \\
s_{41} &= 0 - (0 - 0.5) &= 0.5 \\
s_{43} &= 0 - (0 + 1.0) &= -1.0 \\
s_{44} &= 0 - (0 + 2.0) &= -2.0
\end{aligned}$$

There are three negative shadow costs, therefore, the solution is not optimal. We must re-allocate into the cell with the largest negative shadow cost, that is, cell (Dummy, IV).

The stepping stone circuit is:

	II	III	IV
X	2.5	1.0 350 +	2.0 350 −
Y	3.0	2.0	1.5
Z	1.5 400 +	2.5 0 −	3.0
Dummy	0 100 −	0	0 test +

Look at the − cells, the minimum allocation is 0 loaves, therefore, move 0 loaves around the stepping stone circuit. The new allocation is:

		To shop			
		I	II	III	IV
From bakery	X	1.5 —	2.5 —	1.0 350	2.0 350
	Y	2.0 —	3.0 —	2.0 —	1.5 650
	Z	1.0 400	1.5 400	2.5 —	3.0 —
	Dummy	0 —	0 100	0 —	0 [0]

The allocation is still degenerate but the position of the zero allocation has been moved. Test for optimality using the MODI method:

For the allocated cells:

$$c_{13} = 1.0 = u_1 + v_3 \quad \text{let } u_1 = 0, \text{ then } v_3 = 1.0$$
$$c_{14} = 2.0 = u_1 + v_4 \qquad\qquad\qquad\qquad v_4 = 2.0$$
$$c_{24} = 1.5 = u_2 + v_4 \qquad\qquad\qquad\qquad u_2 = -0.5$$
$$c_{31} = 1.0 = u_3 + v_1 \qquad\qquad\qquad\qquad v_1 = 1.5$$
$$c_{32} = 1.5 = u_3 + v_2 \qquad\qquad\qquad\qquad u_3 = -0.5$$
$$c_{42} = 0 \ \ = u_4 + v_2 \qquad\qquad\qquad\qquad v_2 = 2.0$$
$$c_{44} = 0 \ \ = u_4 + v_4 \qquad\qquad\qquad\qquad u_4 = -2.0$$

For the empty cells, the shadow costs are:

$$s_{ij} = c_{ij} - (u_i + v_j)$$

Therefore:

$$s_{11} = 1.5 - (0 + 1.5) \quad\ \ = 0$$
$$s_{12} = 2.5 - (0 + 2.0) \quad\ \ = 0.5$$
$$s_{21} = 2.0 - (-0.5 + 1.5) = 1.0$$
$$s_{22} = 3.0 - (-0.5 + 2.0) = 1.5$$
$$s_{23} = 2.0 - (-0.5 + 1.0) = 1.5$$
$$s_{33} = 2.5 - (-0.5 + 1.0) = 2.0$$
$$s_{34} = 3.0 - (-0.5 + 2.0) = 1.5$$
$$s_{41} = 0 \ \ - (-2.0 + 1.5) = 0.5$$
$$s_{43} = 0 \ \ - (-2.0 + 1.0) = 1.0$$

SOLUTIONS TO EXERCISES: CHAPTER TEN

All of the shadow costs are now positive or zero, hence the initial allocation was in fact the optimum one.

Bakery X sends 350 loaves to shop III and 350 loaves to shop IV. Bakery Y sends 650 loaves to shop IV. Bakery Z sends 400 loaves to shop I and 400 loaves to shop II. Shop II has a shortfall of 100 loaves which are not supplied. The minimum cost of the transfer is:

$$(350 \times 1.0 + 350 \times 2.0 + 650 \times 1.5 + 400 \times 1.0 + 400 \times 1.5) = 3025 \text{ pence}$$

The total cost is £30.25

$S_{11} = 0$ means there are other allocations, using cell (X, I), which will give a total cost of £30.25. The alternatives are found using a stepping stone circuit for (X, I). Up to 100 loaves can be re-allocated around this circuit.

Exercise 10.4

The problem satisfies the requirements of a transportation problem, but the total number of items available > total number of items required, therefore, a dummy destination is required.

$$\text{Total number of loads available} = 160/\text{day}$$
$$\text{Total number of loads required} = 117/\text{day}$$

The dummy store requires 23 loads per day.

Vogel's method will be used to find the first allocation. The subscripts show the order in which the penalties are used and allocations are made. Consider new warehouse 1 first.

Transport cost £ per load

From warehouse		P	Q	R	S	Dummy	Total available	Penalty cost
		70	85	55	120	0	40	55 15 30$_3$
A		27$_2$	—	13$_3$	—	—	13 0	
		110	90	75	110	0	40	75$_1$ 15 15 15
B		—	17$_4$	—	—	23$_1$	17 0	
		115	115	70	90	0	60	70 20 20 20
1		—	8	17	35	—	43 0	
Total required		2̶7̶ 0	2̶5̶ 8̶ 0	3̶0̶ 1̶7̶ 0	35 0	2̶3̶ 0		
Penalty cost		40 40$_2$	5 5 5 25$_4$	15 15 15 5	20 20 20 20	0		

This initial allocation is basic if there are $(m + n - 1) = (3 + 5 - 1) = 7$ allocations. We can see from the tableau that this is the case.

Test for optimality using the MODI method. For the allocated cells the row, u, and column, v, components are calculated. For the empty cells, the shadow costs are found. This information is given in the tableau below. The shadow costs are ringed.

To store

From warehouse	P	Q	R	S	*Dummy*	
A	70 27	85 (−15)	55 13	120 (+45)	0 (−10)	$u_1 = 0$
B	110 (+50)	90 17	75 (+30)	110 (+45)	0 23	$u_2 = -10$
1	115 (+30)	115 8	70 17	90 35	0 (−25)	$u_3 = 15$
	$v_1 = 70$	$v_2 = 100$	$v_3 = 55$	$v_4 = 75$	$v_5 = 10$	

Three of the shadow costs are negative, therefore, the allocation is not optimum. We must re-allocate. The stepping stone circuit for cell (1, Dummy) is given below.

	Q	R	S	*Dummy*
B	90 17 +	75	110	0 23 −
1	115 8 −	70	90	0 + test

The minimum value in the two − cells is 8, therefore, 8 loads are moved around this circuit. The new allocation is:

To store

From warehouse	P	Q	R	S	*Dummy*	
A	70 / 27	85 / (+10)	55 / 13	120 / (+45)	0 / (+15)	$u_1 = 0$
B	110 / (+25)	90 / 25	75 / (+5)	110 / (+20)	0 / 15	$u_2 = 15$
1	115 / (+30)	115 / (+25)	70 / 17	90 / 35	0 / 8	$u_3 = 15$
	$v_1 = 70$	$v_2 = 75$	$v_3 = 55$	$v_4 = 75$	$v_5 = -15$	

All of the shadow costs are now positive, therefore, this allocation is optimum. The total cost is:

$$£(27 \times 70 + 13 \times 55 + 25 \times 90 + 17 \times 70 + 35 \times 90) = £9195$$

The analysis must now be repeated with the second warehouse. The procedure is identical. The tableau below gives the initial allocation, using Vogel's method, the row and column components and the shadow costs.

To store

From warehouse	P	Q	R	S	*Dummy*	
A	70 / 27	85 / (+15)	55 / 13	120 / (+70)	0 / (+25)	$u_1 = 0$
B	110 / (+20)	90 / 23	75 / 17	110 / (+40)	0 / (+5)	$u_2 = 20$
2	135 / (+40)	95 / 2	80 / (0)	75 / 35	0 / 23	$u_3 = 25$
	$v_1 = 70$	$v_2 = 70$	$v_3 = 55$	$v_4 = 50$	$v_5 = -25$	

All of the shadow costs are positive or zero, therefore, the allocation is optimum. Since the shadow cost in cell (2,R) is actually zero, we know that there is an alternative optimum allocation. We can find out what this is by looking at the stepping stone circuit which includes cell (2,R).

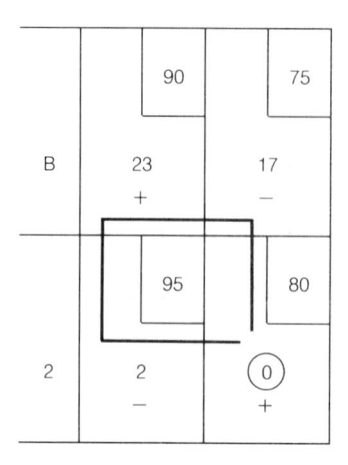

		90	75
B		23 +	17 −
		95	80
2		2 −	0 +

The minimum value in the − cell is 2, therefore if 2 is moved around this circuit, we will find the alternative optimum allocation (see the tableau below).

To store

		P	Q	R	S	*Dummy*
		70	85	55	120	0
From warehouse	A	27	—	13	—	—
		110	90	75	110	0
	B	—	25	15	—	—
		135	95	80	75	0
	2	—	—	2	35	23

The total cost for each of these two allocations is £8765. Therefore, warehouse 2 is the cheaper one to build. The optimum allocation is: warehouse A sends 27 loads per day to store P and 13 loads per day to store R; warehouse B sends 23 loads per day to store Q and 17 loads per day to store R; warehouse 2 sends 2 loads per day to store Q and 35 loads per day to store S; and warehouse 2 has spare capacity for a further 23 loads per day.

Alternatively, warehouse B sends 25 loads per day to store Q and 15 loads per day to store R; warehouse 2 sends 2 loads per day to store R and 35 loads per day to store S. All other allocations remain unchanged.

Exercise 10.5

1 The problem satisfies the requirements of a transportation problem, but the total tonnage available > total tonnage required, therefore, a dummy origin is required.

$$\text{Total tonnage available} = 6500$$
$$\text{Total tonnage required} = 6000$$

Therefore, the dummy freezer plant requires 500 tonnes. We are required to maximise contribution, therefore, we must calculate the contribution per tonne for each combination of farm and plant.

$$\text{Contribution/tonne} = \text{revenue/tonne} - \text{costs/tonne}$$
$$= 200 - (\text{variable costs for farm} + \text{variable costs at plant} + \text{transport costs})$$

Since we are planning for next season, it is assumed that unwanted beans are not grown, therefore, tonnage allocated between a farm and the dummy plant incurs no costs. The unit contributions are shown in the first tableau.

£ /tonne

		To freezer plant			
		Craft	Liver	Dummy	Total available
From farm	Ascent Hill	80	72	0	2000
	Midrow Top	73	70	0	3000
	Alum Up	75	81	0	1500
Total required		2750	3250	500	

Since the objective is to maximise the total contribution, we multiply the unit contributions by -1 before beginning the transportation algorithm.

Vogel's method will be used to find the first allocation. The subscripts show the order in which the penalties are used and the allocations are made.

		To freezer plant				
		Craft	Liver	Dummy	Total available	Penalty costs
From farm	Ascent Hill	−80 / 2000₂	−72 / —	0 / —	~~2000~~ 0	8 8₂
	Midrow Top	−73 / 750	−70 / 1750	0 / 500	~~3000~~ ~~1250~~ ~~500~~ 0	3 3
	Alum Up	−75 / —	−81 / 1500₁	0 / —	~~1500~~ 0	6
Total required		~~2750~~ ~~750~~ 0	~~3250~~ ~~1750~~ 0	~~500~~ 0		
Penalty costs		5 / 7	9₁ / 2	0 / 0		

This initial allocation is basic if there are $(m + n - 1) = (3 + 3 - 1) = 5$ allocations. We can see from the tableau that this is the case.

Test for optimality using the MODI method. For the allocated cells the row, u, and column v, components are calculated. For the empty cells, the shadow costs are found. This information is given in the tableau below. The shadow costs are ringed.

		Craft	Liver	Dummy	
From farm	AH	−80 2000	−72 (+5)	0 (+7)	$u_1 = 0$
	MT	−73 750	−70 1750	0 500	$u_2 = 7$
	AU	−75 (+9)	−81 1500	0 (+11)	$u_3 = −4$
		$v_1 = −80$	$v_2 = −77$	$v_3 = −7$	

All of the shadow costs are positive, therefore, we have the optimum allocation.

Ascent Hill farm sends 2000 tonnes of beans to the Craft plant. Midrow Top farm sends 750 tonnes to Craft plant and 1750 tonnes to Liver plant. The farm will not use 500 tonnes of capacity. Alum Up farm sends 1500 tonnes to Liver plant. The total contribution is:

$$£(2000 \times 80 + 750 \times 73 + 1750 \times 70 + 1500 \times 81) = £458,750$$

2   If Midrow Top is a centre of excellence for bean production, then it should produce at its maximum capacity. The 500 tonnes which are currently allocated between Midrow Top and the dummy should be moved to one of the other dummy cells, with a consequent increase in production at Midrow Top.

This will happen if the costs at Midrow Top reduce sufficiently. Let the decrease in cost be £c per tonne. The contribution will then increase by £c per tonne. The tableau is:

		Craft	Liver	Dummy	
From farm	AH	−80 2000	−72 (+5)	0 (+7 − c)	$u_1 = 0$
	MT	−73 − c 750	−70 − c 1750	0 500	$u_2 = 7 − c$
	AU	−75 (+9)	−81 1500	0 (+11 − c)	$u_3 = −4$
		$v_1 = −80$	$v_2 = −77$	$v_3 = −7 + c$	

The allocation will change when the shadow cost of (AH, dummy) or (AU, dummy) becomes negative. If c takes the value £7, then the shadow cost of (AH, dummy) is zero and we could transfer the surplus 500 tonnes around the following stepping stone circuit:

	Craft	Liver	Dummy
AH	−80 2000 −	−72	0 0 +
MT	−73 − 7 750 +	−70 − 7 1750	0 500 −

The allocation then becomes:

		Craft	Liver	Dummy	
From farm	AH	−80 1500	−72 (+5)	0 500	$u_1 = 0$
	MT	−80 1250	−77 1750	0 (0)	$u_2 = 0$
	AU	−75 (+9)	−81 1500	0 (+4)	$u_3 = -4$
		$v_1 = -80$	$v_2 = -77$	$v_3 = 0$	

To freezer plant

The variable cost at Midrow Top farm must decrease by £7 per tonne to £88 per tonne before this farm is required to produce at its maximum capacity.

Exercise 10.6

The problem satisfies all of the requirements for the assignment problem. We begin the algorithm by subtracting the smallest row element from all of the elements in that row.

568 QUANTITATIVE ANALYSIS

Time per job (hours)

		Job 1	Job 2	Job 3	Job 4	Job 5	Smallest element
	M1	25	16	15	14	13	13
	M2	25	17	18	23	15	15
Employee	M3	30	15	20	19	14	14
	M4	27	20	22	25	12	12
	M5	29	19	17	32	10	10

Subtract the smallest row element from all of the elements in that row.

	Job 1	Job 2	Job 3	Job 4	Job 5
M1	12	3	2	1	0
M2	10	2	3	8	0
M3	16	1	6	5	0
M4	15	8	10	13	0
M5	19	9	7	22	0
	10	1	2	1	0 ← smallest element

Subtract the smallest column elements and make assignments to the zeros.

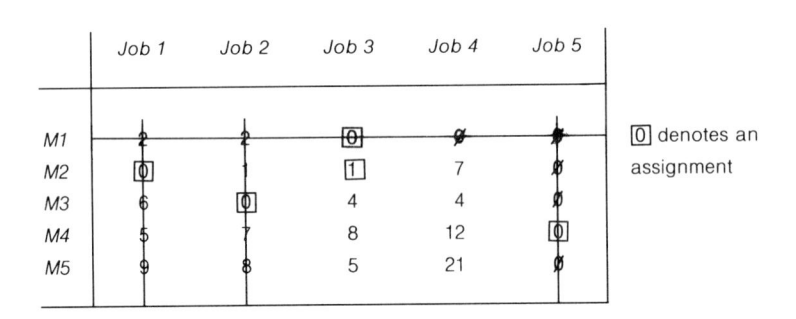

	Job 1	Job 2	Job 3	Job 4	Job 5
M1	2	2	[0]	0	0
M2	[0]	1	1	7	0
M3	6	[0]	4	4	0
M4	5	7	8	12	[0]
M5	9	8	5	21	0

[0] denotes an assignment

It has been possible to make only 4 assignments, therefore, the solution is not feasible.

The next step is to draw the minimum number of straight lines through all of the zeros. See the table above. The smallest number without a line is 1. Subtract 1 from all elements without a line through them and add 1 to all elements at the intersections of two lines. Make new assignments to zeros.

	Job 1	Job 2	Job 3	Job 4	Job 5
M1	3	3	[0]	0	1
M2	[0]	0	0	6	0
M3	5	[0]	3	3	0
M4	5	7	7	11	[0]
M5	9	8	4	20	0

[0] denotes an assignment

The solution is still not feasible, therefore we must repeat the procedure of drawing lines through the zeros. The smallest number without a line is 3. The new tableau is:

	Job 1	Job 2	Job 3	Job 4	Job 5
~~M1~~	~~3~~	~~6~~	~~[0]~~	~~∅~~	~~4~~
~~M2~~	~~[0]~~	~~4~~	~~∅~~	~~6~~	~~3~~
~~M3~~	~~2~~	~~[0]~~	~~∅~~	~~∅~~	~~∅~~
M4	2	7	4	8	[0]
M5	6	8	[1]	17	0

[0] denotes an assignment

The solution is still not feasible, therefore, we must do another iteration. The smallest number without a line is 1. The new tableau is:

	Job 1	Job 2	Job 3	Job 4	Job 5
M1	3	6	∅	[0]	5
M2	[0]	4	∅	6	4
M3	2	[0]	∅	∅	1
M4	1	6	3	7	[0]
M5	5	7	[0]	16	∅

[0] denotes an assignment

This allocation is feasible, therefore, it is the optimum. M1 is assigned to Job 4; M2 is assigned to Job 1; M3 is assigned to Job 2; M4 is assigned to Job 5 and M5 is assigned to Job 3. The total job time is:

$$(25 + 15 + 17 + 14 + 12) = 83 \text{ hours}$$

Exercise 10.7

When the part-timer is added, the tableau is no longer square. A dummy job must be created. It is assumed that this job takes zero time for all of the joiners.

		\multicolumn{5}{c	}{Time per job (hours)}	Dummy	Smallest
		Job 1	Job 2	Job 3	Job 4	Job 5	Job	element
Employee	M1	25	16	15	14	13	0	0
	M2	25	17	18	23	15	0	0
	M3	30	15	20	19	14	0	0
	M4	27	20	22	25	12	0	0
	M5	29	19	17	32	10	0	0
	M6	28	16	19	16	15	0	0

Subtract the smallest row element from all of the elements in that row. Which does not create any new zeros this time.

		Job 1	Job 2	Job 3	Job 4	Job 5	Dummy Job
Employee	M1	25	16	15	14	13	0
	M2	25	17	18	23	15	0
	M3	30	15	20	19	14	0
	M4	27	20	22	25	12	0
	M5	29	19	17	32	10	0
	M6	28	16	19	16	15	0
		25	15	15	14	10	0 ← smallest element

Make assignments to the zeros.

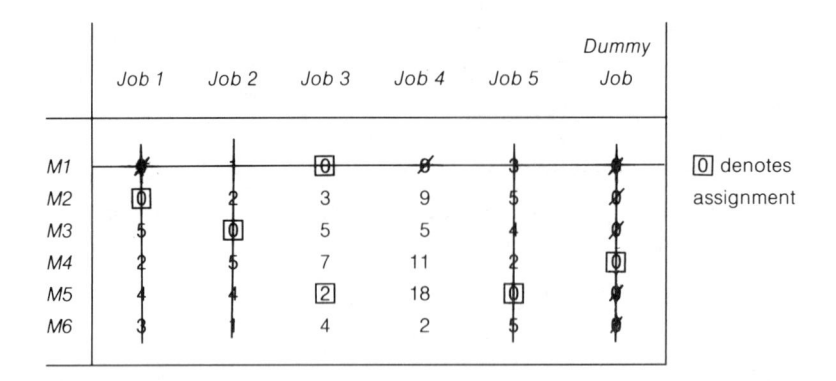

	Job 1	Job 2	Job 3	Job 4	Job 5	Dummy Job
M1	4	4	[0]	∅	3	4
M2	[0]	2	3	9	5	∅
M3	5	[0]	5	5	4	∅
M4	2	5	7	11	2	[0]
M5	4	4	[2]	18	[0]	∅
M6	3	1	4	2	5	∅

[0] denotes assignment

It has been possible to make only 5 assignments, therefore, the solution is not feasible.

The next step is to draw the minimum number of straight lines through all of the zeros. The smallest number without a line is 2. Subtract 2 from all elements without a line through them and add 2 to all elements at the intersections of two lines. Make new assignments to zeros.

	Job 1	Job 2	Job 3	Job 4	Job 5	Dummy Job
M1	2	3	[0]	∅	5	2
M2	[0]	2	1	7	5	∅
M3	5	[0]	3	3	4	∅
M4	2	5	5	9	2	[0]
M5	4	4	∅	16	[0]	∅
M6	3	1	2	[0]	5	∅

[0] denotes assignment

This allocation is feasible, therefore it is the optimum. M1 is assigned to Job 3; M2 is assigned to Job 1; M3 is assigned to Job 2; M4 is assigned to the dummy; M5 is assigned to Job 5 and the part-timer is assigned to Job 4.

The total job time is:

$$(15 + 25 + 15 + 10 + 16) = 81 \text{ hours}$$

Exercise 10.8

Look at the initial tableau which was set up in the text. The total production of the 4 month period is greater than the total demand, therefore, a dummy monthly demand is required. Since items allocated to the dummy represent goods which are not required, then we assume that these items will not be made and the costs are zero.

SOLUTIONS TO EXERCISES: CHAPTER TEN

Production schedule—cost £ /item

		M1	M2	M3	M4	Dummy	Total available
Stock	S1	2	4	6	8	∞	50
production	M1	100	102	104	106	0	300
	M2	∞	100	102	104	0	350
	M3	∞	∞	100	102	0	325
	M4	∞	∞	∞	100	0	375
Total demand		300	275	400	300	125	

Rather than using Vogel's method to make the first allocation, we will guess a sensible loading. We will use the stock, then produce as much as possible in the month when it is needed. This gives the allocation below.

		M1	M2	M3	M4	Dummy	Total available
Stock	S1	2	4	6	8	∞	~~50~~ 0
		50	—	—	—	—	
Production	M1	100	102	104	106	0	~~300~~ ~~50~~ 0
		250	—	—	—	50	
	M2	∞	100	102	104	0	~~350~~ ~~75~~ 0
		—	275	75	—	—	
	M3	∞	∞	100	102	0	~~325~~ 0
		—	—	325	—	—	
	M4	∞	∞	∞	100	0	~~375~~ ~~75~~ 0
		—	—	—	300	75	
Total demand		~~300~~ ~~250~~ 0	~~275~~ 0	~~400~~ ~~75~~ 0	~~300~~ 0	~~125~~ ~~75~~ 0	

For a basic solution there should be $(5 + 5 - 1) = 9$ allocations. We have only 8 allocations, therefore, the solution is degenerate. It will be necessary to make 1 zero allocation during the test for optimality.

The following tableau (see page 572) shows the MODI row and column cost components, the zero allocations and the shadow costs of the empty cells. The shadow costs are ringed.

		M1	M2	M3	M4	Dummy	
Stock	S1	2	4	6	8	∞	$u_1 = 0$
		50	(+6)	(+6)	(+6)	(−98)	
Production	M1	100	102	104	106	0	$u_2 = 98$
		250	(+6)	(+6)	(+6)	50	
	M2	∞	100	102	104	0	$u_3 = 102$
		—	275	75	[0]	(−4)	
	M3	∞	∞	100	102	0	$u_4 = 100$
		—	—	325	(0)	(−2)	
	M4	∞	∞	∞	100	0	$u_5 = 98$
		—	—	—	300	75	
		$v_1 = 2$	$v_2 = -2$	$v_3 = 0$	$v_4 = 2$	$v_5 = -98$	

There are two negative shadow costs, therefore, the allocation may not be optimal. For both (M2, Dummy), −4, and (M3, Dummy), −2, the stepping stone circuits involve taking out of (M2, M4), which contains the zero allocation. As this is not possible, the allocation is optimum. The production schedule is:

Month 1: 250 production + 50 stock = 300 demand (50 not produced).
Month 2: 350 production = 275 demand + 75 stock, used in month 3.
Month 3: 325 production + 75 stock = 400 demand.
Month 4: 300 production = 300 demand (75 not produced).

There is no stock at the end of month 4. The total costs, excluding the cost of producing the initial 50 stock, is £122,750.

Exercise 10.9

Refer to section **10.3.2** for the initial reduction of the tableau. We will begin by making the initial assignments to zeros.

	I	II	III	IV	V	VI
A	8	11	6	[0]	8	14
B	[0]	3	3	0	2	4
C	3	7	3	0	20	18
D	6	11	4	8	[0]	17
E		[0]	0	3		0
F	0	9	[1]	2	0	4

SOLUTIONS TO EXERCISES: CHAPTER TEN

The solution is not feasible. We draw lines through all of the zeros. The smallest number without a line is 1. The extra zeros are created in the usual way. The new tableau is:

	I	II	III	IV	V	VI
A	8	10	5	[0]	8	13
B	[0]	2	2	0	2	3
C	3	6	[2]	0	20	17
D	6	10	3	8	[0]	16
E	2	[0]	0	4	2	0
F	0	8	[0]	2	2	3

The solution is still not feasible. The iterative procedure is repeated. The smallest number without a line through it is 2.

	I	II	III	IV	V	VI
A	6	8	3	[0]	8	11
B	[0]	[2]	2	2	4	3
C	1	4	[0]	0	20	15
D	4	8	1	8	[0]	14
E	2	[0]	0	6	4	0
F	0	8	0	4	2	3

This solution is not feasible. Repeat, using the smallest unlined number, 2.

	I	II	III	IV	V	VI
A	6	6	3	[0]	8	9
B	0	[0]	2	2	4	1
C	1	2	[0]	0	20	13
D	4	6	1	8	[0]	12
E	4	0	2	8	6	[0]
F	[0]	6	0	4	2	1

The solution is feasible, therefore it is optimum. Salesman A goes to area IV; salesman B goes to area II; salesman C goes to area III; salesman D goes to area V; salesman E goes to area VI and salesman F goes to area I. The maximum value of sales is:

$$(83 + 60 + 40 + 53 + 70 + 65) = £371,000$$

Exercise 10.10

1 Five mines supply three preparation plants.

$$\text{The total mines output per day} = 650 \text{ tonnes/day}$$
$$\text{The total preparation plant capacity} = 700 \text{ tonnes/day}$$

Therefore introduce a dummy mine to indicate which plant will not be fully used.
The unit costs for each combination of mine and preparation plant comprise:

unit variable production cost at the mine
+ unit operating cost at the plant
+ unit transport cost

The values are shown in the following tableau. Vogel's penalty cost method is used to find the first allocation and the MODI method to test for optimality.

		To prep plant			Tons/day available	Penalty costs	MODI
		A	B	C			
	1	38	37	50 (+16)	~~120~~ ~~70~~ 0	1 12₅	$u_1 = 0$
		70₅	50₄				
	2	53 (+12)	40	48 (+11)	~~150~~ 0	8₃	$u_2 = 3$
			150₃				
From mine	3	49	49 (+1)	45	~~80~~ 0	4	$u_3 = 11$
		40		40			
	4	54 (+13)	50 (+10)	37	~~160~~ 0	13₂	$u_4 = 3$
				160₂			
	5	42	55 (+14)	42 (+4)	~~140~~ 0	0	$u_5 = 4$
		140					
Dummy		0	0 (+1)	0 (+4)	~~50~~ 0	0	$u_6 = -38$
		50₁					
Tons/day required		~~300~~ ~~250~~ ~~180~~ ~~40~~ 0	~~200~~ ~~50~~ 0	~~200~~ ~~40~~ 0			
Penalty costs		38₁ 4 7	37 3 12₄	37 5 3			
		$v_1 = 38$	$v_2 = 37$	$v_3 = 34$			

There must be (m + n − 1) = 8 entries for a basic solution. There are 8 entries. All the shadow costs are positive, therefore, this is the optimum allocation.

Mine 1 supplies 70 tonnes per day to A and 50 to B; mine 2 supplies 150 tonnes per day to B; mine 3 supplies 40 tonnes per day to A and 40 to C; mine 4 supplies 160 tonnes per day to C; and mine 5 supplies 140 tonnes per day to A.

Preparation plant A has 50 tonnes per day spare capacity even though it has the cheapest operating costs. The total costs of the above allocation are:

$$70 \times 38 + 50 \times 37 + 150 \times 40 + 40 \times 49 + 40 \times 45 + 160 \times 37 + 140 \times 42 = \$26{,}070/\text{day}$$

2. Production costs at Mine 3 fall from $34 to $30 per tonne. All mine output is already taken by the plants and production costs are like a fixed cost and do not affect the allocation, therefore total cost will be reduced by 80 × 4 = $320 per day.

SOLUTIONS TO EXERCISES: CHAPTER TEN   575

3 Mine 5 plans to increase output by 40 tonnes per day from 140 to 180. All of Mine 5's output is allocated to Plant A which has 50 tonnes per day spare capacity. The extra 40 tonnes per day output will go from Mine 5 to Plant A, increasing costs by 40 × 42 = $1,680 per day.

Exercise   10.11

1 Describe two methods of finding an initial feasible solution. We describe two methods in section **10.2.3**. These are the minimum cost and Vogel's penalty cost methods.

2 Braintree Electronics—video tapes. There are six sources of production: October normal (400), October overtime (150), November normal (400), November overtime (150), December normal (400) and December overtime (150). Total capacity is 1650 ('00 tapes) and total demand is 1550 ('00 tapes). Therefore, we need a dummy demand column. The costs in the tableau are production and inventory costs.

	(£/'00 video tapes) To demand				Total	
	October	November	December	Dummy	capacity	MODI
Oct: normal	150 / 300	170 / 50	190 / 50	0 / (+30) —	400 100 50 0	$u_1 = 0$
overtime	180 / (0) —	200 / (0) —	220 / 50	0 / 100	150 100 0	$u_2 = 30$
Nov: normal	∞ / —	150 / 400	170 / (0) —	0 / (+50) —	400 0	$u_3 = -20$
overtime	∞ / —	180 / —	200 / 150	0 / (+20) —	150 0	$u_4 = 10$
Dec: normal	∞ / —	∞ / —	150 / 400	0 / (+70) —	400 0	$u_5 = -40$
overtime	∞ / —	∞ / —	180 / 150	0 / (+40) —	150 0	$u_6 = -10$
Total demand	300 / 0	450 / 50 0	800 400 250 / 200 50 0	100 / 0	1650	
MODI	$v_1 = 150$	$v_2 = 170$	$v_3 = 190$	$v_4 = -30$		

576 QUANTITATIVE ANALYSIS

The initial allocation was done using the minimum cost method. There are 9 entries in the table giving a basic solution. The MODI method is used to test for optimality.

All the shadow costs are positive or zero, therefore, this allocation will give the minimum cost. The total cost is:

$$300 \times 150 + 50 \times 170 \text{ etc} = £251{,}000$$

The zero shadow costs mean that there are other allocations (found using the stepping stone method) that will give this same cost.

Exercise 10.12

1 The sources can be used to supply the destinations like any transportation model, except that all orders must be satisfied on time. The tableau has forbidden cells because February cannot be used to supply January demand and March cannot be used to supply January or February. The costs are:

$$\text{Normal } £80 \quad + £4 \text{ per month, if held in stock}$$
$$\text{Overtime } £90 + £4 \text{ per month, if held in stock}$$

The initial 5000 stock is assumed to have been in stock already for half a month and, therefore, has incurred its production cost plus £2 for half a month of its stocking costs. If it is used in January the cost will therefore be £2 for the extra half month, increasing by £4 per month if it is used later. The tableau is given opposite on page 577, with the required dummy demand. Use the minimum cost method to do the initial allocation.

2 The allocation shown in the tableau is feasible. To test the solution for optimality, we must first determine whether it is basic. For a basic solution there should be:

$$(\text{rows} + \text{columns} - 1) = (7 + 5 - 1) = 11 \text{ allocations}$$

In this case there are only 10 allocations. The solution is degenerate and a zero allocation must be added to an empty cell.

If the MODI procedure is used, the costs in the allocated cells are split into row and column components. These components are then used to calculate the shadow costs for the unallocated cells. If all of the shadow costs are positive or zero, then the solution is optimum. The procedure has been carried out in the tableau above. The required zero allocation was made to cell (January (overtime), March). We can see that the solution is optimum.

Interpretation of the solution:

December stock supplies 5000 units to January demand.
January (normal) production supplies 21,000 units to January demand.
January (normal) production supplies 3000 units to February demand.
February (normal) production supplies 24,000 units to February demand.
February (overtime) production supplies 1000 units to February demand.
February (overtime) production supplies 5000 units to finished goods stock.
March (normal) production supplies 24,000 units to March demand.
March (overtime) production supplies 1000 units to March demand.
March (overtime) production supplies 5000 units to finished goods stock.
The total cost of the solution is £6,902,000.

There are zero shadow costs in some cells. This means that there are other optimum allocations which use these cells.

SOLUTIONS TO EXERCISES: CHAPTER TEN

To demand and stock ('000s)

From ('000s)		Jan	Feb	Mar	Stock end Mar	Dummy	Total available	MODI
	Stock	2 / 5	6 / (+6)	10 / (+10)	12 / (+12)	∞ / —	5 0	$u_1 = 0$
	Jan (normal)	80 / 21	84 / 3	88 / (0)	90 / (0)	0 / (+10)	24 3 0	$u_2 = 78$
	Jan (overtime)	90 / (0)	94 / (0)	98 / (0)	100 / (0)	0 / 6	6 0	$u_3 = 88$
	Feb (normal)	∞ / —	80 / 24	84 / (0)	86 / (0)	0 / (+14)	24 0	$u_4 = 74$
	Feb (overtime)	∞ / —	90 / 1	94 / (0)	96 / 5	0 / (+4)	6 5 0	$u_5 = 84$
	Mar (normal)	∞ / —	∞ / —	80 / 24	82 / (0)	0 / (+18)	24 0	$u_6 = 70$
	Mar (overtime)	∞ / —	∞ / —	90 / 1	92 / 5	0 / (+8)	6 5 0	$u_7 = 80$
Total required		26 / 21 0	28 / 4 1 0	25 / 1 0	10 / 5 0	6 / 0	95	
MODI		$v_1 = 2$	$v_2 = 6$	$v_3 = 10$	$v_4 = 12$	$v_5 = -88$		

3 If orders can be supplied late, eg produced in February for January orders, the ∞ costs in the forbidden cells will be replaced by production costs + £5/month/unit:

	Jan	Feb
Feb (normal)	85	
Feb (overtime)	95	
Mar (normal)	90	85
Mar (overtime)	100	95

578 QUANTITATIVE ANALYSIS

Exercise 10.13

1. An assignment problem. We have 6 aircraft to be assigned to 5 journeys and 1 dummy journey, ie not used. The costs are:

   Total cost = fixed cost/day + (load × variable cost/ton)
   + 2 × (distance × variable cost/mile)   £/day

   Therefore, for example, using aircraft Type A on journey 1, the cost per day is:

   $800 + 10 \times 30 + 2 \times 200 \times 0.6 = £1,340/\text{day}.$

   Notice that load 1 can be carried by Aircraft A only because loads cannot be split and A is the only type with a large enough capacity. Loads 3 and 4 can only be carried by A and B for the same reasons.

2. The assignment tableau is given below. The dummy carries the fixed cost per day of the unused aircraft.

		\multicolumn{6}{c}{Load}	Smallest row element
		1	2	3	4	5	Dummy	
Aircraft	A1	1340	1520	1424	1286	1430	800	800
	A2	1340	1520	1424	1286	1430	800	800
	A3	1340	1520	1424	1286	1430	800	800
	B1	∞	1210	1236	1099	1165	700	700
	B2	∞	1210	1236	1099	1165	700	700
	C1	∞	1650	∞	∞	1475	500	500

Deduct the smallest row element from each row—∞ does not change.

	1	2	3	4	5	Dummy
A1	540	720	624	486	630	0
A2	540	720	624	486	630	0
A3	540	720	624	486	630	0
B1	∞	510	536	399	465	0
B2	∞	510	536	399	465	0
C1	∞	1150	∞	∞	975	0
Smallest column element	540	510	536	399	465	0

Deduct the smallest column element from its column.

	1	2	3	4	5	Dummy
A1	[0]	210	88	87	165	0̸
A2	0̸	210	88	87	165	0̸
A3	0̸	210	88	87	165	0̸
B1	∞	[0]	0̸	0̸	0̸	0̸
B2	∞	0̸	[0]	0̸	0̸	0̸
C1	∞	640	∞	∞	510	[0]

Only four allocations have been made when six are needed. Draw lines through the zeros. The smallest unlined number is 87. Subtract this from all unlined numbers, add to numbers cut by two lines, leave numbers cut by one line.

	1	2	3	4	5	Dummy
A1	[0]	123	1	∅	78	∅
A2	∅	123	1	[0]	78	∅
A3	∅	123	1	∅	78	0
B1	∞	[0]	∅	∅	∅	87
B2	∞	∅	[0]	∅	∅	87
C1	∞	553	∞	∞	423	[0]

Only five assignments have been made when six are needed. Draw lines through the zeros. The smallest unlined number is 1. Add, subtract and ignore as before.

	1	2	3	4	5	Dummy
A1	[0]	122	∅	∅	77	∅
A2	∅	122	[0]	∅	77	∅
A3	∅	122	∅	[0]	77	0
B1	∞	[0]	∅	1	∅	88
B2	∞	∅	∅	1	[0]	88
C1	∞	552	∞	∞	422	[0]

Six assignments have been made as required. This is the optimum allocation.

		£ Cost/day
Aircraft Type A do Journey	1	1340
	3	1424
	4	1286
Aircraft Type B do Journey	2	1210
	5	1165
		6425
Aircraft Type C does nothing		500
Therefore, total cost/day		£6925

Solutions to Chapter Eleven exercises

ANSWERS TO EXERCISES

Exercise 11.1

1 The arrow diagram is:

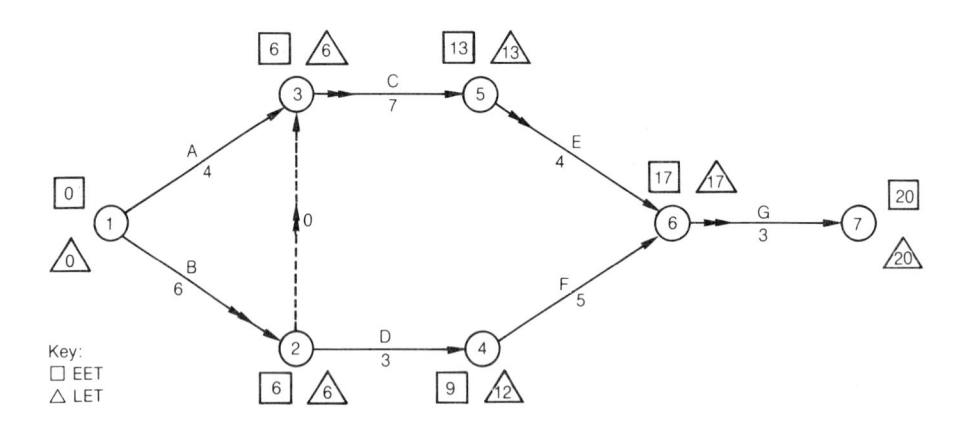

[arrow diagram with nodes 1–7, activities A(4), B(6), C(7), D(3), E(4), F(5), G(3), and dummy activity 0 from node 2 to node 3. EET values in squares: 0, 6, 6, 9, 13, 17, 20. LET values in triangles: 0, 6, 6, 12, 13, 17, 20. Key: □ EET, △ LET]

2

Node	EET, days	Comment
1	0	Assume that the project begins at time zero
2	$EET_1 + 6 = 6$	
3	$EET_1 + 4 = 4$	Choose the later time, 6.
or	$EET_2 + 0 = 6$	
4	$EET_2 + 3 = 9$	
5	$EET_3 + 7 = 13$	
6	$EET_5 + 4 = 17$	Choose the later time, 17.
or	$EET_4 + 5 = 14$	
7	$EET_6 + 3 = 20$	

The overall duration of the project is 20 days.

Node	LET, days	Comment
7	20	Assume that, for the last node, EET = LET
6	$LET_7 - 3 = 17$	
5	$LET_6 - 4 = 13$	
4	$LET_6 - 5 = 12$	
3	$LET_5 - 7 = 6$	
2	$LET_3 - 0 = 6$	Choose the earlier time, 6.
or	$LET_4 - 3 = 9$	
1	$LET_3 - 4 = 2$	Choose the earlier time. For the first node,
or	$LET_2 - 6 = 0$	EET = LET = 0

The EET and LET have been added to the arrow diagram. The critical activities have:

$$\text{EET}_{\text{start}} = \text{LET}_{\text{start}}$$

$$\text{and } \text{EET}_{\text{end}} = \text{LET}_{\text{end}}$$

$$\text{and } \text{EET}_{\text{start}} + \text{duration} = \text{EET}_{\text{end}}$$

From the diagram we can see that the critical activities are B, C, E and G.

Exercise 11.2

1 The activity-on-node diagram is:

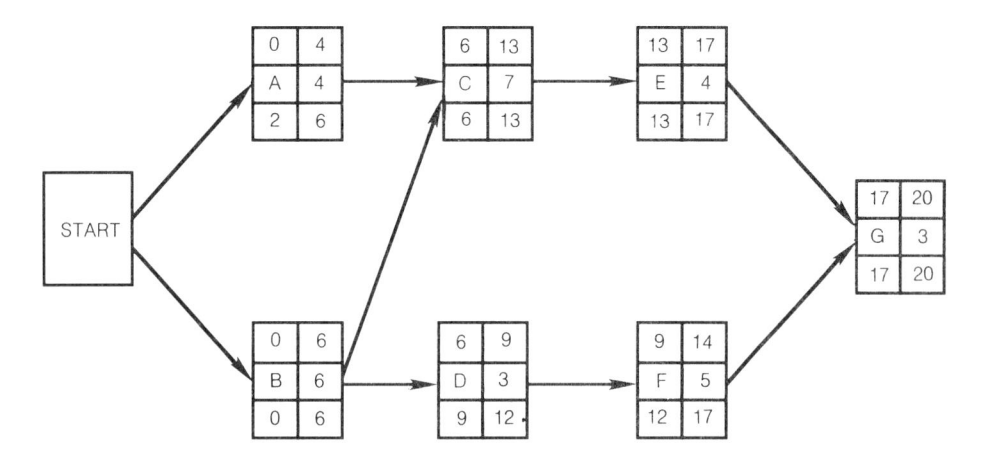

Key:
ES	EF
Act	Dur
LS	LF

2 Activity schedule:

Activity	ES, days	EF, days	Comment
A	0	$\text{ES}_A + 4 = 4$	Assume the project begins at
B	0	$\text{ES}_B + 6 = 6$	time zero.
C	$\text{EF}_A = 4$	$\text{ES}_C + 7 = 13$	Choose the later time, 6.
or	$\text{EF}_B = 6$		
D	$\text{EF}_B = 6$	$\text{ES}_D + 3 = 9$	
E	$\text{EF}_C = 13$	$\text{ES}_E + 4 = 17$	
F	$\text{EF}_D = 9$	$\text{ES}_F + 5 = 14$	
G	$\text{EF}_E = 17$	$\text{ES}_G + 3 = 20$	Choose the later time, 17
or	$\text{EF}_F = 14$		

Activity	LF, days	LS, days	Comment
G	20	$\text{LF}_G - 3 = 17$	Assume, for the last activity, that EF = LF
F	$\text{LS}_G = 17$	$\text{LF}_F - 5 = 12$	
E	$\text{LS}_G = 17$	$\text{LF}_E - 4 = 13$	
D	$\text{LS}_F = 12$	$\text{LF}_D - 3 = 9$	
C	$\text{LS}_E = 13$	$\text{LF}_C - 7 = 6$	
B	$\text{LS}_D = 9$		Choose the earlier time, 6.
or	$\text{LS}_C = 6$	$\text{LF}_B - 6 = 0$	
A	$\text{LS}_C = 6$	$\text{LF}_A - 4 = 2$	

582 QUANTITATIVE ANALYSIS

The critical activities are those for which ES = LS and EF = LF. Activity D is not critical, but if it is delayed by 4 days, then EF_D becomes 13 days, therefore EF_F becomes 18 days and ES_G is delayed to 18 days, hence the whole project is delayed by one day.

Exercise 11.3

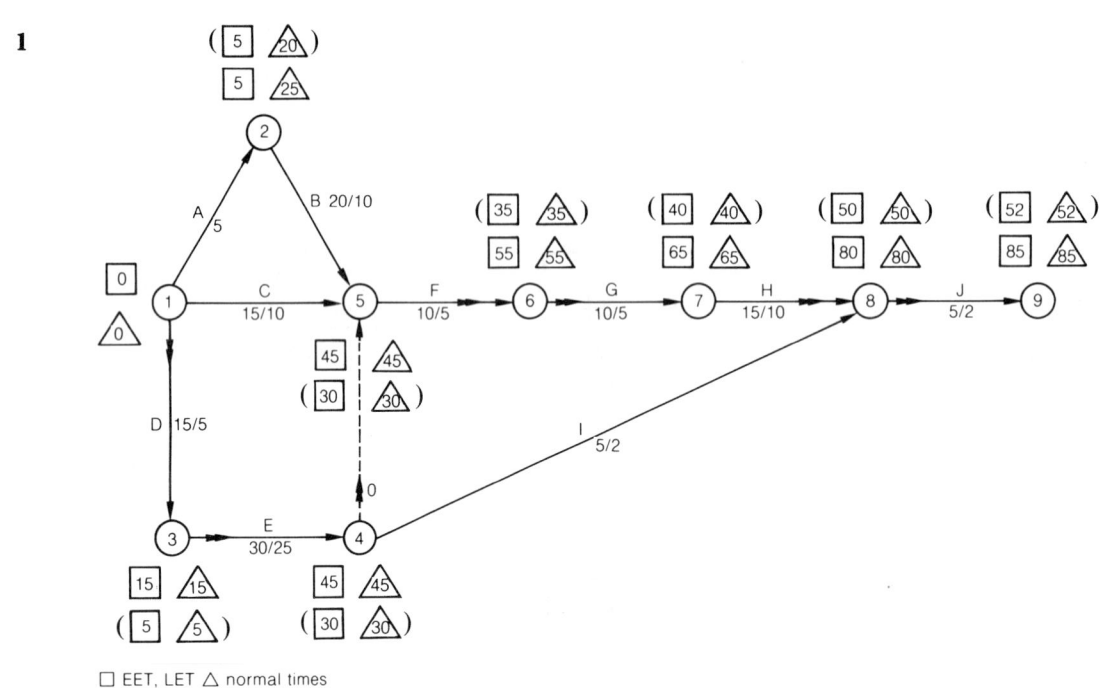

1

[network diagram with nodes 1–9, activities A 5, B 20/10, C 15/10, D 15/5, E 30/25, F 10/5, G 10/5, H 15/10, I 5/2, J 5/2; EET/LET values: node 1 (0,0); node 2 (5,25) normal, (5,20) crash; node 3 (15,15) normal, (5,5) crash; node 4 (45,45) normal, (30,30) crash; node 5 (45,45) normal, (30,30) crash; node 6 (55,55) normal, (35,35) crash; node 7 (65,65) normal, (40,40) crash; node 8 (80,80) normal, (50,50) crash; node 9 (85,85) normal, (52,52) crash]

☐ EET, LET △ normal times
(☐ EET, LET △) crash times
Normal/Crash duration days

2 Calculate the EET and LET for each node. These are shown on the diagram. We can see that the overall project completion time is 85 days. The critical activities are D, E, F, G, H and J.

3 If the duration of C increases to 30 days, there is no effect on the EET at node 5, therefore, there is no effect on the overall project duration.

4 Crash all of the activity times and re-work the EETs and LETs. These are also shown on the diagram above. The minimum completion time is 52 days. The critical path is unchanged. The additional cost incurred when all of the activities are crashed is £825. This cost can be reduced by uncrashing non-critical activities. The non-critical activities are A, B, C and I.

A—no saving possible.
B—the activity can be returned to its normal time without affecting any other activity. This saves £100.
C—the activity can be returned to its normal time without affecting any other activity. This saves £150.
I—the activity can be returned to its normal time without affecting any other activity. This saves £50.

The minimum completion time of 52 days can be achieved for a minimum cost of £(825 − (100 + 150 + 50)) = £525.

Exercise 11.4

1 Assume that the project is suitable for the application of a PERT analysis. The expected duration for each activity can be calculated from:

t = (optimistic + 4 × most likely time + pessimistic)/6

Similarly, the variance of the expected durations is:

$$\sigma^2 = ((\text{pessimistic} - \text{optimistic})/6)^2$$

Activity	Expected duration days	Variance (days)²
A	4	1/9
B	7	1
C	5	1/9
D	6	1/9
E	3	4/9
F	11	4/9
G	4	1/9
H	3	16/9

The expected times are entered on the network diagram below and the EET and LETs calculated. We can see from the diagram that the expected overall completion time is 19 days

Expected cost = cost of the activities + indirect costs + penalty
$$= 10{,}100 + 19 \times 300 + 4 \times 100 = £16{,}200$$

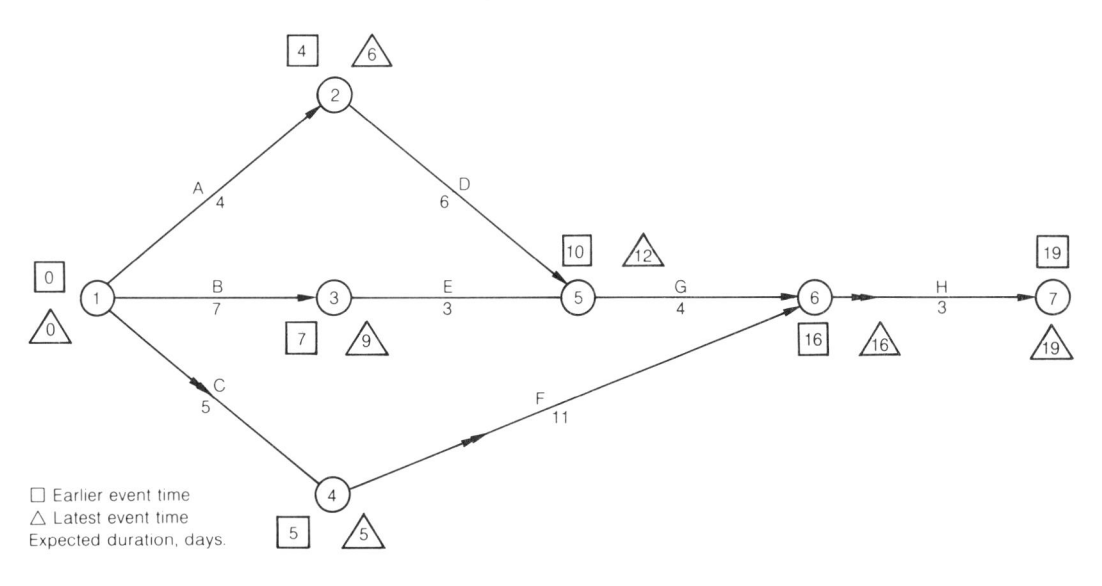

☐ Earlier event time
△ Latest event time
Expected duration, days.

2 The critical path is C–F–H. The non-critical paths are:

A–D–G–H which takes 17 days, with a variance of $2\frac{1}{9}$.
B–E–G–H which also takes 17 days, with a variance of $3\frac{1}{3}$.

The non-critical paths are not much shorter than the critical path, with a mean of 19 days and a variance of $2\frac{1}{3}$.

3 Calculate the probability that the project can be completed in less than 16 days. Assume that the expected overall completion time, T, is normally distributed with mean 19 days, which we assume is the end of the nineteenth day. This is not entirely clear. It could be the middle. The variance of the overall completion time is:

$$\sigma_T^2 = \sigma_C^2 + \sigma_F^2 + \sigma_H^2 = 1/9 + 4/9 + 16/9 = 2.33$$

Therefore,

$$\sigma_T = \sqrt{2.33} = 1.528 \text{ days}$$

584 QUANTITATIVE ANALYSIS

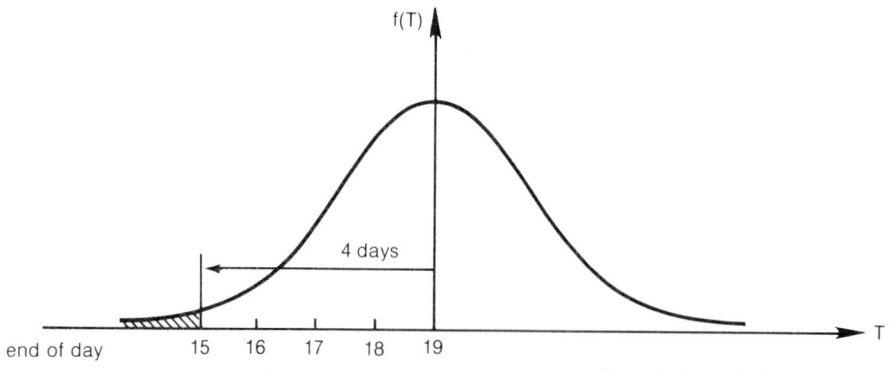

Fifteen days is z standard deviations below the mean, therefore:

$$z = \frac{15 - 19}{1.528} = -2.618$$

From the standard normal tables, $P(z < -2.618) = 0.0044$. Therefore, the probability of not incurring any penalty is 0.44%.

Exercise 11.5

1 The arrow diagram is given below. The EET and LET values have been added.

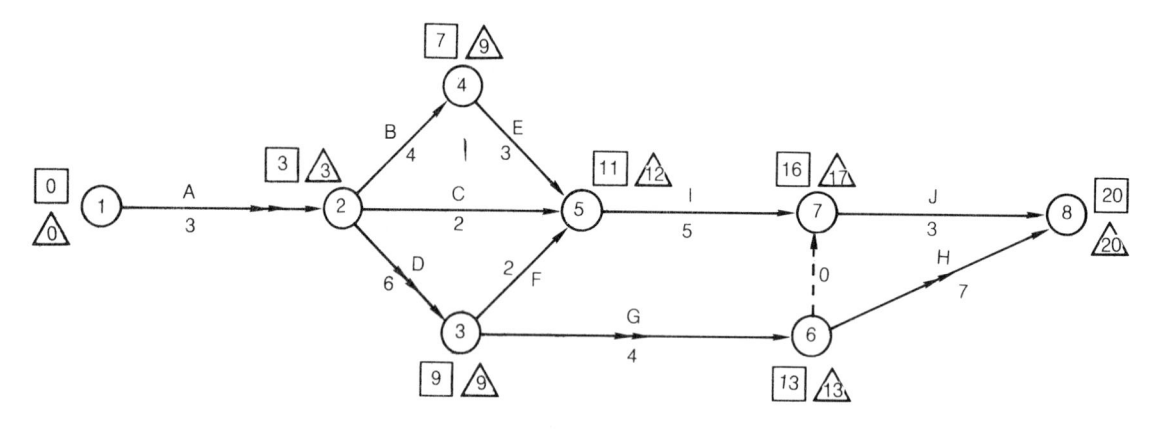

The overall duration is 20 weeks. The critical activities are A, D, G and H.

2 Gantt chart for Rogers plc—all activities at earliest times.

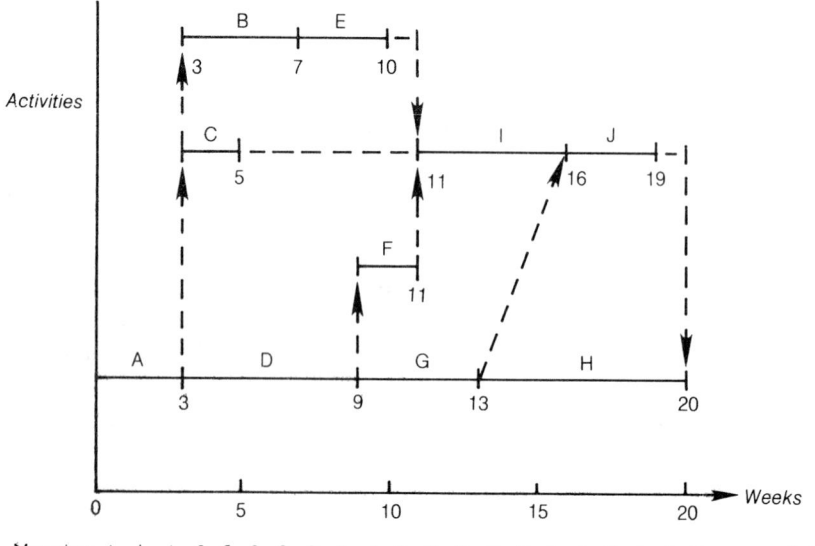

Manning 1 1 1 3 3 2 2 2 2 3 2 2 2 2 2 2 2 2 2 1 (2 men available)
            x x   x

There are 3 weeks (4, 5, and 10) when 3 men are required but only 2 are available. Week 10 can be cleared by moving J, I and then F 1 week later. This does not alter the 20 week duration. The 2 people are then fully used from week 3 onwards. The 3 people requirement in weeks 4–5 can only be reduced by increasing the total duration by 2 weeks from 20 to 22 weeks. I would be started in week 15 and C would be moved into the two weeks 13 to 14.

Exercise 11.6

(a) Arrow diagram for the preparation of year end accounts by the chief accountant of Mercury Manufacturing.

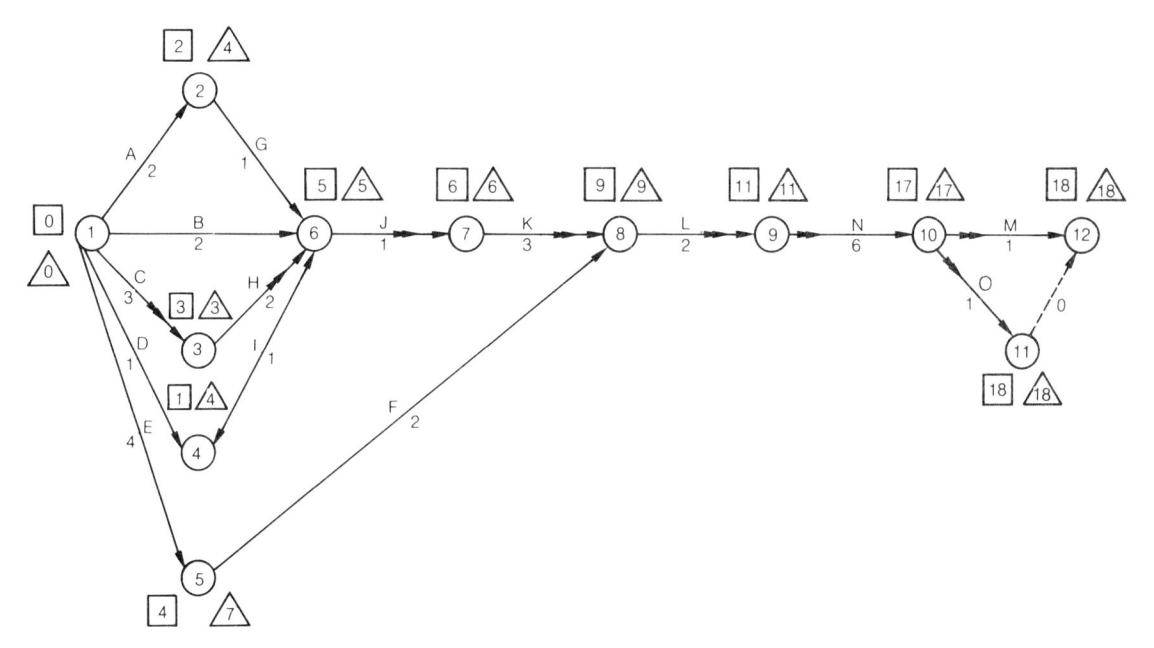

☐ Earliest Event Times
△ Latest Event Times
Duration, Weeks.

The total duration is 18 weeks. The critical activities are C–H–J–K–L–N–M–O.

(b) How many weeks before the final accounts can be submitted to the board, ie before Job O can start? 17 weeks.

(c) The board wants the final accounts approved by the end of week 13, rather than the end of week 18 as the network states. The chief accountant has suggested that Jobs A, C and D could be completed by the end of December and that E could be crashed from 4 to 2 weeks.

If time 0 on the diagram is the first week in January, the early completion of A, C, D can be represented on the diagram by giving them zero duration. Take each activity in turn.

Activity C: is on the critical path. If it is reduced from 3 to 0 weeks, the overall duration will be reduced by 2 weeks (18 to 16 weeks), because A, G, are then critical.

Activity A: is now on the new critical path. If A is reduced from 2 to 0, (C–H–J–K), (D–I–J–K) and (E–F) all become critical, 1 extra week is saved, reducing the time from 16 to 15 weeks. Activity A need only be done 1 week early.

Activity D: although D is now critical, reducing its time from 1 to 0 will have no effect on the overall duration, because of all the other critical jobs.

Activity E: the same applies to E. Any reduction in E will make it non-critical again, but will not reduce the 15 weeks.

The chief accountant's suggestions will not achieve the deadline. He will have to look at other activities, eg N.

(d) Relative advantages and disadvantages of 'activity-on-node' and 'activity-on-arrow' diagram.

The two systems do the same job in different ways. Neither is the better system. Much will depend on the computer package available and what you are used to. If you are used to looking at arrow diagrams and working out the effects of changes, you will find it difficult to get the same feel for a problem with the node diagram.

Node diagrams are easier to draw initially because they do not need logic or identification dummies. These dummies cause problems when people are learning the arrow system. The node system gives all the information about an activity in the node.

The arrow diagrams make it easier to see the effect of changes and to understand the sensitivity analyses. The representation of time by the arrow feels more logical.

Exercise 11.7

(a) Network:

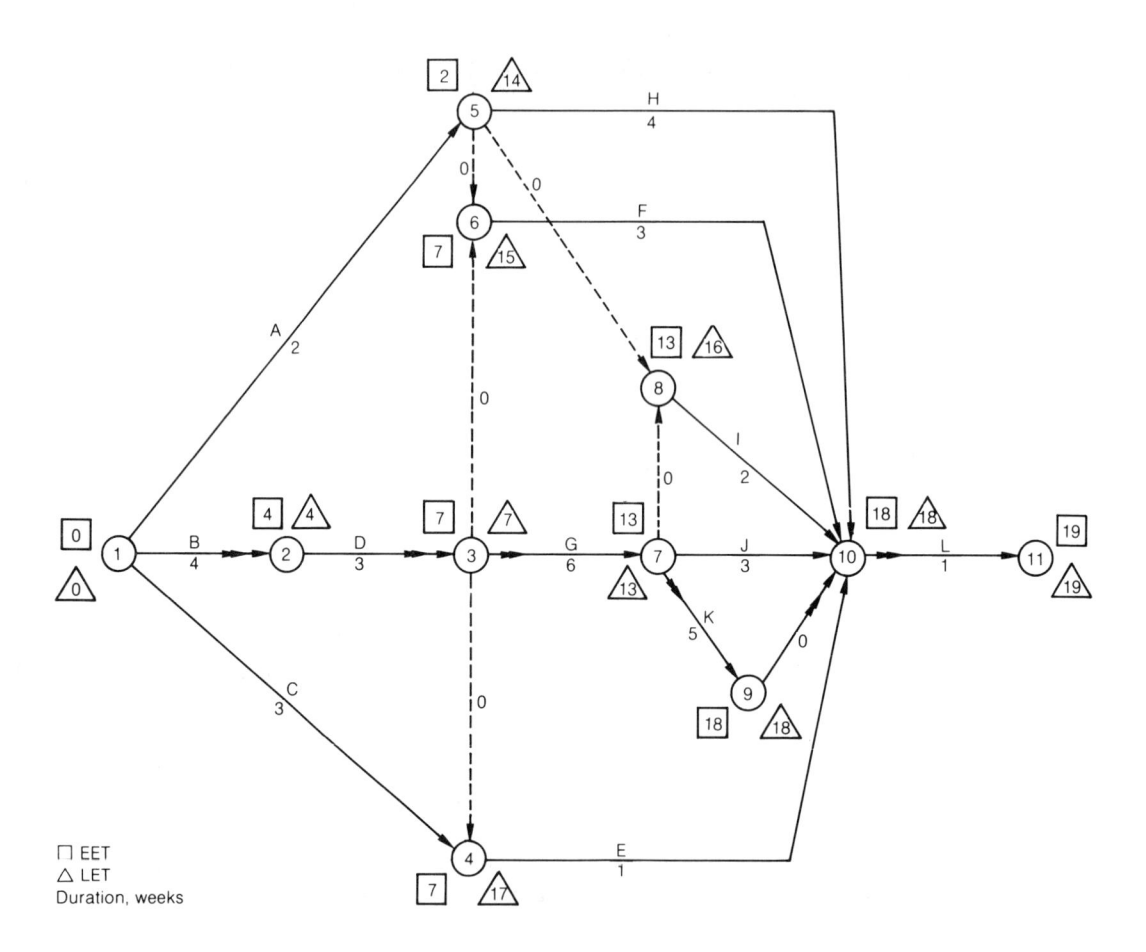

Salemis Ltd begin the preparation of budgets on 1 September.

It must be completed by the end of December—17 working weeks. The network shows that 19 weeks are required. The budget process will not be done on time.

(b) To reduce the time from 19 to 17 weeks, we should look at the critical activities, B–D–G–K–L, to see if any of these can be reduced. As this is done, we must check that no other jobs become critical and prevent the project duration being reduced.

For example, if the critical activity K is reduced below 3 weeks, activity J will become critical and prevent any further reduction in the project duration.

(c) Float: this indicates the 'spare' time on an activity. Total float is the spare time on an activity, assuming that all earlier jobs start as early as possible and all later jobs as late as possible. Free float is the spare time that has no effect on the float of later activities. Independent float is the spare time that has no effect on earlier or later activities. They are defined as:

$$\text{Total float} = \text{LET}_{end} - \text{EET}_{start} - \text{duration}$$
$$\text{Free float} = \text{EET}_{end} - \text{EET}_{start} - \text{duration}$$
$$\text{Independent float} = \text{EET}_{end} - \text{LET}_{start} - \text{duration}$$

For activity I:

$$\text{Total float} = 18 - 13 - 2 = 3 \text{ weeks}$$
$$\text{Free float} = 18 - 13 - 2 = 3 \text{ weeks}$$

Independent float:
from dummy $(5,8) = 18 - 14 - 2 = 2$ weeks
or from dummy $(7,8) = 18 - 13 - 2 = 3$ weeks.
The smaller of these is taken.

Exercise 11.8

(a) Arrow Diagram : incomplete.

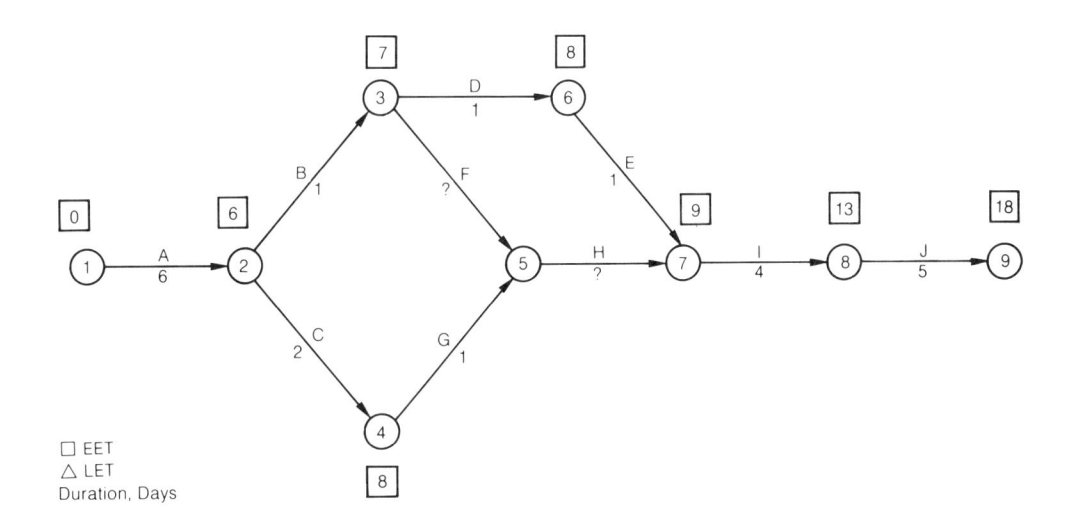

(b) What is the minimum time the project would take, assuming durations of F and H are zero? 18 days.

(c) If the project must be completed in 19 days, what are the restrictions on the durations of F and H?
Activity H: if H is 0 or 1 days, this will give a duration of $\leqslant 19$ days.
Activity F: as long as $F + H \leqslant 3$ days, the project duration is $\leqslant 19$ days.

(d) Poisson distribution: what is the probability project duration $\leqslant 19$ days, if the expected durations of F and H are 2 days and 1 day, respectively.

$$P(\text{project duration} \leqslant 19 \text{ days}) = P(H = 0) \text{ AND } P(F = 0 \text{ or } 1 \text{ or } 2 \text{ or } 3)$$
$$+ P(H = 1) \text{ AND } P(F = 0 \text{ or } 1 \text{ or } 2)$$

From the probabilities provided:

$$P(\text{project duration} \leqslant 19 \text{ days}) = 0.368 \times 0.857 + 0.368 \times 0.677 = 0.565$$

Exercise 11.9

(a) Arrow diagram:

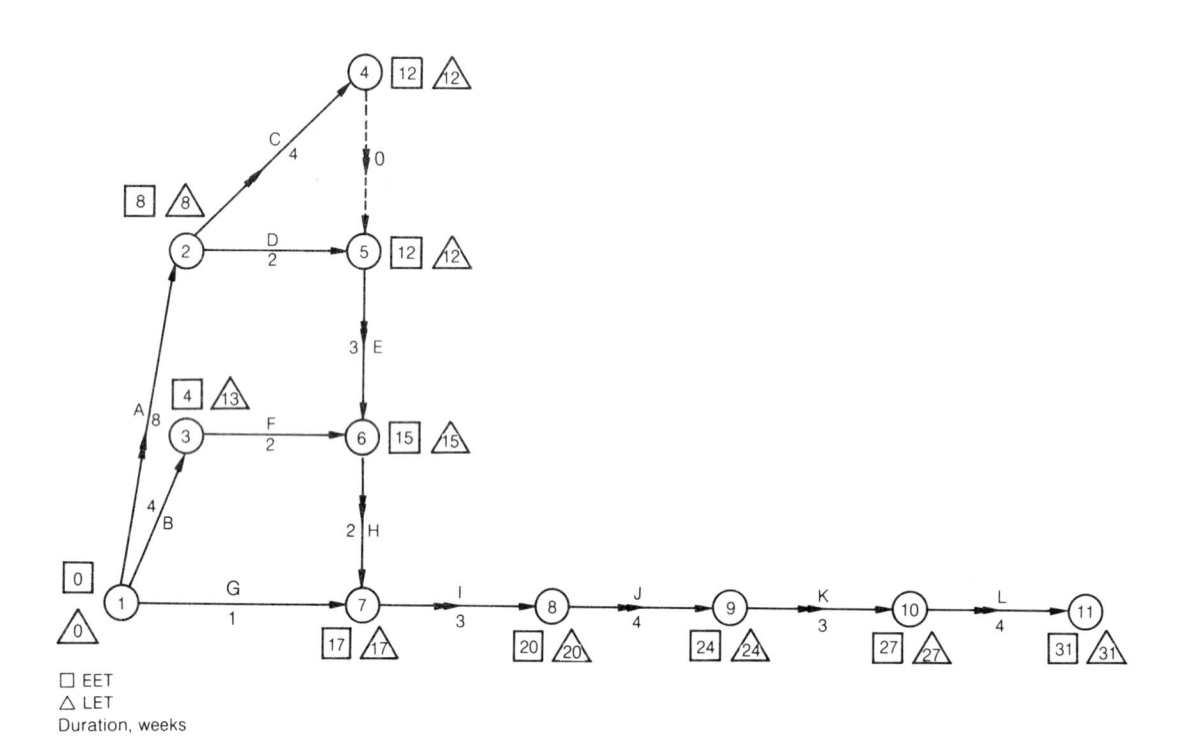

☐ EET
△ LET
Duration, weeks

The new product can be launched in 31 weeks.

(b) Floats—assume free float. The critical jobs are A, C, E, H, I, J, K, L, therefore, the non-critical jobs are B, D, F, G.

The free float = $EET_{end} - EET_{start} - Duration$

B	=	4	−	0	−	4	=	0	can be
F	=	15	−	4	−	2	=	9	shared
D	=	12	−	8	−	2	=	2	
G	=	17	−	0	−	1	=	16	

D has the smallest float, which may be relevant in the next part of the question.

(c) A, B, D, K and L have uncertain times. We are given these optimistic, pessimistic and most likely times and, assuming they follow a beta distribution, we can use the PERT formulae to calculate the expected times and variances.

Activity	a	m	b	$(a + 4m + b)/6$	$((b - a)/6)^2$
		weeks		expected	variance
A	5	8	13	$50/6 = 8\frac{1}{3}$	64/36
B	2	4	6	$24/6 = 4$	16/36
D	1	2	4	$13/6 = 2\frac{1}{6}$	9/36
K	2	3	6	$20/6 = 3\frac{1}{3}$	16/36
L	2	4	8	$26/6 = 4\frac{1}{3}$	36/36

The critical path using these expected times is the same, but the expected project duration increases from 31 weeks to 32 weeks. A, K, L are all 1/3 weeks longer. The variance of the project duration, assuming all jobs are independent, is the sum of the variances of the critical jobs, ie zero, except for A, K, L:

$$\text{Variance} = 64/36 + 16/36 + 36/36 = 3.22 \text{ weeks}^2$$

Therefore, the distribution of the project duration is normal with a mean of 32 weeks and a standard deviation of $\sqrt{3.22} = 1.795$ weeks.

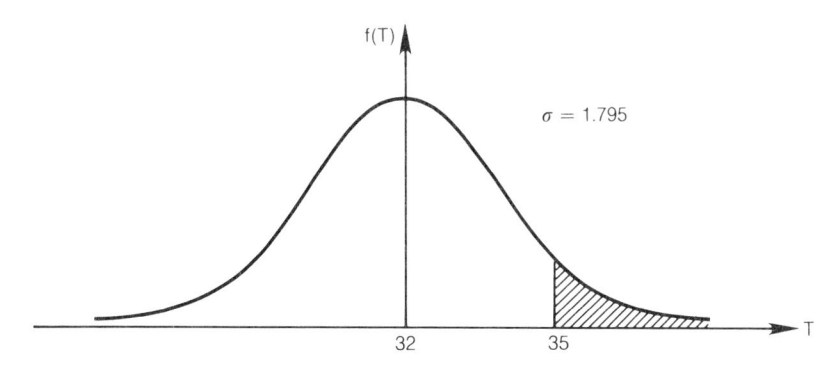

The probability that the project duration > 35 weeks is given by the probability that:

$$z > \frac{35 - 32}{1.795} = 1.67$$

ie approximately 4.7% from the standard normal tables. D is the most likely sub-critical job to affect the project duration, but its duration is always $\leq$ duration of C (4 weeks), therefore variation in D will not affect the overall project time.

Solutions to Chapter Twelve exercises

ANSWERS TO EXERCISES

Exercise 12.1

1 This problem may be modelled by the basic economic order quantity model. The annual demand:

$$D = 25 \times 50 = 1250 \text{ calculators}$$

The cost of ordering:

$$C_o = £15 \text{ per order}$$

The unit holding cost:

$$C_h = £0.50 + 0.15 \times £9 = £1.85 \text{ per calculator per year}$$

The unit cost:

$$C = £9 \text{ per calculator}$$

The optimum order size is given by:

$$q = \sqrt{\frac{2 \times C_o \times D}{C_h}} = \sqrt{\frac{2 \times 15 \times 1250}{1.85}} = 142.4$$

The optimum order quantity is 142 calculators.

2 The total annual variable cost of stocking the calculators is given by:

$$TC = C_o D/q + C_h q/2 \quad £/\text{year}$$

When $q = 142$,

$$TC = 15 \times 1250/142 + 1.85 \times 142/2$$
$$= 132.04 + 131.35 = £263.39 \text{ per year}$$

When $q = 300$,

$$TC = 15 \times 1250/300 + 1.85 \times 300/2$$
$$= 62.5 + 277.5 = £340 \text{ per year}$$

Hence, if Dekkers reduce their order quantity from 300 to 142, they will save:

$$£(340 - 263) = £77 \text{ per year.}$$

3 If C_o is now £5 per order, then the optimum order quantity is reduced to $142/\sqrt{3} = 82$ calculators.

Exercise 12.2

This problem can be treated as a basic EOQ problem.

$$D = 150 \text{ engineers per year}$$
$$C_o = £25{,}000 \text{ per course}$$
$$C_h = £500 \text{ per man per month} = £6000 \text{ per man per year}$$

1 The optimum size for the course is given by:

$$q = \sqrt{\frac{2 \times C_o \times D}{C_h}} = \sqrt{\frac{2 \times 25000 \times 150}{6000}} = 35.4$$

The company should take 35 engineers onto each course.

2 The number of courses per year is $150/35 = 4.3$. The courses should be run every $\frac{(35 \times 12)}{150} = 2.8$ months.

The total annual variable costs of training the engineers is:

$$TC = C_o D/q + C_h q/2$$
$$TC = 25{,}000 \times 150/35 + 6000 \times 35/2$$
$$= 107{,}143 + 105{,}000 = £212{,}143 \text{ per year}$$

3 If the courses are restricted to 25 engineers, then they will have to be run every $\frac{(25 \times 12)}{150} = 2$ months. The total annual variable cost of training the engineers is:

$$TC = 25{,}000 \times 150/25 + 6000 \times 25/2$$
$$= 150{,}000 + 75{,}000 = £225{,}000 \text{ per year}$$

An increase of £12,857 per year.

Exercise 12.3

This problem is suitable for modelling by the EOQ model:

$$D = 20{,}000 \text{ disks per year}$$
$$C_o = £32 \text{ per order}$$
$$C = 0.80 \text{ per disk}$$
$$C_h = 0.01 \times £0.80 = £0.008 \text{ per disk per year}$$

1 The optimum size for the order is given by:

$$q = \sqrt{\frac{2 \times C_o \times D}{C_h}} = \sqrt{\frac{2 \times 32 \times 20{,}000}{0.008}} = 12{,}649.1$$

The company should order 12,650 (for convenience) disks each time. The number of orders per year is $20{,}000/12{,}650 = 1.58$.

2 The total annual variable cost of stocking the disks is:

$$TC = C_o D/q + C_h q/2 \text{ £/year}$$
$$TC = 32 \times 20{,}000/12{,}650 + 0.008 \times 12{,}650/2$$
$$= 50.59 + 50.6 = £101.19 \text{ per year}$$

3 The annual demand is now 24,200 disks per year. The new optimum order quantity is:

$$q = \sqrt{\frac{2 \times 32 \times 24{,}200}{0.008}} = 13{,}914.0$$

592 QUANTITATIVE ANALYSIS

With the increased annual demand the optimum order quantity increases to 13,914 disks. The total annual variable cost of stocking the disks is:

$$TC = C_0 D/q + C_h q/2$$
$$TC = 32 \times 24{,}200/13{,}914 + 0.008 \times 13{,}914/2$$
$$= 55.66 + 55.66 = £111.32 \text{ per year}$$

If the order quantity remains at 12,650 disks per order, then,

$$TC = 32 \times 24{,}200/12{,}650 + 0.008 \times 12{,}650/2$$
$$= 61.22 + 50.6 = £111.82$$

The effect of keeping to the order quantity in **1** is to increase the total annual variable cost by £0.50.

4 The model clearly is not sensitive to changes in the annual demand. The change in the exercise represents more than a 20% increase in the annual demand.

Exercise 12.4

This problem is suitable for modelling by the EOQ model:

$$D = 200 \text{ cars per year}$$
$$C_0 = £500 \text{ per order}$$
$$C = £6000 \text{ per car}$$
$$C_h = 0.3 \times £6000 = £1800 \text{ per car per year}$$

1 The optimum size for the order is given by:

$$q = \sqrt{\frac{2 \times C_0 D}{C_h}} = \sqrt{\frac{2 \times 500 \times 200}{1800}} = 10.54$$

The company should order 10 or 11 cars each time. The total variable cost per year should be examined at both of these order quantities to determine which is the preferred size. However, since we are being offered discounts, we will choose 10 cars as the optimum order size until we find out whether the discounts are worth having. The total annual variable cost of stocking the cars is:

$$TC = C_0 D/q + C_h q/2$$
$$TC = 500 \times 200/10 + 1800 \times 10/2$$
$$= 10{,}000 + 9000 = £19{,}000 \text{ per year}$$

The purchase cost is:

$$200 \times £6000 = £1{,}200{,}000$$

Therefore, the total annual cost of stocking the cars is:

$$£19{,}000 + £1{,}200{,}000 = £1{,}219{,}000 \text{ per year}$$

Now consider the first discount:

$$C = 0.985 \times £6000 = £5910 \text{ per car}$$
$$C_h = 0.3 \times £5910 = £1773 \text{ per car per year}$$

Ignoring the limit on the order size, the optimum order quantity is:

$$q = \sqrt{\frac{2 \times 500 \times 200}{1773}} = 10.6$$

SOLUTIONS TO EXERCISES: CHAPTER TWELVE

The turning point in the total cost curve for the discounted price is outside the range for which the discount is offered, therefore, the optimum order quantity for this price is 50 cars. If q = 50, the total annual cost of stocking the cars is:

$$500 \times 200/50 + 1773 \times 50/2 + 200 \times 5910$$
$$= 2000 + 44325 + 1182000 = £1,228,325 \text{ per year}$$

The discount is not worth having since the increase in holding costs more than off-sets the decrease in the purchase price.

We will now repeat the calculation for the second discount:

$$C = 0.97 \times £6000 = £5820 \text{ per car}$$
$$C_h = 0.3 \times £5820 = £1746 \text{ per car per year}$$

Ignoring the limit on the order size, the optimum order quantity is:

$$q = \sqrt{\frac{2 \times 500 \times 200}{1746}} = 10.7$$

The turning point in the total cost curve for the discounted price is outside the range for which the discount is offered, therefore, the optimum order quantity for this price is 100 cars. If q = 100, the total annual cost of stocking the cars is:

$$500 \times 200/100 + 1746 \times 100/2 + 200 \times 5820$$
$$= 1000 + 87300 + 1164000 = £1,252,300 \text{ per year}$$

The discount is not worth having, since the increase in holding costs more than offsets the decrease in the purchase price.

Since both of the discounts are rejected, it is now worth while determining whether 10 or 11 cars should be ordered. If 10 cars are ordered, the total annual variable costs are £19,000. If 11 cars are ordered, the total annual variable costs are:

$$TC = 500 \times 200/11 + 1800 \times 11/2 = 9090.91 + 9900 = £18,991 \text{ per year}$$

It is marginally cheaper to order the cars in batches of 11.

2 If a discount of 5% is obtained for order quantities above 100 cars, then:

$$C = 0.95 \times £6000 = £5700 \text{ per car}$$
$$C_h = 0.3 \times £5700 = £1710 \text{ per car per year}$$

In view of the previous calculation, we are safe in assuming that optimum order quantity at this price is 100, therefore, the total cost is:

$$500 \times 200/100 + 1710 \times 100/2 + 200 \times 5700 = £1,226,500 \text{ per year}$$

It is still cheaper to order 11 cars at a time. The insensitivity of the model to the discount arises due to the high rate of charging for holding the cars.

Note: The showroom may not be able to take advantage of discounts at large order sizes in any case, because of the problem of space to store the extra cars. Models and fashions also change, so it is undesirable to carry very large stocks.

Exercise 12.5

Initially, we will ignore the variability in the demand.

$$D = 3 \text{ per week} = 3 \times 50 = 150 \text{ per year on average}$$
$$C_o = £50 \text{ per order}$$
$$C = £40 \text{ per tool}$$
$$C_h = 0.3 \times £40 = £12 \text{ per tool per year}$$
$$C_b = £100 \text{ per tool}$$

On the basis of the EOQ model, the optimum order quantity is:

$$q = \sqrt{\frac{2 \times C_o \times D}{C_h}} = \sqrt{\frac{2 \times 50 \times 150}{12}} = 35.4$$

The retailer should order the tool in batches of 35. The number of orders per year is 150/35 = 4.3

Let us now consider the size of the buffer stock which he should hold. If the average demand is for 3 per week, then the average demand during the lead time will be 6 per 2 weeks. Since the weekly demand follows a Poisson distribution, the demand during the lead time may also be assumed to follow a Poisson distribution with a mean of 6. The probability of a lead time demand for r tools is given by:

$$P(r) = \frac{6^r e^{-6}}{r!} \qquad r = 0,1,2,3,\ldots.$$

Demand during the lead time	Probability of this demand	Buffer stock
6	$6^6 e^{-6}/6! = 0.1606$	0
7	$6^7 e^{-6}/7! = 0.1377$	1
8	0.1033	2
9	0.0688	3
10	0.0413	4
11	0.0225	5
12	0.0113	6
13	0.0052	7
14	0.0022	8
15	0.0009	9

Although the unit cost of stock-out is high (£100), we should be safe in assuming that the probability of a lead time demand above 15 is effectively zero.

For each buffer stock, we will calculate the cost of holding it at £12 per unit per year, together with the expected costs of stock-out at this level.

Buffer stock	Demand met	Expected stock-outs /cycle	/year	Stock-out	+ Buffer	= Total
9	15	0	0	0	9 × 12 = 108	108
8	14	1 × 0.0009 = 0.0039	0.0009 × 4.3 = 0.0039	0.0039 × 100 = 0.39	8 × 12 = 96	96.39
7	13	1 × 0.0022 +2 × 0.0009 = 0.004	0.004 × 4.3 = 0.0172	0.0172 × 100 = 1.72	7 × 12 = 84	85.72
6	12	1 × 0.0052 +2 × 0.0022 +3 × 0.0009 = 0.0123	0.0123 × 4.3 = 0.0529	0.0529 × 100 = 5.29	6 × 12 = 72	77.29
5	11	1 × 0.0113 +2 × 0.0052 +3 × 0.0022 +4 × 0.0009 = 0.0319	0.0319 × 4.3 = 0.1372	0.1372 × 100 = 13.72	5 × 12 = 60	73.72
4	10	1 × 0.0225 +2 × 0.0113 +3 × 0.0052 +4 × 0.0022 +5 × 0.0009 = 0.074	0.074 × 4.3 = 0.3182	0.3182 × 100 = 31.82	4 × 12 = 48	79.82

Since the total cost has begun to increase again, we can assume that the best choice for a buffer stock is 5 tools. The re-order level is then $(6 + 5) = 11$.

To minimise the total annual variable cost of stocking this tool, the retailer should order it in batches of 35 whenever the stock level falls to 11 tools.

Exercise 12.6

Initially, we will ignore the variability in the demand.

$$D = 4800 \text{ per year}$$
$$C_o = £30 \text{ per order}$$
$$C_h = £20 \text{ per item per year}$$
$$C_b = £10 \text{ per item}$$

1 On the basis of the EOQ model, the optimum order quantity is:

$$q = \sqrt{\frac{2 \times C_o \times D}{C_h}} = \sqrt{\frac{2 \times 30 \times 4800}{20}} = 120$$

The retailer should order the item in batches of 120.

2 The company requires a 97% level of service, therefore, we must choose a re-order level, R, so that:

$$P(\text{lead time demand} > R) < 0.03$$

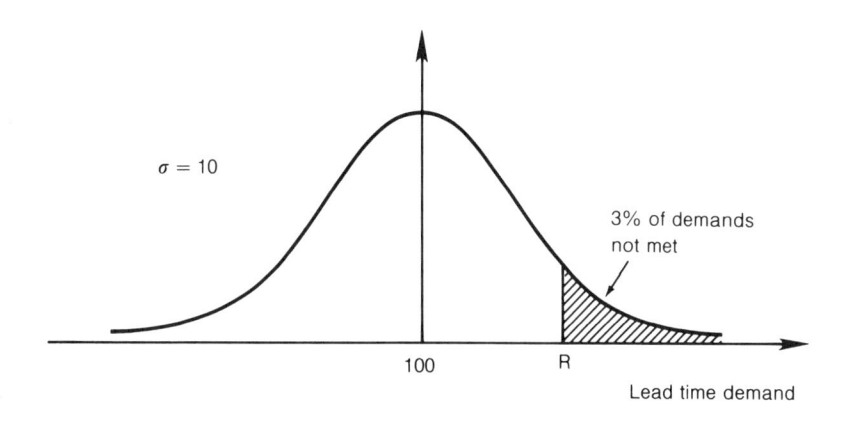

R is z standard deviations above the mean, where:

$$z = \frac{R - 100}{10}$$

If $P(z > (R - 100)/10) = 0.03$, then $z = 1.88$, from the standard normal tables. Therefore:

$$1.88 = \frac{R - 100}{10}$$

and $R = 118.8$.

The company should re-order when the stock level falls to 119 items. This provides for a buffer stock of 19 items.

3 If the re-order level is 115, then there will be a stock-out whenever the lead time demand >115. 115 is z standard deviations above the mean, where:

$$z = \frac{115 - 100}{10} = 1.5$$

From tables, $P(z > 1.5) = 0.0668$. There is a 6.7% probability of a stock-out on any cycle if the re-order level is set at 115 items. The number of stock cycles per year is $4800/120 = 40$. Therefore, the expected number of stock-outs per year is $0.067 \times 40 = 2.68$.

4 We will estimate the best value of buffer stock by calculating the total expected cost of holding the buffer, together with the expected cost of the associated stock-outs. Since the lead time demand is approximated by a continuous distribution, we will examine the cost at buffer stock intervals of 5 items.

Demand during the lead time	Probability of this demand*	Buffer stock
100	0.1974	0
105	0.1747	5
110	0.1210	10
115	0.0655	15
120	0.0279	20
125	0.0092	25
130	0.0024	30

(*Probability that the lead time demand is 105, is approximated by

$$P(102.5 < \text{demand} < 107.5)$$

using the normal distribution for lead time demand.)

We will assume that the probability of a lead time demand above 130 is effectively zero. For each buffer stock, we will calculate the cost of holding it at £20 per unit per year, together with the expected costs of stock-out at this level. The stock-out cost is £10 per item.

Buffer stock	Demand met	Expected stock-outs /cycle	/year	Stock-out	+ Buffer =	Total
30	130	0	0	0	$30 \times 20 = 600$	600
25	125	5×0.0024 $=0.012$	0.012×40 $=0.48$	0.48×10 $=4.8$	$25 \times 20 = 500$	504.8
20	120	5×0.0092 $+10 \times 0.0024$ $=0.07$	0.070×40 $=2.8$	2.8×10 $=28$	$20 \times 20 = 400$	428
15	115	5×0.0279 $+10 \times 0.0092$ $+15 \times 0.0024$ $=0.2675$	0.2675×40 $=10.7$	10.7×10 $=107$	$15 \times 20 = 300$	407
10	110	5×0.0655 $+10 \times 0.0279$ $+15 \times 0.0092$ $+20 \times 0.0024$ $=0.7925$	0.7925×40 $=31.7$	31.7×10 $=317$	$10 \times 20 = 200$	517

Since the total cost has begun to increase again, we can assume that the best choice for a buffer stock is 15 items. The re-order level is then $(100 + 15) = 115$ items.

SOLUTIONS TO EXERCISES: CHAPTER TWELVE

Exercise 12.7

The data given is:

Demand is normal with mean = 800 boxes/week

On the basis of a 50 week year, the mean demand is:

800 × 50 = 40,000 boxes/year

Unit cost, C = £2.50/box; lead time, L = 3 weeks.
Stockholding cost, C_h, = 15% of £2.50 per box/year = £0.375/box/year.
Ordering cost, C_o, = (£160/40) × 12 = £48 per order.

1. The basic principle of an inventory control policy is to organise the size and frequency of the orders so that the total variable cost of the stockholding is minimised, subject to the maintaining of a specified level of service. The costs of holding a lot of stock must be balanced against the cost of ordering frequently and the costs of being out of stock. These latter costs may not be money initially, but loss of customer goodwill will eventually lead to loss of sales.

    If Oxygon holds a year's stock of the typing paper, then they hold an average of 40,000 boxes for which they pay 40,000 × £2.50 = £100,000. Some of this money, which is tied up in the stock, could be used for investment in the business, or, at least, to ease the liquidity problems. Although the demand is very variable, since the lead time is constant, it should be possible to reduce the probability of a stockout to a very low level, with a much reduced stock.

2. Using the average annual demand, the economic order quantity, EOQ, is given by:

$$\text{EOQ} = \sqrt{\frac{2 \times C_o \times D}{C_h}} = \sqrt{\frac{2 \times 48 \times 40000}{0.375}} = 3200 \text{ boxes}$$

The number of orders per year = D/EOQ = 40,000/3200 = 12.5 per year. Therefore, the time interval between orders is 50/12.5 weeks = 4 weeks.

On the basis of the average demand per year, the optimum ordering policy is to order 3200 boxes of the paper, every 4 weeks. This policy makes no allowance for the variable nature of the demand and stockouts are likely to occur on some stock cycles.

3. During the lead time, the average demand is 3 × 800 = 2400 boxes with a standard deviation of $\sqrt{3 \times 250^2} = 433.0$. Choose a re-order level, R, so that P(lead time demand > R) < 0.01.

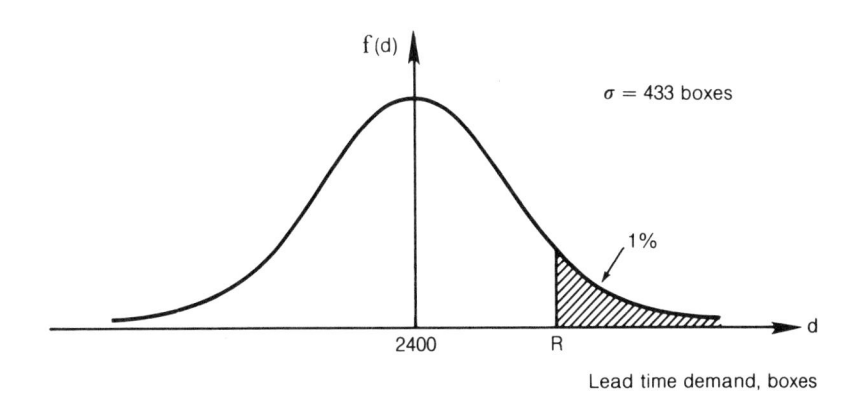

R is z standard deviations above the mean, where:

$$z = \frac{R - 2400}{433} \text{ therefore } R = 2400 + 433z$$

598   QUANTITATIVE ANALYSIS

If $P(z > (R - 2400)/433) = 0.01$, then from the standard normal tables, $z = 2.33$. Therefore:

$$R = 2400 + 2.33 \times 433 = 3408.89$$

The recommended re-order level is 3409 boxes. This means that the company is holding a buffer stock of $(3409 - 2400) = 1009$ boxes.

4   The total stockholding costs:

$$= \text{cost of ordering} + \text{cost of holding normal and buffer stock}$$
$$= C_o \times \frac{D}{q} + C_h\left(\frac{q}{2} + \text{buffer}\right) \quad \text{£/year}$$
$$= 48 \times \frac{40000}{3200} + 0.375\left(\frac{3200}{2} + 1009\right)$$
$$= 600 + 978.375 = 1578.375 \quad \text{£/year}$$

The minimum total variable costs of stockholding is £1578.38 per year.

Exercise 12.8

The data provided are:

$$\text{Demand, } D = 1000 \text{ litres/month} = 12000 \text{ litre/year}$$
$$\text{Order cost, } C_o = £50 \text{ per order (for cleaning etc)}$$
$$\text{Unit cost, } C = £5 \text{ per litre}$$
$$\text{Holding cost, } C_h = 24\% \text{ of £5} = £1.20 \text{ per litre per year}$$

1   The economic order quantity is given by:

$$\text{EOQ} = \sqrt{\frac{2 \times C_o \times D}{C_h}} = \sqrt{\frac{2 \times 50 \times 12000}{1.20}} = 1000 \text{ litres}$$

This is the current order quantity. Therefore, just one storage unit is needed which will be filled each month when a new order arrives. The total annual variable stockholding costs:

$$= C_o \times \frac{D}{q} + C_h \times \frac{q}{2}$$
$$= 50 \times \frac{12000}{1000} + 1.2 \times \frac{1000}{2}$$
$$= £1200/\text{year}$$

Therefore, the total annual cost associated with the storing of the chemical $= £1200 + \text{rent} = £1200 + 12 \times £40 = £1680$ per year.

2   If the monthly demand increases to 2500 litres, then there are two choices:

(a) keep the order size at 1000 litres and one storage unit, but order more often, ie 2.5 times a month or 30 times a year;
(b) increase the number of storage units and order more than 1000 litres at a time.

We will investigate the annual cost for each of these alternatives.

(a) Place 30 orders per year of 1000 litres each: the total annual cost associated with storing the chemical, TC, is given by:

$$\text{TC} = \text{annual cost of cleaning} + \text{cost of holding} + \text{rent}$$
$$= 50 \times 30 + 1.2 \times (1000/2) + 12 \times 40 = £2580/\text{year}$$

SOLUTIONS TO EXERCISES: CHAPTER TWELVE

(b) We do not know how many additional units to rent, therefore, we will look at the effect of the costs of renting 2 units, then 3 and so on. Two units: cost of cleaning = £75 per order. Other costs are unchanged. The EOQ is:

$$\text{EOQ} = \sqrt{\frac{2 \times C_o \times D}{C_h}} = \sqrt{\frac{2 \times 75 \times 30000}{1.20}} = 1936.49 \text{ litres}$$

The optimum order quantity is 1936 litres.

$$\text{TC} = 75 \times \frac{30000}{1936} + 1.2 \times \frac{1936}{2} + 2 \times 12 \times 40$$
$$= £3283.79 \text{ per year}$$

It is, therefore, cheaper to keep one storage unit and to order 1000 litres 2.5 times a month. A similar calculation will show that the cost is even higher with three storage units. Hence, retaining one unit and ordering more frequently is the optimum policy.

3 A second storage unit is more economical if $TC_2 < TC_1$. If:

$$\text{demand} = 7200 \text{ litre/month} = 86,400 \text{ litres/year}$$

then:

$$TC_1 = 50 \times \frac{86400}{1000} + 1.2 \times \frac{1000}{2} + 12 \times 40 = £5400/\text{year}$$

With two units,

$$\text{EOQ} = \sqrt{\frac{2 \times 75 \times 86400}{1.2}} = 3286.3,$$

which is greater than the capacity of 2 units. Hence, if only 2 units are available, the order quantity must be 2000 litres.

$$TC_2 = 75 \times \frac{86400}{2000} + 1.2 \times \frac{2000}{2} + 2 \times 12 \times 40 = £5400/\text{year}$$

Therefore, when demand = 7200 litres per month, $TC_1 = TC_2$. The total cost is linear in demand, therefore, there is only one break even point.

Exercise 12.9

The data given are:

Annual demand = D
Set up cost = £S per run
Holding cost = £H per item per year
Manufacturing cost = £C per item
Batch size = Q

1 If the total annual production cost is £TC per year, then:

TC = annual cost of set up + annual holding cost + cost of manufacture of D items

$$TC = S \times \frac{D}{Q} + H \times \frac{Q}{2} + D \times C \qquad £/\text{year}$$

assuming that the holding cost is based on the average quantity in stock. Therefore, the total annual production cost is given by:

$$TC = \frac{SD}{Q} + \frac{HQ}{2} + DC \qquad \text{£/year}$$

The economic batch quantity is Q_0 where:

$$Q_0 = \sqrt{\frac{2 \times S \times D}{H}}$$

At the optimum, the total annual set up cost is given by:

$$C_s = S \times \frac{D}{Q_0}$$

substituting for Q_0,

$$C_s = \frac{S \times D}{\sqrt{2SD/H}} = \sqrt{\frac{S^2D^2}{2SD/H}} = \sqrt{\frac{SDH}{2}}$$

At the optimum, the total annual holding cost is given by:

$$C_H = H \times \frac{Q_0}{2}$$

substituting for Q_0,

$$C_H = \frac{H\sqrt{2SD/H}}{2} = \sqrt{\frac{H^2 2SD}{4H}} = \sqrt{\frac{SDH}{2}}$$

Therefore, at the optimum the total annual set up costs = total annual holding costs.

2 We are told that manufacturing cost, DC, is 75% of the total annual production cost, when the EBQ is produced. Therefore:

$$DC = 0.75 \times TC$$

Substituting in the total cost equation derived in **1** above:

$$TC = \frac{SD}{Q_0} + \frac{HQ_0}{2} + 0.75TC \qquad \text{£/year}$$

It follows that:

$$\frac{SD}{Q_0} + \frac{HQ_0}{2} = 0.25TC$$

but, at the optimum, set up cost = holding cost, therefore:

$$\frac{SD}{Q_0} = \frac{HQ_0}{2} = 0.125TC \qquad (1)$$

(a) If $Q = 1.5Q_0$, then the total production cost, TC_1, is given by:

$$TC_1 = \frac{SD}{(1.5Q_0)} + \frac{H(1.5Q_0)}{2} + 0.75TC$$

$$= \frac{SD}{Q_0} \times \frac{1}{1.5} + \frac{HQ_0}{2} \times 1.5 + 0.75TC$$

Substituting from (1) above:

$$TC_1 = \frac{0.125TC}{1.5} + (0.125TC) \times 1.5 + 0.75TC$$
$$= 0.0833TC + 0.1875TC + 0.75TC = 1.0208TC$$

Therefore, the total cost has increased by 2.08% compared to the value with the optimum batch quantity.

(b) If $Q = 0.5Q_0$, then the total production cost, TC_2, is given by:

$$TC_2 = \frac{SD}{(0.5Q_0)} + \frac{H(0.5Q_0)}{2} + 0.75TC$$
$$= \frac{SD}{Q_0} \times \frac{1}{0.5} + \frac{HQ_0}{2} \times 0.5 + 0.75TC$$

Substituting from (1) above:

$$TC_2 = \frac{0.125TC}{0.5} + (0.125TC) \times 0.5 + 0.75TC$$
$$= 0.25TC + 0.0625TC + 0.75TC = 1.0625TC$$

Therefore, the total cost has increased by 6.25% compared to the value with the optimum batch quantity.

(c) If $Q = 2Q_0$, then the total production cost, TC_3, is given by:

$$TC_3 = \frac{SD}{(2Q_0)} + \frac{H(2Q_0)}{2} + 0.75TC \times 0.95$$
$$= \frac{SD}{Q_0} \times \frac{1}{2} + \frac{HQ_0}{2} \times 2 + 0.75TC \times 0.95$$

Substituting from (1) above:

$$TC_3 = \frac{0.125TC}{2} + (0.125TC) \times 2 + 0.95 \times 0.75TC$$
$$= 0.0625TC + 0.25TC + 0.7125TC = 1.025TC$$

Therefore, the total cost has increased by 2.5% compared to the value with the optimum batch quantity. The 5% saving in manufacturing costs has not lead to an overall reduction in the total annual production costs.

Let a saving of p% in the manufacturing cost reduce the total annual cost of production to the value with the EBQ, ie TC, then:

$$TC_3 = \frac{0.125TC}{2} + (0.125TC) \times 2 + \left(1 - \frac{P}{100}\right) \times 0.75TC = TC$$

Therefore:

$$0.0625TC + 0.25TC + 0.0075(100 - p)TC = TC$$

Hence:

$$0.3125 + 0.0075(100 - p) = 1$$
$$0.0075(100 - p) = 0.6875$$

Therefore:

$$100 - p = 91.6667$$
$$p = 8.33\%$$

602 QUANTITATIVE ANALYSIS

The manufacturing cost must be reduced by at least 8.3% before it is worthwhile to double the batch size.

Exercise 12.10

(a) Data given are:

Annual demand, D, = 10 × 50 = 500 items per year
Order size, q
Unit cost for order size q, £C_q per item
Unit holding cost, C_h, = 20% of C_q × (q/2) = 0.1C_qq, assuming that the holding cost is based on the average stock level. Total annual cost of purchasing and holding, £TC per annum TC = 0.1C_qq + 500 × C_q £/year

(b) When q in the range 1 − 99,   TC = 0.1 × 10.20 × q + 500 × 10.20
  = 1.02q + 5100   £/year

When q in the range 100 − 249, TC = 0.1 × 9.95 × q + 500 × 9.95
  = 0.995q + 4975   £/year

When q in the range 250 − 499, TC = 0.1 × 9.65 × q + 500 × 9.65
  = 0.965q + 4825   £/year

When q in the range 500 − 999, TC = 0.1 × 9.30 × q + 500 × 9.30
  = 0.93q + 4650   £/year

When q in the range 1000 + ,   TC = 0.1 × 8.80 × q + 500 × 8.80
  = 0.88q + 4400   £/year

These expressions for total cost are linear functions of q over the range of the discount.

q	1	99	100	249	250	499	500	999	1000
TC,£pa	5101	5201	5075	5223	5066	5307	5115	5579	5280

The graph is plotted below:

Graph of total annual purchasing and holding costs for item 53/X2

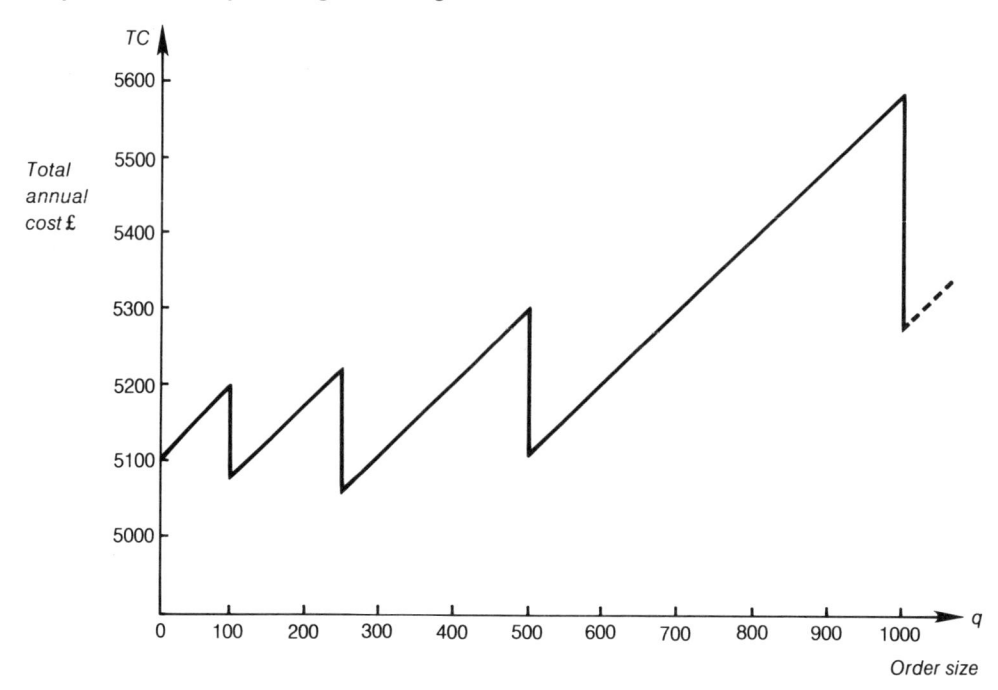

The minimum value of TC occurs for the second discount, that is, when the order size is 250 and the unit cost is £9.65.

(c) Let the unit annual stockholding costs be a proportion, p, of the stock value. Total annual cost of purchasing and holding is given by:

$$TC = pC_q(q/2) + 500C_q \qquad £/year$$

Discount range	q	C_q, £/item	TC, £/year
1 – 99	1	10.20	5.10p + 5100
100 – 249	100	9.95	497.50p + 4975
250 – 499	250	9.65	1206.25p + 4825
500 – 999	500	9.30	2325.00p + 4650
1000 +	1000	8.80	4400.00p + 4400

We require the total cost at q = 1000 items to be less than the total cost at each of the other price breaks. However, we need only compare the last two price breaks. Let us look at the reason for this. When p = 0.2 (current situation), the minimum value of the total cost occurs at q = 250 items, but as the value of p decreases, the unit cost of holding becomes less and the larger order quantities become more attractive due to the savings on the purchase cost. The position of the minimum will move in stages, from q = 250 to q = 500 to q = 1000. We require the value of p which will cause the minimum value of total cost to move from q = 500 to q = 1000. In this case:

$$4400p + 4400 < 2325p + 4650$$
$$2075p < 250$$

therefore p < 0.1205. If the stockholding cost falls below 12.05%, it will be economical to purchase in batches of 1000.

Exercise 12.11

1   The optimum batch size is the one which minimises the total variable costs associated with the stockholding the particular item. We are interested, therefore, only in those costs which depend directly on the size of the batch produced. These costs are the total annual cost of setting up the batch process and the total annual cost of holding the stock.

Set up cost: it takes 6 hours to set up the machine at £10 per hour. There are no other costs which are associated with the setting up of the batch process. Therefore, the total annual set up cost is £60 × the number of batches per year.

Holding cost: the company's cost of capital is 18% per annum, therefore, any capital tied up in stock costs the company 18% of the value of the stock. The value of each item of stock will be the raw materials used to make the item plus the machine time used.

Raw materials cost £7.40 per unit.

Machine time costs 0.5 hours per unit × £1.40 per hour = £0.70 per unit.

(*Note*: it has been assumed that machine time cost is not a fixed cost.)

There are no other variable costs. The storekeeper's wages, the insurance and the overheads are independent of the amount of stock held and, therefore, are regarded as fixed costs. The total annual holding cost is:

0.18 × (7.40 + 0.70) × average stock level = 1.458 × average stock level (£ per unit per year)

The machine operator's wages are assumed to be fixed costs.

2 We will define the following symbols:

D = annual demand for the XT/24 = 50 × 50 = 2500 per year
C_s = set up cost per batch = £10 × 6 hours per batch = £60 per batch
C_h = unit holding cost per year = £1.458 per item per year
R = annual production rate = number made/week × 50 = $\frac{40}{0.5}$ × 50
  = 4000 units per year

The economic batch size, EBQ, is given by:

$$EBQ = Q\sqrt{\frac{R}{R-D}} = \sqrt{\frac{2C_s D}{C_h} \frac{R}{R-D}}$$

(*Note*: R and D are now annual rates in order to be consistent with C_h, which is the annual unit holding cost.)

Therefore:

$$EBQ = \sqrt{\frac{2 \times 60 \times 2500}{1.458} \frac{4000}{(4000 - 2500)}} = 740.7$$

The optimum batch size is 741 units of XT/24. The stock level varies as shown below:

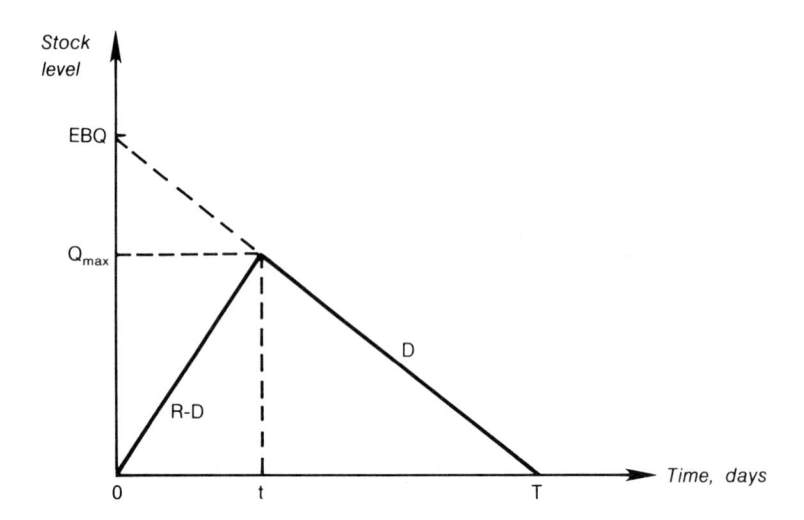

If the production run lasts for t days, then

$$R = \frac{EBQ}{t} \quad (1) \quad \text{and} \quad R - D = \frac{Q_{max}}{t} \quad (2)$$

Dividing (2) by (1) gives:

$$\frac{R - D}{R} = \frac{Q_{max}}{EBQ}$$

Therefore:

$$\frac{4000 - 2500}{4000} = \frac{Q_{max}}{741}$$

$$Q_{max} = \frac{1500 \times 741}{4000} = 277.9$$

The maximum stock level is 278 units of XT/24.

3 Since the new production run begins when the stock level falls to zero, the first unit produced will be used immediately. The minimum time that an item is in stock is zero. The last item is produced t days after the start of the production. If the stock cycle lasts T days, then this item is in stock for (T − t) days. From equation (1) in **2** above,

$$t = EBQ/R = (741/4000) \times 250$$
$$= 46.3 \text{ days}$$

Also, D = EBQ/T, therefore,

$$T = EBQ/D = (741/2500) \times 250$$
$$= 74.1 \text{ days}$$

Hence, the maximum time that an item is in stock is:

$$(74.1 - 46.3) = 27.8 \text{ days, that is, 28 days.}$$

Solutions to Chapter Thirteen exercises

ANSWERS TO EXERCISES

Exercise 13.1

$\lambda = 3$ cars per minute—Poisson $P(r) = \dfrac{\lambda^r e^{-\lambda}}{r!}$.

1 $P(\text{no cars in a one minute interval}) = P(0) = e^{-3} = 0.05$.

2 $P(\text{2 cars in a one minute interval}) = P(2) = \dfrac{3^2 e^{-3}}{2!} = 0.224$.

3 $P(\geqslant 4 \text{ cars}) = 1 - (P(0) + P(1) + P(2) + P(3))$
$= 1 - (0.050 + 0.149 + 0.224 + 0.224) = 0.353$.

Exercise 13.2

Mean = 3 calls per 15 minutes or 1 call per 5 minutes, or 2 calls per 10 minutes.

1 $P(\text{no calls in a 15 minute interval}) = e^{-3} = 0.050$.
2 $P(\text{no calls in a 5 minute interval}) = e^{-1} = 0.368$.
3 $P(\geqslant 2 \text{ calls in a 10 minute interval}) = 1 - (P(0) + P(1))$
$= 1 - (e^{-2} + 2e^{-2}) = 1 - (0.135 + 0.271) = 0.594$.

Exercise 13.3

If it takes exactly 15 minutes to unload a lorry, the loading bay can cope with 0, 1, 2, 3, or 4 lorry arrivals in an hour, but not with 5, 6, 7 ... etc.

$$\lambda = 2.5 \text{ lorries/hour}$$

therefore:

$$P(r \text{ arrivals in an hour}) = \dfrac{\lambda^r e^{-\lambda}}{r!} \qquad r = 0, 1, 2, \ldots$$

$$P(0) + P(1) + P(2) + P(3) + P(4) = 0.082 + 0.205 + 0.257 + 0.214 + 0.134$$
$$= 0.892$$

Therefore:

$$P(\text{bay will cope during an hour}) = 0.892$$

$$P(\text{bay will not cope}) = 0.108$$

In a week there are 24×7 hours = 168 hours per week, so that the unloading bay will be unable to cope for $168 \times 0.108 = 18.1$ hours per week.

Exercise 13.4

Mean is 5 per 5 working days or 1 per working day, ie $\lambda = 1$.
The probability distribution for the time, t, between demands for the item is:

$$P(t) = \lambda e^{-\lambda t}$$

Therefore, the probability that the time between demands < T is:

$$P(t < T) = 1 - e^{-\lambda T}$$

It follows that $P(t > T) = 1 - (1 - e^{-\lambda T})$
$= e^{-\lambda T}$

1 P(time between requests > 2 days) = $e^{-\lambda t} = e^{-2} = 0.135$.
2 P(time 3–4 days) = $e^{-3} - e^{-4} = 0.050 - 0.018 = 0.032$.
3 P(time < 1 day) = $1 - e^{-\lambda t} = 1 - e^{-1} = 0.632$.

Exercise 13.5

Cars pass at rate of 1200 per hour = 20 per minute = λ

1 P(no cars in 30 seconds) = $e^{-20/2} = 0.000045$
2 P(inter-arrival-time between 5 and 15 seconds)
$= e^{-20 \times 1/12} - e^{-20 \times 1/4} = 0.1889 - 0.0067 = 0.1822$

Exercise 13.6

One solicitor dealing with new clients: M/M/1 model. Arrival and service patterns are Poisson:

arrival rate $\lambda = 2$ per hour
service rate $\mu = 3$ per hour

Assume steady state.

1 Mean number in queue $= \dfrac{\lambda^2}{\mu(\mu - \lambda)} = \dfrac{4}{3 \times 1} = 1\tfrac{1}{3}$ clients.

2 Mean waiting time in queue $= \dfrac{\lambda}{\mu(\mu - \lambda)} = \dfrac{2 \text{ hours}}{3 \times 1} = 40$ minutes.

3 P(no one in system) = $P(0) = 1 - \dfrac{\lambda}{\mu} = 1 - \dfrac{2}{3} = \dfrac{1}{3}$.

Exercise 13.7

Enquiries in a library: random—assume Poisson process with $\lambda = 60/5 = 12$ per hour.
M/M/1 model:
With 1 librarian, the service rate $\mu = 60/3.5$ per hour.
Assume steady state:

Mean number of enquiries in queue $= \dfrac{(60/5)^2}{60/3.5 (60/3.5 - 60/5)} = 1.63$.

Mean number in system $= 1.63 + \dfrac{(60/5)}{(60/3.5)} = 2.33$ people.

Mean time in queue $= \dfrac{2.33}{(60/3.5)} = 0.136$ hours = 8.2 minutes.

Mean time in system = 8.2 + 3.5 = 11.7 minutes.

If there are 2 librarians, the service rate drops to $\mu = 60/4$ per hour $= 15$.
M/M/2 model: $\lambda = 12$/hour.

$$P(0) = \frac{1}{1 + \frac{12}{15} + \frac{12^2}{15^2}\frac{15}{(30-12)}} = \frac{1}{2\frac{1}{3}} = 0.429$$

Mean number in queue $= \frac{12^3}{15(30-12)^2} \times 0.429 = 0.1525$.

Mean time waiting $= \frac{0.1525}{12}$ hours $= 0.0127$ hours $= 0.76$ minutes.

Costs: 50 hours per week, with 12 enquiries per hours $= 600$ per week.
M/M/1 system—1 librarian $50 \times 6 = £300$ per week.
Waiting costs per week $= 600 \times 0.136 \times 2 = £163.2$ per week.
Therefore, total M/M/1 costs $= £463.2$ per week.
M/M/2 system—2 librarians $2 \times 50 \times 6 = £600$ per week.
Waiting costs per week $= 600 \times 0.0127 \times 2 = £15.2$ per week
Therefore total M/M/2 costs $= £615.2$ per week.

Comment on costs: the cost of the second librarian is much more than the notional cost of the waiting time—£300 per week against £163 per week—with the M/M/1 system.

Exercise   13.8

Carnucopia—motor dealer.

(a)   The main conditions for pre-delivery inspection to be regarded as an M/M/c system (choose any two):
   (i)   The question implies that the arrival rate of requests for new cars is a Poisson process with $\lambda = 38$ per week.
   (ii) There is a single queue, serviced on a FIFO system by all the channels.
   (iii) The service rate is also Poisson: the service time follows a negative exponential distribution.
(b) Show that $P(0) = 0.0256$.

$$\lambda = 38/\text{week}.$$

For each inspection bay:

$$\text{service rate } \mu = \frac{40}{2} = 20 \text{ cars per week}$$

There are 2 inspection bays, ie $c = 2$. From the question:

$$\rho = \frac{\lambda}{c\mu} = \frac{38}{40}$$

$$P(0) = \left\{ \sum_{i=0}^{c-1} \frac{(\rho c)^i}{i!} + \frac{(\rho c)^c}{c!(1-\rho)} \right\}^{-1}$$

$$= \left\{ \frac{(\rho c)^0}{0!} + \frac{(\rho c)^1}{1!} + \frac{(\rho c)^c}{c!(1-\rho)} \right\}^{-1} = \left\{ 1 + \frac{\lambda}{\mu} + \frac{(\lambda/\mu)^2}{2!(1-\rho)} \right\}^{-1}$$

$$= \left\{ 1 + \frac{38}{20} + \frac{(38/20)^2}{2!(1-38/40)} \right\}^{-1} = (1 + 1.9 + 36.1)^{-1} = (39)^{-1} = 1/39$$

$= 0.0256 =$ proportion of time both bays idle.

SOLUTIONS TO EXERCISES: CHAPTER THIRTEEN

Mean time from placing order to delivery

$$= \text{Mean time in system} + 1 \text{ day}$$
$$= \frac{(\rho c)^c P(0)}{c!(1-\rho)^2 c\mu} + \frac{1}{\mu} = \frac{(38/20)^2 0.0256}{2!(1-38/40)^2 2 \times 20} + \frac{1}{20}$$
$$= 0.462 + 0.05 = 0.512 \text{ weeks}$$
$$= 0.512 \times 5 = 2.56 \text{ days}$$

Therefore, adding 1 day for delivery, mean time from placing order to delivery = 3.56 days.

(c) Proposed system: employ 2 more engineers at a total cost of £220 per week, reduce service time to $1\frac{1}{2}$ hours.

$$\mu = 40/1.5 = 26\tfrac{2}{3} \text{ cars per week}$$

Use this to recalculate P(0) and mean time in system:

$$P(0) = \left\{ 1 + 38/26\tfrac{2}{3} + \frac{(38/26\tfrac{2}{3})^2}{2!(1 - 38/53\tfrac{1}{3})} \right\}^{-1} = \{1 + 1.425 + 3.532\}^{-1}$$
$$= 0.1679$$

Therefore:

$$\text{Mean time in system} = \frac{(38/26\tfrac{2}{3})^2 0.1679}{2!(1 - 38/53\tfrac{1}{3})^2 2 \times 26\tfrac{2}{3}} + \frac{1}{26\tfrac{2}{3}} \text{ weeks}$$
$$= 0.0387 + 0.0375 = 0.0762 \text{ weeks}$$
$$= 0.0762 \times 5 = 0.381 \text{ days}$$

Total time, adding 1 day for delivery = 1.38 days.

So that, we have saved $(3.56 - 1.38) = 2.18$ days on average.

Total savings per week = 2.18 × 38 cars × £2 per day per car.
= £166 per week

On a straight cost basis—net loss of £(220—166) = £54 per week.
However this ignores the benefits to the customer of the 2 days' reduction and any utilisation of the spare inspection time.

Exercise 13.9

(a) Estimate mean weekly cost of idle time. For each machine per hour breakdown we lose 20 components—financially we lose the contribution from these 20 components.

Contribution per component = sales price − direct materials − 0.4 × variable overheads
= 1 − 0.15 − 0.4 × 0.15 = £0.79 per component

Therefore:

$$\text{Contribution per machine hour} = 20 \times 0.78 = £15.8 \text{ per hour}$$
$$\text{Arrival rate (breakdown)} = 3 \text{ machines per week} = \lambda$$
$$\text{Service (repair) rate} = 4 \text{ machines per week} = \mu$$
$$\text{Mean time in system} = \frac{1}{\mu - \lambda} = 1 \text{ week} = 40 \text{ hours}$$

Therefore, since the number of breakdowns is 3, the mean machine down time = 3 × 40 = 120 hours per week. Thus, the cost of down line per week is £1896. *Note*: one of the assumptions of the M/M/1 model is that we have an infinite source from which the arrivals come. Here we have 50 machines, not an infinite number. This will have a small effect on the figures.

(b) Proposed system—employ a second man to help the first, ie still M/M/1 system. Mean service time is reduced from 10 to 8 hours, therefore:

$$\mu = 40/8 = 5 \text{ machines repaired per week}$$
$$\lambda = 3 \text{ machines per week (no change)}$$

Therefore, with the proposal:

$$\text{Mean time in system} = \frac{1}{\mu - \lambda} = \frac{1}{2} \text{ week} = 20 \text{ hours}$$

Thus, the cost of down time is halved from £1896 per week to £948 per week, giving £948 per week contribution from the extra man. Compared with the extra cost of £160 per week for extra engineers, ie net increase in contribution of (948—160) = £788 per week.

If the 2 engineers worked independently, but with identical service rates of 4 machines per week, it would be an M/M/2 model instead of M/M/1 and we would use the M/M/2 equations, which are not given, to calculate the mean time in the system.

(c) Back to (a) and 1 engineer. If there were only 49 operatives, output would be lost when 50 machines were available. This occurs when there are no machines in the repair system and the engineer is idle. The proportion of time for which the engineer is idle is:

$$P(0) = \left(1 - \frac{\lambda}{\mu}\right) = \left(1 - \frac{3}{4}\right) = \frac{1}{4}$$

ie 50 machines are working for 10 hours per week. From (a) each hour is worth £15.8, so that 10 hours, contribute £158 per week. The savings from not employing the fiftieth operative are £140 per week, so that there is a small net loss—it is therefore marginally better to employ 50.

Exercise 13.10

(a) Conditions for using M/M/1 model for the arrival and unloading of lorries at a supermarket:
(i) The lorry arrivals must be random.
(ii) Their actual rate must be constant throughout the day or week, ie it must be a Poisson process—mean λ lorries per hour.
(iii) They must form 1 queue, which has the first-in-first-out queue discipline FIFO, taking lorries from an infinitely large source.
(iv) The single service channel is a Poisson process—with a mean rate μ lorries per hour.
(v) The M/M/1 formulae apply only when the system has reached a steady state, ie not at the start of the day.

Test the data to see if they fit a Poisson process—χ^2 test.

From the observations we calculate the mean arrival rate, λ, and the mean service time, $1/\mu$. Using these figures, we calculate the probabilities based on the null hypothesis that the arrival rate is Poisson or the service rate is negative exponential.

The probabilities are converted to expected frequencies.

$$\chi^2 = \sum \frac{(f_o - f_e)^2}{f_e}$$

The test statistic is compared to $\chi^2_{0.05(n-1-1)}$. If this is not significant we accept H_0 and assume that the Poisson, or negative exponential, fits the data. So long as both arrival and service patterns fit, we can use the M/M/1 model.

(b) Assuming M/M/1: estimate mean number of lorries waiting in queue and mean time in the system. Calculate λ and $1/\mu$ from the data.

Arrivals/hour x	Number of hours (f)	fx	Unloading time mid-point	Number lorries (f)	fx
0	7	0	10	38	380
1	10	10	30	26	780
2	8	16	50	10	500
3	8	24	70	3	210
4	5	20	90	2	180
5	2	10	110	1	110
Totals	40	80	Totals	80	2160

Therefore: $\lambda = \dfrac{80}{40} = 2$ lorries/hour

Therefore: $1/\mu = 27$ minute/lorries

Therefore: $\mu = \dfrac{60}{27} = 2.22$ lorries/hour

Mean number of lorries in queue $= \dfrac{\lambda^2}{\mu(\mu - \lambda)} = \dfrac{4}{2.2\dot{2} \times 0.2\dot{2}} = 8.1$.

Mean time in system $= \dfrac{1}{\mu - \lambda} = \dfrac{1}{0.22\dot{2}}$ hours $= 4.5$ hours.

(c) Costs: 2 employees each paid £100 for a 40 hour week, plus overtime at time and a third. The present system is 9 am to 5 pm, ie an 8 hour day for arrivals—then overtime is worked each evening to clear lorries. We need to estimate overtime. Use mean number of lorries in the system to do this. At steady state:

$$\text{mean number in system} = \dfrac{\lambda}{(\mu - \lambda)} = \dfrac{2}{0.2\dot{2}} = 9 \text{ lorries}$$

Therefore, assume that at 5 pm each evening 9 lorries have to be unloaded. Each takes 27 minutes. Therefore:

$$\text{Overtime/day} = 9 \times 27/60 \text{ hours} = 4.05 \text{ hours}$$

$$\text{Cost/day} = 2 \times \left(8 \times \dfrac{100}{40} + 4.05 \times \dfrac{133\frac{1}{3}}{40}\right) = 2 \times (20 + 13.5)$$
$$= £67/\text{day for an average 12 hour day.}$$

The proposed system is to employ a third man and reduce the unloading time from 27 minutes to 20 minutes per lorry, ie:

$$1/\mu = 20/60 \quad \text{and} \quad \mu = 3 \text{ lorries per hour}$$

Will this reduce the supermarket's costs?
With the proposed system:

Mean number in system $= \dfrac{\lambda}{\mu - \lambda} = \dfrac{2}{3 - 2} = 2$ lorries, and

Mean time in system $= \dfrac{1}{\mu - \lambda} = 1$ hour, ie 40 minutes wait + 20 minutes unload.

A reduction of time in the system, from 4.5 hours to 1 hour, will give a large saving in waiting costs, which are not given in the Exercise. The costs of the proposed system are:

Mean overtime per day = 2 lorries × 20/60 hours = 2/3 hours.

$$\text{Cost/day} = 3 \times \left(8 \times \frac{100}{40} + \frac{2}{3} \times \frac{133\frac{1}{3}}{40}\right) = 3(20 + 2.2\dot{2})$$
$$= £66.67/\text{day}$$

The costs to the supermarket are about equal. The benefits to everyone else are large—reduced waiting time and congestion. Therefore, it is worth employing the third man.

Exercise 13.11

(a) Minicomputer tasks—explain how to test whether arrival pattern is a Poisson distribution. From the data, calculate the observed mean number of tasks arriving per hour, λ. The calculation is needed for part (b).

Observed Number in an hour (r)	Mid point (r_m)	Number of hours f_o	fr_m
0–2	1	8	8
3–5	4	34	136
6–8	7	44	308
9–11	10	12	120
12–14	13	1	13
15	15	1	15
Totals		100	600

Therefore, mean number of tasks arriving per hour = $\frac{600}{100} = 6 = \lambda$. Using the χ^2 test and:

H_0: The distribution is Poisson with $\lambda = 6$ per hour and:

$$P(r \text{ tasks in an hour}) = \frac{6^r e^{-6}}{r!}$$

H_1: the distribution is not Poisson with $\lambda = 6$ per hour.

We can calculate the probabilities of 0,1,2 and 3,4,5 and 6,7,8, etc arrivals in an hour. Multiplying these probabilities the total frequency, 100, gives the expected frequencies.

The test statistic, $\chi^2 = \sum \frac{(f_0 - f_E)^2}{f_E}$, is compared with $\chi^2_{0.05,(n-1-1)}$. If the calculated $\chi^2 < \chi^2_{0.05,(n-1-1)}$, we accept H_0. The distribution is assumed to be Poisson.

(b) Conditions necessary for M/M/1 model.
(i) Arrivals of tasks must be random.
(ii) Arrival rate must be constant throughout the day. The above implies a Poisson process.
(iii) There is 1 queue, drawn from an infinite number of tasks.
(iv) The queue discipline is first-in-first-out, FIFO.
(v) There is 1 server.
(vi) The service times follow a Poisson process, ie random, with a negative exponential distribution.
(vii) To use the M/M/1 equations, we assume that the system reaches a steady state, ie it is not affected by its start-up conditions.

(c) The arrival rate from (a) is $\lambda = 6$ tasks per hour. The service rate from the question is $\mu = 60/7.5 = 8$ per hour

(i) The proportion of time the computer is busy $= \dfrac{\lambda}{\mu} = \dfrac{3}{4}$.

(ii) The mean time a task is in the system

$$= \dfrac{1}{\mu - \lambda} = \dfrac{1}{2} \text{ hours} = 30 \text{ minutes}$$

which is hopeless.

(d) Costs.
Present System: system works 40 hours per week for 50 weeks per year, therefore:

$$\text{Number of tasks in a year} = 40 \times 50 \times 6 = 12{,}000$$

The task system cost is £5 for each hour that a task is in the system, therefore:

$$\text{Annual task system cost £ per year} = 12{,}000 \times 0.5 \times 5 = £30{,}000 \text{ per year}$$

Proposed System: buy an additional processor—assume this still operates as a single server—it just speeds up the service:

$$\mu = 60/(7.5 - 1.5) = 10 \text{ tasks per hour}$$

Therefore, mean time per task in proposed system $= \dfrac{1}{\mu - \lambda} = \dfrac{1}{4}$ hours.

Therefore, annual task system cost $= 12{,}000 \times 0.25 \times 5 = £15{,}000$ per year

However, the company has to pay £8000 per year for the additional processor, so that the cost of the old system is £30,000 per year, the cost of the proposed system is £23,000 per year, giving an overall saving of £7000 per year.

For what value of system cost (at present £5) would it be profitable for the company to purchase an additional processor? Let the task system cost be tsc (£ per hour).

$$\begin{aligned}\text{Cost of old system} &= 6000 \times \text{tsc} \\ \text{Cost of proposed system} &= 8000 + 3000 \times \text{tsc}\end{aligned}$$

Break-even is when old cost = proposed cost, ie:

$$\begin{aligned}6000 \times \text{tsc} &= 8000 + 3000 \times \text{tsc} \\ \text{tsc} &= £8/3 = £2\tfrac{2}{3}\end{aligned}$$

So long as tsc $> £2\tfrac{2}{3}$, therefore, the proposal should be taken up.

Solutions to Chapter Fourteen exercises

ANSWERS TO EXERCISES

Exercise 14.1

The variables in this problem are:

(a) the time between successive people arriving at the bus stop;
(b) the time between successive buses arriving at the bus stop;
(c) the number of empty seats on the bus.

The first step is to allocate random number ranges to the variable values.

Passengers

iat, mins	Probability	Cumulative Probability	Random number
0	0.04	0.04	00–03
1	0.16	0.20	04–19
2	0.24	0.44	20–43
3	0.28	0.72	44–71
4	0.16	0.88	72–87
5	0.10	0.98	88–97
6	0.02	1.00	98–99

Buses

iat, mins	Probability	Cumulative Probability	Random number
8	0.10	0.10	00–09
10	0.38	0.48	10–47
12	0.28	0.76	48–75
14	0.15	0.91	76–90
16	0.09	1.00	91–99

Empty seats

Number of empty seats	Probability	Cumulative Probability	Random number
0	0.06	0.06	00–05
1	0.18	0.24	06–23
2	0.27	0.51	24–50
3	0.34	0.85	51–84
4	0.11	0.96	85–95
5	0.03	0.99	96–98
6	0.01	1.00	99

1 We can now set up the simulation for the arrival of 30 passengers at the bus stop.

	Passenger				Bus			Seats		Passenger		
Number	iat, RN	minute time	Arrival time, mins		iat, RN	minute time	Arrival time, mins	RN	Number	Boards Bus	Queue Size*	Wait time, minute
1	18	1	1	A	26	10	10	23	1	A	1	9
2	18	1	2	B	62	12	22	42	2	B	2	20
3	07	1	3	C	38	10	32	40	2	B	3	19
4	92	5	8	D	97	16	48	64	3	C	4	24
5	46	3	11	E	75	12	60	74	3	C	4	21
6	44	3	14	F	84	14	74	82	3	D	5	34
7	17	1	15	G	16	10	84	97	5	D	6	33
8	16	1	16	H	07	8	92	77	3	D	7	32
9	58	3	19	I	44	10	102	77	3	E	8	41
10	09	1	20	J	99	16	118	81	3	E	9	40
11	79	4	24	K	83	14	132	07	1	E	8	36
12	83	4	28	L	11	10	142	45	2	F	9	46
13	86	4	32							F	8	42
14	19	1	33							F	9	41
15	62	3	36							G	10	48
16	06	1	37							G	11	47
17	76	4	41							G	12	43
18	50	3	44							G	13	40
19	03	0	44							G	14	40
20	10	1	45							H	15	47
21	55	3	48							H	13	44
22	23	2	50							H	14	42
23	64	3	53							I	15	49
24	05	1	54							I	16	48
25	05	1	55							I	17	47
26	71	3	58							J	18	60
27	88	5	63							J	16	55
28	72	4	67							J	17	51
29	93	5	72							K	18	60
30	10	1	73							L	19	69

2 The expected waiting time for the passengers = 1228/30 = 40.9 mins. We should also note that the waiting time is progressively increasing.

$$\text{Mean queue length} = 321/30 = 10.7$$

Again, we should note that the queue is increasing in length.
Note: * the queue length is measured each time a new passenger arrives up to the point at which the thirtieth passenger arrives.
Butterby's mini-bus service does not seem to be a very attractive means of transport. The waiting times are very long.

Exercise 14.2

We will use the same data and random numbers as **Example 14.3**, but run the simulation on the basis of re-ordering when the stock level falls below 1000 batteries at the beginning of the week.

616 QUANTITATIVE ANALYSIS

Week number	Opening stock	Demand RN	Demand Amount	Closing stock	Re-order? Yes/No	Lead time RN	Lead time Weeks	Shortage
1	2000	034	480	1520				
2	1520	743	505	1015				
3	1015	738	505	510				
4	510	636	505	5	Yes	95	4	
5	5	964	520	0				515
6	0	736	505	0				505
7	0	614	505	0				505
8	2500	698	505	1995				
9	1995	637	505	1490				
10	1490	162	490	1000				
11	1000	332	495	505				
12	505	616	505	0	Yes	73	3	
13	0	804	510	0				510
14	0	560	500	0				500
15	2500	111	490	2010				
16	2010	410	500	1510				
17	1510	959	515	995				
18	995	774	510	485	Yes	10	1	
19	2985	246	495	2490				
20	2490	762	505	1985				
		Total	10050	17515				2535

Mean demand = 10,050/20 = 502.5 batteries per week
Mean closing stock = 17,515/20 = 875.75 batteries per week
Mean shortage = 2535/20 = 126.75 batteries per week
Number of orders placed during the 20 week period = 3
ie Mean number of orders/week = 3/20 = 0.15

Expected weekly costs = 875.75 × £0.50 + 126.75 × £20 + 0.15 × £50
= £2980.38

The expected weekly cost has increased compared to the previous model. Savings are made on the holding and ordering costs, but these are more than off-set by the increased cost of shortage. As in all of these simulations, a much longer one is really required before valid conclusions can be drawn.

Exercise 14.3

1 *Advantages*:
  (a) Simulation can be used to investigate the behaviour of problems which are too complex to be modelled mathematically.
  (b) The technique can also be used when the variables in the problem, eg arrival time, service time, do not follow the standard distributions required for the mathematical models, ie Poisson distribution, negative exponential distribution.
  (c) The basic principles of the simulation technique are fairly simple and it is, therefore, more attractive to people who are not expert in quantitative techniques.

  Disadvantages:
  (a) Simulation is not an optimising technique. It simply allows us to select the best of the alternative systems examined.
  (b) Reliable results are possible only if the simulation is continued for a long period.
  (c) A computer is essential to cope with the amount of calculation required in (b).

2 (a) There are three stochastic variables in the problem:
   (i)   time between arrivals;
   (ii) time to deal with the complaint;
   (iii) whether the complaint is 'serious' or not.

   For each variable, a range of random numbers is allocated to each value. The size of the range is determined by the probability of the value occurring.

   (i) Since the time between arrivals is a grouped frequency distribution, we will use the mid-points of the groups to represent the group:

Time between arrivals, mins	Mid-point time, mins	Probability	Cumulative Probability	Random Numbers
0 but < 4	2	0.25	0.25	00–24
4 but < 8	6	0.45	0.70	25–69
8 but < 12	10	0.20	0.90	70–89
12 but < 16	14	0.10	1.00	90–99

   (ii) The time taken to deal with complaints follows a normal distribution with a mean of 7 minutes and a standard deviation of 2 minutes. If each service time, t, is z standard deviations from the mean, then:

   $$z = \frac{t - \mu}{\sigma}$$

   therefore:

   $$t = \mu + \sigma z = 7 + 2z \text{ (minutes)}$$

   We can use the table of random numbers and associated z values, given in the question, to generate a random series of values for z and use these to generate a random series of values for t.

   (iii) Serious complaint.

	Probability	Random number
Serious	0.2	0–1
Not serious	0.8	2–9

(b) Assume that the simulation clock begins at time zero.

Customer number	iat, mins RN	iat, mins Time	Arrival time, mins	Serious complaint RN	Serious complaint Y/No	Service time, mins RN	Service time, mins Time	Service time, mins Start	Service time, mins End	See manager Yes/No
1	09	2	2	5	No	39	7	2	9	No
2	06	2	4	0	Yes	60	7	—	—	Yes
3	51	6	10	7	No	50	7	10	17	No
4	62	6	16	3	No	31	6	17	23	No
5	83	10	26	8	No	02	3	26	29	No
6	61	6	32	2	No	02	3	32	35	No
7	59	6	38	9	No	83	9	38	47	No
8	20	2	40	8	No	90	10	waiting > 5 mins		Yes
9	82	10	50	1	Yes	71	8	—	—	Yes
10	68	6	56	6	No	16	5	56	61	No

(c) Out of the 10 customers in the simulation, 3 saw the manager. We estimate that $3/10 = 30\%$ of all customers see the manager. The time he spends each day dealing with complaints may be estimated by:

30% × average number complaints per hour × average time per complaint × 8 hours
 = 0.3 × number complaints per hour × 7 minutes × 8 hours

To determine the average number of complaints per hour, find the average time between the arrival of complaining customers.

$$\text{Average time between arrivals} = \sum (\text{time} \times \text{probability})$$
$$= 2 \times 0.25 + 6 \times 0.45 + 10 \times 0.2 + 14 \times 0.1$$
$$= 6.6 \text{ minutes}$$

therefore, average number of customers per hour $= 60/6.6 = 9.09$. Time spent by manager dealing with complaints is:

$$0.3 \times 9.09 \times 7 \times 8 = 152.71 \text{ minutes} = 2 \text{ hours } 33 \text{ minutes per day}$$

(d) Repeat the simulation but with a second complaints officer. The proportion of complaints going to the manager can then be calculated and hence, the time he spends each day can be found as in (c). The value of the reduction in the manager's time must be balanced against the cost of providing a second complaints officer.

Exercise 14.4

1 S_n denotes that the account is settled at the end of month n. S_n^* denotes that the account is not settled at the end of month n. The problem may be illustrated by a tree diagram:

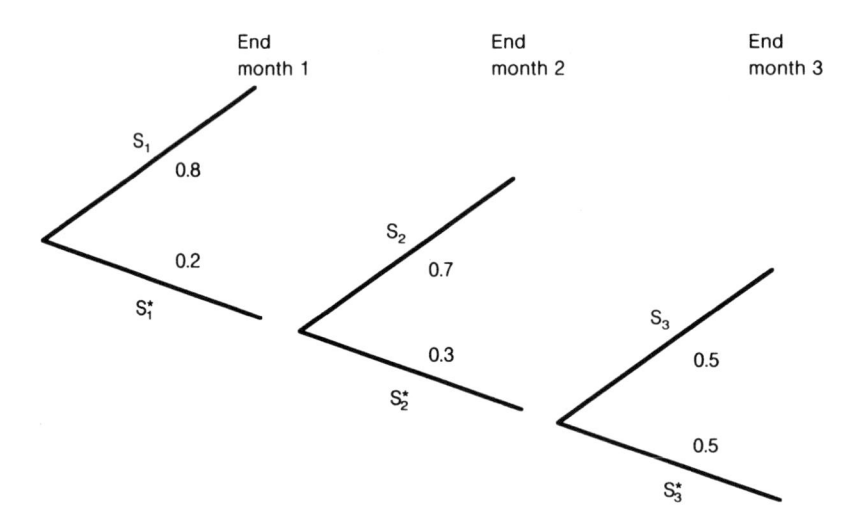

(a) $P(\text{a/c settled at end month 2}) = P(S_1^*) \times P(S_2)$
$$= 0.2 \times 0.7 = 0.14$$

(b) $P(\text{a/c settled at end month 3}) = P(S_1^*) \times P(S_2^*) \times P(S_3)$
$$= 0.2 \times 0.3 \times 0.5 = 0.03$$

(c) Payment is received at the end of month 4 only if the amount is less than £1000 (agency used).
Proportion of accounts $\leqslant £1000$ is $(0.3 + 0.2 + 0.1) = 0.6$
$P(\text{a/c settled at end month 4})$
$$= P(S_1^*) \times P(S_2^*) \times P(S_3^*) \times P(S_4) \times P(\text{a/c} \leqslant £1000)$$
$$= 0.2 \times 0.3 \times 0.5 \times 1 \times 0.6 = 0.018$$

(d) $P(\text{a/c settled at end month 6})$
$$= P(S_1^*) \times P(S_2^*) \times P(S_3^*) \times P(S_6) \times P(\text{a/c} > £1000)$$
$$= 0.2 \times 0.3 \times 0.5 \times 1 \times 0.4 = 0.012$$

2 Expected value of a new account of £2000, A, $= \sum (p_k \times A_k)$ where p_k is the probability that payment A_k is received at the end of month k.

Expected present value of amount outstanding $= \sum p_k \times \dfrac{A_k}{(1 + r/100)^k}$

The value of the debt is £2000, therefore if it is not paid by the end of month 3, legal proceedings are taken. Estimate of the expected proportion recovered if legal action is taken:

Expected proportion recovered $= \sum$ (mid-point proportion) $\times$ probability
$= 0.2 \times 0.1 + 0.5 \times 0.3 + 0.7 \times 0.4 + 0.9 \times 0.2 = 0.63$

therefore the expected amount recovered $= 0.63 \times £2000 = £1260 = A_6$.

$$P(a/c \text{ of } £2000 \text{ settled at end month } 6)$$
$$= P(S_1^*) \times P(S_2^*) \times P(S_3^*) \times P(S_6)$$
$$= 0.2 \times 0.3 \times 0.5 \times 1 = 0.03 = P_6$$

Account settled at end of month	Probability P_k	Amount received, A_k £'000	PV of amount received, £'000	Expected PV, £
1	0.8	2	2 × 0.9852 = 1970.4	1970.4 × 0.8
2	0.14	2	2 × 0.9707 = 1941.4	1941.4 × 0.14
3	0.03	2	2 × 0.9563 = 1912.6	1912.6 × 0.03
6	0.03	1.260	1.26 × 0.9145 = 1152.3	1152.3 × 0.03
			Total	1940.06

The expected PV of the amount recovered is £1940.06

3 The variables in the problem are:

(i) the size of the account;
(ii) the time taken to settle the account;
(iii) if legal proceedings are required, the proportion of the debt recovered.

(i) Size of the account:

Size of account, £	Mid-point, £	Probability	Cumulative Probability	Random number
0 but < 200	100	0.1	0.1	0
200 but < 500	350	0.2	0.3	1–2
500 but < 1000	750	0.3	0.6	3–5
1000 but < 2000	1500	0.3	0.9	6–8
2000 but < 5000	3500	0.1	1.0	9

(ii)

Account settled end of month	Probability	Cumulative Probability	Random number
1	0.80	0.80	00–79
2	0.14	0.94	80–93
3	0.03	0.97	94–96
4 or 6	0.03	1.00	97–99

(iii) If legal proceedings are taken:

Proportion recovered	Mid-point	Probability	Cumulative Probability	Random number
0 but < 0.4	0.2	0.1	0.1	0
0.4 but < 0.6	0.5	0.3	0.4	1–3
0.6 but < 0.8	0.7	0.4	0.8	4–7
0.8 but < 1.0	0.9	0.2	1.0	8–9

Simulation

Account 1: Using the random digits in the order given, the RN 8, produces an account value of £1,500. The next two digit RN, 87, produces a settlement time of end of month 2. Therefore the present value of the amount received is:

$$£1500 \times 0.9707 = £1456.05$$

Account 2: The first RN, 9, produces an account size of £3500. The next 2 random digits, 98, select a settlement time of 4 or 6 months. Since the account is > £1000, legal proceedings are taken and a proportion is recovered at the end of month 6. The next RN is 2 which selects the proportion recovered as 0.5. Therefore the present value of the amount received is:

$$£3500 \times 0.5 \times 0.9145 = £1600.38$$

Exercise 14.5

1 If the total cash flow in years 1, 2 and 3 is less than £42,000, the net cash flow will be negative. The combinations of cash flow which total less than £42,000 are given in the table below:

Cash Flow, £'000

Year 1	Year 2	Year 3	Total	Probability
10	10	10	30	$0.3 \times 0.1 \times 0.3 = 0.009$
10	10	20	40	$0.3 \times 0.1 \times 0.5 = 0.015$
10	20	10	40	$0.3 \times 0.2 \times 0.3 = 0.018$
15	10	10	35	$0.4 \times 0.1 \times 0.3 = 0.012$
20	10	10	40	$0.3 \times 0.1 \times 0.3 = 0.009$
				Total = 0.063

The probability of a negative cash flow is 0.063.

2 Expected cash flow = $\sum$ cash flow × probability.

Year 1: Expected cash flow = $10 \times 0.3 + 15 \times 0.4 + 20 \times 0.3$ = 15 £'000
Year 2: Expected cash flow = $10 \times 0.1 + 20 \times 0.2 + 30 \times 0.4 + 40 \times 0.3$ = 29 £'000
Year 3: Expected cash flow = $10 \times 0.3 + 20 \times 0.5 + 30 \times 0.2$ = 19 £'000

$$\text{Present Value of the cashflow} = 15 \times 0.8696 + 29 \times 0.7561 + 19 \times 0.6575$$
$$= 47.4634 \quad £'000$$

The Net Present Value of the new machine = £(47,463 − 42,000) = £5463

3 Allocate random number ranges to the cash flows for each year.

Year 1

Cash flow £'000	Probability	Random number
10	0.3	0–2
15	0.4	3–6
20	0.3	7–9

Year 2

Cash flow £'000	Probability	Random number
10	0.1	0
20	0.2	1–2
30	0.4	3–6
40	0.3	7–9

SOLUTIONS TO EXERCISES: CHAPTER FOURTEEN

	Year 3	
Cash flow, £'000	Probability	Random number
10	0.3	0–2
20	0.5	3–7
30	0.2	8–9

We can now carry out the simulation.

Number	Year 1, £'000 RN	Cash flow	DCF	Year 2, £'000 RN	Cash flow	DCF	Year 3, £'000 RN	Cash flow	DCF	£'000 Net PV
1	4	15	13.044	2	20	15.122	7	20	13.150	−0.684
2	7	20	17.392	4	30	22.683	9	30	19.725	17.800
3	6	15	13.044	8	40	30.244	4	20	13.150	14.438
4	5	15	13.044	0	10	7.561	0	10	6.575	−14.820
5	0	10	8.696	1	20	15.122	3	20	13.150	−5.032
									Total	11.702

The average net present value of the cash flow = 11.702/5 £'000
= £2340.4

Three out of the five simulations produced negative NPV, therefore we estimate the probability of a negative NPV as 3/5 = 0.6. Since the simulation is small, the estimates are unlikely to be reliable.

Exercise 14.6

1 The policy of the company means that new employees are taken into grade SC1, but vacancies in the other two grades are filled by promotion only. Employees are lost from all three grades by 'wastage', however. The total recruitment into grade SC1 is equal to the total wastage from all three grades.

The variables in this problem are the numbers of people leaving from each of the three grades. Allocate random variable ranges to these values.

Grade SC1

Number leaving	Mid point	Probability	Random number
0–4	2	0.2	0
5–9	7	0.3	1–3
10–14	12	0.4	4–7
15–19	17	0.2	8–9

Grade SC2

Number leaving	Mid point	Probability	Random number
0–4	2	0–2	0–1
5–9	7	0.5	2–6
10–14	12	0.3	7–9

Grade SC3

Number leaving	Mid point	Probability	Random number
0–4	2	0.4	0–3
5–9	7	0.4	4–7
10–14	12	0.2	8–9

We can now run the simulation.

		Wastage/year					
Year	RN	SC1	RN	SC2	RN	SC3	Total/year
1	3	7	5	7	9	12	26
2	2	7	4	7	5	7	21
3	8	17	1	2	2	2	21
4	8	17	6	7	2	2	26
5	2	7	3	7	1	2	16
Total		55		30		25	110

The recruitment each year will be the total wastage for each year.

2 The average number of new employees each year = 110/5 = 22. The number of people promoted from SC1 to SC2 is given by:

$$\text{wastage SC2} + \text{promotion from SC2 to SC3}$$

and promotion from SC2 to SC3 = wastage from SC3. Therefore, the average number of people promoted from SC1 to SC2 per year is given by:

$$(\text{wastage from SC2} + \text{wastage from SC3})/5$$
$$= (30 + 25)/5 = 55/5 = 11$$

Expected wastage/year = $\sum$ (wastage/year) × (probability)
SC1: expected wastage/year = 2 × 0.1 + 7 × 0.3 + 12 × 0.4 + 17 × 0.2 = 10.5
SC2: expected wastage/year = 2 × 0.2 + 7 × 0.5 + 12 × 0.3   = 7.5
SC3: expected wastage/year = 2 × 0.4 + 7 × 0.4 + 12 × 0.2   = 6.0
Total expected wastage/year = 10.5 + 7.5 + 6.0 = 24

This figure is compared to 22 per year from the simulation.

Expected number of promotions/year from SC1 to SC2 = 7.5 + 6.0 = 13.5

This is compared to 11 per year from the simulation.

The two comparisons seem fairly favourable, but as usual we must remember that the simulation is very small.

3 The main difference between the current simulation and one which would enable us to estimate the time taken for a new employee to reach SC3 grade, is that the current simulation does not distinguish between individual employees. A simulation which measured the time required for promotion would have to be set up to identify an individual's movement through the company. It would be a much more complex simulation since we would have to chart each new employee's progress year by year.

Additional data would also be required. We require the probability of an individual leaving each grade in each year. If this probability is assumed to be constant over time, then we could estimate it from the existing data.

Grade	Expected wastage/year	Total in grade	Probability of an individual leaving in year
SC1	10.5	54	10.5/54 = 0.194
SC2	7.5	48	7.5/48 = 0.156
SC3	6.0	32	6.0/32 = 0.188

If the probability of leaving in a given year is thought to be dependent on length of service, then more detailed data must be collected and an appropriate probability distribution constructed. Random numbers can be used to select values of the probabilities and a rule made so that if the probability selected is above a certain level, the employee leaves and if it is below that level the employee stays. Since the promotions are made on the basis of seniority, a record must be kept of each person's position within their grade.

Additional exercises

Students are encouraged to attempt these questions. However, the answers are provided only in the manual issued to lecturers. Lecturers may obtain a free copy of the manual by writing to: Sales and Marketing Department, Longman Group UK Ltd, 21–27 Lamb's Conduit Street, London WC1N 3NJ.

1 (i) Mullins produces STLV components, which are heat treated, turned, and then inspected. Four per cent of components are incorrectly heat treated and, independently, 2% are incorrectly turned. Inspection is difficult because of the high volumes. Of the defective items, the inspection system rejects only 94%. Of the satisfactory items, it actually rejects 1%. What percentage of the components passed by inspection should have been rejected?
What percentage rejected by inspection should have been passed?

(ii) The same STLV components are sold in boxes of 20. What is the probability that all the components in a box are satisfactory?

(iii) Mullins make a different component, HTLV, which is sold to a specification that requires their diameter to be within the range 30mm $\pm$ 0.1mm, and their height to be within the range 50mm $\pm$ 0.2mm. The two dimensions are independent and are normally distributed, the diameter with a mean of 29.95mm and a standard deviation of 0.05mm, and the height with a mean of 50.02mm and a standard deviation of 0.09mm. What proportion of the HTLV components will meet these specifications?

2 A customer orders an item from a mail order catalogue. If the item is in stock the delivery time will have the following distribution:

Time (mail order to customer), days	2	4	6	8
Probability	0.1	0.3	0.4	0.2

There is a 0.9 probability that the item is in stock.

If, however, the mail order company does not have the item in stock, it has to be ordered from the supplier. The delivery time distribution from the supplier is:

Time (supplier to mail order), days	2	4	6
Probability	0.4	0.3	0.3

The company then sends the item by its priority delivery system which has the following distribution:

Priority time (mail order to customer), days	2	3	4	5
Probability	0.1	0.4	0.3	0.2

(a) What is the probability that the customer will receive the item in:

(i)   at least 4 days?
(ii) exactly 6 days?
(iii) more than 6 days?

(b) What is the probability distribution for the delivery time to the customer and what is the expected delivery time?

3 The Crispie Cookie Company produces a range of pre-packed biscuits which are sold in 500g packets. A biscuit is produced by baking a measured quantity of the appropriate mixture, and then a specified number of biscuits is placed into each packet. Because the baking process cannot be completely controlled, there is an unavoidable variation in the weights of biscuits produced. Each packet is therefore automatically weighed, and any packets weighing less than the required 500g are returned for reprocessing. Whenever a packet is rejected in this way, it is estimated to cost £0.1.

The company is about to start selling a new type of biscuit which has an average weight of 40g and a standard deviation of 6g. At full production, the weekly output of this biscuit will be 500,000 packets and the cost per packet (in pounds) is given by:

$$0.05 + 0.01n$$

where n is the number of biscuits in each packet.

Required:
(a) If 13 biscuits are put into each packet, what will be the mean and standard deviation of the weight of a packet?
(b) Explain briefly why the weight of a packet of biscuits will be approximately normally distributed. What is the probability that a packet of 13 biscuits will be rejected as underweight?
(c) Determine the minimum cost number of biscuits per packet by calculating the average weekly production cost for packet sizes of 13, 14 and 15 biscuits.

(ACCA, June 1987)

4 The Titan Tool and Manufacturing Company (TTM) regularly receives large batches of a certain component for further machining. Each batch is inspected on a sampling basis, the procedure employed being to take a sample of 10 components at random, and to accept the batch if the sample contains fewer than 2 defectives. When the contract for the supply of these components was drawn up, it was agreed by the supplier that the proportion defective would not normally exceed 5%. Recently, however, a larger than expected number of batches has been rejected and TTM is beginning to question the adequacy of its sampling scheme. In particular, the production manager of TTM has suggested that the sample size should be increased from 10 to 100 to give a better assessment of the quality of the incoming batches of components.

Required:
(a) On the basis of the current sample of 10 components, what is the probability of accepting a batch which has exactly 5% defective components?
(b) If the sample size were increased to 100 components, the production manager has indicated that a batch which is 5% defective should be accepted with a probability of at least 0.95. If this is to be the case, what is the maximum number of defectives that can be permitted to occur in a sample of 100 if the batch is to be accepted?
(c) As well as increasing the size of the sampling scheme, the production manager of TTM has complained to the company supplying the components about the apparent decrease in quality. The last four batches inspected under the old scheme had respectively 1, 4, 1 and 2 defectives in the 10 components sampled. The first two batches inspected under the new scheme produced respectively 8 and 7 defectives in the sample of 100 components. Is there any evidence of a reduction in the percentage of defective items being supplied following the complaint by the production manager?

(ACCA, June 1983)

5 The South Midlands Pear Growers Federation (SMPGF) markets for a group of orchards. Before the 19X8 season began, SMPGF was approached by Permains, a large retailer who proposed that SMPGF sell 150 tonnes of pears at a fixed price of 17 pence per kg (1 tonne = 1000kg). Permains would pay for all transport costs for the pears to their depot. SMPGF has to decide whether to accept this offer or whether to try to market the entire crop on the open market. In the latter case, SMPGF will have to pay for the transport to the wholesale market. This is likely to cost £10 per tonne for the season.

An uncertainty exists about the size of the crop. If the season is good, the crop is expected to be large (300 tonnes). However, if the weather is bad, the crop will be small (200 tonnes). Based on current forecasts, SMPGF have assessed the probability of a good season as 60%.

The price of the pears on the open market depends on the size of the crop in both the South Midlands and in Kent, the other major producing area. Assume that the crop size in the two areas are independent of each other and that the probability of a good season in Kent has been assessed as 50%. The table below summarises SMPGF's assessment of the market prices for the possible combinations of crop size.

Price, pence/kg

		Kent crop size	
		Large	Small
SMPGF crop size	Large	12	16
	Small	14	30

Required:
(a) Assuming that SMPGF will be able to sell its entire crop in any case, illustrate the situation by a decision tree. Should SMPGF accept Permains' offer if it wishes to maximise its expected revenue?
(b) The Director of SMPGF has proposed an alternative arrangement which could be made with Permains. In this case the contract price paid by Permains would be reduced to 16 pence per kg and SMPGF commits itself to deliver 75 tonnes of pears at this price with the option of supplying the remaining 75 tonnes at the same price after it determines the size of the crop. This decision will be made before the size of the Kent crop is known, however. The drawing up of the necessary contract would involve legal fees of some £500, payable by SMPGF. Draw a new decision tree for this proposal and, on the basis of expected revenues, decide whether SMPGF should propose this new arrangement to Permains?

6   The daily demand for a particular newspaper at Milnes Newsagents has been recorded over the last 30 weekdays.

Number of newspapers demanded per weekday

	100	120	140	160	180
Number of days	3	6	9	6	6

The newsagent buys the papers for 20p each. They sell for 30p. If they are not sold, he can reclaim 5p. If he runs out, he feels he should penalise himself 3p per paper, for loss of future orders.

Required:
(a) Using the criterion of maximin payoff, what would you suggest the newsagent should do?
(b) Using the criterion of minimum expected opportunity loss, what would you now recommend? Compare the effect of the two decision criteria.
(c) How much would it be worth paying for a perfect daily forecasting system?

7   Flora Domestic Appliances Ltd manufactures a range of kitchen equipment, which includes two models of washing machine, A and B. Anyone purchasing either model of machine can take out an annual maintenance contract for which the customer gets free repairs and maintenance in the event of any breakdowns or faults occurring. The cost of any necessary parts is borne by the customer.

Recently, Flora has been evaluating the profitability of these maintenance agreements. To obtain a general indication of the costs involved, a sample of 12 contracts (6 for each model) was analysed, giving rise to the following information on the number of maintenance visits and the total time involved during the previous year.

Contract number	Model	Number of visits	Total time involved (hours)
1	A	1	2.0
2	A	2	5.4
3	A	3	6.6
4	A	5	11.2
5	A	2	3.8
6	A	1	1.8
7	B	1	4.6
8	B	5	14.2
9	B	4	14.2
10	B	3	10.6
11	B	2	6.0
12	B	4	8.4

The company costs routine maintenance and repair work at £10 per visit plus £5 per hour for the time involved.

Required:
(a) Explain why it would be statistically invalid to use a t test to compare the average number of visits for the two models.
(b) Determine the mean and the standard deviation of the total maintenance cost of each model.
(c) Using an appropriate t test, determine whether there is any significant difference between the annual maintenance costs of the two models.

(ACCA, June 1987)

8   The Arcas Appliance Company is currently considering various changes to its pension scheme. In particular, male employees' contributions are to be increased from 5% to 7% of the basic wage, in return for which there will be enhanced pension provision for wives and children in the event of the employee dying before reaching retirement age. It is envisaged that all future employees will be obliged to join the new scheme, but existing employees will be allowed to choose whether or not they will change from the old scheme to the new one. To assess the likely level of acceptance of the new scheme amongst existing employees, a survey was conducted at one of the company's factories, producing the following results according to age and marital status.

Do you intend to change to new scheme?	Age 30 or Under Married	Age 30 or Under Single	Age over 30 Married	Age over 30 Single
Yes	109	153	362	52
No	43	124	207	67
Don't Know	27	102	46	3

Required:
(a) Describe in general terms the main features which seem to be indicated by the data.
(b) Assuming that the results obtained may be regarded as a random sample of the company's employees, test whether there is evidence of a different attitude towards the new scheme depending upon the:

   (i) age of the employee (ignoring marital status);
   (ii) marital status of the employee (ignoring age).

(c) Discuss briefly various ways in which 'Don't Knows' could be dealt with in an analysis of this type and justify the approach that you have adopted.

(ACCA, June 1984)

9   Plutonic Products Ltd is currently involved in wage negotiations with the union representing its 562 manual workers. The union is claiming that on average the earnings of this group of employees in the last 12 months have not increased by as much as the rate of inflation. The company is contesting this claim and, to prove its point, the earnings of a random sample of 10 employees in both 19X1 and 19X2 are presented as follows:

	Annual Earnings (£)	
Employee	19X1	19X2
1	8,694	9,433
2	7,011	7,586
3	6,120	6,708
4	6,761	7,241
5	5,940	6,320
6	6,204	6,738
7	7,177	7,694
8	6,912	7,527
9	7,416	8,002
10	6,581	7,015

During the same period, the value of the Retail Price Index was as follows:

Date	13.1.81	14.4.81	14.7.81	13.10.81	12.1.82	13.4.82	13.7.82	12.10.82	11.1.83
RPI	277.3	292.2	297.1	303.7	310.6	319.7	323.0	322.9	324.6

Due to the fact that the company's business is very seasonal, and also because various different incentive schemes were in operation during the period, both the company and the union agree that it is necessary to take account of a full year's earnings.

Required
(a) Using the Retail Price Index as a measure of inflation, what is the appropriate rate of inflation against which to compare increases in annual earnings between 19X1 and 19X2?
(b) On the basis of a simple arithmetic mean percentage increase for the 10 employees, would you accept the union's claim that the rate of increase in annual earnings has been less than the rate of inflation?

(ACCA, December 1983)

10 (a) Elsi plc manufacture clothing. They have introduced a computerised cutting machine which has not worked consistently to the required tolerances. This has meant that seconds, which are sold at cost, have run at 5% of production. The supplier of the machine has completed modifications. A batch of clothes has been cut, made up and inspected. A sample of 200 contains 6 seconds. Does this indicate that the machine has improved?

(b) The supplier has also suggested that Elsi's standards are too stringent and that he has supplied his machine to customers who are entirely satisfied. He takes the Elsi inspectors to such a customer. They inspect 400 clothes from an identical machine and find 6 seconds. Does this suggest that the two machines produce to different standards?

(c) Another problem with the computerised cutting machine is breakdowns. The manufacturer claimed that it would break down on average once per month. During the last two months it has broken down a total of five times. If the manufacturer's claim were correct, how likely is it that there would have been five or more breakdowns in the two month period? What conclusion do you draw?

11 P G Dips is a restaurant situated in the centre of a city. The owner is trying to expand his business and has introduced a rapid delivery service to the customers' homes. He wants to cost this service more accurately and has decided to collect data on delivery times and distances from the restaurant to the customers. The city is built up for about a five mile radius around the restaurant. The data collected are given below:

Distance, miles	Delivery times, mins	Distance, miles	Delivery times, mins
10.5	27	1.3	10
3.5	16	9.2	26
2.4	12	6.2	21
9.8	27	6.1	22
8.2	23	1.0	8
4.9	19	3.0	14
4.2	17	7.9	24
9.1	25	1.5	10
7.8	23	4.1	17
3.1	13	8.9	25

Required:
(a) Comment briefly on the factors that should be considered when collecting this sample of data.
(b) Establish relationships between time and distance which could form the basis of the owner's costing system for the delivery service. Advise the owner of the likely accuracy of the estimated times.

12 Fast Foods Ltd is a major food retailing company which has recently decided to open several new restaurants. In order to assist with the choice of siting these restaurants the management of Fast Foods Ltd wished to investigate the effect of income on eating habits. As part of their report a marketing agency produced the following table showing the percentage of annual income spent on food, y, for a given annual family income (in pounds), x.

ADDITIONAL EXERCISES   631

x	y
5,000	62
7,500	48
10,000	37
12,500	31
15,000	27
20,000	22
25,000	18

Required:
(a) Plot, on separate scatter diagrams:
  (i) y against x,
  (ii) $\log_{10} y$ against $\log_{10} x$,
  and comment on the relationship between income and percentage of family income spent on food.
(b) Use the method of least squares to fit the relationship:

$$y = ax^b$$

to the data. Estimate a and b.
(c) Estimate the percentage of annual income spent on food by a family with an annual income of £18,000.

(ACCA, December 1987)

13 Neville plc has been trading for about three years in the wholesale fruit and vegetable business. They have been expanding steadily but with the seasonal fluctuations typical of their trade. They have decided that they should formalise their forecasting systems and have gathered the data below on turnover per quarter.

Neville plc turnover in £m per quarter

	Jan–Mar	Apr–Jun	Jul–Sep	Oct–Dec
19X4	—	—	0.5	0.2
19X5	0.5	1.3	1.9	0.7
19X6	1.5	2.0	3.0	1.6

Required:
(a) Use these data to set up a model to forecast turnover for the next four quarters.
(b) If the actual sales in January to March 19X7 were £3.2m, explain, without detailed calculations, how this information would affect your forecasts for the rest of the year.

14 Over the last few years Mr and Mrs North have been saving in a Building Society account. Their balance now stands at £12,700 and they feel that they ought to invest at least £5,000 of this money in an investment scheme which will generate a larger income than the Building Society interest. However, for safety, the Norths want to keep at least £2,000 in the Building Society account.

They consult an accountant friend, Ted West, for advice. He produces two suitable investments for their consideration. Details are given below:

	XYZ Securities 5 Star Plan	ABC Assurance Co Premium Plan
Estimated annual interest rate	16%	20%
Risk measure per £1 invested	0.07	0.10

The risk measure indicates the relative uncertainty associated with the investment in terms of realising the projected annual interest rate (higher values indicate greater risk).

Ted West suggests that a suitable investment portfolio could be constructed by investing funds in both schemes as long as an overall interest rate of at least 18% per annum is achieved. He also advises that for tax purposes the sum invested in ABC should be no more than 70% of the total sum invested in the two schemes.

Required:
(a) If Mr and Mrs North want to minimise the risk in their investment, how should they invest their money?
(b) If the Norths decide to ignore the risk and to use a maximum return-on-investment strategy, how then should they design their portfolio?
(c) What effect would a decrease of 1% in the annual interest rate of the ABC scheme have on the decisions in (a) and (b)?

15 GRM Associates plc is a multi-national conglomerate which owns companies in a variety of industries.

(a) One of these companies is the Toby Pottery Company which specialises in the production of mugs of various kinds. There are five main product lines called A, B, C, D, and E, for convenience. The factory is organised into three departments, 'Forming and Decorating', 'Firing' and 'Packing'. The product requirements per twenty mugs in each of these departments are given in the table below, together with raw material needs, weekly availabilities and selling prices.

	A	B	C	D	E	Availability per week
Raw material, kg	6.0	6.5	6.1	6.1	6.4	35,000
Forming & Decorating (hours)	1.0	0.75	1.25	1.0	1.0	6,000
Firing (hours)	3.0	4.5	6.0	6.0	4.5	30,000
Packing (hours)	0.5	0.5	0.5	0.75	1.0	4,000
Selling price, £ per 20 mugs	40	42	44	48	52	

Product Requirements per 20 mugs

The production costs are as follows:

Raw Material	£2.10 per kg.
Forming and Decorating	£3.00 per hour.
Firing	£1.30 per hour.
Packing	£8.00 per hour.

Formulate the linear programming model which will enable the management of the Toby Pottery Company to determine the product mix which would maximise the weekly contribution to profit. You are not required to solve the model.

What basic assumptions are being made when an LP model is used to solve this problem?

(b) One of GRM's other companies, John Bull plc, produces four similar products in four main departments. Raw materials are unlimited, but in this case there are additional constraints in the shape of maximum predicted demand for each product. The management of John Bull plc has already formulated the contribution maximising linear programming model for these four products and has solved the model using a computer package. The model and the final simplex tableau are given below.

Produce a units of product 1 per month.
Produce b units of product 2 per month.
Produce c units of product 3 per month.
Produce d units of product 4 per month.

Maximise contribution per month, P, in pounds per month, where:

$$P = 40a + 30b + 25c + 80d \quad \text{£ per month}$$

subject to:

Department 1	$3a + 1.5b + 2c + 3.5d$	$\leqslant 1200$ hours per month.
Department 2	$2a + 2b + 3.5c + 4d$	$\leqslant 1200$ hours per month.
Department 3	$3.5a + 3b + 4.5c + 6d$	$\leqslant 1800$ hours per month.
Department 4	$0.5a + 0.5b + c + 1.5d$	$\leqslant 315$ hours per month.
	a	$\leqslant 200$ units per month.
	b	$\leqslant 150$ units per month.
	c	$\leqslant 130$ units per month.
	d	$\leqslant 50$ units per month.
	a,b,c,d	$\geqslant 0$

Final Tableau, with constraints entered in the order given in the model:

	a	b	c	d	s1	s2	s3	s4	s5	s6	s7	s8	rhs
s1	0	0	0	0	1	0	0	−2	−2	−0.5	0	−0.5	70
s2	0	0	0	0	0	1	0	−3.5	−0.25	−0.25	0	1.25	72.5
s3	0	0	0	0	0	0	1	−4.5	−1.25	−0.75	0	0.75	57.5
c	0	0	1	0	0	0	0	1	−0.5	−0.5	0	−1.5	65
a	1	0	0	0	0	0	0	0	1	0	0	0	200
b	0	1	0	0	0	0	0	0	0	1	0	0	150
s7	0	0	0	0	0	0	0	−1	0.5	0.5	1	1.5	65
d	0	0	0	1	0	0	0	0	0	0	0	1	50
z	0	0	0	0	0	0	0	25	27.5	17.5	0	42.5	18125

Advise the management of John Bull plc on the optimum production mix for their products, the consumption of resources and the advisability of increasing resources in any of the departments.

16 The three factories of Cross plc supply its four stores. For one particular product, the variable cost at each of the factories is given below.

Factory—variable cost £ per item

Kenward	Week	Tokay
£20	£22	£21

The transport costs from factory to store, and next month's requirements and availabilities, are given in the following table.

		To Store (£ per item)				
		Beishon	Vinas	Jones	Browns	Total items available
From	Kenward	3	2	1	2	100
	Week	2	2.5	1.5	2.5	150
	Tokay	1	1.5	3.5	3	130
	Total required	80	70	50	150	

Required:
(a) Find the cheapest factory loading and transport system for next month.
(b) What would happen to your allocation if the transport costs from Tokay to Browns were reduced from £3 per item to £1 per item?
(c) If Vinas refused to take any deliveries from Week, what effect would this have on your initial allocation?

17 The government of the Republic of Semele has decided, as a matter of top priority, to build a new road joining the two main cities of Axis and Boreas. Because of the need to complete the project as quickly as possible, the work has been divided into five stages which are to be built simultaneously. Within Semele there are six companies large enough to undertake the construction of any of the five stages and each company has been invited to submit a tender for each stage of the project. The tenders (in millions of Semele dollars) are as follows:

Company	Stage 1	2	3	4	5
A	49	84	63	82	68
B	53	92	62	no bid	67
C	54	86	67	78	68
D	46	86	62	76	no bid
E	57	94	66	83	70
F	50	82	65	80	72

Required:
(a) Assuming that none of the companies is large enough to undertake the work of more than one stage, advise the government how the five contracts should be allocated. What is the minimum total cost for the project?
(b) On speaking to representatives of the six companies it is discovered that A, B, D and F have the capacity to undertake any two stages simultaneously and that C can undertake any three stages simultaneously. Show how the problem may now be formulated and solved using the 'transportation' algorithm. What is now the minimum cost allocation of contracts?

(ACCA, June 1984)

18 Orbona Office Equipment Ltd have recently developed a new colour copying machine and are planning to launch it at a forthcoming trade show. In the meantime, salesmen and service engineers will have to be trained to deal with the new copier, the necessary documentation and publicity material must be got ready, and the preparations for the show itself must be completed. In particular, the product manager has drawn up a list of the activities which will be involved and he has also estimated the order and approximate cost and duration of each activity as shown in the following table.

	Activity description	Preceding activities	Cost (£'00)	Duration (weeks)
A	Train Salesmen	C,J	60	3
B	Train Service Engineers	C,I	80	4
C	Prepare Instruction Manuals	—	30	10
D	Organise Trade Show Display	G,J	20	4
E	Train Staff for Trade Show	D	5	1
F	Arrange Pre-Release Publicity	H,J	60	3
G	Arrange Staff for Trade Show	H	2	2
H	Set up Sales Office	—	45	5
I	Have Service Contract Prepared	—	30	5
J	Produce Promotional Material	I	120	8

Required:
(a) Represent the stages involved in the product launch in the form of an appropriate network of activities. Given that the trade show will take place in six months (i.e. 26 weeks time), can all the necessary activities be completed in time for the show?
(b) Draw a graph showing the week by week schedule of cumulative cash flows, assuming that all activities commence at their earliest time. (You may also assume that, for each activity, the cost is incurred evenly over the duration of the activity.) Comment briefly on the form of your graph.

(ACCA, June 1985)

19 An on-line computer system has recently been installed in the accounts department of a television broadcasting company. The activities involved in introducing a computerised accounting function on this system are listed below together with their normal durations and costs. Since it would be possible to shorten the overall project duration by crashing certain activities at extra cost, the relevant details are also included. In addition there will be a weekly charge of £2,500 to cover overheads.

Activity	Immediate predecessors	Normal Duration (weeks)	Normal Cost (£)	Crash Duration (weeks)	Crash Cost (£)
A	—	3	3,000	2	4,000
B	—	6	6,000	—	—
C	A	4	8,000	1	11,000
D	B	2	1,500	—	—
E	A	8	4,000	5	5,500
F	B	4	3,000	2	5,000
G	C,D	2	2,000	—	—
H	F	3	3,000	1	6,000

The crash time represents the shortest time in which the activity can be completed given the use of more costly methods of completion. Assume that it is possible to reduce the normal time to crash time in steps of one week and that the extra cost will be proportional to the time saved.

Required:
(a) Using the normal durations and costs construct an activity network for the introduction of a computerised system. Determine the critical path and associated total cost.
(b) Activities E and F have to be supervised by the chief accountant who will not be available for the first seven weeks of the project period. Both activities, however, can be supervised simultaneously. Determine whether or not this will affect the completion date and, if so, state how it will be affected.

(c) Assuming that the chief accountant will be available whenever required and that all resources necessary to implement the crashing procedures will also be available, determine the minimum cost of undertaking this project.

(ACCA, December 1987)

20 The purchasing manager of Elstore Ltd, an electrical components retailer, holds a regular stock of, among other things, quasitrons. Over the past year he has sold on average 25 a week and he anticipates that this rate of sale will continue during the next year (which you may take to be 50 weeks). He buys quasitrons from his supplier at the rate of £5 for 10, and every time he places an order it costs on average £10 bearing in mind the necessary secretarial expenses and the time involved in checking the order. As a guide to the stockholding costs involved, the company usually value their cost of capital at 20% and as the storage space required is negligible, he decides that this figure is appropriate in this case. Furthermore, the prices charged to customers are determined by taking the purchasing and stockholding costs and applying a standard mark up of 20%.

Required:
(a) Currently the manager is reviewing his ordering and pricing policies and needs to know how many quasitrons he should order each time and what price he should charge. What would be your advice? (State any assumptions that you make.)
(b) If he now finds out that he can get a discount of 5% for ordering in batches of 1,000 would you advise him to amend the ordering and pricing policy that you have suggested and, if so, to what?
(c) How large would the percentage holding cost have to be for the manager to be indifferent between taking advantage of the quantity discount and maintaining the original ordering policy that you have suggested? Comment on the value that you have obtained.

(ACCA, June 1982)

21 Holders of current accounts at the Countrywide Bank incur quarterly account charges if the balance of their account falls below £100 at any time during the quarter. Recently, however, a scheme has been instituted whereby current account customers who also have a deposit account with the bank may have a fixed amount (specified by the account holder) transferred automatically from their deposit account whenever the balance of their current account falls below £100. In this way, the usual current account charges are avoided but each transfer from the customer's deposit account incurs a fixed transaction charge of £2. The bank currently pays interest at 6% per annum on deposit accounts.

Required:
(a) Explain briefly why this scheme can be thought of as a simple stock control system.
(b) Dick Bond holds both a current and a deposit account with the Countrywide Bank and estimates that he draws, on average, about £400 a month from his current account at a more or less uniform rate. Assuming that transfers from his deposit account would be the only payments into his current account, advise him on how much should be transferred on each occasion.
(c) In the past, Dick Bond has maintained a current account balance of, on average, about £125. As a result, he has usually incurred current account charges of about £7.50 per quarter. Assuming that he continues to use his current account in the same way as he has done in the past, would you advise him to change to the new scheme? By how much would the deposit account interest rate have to change before you would alter your decision?

(ACCA, June 1985)

22 A cost clerk in the sales office of Jupiter Metals is responsible for preparing a sales invoice, following the receipt of an order from a customer. One invoice is required for each order. Orders arrive in the office at an average rate of five per hour, and it takes the clerk, on average, ten minutes to make out an invoice. The office is open for seven hours a day: in the morning from half-past eight until midday, and in the afternoon from half-past one until five o'clock. The outgoing mail is collected from the office at 10 o'clock each morning and 3 o'clock each afternoon. In order that all sales invoices are despatched to customers as quickly as possible, the sales manager has indicated that efforts should be made to ensure that for each order received during the morning, an invoice is sent out that same afternoon, and that an invoice for each order received during the afternoon should be sent the following morning.

Required:
(a) If this situation is to be analysed by means of the basic single server queueing model, identify the components of the queueing system, indicate the assumptions which must be made, and comment on the appropriateness of the assumptions in this particular set of circumstances.
(b) Explain, in detail, how a simulation of the system might be carried out in order to estimate the proportion of invoices which fail to meet the despatch deadlines suggested by the sales manager.

(ACCA, June 1982)

23 A hairdressing salon is run by its owner on her own. She can deal with only one customer at a time, but she has chairs for two other customers to wait. Her customers arrive at the salon with the following inter arrival times, based on 100 observations.

Inter arrival time (minutes)	5	10	15	20	25	30
Frequency	14	12	16	29	18	11

Her service time for a sample of 100 customers is:

Hairdressing time (minutes)	10	15	20	25	30
Frequency	16	23	10	27	14

Required:
(a) Use this information to simulate the first 15 customers arriving at the salon. It opens at 9 00 am. Use the following random numbers for:
Inter arrival times 65 13 29 37 81 54 36 28 17 94 76 79 98 01 42
Hairdressing times 22 83 08 12 15 99 46 75 52 34 07 98 33 93 27
(b) Use the simulation to estimate:
 (i) The total time to dress the hair of 15 customers.
 (ii) The proportion of time the hairdresser is idle.
 (iii) The proportion of time there are 1 or 2 customers waiting.
 (iv) The proportion of customers who cannot wait because there is not sufficient space for them.
(c) From this brief simulation suggest how the salon's owner could improve her services, stating how you would use simulation to check the effects of your suggestions.

The Chartered Association of Certified Accountants

December 1988

Level 2—Professional Examination
Paper 2.6

Quantitative Analysis

Time allowed—3 hours
Number of questions on paper—7
FIVE questions ONLY to be answered
Formulae and Extracts from Tables are on pages 10–17

FIVE questions ONLY to be answered

1   At the end of a financial year the chief accountant of each of ten engineering companies was asked to compute six accounting ratios (A, B, C, D, E, F) to describe his company's performance. The accounting ratios D, E, and F for the ten companies were as follows:

Company	Ratio D	Ratio E	Ratio F
1	1.30	1.30	1.45
2	1.45	1.20	1.20
3	1.30	1.25	1.30
4	0.95	0.80	0.75
5	1.80	1.75	1.90
6	1.50	1.60	1.65
7	1.05	1.35	1.50
8	1.30	1.05	0.90
9	0.90	0.95	0.85
10	0.90	1.10	1.00

Required:
(a) **Calculate Spearman's rank correlation coefficient between:**
    (i)   ratio D and ratio E,
    (ii) ratio E and ratio F,
and hence complete the following Spearman's rank correlation matrix.

	A	B	C	D	E	F
A	1.0	−0.7	0.8	−0.8	−0.9	−0.7
B	−0.7	1.0	−0.8	0.9	0.8	0.7
C	0.8	−0.8	1.0	−0.8	−0.7	−0.6
D	−0.8	0.9	−0.8			
E	−0.9	0.8	−0.7			
F	−0.7	0.7	−0.6			

(8 marks)

(b) Use the correlation matrix described in (a) to divide the six accounting ratios into two distinct groups. Explain your reasoning. (3 marks)

(c) Use the test statistic

$$T = \frac{R\sqrt{n-2}}{\sqrt{1-R^2}} \text{ (which has } t_{(n-2)} \text{ distribution under Ho)}$$

to investigate whether there is a significant rank correlation between ratios **D** and **E**. (5 marks)

(d) Explain why a set of data can give a value of exactly 1.0 for R, Spearman's rank correlation coefficient, and a value of less than 1.0 for r, the product-moment correlation coefficient. Is it possible for a set of data to give r = 1.0 and R < 1.0? Support your arguments with a suitable sketch. (4 marks)

(20 marks)

2 Over the course of each year the Krispy Crisps Company Ltd purchases a large number of wooden pallets for use in the storage and transportation of its products to replace those lost or damaged in transit. The average yearly requirement for the past two years has been 3,000 pallets, a quantity which can be applied realistically to this year as well. The need for replacement pallets is relatively constant and the cost associated with the placing and receipt of an order is £15. The inventory cost policy that Krispy Crisps has traditionally employed is to charge 18% of the purchase cost as the annual inventory holding cost for any item in inventory. The standard price charged by the major manufacturing company is £8 per pallet.

Required:
(a) Determine the optimum order quantity and the consequent time between the orders. (6 marks)

(b) Describe the assumptions you have made in part (a) and assess their likely validity within the context of this question. (4 marks)

(c) The manufacturer offers a discount of 3.125% if Krispy Crisps order 2,000 or more pallets at a time. Show that the discount is not financially beneficial to Krispy Crisps. What percentage discount would be required for Krispy Crisps to order 2,000 or more pallets at a time? (8 marks)

(d) State the effect on the company's inventory policy described in (a) if the supply of pallets has a variable lead time. (2 marks)

(20 marks)

3 (a) Briefly describe the terms:
   (i) additive model,
   (ii) multiplicative model,
   as applied to time series analysis. Explain how to distinguish between the appropriateness of these models. (5 marks)

(b) The table below gives the production figures (in thousands of tonnes) of ceramic goods for 1988.

Jan	Feb	Mar	Apr	May	Jun	Jul	Aug	Sep	Oct	Nov	Dec
335	325	310	354	360	338	333	270	375	395	415	373

As the data exhibit seasonal fluctuations, multiplicative seasonal indices have been calculated using data from several years and are shown below.

Jan	Feb	Mar	Apr	May	Jun	Jul	Aug	Sep	Oct	Nov	Dec
96	93	90	102	105	96	94	78	110	115	120	108

Required:

(i) **Calculate the values of the deseasonalised data for each month of 1988.**
(3 marks)

(ii) **Plot the monthly production figures and the deseasonalised data on the same graph. Comment on any apparent trend of the data.** (5 marks)

(iii) **Use the exponential smoothing model, applied to the deseasonalised data, to produce a forecast of the deseasonalised data for the first month of 1989 using the smoothing constant $\alpha = 0.2$ and starting with a forecast of 350 for August 1988.** (4 marks)

(iv) **Use your answer to part (iii) and the seasonal indices to forecast the production figures of ceramic goods for each of the first three months of 1989, stating any assumptions that you are making.** (3 marks)

(20 marks)

4 (a) **Describe briefly the limitations of the linear programming technique in the solution of real industrial product-mix problems.** (4 marks)

(b) Apollo Products Ltd are capable of manufacturing four products—Alphas, Betas, Gammas, and Deltas—and are currently drawing up their production plans for the forthcoming financial year. Information about the sales price, unit costs and requirements, and maximum annual sales for each of these products are as follows:

	Alpha	Beta	Gamma	Delta
Sales price (£ per unit)	112	120	138	165
Cost of materials (£ per unit)	22	20	25	30
Variable overheads (£ per unit)	8	14	20	25
Assembly labour (hours per unit)	7	5	0	0
Packaging labour (hours per unit)	8	10	0	0
Machine time (hours per unit)	0	0	12	15
Maximum sales (units)	2,500	1,500	1,000	1,500

Fixed overheads of the firm amount to £25,000 per annum. Additional variable costs include assembly labour at £4 per hour, packaging labour at £3 per hour and machine time at £4 per hour. Each year, the resource availabilities are:

Resource	Hours available
Assembly labour	15,000
Packaging labour	21,000
Machine time	12,000

Required:

(i) **Formulate a linear programme from the above information and hence explain how the problem can be formulated as two separate linear programming problems.** (6 marks)

(ii) **By drawing two separate graphs determine the product-mix which will maximise profit for the year and state the amount of profit.** (6 marks)

(iii) **Determine the minimum sales price at which the production of Deltas would be worth while.** (2 marks)

(iv) **Determine the dual price of machine time.** (2 marks)

(20 marks)

5   Each week an internal auditor of a large company checks a random sample of 50 of its financial transactions to find out if the accounting system is working correctly. It is expected that a small number of errors will occur, but the company believes that it is essential to keep the error rate to 2% or below. Hence the internal auditor is dissatisfied if he finds more than one error in his sample.

Required:
(a) **Explain why the Poisson distribution can be used to investigate this situation.** (2 marks)

(b) **Determine the probability that, with the underlying population error rate at 2%, the internal auditor's sample leads to him being dissatisfied.** (3 marks)

(c) **Find the number of errors needed in the sample of 50 transactions for the auditor to be confident (with at least 95% confidence) that the stipulated error rate has been exceeded.** (4 marks)

(d) **If the underlying percentage error rate is x%, show that the probability (P), that the auditor is dissatisfied is given by:**

$$P = 1 - \left(1 + \frac{x}{2}\right)e^{-x/2}$$

Draw a graph of P against x over the range $0 \leqslant x \leqslant 10$ and estimate the value of x that gives P = 0.5. (5 marks)

(e) When the internal auditor is not satisfied with the first sample (that is, he finds more than one error) he takes a further sample of 50 financial transactions. Then, if there is a total of 3 or more errors in the two samples combined, he writes an official letter to the chief accountant informing him of his findings.
If the underlying percentage error rate is 2% determine:
(i) **the probability that the auditor writes such a letter,** (4 marks)
(ii) **the expected number of financial transactions the internal auditor checks each week.** (2 marks)

A selection of Poisson probabilities is given in the following table.

Mean (μ)	0	1	Probabilities of 2	3	4 or more
1	0.368	0.368	0.184	0.061	0.019
2	0.135	0.271	0.271	0.180	0.143
3	0.050	0.149	0.224	0.224	0.328

(20 marks)

6   Saturnite plc are to initiate a project to study the feasibility of a new product. The end result of the feasibility project will be a report recommending the action to be taken for the new product. The activities to be carried out to complete the feasibility project are given below.

Activity	Description	Immediate predecessors	Expected Time (weeks)	Number of Staff required
A	Preliminary design	—	5	3
B	Market research	—	3	2
C	Obtain engineering quotes	A	2	2
D	Construct prototype	A	5	5
E	Prepare marketing material	A	3	3
F	Costing	C	2	2
G	Product testing	D	4	5
H	Pilot survey	B,E	6	4
I	Pricing estimates	H	2	1
J	Final report	F,G,I	6	2

Required:

(a) Draw a network for the scheme of activities set out above. Determine the critical path and the shortest duration of the project. (8 marks)

(b) Assuming the project starts at time zero and that each activity commences at the earliest start date, construct a chart showing the number of staff required at any one time for this project. (6 marks)

(c) The management of Saturnite has decided that it does not want more than 9 staff involved in this project at any one time.
Describe how this can be achieved within the shortest duration time found in (a). How many weeks of this project would require all 9 staff members?
(6 marks)
(20 marks)

7 Buranite Publishing Company Ltd are planning to introduce a new specialist accounting textbook. The company's marketing department estimates that the prior distribution for likely sales is normal with a mean of 10,000 books. In addition it has determined that there is a probability of one half that the likely sales will lie between 8,000 and 12,000 books.

The text-book will sell for £10 per copy but the publishing company pays the author 10% of revenue in royalties and the fixed costs of printing and marketing the book are calculated to be £25,000. Using current printing facilities the variable production costs are £4 per book, however the Buranite Publishing Company has the option of hiring a special machine for £14,000 which will reduce the variable production costs to £2.50 per book.

Required:

(a) **Show that the standard deviation of likely sales is approximately $\sigma = 3{,}000$.** (4 marks)

(b) **Using $\sigma = 3{,}000$ determine the probability that the company will at least break even if**
   (i) **existing printing facilities are used,**
   (ii) **the special machine is hired.** (6 marks)

(c) **By comparing expected profits, decide whether or not the publishing company should hire the special machine.** (3 marks)

(d) By using the normal distribution it can be shown that the following probability distribution may be applied to book sales.

Sales ('000)	0–5	5–8	8–10	10–12	12–15	15–20
Probability	0.05	0.20	0.25	0.25	0.20	0.05

By assuming that the actual sales can only take the midpoints of these classes, determine the expected value of perfect information and interpret its value.
(7 marks)
(20 marks)

Formulae and Extracts from Tables

A. *Sample Statistics*:

    (i)   Arithmetic Mean $= \dfrac{\Sigma x}{n}$

    (ii) Standard Deviation

$$= \sqrt{\dfrac{\Sigma(x - \bar{x})^2}{n - 1}} = \sqrt{\dfrac{\Sigma x^2 - (\Sigma x)^2/n}{n - 1}}$$

B. *Probability Distributions*:

    (i)   Binomial $P(r) = {}^nC_r p^r (1 - p)^{n-r}$

    (ii) Poisson $P(r) = \dfrac{e^{-m} m^r}{r!}$

C. *Standard Errors*:

    (i)   Mean $\dfrac{\sigma}{\sqrt{n}}$

    (ii) Proportion $\sqrt{\dfrac{p(1 - p)}{n}}$

    (iii) Difference between means

$$\sqrt{\dfrac{\sigma_1^2}{n_1} + \dfrac{\sigma_2^2}{n_2}}$$

    (iv) Difference between proportions

$$\sqrt{\dfrac{p_1(1 - p_1)}{n_1} + \dfrac{p_2(1 - p_2)}{n_2}}$$

D. *Test Statistics*:

    (i) $Z = \dfrac{\bar{x} - \mu}{\sigma/\sqrt{n}}$ (one sample)

$$Z = \dfrac{(\bar{x}_1 - \bar{x}_2) - (\mu_1 - \mu_2)}{\sqrt{\dfrac{\sigma_1^2}{n_1} + \dfrac{\sigma_2^2}{n_2}}} \text{ (two samples)}$$

    (ii) $t = \dfrac{\bar{x} - \mu}{\hat{\sigma}/\sqrt{n}}$ (one sample)

$$t = \dfrac{(\bar{x}_1 - \bar{x}_2) - (\mu_1 - \mu_2)}{\hat{\sigma}\sqrt{\dfrac{1}{n_1} + \dfrac{1}{n_2}}} \text{ (two samples)}$$

    (iii) $\chi^2 = \Sigma \dfrac{(O - E)^2}{E}$

E. *Correlation and Regression*:

(i) Product moment correlation coefficient

$$r = \frac{n\Sigma xy - \Sigma x \Sigma y}{\sqrt{[n\Sigma x^2 - (\Sigma x)^2][n\Sigma y^2 - (\Sigma y)^2]}}$$

(ii) Spearman's rank correlation coefficient

$$R = 1 - \frac{6\Sigma d^2}{n(n^2 - 1)}$$

(iii) Least squares regression line

$$y = a + bx$$

$$b = \frac{n\Sigma xy - \Sigma x \Sigma y}{n\Sigma x^2 - (\Sigma x)^2}$$

$$a = \frac{\Sigma y}{n} - \frac{b\Sigma x}{n}$$

(iv) For a single value x, the standard deviation of the estimated value of y

$$\hat{\sigma}\sqrt{1 + \frac{1}{n} + \frac{(x - \bar{x})^2}{\Sigma(x - \bar{x})^2}}$$

where $\hat{\sigma}^2 = \dfrac{\Sigma(y - \hat{y})^2}{n - 2}$

F. *Single Server Queues* $(M/M/1)$:

(i) Average number in system

$$\bar{s} = \frac{\lambda}{\mu - \lambda}$$

(ii) Average number in queue

$$\bar{q} = \bar{s} - \frac{\lambda}{\mu} = \frac{\lambda^2}{\mu(\mu - \lambda)}$$

(iii) Average queueing time

$$\bar{w} = \bar{s} \times \frac{1}{\mu} = \frac{\lambda}{\mu(\mu - \lambda)}$$

(iv) Average time in system

$$\bar{t} = \bar{w} + \frac{1}{\mu} = \frac{1}{\mu - \lambda}$$

The extracts from tables supplied with this examination paper are not reproduced here.

End of Question Paper

Reproduced with permission of the Chartered Association of Certified Accountants

Authors' model answers to this examination paper are on page 647.

Authors' model answers to December 1988 ACCA Quantitative Analysis Paper

1 (a) Calculate Spearman's rank correlation coefficient, r_s:
(i) Between ratio D and ratio E. As we are not told what the ratios are we do not know whether the highest ratio is first or last. Nor do we know if D, E and F ratios all work the same way, ie all highest first. We will assume that all the ratios are consistent and that highest is first. We change the ratios into ranks. When two ranks are equal, say $3 =$, we allocate the rank $(3 + 4)/2 = 3.5$ to both companies. When three ranks are equal, say $5 =$, we allocate the rank $(5 + 6 + 7)/3 = 6$ to all three companies.

Company	Ratio D rank	Ratio E rank	Rank difference(d)	d^2
1	(4 =) 5	4	1	1
2	3	6	3	9
3	(4 =) 5	5	0	0
4	8	10	2	4
5	1	1	0	0
6	2	2	0	0
7	7	3	4	16
8	(4 =) 5	8	3	9
9	(9 =) 9.5	9	0.5	0.25
10	(9 =) 9.5	7	2.5	6.25
				$\sum d^2 = 45.5$

$$r_s = 1 - \frac{6\Sigma d^2}{n(n^2 - 1)} = 1 - \frac{6 \times 45.5}{10 \times 99} = 0.724$$

(ii) Between ratio E and ratio F.

Company	Ratio E rank	Ratio F rank	Differences d
1	4	4	0
2	6	6	0
3	5	5	0
4	10	10	0
5	1	1	0
6	2	2	0
7	3	3	0
8	8	8	0
9	9	9	0
10	7	7	0

Therefore, the rank correlation coefficient between E and F = 1.00. It follows that the rank correlation coefficient between D and F is 0.724. The complete Spearman's rank correlation matrix is:

	A	B	C	D	E	F
A	—	−0.7	0.8	−0.8	−0.9	−0.7
B	−0.7	—	−0.8	0.9	0.8	0.7
C	0.8	−0.8	—	−0.8	−0.7	−0.6
D	−0.8	0.9	−0.8	—	0.724	0.724
E	−0.9	0.8	−0.7	0.724	—	1.0
F	−0.7	0.7	−0.6	0.724	1.0	—

(b) Use the values of r_s in the matrix to divide the six accounting ratios into two distinct groups. All the r_s are reasonably large, but some are positive and some are negative. This allows us to split the group. Ratio A is positively related to Ratio C, but negatively to all the other ratios. This pattern is the same throughout the table. A and C are the first related group; B, D, E, F are the second related group.

(c) Use the test statistic:

$$t = \frac{r_s\sqrt{(n-2)}}{\sqrt{(1-r_s^2)}}$$

to test whether there is a significant rank correlation between D and E.
(This is the test statistic for the Pearson Product moment correlation coefficient. The usual statistic for Spearman's rank correlation coefficient is approximately:

$$z = \frac{r_s}{\sqrt{1/(n-1)}}$$

for $n \geq 10$. Ignore this and do as the question asks).

$H_0: \rho_s = 0$: there is no rank correlation between ratios D and E.

$H_1: \rho_s \neq 0$: there is a rank correlation between the ratios.

This is a two tail test. We will test at the 5% level with (10 − 2) = 8 degrees of freedom. Use the test statistic given.

$$t = \frac{0.724\sqrt{8}}{\sqrt{(1-0.724^2)}} = 2.97$$

This is compared with $t_{0.05/2, 8} = 2.306$. As the calculated t is larger than the boundary value, we reject H_0 at the 5% level. There is a significant correlation between ratios D and E.

(d) Explain why a set of data can have $r_s = 1.0$, but a Pearson product moment correlation r of less than 1. Use Ratios E and F to illustrate.

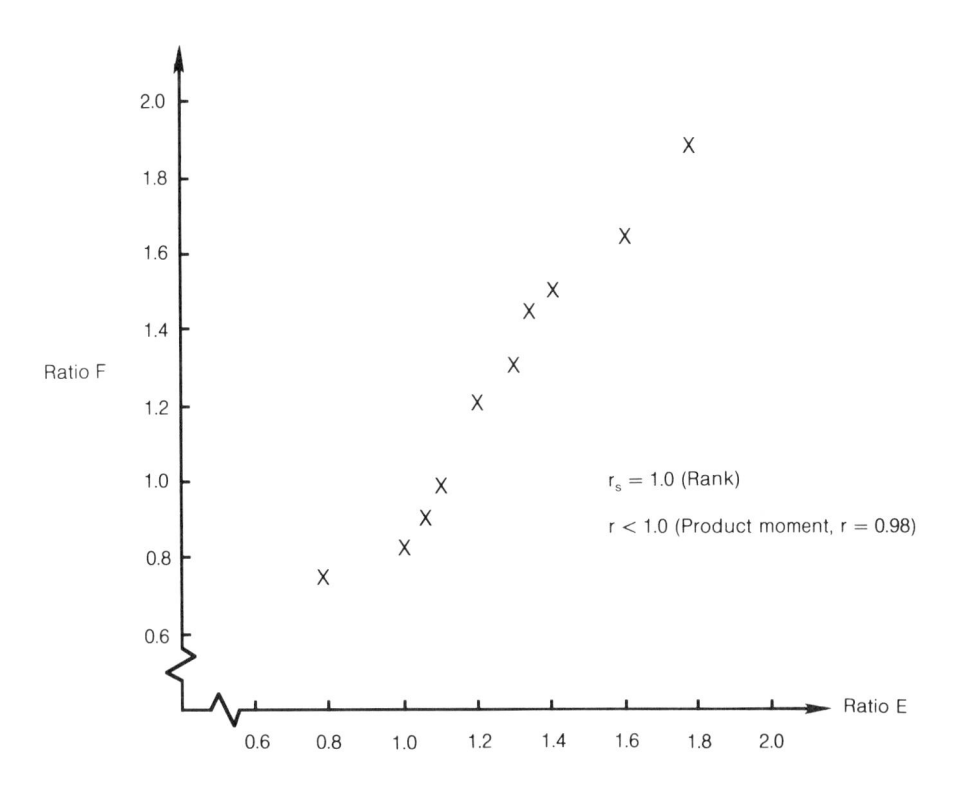

The Spearman rank correlation coefficient for Ratio E and Ratio F is 1.0. It is clear from the diagram that they do not lie exactly on a straight line, which is what must happen for $r = 1.0$.

If, however, Pearson's r does equal exactly 1.0, then the points lie exactly on a straight line and their rank orders must be identical, giving $r_s = 1.0$.

2 (a) Determine the optimum order quantity and the consequent time between the orders.

Annual demand, $D = 3000$ pallets per year.
Cost of order, $C_o = £15$ per order.
Cost of purchase, $C = £8$ per pallet.
Cost of holding, $C_h = 18\%$ of £8 per pallet per year.

The Economic Order Quantity is given by:

$$EOQ = \sqrt{\frac{2C_o D}{C_h}} = \sqrt{\frac{2 \times 15 \times 3000}{0.18 \times 8}} = 250 \text{ pallets per order}$$

The optimum order quantity is 250 pallets.
Number of orders per year $= \dfrac{D}{q} = \dfrac{3000}{250} = 12$.

Therefore the orders should be placed every 1/12 years, that is, once every month approximately.

(b) Assumptions implicit in the EOQ model:
(i) Demand for pallets is spread evenly through the year. This assumes that the pallets are lost or damaged on a regular basis and that production is uniform throughout the year. Both of these are unlikely to be strictly the case in practice.
(ii) Cost of ordering and storage are known and fixed, and independent of the order size. This is probably true for the storage costs, since they are based on the cost of capital. The assumption is probably not reliable for the ordering costs since the order size is likely to affect the cost of delivery, unloading, etc.

650 QUANTITATIVE ANALYSIS

(iii) Lead time is known and fixed so that the new order arrives just as the stock level falls to zero. This assumption may be reliable—it depends on the supplier.

(c) Total annual cost of stockholding and purchase, TC, in pounds per year is:

$$TC = C_o \frac{D}{q} + C_h \frac{q}{2} + CD \text{ £ per year}$$

At q = 250:

$$TC_{250} = 15 \times \frac{3000}{250} + 0.18 \times 8 \times \frac{250}{2} + 8 \times 3000 \text{ £ per year}$$

$$= 180 + 180 + 24000 = £24,360 \text{ per year}$$

With the discount, the purchase price is 96.875% of £8 = £7.75.
Therefore, when q = 2000:

$$TC_{2000} = 15 \times \frac{3000}{2000} + 0.18 \times 7.75 \times \frac{2000}{2} + 7.75 \times 3000 \text{ £ per year}$$

$$= 22.5 + 1395 + 23,250 = £24,667.50 \text{ per year}$$

This is a higher cost. The discount is not worth having, since the saving in the purchase price and ordering costs does not compensate for the increased storage costs.

Consider the purchase price at the point at which the total annual cost is the same, with and without the discount. Suppose at this breakeven point the purchase price is r% of £8. Then:

$$TC_{2000} = 15 \times \frac{3000}{2000} + 0.18 \times \frac{(r \times 8)}{100} \times \frac{2000}{2} + \frac{(r \times 8)}{100} \times 3000 \text{ £ per year}$$

$$= 22.5 + 14.4r + 240r = 22.5 + 254.4r$$

When $TC_{2000} = TC_{250}$:

$$22.5 + 254.4r = 24360$$

therefore:

$$254.4r = 24337.5$$
$$r = 95.6663$$

The discount which must be offered is $(100 - 95.67)\% = 4.33\%$.

(d) If the lead time is variable, then the company may go out of stock of pallets if the re-order level is fixed. This problem can be reduced by the holding of a buffer stock. The size of this buffer stock would be chosen so that the probability of a stock-out is reduced to an acceptable level.

3 (a) Briefly describe the terms 'additive model' and 'multiplicative model'.
One way of modelling time series is to break the actual (A) values into components. Typical components are:

Trend (T)
Seasonal (S)
Cyclical (C)
Error (E)

These components can be combined in a number of ways to model the actual time series. A typical additive model might be:

$$A = T + S + C + E$$

A typical multiplicative model might be:

$$A = T \times S \times C \times E$$

In the additive model we are assuming no connection between, for example, the trend value and the seasonal component. Even if the trend value increases, the seasonal component is constant.

In the multiplicative model if the trend changes, so does the seasonal component. Using a linear trend, the two cases are illustrated below:

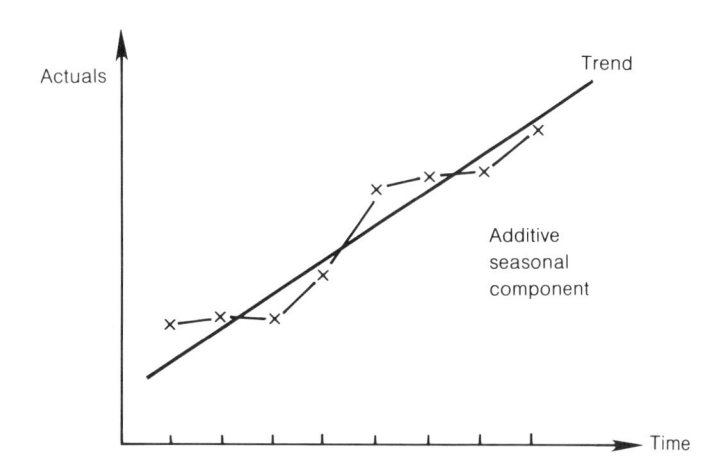

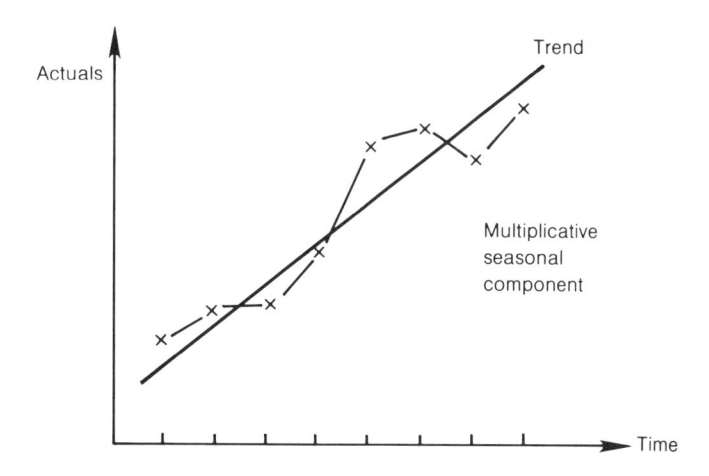

(b) Ceramic goods t(000) per month. The multiplicative model is therefore:

$$A = T \times S \times E$$

Seasonal factors, derived from several years' data are given for each month—except that the figures used are proportions, i.e. 0.96 (not 96).

(i) Calculate the deseasonalised data for each month of 1988:

$$\frac{A}{S} = T \times E = \text{deseasonalised data}$$

	Jan	Feb	Mar	Apr	May	Jun	Jul	Aug	Sep	Oct	Nov	Dec
Actual	335	325	310	354	360	338	333	270	375	395	415	373
A/(S ÷ 100) =	349.0	349.5	344.4	347.1	342.9	352.1	354.3	346.2	340.9	343.5	345.8	345.4

(ii) Plot of monthly production figures and deseasonalised data.

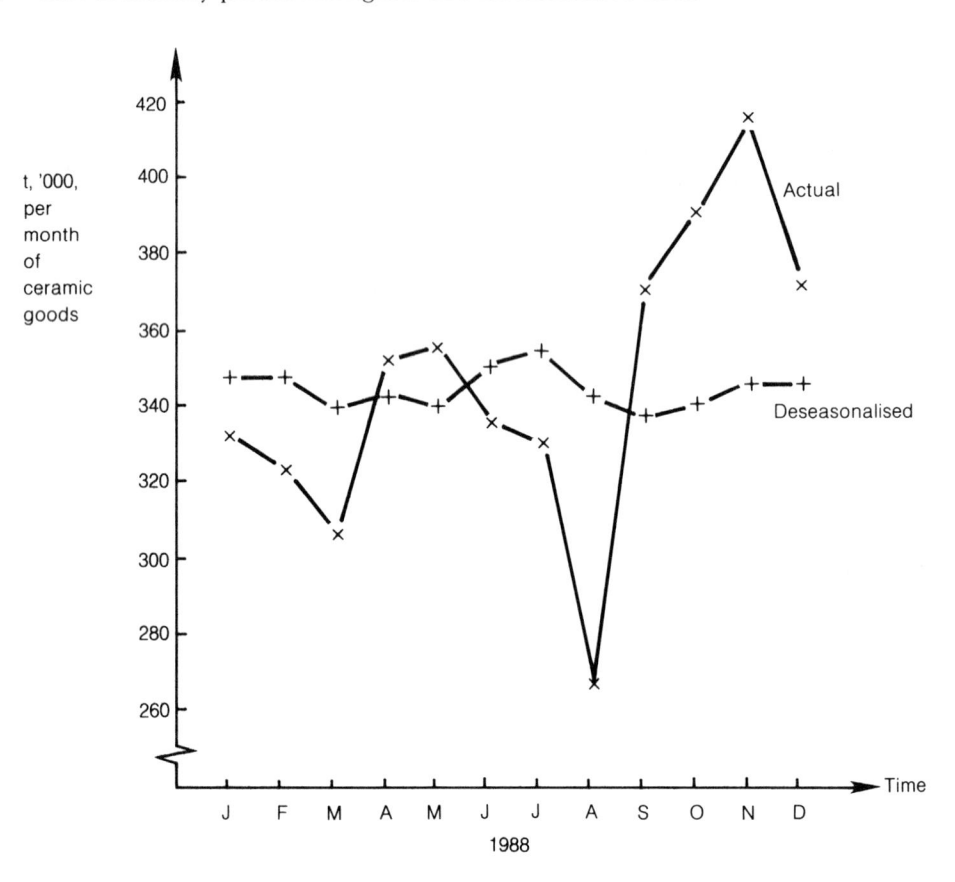

From the graph there appears to be no trend in the deseasonalised data (T × E). It is constant at about 345,000 t per month.

(iii) Use the exponential smoothing model on the deseasonalised data to produce a forecast for January 1989.

$$F_t = \alpha A_{t-1} + (1-\alpha)F_{t-1}$$

where $\alpha = 0.2$ and $F_{August} = 350$.

Month	Actual A_{t-1}	Deseasonalised F_{t-1}	Forecasts $F_t = 0.2A_{t-1} + 0.8F_{t-1}$
Aug 1988	346.2	350	349.2
Sep	340.9	349.2	347.5
Oct	343.5	347.5	346.7
Nov	345.8	346.7	346.7
Dec	345.4	346.7	346.4
Jan 1989		346.4	

Therefore, the forecast for January 1989, deseasonalised, is 346,400 t of ceramic goods per month.

(iv) Use this forecast and the seasonal factors to forecast actual production figures for the first three months of 1989. The model is Forecast Production = T × S where:

$$T = 346.4 \ (000 \ t)$$

AUTHORS' MODEL ANSWERS TO DECEMBER 1988 ACCA QUANTITATIVE ANALYSIS PAPER

Month	T	S	T × S = Forecast (000t per month)
Jan	346.4	0.96	332.5
Feb	346.4	0.93	322.2
Mar	346.4	0.90	311.8

This assumes that the Forecast Trend does not change over the period, that the seasonal factors still apply and that there are no foreseeable effects that should be included in the forecast.

(a) Linear Programming assumes that the objective function and constraints are linear functions of the amount of each product produced. In some situations, the costs of raw materials and the time for manufacture will depend on the quantity produced, hence the coefficients in the objective function and the constraints will be dependent on the production quantity. The objective function and constraints are then no longer linear functions of the decision variables.

Linear Programming requires a knowledge of the amount of resource needed for one unit of each product. In practice there may be uncertainty about the exact values of these figures. Estimates and approximations may limit the usefulness of the linear programme. The linear programming model assumes that there is a single objective to be satisfied. In practice there may be conflicting objectives; overall company strategy may conflict with departmental objectives.

(b) Let the annual production of Alphas be a units; the annual production of Betas be b units; the annual production of Gammas be g units; the annual production of Deltas be d units. Unit contribution for each product:

Unit contribution = selling price − unit variable costs

Alpha: unit contribution = $112 - (22 + 8 + (7 \times 4) + (8 \times 3))$
= $112 - 82 = £30$

Beta: unit contribution = $120 - (20 + 14 + (5 \times 4) + (10 \times 3))$
= $120 - 84 = £36$

Gamma: unit contribution = $138 - (25 + 20 + (12 \times 4))$
= $138 - 93 = £45$

Delta: unit contribution = $165 - (30 + 25 + (15 \times 4))$
= $165 - 115 = £50$

(i) Maximise total annual contribution, £P per year, where:

$$P = 30a + 36b + 45c + 50d \text{ £ per year}$$

Maximise subject to:

Assembly labour:	$7a + 5b$	$\leqslant 15000$ hours per year.
Packaging labour:	$8a + 10b$	$\leqslant 21000$ hours per year.
Machine time:	$12c + 15d$	$\leqslant 12000$ hours per year.
Demand Alpha:	a	$\leqslant 2500$ units per year.
Demand Beta:	b	$\leqslant 1500$ units per year.
Demand Gamma:	c	$\leqslant 1000$ units per year.
Demand Delta:	d	$\leqslant 1500$ units per year.
	a, b, c, d	$\geqslant 0$

Since the constraints on a and b are separate from those on c and d, the linear programme may be split into two 2 variable models.

Model 1: Maximise total annual contribution for Alpha and Beta, £P$_{ab}$ per year, where:

$$P_{ab} = 30a + 36b \text{ £ per year}$$

Maximise subject to:

$$\begin{aligned}\text{Assembly labour:} \quad & 7a + 5b \leq 15000 \text{ hours per year.} \\ \text{Packaging labour:} \quad & 8a + 10b \leq 21000 \text{ hours per year.} \\ \text{Demand Alpha:} \quad & a \leq 2500 \text{ units per year.} \\ \text{Demand Beta:} \quad & b \leq 1500 \text{ units per year.} \\ & a, b \geq 0 \end{aligned}$$

Model 2: Maximise total annual contribution for Gamma and Delta, $£P_{cd}$ per year, where:

$$P_{cd} = 45c + 50d \text{ £ per year}$$

Maximise subject to:

$$\begin{aligned}\text{Machine time:} \quad & 12c + 15d \leq 12000 \text{ hours per year.} \\ \text{Demand Gamma:} \quad & c \leq 1000 \text{ units per year.} \\ \text{Demand Delta:} \quad & d \leq 1500 \text{ units per year.} \\ & c, d \geq 0 \end{aligned}$$

The total annual contribution, $P = P_{ab} + P_{cd}$ £ per year.

(ii) The graph for model 1 is shown below:

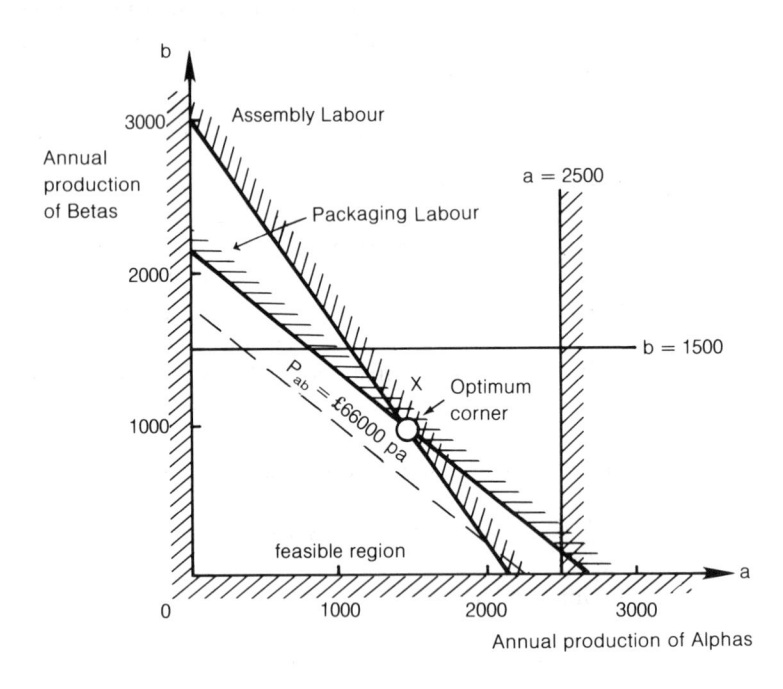

The point $a = b = 1000$ units is a feasible solution. At this point, $P_{ab} = 30 \times 1000 + 36 \times 1000 = £66,000$ per year. Take the trial contribution line as:

$$66000 = 30a + 36b \text{ £ per year}$$

This line is drawn on the graph above.

Moving parallel to this line in the direction of increasing contribution, the optimum corner is found to be X, the intersection of the constraints on assembly labour and packaging labour.

$$\begin{aligned} 7a + 5b &= 15000 \quad (1) \\ 8a + 10b &= 21000 \quad (2) \end{aligned}$$

Multiply (1) by 2 and subtract (2):

$$6a = 9000$$
$$a = 1500$$

therefore:

$$b = 900$$

Maximum value of $P_{ab} = 30 \times 1500 + 36 \times 900 = £77,400$ per year.

The graph for model 2 is shown below:

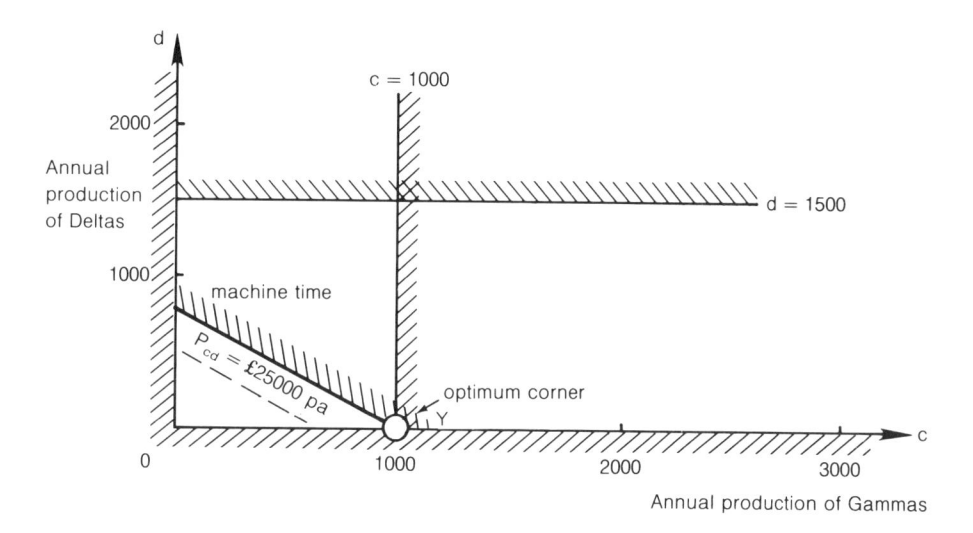

The point $c = 0$, $d = 500$ units is a feasible solution. At this point, $P_{cd} = 45 \times 0 + 50 \times 500 = £25,000$ per year. Take the trial contribution line as:

$$25000 = 45c + 50d \text{ £ per year}$$

This line is drawn on the graph above.

Moving parallel to this line in the direction of increasing contribution, the optimum corner is found to be Y, the intersection of the constraints on machine time, demand for Gamma and the c axis. Therefore the optimum corner is $c = 1000$ units and $d = 0$. Maximum value of:

$$P_{cd} = 45 \times 1000 + 50 \times 0 = £45,000 \text{ per year}$$

The maximum value of the total annual contribution = 77400 + 45000
$$= £122,400 \text{ per year.}$$

The fixed overheads are £25,000 per year, therefore the maximum value of the annual profit is £122,400 − £25,000 = £97,400 per year.

(iii) Let £x be the unit contribution for Deltas.

$$P_{cd} = 45c + xd \text{ £ per year}$$

Delta will be brought into the optimum solution if the objective function is parallel to the constraint on machine time. If the two lines are parallel, the slopes of the two lines will be equal. Hence, Delta will be brought into the optimum solution when:

$$\frac{45}{x} = \frac{12}{15} \text{ therefore } x = 45 \times \frac{15}{12} = 56.25$$

The selling price for Delta = unit contribution + unit variable costs
= 56.25 + 110
= £171.25

(iv) Relax the machine time constraint by 1 hour.

$$12c + 15d <= 12001 \text{ hours per year}$$

The optimum corner will still be the intersection with c = 1000 Therefore the optimum solution is c = 1000, d = 1/15. The new value of the maximum contribution is:

$$P_{cd} = 45 \times 1000 + 50 \times (1/15)$$

This is an increase of £(50/15) compared to the original optimum solution. The shadow price, or dual price, of the machine time is therefore:

$$£(50/15) = £3.33 \text{ per hour}$$

If, however, we reduce the constraint by 1 hour to 11,999 hours per year, the shadow price changes to £(50/12). This happens because we are at the intersection of the three constraints, and the binding constraints change as we pass through the optimum point.

5 (a) The appropriate probability distribution for this problem is the binomial. There are fifty identical, independent trials, each with two possible outcomes, error or not. The probability of an error on any individual trial is constant at 2%.

However, since the number of trials is fairly large, the probability of an error is small and the mean number of errors (np = 50 × 0.02 = 1) is less than five, the Poisson probability distribution may used to approximate the binomial distribution.

(b) Using the Poisson distribution with m = 1 as an approximation:

$$P(r \text{ errors/sample}) = \frac{1^r e^{-1}}{r!} \quad r = 0, 1, 2, \ldots$$

$$P(\text{auditor is dissatisfied}) = P(r > 1) = 1 - P(r \leq 1)$$
$$= 1 - (P(r = 0) + P(r = 1))$$

Using the Poisson tables given in the question:

$$P(\text{auditor dissatisfied}) = 1 - (0.368 + 0.368) = 0.264$$

(c) We require to find the number of errors, k, for which $P(r \geq k) \leq 0.05$.

$$k = 1 \quad P(r > 1) = 0.264 \text{ from (b)}$$
$$k = 2 \quad P(r > 2) = 1 - (P(r = 0) + P(r = 1) + P(r = 2))$$
$$= 1 - (0.368 + 0.368 + 0.184) = 1 - 0.920 = 0.080$$
$$k = 3 \quad P(r > 3) = 1 - (P(r = 0) + P(r = 1) + P(r = 2) + P(r = 3))$$
$$= 1 - (0.368 + 0.368 + 0.184 + 0.061) = 1 - 0.981 = 0.019$$

If the error rate is 2%, then the probability of finding more than three errors in a sample of fifty is less than 5% (1.9% in fact). Therefore, if the auditor finds more than three errors in his sample, it is reasonable for him to assume that the stipulated error has probably been exceeded.

(d) If the error rate is x%, then the mean number of errors per sample is:

$$m = 50 \frac{x}{100} = \frac{x}{2}$$

Assuming that x is small, so that (x/2) < 5, the binomial distribution may be approximated by a Poisson distribution:

$$P(r \text{ errors/sample}) = \frac{(x/2)^r}{r!} e^{-(x/2)} \quad r = 0, 1, 2, \ldots$$

P(auditor dissatisfied), $P = P(r > 1) = 1 - (P(r = 0) + P(r = 1))$

$$P(r = 0) = \frac{(x/2)^0}{0!} e^{-(x/2)} = e^{-(x/2)}$$

$$P(r = 1) = \frac{(x/2)^1}{1!} e^{-(x/2)} = (x/2)e^{-(x/2)}$$

Therefore:

$$P = 1 - (e^{-(x/2)} + (x/2)e^{-(x/2)}) = 1 - \left(1 + \frac{x}{2}\right)e^{-(x/2)}$$

To plot the graph, calculate values of P for given values of x:

x	P	x	P
0	0	5	0.7127
1	0.0902	6	0.8009
2	0.2642	7	0.8641
3	0.4422	8	0.9084
4	0.5940	9	0.9389
		10	0.9596

The graph is plotted below:

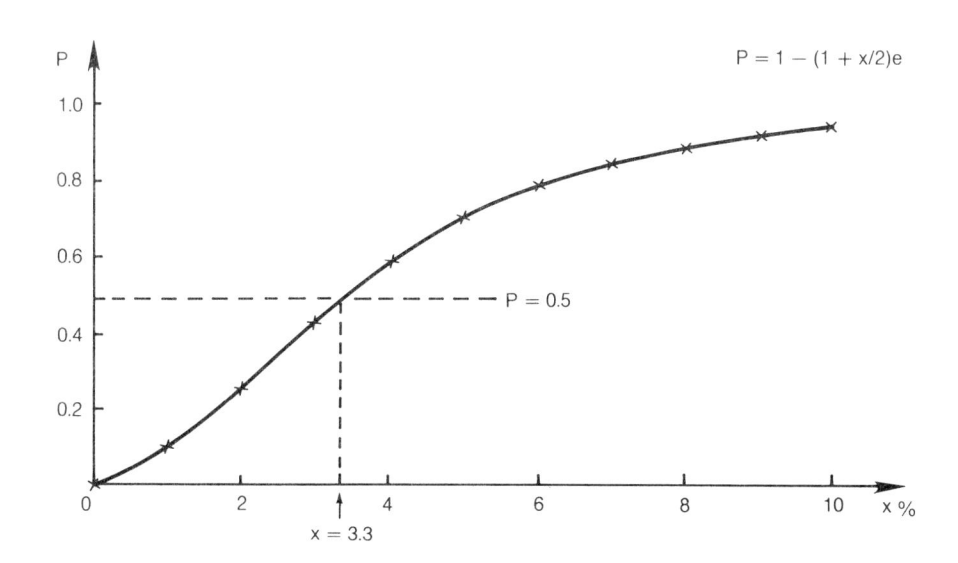

From the graph, when $P = 0.5$, $x = 3.3$.

(e) (i) The various sequences of events are illustrated in the tree diagram below:

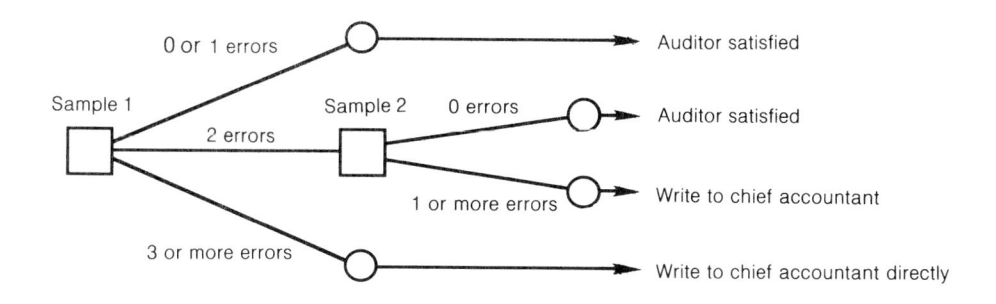

658 QUANTITATIVE ANALYSIS

Note: If there are more than three errors in the first sample, then there is no need to take a second sample. The auditor's criterion for writing to the chief accountant is already satisfied by the first sample.

P_1—refers to the first sample.
P_2—refers to the second sample.

P(auditor writes to the chief accountant) = 1 − P(auditor satisfied)
= 1 − (P_1(0 or 1 error) + P_1(2 errors) × P_2(0 errors))

Using the tables provided in the question:

P(letter sent) = 1 − ((0.368 + 0.368) + (0.184 × 0.368))
= 1 − 0.8037 = 0.1963

P(auditor sends a letter) = 0.196

(ii) Expected number of transactions checked per week is:

50 + 50 × P(second sample needed)
= 50 + 50 × P_1(2 errors) = 50 + 50 × 0.184
= 59.2

Note: In the above calculation it is assumed that the auditor would not take a second sample if the first sample contained three or more errors anyway.

6 (a) The network is illustrated by an arrow diagram. The earliest event times and the latest event times have been added.

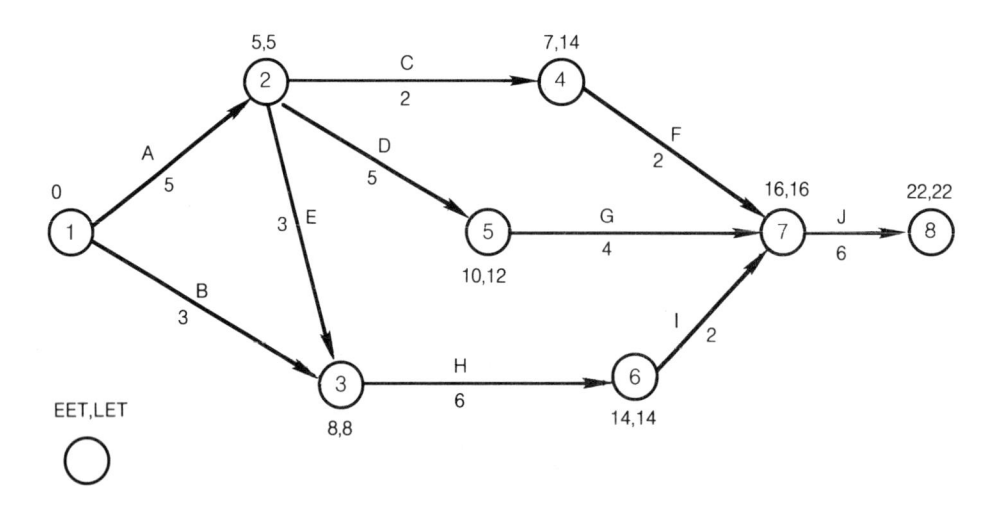

The overall duration of the project is 22 weeks. The critical path is A − E − H − I − J.

(b) A Gantt Chart is constructed which shows all activities beginning at the earliest times and the total float on each path. A resource profile for the number of staff required is then built up. See the diagrams on page 659:

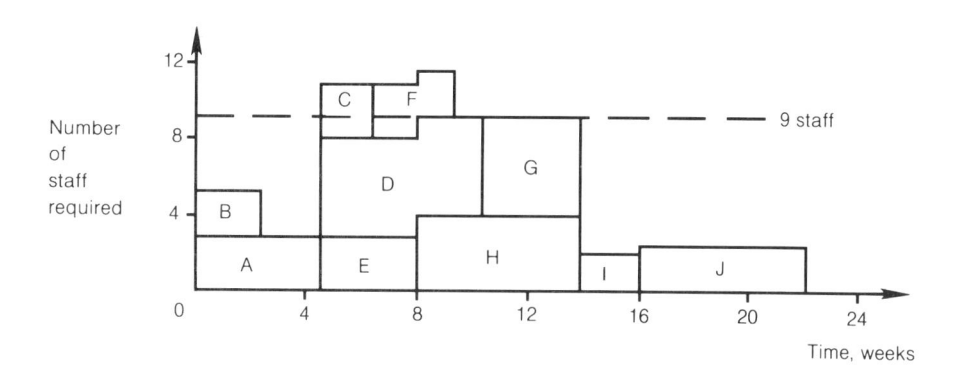

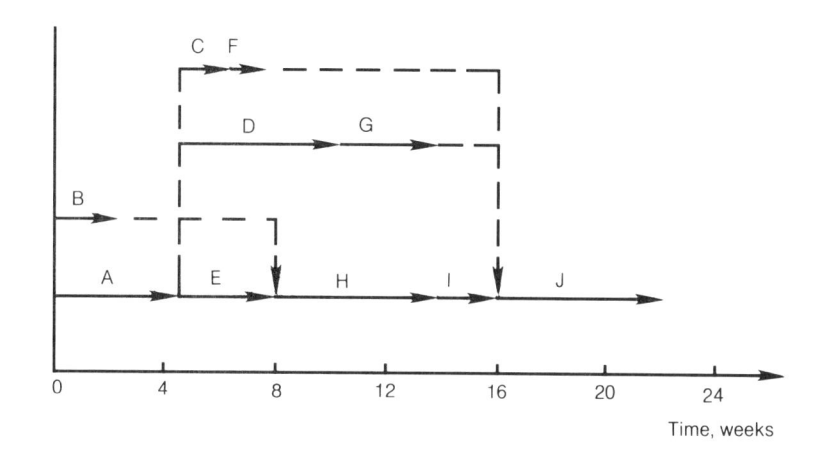

AUTHORS' MODEL ANSWERS TO DECEMBER 1988 ACCA QUANTITATIVE ANALYSIS PAPER

The number of staff required is as follows:

Week	Number of staff
1–3	5
4–5	3
6–8	10
9	11
10–14	9
15–16	1
17–22	2

(c) In order to complete the project with a maximum of 9 staff, some activities must be rescheduled. There are several ways in which this can be achieved, but they all involve activities C, F, D and G. For example:

delay the start of activity D until activity C is complete;
activity G follows immediately after D;
activity F is delayed to begin at the end of week 14.

This schedule means that activities G and F are completed just in time for activity J to begin, therefore there are no delays to the overall completion time. See the diagram below; all 9 members of staff will be required for 6 weeks.

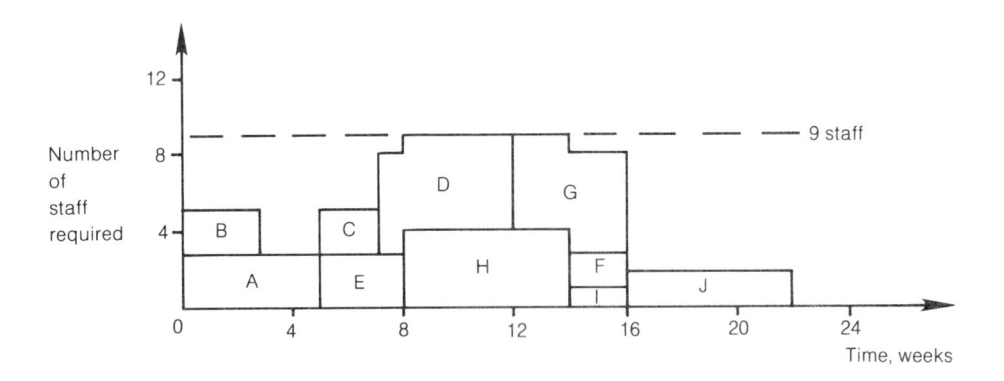

660 QUANTITATIVE ANALYSIS

[Bar chart showing Number of staff required vs Time in weeks, with activities A–J stacked, and a dashed line at 9 staff.]

7 (a) The distribution of the sales is:

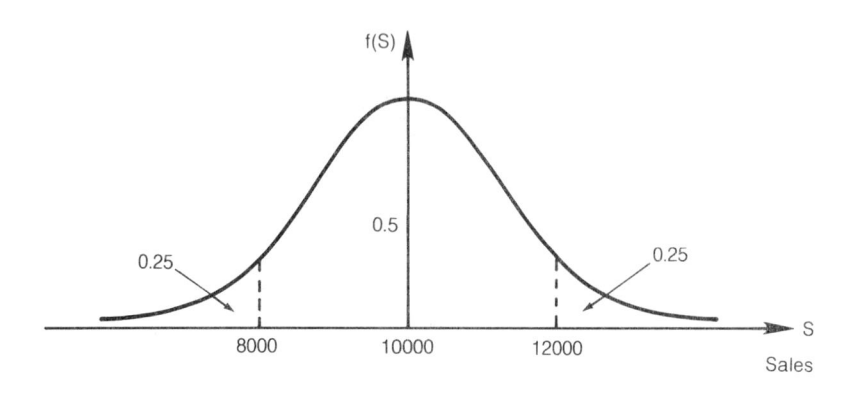

[Normal distribution curve with mean 10000, showing 0.25 probability below 8000 and 0.25 above 12000, peak labelled 0.5.]

Sales of 12,000 is z_1 standard deviations above the mean, where:

$$z_1 = \frac{12{,}000 - 10{,}000}{\sigma} = \frac{2000}{\sigma}$$

P(Sales > 12,000) = 0.25, therefore: $P(z > z_1) = 0.25$.
Using the standard normal tables provided, $z_1 = 0.675$ approximately.
Therefore:

$$0.675 = \frac{2000}{\sigma}$$

$$\sigma = \frac{2000}{0.675} = 2963.0$$

To two significant figures, $\sigma = 3000$ sales.

(b) (i) The company will at least break even if:

$$(\text{revenue} - \text{costs}) \geqslant 0$$

If the company sells b books, they will at least break even if:

$$10b - (4b + 25000 + 0.1 \times 10b) \geqslant 0$$
$$5b \geqslant 25000$$
$$b \geqslant 5000$$

Assuming a normal distribution, 5000 is z standard deviations below the mean, where:

$$z = \frac{5000 - 10{,}000}{3000} = -1.67$$

From the tables:
$$P(z > 1.67) = 0.0475$$

therefore:
$$P(z > -1.67) = 1 - 0.0475 = 0.9525$$

There is 95.25% probability that the company will at least break even if the existing facilities are used.

(ii) If the company sells b books, they will at least break even if:

$$10b - (2.5b + 25000 + 14000 + 0.1 \times 10b) \geq 0$$
$$6.5b \geq 39000$$
$$b \geq 6000$$

Assuming a normal distribution, 6000 is z standard deviations below the mean, where:

$$z = \frac{6000 - 10{,}000}{3000} = -1.33$$

From the tables:
$$P(z > 1.33) = 0.0918$$

Therefore: $P(z > -1.33) = 1 - 0.0918 = 0.9082$.

There is 90.82% probability that the company will at least break even if the special machine is hired.

(c) Expected profit = unit contribution × expected sales − fixed costs.

Expected sales = 10,000 books.

If the existing facilities are used, the unit contribution is £5, therefore, the expected profit = 10,000 × 5 − 25,000 = £25,000.

If the special machine is hired, the unit contribution is £6.5, therefore the expected profit = 10,000 × 6.5 − 39,000 = £26,000. On this basis, the special machine should be hired.

(d) Set up a payoff table for the following three possible decisions:
  (i)   use existing facilities; payoff = £(5 × sales − 25000)
  (ii) hire special machine; payoff = £(6.5 × sales − 39000)
  (iii) do not publish.   payoff = 0

		Payoff, £'000 Possible decision:			
		Existing facilities	Special machine	Do not publish	Probability
Possible outcomes:	2500	−12.5	−22.75	0	0.05
	6500	7.5	3.25	0	0.20
Sales	9000	20.0	19.50	0	0.25
	11000	30.0	32.50	0	0.25
	13500	42.5	48.75	0	0.20
	17500	62.5	74.75	0	0.05

Calculate the expected payoffs by multiplying each payoff by the probability of its occurring.

		Expected Payoff, £'000 Possible decision:	
Possible outcomes: Sales		Existing facilities	Special machine	Do not publish
	2500	−0.625	−1.1375	0
	6500	1.5	0.65	0
	9000	5.0	4.875	0
	11,000	7.5	8.125	0
	13,500	8.5	9.75	0
	17,500	3.125	3.7375	0
Expected payoff		25.00	26.00	0

On the basis of choosing the decision which leads to the maximum expected payoff, the company should hire the special machine. The expected payoff is £26,000.

If the sales are known in advance, the most advantageous decision will be made. In this case, the expected payoff is:

$$(0 \times 0.05) + (7.5 \times 0.2) + (20.0 \times 0.25)$$
$$+ (32.5 \times 0.25) + (48.75 \times 0.2) + (74.75 \times 0.05)$$
$$= 28.1125 \quad (\text{£'000})$$

The expected value of perfect information is £(28,112.5 − 26,000) = £2112.5. This figure is the maximum amount which could be spent on obtaining information about the likely sales of the book. It is the value to the publisher of prior knowledge of the actual sales.

Glossary

Activity. In network analysis, a specific task which forms part of a project.

Activity-on-node diagram. A type of network diagram in which nodes are used to represent the activities and arrows are used to join activities to their immediate predecessors.

Additive time series model. A model of the relationship between the actual variable values and the times series components, in which the components are added.

Algorithm. A solution procedure.

Alternative hypothesis (H_1). The hypothesis which is adopted if the null hypothesis is rejected.

Alternative optima. In linear programming, when the objective function is parallel to a limiting constraint, all solutions on that constraint, between the adjacent corners, are optimum.

Analysis of variance table. A tabulation of the results of an analysis in which the total variation in the dependent variable is split into components. In the context of regression, there are two components; one is associated with the variation due to the regression and the other is the residual variation.

Arrival pattern. The probability distribution which describes the arrival rate, in queuing and simulation.

Arrow diagram. A type of network diagram in which arrows are used to represent activities.

Assignment problem. A special case of the transportation problem when the number of origins and the number of destinations are equal and only one item is transferred in each case.

Association. The relationship between variables in regression analysis.

Attribute. Population characteristic.

Basic variable. In linear programming, a variable which has a non-zero value at a corner of the feasible region.

Beta distribution. A probability distribution which is used to describe the variation in the duration of individual activities in PERT network analysis.

Binomial probability distribution. A standard probability distribution for discrete random variables. It gives the probability of r successes from n identical and independent trials. Each trial has two possible outcomes only and the probability of a success on any one trial is constant.

Bivariate distribution. A distribution of two variables.

Boundary value. In hypothesis testing, the value of the standard statistic at the specified significance level.

Buffer stock. Additional stock which is held as an insurance against above average demand during the lead time. Its purpose is to reduce the probability of a stock-out occurring.

Central limit theorem. The sample distribution of sample mean, $\bar{x}$, is approximately normal even if the parent population is non-normal, as long as the sample size is large, eg at least 30.

Chi-squared (χ^2) test. A non-parametric test which compares a set of observed frequencies with a set of expected frequencies.

Coefficient of determination. A measure of the degree to which the regression model explains the variation in the dependent variable.

Combination. The number of ways of selecting r items from a group of n items, when the order of selection is not important.

Compound event. This term specifically refers to a sequence of outcomes.

Conditional probability. The probability of event B occurring, given that event A has already occurred, is called the conditional probability of B.

Confidence interval. A range of values about a sample statistic, within which the corresponding population parameter is expected to lie with a stated probability.

Confidence level. The proportion of samples for which the population parameter can be expected to lie in the confidence interval. For example, for a 95% confidence level, the resulting confidence interval will contain the population parameter for 95% of all samples.

Constraint. In linear programming, a limitation on the allocation of a resource.

Contingency table. A table used to display the observed or expected frequencies for a χ^2 test of the association of attributes.

Continuity correction. When a continuous probability distribution is used to approximate a discrete probability distribution, it is necessary to adjust the value of the discrete variable by ± 0.5.

Continuous probability distribution. The probability that a continuous random variable will take a value within a given interval.

Continuous random variable. A random variable which can take any value.

Corner of the feasible region. In linear programming, a feasible solution which is the intersection of two or more boundaries of the feasible region.

Correlation. The strength of the linear relationship between variables.

Correlation coefficient. A measure of the strength of the linear relationship between the dependent and independent variables. Referred to also as Pearson's product moment correlation coefficient.

Crash time. In network analysis, the shortest possible time in which an activity can be completed.

Critical activity. In network analysis, an activity for which any delay will cause delay in the whole project.

Critical path. The path through the network which links the critical activities. This path takes the longest time.

Cyclical variation. Long-term cyclical variation in the value of the time series variable. The cycle takes several years to repeat.

Decision rule. A criterion for selecting a particular decision from a set of uncertain alternatives.

Decision tree. A branching diagram which illustrates all possible sequences of decision alternatives and decision outcomes for a problem.

Decision variable. The variables in a programming problem for which values must be chosen to satisfy a specific objective.

Degenerate solution. Occurs in transportation or assignment problems, when one or more of the basic variables in a feasible allocation has the value zero. This results in fewer cells than expected which contain an allocation.

Degrees of freedom. A parameter of the t, χ^2 and F distributions which depends on the sample size, n, of the simple random sample(s).

Delphi method. A method of forecasting when there is little or no past data. Use is made of expert guesswork.

Dependent variable. A variable for which the values depend on the values of one or more independent variable. This is the variable which is explained or predicted by the regression model.

Deseasonalise. Remove the effect of the seasonal variation from the time data by subtracting an estimate of the seasonal component from the actual values of the variable.

Deterministic system. A (stock) system for which all of the parameters are known and fixed. There is no uncertainty about the way in which the system will behave.

Detrend. Remove the effect of the trend in the data by subtracting an estimate of the trend component from the actual values of the variable.

Deviation. In regression analysis, residual, or error.

Discrete probability distribution. The values of the random variable and their associated probabilities, for a particular experiment.

Discrete random variable. A random variable which takes integer values only.

Dual problem. A re-formulation of the linear programming model in terms of the shadow prices of the primal problem.

Dual variable. The variables used in the dual linear programme. These are the shadow prices of the primal problem.

Dummy. In transportation or assignment, an additional origin or destination which is included in the tableau so that total supply is the same as total demand. This is a condition required by the transportation and assignment algorithms.

Dummy. In network analysis, an imaginary activity with zero activity time which is used in arrow diagrams.

Earliest event time. In network analysis, the earliest time by which an event can be completed.

Earliest finish time. The earliest time by which an activity can be completed.

Earliest start time. The earliest time at which an activity can begin.

Economic batch quantity. The optimum production batch size if the objective is to minimise the total variable cost associated with the production and storage of the item.

Economic order quantity. The optimum order quantity if the objective is to minimise the total variable cost associated with stockholding.

Error. In regression analsysis, residual, or deviation.

Event. One or more outcomes.

Event. This is represented by a node in an arrow network diagram. The event is complete when all of the activities leading into the node have been completed.

Expected value. The average value of the outcome of an experiment. The possible outcomes of the experiment all have numerical values.

Expected value of a probability distribution. The mean value of the probability distribution.

Expected value of perfect information. The monetary value of information about the outcome of a decision problem. It is the difference between the expected payoff when the decision outcome is known and the best expected payoff when there is no information about the decision outcome.

Experiment. One or more trials.

Explained variation. In regression analysis, the amount of the variation in the dependent variable which can be attributed to the relationship with the independent variable(s).

Exponential smoothing. The use of weighted averages of past time series data to smooth the time series so that forecasts can be made.

F distribution. The probability distribution for the ratio of two sample variances, from assumed normal populations.

Feasible region. In linear programming, the set of all of the feasible solutions.

Feasible solution. A set of values for the decision variables which satisfy all of the constraints in the problem.

Finite population. A population which is limited in size.

Float. In network analysis, the spare time associated with an activity.

Forbidden allocation. In transportation or assignment, a combination of origin and destination which is not allowable.

Free float. The amount of time by which an activity can be delayed or re-scheduled without affecting the earliest start time of any following activity.

F test. Variance ratio test.

Gantt chart. A type of network diagram in which a time axis is used. The total float is illustrated.

Goodness of fit test. A hypothesis test, using χ^2, to determine the likelihood that a set of observed data follows a specified standard probability distribution.

Holding cost. In stock control, the variable cost incurred only when an item is being stored.

Hungarian algorithm. A solution procedure for the assignment problem.

Hypothesis. In hypothesis testing, a statement about the value of a population parameter.

Hypothesis test. A statistical test which examines a hypothesis in the light of sample evidence.

Identification dummy. A dummy activity which is inserted in an arrow network diagram if two activities have the same start and end nodes. The purpose is to avoid ambiguity.

Idle time. In queuing, the time for which the service channel is empty.

Immediate predecessor. In network analysis, an activity which must be completed before another specific activity can begin.

Independent. Events are independent when the occurrence of one event does not influence the probability of the occurrence of any of the others.

Independent float. In network analysis, the float which is associated entirely with one activity and is not shared in any way with preceding or succeeding activities.

Independent samples. Samples which are selected independently of one another, from two or more populations.

Independent variable. A variable for which the values do not depend on the value of any other variable. This is the variable which is used to explain or predict the values of the dependent variable.

Infeasible solution. A set of values of the decision variables which do not satisfy all of the constraints in the linear programme. The constraints on the problem may be such that there is no feasible region.

Infinite population. A population which is not limited in size.

Interval data. A data set in which the values are measurements of some property of the population or sample members.

Just-in-time. A manufacturing system which requires the items to be provided just before they are needed.

Latest event time. In network analysis, the latest time by which an event must be completed if there is to be no delay to the whole project.

Latest finish time. The latest time by which an activity must be complete if there is to be no delay to the whole project.

Latest start time. The latest time at which an activity can begin without causing delay to the whole project.

Lead time. The time between the placing of an order and its delivery.

Level of service. In stock control, the maximum acceptable probability of a stock-out.

Limiting constraint. In linear programming, a constraint which is limiting the value of the objective function. The intersection of the limiting constraints form the optimum corner.

Linear programme. A mathematical model in which the objective function and constraints are all linear functions of the decision variables.

Linear regression. A linear model is used to describe the relationship between one variable and one or more other variables.

Linear transformation. The variables in a non-linear model are re-defined so that the form of the model becomes linear.

Logic dummy. A dummy activity which is used to create the correct logic in an arrow network diagram.

Mathematical programming. The use of mathematical techniques and models in programming activities, eg linear programming.

Maximax rule. A particular decision rule in which criterion of maximising the maximum payoff is used.

Maximin rule. A particular decision rule in which criterion of maximising the minimum payoff is used.

Maximise expected payoff. A particular decision rule in which the criterion of maximising the expected payoff is used.

Mean absolute difference. A method of measuring the reliability of a time series model. It is the average of the size of the differences between the actual values of the variable and the values forecast from the model.

Mean square error. A method of measuring the reliability of a time series model. It is the average of the squared differences between the actual values of the variable and the values forecast from the model.

Minimax rule. A particular decision rule in which criterion of minimising the maximum opportunity loss is used.

Minimise expected opportunity loss or regret. A particular decision rule in which the criterion of minimising the expected opportunity loss is used.

Minimum cost method. In transportation, a method of making an initial feasible allocation in the transportation algorithm.

M/M/1 model. A queueing model in which both the arrival and service patterns are Poisson processes and there is only one server.

M/M/2 model. A queueing model in which both the arrival and service patterns are Poisson processes and there are two servers.

Model. A mathematical equation(s) used to describe the relationship between one variable and one or more other variable.

MODI Method. In transportation, a procedure for testing a feasible transportation solution for optimality.

Monte Carlo. A particular type of simulation in which the variables are discrete and random numbers are used to generate a series of values for the variables.

Most probable time. In PERT network analysis, the most likely time needed to complete the activity.

Moving average. In time series, a method of smoothing out seasonal variation from a data set, by averaging successive small groups of values. The group size corresponds to the cycle length for the seasonal variation.

Multiplicative time series model. A model of the relationship between the actual variable values and the times series components, in which the components are multiplied.

Mutually exclusive. Events are mutually exclusive if they cannot occur at the same time.

Network diagram. A diagram which illustrates the sequences in which activities must be done in a given project.

Non-basic variable. In linear programming, a variable which has a zero value at a corner of the feasible region.

Non-parametric test. A hypothesis test which does not require the use of population parameters, or assumptions about the population distribution, eg χ^2 test.

Normal probability distribution. A standard probability distribution for continuous random variables. The probability density function is defined in terms of the mean and standard deviation of the variable, and it generates a symmetrical bell-shaped distribution.

Normal (z) test. A hypothesis test of a population mean when the population variance is known and the population is normal, or the sample size is large ($n \geq 30$).

Null hypothesis (H_0). The hypothesis to be examined in the hypothesis test. The sample evidence leads to the acceptance or rejection of this hypothesis.

Objective function. An expression of the objective of a programming problem in terms of the decision variables.

One tail test. In hypothesis testing, a test in which the null hypothesis is rejected if the value of the test statistic falls into one tail of the sampling distribution. The alternative hypothesis states that the population parameter is greater than or less than a specified value.

Opportunity loss. The shortfall in the payoff which arises if the wrong decision is made.

Optimistic time. In PERT network analysis the minimum time in which the activity can be completed when all conditions are favourable.

Optimum solution. In linear programming, the particular feasible solution(s) which meets the objective.

Ordering cost. In stock control, the variable cost incurred only when an order is placed.

Ordinal data. A data set in which the members are identified in terms of their rank position, rather than in terms of a measurement of a characteristic.

Outcome. A result of a trial.

Paired samples. In hypothesis testing, each item in one sample is paired with an item in the other sample.

Parameter. A population measure, eg population mean, range.

Payoff. The result, in terms of profit, cost, etc, of a particular combination of a decision alternative and an outcome from an experiment.

Payoff table. A tabulation of the payoffs for every combination of decision and outcome in a problem.

Permutation. The number of ways of selecting r items from a group of n items, when the order of selection is important.

PERT. A procedure for analysing a project network when the activity times are uncertain.

Pessimistic time. In PERT network analysis, the longest time needed to complete the activity when conditions are unfavourable.

Pivotal column. In linear programming, the column in the Simplex tableau which corresponds to the non-basic variable which it is most advantageous to introduce into the solution.

Pivotal element. The cell in the Simplex tableau at the intersection of the pivotal row and column.

Pivotal row. The row in the Simplex tableau which corresponds to the basic variable which it is most advantageous to remove from the solution.

(Point) estimator. A sample statistic, such as $\bar{x}$, p, s, which is used to estimate the corresponding population parameter.

Poisson probability distribution. A standard probability distribution for discrete random variables. It gives the probability of r successes per unit interval from an infinite number of identical and independent trials. Each trial has two possible outcomes only and the probability of a success on any one trial is very small and constant.

Poisson process. The rate of occurrence of an event is described by the discrete Poisson probability distribution, and the time between events is described by the negative exponential distribution.

Population. The entire group of items to which the statistical investigation relates.

Primal problem. The initial formulation of the linear programming model.

Probability. A numerical measure of the likelihood that a particular outcome will occur when an experiment is done.

Probability density function. The mathematical function which describes the probability distribution of a continuous random variable.

Probability theory. A mathematical framework which enables us to evaluate and use probability measures.

Probability tree. A diagram which illustrates all of the possible sequences of outcomes from an experiment. It can be used to determine the probabilities of compound events.

Programming. The allocation of resources.

Quantity discount. A discount on unit purchase price, offered by a supplier, if a large order is placed.

Queue. An ordered group of items waiting for some activity.

Queue discipline. The order in which the items are selected from the queue for service.

Queueing system. A combination of a queue and a service mechanism.

Random number. A number in which each digit, $(0 - 9)$, is generated randomly.

Random sample. Each item in the sample is selected independently and has the same chance of being selected.

Random variable. A numerical description of the outcome of an experiment. The values of the variable arise entirely by chance.

Rank correlation coefficient. A measure of the degree of agreement between two independent rankings of a set of items. It is also referred to as Spearman's rank correlation coefficient.

Regression. The nature or shape of the relationship between variables.

Regression coefficient. The coefficient of the independent variable in a linear regression model. In the two variable case, it is the slope of the regression line.

Regret. In decision theory, opportunity loss.

Re-order cycle model. A stock model in which an order is placed at a regular interval of time. The order size is adjusted according to the existing stock level at the re-order time.

Re-order interval. The time interval between the placing of successive orders.

Re-order level. The stock level at which a new order is placed.

Re-order level model. A stock model in which an order is placed whenever the stock falls to a pre-determined level. The order size is fixed.

Residual. In regression or time series analysis, the difference between the actual value of the dependent variable and the value predicted by the model.

Resource profile. In network analysis, a graph of resource requirements against time.

Sample. A small group selected from a population.

Sampling distribution. A probability distribution for all possible values of a sample statistic, using samples of a given size.

Sampling frame. A list of all of the members of a population.

Scatter diagram. A plot of corresponding pairs of values from a bivariate distribution.

Scenario Writing Method. A method of forecasting when there is little or no past data. Use is made of expert guesswork.

Seasonal variation. Short-term cyclical variation in the value of a time series variable. The cycle is repeated over periods of a year or less.

Semi-average. The average of a group of values from the data set, when the set is divided into two matching groups.

Set up cost. In stock control, the variable cost incurred only when a batch production process is being set up for a production run.

Service. An activity or process in queuing or simulation.

Service channel. The point at which the items in the queue are serviced.

Service pattern. The probability distribution which describes the service rate.

Shadow price. In linear programming, the change in the value of the objective function when one additional unit of a limiting resource is provided.

Significance level. In hypothesis testing, the probability at which a decision is made to accept or reject the null hypothesis.

Significant result. The probability of exceeding the value of the test statistic is less than the chosen significance level. The result is therefore unlikely to occur due to random sampling.

Simplex method. A matrix algebra method of solving a linear programme.

Simplex tableau. A tabulation of the coefficients of the variables in the linear programme, including those in the objective function and the slack or surplus variables, together with the right hand side values of the constraints.

Simulation. A technique which replicates the behaviour over time of a real system.

Slack variable. A variable which is added to the left hand side of ⩽ constraints. It represents the spare capacity of a particular resource for a given feasible solution.

Smoothing constant. The weight given to the most recent value of the variable in the time series. This is used in the exponential smoothing model to calculate forecasts.

Standard deviation of a probability distribution. A measure of the spread of the distribution relative to the expected value.

Standard error. Standard deviation of a sampling distribution.

Standard normal probability distribution. A normal distribution with a mean of zero and a variance of 1.

Statistic. A sample measure, eg sample mean.

Stepping stone method. A procedure for testing a feasible transportation solution for optimality; also a procedure for re-allocating in a non-optimal solution.

Stock-out. Demand exceeds available stock.

Surplus variable. A variable which is subtracted from the left hand side of ⩾ constraints. It represents the extra amount provided by a particular feasible solution, over and above the minimum level specified by the constraint.

Tableau (transportation or assignment). A tabulation of the unit cost of transfer between each combination of origin and destination.

t distribution. The probability distribution for the sample mean, $\bar{x}$, when the population variance is unknown, but the population is assumed to be normal.

Test of significance. Hypothesis test.

Test statistic. A value of a standard statistic (z, t, F, χ^2) which is calculated using the sample and the null hypothesis.

Time series. A data set in which the values of the variable are recorded at successive time intervals.

Total float. In network analysis, the amount of time by which an activity can be delayed or re-scheduled without affecting the overall completion time of the project.

Traffic intensity. In queuing, the ratio of mean arrival rate to mean service rate.

Transportation problem. An allocation problem which concerns the transfer of items from a set of origins to a set of destinations.

Trend. The long-run change in the value of the variable over time.

Trial. An activity or process for which there are several possible results. It is not known which of the possible results will actually occur until the activity is complete.

t test. A test of a sample mean when the population variance is unknown but the population is assumed to be normal.

Two tail test. A test in which the null hypothesis is rejected if the value of the test statistic falls into either tail of the sampling distribution. The alternative hypothesis states that the population parameter is not equal to a specified value.

Unbiased estimator. An estimator for which its expected value is equal to the population parameter being estimated.

Unbounded problem. This arises if the feasible region is open on one side so that the value of the objective function could be increased indefinitely. This usually arises because the linear programme is incomplete or there is an error.

Uniform probability distribution. A continuous probability distribution for which the probability density function is a constant. The probability that the variable will take a value in an interval of a given size, is the same for all intervals of that size, within the range of the variable.

Utility. In decision analysis, the payoff for a decision alternative which takes into account the risk associated with that decision.

Utility value. A numerical measure of the worth of the payoff to the decision maker.

Variance. Square of the standard deviation.

Variance ratio test. A test which compares two sample variances, assuming that the populations are normal.

Venn diagram. A pictorial representation of the outcomes of a trial, together with the probabilities of occurrence.

Vogel's penalty cost method. A method of making an initial feasible allocation in the transportation algorithm.

χ^2 distribution. The probability distribution for the sample variance, s^2, for an assumed normal population.

Yates' correction. A continuity correction used in the χ^2 test when the contingency table is small, eg 2×2.

Mathematical formulae

Chapter One

Addition Rule:

$$P(A \text{ or } B \text{ or both}) = P(A) + P(B) - P(A \text{ and } B)$$

Addition Rule for mutually exclusive events:

$$P(A \text{ or } B) = P(A) + P(B)$$

Multiplication Rule:

$$P(A \text{ and } B) = P(A)P(B \text{ given } A)$$

Multiplication Rule for independent events:

$$P(A \text{ and } B) = P(A)P(B)$$

Bayes' Rule:

$$P(A \text{ given } B) = \frac{P(A \text{ and } B)}{P(B)}$$

Expected value:

$$E(r) = \sum pr$$

Permutation:

$$^nP_r = \frac{n!}{(n-r)!} \qquad \text{where } n! = n \times (n-1) \times (n-2) \times \cdots \times 2 \times 1$$
$$\text{and } 0! = 1$$

Combination:

$$^nC_r = \frac{n!}{r!(n-r)!}$$

Chapter Two

Expected value of a discrete probability distribution:

$$E(r) = \sum rP(r)$$

Variance of a discrete probability distribution:

$$\sigma^2 = \sum r^2 P(r) - (E(r))^2$$

Standard deviation of a discrete probability distribution:

$$\sigma = \sqrt{\sum r^2 P(r) - (E(r))^2}$$

Binomial probability distribution:

$$P(r \text{ successes in n trials}) = {}^nC_r p^r q^{n-r} \qquad r = 0, 1, 2, \ldots, n$$

$$\text{where } {}^nC_r = \frac{n!}{r!\,(n-r)!}$$

Expected value for a binomial probability distribution:

$$\text{Expected number of successes} = np$$
$$\text{Expected proportion of successes} = p$$

Standard deviation for a binomial probability distribution

$$\text{Standard deviation of number of successes} = \sqrt{npq}$$

$$\text{Standard deviation of proportion of successes} = \sqrt{\frac{pq}{n}}$$

Poisson probability distribution

$$P(r \text{ successes/unit interval}) = \frac{m^r}{r!} e^{-m} \qquad r = 0, 1, 2, \ldots$$

where m is the mean number of successes per unit interval

Expected value and standard deviation of a Poisson probability distribution:

$$\text{Expected value, } E(r) = m$$
$$\text{Standard deviation, } \sigma = \sqrt{m}$$

Normal probability distribution:

x-values of a normal variable with mean, μ and standard deviation, σ

z-values of the standard normal variable with mean, 0 and standard deviation, 1.

$$z = \frac{x - \mu}{\sigma} \text{ standard deviations from the mean}$$

Combinations of independent normal variables:
x and y are independent and normally distributed. If $z = x \pm y$, then z is normally distributed:

$$\mu_z = \mu_x \pm \mu_y \quad \text{and} \quad \sigma_z^2 = \sigma_x^2 + \sigma_y^2$$

If:

$$z = ax \quad \text{where} \quad a \text{ is a constant}$$
$$\mu_z = a\mu_x \quad \text{and} \quad \sigma_z^2 = a^2 \sigma_x^2$$

Chapter Four

Sampling distribution of sample means:

$$E(\bar{x}) = \mu$$

$$\bar{x} = \hat{\mu} \quad \text{unbiased estimate}$$

$$SE_{\bar{x}} = \sqrt{\frac{(N-n)}{(N-1)} \frac{\sigma^2}{n}}$$

$$SE_{\bar{x}} = \frac{\sigma}{\sqrt{n}} \quad \text{if N large}$$

Sampling distribution of sample variance:

$$E(s^2) = \frac{N}{(N-1)} \frac{(n-1)}{n} \sigma^2$$

$$\hat{\sigma}^2 = \frac{(N-1)}{N} \frac{n}{(n-1)} s^2 \quad \text{unbiased estimate}$$

where
$$s^2 = \frac{\sum (x - \bar{x})^2}{n}$$

$$\hat{\sigma}^2 = \frac{n}{(n-1)} s^2 \quad \text{if N large}$$

$$\widehat{SE}_{\bar{x}} = \frac{s}{\sqrt{n-1}} = \frac{\hat{\sigma}}{\sqrt{n}}$$

Standard normal distribution for sample means:
σ^2 known, population normal:

$$z = \frac{\bar{x} - \mu}{SE_{\bar{x}}} = \frac{\bar{x} - \mu}{\sigma/\sqrt{n}}$$

t distribution for sample means:
σ^2 unknown, population normal:

$$t = \frac{\bar{x} - \mu}{\widehat{SE}_{\bar{x}}} = \frac{\bar{x} - \mu}{\hat{\sigma}/\sqrt{n}} = \frac{\bar{x} - \mu}{s/\sqrt{n-1}} \quad \text{with } (n-1) \text{ degrees of freedom}$$

χ^2 sampling distribution for sample variance:
Population normal:

$$\chi^2 = \frac{ns^2}{\sigma^2} \quad \text{with } (n-1) \text{ degrees of freedom}$$

F sampling distribution for two sample variances:
Populations normal:

$$F = \frac{n_1 s_1^2}{(n_1-1)\sigma_1^2} \bigg/ \frac{n_2 s_2^2}{(n_2-1)\sigma_2^2} \quad \text{with } (n_1-1) \text{ and } (n_2-1) \text{ degrees of freedom}$$

Chapter Five

$(1-\alpha) \times 100\%$ confidence interval for a population mean, μ:

$$\bar{x} \pm z_{\alpha/2} SE_{\bar{x}}$$

if σ is known, CI is:

$$\bar{x} \pm z_{\alpha/2} \frac{\sigma}{\sqrt{n}}$$

if σ is unknown, CI is:

$$\bar{x} \pm t_{\alpha/2,(n-1)} \frac{s}{\sqrt{n-1}}$$

678 QUANTITATIVE ANALYSIS

$(1 - \alpha) \times 100\%$ confidence interval for a population proportion, p:

$$\hat{p} \pm z_{\alpha/2} \sqrt{\frac{\hat{p}(1 - \hat{p})}{n}}$$

$(1 - \alpha) \times 100\%$ confidence interval for a population variance, σ^2:

$$\frac{ns^2}{\chi^2_{\alpha/2,(n-1)}} \quad \text{to} \quad \frac{ns^2}{\chi^2_{(1-\alpha/2),(n-1)}}$$

Chapter Six

Hypothesis test on sample means:
Single sample; σ^2 known, or $n \geq 30$:

$$z = \frac{\bar{x} - \mu}{SE_{\bar{x}}} = \frac{\bar{x} - \mu}{\sigma/\sqrt{n}}$$

σ^2 unknown:

$$t = \frac{\bar{x} - \mu}{\widehat{SE}_{\bar{x}}} = \frac{\bar{x} - \mu}{\hat{\sigma}/\sqrt{n}} = \frac{\bar{x} - \mu}{s/\sqrt{n-1}} \text{ with } (n-1) \text{ degrees of freedom}$$

Two samples; σ_1^2, σ_2^2 known:

$$z = \frac{(\bar{x}_1 - \bar{x}_2) - (\mu_1 - \mu_2)}{SE_{\bar{x}_1 - \bar{x}_2}}$$

where

$$SE_{\bar{x}_1 - \bar{x}_2} = \sqrt{\frac{\sigma_1^2}{n_1} + \frac{\sigma_2^2}{n_2}}$$

σ_1^2, σ_2^2 unknown but assume $\sigma_1^2 = \sigma_2^2$:

$$t = \frac{(\bar{x}_1 - \bar{x}_2) - (\mu_1 - \mu_2)}{\widehat{SE}_{\bar{x}_1 - \bar{x}_2}} \text{ with } (n_1 + n_2 - 2) \text{ degrees of freedom}$$

where

$$\widehat{SE}_{\bar{x}_1 - \bar{x}_2} = \hat{\sigma}\sqrt{\frac{1}{n_1} + \frac{1}{n_2}}$$

and

$$\hat{\sigma} = \sqrt{\frac{n_1 s_1^2 + n_2 s_2^2}{n_1 + n_2 - 2}}$$

Variance ratio, F, test:

$$F = \frac{\hat{\sigma}^2_{larger}}{\hat{\sigma}^2_{smaller}}$$

where

$$\hat{\sigma}^2_{larger} = \frac{n_1 s_1^2}{(n_1 - 1)}$$

$$\hat{\sigma}^2_{smaller} = \frac{n_2 s_2^2}{(n_2 - 1)} \text{ with } (n_1 - 1) \text{ and } (n_2 - 1) \text{ degrees of freedom}$$

Hypothesis test on sample proportions
$n \geq 30$
Single sample:

$$z = \frac{\hat{p} - p}{SE_{\hat{p}}} = \frac{\hat{p} - p}{\sqrt{\frac{p(1-p)}{n}}}$$

Two samples:

$$z = \frac{(\hat{p}_1 - \hat{p}_2) - (p_1 - p_2)}{SE_{\hat{p}_1 - \hat{p}_2}}$$

where $\quad SE_{\hat{p}_1 - \hat{p}_2} = \sqrt{\dfrac{p_1(1-p_1)}{n_1} + \dfrac{p_2(1-p_2)}{n_2}} \quad$ if p_1 and p_2 given

or $\quad \widehat{SE}_{\hat{p}_1 - \hat{p}_2} = \sqrt{\bar{p}(1-\bar{p})\left(\dfrac{1}{n_1} + \dfrac{1}{n_2}\right)} \quad$ if p_1 and p_2 are unknown but assumed equal
$p_1 = p_2 = p$ is estimated by pooling $\hat{p}_1$ and $\hat{p}_2$

χ^2 *test*

$$\chi^2 = \sum \left\{ \frac{(f_E - f_0)^2}{f_E} \right\}$$

Degrees of freedom:
Contingency table: $(r-1)(c-1)$
Goodness of fit: $n - 1 - k \quad$ k is the number of population parameters which are estimated from the sample.

Chapter Seven

Simple linear regression:

$$\hat{y} = a + bx \quad \text{estimates} \quad y = \alpha + \beta x + \varepsilon$$

For the 'least squares' line of best fit:

$$b = \frac{n \sum xy - \sum x \sum y}{n \sum x^2 - (\sum x)^2}$$

$$a = \frac{\sum y}{n} - \frac{b \sum x}{n}$$

$$r = \frac{n \sum xy - \sum x \sum y}{\sqrt{(n \sum x^2 - (\sum x)^2)(n \sum y^2 - (\sum y)^2)}}$$

Statistical inference in regression analysis—two variable models:
Test correlation coefficient:

$$t = \sqrt{\frac{r^2(n-2)}{(1-r^2)}} \quad \text{with } (n-2) \text{ degrees of freedom}$$

Test regression coefficient:

$$t = \frac{b}{SE_b} \quad \text{with } (n-2) \text{ degrees of freedom}$$

where: $\quad SE_b = \dfrac{\hat{\sigma}_e}{\sqrt{\sum (x - \bar{x})^2}}$

and: $\quad \hat{\sigma}_e = \dfrac{\sum (y - \hat{y})^2}{(n-2)}$

$$= \frac{(\sum y^2 - \sum y - b \sum xy)}{n - 2}$$

Confidence interval for the mean value of y, for a given x (x_o), $\mu_{y/x}$:

$$\hat{y} \pm t_{p/2,(n-2)}\hat{\sigma}_e\sqrt{\frac{1}{n} + \frac{(x_0 - \bar{x})^2}{\sum(x - \bar{x})^2}}$$

Confidence interval for individual y values, for a given x, (x_o):

$$\hat{y} \pm t_{p/2,(n-2)}\hat{\sigma}_e\sqrt{1 + \frac{1}{n} + \frac{(x_0 - \bar{x})^2}{\sum(x - \bar{x})^2}}$$

Statistical inference in regression analysis—multi-variable models:
Test of the model as a whole:

$$F = \frac{\text{Mean squares due to regression}}{\text{Mean squares due to residual}}$$

$$= \frac{\sum(\hat{y} - \bar{y})^2/df_{reg}}{\sum(y - \hat{y})^2/df_{resid}}$$

d_{freg} = number of independent variables in the model, k

$df_{resid} = n - 1 - k$

The F statistic has k and $(n - 1 - k)$ degrees of freedom.
Partial F test:

$$F = \frac{(r^2_{larger} - r^2_{smaller})/(k_{larger} - k_{smaller})}{(1 - r^2_{larger})/(n - 1 - k_{larger})}$$

Rank correlation:

$$r_s = 1 - \frac{6\sum d^2}{n(n^2 - 1)}$$

Test of significance:

$$z = \frac{(r_s - 0)}{1/\sqrt{(n-1)}} \qquad n \geqslant 10$$

Chapter Eight

Time series:

$$A = T + S + E: \text{additive model}$$

or

$$A = T \times S \times E: \text{multiplicative model}$$

$$\text{MAD} = \frac{\sum|\text{Actual} - \text{Forecast}|}{n} = \frac{\sum|E_t|}{n}$$

$$\text{MSE} = \frac{\sum(E_t)^2}{n}$$

$$F_t = aA_t + (1 - a)F_{t-1} \qquad 0 < a < 1$$

Chapter Seven

Network analysis:
PERT:

$$\text{Expected activity duration, } t = \frac{a + 4m + b}{6}$$

$$\text{Variance, } \sigma_t^2 = \left(\frac{b - a}{6}\right)^2$$

Chapter Twelve

Inventory control:
Basic model:

$$\text{Total annual variable cost of stockholding, TC} = \frac{C_o D}{q} + \frac{C_h q}{2} \quad \text{£ per year}$$

$$\text{EOQ} = \sqrt{\frac{2C_o D}{C_h}}$$

$$\text{Re-order interval} = \frac{q}{D}$$

$$\text{Simple EBQ} = \sqrt{\frac{2C_s D}{C_h}}$$

Batch production model—goods used while production is in progress:

$$\text{EBQ} = \sqrt{\frac{2C_s D}{C_h} \frac{P}{(P - D)}}$$

$$\text{TC} = \frac{C_s D}{q} + \frac{C_h (P - D) q}{2P} \quad \text{£ per year}$$

Planned shortages model:

1 Orders are filled from the new stock:

$$\text{TC} = \frac{C_o D}{q} + \frac{C_h (q - S)^2}{2q} + \frac{C_b S^2}{2q} \text{ £ per year}$$

$$\text{Optimum order size, } q_o = \sqrt{\frac{2C_o D}{C_h} \frac{(C_h + C_b)}{C_b}} = \text{EOQ} \sqrt{\frac{C_h + C_b}{C_b}}$$

$$S = \sqrt{\frac{2C_o D}{C_b} \frac{C_h}{(C_h + C_b)}}$$

2 Orders are not filled from the new stock:

$$\text{TC} = \frac{C_o D}{(q + S)} + \frac{C_h q^2}{2(q + S)} + \frac{C_b S^2}{2(q + S)} \text{ £ per year}$$

$$q_o = \sqrt{\frac{2C_o D}{C_h} \frac{C_b}{(C_h + C_b)}} = \text{EOQ} \sqrt{\frac{C_b}{(C_h + C_b)}}$$

$$S = \sqrt{\frac{2C_o D}{C_b} \frac{C_h}{(C_h + C_b)}}$$

Re-order interval model:

$$\text{Optimum interval, } T = \sqrt{\frac{2C_o}{C_h D}}$$

Chapter Thirteen

Queueing:

$$P(r \text{ arrivals/unit time}) = \frac{\lambda^r}{r!} e^{-\lambda} \quad r = 0, 1, 2, \ldots$$

$$P(\text{time, t, between arrivals}) = \lambda e^{-\lambda t}$$

M/M/1 model—steady state:

$$\rho = \frac{\lambda}{\mu}$$

$$P(o) = 1 - \frac{\lambda}{\mu}$$

$$P(n) = \left(\frac{\lambda}{\mu}\right)^n P(o)$$

$$\text{Mean number in queue} = \frac{\lambda^2}{\mu(\mu - \lambda)}$$

$$\text{Mean number in system} = \frac{\lambda}{\mu - \lambda}$$

$$\text{Mean time in queue} = \frac{\lambda}{\mu(\mu - \lambda)}$$

$$\text{Mean time in system} = \frac{1}{\mu - \lambda}$$

M/M/2 model—steady state:

$$P(o) = \frac{1}{1 + \frac{\lambda}{\mu} + \left(\frac{\lambda}{\mu}\right)^2 \frac{\mu}{(2\mu - \lambda)}}$$

$$P(n) = \frac{\left(\frac{\lambda}{\mu}\right)^n}{n!} P(o) \quad 0 \leqslant n \leqslant 2$$

$$P(n) = \frac{\left(\frac{\lambda}{\mu}\right)^n}{2!\, 2^{n-2}} P(o) \quad n \geqslant 2$$

$$\text{Mean number in queue} = \frac{\lambda^3}{\mu(2\mu - \lambda)^2} P(o)$$

$$\text{Mean number in system} = \text{mean number in queue} + \left(\frac{\lambda}{\mu}\right)$$

$$\text{Mean time in queue} = \text{mean number in queue}/\lambda$$

$$\text{Mean time in system} = \text{mean time in queue} + \frac{1}{\mu}$$

APPENDIX TWO

Statistical tables

Normal distribution (areas)

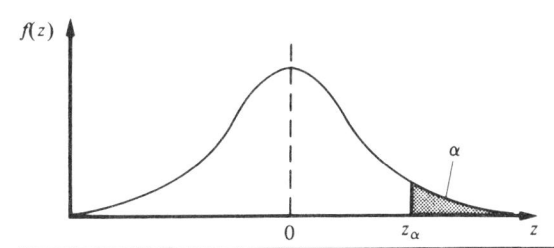

Area (α) in the tail of the standardised Normal curve, $N(0, 1)$, for different values of z. **Example**: Area beyond $z = 1.96$ (or below $z = -1.96$) is $\alpha = 0.02500$. For Normal curve with $\mu = 10$ and $\sigma = 2$, area beyond $x = 12$, say, is the same as area beyond $z = \dfrac{x - \mu}{\sigma} = \dfrac{12 - 10}{2} = 1$, i.e. $\alpha = 0.15866$.

$z \rightarrow$ $\downarrow$	0.00	0.01	0.02	0.03	0.04	0.05	0.06	0.07	0.08	0.09
0.0	.50000	.49601	.49202	.48803	.48405	.48006	.47608	.47210	.46812	.46414
0.1	.46017	.45620	.45224	.44828	.44433	.44038	.43644	.43251	.42858	.42465
0.2	.42074	.41683	.41294	.40905	.40517	.40129	.39743	.39358	.38974	.38591
0.3	.38209	.37828	.37448	.37070	.36693	.36317	.35942	.35569	.35197	.34827
0.4	.34458	.34090	.33724	.33360	.32997	.32636	.32276	.31918	.31561	.31207
0.5	.30854	.30503	.30153	.29806	.29460	.29116	.28774	.28434	.28096	.27760
0.6	.27425	.27093	.26763	.26435	.26109	.25785	.25463	.25143	.24825	.24510
0.7	.24196	.23885	.23576	.23270	.22965	.22663	.22363	.22065	.21770	.21476
0.8	.21186	.20897	.20611	.20327	.20045	.19766	.19489	.19215	.18943	.18673
0.9	.18406	.18141	.17879	.17619	.17361	.17106	.16853	.16602	.16354	.16109
1.0	.15866	.15625	.15386	.15150	.14917	.14686	.14457	.14231	.14007	.13786
1.1	.13567	.13350	.13136	.12924	.12714	.12507	.12302	.12100	.11900	.11702
1.2	.11507	.11314	.11123	.10935	.10749	.10565	.10383	.10204	.10027	.09853
1.3	.09680	.09510	.09342	.09176	.09012	.08851	.08692	.08534	.08379	.08226
1.4	.08076	.07927	.07780	.07636	.07493	.07353	.07214	.07078	.06944	.06811
1.5	.06681	.06552	.06426	.06301	.06178	.06057	.05938	.05821	.05705	.05592
1.6	.05480	.05370	.05262	.05155	.05050	.04947	.04846	.04746	.04648	.04551
1.7	.04457	.04363	.04272	.04182	.04093	.04006	.03920	.03836	.03754	.03673
1.8	.03593	.03515	.03438	.03362	.03288	.03216	.03144	.03074	.03005	.02938
1.9	.02872	.02807	.02743	.02680	.02619	.02559	.02500	.02442	.02385	.02330
2.0	.02275	.02222	.02169	.02118	.02068	.02018	.01970	.01923	.01876	.01831
2.1	.01786	.01743	.01700	.01659	.01618	.01578	.01539	.01500	.01463	.01426
2.2	.01390	.01355	.01321	.01287	.01254	.01222	.01191	.01160	.01130	.01101
2.3	.01072	.01044	.01017	.00990	.00964	.00939	.00914	.00889	.00866	.00842
2.4	.00820	.00798	.00776	.00755	.00734	.00714	.00695	.00676	.00657	.00639
2.5	.00621	.00604	.00587	.00570	.00554	.00539	.00523	.00509	.00494	.00480
2.6	.00466	.00453	.00440	.00427	.00415	.00403	.00391	.00379	.00368	.00357
2.7	.00347	.00336	.00326	.00317	.00307	.00298	.00289	.00280	.00272	.00263
2.8	.00256	.00248	.00240	.00233	.00226	.00219	.00212	.00205	.00199	.00193
2.9	.00187	.00181	.00175	.00169	.00164	.00159	.00154	.00149	.00144	.00139
3.0	.00135	.00131	.00126	.00122	.00118	.00114	.00111	.00107	.00104	.00100
3.1	.00097	.00094	.00090	.00087	.00085	.00082	.00079	.00076	.00074	.00071
3.2	.00069	.00066	.00064	.00062	.00060	.00058	.00056	.00054	.00052	.00050
3.3	.00048	.00047	.00045	.00043	.00042	.00040	.00039	.00038	.00036	.00035
3.4	.00034	.00032	.00031	.00030	.00029	.00028	.00027	.00026	.00025	.00024
3.5	.00023	.00022	.00022	.00021	.00020	.00019	.00019	.00018	.00017	.00017
3.6	.00016	.00015	.00015	.00014	.00014	.00013	.00013	.00012	.00012	.00011
3.7	.00011	.00010	.00010	.00010	.00009	.00009	.00009	.00008	.00008	.00008
3.8	.00007	.00007	.00007	.00006	.00006	.00006	.00006	.00005	.00005	.00005
3.9	.00005	.00005	.00004	.00004	.00004	.00004	.00004	.00004	.00004	.00003
4.0	.00003	.00003	.00003	.00003	.00003	.00002	.00002	.00002	.00002	.00002

α	0.4	0.25	0.2	0.15	0.1	0.05	0.025	0.01	0.005	0.001
z_α	.2533	.6745	.8416	1.0364	1.2816	1.6449	1.9600	2.3263	2.5758	3.0902

χ^2 (Chi-squared)-distribution

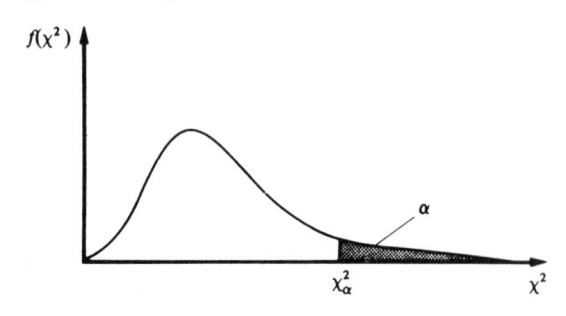

Values of χ_α^2 giving area (α) in the right-hand tail for different number of degrees of freedom (v). **Example:** For $v = 15$ area beyond $\chi_{0.95}^2 = 7.261$ is 0.950 and beyond $\chi_{0.10}^2 = 22.307$ is 0.100.

v \ α	0.995	0.990	0.975	0.950	0.900	0.750	0.500	0.250	0.100	0.050	0.025	0.010	0.005
1	0.0⁴3927*	0.0³1571*	0.0³9821*	0.0²3932*	0.01579	0.1015	0.4549	1.323	2.706	3.841	5.024	6.635	7.879
2	0.01003	0.02010	0.05065	0.1026	0.2107	0.5754	1.386	2.773	4.605	5.991	7.378	9.210	10.597
3	0.07172	0.1148	0.2158	0.3518	0.5844	1.213	2.366	4.108	6.251	7.815	9.348	11.345	12.838
4	0.2070	0.2971	0.4844	0.7107	1.064	1.923	3.357	5.385	7.779	9.488	11.143	13.277	14.860
5	0.4117	0.5543	0.8312	1.145	1.610	2.675	4.351	6.626	9.236	11.070	12.833	15.086	16.750
6	0.6757	0.8721	1.237	1.635	2.204	3.455	5.348	7.841	10.645	12.592	14.449	16.812	18.548
7	0.9893	1.239	1.690	2.167	2.833	4.255	6.346	9.037	12.017	14.067	16.013	18.475	20.278
8	1.344	1.646	2.180	2.733	3.490	5.071	7.344	10.219	13.362	15.507	17.535	20.090	21.955
9	1.735	2.088	2.700	3.325	4.168	5.899	8.343	11.389	14.684	16.919	19.023	21.666	23.589
10	2.156	2.558	3.247	3.940	4.865	6.737	9.342	12.549	15.987	18.307	20.483	23.209	25.188
11	2.603	3.053	3.816	4.575	5.578	7.584	10.341	13.701	17.275	19.675	21.920	24.725	26.757
12	3.074	3.571	4.404	5.226	6.304	8.438	11.340	14.845	18.549	21.026	23.337	26.217	28.300
13	3.565	4.107	5.009	5.892	7.041	9.299	12.340	15.984	19.812	22.362	24.736	27.688	29.819
14	4.075	4.660	5.629	6.571	7.790	10.165	13.339	17.117	21.064	23.685	26.119	29.141	31.319
15	4.601	5.229	6.262	7.261	8.547	11.036	14.339	18.245	22.307	24.996	27.488	30.578	32.801
16	5.142	5.812	6.908	7.962	9.312	11.912	15.338	19.369	23.542	26.296	28.845	32.000	34.267
17	5.697	6.408	7.564	8.672	10.085	12.792	16.338	20.489	24.769	27.587	30.191	33.409	35.718
18	6.265	7.015	8.231	9.390	10.865	13.675	17.338	21.605	25.989	28.869	31.526	34.805	37.156
19	6.844	7.633	8.907	10.117	11.651	14.562	18.338	22.718	27.204	30.143	32.852	36.191	38.582
20	7.434	8.260	9.591	10.851	12.443	15.452	19.337	23.828	28.412	31.410	34.170	37.566	39.997
21	8.034	8.897	10.283	11.591	13.240	16.344	20.337	24.935	29.615	32.670	35.479	38.932	41.401
22	8.643	9.542	10.982	12.338	14.041	17.240	21.337	26.039	30.813	33.924	36.781	40.289	42.796
23	9.260	10.196	11.688	13.090	14.848	18.137	22.337	27.141	32.007	35.172	38.076	41.638	44.181
24	9.886	10.856	12.401	13.848	15.659	19.037	23.337	28.241	33.196	36.415	39.364	42.080	45.558
25	10.520	11.524	13.120	14.611	16.473	19.939	24.337	29.339	34.382	37.652	40.646	44.314	46.928
26	11.160	12.198	13.844	15.379	17.292	20.843	25.336	30.434	35.563	38.885	41.923	45.642	48.290
27	11.808	12.879	14.573	16.151	18.114	21.749	26.336	31.528	36.741	40.113	43.194	46.963	49.645
28	12.461	13.565	15.308	16.928	18.939	22.657	27.336	32.620	37.916	41.337	44.461	48.278	50.993
29	13.121	14.256	16.047	17.708	19.768	23.567	28.336	33.711	39.087	42.557	45.722	49.588	52.336
30	13.787	14.954	16.791	18.493	20.599	24.478	29.336	34.800	40.256	43.773	46.979	50.892	53.672
35	17.192	18.509	20.569	22.465	24.797	29.054	34.336	40.223	46.059	49.802	53.203	57.342	60.275
40	20.707	22.164	24.433	26.509	29.050	33.660	39.335	45.616	51.805	55.758	59.342	63.691	66.766
45	24.311	25.901	28.366	30.612	33.350	38.291	44.335	50.985	57.505	61.656	65.410	69.957	73.166
50	27.991	29.707	32.357	34.764	37.689	42.942	49.335	56.334	63.167	67.505	71.420	76.154	79.490
55	31.735	33.571	36.398	38.958	42.060	47.611	54.335	61.665	68.796	73.311	77.381	82.292	85.749
60	35.535	37.485	40.482	43.188	46.459	52.294	59.335	66.981	74.397	79.082	83.298	88.379	91.952
70	43.275	45.442	48.758	51.739	55.329	61.698	69.334	77.577	85.527	90.531	95.023	100.425	104.215
80	51.172	53.540	57.153	60.391	64.278	71.144	79.334	88.130	96.578	101.879	106.629	112.329	116.321
90	59.196	61.754	65.647	69.126	73.291	80.625	89.334	98.650	107.565	113.145	118.136	124.116	128.299
100	67.328	70.065	74.222	77.929	82.358	90.133	99.334	109.141	118.498	124.342	129.561	135.807	140.169
120	83.829	86.909	91.568	95.705	100.627	109.224	119.335	130.051	140.228	146.565	152.214	158.963	163.670
150	109.122	112.655	117.980	122.692	128.278	137.987	149.334	161.288	172.577	179.579	185.803	193.219	198.380
200	152.224	156.421	162.724	168.279	174.828	186.175	199.334	213.099	226.018	233.993	241.060	249.455	255.281
250	196.145	200.929	208.095	214.392	221.809	234.580	249.334	264.694	279.947	287.889	295.691	304.948	311.361
z_α	−2.5758	−2.3263	−1.9600	−1.6449	−1.2816	−0.6745	0.0000	0.6745	1.2816	1.6449	1.9600	2.3263	2.5758

* e.g. $0.0^43927 = 0.00003927$

Interpolation: For $v > 100$, $\chi_\alpha^2 = \frac{1}{2}(z_\alpha + \sqrt{2v-1})^2$ where z_α is the standardised Normal variable shown in the bottom line of the table.

t-distribution

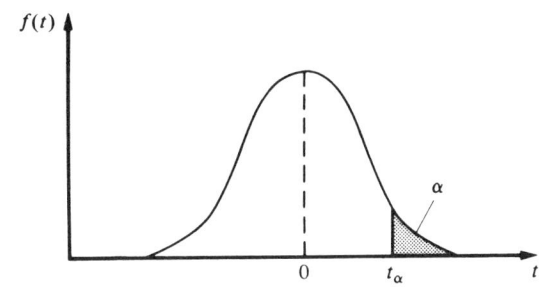

Critical points (t_α) for different probability levels (α) and different number of degrees of freedom (ν). **Example:** For $\nu = 19$, $P(t > 2.0930) = 0.025$ and $P(|t| > 2.0930) = 0.05$.

ν \ α	0.4	0.25	0.15	0.1	0.05	0.025	0.01	0.005	0.001	0.0005
1	0.3249	1.0000	1.9626	3.0777	6.3138	12.7062	31.8205	63.6567	318.3087	636.6189
2	0.2887	0.8165	1.3862	1.8856	2.9200	4.3027	6.9646	9.9248	22.3271	31.5991
3	0.2767	0.7649	1.2498	1.6377	2.3534	3.1824	4.5407	5.8409	10.2145	12.9240
4	0.2707	0.7407	1.1896	1.5332	2.1318	2.7764	3.7469	4.6041	7.1732	8.6103
5	0.2672	0.7267	1.1558	1.4759	2.0150	2.5706	3.3649	4.0321	5.8934	6.8688
6	0.2648	0.7176	1.1342	1.4398	1.9432	2.4469	3.1427	3.7074	5.2076	5.9588
7	0.2632	0.7111	1.1192	1.4149	1.8946	2.3646	2.9980	3.4995	4.7853	5.4079
8	0.2619	0.7064	1.1081	1.3968	1.8595	2.3060	2.8965	3.3554	4.5008	5.0413
9	0.2610	0.7027	1.0997	1.3830	1.8331	2.2622	2.8214	3.2498	4.2968	4.7809
10	0.2602	0.6998	1.0931	1.3722	1.8125	2.2281	2.7638	3.1693	4.1437	4.5869
11	0.2596	0.6974	1.0877	1.3634	1.7959	2.2010	2.7181	3.1058	4.0247	4.4370
12	0.2590	0.6955	1.0832	1.3562	1.7823	2.1788	2.6810	3.0545	3.9296	4.3178
13	0.2586	0.6938	1.0795	1.3502	1.7709	2.1604	2.6503	3.0123	3.8520	4.2208
14	0.2582	0.6924	1.0763	1.3450	1.7613	2.1448	2.6245	2.9768	3.7874	4.1405
15	0.2579	0.6912	1.0735	1.3406	1.7531	2.1314	2.6025	2.9467	3.7328	4.0728
16	0.2576	0.6901	1.0711	1.3368	1.7459	2.1199	2.5835	2.9208	3.6862	4.0150
17	0.2573	0.6892	1.0690	1.3334	1.7396	2.1098	2.5669	2.8982	3.6458	3.9651
18	0.2571	0.6884	1.0672	1.3304	1.7341	2.1009	2.5524	2.8784	3.6105	3.9216
19	0.2569	0.6876	1.0655	1.3277	1.7291	2.0930	2.5395	2.8609	3.5794	3.8834
20	0.2567	0.6870	1.0640	1.3253	1.7247	2.0860	2.5280	2.8453	3.5518	3.8495
21	0.2566	0.6864	1.0627	1.3232	1.7207	2.0796	2.5176	2.8314	3.5272	3.8193
22	0.2564	0.6858	1.0614	1.3212	1.7171	2.0739	2.5083	2.8188	3.5050	3.7921
23	0.2563	0.6853	1.0603	1.3195	1.7139	2.0687	2.4999	2.8073	3.4850	3.7676
24	0.2562	0.6848	1.0593	1.3178	1.7109	2.0639	2.4922	2.7969	3.4668	3.7454
25	0.2561	0.6844	1.0584	1.3163	1.7081	2.0595	2.4851	2.7874	3.4502	3.7251
26	0.2560	0.6840	1.0575	1.3150	1.7056	2.0555	2.4786	2.7787	3.4350	3.7066
27	0.2559	0.6837	1.0567	1.3137	1.7033	2.0518	2.4727	2.7707	3.4210	3.6896
28	0.2558	0.6834	1.0560	1.3125	1.7011	2.0484	2.4671	2.7633	3.4082	3.6739
29	0.2557	0.6830	1.0553	1.3114	1.6991	2.0452	2.4620	2.7564	3.3962	3.6594
30	0.2556	0.6828	1.0547	1.3104	1.6973	2.0423	2.4573	2.7500	3.3852	3.6460
35	0.2553	0.6816	1.0520	1.3062	1.6896	2.0301	2.4377	2.7238	3.3400	3.5911
40	0.2550	0.6807	1.0500	1.3031	1.6839	2.0211	2.4233	2.7045	3.3069	3.5510
45	0.2549	0.6800	1.0485	1.3006	1.6794	2.0141	2.4121	2.6896	3.2815	3.5203
50	0.2547	0.6794	1.0473	1.2987	1.6759	2.0086	2.4033	2.6778	3.2614	3.4960
60	0.2545	0.6786	1.0455	1.2958	1.6706	2.0003	2.3901	2.6603	3.2317	3.4602
70	0.2543	0.6780	1.0442	1.2938	1.6669	1.9944	2.3808	2.6479	3.2108	3.4350
80	0.2542	0.6776	1.0432	1.2922	1.6641	1.9901	2.3739	2.6387	3.1953	3.4163
90	0.2541	0.6772	1.0424	1.2910	1.6620	1.9867	2.3685	2.6316	3.1833	3.4019
100	0.2540	0.6770	1.0418	1.2901	1.6602	1.9840	2.3642	2.6259	3.1737	3.3905
120	0.2539	0.6765	1.0409	1.2886	1.6577	1.9799	2.3578	2.6174	3.1595	3.3735
150	0.2538	0.6761	1.0400	1.2872	1.6551	1.9759	2.3515	2.6090	3.1455	3.3566
200	0.2537	0.6757	1.0391	1.2858	1.6525	1.9719	2.3451	2.6006	3.1315	3.3398
300	0.2536	0.6753	1.0382	1.2844	1.6499	1.9679	2.3388	2.5923	3.1176	3.3233
∞	0.2533	0.6745	1.0364	1.2816	1.6449	1.9600	2.3263	2.5758	3.0902	3.2905

F-distribution

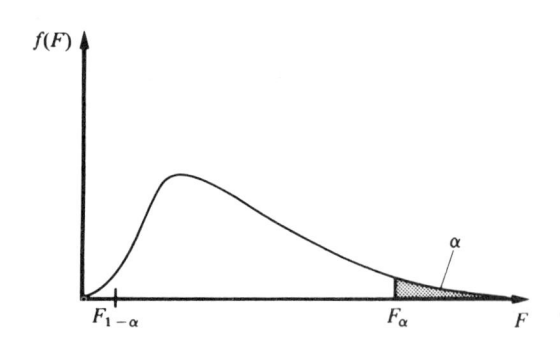

Values of F_α ($\alpha = 0.1, 0.05, 0.025, 0.01$ and 0.005) for different combinations of degrees of freedom in the numerator, v_1, and denominator, v_2. **Example:** When $v_1 = 10$ and $v_2 = 2$, area (α) to the right of $F_{0.05} = 19.40$ is 0.05. To find $F_{1-\alpha}$, leaving an area α in the left-hand tail, use the relation: $F_{1-\alpha}(v_1, v_2) = 1/F_\alpha(v_2, v_1)$. **Example:** $F_{0.95}(2, 10) = 1/19.40 = 0.05155$.

v_2	α	1	2	3	4	5	6	7	8	9	10	12	15	20	25	30	50	100	∞
1	0.100	39.86	49.50	53.59	55.83	57.24	58.20	58.91	59.44	59.86	60.19	60.71	61.22	61.74	62.05	62.26	62.69	63.01	63.33
	0.050	161.4	199.5	215.7	224.6	230.2	234.0	236.8	238.9	240.5	241.9	243.9	245.9	248.0	249.3	250.1	251.8	253.0	254.3
	0.025	647.8	799.5	864.2	899.0	921.8	937.1	948.2	956.7	963.3	968.6	976.7	984.9	993.1	998.1	1001	1008	1013	1018
	0.010	4052	4999	5403	5625	5764	5859	5928	5981	6022	6056	6106	6157	6209	6240	6261	6303	6334	6366
	0.005	16211	20000	21615	22500	23056	23437	23715	23925	24091	24224	24426	24630	24836	24960	25044	25211	25337	25464
2	0.100	8.526	9.000	9.162	9.243	9.293	9.326	9.349	9.367	9.381	9.392	9.408	9.425	9.441	9.451	9.458	9.471	9.481	9.491
	0.050	18.51	19.00	19.16	19.25	19.30	19.33	19.35	19.37	19.38	19.40	19.41	19.43	19.45	19.46	19.46	19.48	19.49	19.50
	0.025	38.51	39.00	39.17	39.25	39.30	39.33	39.36	39.37	39.39	39.40	39.41	39.43	39.45	39.46	39.46	39.48	39.49	39.50
	0.010	98.50	99.00	99.17	99.25	99.30	99.33	99.36	99.37	99.39	99.40	99.42	99.43	99.45	99.46	99.47	99.48	99.49	99.50
	0.005	198.5	199.0	199.2	199.2	199.3	199.3	199.4	199.4	199.4	199.4	199.4	199.4	199.4	199.5	199.5	199.5	199.5	199.5
3	0.100	5.538	5.462	5.391	5.343	5.309	5.285	5.266	5.252	5.240	5.230	5.216	5.200	5.184	5.175	5.168	5.155	5.144	5.134
	0.050	10.13	9.552	9.277	9.117	9.013	8.941	8.887	8.845	8.812	8.786	8.745	8.703	8.660	8.634	8.617	8.581	8.554	8.526
	0.025	17.44	16.04	15.44	15.10	14.88	14.73	14.62	14.54	14.47	14.42	14.34	14.25	14.17	14.12	14.08	14.01	13.96	13.90
	0.010	34.12	30.82	29.46	28.71	28.24	27.91	27.67	27.49	27.35	27.23	27.05	26.87	26.69	26.58	26.50	26.35	26.24	26.13
	0.005	55.55	49.80	47.47	46.19	45.39	44.84	44.43	44.13	43.88	43.69	43.39	43.08	42.78	42.59	42.47	42.21	42.02	41.83
4	0.100	4.545	4.325	4.191	4.107	4.051	4.010	3.979	3.955	3.936	3.920	3.896	3.870	3.844	3.828	3.817	3.795	3.778	3.761
	0.050	7.709	6.944	6.591	6.388	6.256	6.163	6.094	6.041	5.999	5.964	5.912	5.858	5.803	5.769	5.746	5.699	5.664	5.628
	0.025	12.22	10.65	9.979	9.605	9.364	9.197	9.074	8.980	8.905	8.844	8.751	8.657	8.560	8.501	8.461	8.381	8.319	8.257
	0.010	21.20	18.00	16.69	15.98	15.52	15.21	14.98	14.80	14.66	14.55	14.37	14.20	14.02	13.91	13.84	13.69	13.58	13.46
	0.005	31.33	26.28	24.26	23.15	22.46	21.97	21.62	21.35	21.14	20.97	20.70	20.44	20.17	20.00	19.89	19.67	19.50	19.32
5	0.100	4.060	3.780	3.619	3.520	3.453	3.405	3.368	3.339	3.316	3.297	3.268	3.238	3.207	3.187	3.174	3.147	3.126	3.105
	0.050	6.608	5.786	5.409	5.192	5.050	4.950	4.876	4.818	4.772	4.735	4.678	4.619	4.558	4.521	4.496	4.444	4.405	4.365
	0.025	10.01	8.434	7.764	7.388	7.146	6.978	6.853	6.757	6.681	6.619	6.525	6.428	6.329	6.268	6.227	6.144	6.080	6.015
	0.010	16.26	13.27	12.06	11.39	10.97	10.67	10.46	10.29	10.16	10.05	9.888	9.722	9.553	9.449	9.379	9.238	9.130	9.020
	0.005	22.78	18.31	16.53	15.56	14.94	14.51	14.20	13.96	13.77	13.62	13.38	13.15	12.90	12.76	12.66	12.45	12.30	12.14
6	0.100	3.776	3.463	3.289	3.181	3.108	3.055	3.014	2.983	2.958	2.937	2.905	2.871	2.836	2.815	2.800	2.770	2.746	2.722
	0.050	5.987	5.143	4.757	4.534	4.387	4.284	4.207	4.147	4.099	4.060	4.000	3.938	3.874	3.835	3.808	3.754	3.712	3.669
	0.025	8.813	7.260	6.599	6.227	5.988	5.820	5.695	5.600	5.523	5.461	5.366	5.269	5.168	5.107	5.065	4.980	4.915	4.849
	0.010	13.75	10.92	9.780	9.148	8.746	8.466	8.260	8.102	7.976	7.874	7.718	7.559	7.396	7.296	7.229	7.091	6.987	6.880
	0.005	18.63	14.54	12.92	12.03	11.46	11.07	10.79	10.57	10.39	10.25	10.03	9.814	9.589	9.451	9.358	9.170	9.026	8.879
7	0.100	3.589	3.257	3.074	2.961	2.883	2.827	2.785	2.752	2.725	2.703	2.668	2.632	2.595	2.571	2.555	2.523	2.497	2.471
	0.050	5.591	4.737	4.347	4.120	3.972	3.886	3.787	3.726	3.677	3.637	3.575	3.511	3.445	3.404	3.376	3.319	3.275	3.230
	0.025	8.073	6.542	5.890	5.523	5.285	5.119	4.995	4.899	4.823	4.761	4.666	4.568	4.467	4.405	4.362	4.276	4.210	4.142
	0.010	12.25	9.547	8.451	7.847	7.460	7.191	6.993	6.840	6.719	6.620	6.469	6.314	6.155	6.058	5.992	5.858	5.755	5.650
	0.005	16.24	12.40	10.88	10.05	9.522	9.155	8.885	8.678	8.514	8.380	8.176	7.968	7.754	7.623	7.534	7.354	7.217	7.076
8	0.100	3.458	3.113	2.924	2.806	2.726	2.668	2.624	2.589	2.561	2.538	2.502	2.464	2.425	2.400	2.383	2.348	2.321	2.293
	0.050	5.318	4.459	4.066	3.838	3.687	3.581	3.500	3.438	3.388	3.347	3.284	3.218	3.150	3.108	3.079	3.020	2.975	2.928
	0.025	7.571	6.059	5.416	5.055	4.817	4.652	4.529	4.433	4.357	4.295	4.200	4.101	3.999	3.937	3.894	3.807	3.739	3.670
	0.010	11.26	8.649	7.591	7.006	6.632	6.371	6.178	6.029	5.911	5.814	5.667	5.515	5.359	5.263	5.198	5.065	4.963	4.859
	0.005	14.69	11.04	9.596	8.805	8.302	7.952	7.694	7.496	7.339	7.211	7.015	6.814	6.608	6.482	6.396	6.222	6.088	5.951
9	0.100	3.360	3.006	2.813	2.693	2.611	2.551	2.505	2.469	2.440	2.416	2.379	2.340	2.298	2.272	2.255	2.218	2.189	2.159
	0.050	5.117	4.256	3.863	3.633	3.482	3.374	3.293	3.230	3.179	3.137	3.073	3.006	2.936	2.893	2.864	2.803	2.756	2.707
	0.025	7.209	5.715	5.078	4.718	4.484	4.320	4.197	4.102	4.026	3.964	3.868	3.769	3.667	3.604	3.560	3.472	3.403	3.333
	0.010	10.56	8.022	6.992	6.422	6.057	5.802	5.613	5.467	5.351	5.257	5.111	4.962	4.808	4.713	4.649	4.517	4.415	4.311
	0.005	13.61	10.11	8.717	7.956	7.471	7.134	6.885	6.693	6.541	6.417	6.227	6.032	5.832	5.708	5.625	5.454	5.322	5.188
10	0.100	3.285	2.924	2.728	2.605	2.522	2.461	2.414	2.377	2.347	2.323	2.284	2.244	2.201	2.174	2.155	2.117	2.087	2.055
	0.050	4.965	4.103	3.708	3.478	3.326	3.217	3.135	3.072	3.020	2.978	2.913	2.845	2.774	2.730	2.700	2.637	2.588	2.538
	0.025	6.937	5.456	4.826	4.468	4.236	4.072	3.950	3.855	3.779	3.717	3.621	3.522	3.419	3.355	3.311	3.221	3.152	3.080
	0.010	10.04	7.559	6.552	5.994	5.636	5.386	5.200	5.057	4.942	4.849	4.706	4.558	4.405	4.311	4.247	4.115	4.014	3.909
	0.005	12.83	9.427	8.081	7.343	6.872	6.545	6.302	6.116	5.968	5.847	5.661	5.471	5.274	5.153	5.071	4.902	4.772	4.639
11	0.100	3.225	2.860	2.660	2.536	2.451	2.389	2.342	2.304	2.274	2.248	2.209	2.167	2.123	2.095	2.076	2.036	2.005	1.972
	0.050	4.884	3.982	3.587	3.357	3.204	3.095	3.012	2.948	2.896	2.854	2.788	2.719	2.646	2.601	2.570	2.507	2.457	2.404
	0.025	6.724	5.256	4.630	4.275	4.044	3.881	3.759	3.664	3.588	3.526	3.430	3.330	3.226	3.162	3.118	3.027	2.956	2.883
	0.010	9.646	7.206	6.217	5.668	5.316	5.069	4.886	4.744	4.632	4.539	4.397	4.251	4.099	4.005	3.941	3.810	3.708	3.602
	0.005	12.23	8.912	7.600	6.881	6.422	6.102	5.865	5.682	5.537	5.418	5.236	5.049	4.855	4.736	4.654	4.486	4.359	4.226

STATISTICAL TABLES

F-distribution—*continued*

v_2	α \ v_1	1	2	3	4	5	6	7	8	9	10	12	15	20	25	30	50	100	∞
12	0.100	3.177	2.807	2.606	2.480	2.394	2.331	2.283	2.245	2.214	2.188	2.147	2.105	2.060	2.031	2.011	1.970	1.938	1.904
	0.050	4.747	3.885	3.490	3.259	3.106	2.996	2.913	2.849	2.796	2.753	2.687	2.617	2.544	2.498	2.466	2.401	2.350	2.296
	0.025	6.554	5.096	4.474	4.121	3.891	3.728	3.607	3.512	3.436	3.374	3.277	3.177	3.073	3.008	2.963	2.871	2.800	2.725
	0.010	9.330	6.927	5.953	5.412	5.064	4.821	4.640	4.499	4.388	4.296	4.155	4.010	3.858	3.765	3.701	3.569	3.467	3.361
	0.005	11.75	8.510	7.226	6.521	6.071	5.757	5.525	5.345	5.202	5.085	4.906	4.721	4.530	4.412	4.331	4.165	4.037	3.904
13	0.100	3.136	2.763	2.560	2.434	2.347	2.283	2.234	2.195	2.164	2.138	2.097	2.053	2.007	1.978	1.958	1.915	1.882	1.846
	0.050	4.667	3.806	3.411	3.179	3.025	2.915	2.832	2.767	2.714	2.671	2.604	2.533	2.459	2.412	2.380	2.314	2.261	2.206
	0.025	6.414	4.965	4.327	3.996	3.707	3.604	3.483	3.388	3.312	3.250	3.153	3.053	2.948	2.882	2.837	2.744	2.671	2.595
	0.010	9.074	6.701	5.739	5.205	4.862	4.620	4.441	4.302	4.191	4.100	3.960	3.815	3.665	3.571	3.507	3.375	3.272	3.165
	0.005	11.37	8.186	6.926	6.233	5.791	5.482	5.253	5.076	4.935	4.820	4.643	4.460	4.270	4.153	4.073	3.908	3.780	3.647
14	0.100	3.102	2.726	2.522	2.395	2.307	2.243	2.193	2.154	2.122	2.095	2.054	2.010	1.962	1.933	1.912	1.869	1.834	1.797
	0.050	4.600	3.739	3.344	3.112	2.958	2.848	2.764	2.699	2.646	2.602	2.534	2.463	2.388	2.341	2.308	2.241	2.187	2.131
	0.025	6.298	4.857	4.222	3.892	3.663	3.501	3.380	3.285	3.209	3.147	3.050	2.949	2.844	2.778	2.732	2.638	2.565	2.487
	0.010	8.862	6.515	5.564	5.035	4.695	4.456	4.278	4.140	4.030	3.939	3.800	3.656	3.505	3.412	3.348	3.215	3.112	3.004
	0.005	11.06	7.922	6.680	5.998	5.562	5.257	5.031	4.857	4.717	4.603	4.428	4.247	4.059	3.942	3.862	3.698	3.569	3.436
15	0.100	3.073	2.695	2.490	2.361	2.273	2.208	2.158	2.119	2.086	2.059	2.017	1.972	1.924	1.894	1.873	1.828	1.793	1.755
	0.050	4.543	3.682	3.287	3.056	2.901	2.790	2.707	2.641	2.588	2.544	2.475	2.403	2.328	2.280	2.247	2.178	2.123	2.066
	0.025	6.200	4.765	4.153	3.804	3.576	3.415	3.293	3.199	3.123	3.060	2.963	2.862	2.756	2.689	2.644	2.540	2.474	2.395
	0.010	8.683	6.359	5.417	4.893	4.556	4.318	4.142	4.004	3.895	3.805	3.666	3.522	3.372	3.278	3.214	3.081	2.977	2.868
	0.005	10.80	7.701	6.476	5.805	5.372	5.071	4.847	4.674	4.536	4.424	4.250	4.070	3.883	3.766	3.687	3.523	3.394	3.260
16	0.100	3.048	2.668	2.462	2.333	2.244	2.178	2.128	2.088	2.055	2.028	1.985	1.940	1.891	1.860	1.839	1.793	1.757	1.718
	0.050	4.494	3.634	3.239	3.007	2.852	2.741	2.657	2.591	2.538	2.494	2.425	2.352	2.276	2.227	2.194	2.124	2.068	2.010
	0.025	6.115	4.687	4.077	3.729	3.502	3.341	3.219	3.125	3.049	2.986	2.889	2.788	2.681	2.614	2.568	2.472	2.396	2.316
	0.010	8.531	6.226	5.292	4.773	4.437	4.202	4.026	3.890	3.780	3.691	3.553	3.409	3.259	3.165	3.101	2.967	2.863	2.753
	0.005	10.58	7.514	6.303	5.638	5.212	4.913	4.692	4.521	4.384	4.272	4.099	3.920	3.734	3.618	3.539	3.375	3.246	3.112
17	0.100	3.026	2.645	2.437	2.308	2.218	2.152	2.102	2.061	2.028	2.001	1.958	1.912	1.862	1.831	1.809	1.763	1.726	1.686
	0.050	4.451	3.592	3.197	2.965	2.810	2.699	2.614	2.548	2.494	2.450	2.381	2.308	2.230	2.181	2.148	2.077	2.020	1.960
	0.025	6.042	4.619	4.011	3.665	3.438	3.277	3.156	3.061	2.985	2.922	2.825	2.723	2.616	2.548	2.502	2.405	2.329	2.247
	0.010	8.400	6.112	5.185	4.669	4.336	4.102	3.927	3.791	3.682	3.593	3.455	3.312	3.162	3.068	3.003	2.869	2.764	2.653
	0.005	10.38	7.354	6.156	5.497	5.075	4.779	4.559	4.389	4.254	4.142	3.971	3.793	3.607	3.492	3.412	3.248	3.119	2.984
18	0.100	3.007	2.624	2.416	2.286	2.196	2.130	2.079	2.038	2.005	1.977	1.933	1.887	1.837	1.805	1.783	1.736	1.698	1.657
	0.050	4.414	3.555	3.160	2.928	2.773	2.661	2.577	2.510	2.456	2.412	2.342	2.269	2.191	2.141	2.107	2.035	1.978	1.917
	0.025	5.978	4.560	3.954	3.608	3.382	3.221	3.100	3.005	2.929	2.866	2.769	2.667	2.559	2.491	2.445	2.347	2.269	2.187
	0.010	8.285	6.013	5.092	4.579	4.248	4.015	3.841	3.705	3.597	3.508	3.371	3.227	3.077	2.983	2.919	2.784	2.678	2.566
	0.005	10.22	7.215	6.028	5.375	4.956	4.663	4.445	4.276	4.141	4.030	3.860	3.683	3.498	3.382	3.303	3.130	3.009	2.873
19	0.100	2.990	2.606	2.397	2.266	2.176	2.109	2.058	2.017	1.984	1.956	1.912	1.865	1.814	1.782	1.759	1.711	1.673	1.631
	0.050	4.381	3.522	3.127	2.895	2.740	2.628	2.544	2.477	2.423	2.378	2.308	2.234	2.155	2.106	2.071	1.999	1.940	1.878
	0.025	5.922	4.508	3.903	3.559	3.333	3.172	3.051	2.956	2.880	2.817	2.720	2.617	2.509	2.441	2.394	2.295	2.217	2.133
	0.010	8.185	5.926	5.010	4.500	4.171	3.939	3.765	3.631	3.523	3.434	3.297	3.153	3.003	2.909	2.844	2.709	2.602	2.489
	0.005	10.07	7.093	5.916	5.268	4.853	4.561	4.345	4.177	4.043	3.933	3.763	3.587	3.402	3.287	3.208	3.043	2.913	2.776
20	0.100	2.975	2.589	2.380	2.249	2.158	2.091	2.040	1.999	1.965	1.937	1.892	1.845	1.794	1.761	1.738	1.690	1.650	1.607
	0.050	4.351	3.493	3.098	2.866	2.711	2.599	2.514	2.447	2.393	2.348	2.278	2.203	2.124	2.074	2.039	1.966	1.907	1.843
	0.025	5.871	4.461	3.859	3.515	3.289	3.128	3.007	2.913	2.837	2.774	2.676	2.573	2.464	2.396	2.349	2.249	2.170	2.085
	0.010	8.096	5.849	4.938	4.431	4.103	3.871	3.699	3.564	3.457	3.368	3.231	3.088	2.938	2.843	2.778	2.643	2.535	2.421
	0.005	9.944	6.986	5.818	5.174	4.762	4.472	4.257	4.090	3.956	3.847	3.678	3.502	3.318	3.203	3.123	2.950	2.828	2.690
21	0.100	2.961	2.575	2.365	2.233	2.142	2.075	2.023	1.982	1.948	1.920	1.875	1.827	1.776	1.742	1.719	1.670	1.630	1.586
	0.050	4.325	3.467	3.072	2.840	2.685	2.573	2.488	2.420	2.366	2.321	2.250	2.176	2.096	2.045	2.010	1.936	1.876	1.812
	0.025	5.827	4.420	3.819	3.475	3.250	3.090	2.969	2.874	2.798	2.735	2.637	2.534	2.425	2.356	2.308	2.208	2.128	2.042
	0.010	8.017	5.780	4.874	4.369	4.042	3.812	3.640	3.506	3.398	3.310	3.173	3.030	2.880	2.785	2.720	2.584	2.475	2.360
	0.005	9.830	6.891	5.730	5.091	4.681	4.393	4.179	4.013	3.880	3.771	3.602	3.427	3.243	3.128	3.049	2.884	2.753	2.614
22	0.100	2.949	2.561	2.351	2.219	2.128	2.060	2.008	1.967	1.933	1.904	1.859	1.811	1.759	1.726	1.702	1.652	1.611	1.567
	0.050	4.301	3.443	3.049	2.817	2.661	2.549	2.464	2.397	2.342	2.297	2.226	2.151	2.071	2.020	1.984	1.909	1.849	1.783
	0.025	5.786	4.383	3.783	3.440	3.215	3.055	2.934	2.839	2.763	2.700	2.602	2.498	2.389	2.320	2.272	2.171	2.090	2.003
	0.010	7.945	5.719	4.817	4.313	3.988	3.758	3.587	3.453	3.346	3.258	3.121	2.978	2.827	2.733	2.667	2.531	2.422	2.305
	0.005	9.727	6.806	5.652	5.017	4.609	4.322	4.109	3.944	3.812	3.703	3.535	3.360	3.176	3.061	2.982	2.817	2.685	2.545
23	0.100	2.937	2.549	2.339	2.207	2.115	2.047	1.995	1.953	1.919	1.890	1.845	1.796	1.744	1.710	1.686	1.636	1.594	1.549
	0.050	4.279	3.422	3.028	2.796	2.640	2.528	2.442	2.375	2.320	2.275	2.204	2.128	2.048	1.996	1.961	1.885	1.823	1.757
	0.025	5.750	4.349	3.750	3.408	3.183	3.023	2.902	2.808	2.731	2.668	2.570	2.466	2.357	2.287	2.239	2.137	2.056	1.968
	0.010	7.881	5.664	4.765	4.264	3.939	3.710	3.539	3.406	3.299	3.211	3.074	2.931	2.781	2.686	2.620	2.483	2.373	2.256
	0.005	9.635	6.730	5.582	4.950	4.544	4.259	4.047	3.882	3.750	3.642	3.475	3.300	3.116	3.001	2.922	2.756	2.624	2.484
24	0.100	2.927	2.538	2.327	2.195	2.103	2.035	1.983	1.941	1.906	1.877	1.832	1.783	1.730	1.696	1.672	1.621	1.579	1.533
	0.050	4.260	3.403	3.009	2.776	2.621	2.508	2.423	2.355	2.300	2.255	2.183	2.108	2.027	1.975	1.939	1.863	1.800	1.733
	0.025	5.717	4.319	3.721	3.379	3.155	2.995	2.874	2.779	2.703	2.640	2.541	2.437	2.327	2.257	2.209	2.107	2.024	1.935
	0.010	7.823	5.614	4.718	4.218	3.895	3.667	3.496	3.363	3.256	3.168	3.032	2.889	2.738	2.643	2.577	2.440	2.329	2.211
	0.005	9.551	6.661	5.519	4.890	4.486	4.202	3.991	3.826	3.695	3.587	3.420	3.246	3.062	2.947	2.868	2.702	2.569	2.428

F-distribution—*continued*

v_2	α \ v_1	1	2	3	4	5	6	7	8	9	10	12	15	20	25	30	50	100	∞
25	0.100	2.918	2.528	2.317	2.184	2.092	2.024	1.971	1.929	1.895	1.866	1.820	1.771	1.718	1.683	1.659	1.607	1.565	1.518
	0.050	4.242	3.385	2.991	2.759	2.603	2.490	2.405	2.337	2.282	2.236	2.165	2.089	2.007	1.955	1.919	1.842	1.779	1.711
	0.025	5.686	4.291	3.694	3.353	3.129	2.969	2.848	2.753	2.677	2.613	2.515	2.411	2.300	2.230	2.182	2.070	1.996	1.906
	0.010	7.770	5.568	4.675	4.177	3.855	3.627	3.457	3.324	3.217	3.129	2.993	2.850	2.699	2.604	2.538	2.400	2.289	2.169
	0.005	9.475	6.598	5.462	4.835	4.433	4.150	3.939	3.776	3.645	3.537	3.370	3.196	3.013	2.898	2.819	2.652	2.519	2.377
26	0.100	2.909	2.519	2.307	2.174	2.082	2.014	1.961	1.919	1.884	1.855	1.809	1.760	1.706	1.671	1.647	1.594	1.551	1.504
	0.050	4.225	3.369	2.975	2.743	2.587	2.474	2.388	2.321	2.265	2.220	2.148	2.072	1.990	1.938	1.901	1.823	1.760	1.691
	0.025	5.659	4.265	3.670	3.329	3.105	2.945	2.824	2.729	2.653	2.590	2.491	2.387	2.276	2.205	2.157	2.053	1.969	1.878
	0.010	7.721	5.526	4.637	4.140	3.818	3.591	3.421	3.288	3.182	3.094	2.958	2.815	2.664	2.569	2.503	2.364	2.252	2.131
	0.005	9.406	6.541	5.409	4.785	4.384	4.103	3.893	3.730	3.599	3.492	3.325	3.151	2.968	2.853	2.774	2.607	2.473	2.330
27	0.100	2.901	2.511	2.299	2.165	2.073	2.005	1.952	1.909	1.874	1.845	1.799	1.749	1.695	1.660	1.636	1.583	1.539	1.491
	0.050	4.210	3.354	2.960	2.728	2.572	2.459	2.373	2.305	2.250	2.204	2.132	2.056	1.974	1.921	1.884	1.806	1.742	1.672
	0.025	5.633	4.242	3.647	3.307	3.083	2.923	2.802	2.707	2.631	2.568	2.469	2.364	2.253	2.183	2.133	2.028	1.945	1.853
	0.010	7.677	5.488	4.601	4.106	3.785	3.558	3.388	3.256	3.149	3.062	2.926	2.783	2.632	2.536	2.470	2.330	2.218	2.097
	0.005	9.342	6.489	5.361	4.740	4.340	4.059	3.850	3.687	3.557	3.450	3.284	3.110	2.928	2.812	2.733	2.565	2.431	2.287
28	0.100	2.894	2.503	2.291	2.157	2.064	1.996	1.943	1.900	1.865	1.836	1.790	1.740	1.685	1.650	1.625	1.572	1.528	1.478
	0.050	4.196	3.340	2.947	2.714	2.558	2.445	2.359	2.291	2.236	2.190	2.118	2.041	1.959	1.906	1.869	1.790	1.725	1.654
	0.025	5.610	4.221	3.626	3.286	3.063	2.903	2.782	2.687	2.611	2.547	2.448	2.344	2.232	2.161	2.112	2.007	1.922	1.829
	0.010	7.636	5.453	4.568	4.074	3.754	3.528	3.358	3.226	3.120	3.032	2.896	2.753	2.602	2.506	2.440	2.300	2.187	2.064
	0.005	9.284	6.440	5.317	4.698	4.300	4.020	3.811	3.649	3.519	3.412	3.246	3.073	2.890	2.775	2.695	2.527	2.392	2.247
29	0.100	2.887	2.495	2.283	2.149	2.057	1.988	1.935	1.892	1.857	1.827	1.781	1.731	1.676	1.640	1.616	1.562	1.517	1.467
	0.050	4.183	3.328	2.934	2.701	2.545	2.432	2.346	2.278	2.223	2.177	2.104	2.027	1.945	1.891	1.854	1.775	1.710	1.638
	0.025	5.588	4.201	3.607	3.267	3.044	2.884	2.763	2.669	2.592	2.529	2.430	2.325	2.213	2.142	2.092	1.987	1.901	1.807
	0.010	7.598	5.420	4.538	4.045	3.725	3.499	3.330	3.198	3.092	3.005	2.868	2.726	2.574	2.478	2.412	2.271	2.158	2.034
	0.005	9.230	6.396	5.276	4.659	4.262	3.983	3.775	3.613	3.483	3.377	3.211	3.038	2.855	2.740	2.660	2.492	2.357	2.210
30	0.100	2.881	2.489	2.276	2.142	2.049	1.980	1.927	1.884	1.849	1.819	1.773	1.722	1.667	1.632	1.606	1.552	1.507	1.456
	0.050	4.171	3.316	2.922	2.690	2.534	2.421	2.334	2.266	2.211	2.165	2.092	2.015	1.932	1.878	1.841	1.761	1.695	1.622
	0.025	5.568	4.182	3.589	3.250	3.026	2.867	2.746	2.651	2.575	2.511	2.412	2.307	2.195	2.124	2.074	1.968	1.882	1.787
	0.010	7.562	5.390	4.510	4.018	3.699	3.473	3.304	3.173	3.067	2.979	2.843	2.700	2.549	2.453	2.386	2.245	2.131	2.006
	0.005	9.180	6.355	5.239	4.623	4.228	3.949	3.742	3.580	3.450	3.344	3.179	3.006	2.823	2.708	2.628	2.459	2.323	2.176
35	0.100	2.855	2.461	2.247	2.113	2.019	1.950	1.896	1.852	1.817	1.787	1.739	1.688	1.632	1.595	1.569	1.513	1.465	1.411
	0.050	4.121	3.267	2.874	2.641	2.485	2.372	2.285	2.217	2.161	2.114	2.041	1.963	1.878	1.824	1.786	1.703	1.635	1.558
	0.025	5.485	4.106	3.517	3.179	2.956	2.796	2.676	2.581	2.504	2.440	2.341	2.235	2.122	2.049	1.999	1.890	1.801	1.702
	0.010	7.419	5.268	4.396	3.908	3.592	3.368	3.200	3.069	2.963	2.876	2.740	2.597	2.445	2.348	2.281	2.137	2.020	1.891
	0.005	8.976	6.188	5.086	4.479	4.088	3.812	3.607	3.447	3.318	3.212	3.048	2.876	2.693	2.577	2.497	2.327	2.188	2.036
40	0.100	2.835	2.440	2.226	2.091	1.997	1.927	1.873	1.829	1.793	1.763	1.715	1.662	1.605	1.568	1.541	1.483	1.434	1.377
	0.050	4.085	3.232	2.839	2.606	2.449	2.336	2.249	2.180	2.124	2.077	2.003	1.924	1.839	1.783	1.744	1.660	1.589	1.509
	0.025	5.424	4.051	3.463	3.126	2.904	2.744	2.624	2.529	2.452	2.388	2.288	2.182	2.068	1.994	1.943	1.832	1.741	1.637
	0.010	7.314	5.170	4.313	3.828	3.514	3.291	3.124	2.993	2.888	2.801	2.665	2.522	2.369	2.271	2.203	2.058	1.938	1.805
	0.005	8.828	6.066	4.976	4.374	3.986	3.713	3.509	3.350	3.222	3.117	2.953	2.781	2.598	2.482	2.401	2.230	2.088	1.932
45	0.100	2.820	2.425	2.210	2.074	1.980	1.909	1.855	1.811	1.774	1.744	1.695	1.643	1.585	1.546	1.519	1.460	1.409	1.349
	0.050	4.057	3.204	2.812	2.579	2.422	2.308	2.221	2.152	2.096	2.049	1.974	1.895	1.808	1.752	1.713	1.626	1.554	1.470
	0.025	5.377	4.009	3.422	3.086	2.864	2.705	2.584	2.489	2.412	2.348	2.248	2.141	2.026	1.952	1.900	1.788	1.694	1.586
	0.010	7.234	5.110	4.249	3.767	3.454	3.232	3.066	2.935	2.830	2.743	2.608	2.464	2.311	2.213	2.144	1.997	1.875	1.737
	0.005	8.715	5.974	4.892	4.294	3.909	3.638	3.435	3.276	3.149	3.044	2.881	2.709	2.527	2.410	2.329	2.155	2.012	1.851
50	0.100	2.809	2.412	2.197	2.061	1.966	1.895	1.840	1.796	1.760	1.729	1.680	1.627	1.568	1.529	1.502	1.441	1.388	1.327
	0.050	4.034	3.183	2.790	2.557	2.400	2.286	2.199	2.130	2.073	2.026	1.952	1.871	1.784	1.727	1.687	1.599	1.525	1.438
	0.025	5.340	3.975	3.390	3.054	2.833	2.674	2.553	2.458	2.381	2.317	2.216	2.109	1.993	1.919	1.866	1.752	1.656	1.545
	0.010	7.171	5.057	4.199	3.720	3.408	3.186	3.020	2.890	2.785	2.698	2.562	2.419	2.265	2.167	2.098	1.949	1.825	1.683
	0.005	8.626	5.902	4.826	4.232	3.849	3.579	3.376	3.219	3.092	2.988	2.825	2.653	2.470	2.353	2.272	2.097	1.951	1.786
60	0.100	2.791	2.393	2.177	2.041	1.946	1.875	1.819	1.775	1.738	1.707	1.657	1.603	1.543	1.504	1.476	1.413	1.358	1.291
	0.50	4.001	3.150	2.758	2.525	2.368	2.254	2.167	2.097	2.040	1.993	1.917	1.836	1.748	1.690	1.649	1.559	1.481	1.389
	0.025	5.286	3.925	3.343	3.008	2.786	2.627	2.507	2.412	2.334	2.270	2.169	2.061	1.944	1.869	1.815	1.690	1.599	1.482
	0.010	7.077	4.977	4.126	3.649	3.339	3.119	2.953	2.823	2.718	2.632	2.496	2.352	2.198	2.098	2.028	1.877	1.749	1.601
	0.005	8.495	5.795	4.729	4.140	3.760	3.492	3.291	3.134	3.008	2.904	2.742	2.570	2.387	2.270	2.187	2.010	1.861	1.689
70	0.100	2.779	2.380	2.164	2.027	1.931	1.860	1.804	1.760	1.723	1.691	1.641	1.587	1.526	1.486	1.457	1.392	1.335	1.265
	0.050	3.978	3.128	2.736	2.503	2.346	2.231	2.143	2.074	2.017	1.969	1.893	1.812	1.722	1.664	1.622	1.530	1.450	1.353
	0.025	5.247	3.890	3.309	2.975	2.754	2.595	2.474	2.379	2.302	2.237	2.136	2.028	1.910	1.833	1.779	1.660	1.558	1.436
	0.010	7.011	4.922	4.074	3.600	3.291	3.071	2.906	2.777	2.672	2.585	2.450	2.306	2.150	2.050	1.980	1.826	1.695	1.540
	0.005	8.403	5.720	4.661	4.076	3.698	3.431	3.232	3.076	2.950	2.846	2.684	2.513	2.329	2.211	2.128	1.940	1.797	1.618
80	0.100	2.769	2.370	2.154	2.016	1.921	1.849	1.793	1.748	1.711	1.680	1.629	1.574	1.513	1.472	1.443	1.377	1.318	1.245
	0.050	3.960	3.111	2.719	2.486	2.329	2.214	2.126	2.056	1.999	1.951	1.875	1.793	1.703	1.644	1.602	1.508	1.426	1.325
	0.025	5.218	3.864	3.284	2.950	2.730	2.571	2.450	2.355	2.277	2.213	2.111	2.003	1.884	1.807	1.752	1.632	1.527	1.400
	0.010	6.963	4.881	4.036	3.563	3.255	3.036	2.871	2.742	2.637	2.551	2.415	2.271	2.115	2.015	1.944	1.788	1.655	1.494
	0.005	8.335	5.665	4.611	4.029	3.652	3.387	3.188	3.032	2.907	2.803	2.641	2.470	2.286	2.168	2.084	1.903	1.748	1.563

F-distribution—continued

v_2	α	1	2	3	4	5	6	7	8	9	10	12	15	20	25	30	50	100	∞
90	0.100	2.762	2.363	2.146	2.008	1.912	1.841	1.785	1.739	1.702	1.670	1.620	1.564	1.503	1.461	1.432	1.365	1.304	1.228
	0.050	3.947	3.098	2.706	2.473	2.316	2.201	2.113	2.043	1.986	1.938	1.861	1.779	1.688	1.629	1.586	1.491	1.407	1.302
	0.025	5.196	3.844	3.265	2.932	2.711	2.552	2.432	2.336	2.259	2.194	2.092	1.983	1.864	1.787	1.731	1.610	1.503	1.371
	0.010	6.925	4.849	4.007	3.535	3.228	3.009	2.845	2.715	2.611	2.524	2.389	2.244	2.088	1.987	1.916	1.759	1.623	1.457
	0.005	8.282	5.623	4.573	3.992	3.617	3.352	3.154	2.999	2.873	2.770	2.608	2.437	2.253	2.134	2.051	1.868	1.711	1.520
100	0.100	2.756	2.356	2.139	2.002	1.906	1.834	1.778	1.732	1.695	1.663	1.612	1.557	1.494	1.453	1.423	1.355	1.293	1.214
	0.050	3.936	3.087	2.696	2.463	2.305	2.191	2.103	2.032	1.975	1.927	1.850	1.768	1.676	1.616	1.573	1.477	1.392	1.283
	0.025	5.179	3.828	3.250	2.917	2.696	2.537	2.417	2.321	2.244	2.179	2.077	1.968	1.849	1.770	1.715	1.592	1.483	1.347
	0.010	6.895	4.824	3.984	3.513	3.206	2.988	2.823	2.694	2.590	2.503	2.368	2.223	2.067	1.965	1.893	1.735	1.598	1.427
	0.005	8.241	5.589	4.542	3.963	3.589	3.325	3.127	2.972	2.847	2.744	2.583	2.411	2.227	2.108	2.024	1.840	1.681	1.485
120	0.100	2.748	2.347	2.130	1.992	1.896	1.824	1.767	1.722	1.684	1.652	1.601	1.545	1.482	1.440	1.409	1.340	1.277	1.193
	0.050	3.920	3.072	2.680	2.447	2.290	2.175	2.087	2.016	1.959	1.910	1.834	1.750	1.659	1.598	1.554	1.457	1.369	1.254
	0.025	5.152	3.805	3.227	2.894	2.674	2.515	2.395	2.299	2.222	2.157	2.055	1.945	1.825	1.746	1.690	1.565	1.454	1.310
	0.010	6.851	4.787	3.949	3.480	3.174	2.956	2.792	2.663	2.559	2.472	2.336	2.192	2.035	1.932	1.860	1.700	1.559	1.381
	0.005	8.179	5.539	4.497	3.921	3.548	3.285	3.087	2.933	2.808	2.705	2.544	2.373	2.188	2.069	1.984	1.798	1.636	1.431
150	0.100	2.739	2.338	2.121	1.983	1.886	1.814	1.757	1.712	1.674	1.642	1.590	1.533	1.470	1.427	1.396	1.325	1.259	1.169
	0.050	3.904	3.056	2.665	2.432	2.274	2.160	2.071	2.001	1.943	1.894	1.817	1.734	1.641	1.580	1.535	1.436	1.345	1.223
	0.025	5.126	3.781	3.204	2.87˜	2.652	2.494	2.373	2.278	2.200	2.135	2.032	1.922	1.801	1.722	1.665	1.538	1.423	1.271
	0.010	6.807	4.749	3.915	3.447	3.142	2.924	2.761	2.632	2.528	2.441	2.305	2.160	2.003	1.900	1.827	1.665	1.520	1.331
	0.005	8.118	5.490	4.453	3.878	3.508	3.245	3.048	2.894	2.770	2.667	2.506	2.335	2.150	2.030	1.944	1.756	1.590	1.374
200	0.100	2.731	2.329	2.111	1.973	1.876	1.804	1.747	1.701	1.663	1.631	1.579	1.522	1.458	1.414	1.383	1.310	1.242	1.144
	0.050	3.888	3.041	2.650	2.417	2.259	2.144	2.056	1.985	1.927	1.878	1.801	1.717	1.623	1.561	1.516	1.415	1.321	1.189
	0.025	5.100	3.758	3.182	2.850	2.630	2.472	2.351	2.256	2.178	2.113	2.010	1.900	1.778	1.698	1.640	1.511	1.393	1.229
	0.010	6.763	4.713	3.881	3.414	3.110	2.893	2.730	2.601	2.497	2.411	2.275	2.129	1.971	1.868	1.794	1.629	1.481	1.279
	0.005	8.057	5.441	4.408	3.837	3.467	3.206	3.010	2.856	2.732	2.629	2.468	2.297	2.112	1.991	1.905	1.715	1.544	1.314
500	0.100	2.716	2.313	2.095	1.956	1.859	1.786	1.729	1.683	1.644	1.612	1.559	1.501	1.435	1.391	1.358	1.282	1.209	1.087
	0.050	3.860	3.014	2.623	2.390	2.232	2.117	2.028	1.957	1.899	1.850	1.772	1.686	1.592	1.528	1.482	1.376	1.275	1.113
	0.025	5.054	3.716	3.142	2.811	2.592	2.434	2.313	2.217	2.139	2.074	1.971	1.859	1.736	1.655	1.596	1.462	1.336	1.137
	0.010	6.686	4.648	3.821	3.357	3.054	2.838	2.675	2.547	2.443	2.356	2.220	2.075	1.915	1.810	1.735	1.566	1.408	1.164
	0.005	7.950	5.355	4.330	3.763	3.396	3.137	2.941	2.789	2.665	2.562	2.402	2.230	2.044	1.922	1.835	1.640	1.460	1.184
1000	0.100	2.711	2.308	2.089	1.950	1.853	1.780	1.723	1.676	1.638	1.605	1.552	1.494	1.428	1.383	1.350	1.273	1.197	1.060
	0.050	3.851	3.005	2.614	2.381	2.223	2.108	2.019	1.948	1.889	1.840	1.762	1.676	1.581	1.517	1.471	1.363	1.260	1.078
	0.025	5.039	3.703	3.129	2.799	2.579	2.421	2.300	2.204	2.126	2.061	1.958	1.846	1.722	1.640	1.581	1.445	1.316	1.094
	0.010	6.660	4.625	3.801	3.338	3.036	2.820	2.657	2.529	2.425	2.339	2.203	2.056	1.897	1.791	1.716	1.544	1.383	1.112
	0.005	7.915	5.326	4.305	3.739	3.373	3.114	2.919	2.766	2.643	2.541	2.380	2.208	2.022	1.900	1.812	1.615	1.431	1.125
∞	0.100	2.706	2.303	2.084	1.945	1.847	1.774	1.717	1.670	1.632	1.599	1.546	1.487	1.421	1.375	1.342	1.263	1.185	1.000
	0.050	3.841	2.996	2.605	2.372	2.214	2.099	2.010	1.938	1.880	1.831	1.752	1.666	1.571	1.506	1.459	1.350	1.243	1.000
	0.025	5.024	3.689	3.116	2.786	2.567	2.408	2.288	2.192	2.114	2.048	1.945	1.833	1.708	1.626	1.566	1.428	1.296	1.000
	0.010	6.635	4.605	3.782	3.319	3.017	2.802	2.639	2.511	2.407	2.321	2.185	2.039	1.878	1.773	1.696	1.523	1.358	1.000
	0.005	7.879	5.298	4.279	3.715	3.350	3.091	2.897	2.744	2.621	2.519	2.358	2.187	2.000	1.877	1.789	1.590	1.402	1.000

Random digits

Random digits 0 to 9, which are 'blocked' for convenience, may be used in any systematic way; e.g. if a random sample of 5 is required from population consisting of 83 members, the first two columns may be used to identify the members of the population, i.e. numbers 01 to 83, and the selected numbers are: 29, 12, 02, 69 and 11. Numbers greater than 83 may be ignored. When the first two columns are exhausted, columns 3 and 4 may be used, etc.

29	32	95	99	57	98	08	36	97	08	65	30	47	22	00	38	60	10	01	10
12	11	80	16	17	01	03	97	59	73	74	98	73	65	85	59	74	66	37	58
87	58	22	25	55	35	72	79	28	15	69	17	42	98	72	05	47	12	40	99
02	92	42	87	57	53	53	34	55	75	83	64	09	10	19	33	29	57	62	98
69	28	63	73	98	45	61	10	43	20	08	10	43	16	81	17	62	99	09	16
11	95	68	77	86	91	76	11	63	34	15	08	35	39	37	12	74	15	00	10
06	43	41	02	13	65	23	94	48	88	88	87	03	90	77	68	98	09	17	22
68	55	98	08	39	59	85	46	66	13	42	90	86	13	29	12	38	48	27	54
41	01	06	65	10	29	29	91	86	24	45	59	04	88	17	68	31	01	91	13
46	75	71	76	88	04	42	94	41	42	39	79	14	46	13	49	37	18	28	08
80	14	13	43	24	47	61	47	42	24	24	82	12	23	54	81	33	18	96	89
30	56	60	77	80	33	67	68	31	67	73	23	45	30	55	81	51	87	68	58
53	50	41	02	98	49	97	32	43	55	75	33	51	20	99	64	76	20	80	98
84	14	75	87	37	58	51	94	06	73	27	94	23	76	77	81	72	90	45	41
08	27	89	33	87	52	24	57	50	22	22	76	60	05	79	86	58	83	88	41
97	08	50	16	41	67	40	56	13	12	68	67	36	22	08	55	76	86	45	67
97	08	37	42	48	95	90	48	34	88	19	66	38	94	64	95	07	78	23	86
70	15	04	10	34	95	57	63	75	82	88	74	28	24	66	99	52	65	36	98
06	38	31	17	38	24	98	52	67	04	95	54	89	79	45	28	05	18	60	17
63	87	79	25	86	56	74	17	45	32	53	62	09	04	86	65	87	48	82	02
17	00	56	31	14	18	56	97	91	78	85	82	06	24	88	49	17	68	51	50
17	76	35	38	19	24	47	21	09	43	09	72	02	64	66	06	78	21	70	41
57	77	32	13	60	37	68	66	11	23	30	62	97	71	02	20	13	22	00	40
35	86	97	84	91	77	73	03	37	77	50	24	54	51	40	20	66	16	34	84
72	68	64	77	89	72	77	67	45	72	25	56	78	69	72	63	86	52	07	43
91	01	78	50	50	91	99	15	36	02	74	42	55	33	19	88	35	17	58	37
70	37	55	94	53	05	78	53	23	29	15	57	70	30	88	63	20	12	64	38
11	06	17	48	24	57	50	76	81	77	30	12	92	27	19	32	63	70	97	80
60	37	89	98	61	05	51	89	47	28	34	83	98	44	66	96	84	64	64	92
37	41	11	09	04	84	38	51	91	49	23	78	53	95	40	17	73	23	04	70
28	97	38	27	97	54	95	94	54	79	93	88	00	82	39	61	93	78	07	88
14	29	17	18	84	03	10	62	15	70	01	15	06	30	97	79	55	98	79	39
81	70	53	83	20	25	26	56	55	56	33	58	74	21	76	94	24	80	12	50
08	20	90	25	43	22	81	74	51	76	53	39	59	35	34	46	55	54	73	50
61	95	25	85	66	34	76	39	98	88	45	57	64	11	17	06	43	35	27	09
64	58	31	05	45	77	25	20	02	09	36	87	63	01	10	08	01	19	19	06
75	49	97	87	79	31	66	57	89	56	56	97	71	43	65	62	36	77	50	87
66	95	10	78	42	24	91	82	74	29	00	53	44	70	18	23	48	09	90	99
85	37	61	48	07	99	13	01	16	94	37	31	28	96	59	77	62	24	95	84
06	87	15	09	48	31	18	66	87	11	19	71	67	20	93	92	02	96	15	65
11	15	95	59	69	81	75	75	88	69	95	12	75	69	18	10	60	35	31	47
03	64	44	33	46	16	02	28	14	33	61	57	28	33	96	47	49	86	85	83
68	89	57	51	94	84	09	80	37	90	52	99	85	52	49	66	63	69	11	31
43	13	09	12	00	65	69	54	11	00	20	94	22	93	90	16	82	64	27	46
42	68	71	56	74	17	71	63	80	81	02	41	49	27	92	44	44	13	45	21
12	55	09	80	30	50	34	96	31	71	19	21	79	42	17	57	04	04	19	00
88	84	87	74	01	39	99	02	75	76	61	88	97	89	06	97	15	70	26	27
49	27	92	08	87	65	12	32	27	96	11	26	30	88	48	89	29	73	50	47
46	51	54	92	06	44	85	83	14	78	68	83	33	17	03	10	99	10	17	34
34	96	78	90	18	41	44	69	10	30	48	98	32	76	12	81	29	83	02	87

Random digits—*continued*

80	07	15	41	15	37	42	39	24	45	48	73	61	15	44	74	40	27	26	47
39	08	51	67	63	03	76	76	86	09	39	32	62	77	60	85	37	14	69	76
51	32	57	06	49	13	01	25	98	83	44	96	92	78	37	24	49	35	54	52
84	46	17	46	71	53	88	78	30	71	53	85	55	10	93	40	05	66	72	38
04	88	20	78	89	94	31	36	83	74	51	25	28	43	54	76	57	08	21	23
21	45	86	26	12	21	28	37	56	47	86	18	38	39	18	89	99	62	81	98
71	38	27	31	40	52	36	03	51	54	83	14	51	17	86	77	66	84	50	84
78	50	39	32	55	17	25	06	90	90	69	48	70	68	22	07	85	07	95	84
22	76	93	40	26	30	77	61	71	74	81	13	73	21	99	00	47	52	43	18
25	21	70	62	69	05	05	58	75	92	85	60	50	87	81	35	80	83	42	16
96	79	06	87	51	04	17	61	42	12	64	77	45	06	55	68	19	39	17	22
97	76	01	89	33	70	46	23	44	83	99	55	95	03	41	89	33	49	89	86
78	03	18	58	00	47	18	01	33	49	99	55	54	70	65	34	76	58	86	20
09	63	31	80	30	17	11	75	34	81	25	45	91	80	50	25	64	70	05	48
61	33	89	72	78	98	26	56	88	66	51	69	71	48	13	71	40	57	31	22
64	83	61	76	37	68	22	25	09	82	53	59	78	66	81	66	45	56	64	78
18	93	65	67	39	81	96	44	68	46	96	50	08	71	70	81	23	32	89	61
86	84	70	40	22	89	25	42	62	69	95	98	59	26	69	55	33	62	91	88
96	57	56	48	81	92	77	95	43	50	29	89	07	58	10	83	66	04	15	74
54	35	65	28	09	99	04	41	86	60	69	54	82	74	49	86	82	25	07	29
18	79	09	01	55	60	31	19	19	48	01	89	54	63	96	70	99	15	71	84
19	78	77	63	36	52	38	88	16	92	23	42	49	79	27	15	09	94	49	35
55	71	79	75	30	29	13	32	60	07	33	73	61	89	63	64	17	15	21	39
38	58	83	62	94	73	84	48	95	17	79	74	78	38	09	37	35	75	74	70
78	29	66	85	65	45	79	70	88	92	73	24	71	71	63	70	47	56	70	28
87	55	81	22	04	62	21	45	81	82	43	96	17	70	61	80	59	10	59	00
06	98	70	24	03	20	67	45	67	65	04	61	76	89	25	13	73	06	41	16
33	08	62	21	90	70	72	16	01	23	26	05	10	33	23	23	03	07	46	08
54	03	25	45	50	40	58	15	41	07	16	24	16	63	46	64	27	85	27	47
68	90	88	08	25	70	23	82	53	40	51	91	84	67	84	08	09	76	19	19
90	18	00	18	76	88	55	07	52	00	30	04	83	72	04	74	87	56	90	80
70	07	33	78	52	59	92	46	58	33	61	42	31	47	58	89	32	02	55	36
19	13	05	69	12	74	49	85	21	49	18	11	60	96	94	04	74	26	23	44
95	70	86	00	19	44	74	51	22	34	63	14	11	30	48	54	71	78	97	12
65	12	41	20	32	33	72	70	71	24	51	39	43	28	90	51	14	46	17	40
15	53	57	75	61	54	95	63	75	51	28	43	39	55	90	58	01	50	31	88
60	27	72	94	00	25	71	09	76	19	66	69	44	09	39	12	60	43	02	52
57	91	58	68	24	78	33	54	25	46	08	87	72	85	28	98	89	67	68	92
40	15	42	80	71	35	81	75	95	40	04	85	70	88	19	44	75	50	63	41
23	97	89	48	74	96	60	10	40	24	33	88	86	93	30	79	96	32	25	34
48	25	55	19	87	97	39	79	66	73	50	78	72	75	08	78	66	69	13	35
24	58	57	51	61	90	39	52	91	33	77	67	76	78	40	42	05	70	73	08
60	22	38	11	98	95	66	00	95	19	32	99	90	77	55	50	86	94	41	83
84	89	06	96	10	47	83	22	11	81	19	13	48	21	71	99	16	81	88	56
30	80	70	60	93	09	74	04	99	72	67	91	91	75	20	36	08	45	28	35
23	95	78	32	20	71	90	24	20	66	09	27	14	97	94	78	67	45	20	62
48	52	58	73	69	63	54	77	76	89	09	15	50	05	85	91	12	10	12	29
33	69	72	87	15	96	24	09	14	84	41	57	16	17	78	18	46	46	23	04
71	71	53	72	84	65	86	16	70	43	62	10	33	15	61	60	80	73	18	21
29	53	27	21	49	53	31	68	21	10	17	47	35	74	84	18	58	07	17	32
17	70	60	84	24	50	82	33	67	40	15	88	50	22	54	28	39	46	14	28
98	37	60	93	52	27	20	93	10	62	90	69	27	96	44	54	01	13	81	14
16	39	86	14	17	56	74	44	76	20	77	74	52	52	56	06	99	78	52	67
53	17	93	61	99	15	08	47	04	09	46	95	53	02	57	60	02	02	99	83
05	38	06	80	55	75	49	12	95	96	98	63	46	51	49	74	97	71	95	88

Index

Assignment—
   algorithm, 311–14
   forbidden allocations, 316
   problem, 311
      special cases, 315–16
Association—
   meaning, 181
Attributes—
   association, testing, 158–64

Bivariate distribution—
   meaning, 182

Central Limit Theorem—
   meaning, 101
Combinations—
   illustration of, 18
   meaning, 18
   probabilities, use to calculate, 19
Confidence interval—
   linear regression analysis, in, 197–9
   meaning, 115
   population mean, for, 115–19
      sample size, 123–4
   population proportion, for, 119–21
      sample size, 125
   population variance, for, 121–3
      sample size, 125–6
Contingency table—
   example of, 158
   meaning, 158
   Yates' correction, 159
Continuous random variables—
   meaning, 27
   observed frequency distribution, and, 168–71
   probability distributions for, 43–6
      uniform distribution, 45–6
Correlation—
   meaning, 181

Decisions—
   best, meaning, 67
   changes in probabilities, sensitivity to, 73
   risk, *see* Risk assessment
   rules—
      probability values, not using, 68–71
      probability values, using, 71–3
   significant, 133
   trees—
      sensitivity analysis, and, 87–9
      single decision, illustrating, 79–80
      time value of money and two-stage outcomes, involving, 83–6
      two stage decision process, 80–3
      value of perfect information, 73–4
Discrete random variables—
   observed frequency distribution, and, 164–5
   outcomes represented by, 27–8
   probability distributions for, 27–32
      graphical representations, 29–30

Estimates—
   subjective, probability evaluated from, 5

Event—
   compound, probability of, 5
   independent, 6
   meaning, 1
   mutually exclusive, 6
Experiment—
   expected values, 15–16
   meaning, 1
   outcome, meaning, 1
   pattern in outcomes of, 1
   permutations and combinations, 16–19
   probability, to evaluate, 4
Exponential smoothing—
   forecasting using, 235–9
   meaning, 235

Forecasting—
   additive model, using, 230
   exponential smoothing, using, 235–9
   multiplicative component model, with 234–5

Games—
   mathematical theory of, 1
Gantt Chart—
   construction of, 334

Hungarian algorithm, 311–12
Hypothesis—
   alternative, 132
   null, 131
   test, *see* Hypothesis test
Hypothesis test—
   Chi squared test, 134, 158–72
   F test, 134, 144–6
   illustration of principle, 132–3
   non-parametric, 158–72
   normal, 134
   one and two tailed, 134–5
   overall linearity, assessing, 194–7
   paired data, on, 157–8
   procedure for, 131–5, 172
   sample mean, on—
      population variance known, 135–9, 147–9
      population variance unknown, 140–2, 149–55
      two, on, 147–55
   sample proportion, on, 142–3
   standard, 133–4
   t test, 134
   two population variances, on, 144–6
   two sample proportions, on, 155–7
   use of, 131
   variance ratio test, 134, 144–6

Information—
   perfect, value of, 73–4

Linear programming—
   algebraic solution, 269
   coefficients of objective function, changes in, 265–9
   dual model, 279–83
   graphical solution, 253–62
   limiting resource, change in provision of, 263

696 INDEX

Linear programming—*contd*
  meaning, 247
  multi-variable, simplex solution of, 269–75
  non-limiting resource, changes in, 265
  problem formulation, 247–52
  sensitivity analysis, 262–9
  simplex method, 270–9
  solving, 252–62
  use of, 247
Linear regression—
  confidence intervals, 197–9
  correlation coefficient r, 186–9
  hypothesis tests, 194–7
  linear relationship, 181
    strength of, 186–9
  meaning, 181
  models—
    multiple, 199–207
    non-linear relationships, 207–12
    prediction and estimation using, 189–92
    simple model, 182–6
    use of, 182
  Spearman's rank correlation—r_s, 212–14
  statistical inferences, 192–99
Linear transformation—
  meaning, 207

Model—
  linear programming *see* Linear programming
  linear regression, *see* Linear regression
  non-linear regression, 209
  purpose of, 182
  queueing, *see* Queueing
  simulation, *see* Simulation
  stock, *see* Stock
  time series, in, *see* Time series
  use of, 181
Monte Carlo simulation, 403

Network analysis—
  activity-on-node diagrams, 328–31
  arrow diagrams, 325–8, 332–4
  critical path analysis, 329–34
  Gantt Chart, construction of, 334
  network diagrams, 325–9
  stages of, 325
  use of, 325

Operational research—
  meaning, 245
Outcome—
  meaning, 1

Parameter—
  meaning, 95
Pareto effect, 379
Permutations—
  illustration, 16–18
  probabilities, use to calculate, 19
Population—
  finite, 95
  infinite, 95
  mean—
    confidence interval for, 115–19
    estimating, sample size for, 123–4
  meaning, 95
  parameter, 95
  proportion—
    calculating, sample size for, 125
    confidence interval for, 119–21
  variance—
    confidence interval for, 121–3
    estimating, sample size for, 125–6
    hypothesis test on, 144–6

Programming—
  allocation of resources as, 247
  linear, *see* Linear programming
Project—
  activity times, uncertainty in, 339–42
  analysis of, 325
  costs, 335–9
  duration, minimising, 335–8
  minimum cost, completing at, 338–9
  network analysis, *see* Network analysis
  resource allocation, 342–5
Project Evaluation and Review Technique (PERT)—
  application of, 340–42
  use of, 339–40
Probability—
  basic idea of, 2
  compound events, of, 5
  conditional, 8–9
  definition, 2
  distributions, *see* Probability distributions
  evaluation—
    recorded information, use of, 4
    repeated experiment, by, 4–5
    subjective estimates, from, 5
    symmetry, from, 3–4
  measure, properties of, 2
  measurement, evaluated by, 4–5
  permutations and combinations, use to calculate, 19
  rules of—
    addition, 6–8
    Bayes' rules, 12–15
    conditional, 8–9
    more than two events, for, 10
    multiplication, 9–12
    terminology, 6
  theory, *see* Probability theory
  trees, 5–6, 10–11
  value of, 2–3
Probability distributions—
  binomial, 27, 32–8, 57
    expected value and standard deviation, 36–8
    normal distribution as approximation to, 51–4
    observed distribution, goodness of fit, 165–6
    Poisson distribution as approximation to, 41–3
  combined variables, of, 56–7
  continuous random variables, for, 43–6
  discrete random variables, for, 27–32
    graphical representations, 29–30
  expected value and standard deviation, 30–2
  normal, 27, 45–7
    approximation for Poisson distribution, as, 54
    approximation to binomial distribution, as, 51–4
    standard, 47–51
  Poisson, 27, 38–41, 57
    approximation to binomial distribution, as, 41–3
    expected value and variance, 41
    normal distribution as approximation for, 54
    observed distribution, goodness of fit, 166–8
Probability theory—
  origins of, 1
  use of, 1

Quantity discounts, 360–3
Queueing—
  discipline, 388
  models—
    M/M/1, 392–5
    Poisson process, 389–92, 395
  servers, effect of, 389
  simulation model, application of, 407–12
  system—
    arrivals and arrival patterns, 387–8
    components of, 387–9
    costs of, 396
    good, judging, 389

Random variables—
    combination of, 55–7
    continuous, see Continuous random variables
    discrete, see Discrete random variables
    independent, 55–6
    non-independent, 56
    numerical outcomes of experiment, representing, 57
Regression—
    linear, see Linear regression
    meaning, 181
Resources—
    allocation, 342–5
    profiles, 343
Risk assessment—
    standard deviations, use of, 74–6
    utility, use of, 76–9
Sample—
    advantages of using, 96
    appropriate size, choice of, 123–6
    meaning, 95
    non-random designs, 98
    random designs, 97–8
    selection of, 96–9
    statistic, 95
Sampling—
    distributions—
        illustration of, 99–100
        meaning, 99
        sample means, of, 101, 106
        sample variance, of, 103–6
        standard 107–9
    frame, 97
    judgement, 98
    hypothesis test, see Hypothesis test
    non-random, 98
    quota, 98
    random, 96–8
    reasons for, 95–6
    terminology, 95
Sensitivity analysis—
    decision trees, and, 87–9
    limiting constraints, of, 263–5
    linear programming, in, 262–9
    simplex method, and, 270–9
    transportation problem, of, 305–7
Simulation—
    models—
        discrete, principles of, 403–7
        queuing problem, application to, 407–12
        stock control problem, application to, 412–14
    Monte Carlo, 403
    use of, 403
Spearman's rank correlation—r_s, 212–14
Statistic—
    meaning, 95

Stock—
    basic model—
        assumptions, 353
        costs of stocking, 354–5
        economic batch quantity, 358–9
        re-order level and interval, 357–8
        total cost equation, 355–7
        uncertainty, and, 370–9
    batch production model, 363–6
    control, 379
    economic batch quantity model, 358–9
    optimum order quantity, 356–7
    Pareto effect, 379
    planned shortages model, 367–70
    quantity discounts, 360–3
    re-order cycle system, 377–9
    re-order level system, 371–5
    simulation model, application of, 412–14
Symmetry—
    probability evolved from, 3–4

Test of significance, see Hypothesis test
Time series—
    components, 223–6
    cyclical variation, 234
    exponential smoothing, 235–9
    mean absolute deviation, 224
    mean square error, 225
    meaning, 223
    model—
        additive component, analysis of 226–30
        choice of, 225
        multiplicative, analysis of, 230–5
    moving average technique, 223
    seasonal variation, 224
    trend, 223
Transportation—
    algorithm, 293
    degenerate solutions, 308–11
    forbidden allocations, 307
    initial allocation, 294–7
    optimality, testing for, 297–303
    optimum solution, finding, 303–5
    problem, illustration of, 291–3
    sensitivity analysis, 305–7
    variation in problem, 307–11
Trial—
    meaning, 1

Utility—
    assessment of risk, use in incorporating, 76–9
    values, use of, 77–9